REGULATION OF GENE EXPRESSION IN *ESCHERICHIA COLI*

E. C. C. Lin, Ph.D.

A. Simon Lynch, Ph.D.

Harvard Medical School
Boston, Massachusetts, U.S.A.

CHAPMAN & HALL
I(T)P An International Thomson Publishing Company

New York • Albany • Bonn • Boston • Cincinnati • Detroit • London • Madrid • Melbourne •
Mexico City • Pacific Grove • Paris • San Francisco • Singapore • Tokyo • Toronto • Washington

R.G. LANDES COMPANY
AUSTIN

REGULATION OF GENE EXPRESSION IN *ESCHERICHIA COLI*

R.G. LANDES COMPANY
Austin, Texas, U.S.A.

U.S. and International Copyright © 1996 R.G. Landes Company and Chapman & Hall

Please address all inquiries to the Publishers:
R.G. Landes Company, 909 Pine Street, Georgetown, Texas, U.S.A. 78626
Phone: 512/ 863 7762; FAX: 512/ 863 0081

Chapman & Hall, 115 Fifth Avenue, New York, New York, U.S.A. 10003
Chapman & Hall, 2-6 Boundary Row, London SE1 8HN, United Kingdom

ISBN number: 0-412-10291-9

Library of Congress Cataloging-in-Publication Data

Regulation of gene expression in Escherichia coli / [edited by] E.C.C. Lin, A. Simon Lynch.
    p. cm.
   Includes bibliographical references and index.
   ISBN 0-412-10291-9 (alk. paper)
   1. Escherichia coli--Genetics. 2. Genetic regulation. I. Lin, E.C.C. II. Lynch, A. Simon, 1964- .

    [DNLM: 1. Escherichia coli--genetics. 2. Gene Expression Regulation, Bacterial. QW 138.5.E8
R344 1995]
QH434.R44 1995
589.9'5--dc20
DNLM/DLC
for Library of Congress
                95-43432
                CIP

CONTENTS

1. **Introduction: From Physiology to DNA and Back**.....................1
   Jon Beckwith

2. **RNA Chain Initiation and Promoter Escape by RNA Polymerase** 7
   Michael J. Chamberlin and Lilian M. Hsu
   Promoter Function is Regulated at Two Distinct Phases of Transcription:
      Promoter Binding and RNA Chain Initiation 7
   The Biochemistry of the RNA Chain Initiation Phase of Transcription 8
   Parameters That Describe the RNA Chain Initiation Reaction at Different Promoters 11
   Factors That Affect the Initiation Reaction: Intrinsic Factors 13
   Factors That Affect the Initiation Reaction: Extrinsic Factors 15
   Models for the Mechanism of RNA Chain Initiation:
      Some Simple Models Do Not Account for What Is Known.................. 15
   Models for the Mechanism of RNA Chain Initiation-Models Based on Recent
      Models for RNA Chain Elongation 16

3. **Transcription Termination and Its Control**...........................27
   Jeffrey W. Roberts
   Termination 27
   Antitermination 34

4. **Codon Context, Translational Step-Times and Attenuation**...............47
   G. Wesley Hatfield
   Overview... 47
   Attenuation 48
   Codon Context and Translational Efficiency 52
   Effects of Codon Pair Bias on Translational Step-Times 54
   Discussion 60

5. **Control by Antisense RNA** 67
   Brian N. Zeiler and Robert W. Simons
   Introduction 67
   Antisense RNAs Control Diverse Biological Functions 67
   Antisense RNAs Control Gene Expression at Many Different
      Post-Transcriptional Levels 71
   Antisense RNAs Pair to Their Target RNAs by Defined Mechanisms 76
   Overview... 78

6. **Translational Control of Gene Expression in *E. Coli* and Bacteriophage** 85
   Mathias Springer
   Introduction 85
   Translation Initiation 88
   Translational Operators 89
   Translational Repressors 98
   Mechanisms of Control 102
   Translational Control and mRNA Processing and/or Degradation............ 108
   The Role of Translational Control in Growth Rate Regulation 112
   Conclusions and Perspectives 113

7. **Effects of DNA Supercoiling on Gene Expression** 127
   James C. Wang and A. Simon Lynch
   Synopsis 127
   Introduction 128
   The Dependence of Transcription on the Cellular Level of DNA Gyrase
      and DNA Topoisomerase I 129
   Mechanistic Considerations 130
   Supercoiling of the DNA Template by Transcription 136
   Concluding Remarks 139

8. **The HU and IHF Proteins:**
   Accessory Factors for Complex Protein-DNA Assemblies 149
   Howard A. Nash
   Perspective 149
   Structure 150
   Interaction with Nucleic Acids 152
   Control of Intracellular Concentration and Activity 156
   Participation of IHF and HU in Well-Characterized Biochemical Processes 158
   Unfinished Business 166

9. **The *lac* and *gal* Operons Today** 181
   Sankar Adhya
   Introduction 181
   The *lac* and *gal* Operons Encode Enzymes of a Continuous Biochemical Pathway 182
   The Regulatory Circuits and Their Components 182
   Modulation of Promoters by cAMP•CRP 187
   Control of *P2* by UTP in *gal* 191
   Natural Polarity 191
   Negative Control by Repressor-Operator Interactions 192
   Epilogue 194

10. **The Maltose System** 201
    Winfried Boos, Ralf Peist, Katja Decker and Eva Zdych
    Introduction and Scope 201
    The Positive Transcriptional Activator MalT 203
    The Maltose/Maltodextrin Transport System 206
    The Enzymes of the Maltose System 209
    Nonclassical Regulatory Phenomena 212
    Perspectives 219

11. **The Phosphoenolpyruvate-Dependent Carbohydrate:**
    Phosphotransferase System (PTS) and Control of Carbon Source Utilization 231
    Joseph W. Lengeler
    Introduction 231
    Regulatory Phenomena Related to Carbon Source Utilization 232
    Bacterial Transport Systems and Global Regulatory Networks Form a Unit 234
    The Bacterial PTS Is a Transport and Signal Transduction System 237
    IIA^{Glc} of the PTS Is Central to Carbon Catabolite Repression 239
    IIA^{Glc}, the Regulation of Adenylate Cyclase Activity and of Intracellular cAMP Levels 241

Not Only cAMP Levels, but Also CRP Levels Are Essential in Catabolite Repression 242
IIA^{Glc} and Inducer Exclusion 243
Catabolite Repression and Inducer Exclusion Act in Concert 244
Carbon Catabolite Repression through PTS-Control Is Part of a Stimulon 245
Concluding Remarks 247

12. **The Cap Modulon** 255
    Stephen Busby and Annie Kolb
    The Long History of CAP 255
    Cyclic AMP and Gene Expression 255
    CAP as a Global Regulator: The CAP Modulon 256
    CAP Binding at Target Promoters and Structural Studies 259
    Activation by CAP at "Simple" Promoters 265
    Activation by CAP at Complex Promoters 266
    CAP as a Repressor and a Co-Repressor 269
    CAP: Paradigm or Artifact? 270

13. **Regulation of Nitrogen Assimilation** 281
    Boris Magasanik
    The *glnALG(glnA ntrBC)* Operon 282
    The σ^{54}-Dependent Promoter 282
    Transcriptional Enhancers 283
    Phosphorylation of NRI 283
    NR$_I$/NR$_{II}$ as Two-Component Paradigm 284
    Activation of Transcription 285
    Response to Nitrogen Availability 287

14. **History of the Pho System** 291
    Annamaria Torriani-Gorini

15. **Are the Multiple Signal Transduction Pathways of the Pho Regulon
    Due to Cross Talk or Cross Regulation?** 297
    *Barry L. Wanner, Weihong Jiang, Soo-Ki Kim, Sayaka Yamagata, Andreas Haldimann
        and Larry L. Daniels*
    Introduction 297
    Genes for P$_i$ Control of the Pho Regulon 299
    Transmembrane Signaling by Environmental P$_i$ 300
    Genes for P$_i$ Independent Controls of the Pho Regulon 303
    Activation by CreC and Acetyl Phosphate 306
    Cross Talk, Cross Regulation and a Hypothesis 308
    Is There Evidence for Cross Regulation? 309
    Overview and Prospects for Future Studies 311

16. **The FNR Modulon and FNR-Regulated Gene Expression** 317
    John R. Guest, Jeffrey Green, Alistair S. Irvine and Stephen Spiro
    Introduction 317
    The Metabolic Arena 318
    The FNR Modulon 321
    The FNR Protein and Relationships with CAP 323

The DNA-Binding Specificity of FNR 324
In Vitro Transcription Activation and Repression 326
Transcriptional Organization of Representative Promoters 327
Potential FNR Contacts with RNA Polymerase and DNA-Bending 330
The Mystery of Redox-Sensing 331
Structural and Functional Homologs of FNR 335
Concluding Remarks 337

17. The NAR Modulon Systems:
Nitrate and Nitrite Regulation of Anaerobic Gene Expression 343
Andrew J. Darwin and Valley Stewart
Introduction 343
Anaerobic Respiration 344
The Characterization of the Nar Regulatory System 345
Dual Two-Component Regulatory Systems 346
The Role of Nitrite in the Nar Regulatory System 347
The Sensor Proteins 347
The Response Regulators 349
Indirect Nitrate Regulation of Gene Expression 353
Concluding Remarks 354

18. Regulation of Aerobic and Anaerobic Metabolism by the Arc System 361
A. Simon Lynch and Edmund C. C. Lin
Introduction 361
Identification of the *arc* Genes 361
In Vivo Studies of *arc* Mutants 365
In Vitro Phosphorylation Studies 366
The Arc Modulon 368
ArcA DNA Binding 370
The Arc Stimulus 372
Future Studies 373

19. The Porin Regulon: A Paradigm for the Two-Component Regulatory Systems ... 383
James M. Slauch and Thomas J. Silhavy
Introduction 383
Background 384
The History of Porin Regulation 385
The Structure of the *ompB* Locus 389
The Structure of the *ompF* and *ompC* Genes 390
The Roles of OmpR and EnvZ 393
Phosphorylation and Signal Transduction 397
Summary and Conclusions 404

20. The Leucine\Lrp Regulon 419
Elaine B. Newman and Rongtuan Lin
Introduction 419
The Leucine-Responsive Regulatory Protein 420
Regulation of Lrp Synthesis 420
Target Operons of Lrp and Mutant Phenotypes 421

Lrp as a Chromosome Organizer 424
Molecular Aspects of Lrp Interactions at Individual Promoters 425
Appendix 428

21. **Adaptive responses to Oxidative Stress: The *soxRS* and *oxyR* Regulons** 435
Elena Hidalgo and Bruce Demple
Reactive Oxygen Species 435
Antioxidant Defenses 436
Oxidative Stress 436
Global Responses to Oxidative Stress 437
The *soxRS* Regulon 437
The *oxyR* Regulon 442
Control of Antibiotic Resistance Genes 445

22. **The SOS Regulatory System** 453
John W. Little
Introduction and Current Regulatory Model 453
Development of the SOS Model 454
Recent Developments 458
Behavior of the SOS Gene Regulatory Circuitry 469
Future Prospects 474

23. **Heat Shock Regulation** 481
Dominique Missiakas, Satish Raina and Costa Georgopoulos
Introduction 481
Properties of Important Heat Shock Proteins 482
Regulation of the σ^{32}-Promoted Heat Shock Response 487
A Second Heat Shock Regulon 491
The σ^{54}-Promoted Stress Response 495
Heat Shock or Stress Responses in Other Eubacteria 495

24. **Roles for Energy-Dependent Proteases in Regulatory Cascades** 503
Susan Gottesman
Introduction 503
The Proteases and Their Targets 504
Summary and General Conclusions 513

25. **Control of rRNA and Ribosome Synthesis** 521
Richard L. Gourse and Wilma Ross
Introduction 521
rRNA Gene Organization 522
High Activity of rRNA Synthesis Rates 523
Stringent Control 530
Growth Rate Dependent Control 531
Additional Considerations 535
Conclusion and Future Prospects 536

26. **Cell Division** 547
    Lawrence I. Rothfield and Jorge Garcia-Lara
    Introduction 547
    The Cell Division Process 548
    Essential Cell Division Genes 548
    Transcriptional Regulation of Cell Division Genes 552
    Translational Control 558
    Division Inhibitors 558
    Cell Division Inhibition by Mutations in Genes That Do Not Code
       for Cell Division Proteins 561
    Past, Present and Future.. 562

27. **Regulation of Gene Expression in Stationary Phase** 571
    Heidi Goodrich-Blair, María Uría-Nickelsen and Roberto Kolter
    Introduction 571
    The σ^s Regulon 573
    RpoS Regulation 575

Index 585

EDITORS

Edmund C. C. Lin, Ph.D.
Department of Microbiology and Molecular Genetics
Harvard Medical School
Boston, Massachusetts, U.S.A.
Chapter 18

A. Simon Lynch, Ph.D.
Department of Microbiology and Molecular Genetics
Harvard Medical School
Boston, Massachusetts, U.S.A.
Chapters 7,18

CONTRIBUTORS

Sankar Adhya, Ph.D.
Laboratory of Molecular Biology
National Cancer Institute
National Institutes of Health
Bethesda, Maryland, U.S.A.
Chapter 9

Jon Beckwith, Ph.D.
Department of Microbiology
  and Molecular Genetics
Harvard Medical School
Boston, Massachusetts, U.S.A.
Chapter 1

Winfried Boos, Ph.D.
Department of Biology
University of Konstanz
Konstanz, Germany
Chapter 10

Stephen Busby, D.Phil.
School of Biochemistry
University of Birmingham
Birmingham, U.K.
Chapter 12

Michael J. Chamberlin, Ph.D.
Division of Biochemistry and Molecular Biology
University of California, Berkeley
Berkeley, California, U.S.A.
Chapter 2

Larry L. Daniels, Ph.D.
Department of Biological Sciences
Purdue University
West Lafayette, Indiana, U.S.A.
Chapter 15

Andrew J. Darwin, Ph.D.
Section of Microbiology
Cornell University
Ithaca, New York, U.S.A.
Chapter 17

Katja Decker
Department of Biology
University of Konstanz
Konstanz, Germany
Chapter 10

Bruce Demple, Ph.D.
Department of Molecular
  and Cellular Toxicology
Harvard University, School of Public Health
Boston, Massachusetts, U.S.A.
Chapter 21

Jorge Garcia-Lara, Ph.D.
Department of Microbiology
University of Connecticut Health Center
Farmington, Connecticut, U.S.A.
Chapter 26

CONTRIBUTORS

Costa Georgopoulos, Ph.D.
Département de Biochimie Médicale
Centre Médical Universitaire
Genève, Switzerland
Chapter 23

Heidi Goodrich-Blair, Ph.D.
Department of Microbiology
   and Molecular Genetics
Harvard Medical School
Boston, Massachusetts, U.S.A.
Chapter 27

Susan Gottesman, Ph.D.
Laboratory of Molecular Biology
National Cancer Institute
Bethesda, Maryland, U.S.A.
Chapter 24

Richard L. Gourse, Ph.D.
Department of Bacteriology
University of Wisconsin
Madison, Wisconsin, U.S.A.
Chapter 25

Jeffrey Green, Ph.D.
The Krebs Institute
Department of Molecular Biology
   and Biotechnology
University of Sheffield
Sheffield, United Kingdom
Chapter 16

John R. Guest, D.Phil.
The Krebs Institute
Department of Molecular Biology
   and Biotechnology
University of Sheffield
Sheffield, United Kingdom
Chapter 16

Andreas Haldimann, Ph.D.
Department of Biological Sciences
Purdue University
West Lafayette, Indiana, U.S.A.
Chapter 15

G. Wesley Hatfield, Ph.D.
Department of Microbiology
   and Molecular Genetics
University of California
College of Medicine
Irvine, California, U.S.A.
Chapter 4

Elena Hidalgo, Ph.D.
Department of Molecular
   and Cellular Toxicology
Harvard University, School of Public Health
Boston, Massachusetts, U.S.A.
Chapter 21

Lilian M. Hsu, Ph.D.
Program in Biochemistry
Mt. Holyoke College
South Hadley, Massachusetts, U.S.A.
Chapter 2

Alistair S. Irvine, B.Sc.
The Krebs Institute
Department of Molecular Biology
   and Biotechnology
University of Sheffield
Sheffield, United Kingdom
Chapter 16

Weihong Jiang, Ph.D.
Department of Biological Sciences
Purdue University
West Lafayette, Indiana, U.S.A.
Chapter 15

Soo-Ki Kim, Ph.D.
Department of Biological Sciences
Purdue University
West Lafayette, Indiana, U.S.A.
Chapter 15

Annie Kolb, Ph.D.
Département de Biologie Moléculaire
Institut Pasteur
Paris, France
Chapter 12

CONTRIBUTORS

Roberto Kolter, Ph.D.
Department of Microbiology
 and Molecular Genetics
Harvard Medical School
Boston, Massachusetts, U.S.A.
Chapter 27

Joseph W. Lengeler, Ph.D.
Fachbereich Biologie/Chemie
Universität Osnabrück
Osnabrück, Germany
Chapter 11

Rongtuan Lin, Ph.D.
Concordia University and
Lady Davis Institute
Montreal, Canada
Chapter 20

John W. Little, Ph.D.
Department of Biochemistry and Department
 of Molecular and Cellular Biology
University of Arizona
Tucson, Arizona, U.S.A.
Chapter 22

Boris Magasanik, Ph.D.
Department of Biology
Massachusetts Institute of Technology
Cambridge, Massachusetts, U.S.A
Chapter 13

Dominique Missiakas, Ph.D.
Département de Biochimie Médicale
Centre Médical Universitaire
Genève, Switzerland
Chapter 23

Howard A. Nash, M.D., Ph.D.
Section on Molecular Genetics
Laboratory of Molecular Biology
National Institute of Mental Health
Bethesda, Maryland, U.S.A.
Chapter 8

Elaine B. Newman, Ph.D.
Concordia University
Montreal, Canada
Chapter 20

Ralf Peist
Department of Biology
University of Konstanz
Konstanz, Germany
Chapter 10

Satish Raina, Ph.D.
Département de Biochimie Médicale
Centre Médical Universitaire
Genève, Switzerland
Chapter 23

Jeffrey W. Roberts, Ph.D.
Section of Biochemistry,
 Molecular and Cell Biology
Cornell University
Ithaca, New York, U.S.A.
Chapter 3

Wilma Ross, Ph.D.
Department of Bacteriology
University of Wisconsin
Madison, Wisconsin, U.S.A.
Chapter 25

Lawrence I. Rothfield, M.D.
Department of Microbiology
University of Connecticut Health Center
Farmington, Connecticut, U.S.A.
Chapter 26

Thomas J. Silhavy, Ph.D.
Department of Molecular Biology
Princeton University
Princeton, New Jersey, U.S.A.
Chapter 19

CONTRIBUTORS

Robert W. Simons, Ph.D.
Department of Microbiology
  and Molecular Genetics
University of California
Los Angeles, California, U.S.A.
Chapter 5

James M. Slauch, Ph.D.
Department of Microbiology
University of Illinois
Urbana, Illinois, U.S.A.
Chapter 19

Stephen Spiro, Ph.D.
School of Biological Sciences
University of East Anglia
Norwich, United Kingdom
Chapter 16

Mathias Springer, Ph.D.
Institut de Biologie Physico-Chimique
Paris, France
Chapter 6

Valley Stewart, Ph.D.
Section of Microbiology
Section of Genetics and Development
Cornell University
Ithaca, New York, U.S.A.
Chapter 17

Annamaria Torriani-Gorini, Ph.D.
Massachusetts Institute of Technology
Cambridge, Massachusetts, U.S.A.
Chapter 14

María Uría-Nickelsen, Ph.D.
Department of Microbiology
  and Molecular Genetics
Harvard Medical School
Boston, Massachusetts, U.S.A.
Chapter 27

James C. Wang, Ph.D.
Department of Molecular
  and Cellular Biology
Harvard University
Cambridge, Massachusetts, U.S.A.
Chapter 7

Barry L. Wanner, Ph.D.
Department of Biological Sciences
Purdue University
West Lafayette, Indiana, U.S.A.
Chapter 15

Sayaka Yamagata, B.S.
Department of Biological Sciences
Purdue University
West Lafayette, Indiana, U.S.A.
Chapter 15

Eva Zdych
Department of Biology
University of Konstanz
Konstanz, Germany
Chapter 10

Brian N. Zeiler, B.S.
Department of Microbiology
  and Molecular Genetics
University of California
Los Angeles, California, U.S.A.
Chapter 5

The aim of this monograph is to give an up-to-date presentation of our understanding of key regulatory mechanisms governing gene expression in *Escherichia coli*. Although regulation at the level of transcription probably occurs most extensively, the important roles played by post-transcriptional, translational, and post-translational controls have become increasingly clear. The coverage is not intended to be comprehensive. Instead, we have selected several well-defined systems to serve as paradigms, as well as a number of complex systems that are in the early phase of being investigated. To give the book a personal flavor, we encouraged authors to present the topics with their special perspectives, instead of only emphasizing the latest findings. In addition, authors were urged to highlight some of the elegant experimental strategies employed in their field, as well as their individual research styles. We wish to convince the readers that *E coli*, from which so much fundamental knowledge of molecular biology has been harvested, remains a highly rewarding organism for further study because of the large areas of its biology still to be explored despite the bank of data that have already accumulated, and because of the relatively simple and yet powerful investigative tools made available by so many years of skillful efforts. The frontiers of knowledge need endless updating, but it is hoped that parts of this book will have some heuristic value and therefore will be more lasting.

Although the analysis of genetic expression was initially focused on the *lac* operon under inducer/repressor control, the scope of studies expanded swiftly to cover many other operons. As a reward, a rich variety of mechanisms governing gene expression was discovered, including transcriptional activation, attenuation, and feedback control of translation, none of which were anticipated by the first pioneers. Soon it became evident that even in the expression of a single operon, like *lac*, multiple regulatory mechanisms are involved. For instance, the transport of glucose causes exclusion of the inducer of the *lac* operon, and the metabolism of glucose results in the lowering of cellular concentration of cAMP, thereby curtailing transcriptional activation of *lac* by the global regulator protein, CAP.

It was also learned that genes concerned with a particular pathway are sometimes contained in more than one operon which are often not genetically linked. The study of arginine biosynthesis uncovered several *arg* operons dispersed around the chromosome, even though their expression is under the control of a common repressor protein. The term regulon is therefore used to designate such a set of operons (whether linked or not). The advantage of placing functionally connected genes in different operons is that each promoter can have its own dissociation constant for the specific regulator, maximal rate of transcription, and perhaps additional sites for modulation of expression by global regulators. In this way, each promoter can be specially tailored for expression under different combinations of environmental conditions. As increasing numbers of global regulators were identified, the term modulon was proposed to designate a superset of target operons under the control of a broadly acting regulator. In contrast to regulons, modulons comprise target operons that do not all share the same specific regulator. Hence not all members of a modulon need to belong to the same regulon. Also, the global regulator of a modulon can in some cases prevail over the action of a specific regulator. For such integrative control of gene expression to be possible, complex promoters have evolved which include binding sites for multiple regulators. For example, the expression of the *sodA* operon is regulatable by seven global regulators.

The integrative control networks that have become apparent through the uncovering of successive layers of regulatory systems lead us to consider gene regulation as a whole in terms of molecular economics. Also, as complete bacterial genomic DNA sequences become available, physiological studies of diverse bacterial species can be expected to expand rapidly, at least in part driven by the need to understand pathogenesis and to discover new antibiotic targets. Perhaps the book will serve to stimulate some thought on the challenging tasks ahead and to guide the development of suitable experimental strategies. In the meantime, we hope that the book will be a useful reference tool to teachers and advanced students in the prokaryotic field.

E. C. C. Lin and A. Simon Lynch

ACKNOWLEDGMENTS

We thank Sankar Adhya and Jon Beckwith for advice and encouragement, Shiro Iuchi for reading certain chapters, and Alison Ulrich for countless patient hours of assistance in the editing process and correspondence. Support by the Institute of General Medical Science of NIH is also gratefully acknowledged.

RG Landes would like to thank Joe Pogliano and Kit Pogliano for the generous use of their photographs for this book's cover illustrations.

Front cover:

A progression through the *Escherichia coli* cell cycle showing the localization of the FtsZ protein (green) and the nucleoids (red), (left half of panel), or the nucleoids alone (right half of panel). FtsZ assembles from a cytoplasmic pool into a contractile ring which remains at the leading edge of the invaginating septum during cytokinesis. FtsZ was visualized by immunofluorescence microscopy using FtsZ-specific antibodies and fluorescein-conjugated secondary antibodies and the nucleoids were visualized by propidium iodide staining. Courtesy of Joe Pogliano and Kit Pogliano, Harvard University.

Back cover:

A micrograph showing a field of exponentially growing *Escherichia coli* cells after immunolocalization of the cell division protein FtsZ (green) and the chromosomes (red). Courtesy of Kit Pogliano, Harvard University.

CHAPTER 1

INTRODUCTION: FROM PHYSIOLOGY TO DNA AND BACK

Jon Beckwith

I t all began with physiology. A century ago we learned that microbes can adapt to their environment by expressing new enzymatic activities. One of the earliest findings was in yeast, where the enzymes for galactose metabolism were found to appear only when cells are exposed to lactose or galactose as carbon and energy sources. Similar findings were made subsequently with bacteria. However, in a sense, the focus on regulatory mechanisms as a major scientific problem began only in the 1930s when Karstrom coined the term "enzyme adaptation." Giving a name to the phenomenon focused greater attention on it and sparked debate about the nature of the process. In 1944, Jacques Monod, working by night with the French Resistance and by day at the Institut Pasteur in Paris, initiated his studies on "adaptation" to lactose by *E. coli*. Over the next decade, Monod and his co-workers described fundamental properties of this system that were ultimately to have a profound generative effect on the evolution of studies on gene regulation. First, he described diauxie, a phenomenon in which growth on glucose as carbon source prevents lactose from inducing the appearance of β-galactosidase activity (which is ordinarily induced by lactose). Second, with his co-worker Alice Audureau, he obtained constitutive mutants that no longer required the presence of lactose for the appearance of β-galactosidase activity. Finally, using newly developed tools that allowed radioactive labeling of proteins, he and his colleagues demonstrated that β-galactosidase was synthesized de novo after the addition of lactose. This finding put an end to proposals that adaptation involved activation of a pre-existing protein.

The 1950s saw the flowering of physiological studies and the beginning of the genetic approaches that were to provide persuasive evidence for a mechanism of genetic regulation. In addition to the inducible systems such as lactose, several laboratories discovered end-product repression, in which a small molecule (e.g., an amino acid or a purine or pyrimidine) when added to the growth medium would shut off the expression of the enzymes responsible for the biosynthesis of that molecule. Detailed physiological studies on the properties of *E. coli*

Regulation of Gene Expression in Escherichia coli, edited by E. C. C. Lin and A. Simon Lynch. © 1996 R.G. Landes Company.

cells growing at different rates led to the finding that ribosome synthesis was also a regulated process.

The proposal of the repressor model for control of the lactose operon and the elaboration of a genetic approach to analyzing regulatory systems offered a paradigm for studies in genetic control. The paradigm was taken up with enthusiasm, as the 1960s witnessed the elaboration of numerous metabolic regulatory pathways as well as the detailed analysis of genetic regulatory steps involved in bacteriophage development and lysogeny. Even though attempts to fit all of these phenomena into the paradigmatic repressor model began to fall apart, the paradigm had proved enormously valuable in the development of a field which flourishes to this day and which is central to the study of a plethora of biological issues from development to cancer.[1] By 1970, numerous novel regulatory mechanisms had been discovered ranging from the establishment by Ellis Englesberg and co-workers of positive control by the activator of the genes determining arabinose catabolism, to regulation by antitermination of transcription and to repression of translation initiation at the RNA level.

While physiological and genetic studies continued to contribute novel regulatory phenomena and mechanisms in the 1970s, the advent of the new technologies for manipulating DNA focused attention on the molecular details of regulatory protein-DNA interactions. On the one hand, new regulatory problems were presented; these included heat-shock control, the SOS system responding to DNA damage, regulatory responses to turgor pressure or osmotic pressure, responses to different ions in the media, differential gene expression under aerobic vs. anaerobic conditions, stationary phase gene expression, challenges by alterations of the redox environment of cells, and others. On the other hand, regulatory proteins were purified and their interactions with DNA studied in detail. These latter studies have revealed unanticipated aspects of gene regulation and DNA structure, including DNA bending and looping. They have also been an important component of studies of protein structure and function. The crystal structures of DNA binding proteins, often in complex with their specific DNA binding site have contributed to a detailed understanding of how these proteins interact with DNA and with the bacterial transcription apparatus.

These regulatory studies have led to the unexpected finding that a specific regulatory protein can function in a variety of different ways. For instance, many proteins originally identified as repressors, can also act as activators and vice versa. Furthermore, activators can promote transcription by a number of different mechanisms. For instance, Busby and Kolb, in their chapter on the cyclic-AMP receptor protein (CRP), describe several modes by which this regulatory molecule activates genes encoding products involved in the catabolism of various sugars. According to them, "The history of CAP is a story of increasing complexity." In the *lac* operon, CAP is thought to interact directly with RNA polymerase to promote high levels of transcription. In the *mel* system, CAP is required for the expression of the positive regulatory gene, *melR*, and does not act directly on the *mel* operon itself. In one of the *mal* operons, CAP acts to reposition the positive activator of the system, MalT, thus allowing effective activation of RNA polymerase. In the *pap* regulon, CAP may activate gene expression simply by displacing a negative regulator. Finally, CAP can repress the transcription of certain genes. As Adhya puts it in his chapter on the *lac* and *gal* operons, "Even studying the regulatory mechanisms in the classical operons three decades later continues to provide new and fascinating insights about both positive and negative controls."

Throughout this period, the models and mechanisms of gene regulation worked out in bacteria were used as a starting point for analogous studies in eukaryotic systems. However, in recent years, there has been significant cross-fertilization in the opposite direction. Perhaps the best example of this direction of information transfer is the field of protein phosphorylation and its importance in gene regulation. It has become clear that kinases acting on histidine, serine/threonine, aspartate and tyrosine residues are central players in many developmental pathways, in the eukaryotic cell cycle and in pathways leading to tumor formation. It is only recently that analogous activities have been found in bacteria. Protein phosphorylation is central to the functioning of the two-component signal transduction pathways of gene regulation described in several chapters of this volume. More recently, the existence of a serine/threonine kinase important in regulation has been described for *Myxococcus xanthus*.[2] It seems likely that such proteins will be found in other bacteria and will participate in regulatory pathways.

Historically, as we have seen, the field of gene regulation began with the study of the effect of small molecules on gene expression. It has ended up largely today as a study of protein-DNA and protein-protein interactions at the molecular level. This volume not only gives a sense of the history and accomplishments of the field of bacterial gene regulation, but also points towards the problems of the future. Clearly, there is a lot to be learned by probing ever more deeply into the ways in which proteins interact with DNA, the ability of proteins to alter DNA structure and the interactions between regulatory proteins and RNA polymerase when bound to DNA. Similar issues exist for the association of proteins with RNA for effects on translation, degradation and transcription termination. In addition, it seems likely that there will be more broad regulatory mechanisms discovered. The interest in bacterial response to stresses such as heat-shock or oxidative damage is a relatively recent phenomenon. Even more recently, attention has focused on gene expression during starvation or the stationary phase of the bacterial life cycle (see chapter by Kolter). In all these cases major regulatory phenomena affecting large number of genes had been ignored until people happened to ask the right questions. Whether deeper analyses of such processes as bacterial cell division and the cell cycle will reveal additional novel regulatory mechanisms is not clear. To significantly probe these questions may require novel or improved cell biological techniques for studying the growth of the bacterial cell and for examining the formation of structures such as the cell septum.

But, at the end, ironically, a major body of gaps in our understanding of regulation relates to questions of physiology. We started out looking at gene regulation because we knew that lactose would induce the synthesis of the proteins required for its utilization, and because we knew that tryptophan added to the growth medium would repress the synthesis of the enzymes involved in its biosynthesis. The focus in the first couple of decades of research in gene regulation was largely on such systems where one knew the physiological regulatory signal (or a close relative of it). The chapters in this volume on the *lac*, *gal*, *mal*, *glp*, *pho*, and *leu* systems cover pathways where studies were initiated in that era. For the most part, these operons or regulons for specific biosynthetic or catabolic pathways were typical of the ones under study. Exceptions included the regulation of ribosomal synthesis with growth rate, the phenomenon of catabolite repression and, somewhat later, the SOS regulatory system, responding to DNA damage.

Now we face a host of interesting regulatory phenomena where the physiological effectors are not at all obvious. As John Little puts it in his chapter on SOS, "Still the least understood aspect...are the...inducing signals." Examples of systems where the effector may well be "the least understood" feature of the regulation include the regulatory mechanisms for controlling aerobic vs. anaerobically expressed genes, regulation in response to changes in osmotic pressure, gene expression compensating for oxidative challenges, the heat-shock system, and still today the growth rate regulation of ribosomal synthesis. The problem may become even more acute as *E. coli* genome sequencing reveals new genes for which the function of the proteins is not apparent from the predicted sequences of the products encoded.

How will we proceed to define the signals that control these regulatory pathways? In many cases, the problem is effectively one of cell physiology. For instance, the sensing of aerobic vs. anaerobic conditions may require components of electron transport pathways or other factors involved in cellular energetics. Response to oxidative challenge may use one of the same components, or, alternatively, may detect an altered metabolite. Experience does not teach us how to pinpoint the regulatory molecules or alterations. One of the shining examples of a signal identified for a global control mechanism is 3',5' cyclic AMP. However, this molecule was not identified through a direct search for an effector. Rather, Sutherland's study of the physiological regulation of cyclic-AMP synthesis had led to the finding that glucose lowers cellular levels of the compound. Sutherland himself was not working on catabolite repression. But this finding was then noticed by those who were working on the subject, and the effect of the metabolite on the regulatory mechanism subsequently tested.

More recently, another class of signals has been identified, the homoserine lactones, that appear to be used by many bacteria for sensing cell density. The first of this class of molecules to be characterized was detected as a regulatory signal in the case of bacterial luminescence exhibited by *Vibrio fischeri*. At high densities of bacteria, this molecule was found in the growth media and its structure determined. Both the gene coding for the regulatory protein involved (*luxR*) and the genes for biosynthesis of the homoserine lactone were sequenced. The existence of these sequences in DNA databanks allowed the identification of homologous genes involved in other regulatory pathways. In recent years, homoserine lactones have been suggested as key signaling molecules regulating conjugal transfer of Ti plasmids of *Agrobacterium tumefaciens*, the synthesis of extracellular proteases involved in the virulence of *Pseudomonas aeruginosa*, expression of extracellular proteases in *Erwinia carotovora* and expression of σ[s], a regulator of genes induced under starvation conditions in *Escherichia coli*.[3,4] In the latter case, the evidence for involvement of a homoserine lactone first came from the sequencing of a regulatory gene whose product shared homology with a known lactonizing enzyme. In this book, Rothfield's chapter describes a tantalizing possible connection between a homoserine lactone and the process of cell division.

Finally, the metabolic intermediate acetyl-phosphate may play a key role in signaling for a number of regulatory pathways. This molecule was first recognized as such in studies on the phosphate modulon (see chapter by Wanner). Suppressor mutations that restored expression of phosphate-regulated genes in certain mutant backgrounds turned out to be in the gene for acetate kinase. Further physiological and genetic studies have since strengthened the notion that acetyl phosphate serves as a signaling molecule.[5]

Nevertheless, a number of other systems have resisted such approaches to identifying regulatory signals. It seems likely that many of these will require a deeper knowledge of bacterial cell physiology than we now have. For some pathways, a more detailed analysis of electron transport, bacterial energetics and redox factors in different cellular compartment will be important. In other cases, a greater understanding of membrane structure, membrane function and its responses to environmental stresses may be necessary to fully understand how the organism regulates gene expression. Understanding the basis of signaling in the SOS and heat-shock systems may require fuller characterization of DNA and protein metabolism, respectively.

Finally, understanding many of these regulatory phenomena will require an integration of existing knowledge and knowledge to be acquired about metabolic and other physiological pathways. Many of the newer regulatory systems described here are integrated into the larger framework of cellular structure, membrane function, metabolism and energetics. The broad field of gene regulation is truly becoming the cell biology of bacteria.

ACKNOWLEDGMENTS

Jon Beckwith is an American Cancer Society Research Professor. His research is supported by grants from the National Institutes of Health, the National Science Foundation and the American Cancer Society.

REFERENCES

1. Beckwith J. Genetics at the Institut Pasteur: substance and style. ASM News 1987; 53:551-555.
2. Muñoz-Dorado J, Inouye S, Inouye M. A gene encoding a protein serine/threonine kinase is required for normal development of *M. xanthus*, a gran-negative bacterium. Cell 1991; 67:995-1006.
3. Fuqua WC, Winans SC, Greenberg EP. Quorum sensing in bacteria: the LuxR-LuxI family of cell density-responsive transcriptional regulators. J Bacteriol 1994; 176:269-275.
4. Huisman GW, Kolter R. Sensing starvation: a homoserine lactone-dependent signaling pathway in *Escherichia coli*. Science 1994; 265:537-539.
5. McCleary WR, Stock JB, Ninfa AJ. Is acetyl phosphate a global signal in *Escherichia coli*? J Bacteriol 1993; 175:2793-2798.

RNA CHAIN INITIATION AND PROMOTER ESCAPE BY RNA POLYMERASE

Michael J. Chamberlin and Lilian M. Hsu

PROMOTER FUNCTION IS REGULATED AT TWO DISTINCT PHASES OF TRANSCRIPTION: PROMOTER BINDING AND RNA CHAIN INITIATION

The efficiency of promoter function is determined by two distinct phases of transcription, promoter binding and activation, and RNA chain initiation and promoter escape. The former process has been extensively studied in both prokaryotic and eukaryotic organisms, and involves interaction of the RNA polymerase with general transcription factors, promoter specific factors, and with the DNA sequences in the recognition region of the promoter (for reviews see refs. 1-4). The latter process has generally been overlooked, and is often subtended into the promoter binding phase by use of the general term "initiation" to include both promoter binding and RNA chain initiation. However, it has been recognized for over 25 years that the two phases are distinct. Hence, true RNA chain initiation has, in some ways, become the lost step in transcription. The promoter binding and activation process can normally be completed in the absence of the nucleoside triphosphate substrates.[5] Similarly, the RNA chain initiation and promoter escape phase is a biochemically distinct process, and is controlled and regulated by DNA sequences and factors that have no similar role in promoter binding. These sequences and factors can affect promoter strength considerably,[6,7] and can lead to regulation that is independent of the promoter binding step.[8-11] Hence, RNA chain initiation is a central and regulated step in the transcription reaction.

This review will discuss the RNA chain initiation and promoter escape reaction primarily as it has been studied with the *E. coli* RNA polymerase. However, it is quite clear that the general features of this reaction are shared by all of the known studied RNA polymerases that carry out de novo initiation of RNA chains.

Regulation of Gene Expression in Escherichia coli, edited by E. C. C. Lin and A. Simon Lynch. © 1996 R.G. Landes Company.

THE BIOCHEMISTRY OF THE RNA CHAIN INITIATION PHASE OF TRANSCRIPTION

The RNA chain initiation phase of transcription involves the stepwise synthesis of short oligoribonucleotides. The majority of these nascent transcripts appear to be released from the RNA polymerase-promoter complex in a process termed abortive initiation.[12-15] The abortive products normally range from 2-10 nt in length, although released products of up to 16 nt are synthesized from some promoters as part of the initiation process. During this process, the RNA polymerase holoenzyme can remain bound to the promoter,[13,16] although this depends entirely on the equilibrium between closed- and open-promoter complexes, as well as the actual binding constant of polymerase to the particular promoter in question.[17,18] The completion of the initiation phase for *E. coli* RNA polymerase is signaled by three distinct biochemical changes in the complex: the formation of a stable ternary complex in which the RNA is now tightly bound, the release of the sigma specificity subunit, and the initial translocation of the RNA polymerase away from the promoter. It is not known whether these reactions are concerted or sequential, and if sequential, what the order might be.

The initial translocation of the RNA polymerase away from the promoter is often referred to as the promoter clearance reaction. However, it is becoming quite evident that this is not a correct view, and we shall use the term promoter escape in its place. While the initial translocation does expose most of the upstream sequences needed for promoter recognition, the enzyme, just after the initial translocation away from the promoter, remains bound to initial transcribed sequences (ITS) out to about +20, that are also contacted in the open promoter complex. It would be expected then, that if the enzyme should pause in this region, it would still block use of the promoter by a subsequent enzyme. In fact, this kind of situation has been shown to occur with a mutant form of the lambda P_R' promoter, when the RNA polymerase pauses at about position +16.[19] Under these conditions, promoter clearance actually involves movement of the enzyme away from this pause site, which takes place during the chain elongation phase of transcription. A similar block to promoter clearance in the region of +6 to +12 nt has also been described for synthetic promoters that bind RNA polymerase rapidly, but have low rates of RNA synthesis, due to limitations in the clearance reaction.[20] Hence, the use of the term promoter clearance involves steps subsequent to those needed to complete the RNA chain initiation reaction and should be avoided in discussing initiation itself. Promoter escape will be used to refer to the initial translocation, since it focuses only on steps in initiation.

The combined effects of the release of the sigma factor, together with the translocation away from the promoter, precludes "abortive elongation", in which elongating transcripts would be synthesized, released, and resynthesized by the elongating core RNA polymerase. There is no evidence for such a reaction with duplex DNA templates. During elongation, release of the transcript leads instead to termination with the requirement for release of the RNA polymerase from the DNA template, and reentry into the sigma binding and promoter binding phase of transcription. It seems likely that this is a general feature of transcription by RNA polymerases and is not limited to the bacterial paradigm.

Early structural studies of the intermediates in the transcription initiation process were handicapped by the general instability of these initial transcribing complexes (ITC), which precluded their direct biochemical study.[21-23] However, at a particular *tac* promoter, ITC complexes are formed that are much more stable, and this has led to the isolation of such complexes and their biochemical characterization.[16] Footprinting of such ITC complexes with DNase I showed that the footprint is almost indistinguishable from that of the open promoter complex with the exception of a slight translocation downstream by about four base pairs, and the appearance of a hypersensitive site at -22. This footprint appears to be independent of nascent RNA chain length for the ITC complexes studied bearing chains from 5-8 nt.[16] All of the stable ITC complexes studied retained the sigma subunit, and hence involved complexes with the RNA polymerase holoenzyme.

In contrast, complexes at this same *tac* promoter bearing chains of 11 nt had lost the sigma subunit, and had a DNase I footprint distinctly different from that of the open promoter complex that had moved away from the upstream promoter sequences. This result is consistent with other studies that had shown sigma release to occur with most promoters after synthesis of RNA chains of about 8-10 nt,[21-24] and that had footprinted complexes in which the RNA polymerase had moved away from the promoter.[22,23]

It is clear in retrospect that the early report by Shultz and Zillig[25] of a stable complex formed with a poly d(A-T) template, using the trinucleotide ApUpA, was probably the first example of formation of a stable ITC complex. Since that time, additional examples of relatively stable ITC complexes have been reported.[26-28] However, early reports describing the purported physical properties of similar complexes using promoters that cannot form stable ITC complexes are clearly in error and should be disregarded, since there was no direct demonstration that ITC complexes were formed and characterized.[29]

Mechanistically, it has been suggested that abortive initiation is a necessary stage of the RNA polymerase initiation reaction; it leads to the formation of a stably "primed" enzyme-template complex (i.e., an elongation ternary complex) and appears to be uniquely required of RNA polymerases that perform de novo synthesis.[22] In keeping with this view, every RNA polymerase examined thus far that can initiate chains de novo has been found to undergo abortive initiation; these include RNA polymerases from phages T3, T7, and SP6,[17,30-32] various eukaryotic RNA polymerases,[33-36] and prokaryotic RNA polymerases.[14,18,21,37,38] Despite this ubiquity, it is still not clear what features of the initiation process require that short transcripts be released, and it is also unclear what the relationship is between promoter structure, promoter strength, and the yield of abortive transcripts as compared to productive transcripts.[14,38] We will discuss this matter further in the sections that follow.

A final reaction that can occur during the initiation phase has been termed slippage or reiterative transcription. In this reaction, a sequence of at least three identical nucleotides within the initiation region of the template generates runs of complementary nucleotides in the RNA transcript that can be much longer than three.[39-42] A similar reaction involving synthesis of long poly (rA) sequences from much shorter, single-stranded runs of oligo (dT) residues in DNA templates was discovered very early in the study of *E. coli* RNA polymerase.[43-45] Such template-directed slippage during the initiation reaction often leads to regularly spaced ladders of products on PAGE gels, that are generated by abortive release of a fraction of the chains after each nucleotide is added. Thus, although very long chains are formed, they retain the characteristic of ITC complexes in abortive releasing, and almost certainly also remain at the promoter with sigma subunit in the complex, although this is difficult to show directly.

The sequence of 3+ nucleotides that generates the slippage product may be at the 5' terminus of the template coding sequence,[39,40] or may initiate at positions out to about +3; slipping from sequences farther out in the initiation region has not yet been observed. It is interesting that, in a collection of 112 ITS regions of promoters, almost half have three or more identical nucleotides in the first nine bases of the sequence.[46,47] The slippage reaction during initiation appears to be exploited as part of a regulatory process in the control of transcription from the *pyrB* promoter[9] (also see below).

A reaction that may represent an early part of the slippage process is termed "primer shifting", and has been shown to occur with the *rrnB* P$_L$ promoter[27] and with the lambda phage P$_L$ promoter.[28] In the former case, the sequence of the nontranscribed strand at the start site of transcription is CACCACTG in which the underlined A is the normal start site reading from left to right. When such a promoter is incubated with ATP and CTP it forms an ITC complex, bearing a chain with the sequence pppACCAC which is relatively stable. It appears that the formation of this transcript involves first synthesis of pppAC at the normal start site, followed by partial release and transfer to base pairing with the earlier AC sequence at positions -3 and -2, followed by resynthesis of the start site sequence beginning at nucleotide -1, to give the stated product. A similar reaction is seen with a T7 A1 promoter bearing an altered ITS sequence[48] in which the initial transcribed nascent RNA, AUCC or AUCCC, can slip back to pair with two G residues at positions -2 and -1 on the template strand, leading to synthesis of an RNA chain with the 5'-terminal sequence AUCCAUCCC or AUCCCAUCCC. Since such shifting almost certainly must involve movement of the transcript backward along the DNA template, it is interesting that the cleavage factor GreA is able to stimulate primer shifting with this T7 A1 promoter;[48] this will be discussed below.

Goldfarb and his colleagues[27,28] suggest that the stability of primer shifted transcripts is due to positioning of the 5 nt product in a tight RNA binding site not used by other ITC complexes (this would correspond to RNA binding site II; see below). However, the stability of this product is quite comparable to the stable 5 and 6 nt ITC complexes formed on the variant *tac* promoter[16] and there is no reason to assume that these *rrnB* P1 complexes require special features in the RNA binding of ITC complexes.

An important element of both the slippage and primer shifting reactions is that they lead to a change in the register of the transcript; that is, the correspondence between the nucleotide sequence of the DNA template and the RNA product is altered. The slippage products are often called "noncognate" or "pseudotemplated" bases. These appellations are misnomers, since the reactions leading to their formation require an exact sequence of complementary template bases in the DNA template strand, and such bases must be incorporated by templating using complementary Watson-Crick base pairs.

It has been shown that slippage can occur during normal elongation by both *E. coli* and T7 RNA polymerases, if more than 10 identical residues are present, leading to incorporation of additional RNA nucleotides in the transcript.[49,50] Interestingly, in the case of T7 RNA polymerase, products are found in which RNA nucleotides have been deleted as well, leading to RNAs shorter than the sequence expected from the template. Another example of a register change has recently been discovered by Zhou and Doetsch,[51] who have shown that T7 RNA polymerase can elongate past a substantial gap in the DNA template strand (up to 24 nt), giving an RNA in which the gap in the template strand is deleted from the RNA sequence in the product. Such register changes pose important problems for models of elongation and initiation, as we discuss below.

PARAMETERS THAT DESCRIBE THE RNA CHAIN INITIATION REACTION AT DIFFERENT PROMOTERS

Some of the earliest studies of the RNA chain initiation reaction revealed that there are quantitative and qualitative differences between different promoters inherent in this reaction.[6,7,13,38,52] We suggest that there are at least four distinct quantitative parameters that can be used to describe these differences between promoters and to characterize the effects of other factors that affect the reaction. One of the most important differences lies in the rate with which the RNA polymerase escapes from the promoter once an open promoter complex has been formed (which we will call simply *the rate of promoter escape*), since this parameter directly affects the *strength of the promoter*, measured in the steady state rate of transcripts produced per second.[6] Note that this definition of promoter strength is not universal, and other definitions have been used. However, given the fact that the rate of promoter function is the product of the rates of all of the different processes up to and including escape, ours would seem to be the most general definition.

When the percentage yield of abortive products (which we will call *abortive yield*) is compared to total initiated products (total initiated products = abortive products + productive products), it is clear that this ratio varies for different promoters as well.[7,15,52] It is sometimes useful to express abortive yield as the number of abortive transcripts produced for each productive transcript, since this gives meaningful numbers with which different promoters can be compared. However, this number can become infinite, hence meaningless, at certain promoters, and with certain RNA polymerase mutants, where there is little or no formation of productive transcripts. The important question of whether there is a simple relationship between the rate of promoter escape and abortive yield will be discussed below.

Finally, the molar yield of each size of abortive transcript (which we will call *abortive pattern*) also varies for different promoters, as does the maximum size of aborted transcripts.[15,38] The abortive pattern can be used to calculate the *probability of dissociation* of the RNA chain for each nucleotide added to the nascent ITC complex.[15] In this calculation, the amount of abortive product at each step reduces the mole fraction of RNA polymerase that proceeds on to addition of the next nucleotide. Hence, the probability of dissociation at a position well downstream, say at 9 nt, will be much higher than at a position near the promoter, say at 3 nt, which gives an equal molar concentration of aborted product.

The variation of the maximum size of aborted transcript is most likely due to differences in the length of RNA that must be synthesized before the transition of the complex into the elongation mode. Earlier estimates suggested that this transition, together with sigma release, occurred at about 9-10 nt.[21,23,24] It has been shown biochemically that for the modified *tac* promoter studied by Krummel and Chamberlin,[16] the transition into elongation with release of sigma subunit is complete by 11 nt.

However, for the T5 N25$_{antiDSR}$ promoter, and also with some other promoters abortive transcripts up to 16 nt are formed (Nam Vo, Hsu and Chamberlin, unpublished).[15] For these promoters it is most plausible to assume that sigma is not released until about 16 nt. Earlier evidence from crosslinking studies had suggested that the position of sigma release might vary, out to about 15 nt.[53] The structural features of a promoter that lead to these differences in the point of sigma release are not yet understood.

There are a number of different experimental systems in which these four quantitative parameters have been studied. Some of the earliest studies of the rate of promoter escape employed $[\gamma\text{-}^{32}P]$-ATP or $[\gamma\text{-}^{32}P]$-GTP to study the rate of productive initiation at the T7 A1 and A2 promoters, respectively, using a single cycle transcription reaction.[54,55] By studying the rate with which end labeled RNA chains became acid insoluble and resistant to dilution by excess unlabeled nucleotide, one should obtain a measure of the rate of promoter escape. Nierman and Chamberlin[54,55] found that the rate of escape was inversely proportional to the concentration of the initial nucleotide in the reaction. For comparison, at 100 µM NTP concentrations, promoter escape occurred with a half time of 6 seconds, while at 10 µM NTP, escape required 60 seconds. At the time of these studies it was not known whether there were stable intermediates in the initiation reaction, nor what the abortive yield might be. However, it is now clear that, at least for T7 A1, fewer than 8% of the initiated chains become productive chains, no stable ITC complexes are formed, and no strong pauses occur in the first 20 nt of the initial transcribed sequence. Hence these procedures should give valid measurements of the rate of promoter escape.

A similar method has been described by Gralla and his collaborators[38,56] that is primarily suited to promoters that show a slow rate of escape; in the case of the *lac* promoter about 60 seconds are required. In this method shorter DNA templates can be used, and the reaction is limited to a single round by addition of heparin. The authors actually used rifampicin to block further productive initiation. For rapidly binding and initiating promoters, it is normally difficult to limit the reaction to a single round of transcription with these methods.

Both of these methods require that there be no highly stable ITC complexes, or paused elongation complexes near the promoter, since these complexes are resistant to rifampicin and to dilution of end-labeled nucleotides by unlabeled nucleotides. Hence, kinetics experiments are needed to address these issues for any promoter under study. It would be quite desirable to develop a better assay for the rate of escape that is applicable to short DNA templates and works for either rapidly clearing or slowly clearing promoters.

In principle, determination of the other parameters for the initiation process can be carried out rather simply by determining the molar yields and sizes of each of the abortive products and comparing these quantitatively to the molar yield of productive products. However, there are numerous technical problems to consider. In most cases it is difficult to limit transcription to a single round with rapidly initiating promoters, and single round transcription reactions will also give very low concentrations of terminally labeled products. When $[\alpha\text{-}^{32}P]$-labeled nucleotides are used one can get somewhat more labeling, but there are other complications that set in with such experiments. First, the radioactivity in each transcript must be corrected for the number of residues of that base in the transcript. Second, there will be interference in such an assay by RNA cleavage products produced during the initiation reaction;[15,48,57] these products are not true abortive transcripts. Finally, depending on the exact $[\alpha\text{-}^{32}P]$-labeled nucleotide chosen, some of the abortive products may not be labeled, and hence will not be seen by autoradiography.

These experimental problems are especially true when di- or tri-nucleotide primers are used to follow abortive initiation. Because high concentrations of these oligomers are needed for effective initiation, it is difficult to label the oligomer itself to a sufficiently high specific activity. Hence, all experiments must be done with [α-^{32}P]-labeled nucleotides. It should also be specifically noted that the abortive yield and the abortive pattern both appear to change when dinucleotides are used as initiators as compared with nucleoside triphosphates.[15,58]

One solution to these problems is the use of a steady state reaction, in which the RNA polymerase is permitted to initiate multiple rounds of transcription.[15,59,62] Under these conditions, after an appropriate amount of time has passed, kinetic intermediates disappear as significant products, and the only products are productive transcripts and abortive transcripts. If a template with a short productive transcription unit is used (30-60 nt), for example from a PCR-produced template, all of these products may be visualized on a single denaturing PAGE system. Finally, if the chains are terminally labeled with a [γ-^{32}P]-nucleotide, all of the abortive products will be labeled, and the molar concentration of each transcript will be directly proportional to the radioactivity of each band. This assay is probably the simplest way to measure the three parameters in question: abortive yield, abortive pattern and the maximum length of abortive transcript. Furthermore, because of the multiple rounds of initiation that occur, it is usually not difficult to get highly labeled products with relatively low specific activity [γ-^{32}P]-NTP.

FACTORS THAT AFFECT THE INITIATION REACTION: INTRINSIC FACTORS

It is convenient to characterize the factors that affect different aspects of the initiation reaction as intrinsic factors and extrinsic factors. Intrinsic factors would include the sequence of the DNA template, as well as the nature of the RNA polymerase involved including RNA polymerase mutants, together with promoter-specific activators or repressors. Extrinsic factors would include solution conditions such as the concentration of nucleoside triphosphates, the concentration and nature of salts and metals, the temperature and so forth.

Of the intrinsic factors affecting RNA chain initiation, the most studied has been the effect of promoter sequences themselves. However, there have still been relatively few studies in which the effects of varying different promoter regions have been systematically examined. It is clear that different promoters can vary greatly in the rate of promoter escape.[6,14,38,54,55] These differences can be due to alterations of the ITS region, or of the nontranscribed promoter recognition region (PRR) that primarily involves the consensus -10 and -35 conserved regions.[7,15,16,57]

Alteration of ITS sequences and PRR sequences can also change the abortive yield and the abortive pattern quite dramatically, as well as the maximum size of aborted transcripts.[7,15,16,57] However, only recently have systematic studies of the effects of sequence changes on the different initiation parameters started.

The role of the abortive yield in determining the rate of promoter escape was first raised by Gralla and his co-workers,[13,38,56] who showed convincingly that the strength of the *lac* promoter was limited by the rate of the escape reaction. Because of the large yield of abortive products, their attention was drawn to the possibility that there was a causal relationship. In comparison to T7 and lambda phage promoters, the *lac* promoter completes the initiation reaction very slowly, and it was believed at that time that there was little abortive initiation at T7 promoters A1 and A2.[38,54,55] This confusion arose because the studies with T7 DNA were done analyzing for dinucleotides using thin layer chromatography, which did not reveal longer abortive products. These comparisons led to the idea that the rate of promoter escape was inversely related to the abortive yield, that is, promoters from which escape occurs rapidly produce little or no abortive transcripts.[14,38] This idea received further support from the demonstration that the initial transcribed sequence of promoters (ITS), roughly defined as the first 20 bp downstream of the transcription start site, can dramatically affect promoter strength.[6] In fact, one of the very strongest *E. coli* promoters was found to be the N25 promoter from bacteriophage T;5[60,61] its activity could be reduced about 10-fold in vivo by replacing the ITS sequence with a modified sequence, to produce the N25$_{antiDSR}$ promoter.[6,62]

Support for these ideas was weakened somewhat by the finding that the T7A1 promoter, despite its rapid escape rate, actually makes a substantial amount of abortive transcripts for each productive transcript when compared to the *rrnB* P1 and lambda P$_L$ promoters.[58] Furthermore, it was found that the abortive yield and abortive pattern were quite different for the three promoters. However, this study did not analyze these parameters quantitatively, nor measure the effect of elevated NTP concentrations.

The studies of Hsu and Chamberlin[15,62] have resolved several of these questions, and raised significant other questions. Using a steady state assay for labeling of RNA chains with γ-^{32}P-ATP, and transcribing short PCR-amplified templates, they were able to follow the formation of both kinds of products, and quantitatively analyze both. This allowed them to determine the actual abortive yields and abortive patterns, and to express the latter as the probability of abortive escape for each nascent RNA in the initiation reaction. The abortive yields for the T7 A1 promoter, and the T5 N25 promoter were 92% and 98%, respectively; these correspond to synthesis of 11.5 and 49 abortive transcripts for each productive transcript (these are calculated using the definitions given earlier). These values are much higher than one might have expected, considering that these are two of the strongest promoters known for *E. coli* RNA polymerase. Furthermore, since the N25 promoter appears to permit escape more rapidly than T7 A1,[6] this suggests that there is no simple correlation between abortive yield and promoter escape rate, at least for the stronger promoters.

For the T5 N25$_{antiDSR}$ promoter, that is limited in promoter escape by the ITS sequence, an abortive yield of 99.7% is obtained, or about 330 abortive transcripts per productive transcript. This increase of about 7-fold in abortive transcripts per productive transcript is close to the decrease in promoter strength brought about by replacing the ITS sequence, about 10-fold in vitro,[6] this is well within the error of the measurements. However, more data are needed on many different promoters to determine the extent to which these two parameters may be related for weaker promoters.

Several RNA polymerase mutants are known that affect the efficiency of the initiation reaction. Goldfarb and his collaborators have described several kinds of mutations in the *rpoB* gene that block, or reduce, the rate of promoter escape, leading to a large increase in abortive yield.[63-65] Both mutations in the protein and proteolytic cleavage of

T7 RNA polymerase can reduce or eliminate the ability of this enzyme to escape from the promoter.[66] Several groups are now carrying out site-directed mutagenesis of both the *rpoB* and *rpoC* genes, which should lead to additional information about regions that affect the initiation process.

Some promoters require accessory factors for promoter escape to occur. The best studied case of this involves the *malT* promoter, where binding of the CAP protein is required for promoter escape.[8] It is thus clear that CAP can act at several distinct steps in the promoter recognition and RNA chain initiation reactions.

In the case of the lac and gal repressors, it appears that the binding of the repressor to its operator need not block binding of RNA polymerase[67,68] (Sankar Adya, personal comm.). In the former case, repressor binding appears to affect the escape reaction, trapping the RNA polymerase at the promoter in a mode that permits abortive initiation.[68]

FACTORS THAT AFFECT THE INITIATION REACTION: EXTRINSIC FACTORS

Solution factors such as the cation and anion concentrations, the concentration and nature of the divalent cation, and the temperature, can affect several of the steps of the transcription reaction (for reviews see refs. 69-71). Modest changes in the concentrations of KCl and Mg^{++} ion in the reaction, that can affect the efficiency of promoter utilization and transcription from DNA ends, have no profound effects on either abortive yields or abortive patterns with T7 A1, and T5 N25 promoters.[15] However, for the most part there have been few quantitative studies of the effects of these parameters on the RNA chain initiation reaction itself.

One instance in which extrinsic factors play a central role in initiation is the *pyrB* promoter, mentioned earlier, where in the presence of elevated UTP levels, transcript slippage occurs generating tracks of oligoU residues during the reaction. At lower UTP concentrations, normal initiation and promoter escape takes place.[9] Hence, this is a promoter that is regulated in part at the escape stage by the levels of UTP, and where UTP is the final product of the genes of the regulated operon.

Another example of extrinsic factors capable of affecting promoter escape was shown with the T5 N25$_{antiDSR}$ promoter. The addition of moderate amounts (approximately 10-fold molar excess over RNA polymerase concentration) of the transcription cleavage factors GreA and/or GreB to the reaction can stimulate promoter escape and reduce the abortive yield by up to 10-fold.[62] The effect is promoter specific; thus there is a 10-fold effect on the T5 N25$_{antiDSR}$ promoter, but no detectable effects on T7 A1 or T5 N25. Similarly Feng et al[48] have seen a small stimulation with a mutant T7 A1 promoter. The mechanism of this stimulation is not known; however, there is a considerable amount of cleavage of transcripts in the abortive size range in such experiments, whether or not any stimulation is seen.[48,62] A discussion of possible mechanisms is found below.

MODELS FOR THE MECHANISM OF RNA CHAIN INITIATION: SOME SIMPLE MODELS DO NOT ACCOUNT FOR WHAT IS KNOWN

To our knowledge, there have been no formal models proposed for the RNA chain initiation process. Rather, the classical model for RNA chain elongation has often been applied to initiation. In this model[72-74] it is assumed that transcription proceeds through formation of a bubble in the DNA, and that the RNA chain is engaged in a DNA-RNA hybrid that normally spans 12 base pairs. For such a model, one might predict that formation of short RNA oligomers, having melting temperatures below 37°C, would lead to loss of the oligomer from the template and hence give rise to abortive products. This model would predict that: (1) the pattern of abortive products would decrease as the length of the transcript increases, due to the enhanced stabilization of the hybrid; and (2) that the GC content of the transcript would play a major role in determining the abortive probability. Neither of these predictions is met[15,58] and the abortive release of transcripts out to 16 nt fairly rules out the idea that the binding of the transcript to the transcription complex is mediated primarily through base pairing to DNA. There are also serious reservations as to the applicability of this model for the normal elongation process as well (for additional references addressing this point see refs. 75-77).

A second model for the initiation reaction was proposed by McClure,[1] who suggested that the rate of passage through the initiation region was determined by a competition between addition of each nucleotide and the abortive release of each oligomer. This idea was furthered by discussions of a "stressed intermediate" in the initiation reaction, which presumably means that the intermediates were quite unstable, and must either be elongated or dissociate.[23] We will call this the "kinetic competition" model for initiation, since it is similar to a model for elongation and termination by the same name.[78] If this model is correct, then at sufficiently high NTP concentrations, abortive initiation should decrease, and possibly vanish completely.

This model does not explain the results of Hsu and Chamberlin[15] who showed that at very high NTP concentrations, the abortive yield does not necessarily decrease, but can actually increase. In addition, experiments involving varying concentrations of single NTP showed that for most positions in the initiation process, the yield of a particular sized abortive product was not determined by the concentration of the next nucleotide to be added. Obviously, such experiments can be manipulated at very low NTP concentrations; the complete omission of a particular NTP needed in initiation often, but not always,[54] leads to accumulation of the intermediate just prior to addition of that nucleotide.

MODELS FOR THE MECHANISM OF RNA CHAIN INITIATION-MODELS BASED ON RECENT MODELS FOR RNA CHAIN ELONGATION

In the past few years, several novel models have been proposed for the mechanism of RNA chain elongation. These models originated with a proposal by one of us at a Harvey Lecture in 1992 that was circulated widely and ultimately published.[76,79] Several variations on this model have now been described.[48,80-82] The rationale for choosing this model over the older model of Hearst and von Hippel has been presented in detail previously,[76] and will not be reviewed here.

In this model, the elongation complex of RNA polymerase is a dynamic structure that changes continuously during elongation (Fig. 2.1). There are two major DNA binding sites, a leading site, designated I, and a trailing site, designated II. These move discontinuously along the DNA, in an inchworm-like movement, with one of the sites always being tightly bound, while the other site can slide along the DNA. By alternating which site is fixed and which can slide, the enzyme can move along DNA with no possibility of accidental termination;

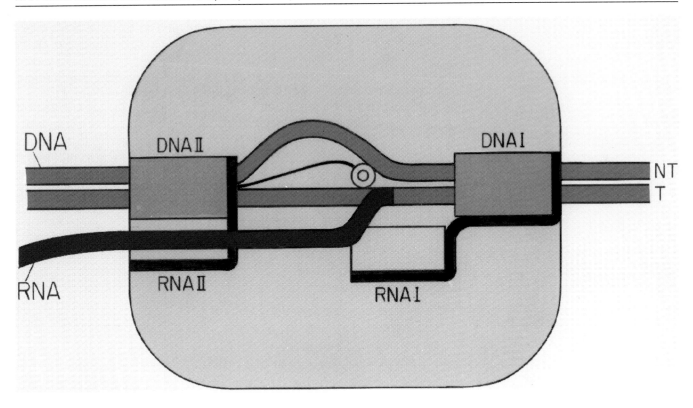

this is an important feature of any model for elongation by RNA polymerases, which are intrinsically unable to function distributively.

RNA is bound to the complex, not through a stable 12 bp DNA-RNA hybrid, but through interaction with two large (8-12 nt) RNA binding sites, that bind single-stranded RNA preferentially. These sites, like the DNA binding sites, are designated I (leading) and II (trailing), and alternately are locked and can slide, in parallel with the DNA binding sites. Transcription elongation takes place through DNA template-nucleoside triphosphate base pairing at the catalytic site of the enzyme; however, the pairing between DNA template and the transcript is broken after a few nucleotides, and the transcript is then bound in RNA binding site I.

The elongation process as it is envisioned for this model can be broken down into two stages. The first stage involves nucleotide addition, and begins with DNA and RNA sites I, locked or in a tight binding mode, with the 3'-OH of the nascent transcript positioned at the upstream edge of the RNA binding site I. DNA and RNA sites II are unlocked, and can slide along the DNA and RNA, respectively. The action of the catalytic site adds nucleotides to the end of the transcript, moving the catalytic site forward along the DNA template strand, and filling RNA site I. As the catalytic site moves forward, it pulls DNA and RNA binding sites forward in concert, until RNA site I is filled.

The second stage of this process, called translocation, involves locking of DNA and RNA sites II, unlocking of DNA and RNA sites I, and the subsequent translocation of the forward sites along the DNA strands, leading to the emptying of RNA site I. Note that the DNA binding site I tracks smoothly along the DNA helix, and hence the RNA polymerase carries out one full rotation about the helix axis for every 10-10.5 bp of DNA duplex.

Fig. 2.1. Diagrammatic model of elongating E. coli RNA polymerase. Model is taken from Chamberlin.[76] The figure shows the elongating RNA polymerase at the beginning of the nucleotide addition phase of transcription; transcription is from right to left.

The black shaded outline shows the approximate limits of the body of the RNA polymerase as measured from DNA footprinting studies (see text). The DNA strands are shown in orange, and are labeled "T" for template strand and "NT" for nontemplate strand. The DNA strands are shown as separated for about 12 bp just up to the catalytic site.[76] The RNA product is shown in red extending from the 5' end at the left to the 3'-OH terminus at the right.

DNA I and DNA II are the leading and trailing DNA binding sites, respectively. These are shown in blue. RNA I and RNA II are the leading and trailing RNA binding sites, respectively. These are shown in green. The DNA and RNA sites are arbitrarily assumed to bind 8-10 nt or bp, hence the complex covers about 40 bp of DNA.

The catalytic center for nucleotide addition is envisioned as attached to the DNA and RNA binding sites II; this is shown diagramatically by placing it at the end of a black rod attached to these sites. It is designated by a small green circle. The DNA helical structure has been unwound for simplicity of representation.

In a model for initiation, RNA binding site II is either absent, or blocked by sigma subunit, and initiation might occur by alignment of the initiating NTP or the primer dinucleotide at the left side of RNA site I.

A great deal of evidence supports the general form of the model, including strong evidence that there is no stable formation of a 12 bp DNA-RNA hybrid,[75,77] direct evidence for both *E. coli* RNA polymerase and yeast RNA polymerase II that there are two large RNA binding sites on the polymerase,[76,80] and studies of the transcript cleavage reaction in which up to 17 nt at the 3'-terminus of the nascent RNA can be cleaved from the transcript, and rapidly dissociated into the solution, leaving the 5'-terminal fragment still bound tightly to the RNA polymerase.[57,84-86] Evidence that the enzyme translocates discontinuously, as proposed in the model, has now been presented by a number of groups; however, studies on halted RNA polymerase complexes have theoretical difficulties that preclude an unambiguous interpretation of structural studies of these complexes.[87,88]

This model has been extended to the initiation reaction by several groups.[48,64,65,89,90] Indeed, if a plausible model for elongation were unable to provide information about initiation events, that would weigh against it. However, it should be noted that the changes in structure that accompany the transition out of the initiation phase are so great, that there may be no direct homology between the two complexes, save perhaps the existence of the single nucleotide binding and catalysis site that is probably common to the two complexes.

What changes are needed to extend this model to the initiation phase of transcription? Of course, the open promoter complex is one in which direct interactions between the sigma subunit and nucleotides in the -35 and -10 regions take place, as well as interactions upstream of -35 between the α subunit and the DNA.[1,91] However, there is reasonably good evidence to suggest that sigma may also block the second (trailing) RNA binding site II, postulated in the above model. First, binding of small RNAs (less than 8-10 nt) is possible for both open promoter complexes and initial transcribing complexes while sigma remains bound to the RNA polymerase.[87,88] However, after the initial translocation away from the promoter, sigma is released and at the same time, the nascent RNA becomes stably bound in the ternary elongation complex. Since, in the model for elongation, this stability reflects the binding of RNA in site II, it is tempting to equate sigma release with occupancy of the second site.

However, there is additional evidence that supports this view. It has been known for some time[92,93] that binding of larger RNAs to RNA polymerase leads immediately to sigma release; that is, the stable form of the binary RNA-RNA polymerase contains core polymerase, not the holoenzyme. For these reasons it is attractive to assume that, by direct steric hindrance, or by conformational effects, RNA binding site II is not available in the open promoter complex.

This assumption then leads to the notion that the chain initiation reaction, together with abortion of nascent RNAs, takes place at RNA binding site I. Indeed, this view was strongly supported by the original discovery that RNA polymerase synthesizes transcripts out to about 9nt without significant movement away from the promoter.[13,14,16,26] Additional support for this view has come from studies in which the initiating triphosphate is covalently coupled to residues in the active site of the RNA polymerase, thus locking it to the site.[89,94] In a modified form of this experiment, the locking is accomplished by binding rifampicin to the terminal phosphate of the triphosphate.[90] The results in all cases support the idea that the RNA polymerase, in this initial phase of synthesis, can form about 9nt transcripts prior to executing a structural transition. Thus, in these crosslinked complexes, up to 9nt transcripts can be formed, but longer chains are not synthesized.

The addition of successive nucleotide residues to the nascent RNA in site I cannot, however, simply involve addition of successive residues to a tight RNA binding site. Such a mechanism would give a pattern of abortive transcripts much like that expected from a base paired hybrid, with abortion of short, but not of long, transcripts. The existence of the primer shifting reaction[27,48] also testifies to the occurrence of relatively weak binding to site I during the initiation reaction in all but rare cases.

A plausible explanation for this weak binding and abortion may lie in the interaction of the catalytic, nucleotide binding site with the nascent oligomers. It has always been assumed that the catalytic site is fixed at the 3'-OH terminus of the RNA transcript. However, there is good evidence from studies of the cleavage reaction that this reaction is actually carried out by the catalytic site, and represents a form of the reverse reaction in which water replaces inorganic pyrophosphate[15,84] although this has yet to be shown for the *E. coli* RNA polymerase. This view of the cleavage reaction readily explains why cleavage can take place rapidly and quantitatively in the complete absence of either GreA or GreB factors (Solow-Cordero and Milan, personal communication; Landick, personal communication).[95]

If this view of the cleavage reaction is correct, then we must view GreA and GreB not simply as cleavage factors, but as factors that can facilitate repositioning of the catalytic site. Since the catalytic site can move back along the RNA for at least 11 nt in the absence of Gre factors (Milan and Solow-Cordero, personal communication), this provides a clear model for how a transcript can be aborted, yet show no evidence of kinetic competition due to the presence of high levels of the next nucleotide. Since the template base must make up the specific part of the active site in binding a nucleotide, movement away from that template base means in effect that the nucleotide binding site does not exist on this form of the polymerase! In this model, slippage of the catalytic site back along the nascent RNA can either lead to abortion of that oligomer, primer slippage through the RNA site to rejoin the 3'-OH of the RNA with the site, or to cleavage of the RNA, followed either by resynthesis or release of the RNA.[48,62]

It is not clear at this point what determines the movement of the catalytic site away from the RNA terminus in the absence of Gre factors. This event may be a central determining factor in the establishment of the abortive pattern for a promoter. However, the addition of a Gre factor to the reaction can either reposition the catalytic site backwards, leading to abortion or cleavage, or hold the catalytic site at the RNA terminus, leading to continued synthesis. This latter possibility is suggested by Hsu et al[62] as the mechanism of stimulation of initiation from T5 N25$_{antiDSR}$ by these factors, and has also been suggested by Goldfarb and his collaborators to explain parallel effects of RNA polymerase mutants on abortive initiation and pausing.[63-65] Alternatively, it appears that Gre factors can also enhance the primer shifting reaction, which involves movement of the nascent RNA along the RNA binding site.[48]

Although these possibilities are intriguing, there are several elements of the elongation model that do not fit well with certain experimental aspects of initiation. First, in the model of Chamberlin[76] the distance between the two RNA binding sites at the beginning of the nucleotide addition stage of elongation is postulated to be about

10 nt. However, this would predict that translocation out of initiation would require about 16-20 nt of synthesis, rather than about 10nt. The finding that some promoters may actually require 16 nt for escape is interesting and needs further study.

Second, the pattern of the cleavage products formed by GreA and GreB suggests that these factors prefer to cleave RNAs that extend just out of the RNA binding site I at the 3'-end of the RNA (GreA) or that are just leaving that site at its trailing edge (GreB).[48] However, the preferential cleavage of 7 nt and 8 nt RNAs to give stable 5 nt RNAs[57] would require that the 5 nt RNAs be bound just outside of site I, rather than in site I as the model predicts. One possible solution to both of these problems is that the RNA binding site for initiation is unique, and is located between sites I and II. In this case it may be lost in the promoter clearance reaction, transferring the nascent RNA into site I, where it is now tightly bound in the elongation mode.

Despite these deficiencies in the different models for the RNA chain initiation process, it is clear that this phase of transcription is an important regulatory step and will continue to lead to new discoveries. Certainly, as the mechanism of this phase of transcription becomes better understood, new examples of the regulation at this step will be identified that are now hidden by the mystery that surrounds this process.

REFERENCES

1. McClure WR. Mechanism and control of transcription initiation in prokaryotes. Annu Rev Biochem 1985; 54:171-204.
2. Hoopes BC, McClure WR. Strategy in Regulation of Transcription Initiation. In: Neidhardt FC et al, eds. *Escherichia coli* and *Salmonella typhimurium.* ASM Press 1987:1231-1240.
3. Tjian R, Maniatis T. Transcriptional activation: a complex puzzle with few easy pieces. Cell 1994; 77:5-8.
4. Gralla JD. Promoter recognition and mRNA initiation by *Escherichia coli* σ^{70}. Methods in Enzymol 1990; 185:37-54.
5. Chamberlin MJ. Transcription 1970: a summary. CSHSQB 1970; 35:851-873.
6. Kammerer W, Deuschle U, Gentz R, Bujard H. Functional dissection of *Escherichia coli* promoters: information in the transcribed region is involved in late steps of the overall process. EMBO J 1986; 11:2995-3000.
7. Knaus R, Bujard H. P_L of coliphage lambda: an alternative solution for an efficient promoter. EMBO J 1988; 7:2919-2923.
8. Menendez M, Kolb A, Buc H. A new target for CRP action at the *malT* promoter. EMBO J 1987; 6:4227-4234.
9. Jin DJ, Turnbough CL, Jr. An *Escherichia coli* RNA polymerase defective in transcription due to its overproduction of abortive initiation products. J Mol Biol 1994; 236:72-80.
10. Jin DJ. Slippage synthesis at the *gal* P2 promoter of *Escherichia coli* and its regulation by UTP concentration and cAMP.cAMP receptor protein. J Biol Chem 1994; 269:17221-17227.
11. Maxon ME, Goodrich JA, Tjian R. Transcription factor IIE binds preferentially to RNA polymerase IIa and recruits TFIIH: a model for promoter clearance. Genes and Dev 1994; 8:515-524.
12. McClure WR, Cech CL. On the mechanism of rifampicin inhibition of RNA synthesis. J Biol Chem 1978; 253:8949-8956.
13. Carpousis AJ, Gralla JD. Cycling of ribonucleic acid polymerase to produce oligonucleotides during initiation in vitro at the *lac* UV5 promoter. Biochemistry 1980; 19:3245-3253.

14. Munson LM, Reznikoff WS. Abortive initiation and long ribonucleic acid synthesis. Biochemistry 1981; 20:2081-2085.

15. Hsu LM, Chamberlin MJ. Abortive initiation, productive initiation and promoter escape by *E. coli* RNA polymerase. (In preparation).

16. Krummel B, Chamberlin MJ. RNA chain initiation by *Escherichia coli* RNA polymerase. Structural transitions of the enzyme in early ternary complexes. Biochemistry 1989; 28:7829-7842.

17. Martin CT, Muller DK, Coleman JE. Processivity in early stages of transcription by T7 RNA polymerase. Biochemistry 1988; 27:3966-3974.

18. Whipple FW, Sonenshein AL. Mechanism of initiation of transcription by *Bacillus subtilis* RNA polymerase at several promoters. J Mol Biol 1992; 223:399-414.

19. Kainz M, Roberts JW. Initiation and pausing at the lambda late gene promoter in vivo. J Mol Biol 1995; 254:808-814.

20. Ellinger T, Behnke D, Bujard H, Gralla JD. Stalling of *E. coli* RNA polymerase in the +6 to +12 region in vivo is associated with tight binding to consensus promoter elements. J Mol Biol 1994; 239:455-465.

21. Grachev MA, Zaychikov EF. Initiation by *Escherichia coli* RNA polymerase: transformation of abortive to productive complex. FEBS Lett 1980; 115:23-26.

22. Carpousis AJ, Gralla JD. Interaction of RNA polymerase with *lac* UV5 promoter DNA during mRNA initiation and elongation. Footprinting, methylation and rifampicin-sensitivity changes accompanying transcription initiation. J Mol Biol 1985; 183:165-177.

23. Straney DC, Crothers DM. A stressed intermediate in the formation of stably initiated RNA chains at the *Escherichia coli lac* UV5 promoter. J Mol Biol 1987; 193:279-292.

24. Hansen UM, McClure WR. Role of σ subunit of *Escherichia coli* RNA polymerase in initiation. II. Release of σ from ternary complexes. J Biol Chem 1980; 255:9564-9570.

25. Schultz W, Zillig W. Nucleic Acids Res 1981; 9:6889-6906.

26. Metzger W, Schickor P, Meier T et al. Nucleation of RNA chain formation by *Escherichia coli* DNA-dependent RNA polymerase. J Mol Biol 1993; 232:35-49.

27. Borukhov S, Sagitov V, Josaitis CA et al. Two modes of transcription initiation in vitro at the *rrnB* P1 promoter of *Escherichia coli*. J Biol Chem 1993; 268:23477-23482.

28. Severinov K, Goldfarb A. Topology of the product binding site in RNA polymerase revealed by transcript slippage at the phage λ PL promoter. J Biol Chem 1994; 269:31701-31705.

29. Ruetsch NR, Dennis D. RNA polymerase. Limit cognate primer for initiation and stable ternary complex fromation. J Biol Chem 1987; 262:1674-1679.

30. Milligan JF, Groebe DR, Witherell GW, Uhlenbeck OC. Oligonucleotide synthesis using T7 RNA polymerase and synthetic DNA templates. Nucleic Acids Res 1987; 15:8783-8798.

31. Nam S-C, Kang C. Transcription initiation site selection and abortive initiation: cycling of phage SP6 RNA polymerase. J Biol Chem 1988; 263:18123-18127.

32. Ling M-L, Risman SS, Klement JF et al. Abortive initiation by bacteriophage T3 and T7 RNA polymerases under conditions of limiting substrate. Nucleic Acids Res 1989; 17:1605-1618.

33. Hinrichsen AI, Ortner I, Hartmann GR. Synthesis of dinucleoside tetraphosphates by RNA polymerase B (II) from calf thymus. FEBS Lett 1985; 193:199-202.

34. Mosig H, Schaffner AR, Sieber H, Hartmann GR. Primer-independent abortive initiation by wheat germ RNA polymerase B (II). Eur J Biochem 1985; 149:337-343.

35. Luse DS, Jacob GA. Abortive initiation by RNA polymerase II in vitro at the adenovirus 2 major late promoter. J Biol Chem 1987; 262:14990-14997.

36. Goodrich JA, Tjian R. Transcription factor IIE, IIH and ATP hydrolysis direct promoter clearance by RNA polymerase II. Cell 1994; 77:145-156.

37. Johnston DE, McClure WR. Abortive initiation of in vitro RNA synthesis on bacteriophage λ DNA. In: Losick RR, Chamberlin MJ, eds. *RNA Polymerase*. Cold Spring Harbor Laboratory, 1976:101-126.

38. Gralla JD, Carpousis AJ, Stefano JE. Productive and abortive initiation of transcription in vitro at the *lac* UV5 promoter. Biochemistry 1980; 19:5864-5869.

39. Harley CB, Lawrie J, Boyer HW, Hedgpeth J. Reiterative copying by *E. coli* RNA polymerase during transcription initiation of mutant pBR322 *tet* promoters. Nucleic Acids Res 1990; 18:547-552.

40. Jacques J-P, Susskind MM. Pseudo-templated transcription by *Escherichia coli* RNA polymerase at a mutant promoter. Genes Dev 1990; 4:1801-1810.

41. Guo H-C, Roberts JW. Heterogeneous initiation due to slippage at the bacteriophage 82 late gene promoter in vitro. Biochemistry 1990; 29:10702-10709.

42. Xiong XF, Reznikoff WS. Transcriptional slippage during the transcription initiation process at a mutant lac promoter in vivo. J Mol Biol 1993; 231:569-580.

43. Chamberlin M, Berg P. Deoxyribonucleic acid-directed synthesis of ribonucleic acid by an enzyme from *Escherichia coli*. Proc Natl Acad Sci USA 1962; 48:81-94.

44. Chamberlin M, Berg P. Mechanism of RNA polymerase action: characterization of the DNA-dependent synthesis of polyadenylic acid. J Mol Biol 1964; 8:708-726.

45. Falaschi A, Adler J, Khorana HG. Chemically synthesized deoxypolynucleotides as templates for ribonucleic acid polymerase. J Biol Chem 1963; 238:3080-3085.

46. Hawley DK, McClure WR. Compilation and analysis of *Escherichia coli* promoter DNA sequences. Nucleic Acids Res 1983; 8:2237-2255.

47. Lisser S, Margalit H. Compilation of *E. coli* mRNA promoter sequences. Nucleic Acids Res 1993; 21:1507-1516.

48. Feng GH, Lee DN, Wang D et al. Gre-A induced transcript cleavage in transcription complexes containing *Escherichia coli* RNA polymerase is controlled by multiple factors, including nascent transcript location and structure. J Biol Chem 1994; 269:22282-22294.

49. MacDonald LE, Zhou Y, McAllister WT. Termination and slippage by bacteriophage T7 RNA polymerase. J Mol Biol 1993; 232:1030-1047.

50. Wagner LA, Weiss RB, Driscoll R et al. Transcription slippage occurs during elongation at runs of adenine or thymine in *Escherichia coli*. Nucleic Acids Res 1990; 18:3529-3535.

51. Zhou, Reines, Doetsch. T7 RNA polymerase bypass of large gaps on the template strand reveals a critical role of the nontemplate strand in elongation. Cell 1995; 82:577-585.

52. Levin JR, Chamberlin MJ. Mapping and characterization of transcriptional pause sites in the early genetic region of bacteriophage T7. J Mol Biol 1987; 196:61-84.

53. Bernhard SL, Meares CF. The σ subunit of RNA polymerase contacts the leading ends of transcripts 9-13 bases long on the λ P_R promoter but not on T7 A1. Biochemistry 1986; 25:5914-5919.

54. Nierman WC, Chamberlin MJ. Studies of RNA chain initiation by *Escherichia coli* RNA polymerase bound to T7 DNA. Direct analysis of the kinetics of RNA chain initiation at T7 promoter A_1. J Biol Chem 1979; 254:7921-7926.

55. Nierman WC, Chamberlin MJ. Studies of RNA chain initiation by *Escherichia coli* RNA polymerase bound to T7 DNA. Direct analysis of the kinetics of RNA chain initiation at T7 promoter A_2. J Biol Chem 1980; 255:1819-1823.

56. Stefano JE, Gralla J. *Lac* UV5 transcription in vitro. Rate limitation subsequent to formation of an RNA polymerase-DNA complex. Biochemistry 1979; 18:1063-1067.

57. Surratt CK, Milan SC, Chamberlin MJ. Spontaneous cleavage of RNA in ternary complexes of *Escherichia coli* RNA polymerase and its significance for the mechanism of transcription. Proc Natl Acad Sci USA 1991; 88:7983-7987.

58. Levin JR, Krummel B, Chamberlin MJ. Isolation and properties of transcribing ternary complexes of *Escherichia coli* RNA polymerase positioned at a single template bases. J Mol Biol 1987; 196:85-100.

59. Arndt KM, Chamberlin MJ. Transcription termination in *Escherichia coli*. Measurement of the rate of enzyme release from Rho-independent terminators. J Mol Biol 1988; 202:271-285.

60. Deuschle U, Kammerer W, Gentz R, Bujard H. EMBO J 1986; 5:2987-2994.

61. Brunner M, Bujard H. Promoter recognition and promoter strength in the *Escherichia coli* system. EMBO J 1987; 6:3139-3144.

62. Hsu LM, Vo NV, Chamberlin MJ. *Escherichia coli* transcript cleavage factors GreA and GreB stimulate promoter clearance and gene expression in vitro and in vivo. Proc Natl Acad Sci USA 1995; 92:11588-11592.

63. Kashlev M, Lee J, Zalenskaya K et al. Blocking of the initiation-to-elongation transition by a transdominant RNA polymerase mutation. Science 1990; 248:1006-1009.

64. Lee J, Kashlev M, Borukhov S, Goldfarb A. A beta subunit mutation disrupting the catalytic function of *Escherichia coli* RNA polymerase. Proc Natl Acad Sci USA 1991; 88:6018-6022.

65. Sagitov V, Nikiforov V, Goldfarb A. Dominant lethal mutations near the 5' substrate binding site affect RNA polymerase propagation. J Biol Chem 1993; 268:2195-2202.

66. Straney S, Crothers DM. Lac repressor is a transient gene-activating protein. Cell 1987; 51:699-707.

67. Lee J, Goldfarb A. Lac repressor acts by modifying the initial transcribing complex so that it cannot leave the promoter. Cell 1991; 66:793-798.

68. Muller DK, Martin CT, Coleman JE. Processivity of proteolytically modified forms of T7 RNA polymerase. Biochemistry 1988; 27:5763-5771.

69. Chamberlin M. The selectivity of transcription. Annu Rev Biochem 1974; 43:721-775.

70. Reynolds R, Bermudez-Cruz RM, Chamberlin MJ. Parameters affecting transcription termination by *Escherichia coli* RNA polymerase. I. Analysis of 13 rho-independent terminators. J Mol Biol 1992; 224:31-51.

71. Leirmo S, Harrison C, Cayley DS, Burgess RR, Records MT, Jr. Replacement of potassium chloride by potassium glutamate dramatically enhances protein-DNA interactions in vitro. Biochemistry 1987; 26: 2095-2101.

72. Gamper HB, Hearst JE. A topological model for transcription based on unwinding angle analysis of *E. coli* RNA polymerase binary, initiation and ternary complexes. Cell 1982; 29:81-90.

73. von Hippel PH, Bear DG, Morgan WD, McSwiggen JA. Protein nucleic acid interaction in transcription: a molecular analysis. Annu Rev Biochem 1984; 53:389-446.

74. Yager TD, von Hippel PH. Transcription elongation and termination in *Escherichia coli*. In: Neidhardt FC et al, eds. *Escherichia coli* and *Salmonella typhimurium*. ASM Press 1987:1241-1275.

75. Rice GA, Kane CM, Chamberlin MJ. Footprinting analysis of mammalian RNA polymerase II along its transcript: an alternative view of transcript elongation. Proc Natl Acad Sci USA 1991; 88:4245-4249.

76. Chamberlin MJ. New models for the mechanism of transcription elongation and its regulation. Harvey Lectures, Series 88, New York: Wiley-Lis, 1995:1-21.

77. Milan S, Chamberlin MJ. Structural analysis of ternary complexes of Escherichia coli RNA polymerase. Ribonuclease footprinting of nascent transcripts. 1995; (in preparation).

78. von Hippel PH, Yager TD. Transcript elongation and termination are competitive kinetic processes. Proc Natl Acad Sci USA 1991; 88: 2307-2311.

79. Johnson TL, Chamberlin MJ. Complexes of yeast RNA Polymerase II and RNA are substrates for TFIIS-induced RNA cleavage. Cell 1994; 77:217-224.

80. Nudler E, Goldfarb A, Kashlev M. Discontinuous mechanism of transcription elongation. Science 1994; 265:793-796.

81. Chan CL, Landick R. New perspectives on RNA chain elongation and termination by *E. coli* RNA polymerase. In: Conaway RC, Conaway JW, eds. Transcription: Mechanisms and Regulation. New York: Raven Press, Ltd., 1994:297-321.

82. Wang D, Meier T, Chan C et al. Discontinuous movements of DNA and RNA in RNA polymerase accompany formation of a paused transcription complex. Cell 1995; 81:341-350.

83. Altmann CR, Solow-Cordero DE, Chamberlin MJ. RNA cleavage and chain elongation by *Escherichia coli* DNA-dependent RNA polymerase in a binary enzyme-RNA complex. Proc Natl Acad Sci USA 1994; 91:3784-3788.

84. Rudd MD, Izban MG, Luse DL. The active site of RNA polymerase II participates in transcript cleavage within arrested ternary complexes. Proc Natl Acad Sci USA 1994; 91:8057-8061.

85. Borukhov S, Polyakov A, Nikiforov V, Goldfarb A. GreA protein: a transcription elongation factor from *Escherichia coli*. Proc Natl Acad Sci USA 1992; 89:8899-8902.

86. Borukhov S, Sagitov V, Goldfarb A. Transcript cleavage factor from *E. coli*. Cell 1993; 72:459-466.

87. Krummel B, Chamberlin MJ. Structural analysis of ternary complexes of *Escherichia coli* RNA polymerase. Individual complexes halted along different transcription units have distinct and unexpected biochemical properties. J Mol Biol 1992; 225:221-237.

88. Krummel B, Chamberlin MJ. Structural analysis of ternary complexes of *Escherichia coli* RNA polymerase. Deoxyribonuclease I footprint of defined complexes. J Mol Biol 1992; 225:239-250.

89. Mustaev A, Kashlev M, Zaychikov et al. Active center rearrangement in RNA polymerase initiation complex. J Biol Chem 1993; 268:19185-19187.

90. Mustaev A, Zaychikov E, Severinov K et al. Topology of the RNA polymerase active center probed by chimeric rifampicin-nucleotide compounds. Proc Natl Acad Sci USA 1994; 91:12036-12040.

91. Ross W, Gosink KK, Salomon J et al. A third recognition element in bacterial promoters: DNA binding by the α subunit of RNA polymerase. Science 1993; 262:1407-1413.

92. Krakow JS, von der Helm K. *Azotobacter* RNA polymerase transitions and the release of sigma. Cold Spring Harbor Symp. Quant Biol 1970; 35:73-82.

93. Busby S, Spaasky A, Buc H. On the binding of tRNA to *Escherichia coli* RNA polymerase. Interactions between the core enzyme, DNA and tRNA. Eur J Biochem 1981; 118:443-451.

94. Grachev M, Kolocheva T, Lukhtanov E et al. Studies on the functional topography of *Escherichia coli* RNA polymerase. Highly selective affinity labeling by analogs of initiating substrates. Eur J Biochem 1987; 163:113-121

95. Orlova M, Newlands J, Das A et al. Intrinsic cleavage activity of RNA polymerase. Proc Natl Acad Sci USA 1995; 92: 4596-4600.

TRANSCRIPTION TERMINATION AND ITS CONTROL

Jeffrey W. Roberts

Transcription terminators form the natural boundaries of gene expression as well as act as sites of genetic regulation. *E. coli* and its phages have provided the central models to understand the biochemical mechanisms of these processes. The notable conservation of the two large major subunits of RNA polymerase between bacterial RNAP and all three eukaryotic enzymes,[1] and the fact that termination is at least partly a function of these subunits,[2] suggest that results from studies of *E. coli* should be of universal significance.

This chapter will discuss mechanisms of transcription termination and the regulatory processes that involve termination, particularly those which affect the general termination ability of RNA polymerase.

TERMINATION

There are two distinct mechanisms by which *E. coli* RNA polymerase terminates: direct recognition of sites in the DNA and transcript, a process called "intrinsic" or rho-independent termination; and through the intervention of the protein termination factor rho, which acts at specific sites distinct from intrinsic terminators. There is considerable understanding of both the sites and how the proteins interact with them.

INTRINSIC TERMINATORS

Intrinsic terminators are found at the boundaries of most transcription units in *E. coli*. By definition, they function with core enzyme alone—the β, β', and α subunits of RNA polymerase—although the NusA protein can significantly enhance their efficiency.[3] Essential features of intrinsic terminators have been well characterized:[2,4,5] a DNA sequence of inverted symmetry that specifies a hairpin in the transcript, and a following segment typically rich in encoded uridines that appear at or near the end of the released RNA (Fig. 3.1). In addition to these more defined elements, regions both downstream and upstream also have important effects on termination efficiency.[6,7]

ROLE OF TERMINATOR HAIRPIN

There is clear evidence that the RNA hairpin, rather than the symmetrical DNA that specifies it, is the important structure. Experiments

Regulation of Gene Expression in Escherichia coli, edited by E. C. C. Lin and A. Simon Lynch. © 1996 R.G. Landes Company.

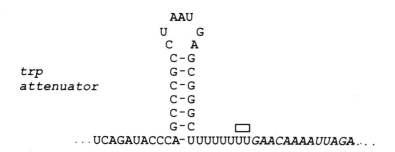

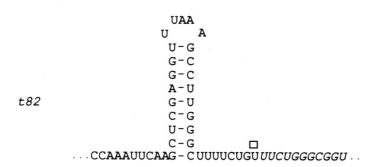

Fig. 3.1. RNA structures of two intrinsic terminators: the tryptophan operon attenuator (Landick and Yanofsky, 1987) and the bacteriophage 82 late gene terminator (McDowell et al, 1994). The trp attenuator has the more typical long uninterrupted uridine-encoding segment, whereas t82 is an example of an efficient terminator that lacks this feature; possibly other surrounding sequences replace its function. Arbitrary amounts of sequence adjacent to the hairpin and U-encoding region are shown; it is not clear how far regions important to promoter function may extend. The hatched box marks the ends of the terminated RNA, and nucleotides of RNA that reads through the terminator are shown in italic.

using heteroduplexes show that the DNA template strand in the symmetrical segment, and thus presumably its encoded RNA, specifies the active terminator.[8] Mutational analysis shows that weakening the potential hairpin by mutation in the stem reduces or eliminates termination, whereas compensating mutations that restore the stem but change the base pair can restore termination.[4,9,10] Incorporation into the transcript of base analogs that weaken (inosine for guanosine) or strengthen (iodocytosine for cytosine) the hairpin has the predicted effects: less and more termination, respectively.[4] Thus, to a first approximation at least, it is the secondary structure and not base-specific features that are important. However, not any designed hairpin sequence will work at the naively predicted efficiency.[10] The failure of theoretically stable hairpins to replace natural structures could have several origins: the rate of formation during transcription, if this is important, might be deficient; unapparent adventitious pairings might interfere with hairpin formation; or base-specific or other subtle structural elements might be involved, if, for example, the hairpin binds the enzyme.[11]

ROLE OF U-RICH REGION

The region encoding uridine-rich RNA is nearly universal in intrinsic terminators. Progressive deletion of the U-rich region from the downstream end—thereby substituting nonuridine bases in these positions—also progressively reduces termination in one model system[12] (although not in proportion to the uridine content), and another natural defective terminator mutant has deleted four of eight successive uridines.[13,14] Physical chemical measurements show that heteroduplexes of oligouridine and oligodeoxyadenosine are particularly unstable.[15] Furthermore, substitution of bromouridine for uridine in the transcript, which should stabilize base-pairing with the dA template segment, decreases termination.[14] These results have led to the notion that termination efficiency is directly related to the stability of an RNA/DNA hybrid at or near the end of the released RNA.

When RNA synthesis by purified RNA polymerase is stalled artificially within a few nucleotides of the natural release site, the complex is much less stable than are elongation complexes stopped at other sites.[16] The region of instability extends beyond the actual release sites, which generally are only a few adjacent nucleotides. Furthermore, sites of natural instability that are not terminators occur just after an RNA hairpin could form in the transcript, implying that the hairpin even in the absence of an adjacent U-encoding segment confers instability.[11] These results also imply that destabilization does not require active movement of the complex, even though release occurs naturally during transcriptional advance, so that stopped complexes must retain some of the essential properties that lead to termination.

Two Models for Intrinsic Termination

The long standing model derived from these observations—originally presented qualitatively,[14] and then given in a thermodynamic framework[17]—attributed the stability of the transcription complex primarily to a region of RNA/DNA base pairing (a "hybrid") at the growing point, nominally about 12 nucleotides long. Intrinsic termination was ascribed to extraction of part of this hybrid through formation of the hairpin, followed by dissociation of the remaining unstable rU/dA segment.

An alternative model questions the existence of an extended RNA/DNA hybrid (more than ~3 base pairs), and proposes instead that the stability of the elongating complex derives from tight binding of the product RNA to RNA polymerase (see Chamberlin and Hsu, this volume).[18] An important motivation was the observation that the U-encoding segment is not invariably as extended or uniform as would be predicted if the strength of the rU/dA base pairs directly determined the stability of the complex, as well as evidence that sequences before and after the hairpin and release site can have dominant effects on termination efficiency.[19] These and other considerations led to the "inchworm" model of elongation and termination,[18] in which the emerging transcript is held successively in two RNA binding sites as the enzyme advances discontinuously. The role of the hairpin is to remove RNA from a binding site required for integrity of the complex, thereby promoting release of the RNA. An essential part of the model is that the elongating complex exists in different conformations, only certain of which favor termination; these conformations reflect different positions of the catalytic center relative to the RNA binding sites during inchworming. Several phenomena suggested that distinct conformations exist, including the susceptibility of the emerging transcript to cleavage near the 3' end,[20] particularly promoted by the factors GreA and GreB,[21] and variable footprints of stopped elongation complexes.[18,22,23] The model as further elaborated[24-26] proposes that sequences besides those encoding the hairpin are required to set the correct conformation for RNA release when the hairpin forms. These signals might include the U-encoding segment (see below) as well as adjacent sequences.[6]

There is strong evidence for certain aspects of this model. First, distinct conformations of elongating complexes in vitro, measured as different separations between the growing RNA end and the front of the enzyme on DNA, have been found.[22-24,27] Furthermore, the acquisition of a particular "strained" conformation was shown to occur during a hairpin-induced transcription pause,[26] and also while RNA polymerase transited the U-encoding segment of a terminator.[25] Second, RNA binding sites of RNA polymerase have been characterized (see chapter 2). Third, it has been shown that DNA polymerase of phage T4 can pass an elongating transcription complex from either direction without displacing RNA polymerase from the DNA or destroying its register on the template.[28,29] This result surely implies that some mechanism other than the strength of the RNA/DNA hybrid maintains the integrity of the complex. It does not mean that no extended hybrid exists, or might not exist during some stages of transcription; there is evidence for hybrid from enzymatic probing of stalled purified complexes[30] and chemical probing of complexes in vivo,[31] and it is difficult otherwise to explain the loss of register between transcript and template during transcription of homopolymeric template segments of greater than (but not less than) 10 nucleotides.[32] However it is clear that RNA polymerase has activities far more complex than simply holding an elongating bubble, the structure of which is maintained by RNA/DNA interactions during monotonous addition of nucleotides to the growing RNA end.

Even though rU/dA interactions may not determine the stability of the complex exactly as originally thought, they certainly could be important to the termination process. One interpretation consistent with evidence implicating the instability of the rU/dA segment in termination is that its weakness favors displacement of the growing RNA end from the template and the catalytic site of the enzyme; displacement, in turn, might be essential to RNA release, as in the earlier model, or displacement might promote or contribute to a conformation of RNA polymerase that allows the hairpin to induce release.

It is important that U-rich segments are generally involved in pausing and termination phenomena of RNA polymerases. For example, the much smaller phage T7 RNA polymerase and its relatives (~100 kDa versus ~400 kDa for the bacterial RNAP family) recognize hairpin-rU rich terminators like those discussed above, and, in addition, T7 RNAP recognizes a terminator with a rU-rich segment and no discernible hairpin.[33] Eukaryotic RNA polymerase I and III terminators include essential U rich-encoding segments just preceding the site of release.[34,35] The mechanism of RNAP II termination is unknown, although it likely is an enzymatically catalyzed process linked to mRNA 3' end formation, conceivably modeled by the activity of *E. coli* rho protein.[36] Pol II does pause and arrest at sequence tracts encoding rU-rich segments,[37,38] consistent with a common influence of these segments on elongation of all RNA polymerases.

PAUSING, TERMINATION, AND ELONGATION KINETICS

Purified RNA polymerase does not transcribe DNA at a uniform rate, but instead displays distinct pauses as it elongates, remaining at discrete sites for seconds to minutes.[2] It is likely that such pausing is involved in the function of terminators, as discussed below. Different template elements can induce pausing: G/C rich segments about 10 nucleotides upstream of the RNA end;[39] RNA hairpins that form just upstream of the pause site;[40] a nontemplate DNA strand sequence near the lambda late gene promoter where σ^{70} is thought to bind to induce an early elongation pause;[41] and elements of unknown character.[42] Complex pause sites, the best characterized of which are in the leader regions of the *trp* and *his* operons, involve both hairpins and adjacent sequence elements.[7]

The involvement of RNA hairpins in pausing may imply a similar function of the terminator hairpin in intrinsic termination, although this has not been shown. The most evident distinction between terminators and hairpin-associated pauses is the lack of a U-encoding segment after a pause hairpin. However, since sequences surrounding these elements contribute to both pausing and termination in unknown ways, it is not clear if this is the most significant difference; in fact the combination of an effective pause hairpin and a U-encoding segment does not necessarily make a terminator.[43]

Some pausing clearly has a kinetic function: RNAP waits for something to happen, such as translation of leader mRNA in the regulatory regions of operons like *trp* and *his* that are controlled by attenuators.[44] Although this function has not been established for pausing elements in termination, it is clear that pausing is correlated in an important way with transcription termination and its control. For example, Rho-dependent termination (see below) occurs at sites that in the absence of rho are pause sites, at least in vitro.[45] The pause at the beginning of the lambda late gene transcript acts to present RNA polymerase for modification by the lambda gene Q antiterminator, and the Q protein in turn releases RNA polymerase from the pause (see below). Q protein and N protein both suppress pausing during downstream transcription.[46,47] Furthermore, there exist mutationally altered RNA polymerases that change pausing and termination in a corresponding way: less termination correlates with less pausing, and more termination with more efficient pausing.[48] The simplest interpretation of this relation is that the enzyme undergoes the same change at terminators as it does at pauses, and that the probability or duration of this change determines the efficiency of termination.

The detailed behavior of these mutationally altered RNA polymerases at a particular intrinsic terminator, named t21, implies a competition between elongation and release of the transcript.[49] Release of the transcript occurs at several adjacent sites at the end of the U-encoding segment. Changing the NTP concentration corresponding to the inserted nucleotides also moves the distribution of release sites, more promoter distal for higher concentrations and more promoter proximal for lower. For a given nucleotide concentration, the effect of the RNA polymerase mutations is to move the distribution of RNAs in a promoter distal direction for the hypoterminating mutant (thus moving through the region of release and into further elongation), and to move the released ends promoter proximal for the hyperterminating mutant. It is as though the tendency to pause—perhaps reflecting the same conformational shift that is required for RNA release—determines how efficiently the enzyme progresses through the area of potential release, so that the actual release is controlled by elongation rate. The result is a higher or lower apparent K_s for NTPs in elongation; this value may reflect a variety of reaction steps, and not necessarily actual binding of the NTP in the active site. An effect of changing NTP concentration on termination efficiency of wild-type RNA polymerase also has been found.[19]

LOCATION OF HAIRPIN STEM RELATIVE TO RELEASE SITE

An important dimensional relation between the hairpin stem and the release sites has been noted. As shown in detail for one terminator and inferred for others, there is a distinct boundary downstream of the hairpin beyond which no release occurs.[49] This site occurs at a particular number of nucleotides, 9-10, beyond the last G/C base pair of the stem, although not in a specific location relative to any sequence in the U-rich segment; this distance is consistent with other determinations of terminator release sites.[7,19] Regions just beyond the boundary may be as or more U-rich than the actual end of the transcript. This result suggests that the release potential is limited by some structural parameter of the enzyme—e.g., a defined extension of the transcript from a hairpin binding site—rather than a particular nucleotide sequence at the end of the transcript.

Studies of conformational changes of RNA polymerase during elongation through both a pause site and an intrinsic terminator suggest that a particular inchworming transition is correlated with both events: reduction of the distance from the front of the enzyme to the catalytic center at the growing end of the RNA—also described as change from a "relaxed" to "strained" state of the enzyme.[25,26] It was speculated that release of RNA at a terminator is induced by a jump from a strained state back to a relaxed state, perhaps in coincidence with formation of the hairpin.[25,26] No clear mechanistic description yet exists, and it is unclear in what important way the termination and pause sites differ; however, it seems very likely that these conformational changes are essential to both pausing and termination.

RNA polymerase clearly does pause at rho-dependent terminators in vitro, where continued elongation occurs in the absence of rho. A competition between elongation and release at a rho-dependent terminator was inferred from the ability of an RNA polymerase mutant expressing slow elongation and enhanced pausing (the same that hyperterminates at intrinsic terminators) to restore activity to a defective rho protein that is inactive with wild-type RNA polymerase.[50] This hypothetical relation between rates of elongation and rho function was called "kinetic coupling."

EFFECTS OF UPSTREAM SEQUENCES ON TERMINATION

Finally, there exists an important influence on intrinsic termination efficiency that might have general implications for the activity of RNA, but has received little notice. The fusion of novel transcribed sequences several hundred nucleotides upstream of well characterized intrinsic terminators gives partial relief of termination both in vivo and in vitro, in a fashion sensitive even to single base changes in the new upstream sequences.[51,52] One view is that the novel RNA segments induce termination-resistant states of RNAP, conceivably related to conformational variants of inchworming;[52] this potential promoter effect on terminator function has not been carefully explored. An alternative interpretation is that these novel sequences interact with RNA segments of the hairpin, impairing its stability or delaying its formation.[51] The abundance of these effects in various fusions may reflect the promiscuity of RNA-RNA interactions, and, perhaps more important, may imply that each natural RNA segment has evolved to exist and function in a natural context that extends considerable distances along the RNA sequence. An implication is that any novel fusion of a transcribed segment (e.g., one associated with a promoter) to a terminator may yield an unpredictable terminator activity. Similar caution might apply to fusions testing other activities of RNA such as splicing, processing, or enzymatic catalysis.

RHO-DEPENDENT TERMINATION[53]

Rho was detected and purified from bacterial extracts as an activity that modifies, and, as it turned out, terminates transcription from the bacteriophage lambda early operons.[54] Its function in inducing genetic polarity was suggested by the finding of rho-dependent terminators in polar insertions,[55] and was established by the identification of rho as the product of the polarity suppressor gene *suA* (now named *rho*);[56] many rho mutants have been found, and the gene is known to be essential. The active protein is a hexamer of identical subunits of 46,000 Da, containing RNA binding and nucleoside triphosphatase activities.[57]

Rho-dependent terminators have been found at numerous specific sites, including the ends of *E. coli* operons,[58] the end of a tRNA transcription unit,[59] the regulatory region of the *E. coli* tryptophanase operon,[60] and at sites in the lambda and T4 phage genomes. Operons are sometimes bounded by both an intrinsic terminator and a rho-dependent terminator in tandem.[61,62] Besides these particular sites, rho-dependent terminators appear in many operons when nonsense mutations interrupt translation—the phenomenon of genetic polarity. It would appear that the bacterial cell has evolved so that routine termination is mainly the business of intrinsic terminators, whereas rho often functions in cases where particular conditions have rendered transcription futile and wasteful, e.g., when translation of an emerging operon transcript fails to occur.[63]

The primary requirement of rho activity, which also explains its ability to recognize untranslated transcripts, is for free RNA upstream of the site or sites of RNA release; rho must interact with this RNA in order for termination to occur.[57] The RNA requirement was discovered through its role as an essential cofactor for an ATPase (or NTPase) activity of rho;[64] ATP hydrolysis also is essential for termination activity of rho, because the substitution for ATP of an analog with a nonhydrolizable β–γ bond (e.g., AMP-PNP) that still serves as a substrate for RNA synthesis prevents rho-dependent termination. In accord with the strong preference of cytidine containing polymers to activate the ATPase, a substantial content of cytidine is required in the upstream RNA.[65,66] However, no particular required RNA structure or sequence has been found from analysis of either authentic termination sites or RNAs active in vitro;[66-69] possibly any RNA relatively free of secondary structure with some cytidine content is active,[70,71] as is an entirely artificial transcript (e.g., of yeast telomeric DNA[72]).

The pauses displayed by RNA polymerase during transcription in vitro through rho-dependent terminators in the absence of rho occur exactly where RNAs are released by rho,[73] as would be expected if the pauses are essential for rho-dependent release. Rho also releases RNA polymerase stopped in vitro by a bound protein at a site where pausing does not otherwise occur,[74] suggesting that the kinetic element of the pause is important—time for rho to interact with RNA polymerase— rather than any more specific response to the pausing signal. It is not known exactly what signals are pausing at rho-dependent terminators, although, as for the *his* leader pause site[7] there likely exist distinct elements. As is true for *his*, sequences that could encode a hairpin do contribute to the pause at the site of the major RNA release at the lambda tR1 rho-dependent terminator,[45] but much of the pause remains if the hairpin is removed.[72]

The mechanism of rho function has been approached through structural studies of the rho polypeptide and hexamer, and biochemical studies of its interaction with RNA.[53,75,76] The rho monomer has several domains, including an amino terminal segment that binds RNA and a carboxy terminal domain containing ATPase motifs. There exist functionally separate sites in the hexamer that bind polynucleotide strongly (primary sites) and weakly (secondary sites), although these could be alternate manifestations of the same RNA binding domain; both primary and secondary sites must be occupied for ATPase activity,[77,78] a situation that presumably models the active configuration for termination. A substantial length of transcribed RNA, greater than 70 nucleotides, is required for termination activity.[68,79] This is similar to the total binding capacity of rho for polynucleotides,[80] and clearly represents binding of RNA to much of the hexamer. One model of rho structure and function suggests an alternating of strong and weak binding by asymmetrical dimers within the hexamer,[81] a transition that could correspond to essential movements between RNA and protein that accompany ATP hydrolysis and RNA release.

The function of these movements in termination may be reflected by the ability of rho, in the absence of RNA polymerase, to extract RNA from an RNA/DNA hybrid bearing a single-stranded 5' RNA extension, in an ATP-dependent activity termed a helicase.[82] This activity has the same requirements for RNA structure and sequence in the single-stranded portion as does the termination activity. Rho helicase acts by a stoichiometric reaction in which rho binds the released RNA in an ATP-dependent fashion, rather than the catalytic unwinding characteristic of DNA helicases. It is not clear that there is direct interaction with DNA/RNA hybrid, or if, instead, rho wedges into the hybrid by binding RNA as the hybrid breathes. The fact that rho can remove RNA from a binding site suggests that the reaction is important even if RNA is bound to protein rather than the DNA template strand in the transcription complex. It has been speculated that rho translocates 5' to 3' from an initial binding site to the site of termination; however, in contrast to this notion, there is evidence that rho retains grip on the initial binding site,[83] and although some movements clearly occur, it has not been shown that there is actual translocation.

Rho appears to be absolutely essential for termination at sites that have been characterized, although other proteins modulate its activity. NusA decreases the activity of rho in vivo and in vitro; this appears to be the essential function of NusA, because a *rho* mutation that decreases the activity of rho allows the cell to survive without NusA.[84] The essential protein NusG is required for efficient rho function at certain sites in vivo,[85] and NusG enhances rho activity in vitro by binding cooperatively with rho in the transcription complex.[86] Both NusA and NusG are also essential accessory factors in lambda gene *N*-mediated antitermination (see below), for which they partake in a multi-subunit complex that also includes other Nus proteins and RNA polymerase, N protein, and the RNA site that binds N.[87,88] It has been suggested that rho also might be incorporated into this antitermination complex, and that there are direct parallels between these protein ensembles required for termination and antitermination.[57]

ANTITERMINATION

The existence of a regulatory mechanism in *E. coli* targeted to transcription termination sites was proposed in 1969,[54] when evidence was presented that a cellular termination mechanism involving rho factor acts in vitro to prevent gene expression that depends in vivo upon the regulatory activity of the phage lambda gene *N* protein, and that arises from promoters upstream of the termination sites. Substantial confirmation was provided by proof that rho factor is responsible for genetic polarity—which thus is due to transcription termination—and the demonstration that N protein prevents genetic polarity.[89,90] The activities of both Q and N proteins in defined transcription systems[87,88,91] have established that their function is to prevent the recognition of transcription terminators by RNA polymerase, and that they act by changing the behavior of RNA polymerase itself rather than modifying the terminators involved. A cellular antitermination mechanism closely related to that of lambda N protein acts during transcription of the ribosomal RNA operons of *E. coli*.[92]

SITE-SPECIFIC ANTITERMINATION (ATTENUATION)

A transcription antitermination mechanism of a different sort, *attenuation*, controls expression of several operons involved in amino acid biosynthesis, including those for trytophan, leucine, histidine, and threonine, and also the *pyrBI* operon encoding aspartate transcarbamylase. In contrast to the systems described above that alter the response of RNA polymerase to terminators, attenuation prevents termination only at a single terminator before the relevant genes, acting through changes in localized RNA structures that affect the formation of the terminator. The molecular mechanism of attenuation control has been explicated in satisfying detail, and has been well reviewed.[44,93] For the amino acid operons, impaired translation of mRNA encoding a *leader peptide* featuring repeated tandem codons for the scarce amino acid prevents function of a downstream intrinsic terminator at the attenuation site, acting by exposing structures in the mRNA that interfere with formation of the essential terminator hairpin. In *pyrBI*, slowed transcription through a region encoding multiple uridines in conditions of pyrimidine limitation (and thus low UTP) allows ribosomes to block formation of a transcription termination hairpin, thus allowing downstream gene expression.[94]

Several variants of this site-specific antitermination mechanism are known in *E. coli* and other bacteria: 1. In the regulation of the tryptophan biosynthetic genes of *Bacillus*, a separate regulatory protein (TRAP) whose activity is controlled by tryptophan binds RNA to prevent formation of the termination hairpin.[95] 2. In the tryptophanase operon of *E. coli* attenuation involves both rho-dependent termination and the act of translation of a leader peptide, whose processes interact through a still obscure mechanism to sense the availability of tryptophan.[96] 3. In the *bgl* operon of *E. coli* a specific regulatory protein obstructs the terminator hairpin, and its state of phosphorylation and consequent activity are controlled by the availability of a β-glucoside.[97] 4. For operons of various gram-positive bacteria encoding mostly aminoacyl tRNA synthetases, some structure involving both upstream RNA sequences and the uncharged tRNA cognate to the synthetase interferes with formation of the terminator stem and thus allows readthrough into the regulated genes.[98]

Persistent Antitermination through Control of RNA Polymerase Function: The Phage Lambda Gene N and Q Proteins

The lambda gene *N* and *Q* proteins invoke phage gene expression at, respectively, early and late times of phage growth. Their common property is to modify RNA polymerase at a defined locus in the genome to provide antitermination at any terminator encountered downstream. The existence of distinct sites of recognition and modification was shown originally for N, whose genome-specific recognition site (named *nut* for *N u*tilization) is near the phage early promoters, but which causes readthrough of distal terminators.[99] Similarly, the recognition site of Q protein (named *qut* for *Q u*tilization) is within the single promoter that serves the phage late genes,[100,101] but the enzyme again is modified for downstream terminators. An essential difference between Q and N is the nature of this recognition site: for Q, it is DNA, to which Q binds specifically,[102] whereas N binds a structure in the transcribed mRNA.[103,104] Most of the progress in understanding how these proteins work has been to determine the structures that form at their respective recognition sites. In addition, however, the way antitermination occurs is suggested by the fact that these proteins suppress pausing by RNA polymerase during transcription downstream.[46,47]

Lambda Q Protein

Characterization of the recognition complex between Q protein and RNA polymerase was simplified by the fact that purified Q is active in reactions containing only purified RNA polymerase, NusA protein, and DNA and small molecules.[105,106] Deletion analysis showed that sequences necessary for Q function are within and adjacent to the late gene promoter, but separate from the promoter elements themselves. Q binds an essential sequence between the -10 and -35 promoter elements (Fig. 3.2). Its function, however, also requires some DNA in the early transcribed segment, particularly nucleotides 1 through 9.[100,101] This segment acts, in the absence of Q, to induce a prolonged pause in RNA synthesis at +16 and +17 that is detectable both in vivo[31] and in vitro; during the pause Q binds RNA polymerase. A consequence of Q binding the paused complex is to reduce the half-life of the pause about 5-fold: it "chases" RNAP from the pause site.[106]

This pause is induced by σ^{70}, acting in a completely unexpected way.[107] Characterization of sequences required for the pause, and for Q function, identified particularly an A residue at +2 and a T residue at +6;[108,109] it was further shown that these act almost exclusively through the *nontemplate* strand of DNA.[41] The realization that these bases are congruent with the most highly conserved A and T of the -10 consensus (TATAAT), and furthermore occupy the same position of the open DNA bubble in pause and in the open complex, suggested that the pause actually is induced at least in part by the same sequence that constitutes the -10 recognition sequence.[107] In fact, sigma is required for the pause and is present in the paused complex.[107,110] This is true although the paused complex is an authentic elongation complex in most ways: it has broken the original σ^{70}-promoter bonds, it is stable, and its footprint is about 30 base pairs in extent, like elongation complexes, except for some extra upstream contacts.[102] Presumably the same σ^{70} molecule that was fixed to -10 remains nonspecifically bound to core enzyme, and then "hops" to the pause site.[107] An incidental but important implication of this result is that base-specific recognition of the -10 region by sigma in initiation occurs in the open complex, and that this recognition is also to the nontemplate DNA strand.[111]

The most important observation about the activity of Q, and probably the only indication how antiterminators affect the properties of

Fig. 3.2. A proposed structure of the paused transcription complex at nucleotides 16 and 17 of the phage lambda late gene promoter. Q protein is shown bound to the qut *site. The highly conserved A and T of the nontemplate DNA strand of both the -10 consensus sequence of the promoter and the pause-inducing sequence are shown enlarged. As described in the text, recent evidence indicates that region 2 of σ^{70} is bound to the nontemplate DNA strand at the pause-inducing sequence in the early transcribed region that is related to the -10 promoter consensus TATAAT. The positions of both Q and RNA polymerase approximately reflect boundaries found by exonuclease digestion. It is thought that NusA protein also is present in the complex.*

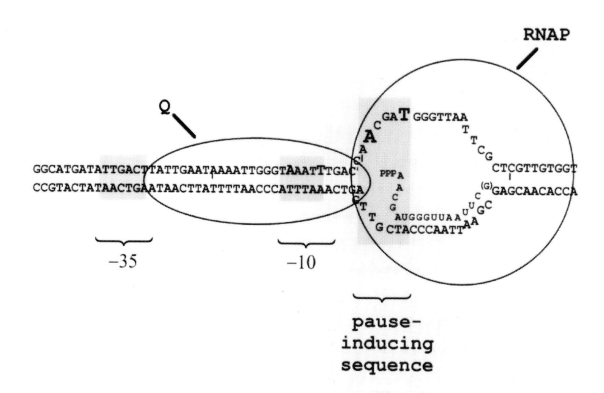

RNAP in elongation, is that transcription pausing at sites far downstream from the terminator is reduced by the Q modification.[46] Furthermore, this effect is related to the pausing capacity of the polymerase: pausing is reduced more dramatically for a hyperterminating mutant of RNA polymerase that also pauses more than wild type, whereas there is only a slight effect on pausing by a hypoterminating polymerase that naturally pauses less than wild-type.[112] These results suggest that Q affects a specialized pausing state, and not the elongation process in general.

It is not understood how this pausing state of transcribing RNA polymerase relates to termination or to structures of the transcription complex described above, nor is it understood how Q affects them. As proposed previously,[91] there would seem to be two broad possibilities for the function of an antiterminator. First, it could interfere directly with the putative essential mechanics of termination—e.g., formation or binding of an RNA hairpin, or inchworming movements of the enzyme, or transfer of the RNA end out of the active site. If pausing necessarily accompanies these movements, the antipausing activity of Q could reflect its inhibition of an essential step of termination. Second, an antiterminator could affect a signal distinct from these elements, but associated with the terminator, that pauses the enzyme long enough for the mechanics of release to occur. According to this model, the antiterminator would act simply by affecting the kinetics of elongation.

It is presumed that Q forms a complex that travels with RNA polymerase through elongation, although this has not been shown.

LAMBDA N PROTEIN

Understanding of N function was based originally on genetic experiments that characterized its activity and identified four cellular factors—NusA, NusB, NusE, and NusG—required for its function.[113-117] These experiments (1) revealed polarity suppression by N; (2) showed that the site of N recognition differs from its site of action; (3) identified the N recognition (*nut* for *N ut*ilization) site;[118] and (4) showed that translation of the RNA message of *nut* prevents its function, thus suggesting strongly that the site of action is the RNA. The first important cellular protein characterized through its interaction with N was NusA, which was shown to bind both N and the core RNA polymerase in vitro.[119,120] Subsequent experiments established a structure of the active complex of N-modified RNA polymerase that contains the N site of action *nut*, and the cellular factors NusA, NusB, NusE (identical to the ribosomal protein S10), and NusG (Fig. 3.3).[87,88] These factors are believed to bind cooperatively and tightly to the transcribing RNAP.

A network of connections, centered on the RNA, stabilizes the complex (Fig. 3.3). The *nut* site has two parts, *boxB*, to which N itself binds;[121] and *boxA*, to which a heterodimer of NusB and NusE (S10) binds.[122] Mutagenesis shows that bases of the loop are mostly responsible for N function and binding.[123] N binds NusA, and at least the three proteins NusA, S10, and NusG make important contacts with RNA polymerase. This entire array can be detected by bandshift experiments, and it may include yet more complex interactions among its components.[124]

Even though the active complex contains all of these factors, the essential activity is carried by N itself. Over short distances, N, NusA, and the *nut* site alone suffice for in vitro antitermination.[125,126] It is

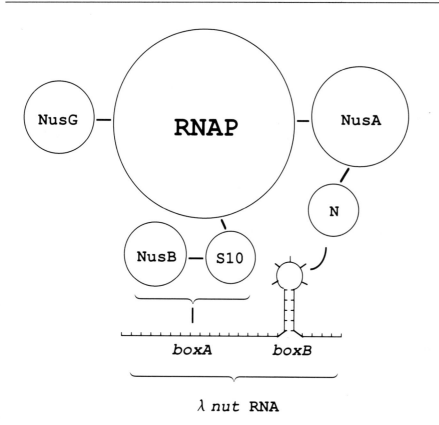

λ *nut* RNA

Fig. 3.3. Protein and RNA components of the lambda N protein antitermination complex (see refs. 104, 118-123). Thick lines represent interactions that have been shown clearly in genetic or biochemical experiments. Spherical volumes represent approximate masses of protein monomers.

believed that the other factors act to stabilize the complex so that it persists over kilobases, for example during transcription of the entire length of the phage lambda genome.[47,126]

ANTITERMINATION IN THE RIBOSOMAL OPERONS

An antitermination system closely related to that of N protein acts in the ribosomal RNA operons of *E. coli*. It was first recognized through the failure of a transposon to exert its expected polarity in ribosomal genes, suggesting a polarity suppression mechanism.[127] Genetic studies, and particularly detailed mutagenesis experiments, established that sequences near the beginning of the transcript closely related to lambda *boxA* endow antitermination properties on downstream transcription.[128] The relation of ribosomal antitermination to the lambda system and the role of proteins NusB and NusE (S10) was clarified gratifyingly by the demonstration that these proteins bind the *boxA*-like sequence; in fact they bind it more strongly than *boxA* in the *nut* segment, implying that the ribosomal RNA sequence is the better one.[122] One interpretation is that the weaker interactions in the lambda system are importantly strengthened by the involvement of N in the complex, by which means lambda therefore controls the assembly of the complex. Despite this view, however, there is evidence for yet another factor, detectable in cell extracts, that is required besides the defined set already mentioned for the ribosomal antitermination system.[129]

HK022 NUN PROTEIN AND THE HK022 ANTITERMINATION SYSTEM

The close lambda relative HK022 has provided two surprise variants of termination and its control. First, HK022 has a protein (Nun) that is similar in sequence and many properties to lambda N, is placed identically in the genome, and also is targeted to *nut* sites. However, the target of Nun is foreign and its function is reversed: it binds lambda (not HK022) *nut* sites to *induce* termination rather than to prevent it, and thus lets HK022 poison its relative.[130,131] And then, in addition to having this N-like protein of switched function, HK022 has its own unique early antitermination system, which appears to require no phage encoded protein, but does have essential phage genomic sites. The specificity of this system is shown clearly by the isolation of mutations altering the β' subunit of RNA polymerase that prevent its operation, and mutations in the HK022 genetic site.[132,133] There is promise these systems will give yet new insights into the regulation of transcription termination.

ACKNOWLEDGMENTS

I thank Robert Landick, Robert Weisberg, and members of the laboratory for discussing the manuscript. Research in the author's laboratory is supported by the National Institutes of Health.

REFERENCES

1. Sweetser D, Nonet M, Young RA. Prokaryotic and eukaryotic RNA polymerases have homologous core subunits. Proc Natl Acad Sci USA 1987; 84:1192-1196.
2. Chan CL, Landick R. New perspectives on RNA chain elongation and termination by *E. coli* RNA polymerase. In: Conaway RC, Conaway JW, eds. Transcription: Mechanisms and Regulation. New York: Raven Press, 1994:297-321.
3. Schmidt MC, Chamberlin MJ. NusA protein of *Escherichia coli* is an efficient transcription termination factor for certain terminator sites. J Mol Biol 1987; 195:809-818.
4. Yager T, von Hippel P. Transcript elongation and termination in *Escherichia coli*. In: Neidhardt F et al, eds. *Escherichia coli* and Salmonella typhimurium: cellular and molecular biology, Washington, DC: Amer Soc Microbiol, 1987:1241-1275.
5. Platt T. Transcription termination and the regulation of gene expression. Ann Rev Biochem 1986; 55:339-372.
6. Reynolds R, Chamberlin MJ. Parameters affecting transcription termination by *Escherichia coli* RNA polymerase I: Construction and analysis of hybrid terminators. J Mol Biol 1992; 224:53-63.
7. Chan CL, Landick R. Dissection of the his leader pause site by base substitution reveals a multipartite signal that includes a pause RNA hairpin. J Mol Biol 1993; 233:25-42.
8. Ryan T, Chamberlin M. Transcription analysis with heteroduplex trp attenuator templates indicates that the transcript stem and loop structure serve as the termination signal. J Biol Chem 1983; 258:4690.
9. Yanofsky C. Attenuation in the control of expression of bacterial operons. Nature 1981; 289:751-758.
10. Cheng S-W et al. Functional importance of sequence in the stem-loop of a transcription terminator. Science 1992; 254:1205-1207.
11. Arndt KM, Chamberlin MJ. RNA chain elongation by *Escherichia-Coli* RNA polymerase: factors affecting the stability of elongating ternary complexes. J Mol Biol 1990; 213:79-108.
12. Lynn SP, Kasper LM, Gardner JF. Contributions of RNA secondary structure and length of the thymidine tract to transcription termination at the thr operon attenuator. J Biol Chem 1988; 263:472-479.
13. Bertrand K et al. The attenuator of the tryptophan operon of *Escherichia coli*: heterogeneous 3'-OH termini in vivo and deletion mapping of functions. J Mol Biol 1977; 227-247.
14. Farnham PJ, Platt T. A model for transcription termination suggested by studies on the trp attenuator in vitro using base analogs. Cell 1980; 20:739-748.
15. Martin FH, Tinoco I Jr. DNA-RNA hybrid duplexes containing oligo(dA:rU) sequences are exceptionally unstable and may facilitate termination of transcription. Nucleic Acids Res 1980; 8:2295-2299.
16. Wilson, von Hippel PH. Stability of Escherichia coli transcription complexes near an intrinsic terminator. J Mol Biol 1994; 244:36-51.
17. Yager TD, von Hippel PH. A thermodynamic analysis of RNA transcript elongation and termination in *Escherichia coli*. Biochemistry 1991; 30:1097-1118.

18. Chamberlin M. New models for the mechanism of transcription elongation and its regulation. In: Harvey Lectures. New York: Wiley-Liss, 1994:1-21.

19. Reynolds R, Bermudez-Cruz RM, Chamberlin MJ. Parameters affecting transcription termination by *Escherichia coli* RNA polymerase I. Analysis of 13 Rho-independent terminators. J Mol Biol 1992; 224:31-51.

20. Surratt CK, Milan SC, Chamberlin MJ. Spontaneous cleavage of RNA in ternary complexes of *Escherichia coli* RNA polymerase and its signifiance for the mechanism of transcription. Proc Natl Acad Sci USA 1991; 88:7983-7987.

21. Borukhov S, Sagitov V, Goldfarb A. Transcript cleavage factors from *E. coli*. Cell 1993; 72:459-466.

22. Krummel B, Chamberlin M. Structural analysis of ternary complexes of *Escherichia coli* RNA polymerase: Deoxyribonuclease I footprinting of defined complexes. J Mol Biol 1992; 225:239-250.

23. Krummel B, Chamberlin M. Structural analysis of ternary complexes of *Escherichia coli* RNA polymerase: Individual complexes halted along different transcription units have distinct and unexpected biochemical properties. J Mol Biol 1992; 225:221-237.

24. Nudler E, Goldfarb A, Kashlev M. Discontinuous mechanism of transcription elongation. Science 1994; 265:793-796.

25. Nudler E et al. Coupling between transcription termination and RNA polymerase inchworming. Cell 1995; 81:351-357.

26. Wang D et al. Discontinuous movements of DNA and RNA in RNA polymerase accompany formation of a paused transcription complex. Cell 1995; 81:341-350.

27. Feng G et al. GreA-induced transcript cleavage in transcription complexes containing *Escherichia coli* RNA polymerase is controlled by multiple factors, including nascent transcript location and structure. J of Biol Chem 1994; 269:22282-22294.

28. Liu B et al. The DNA replication fork can pass RNA polymerase without displacing the nascent transcript. Nature (London) 1993; 366:33-39.

29. Liu B, Alberts BM. Head-on collision between a DNA replication apparatus and RNA polymerase transcription complex. Science (Washington DC) 1995; 267:1131-1137.

30. Lee DN, Landick R. Structure of RNA and DNA chains in paused transcription complexes containing *Escherichia-coli* RNA polymerase. J Mol Biol 1992; 228:759-777.

31. Kainz M, Roberts JW. Structure of transcription elongation complexes in vivo. Science 1992; 255:838-841.

32. Wagner LA et al. Transcriptional slippage occurs during elongation at runs of adenine or thymine in *Escherichia coli*. Nucleic Acids Res 1990; 18:3529-3536.

33. MacDonald LE et al. Characterization of two types of termination signal for bacteriophage T7 RNA polymerase. J Mol Bio 1994; 238:145-158.

34. Bogenhagen D, Brown DB. Nucleotide sequences in Xenopus 5S DNA required for transcription termination. Cell 1981; 24:261.

35. Lang WH et al. A model for transcription termination by RNA polymerase I. Cell 1994; 79:527-534.

36. Wu SY, Platt T. Transcriptional arrest of yeast RNA polymerase II by *Escherichia coli* Rho protein in vitro. Proc Natl Acad Sci USA 1993; 90:6606-6610.

37. Wiest K, Hawley DK. In vitro analysis of a transcription termination site for RNA polymerase II. Mol Cell Biol 1990; 10:5782-5795.

38. Reines D et al. Identification of intrinsic termination sites in-vitro for RNA polymerase II within eukaryotic gene sequences. J Mol Biol 1987; 196:299-312.

39. Maizels N. The nucleotide sequence of the lactose messenger ribonucleic acid transcribed from the UV5 promoter mutant of *E. coli*. Proc Nat Acad Sci USA 1973; 70:3585-3589.

40. Farnham PJ, Platt T. Rho-independent termination: dyad symmetry in DNA causes RNA polymerase to pause during transcription in vitro. Nucleic Acids Res 1981; 9:563-577.

41. Ring BZ, Roberts JW. Function of a nontranscribed DNA strand site in transcription elongation. Cell 1994; 78:317-324.

42. Levin JR, Chamberlin MJ. Mapping and characterization of transcriptional pause sites in the early genetic region of bacteriophage T7. J Mol Biol 1987; 196:61-84.

43. Lee DN et al. Transcription pausing by *Escherichia-coli* RNA polymerase is modulated by downstream DNA sequences. J Biol Chem 1990; 265:15145-15153.

44. Landick R, Yanofsky C. Transcription attenuation. In: Neidhardt F et al, eds. *Escherichia coli* and salmonella typhimurium: cellular and molecular biology. Washington, DC: Amer Soc Microbiol, 1987:1276-1301.

45. Rosenberg M et al. The relationship between function and DNA sequences in an intercistronic regulatory region of phage lambda. Nature 1978; 272:414-423.

46. Yang X, Roberts JW. Gene Q Antiterminator proteins of *Escherichia-coli* phages 82 and lambda suppress pausing by RNA polymerase at a Rho-dependent terminator and at other sites. Proc Natl Acad Sci USA 1989; 86:5301-5305.

47. Mason SW, Li J, Greenblatt J. Host factor requirements for processive antitermination of transcription and suppression of pausing by the N protein of bacteriophage lambda. J Biol Chem 1992; 267:19418-19426.

48. Fisher R, Yanofsky C. Mutations of the beta subunit of RNA polymerase alter both transcription pausing and transcription termination in the trp operon leader region in vitro. J Biol Chem 1983; 258:8146-8150.

49. McDowell JC et al. Determination of intrinsic transcription termination efficiency by RNA polymerase elongation rate. Science (Washington DC) 1994; 266:822-825.

50. Jin DJ et al. Termination efficiency at Rho-dependent terminators depends on kinetic coupling between RNA polymerase and Rho. Proc Natl Acad Sci USA 1992; 89:1453-1457.

51. Goliger JA et al. Early transcribed sequences affect termination efficiency of Escherichia-coli RNA polymerase. J Mol Biol 1989; 205:331-342.

52. Telesnitsky APW, Chamberlin MJ. Sequences linked to prokaryotic promoters can affect the efficiency of downstream termination sites. J Mol Biol 1989; 205:315-330.

53. Platt T, Richardson JP. Escherichia coli rho factor: Protein and enzyme of transcription termination. In: McKnight SL, Yamamoto KR, eds. Transcriptional Regulation, Book 1. 1992:365-388.

54. Roberts JW. Termination factor for RNA synthesis. Nature 1969; 224:1168-1174.

55. de Crombrugghe G et al. Effect of rho on transcription of bacterial operons. Nature New Biology 1973; 241:260.

56. Richardson JP, Grimley C, Lowery C. Transcription termination factor rho activity is altered in *Escherichia coli* with suA gene mutations. Proc Nat Acad Sci 1975; 72:1725-.

57. Platt T. Rho and RNA: Models for recognition and response. Molecular Microbiology 1994; 11:983-990.

58. Galloway JL, Platt T. Signals sufficient for rho-dependent transcription termination at Trp T' span a region centered 60 base pairs upstream of the earliest 3' end point. J Biol Chem 1988; 263:1761-1767.

59. Kupper H et al. A rho-dependent termination site in the gene coding for tyrosine tRNA su3 of *Escherichia coli.* Nature 1978; 272:423-428.

60. Stewart V, Landick R, Yanofsky C. Rho-dependent transcription termination in the tryptophanase operon leader region of *Escherichia-coli* K-12. J Bacteriol 1986; 166:217-223.

61. Platt T. Termination of transcription and its regulation in the tryptophan operon of E. coli. Cell 1981; 24:10-23.

62. Albrechtsen B et al. Transcriptional termination sequence at the end of the *Escherichia-coli* ribosomal rrnG operon complex terminators and antitermination. Nucleic Acids Res 1991; 19:1845-1852.

63. Richardson JP. Preventing the synthesis of unused transcripts by rho factor. Cell 1991; 64:1047-1050.

64. Lowery-Goldhammer C, Richardson JP. An RNA-dependent nucleoside triphosphate phosphohydrolase (ATPase) activity associated with rho termination factor. Proc Natl Acad Sci USA 1974; 71:2003-2007.

65. Zalatan F, Platt T. Effects of decreased cytosine content on rho interaction with the rho-dependent terminator trp t' in *Escherichia coli.* J Biol Chem 1992; 267:19082-19088.

66. Hart CDM, Roberts JW. Rho-dependent transcription termination characterization of the requirement for cytidine in the nascent transcript. J Biol Chem 1991; 266:24140-24148.

67. Chen CYA, Richardson JP. Sequence elements essential for rho-dependent transcription termination at lambda tR1. J Biol Chem 1987; 262: 11292-11299.

68. Lau LF, Roberts J. ρ-Dependent transcription termination at lambda tR1 requires upstream sequences. J Biol Chem 1985; 260:574-584.

69. Zalatan F, Galloway Salvo- J, Platt T. Deletion analysis of the *Escherichia coli* rho dependent transcription terminator trp t'. J Biol Chem 1993; 268:17051-17056.

70. Morgan WD et al. RNA sequence and secondary structure requirements for rho-dependent trancription termination. Nucleic Acids Research 1985; 13:3739-3754.

71. Alifano P et al. A consensus motif common to all rho-dependent prokaryotic transcription terminators. Cell 1991; 64:553-563.

72. Hart CM, Roberts JW. Deletion analysis of the lambda tR1 termination region. Effect of sequences near the transcript release sites, and the minimum length of rho-dependent transcripts. J of Mol Biol 1994; 237:255-265.

73. Lau LF, JW Roberts, Wu R. RNA polymerase pausing and transcript release at the lambda tR1 terminator in vitro. J Biol Chem 1983; 258:9391-9397.

74. Pavco PA, Steege DA. Elongation by *Escherichia coli* RNA polymerase is blocked in vitro by a site-specific DNA binding protein. J Biol Chem 1990; 265:9960-9969.

75. Dolan JW, Marshall NF, Richardson JP. Transcription termination factor rho has three distinct structural domains. J Biol Chem 1990; 265: 5747-5754.

76. Dombroski AJ, Platt T. Structure of rho factor an RNA-binding domain and a separate region with strong similarity to proven ATP-binding domains. Proc Natl Acad Sci USA 1988; 85:2538-2542.

77. Richardson JP. Activation of rho protein ATPase requires simultaneous interaction at two kinds of nucleic acid-binding sites. J Biol Chem 1982; 257:5760-5766.

78. Seifried SE, Easton JB, Hippel von PH. ATPase activity of transcription termination factor rho: Functional dimer model. Proc Natl Acad Sci USA 1992; 89:10454-10458.

79. Morgan WD, Bear DG, von Hippel PH. Rho-dependent termination of transcription. I. Identification and characterization of termination sites for transcription from the bacteriophage Lambda pR promoter. J Biol Chem 1983; 258:9553-9564.

80. McSwiggen JA, Bear DG, von Hippel PH. Interactions of *Escherichia-coli* transcription termination factor rho with RNA I. Binding stoichiometries and free energies. J Mol Biol 1988; 199:609-622.

81. Geiselmann J et al. A physical model for the translocation and helicase activities of *Escherichia-coli* transcription termination protein rho. Proc Natl Acad Sci USA 1993; 90:7754-7758.

82. Brennan CA, Dombroski AJ, Platt T. Transcription termination factor rho is an RNA-DNA helicase. Cell 1987; 48:945-952.

83. Steinmetz EJ, Platt T. Evidence supporting a tethered tracking model for helicase activity of *Escherichia coli* Rho factor. Proc Nat Acad Sci USA 1994; 91:1401-1405.

84. Zheng C, Friedman DI. Reduced Rho-dependent transcription termination permits NusA-independent growth of *Escherichia coli*. Proc Nat Acad Sci USA 1994; 91:7543-7547.

85. Sullivan SL, Gottesman ME. Requirement for *Escherichia coli* NusG protein in factor-dependent transcription termination. Cell 1992; 68:989-994.

86. Nehrke KW, Platt T. A quaternary transcription termination complex: Reciprocal stabilization by Rho factor and NusG protein. J Mol Biol 1994; 243:830-839.

87. Greenblatt J, Nodwell JR, Mason SW. Transcriptional antitermination. Nature (London) 1993; 364:401-406.

88. Das A. How the phage lambda N gene product suppresses transcription termination: Communication of RNA polymerase with regulatory proteins mediated by signals in nascent RNA. J Bact 1992; 174:6711-6716.

89. Adhya S, Gottesman M, Crombrugghe B. Release of polarity in *Escherichia coli* by gene *N* of phage λ: Termination and antitermination of transcription. Proc Natl Acad Sci USA 1974; 71:2534-2538.

90. Franklin NC. Altered reading of genetic signals fused to the N operon of bacteriophage λ: Genetic evidence for the modification of polymerase by the protein product of the *N* gene. J Mol Biol 1974; 89:33-48.

91. Roberts JW. Antitermination and the control of transcription elongation. In: McKnight SL, Yamamoto KR, eds. Transcriptional regulation, Book I. 1992:389-406.

92. Berg KL, Squires C, Squires CL. Ribosomal RNA operon anti-termination function of leader and spacer region box B-box A sequences and their conservation in diverse microorganisms. J Mol Biol 1989; 209:345-358.

93. Landick R, Turnbough CL Jr. Transcriptional attenuation. In: McKnight SL, Yamamoto KR, eds. Transcriptional regulation. Cold Spring Harbor: Cold Spring Harbor Press, 1992:407-446.

94. Turnbough CL Jr, Hicks KL, Donahue JP. Attenuation control of pyrBI operon expression in *Escherichia coli* K12. Proc Natl Acad Sci 1983; 80:368-372.

95. Antson AA et al. The structure of trp RNA-binding attenuation protein. Nature (London) 1995; 374:693-700.

96. Stewart V, Yanofsky C. Evidence for transcription antitermination control of tryptophanase operon expression in *Escherichia coli* K-12. J Bacteriol 1985; 164:731-740.

97. Choder-Amster O, Wright A. Transcriptional regulation of the bgl operon of *Escherichia-coli* involves phosphotransferase system-mediated phosphorylation of a transcriptional antiterminator. J Cell Biochem 1993; 51:83-90.

98. Grundy FJ, Henkin TM. tRNA as a positive regulator of transcription antitermination in bacillus subtilis. Cell 1993; 74:475-482.

99. Friedman DO, Wilgus GS, Mural RJ. Gene N regulator function of phage λimm21: evidence that a site of action of N differs from a site of recognition. J Mol Biol 1973; 81:505-516.

100. Somasekhar G, Szybalski W. The functional boundaries of the Q utilization site required for antitermination of late transcription in bacteriophage lambda. Virology 1987; 158:414-426.

101. Yang X et al. Transcription antitermination by phage lambda gene Q protein requires a DNA segment spanning the RNA start site. Genes Dev 1987; 1:217-226.

102. Yarnell WS, Roberts JW. The phage lambda gene Q transcription antiterminator binds DNA in the late gene promoter as it modifies RNA polymerase. Cell 1992; 69:1181-1189.

103. Lazinski D, Grzadzielska E, Das A. Sequence-specific recognition of RNA hairpins by bacteriophage antiterminators requires a conserved arginine-rich motif. Cell 1989; 59:207-218.

104. Nodwell JR, Greenblatt J. The nut site of bacteriophage lambda is made of RNA and is bound by transcription antitermination factors on the surface of RNA polymerase. Genes Dev 1991; 5:2141.

105. Grayhack EJ, Roberts JW. The phage λ gene Q product: activity of a transcription antiterminator in vitro. Cell 1982; 30:637-648.

106. Grayhack EJ et al. Phage lambda gene Q antiterminator recognizes RNA polymerase near the promoter and accelerates it through a pause site. Cell 1985; 42:259-270.

107. Ring BZ, Roberts JW. Unpublished experiments. 1995.

108. Guo H-C. Mutational analysis of transcription antitermination mediated by lambdoid phage gene Q products. Ph.D. thesis. Ithaca, NY: Cornell University, 1990.

109. Yang X. Transcription antitermination mediated by lambdoid phage Q proteins. Ph.D. thesis. Ithaca, NY: Cornell University, 1988.

110. Yarnell W, Roberts JW. Unpublished experiments. 1995.

111. Roberts CW, Roberts JW. Unpublished experiments. 1995.

112. McDowell JC. Relation of transcription termination and antitermination to the kinetics of transcription elongation. Ph.D. thesis. Ithaca, NY: Cornell University, 1994.

113. Friedman DI. A bacterial mutant affecting lambda development. In: Hershey AD, ed. The Bacteriophage Lambda. Cold Spring Harbor: Cold Spring Harbor Laboratory, 1971:733.

114. Georgopoulos CP. A bacterial mutation affecting N function. In: Hershey AD, ed. The Bacteriophage Lambda. Cold Spring Harbor: Cold Spring Harbor Laboratory, 1971.

115. Friedman DI, Gottesman ME. Lytic mode of lambda development. In: Hendrix et al, eds. Lambda II. 1983.

116. Li J et al. Nusg a new *Escherichia-coli* elongation factor involved in transcriptional antitermination by the N protein of phage lambda. J Biol Chem 1992; 267:6012-6019.

117. Sullivan SL, Ward DF, Gottesman ME. Effect of *Escherichia coli* NusG function on lambda N-mediated transcription antitermination. J Bacteriol 1992; 174:1339-1344.

118. Salstrom JS, Szybalski W. Coliphage λ*nut*L–: a unique class of mutants defective in the site of gene N product utilization for antitermination of leftward transcription. J Mol Biol 1978; 134:195-221.

119. Greenblatt J, Li J. Interaction of the sigma factor and the *nus*A gene protein of *E. coli* with RNA polymerase in the initiation-termination cycle of transcription. Cell 1981; 24:421-428.

120. Greenblatt J, Li J. The *nus*A gene product of *Escherichia coli*: Its identification and a demonstration that it interacts with the gene N transcription antitermination protein of bacteriophage lambda. J Mol Biol 1981; 147:11-23.

121. Chattopadhyay S et al. Bipartite function of a small RNA hairpin in transcription antitermination in bacteriophage lambda. Proc Nat Acad Sci USA 1995; 92:4061-4065.

122. Nodwell JR, Greenblatt J. Recognition of boxA antiterminator RNA by the *Escherichia coli* antitermination factors NusB and ribosomal protein S10. Cell 1993; 72:261-268.

123. Doelling JH, Franklin NC. Effects of all single base substitutions in the loop of boxB on antitermination of transcription by bacteriophage lambda's N protein. Nucleic Acids Res 1989; 17:5565-5578.

124. Mogridge J, Mah T-F, Greenblatt J. A protein-RNa interaction network facilitates the template-independent cooperative assembly on RNA polymerase of a stable antitermination complex containing the N protein of phage λ. Genes and Development 1995; (in press).

125. Whalen W, Ghosh B, Das A. NusA protein is necessary and sufficient in vitro for phage lambda N gene product to suppress a rho-independent terminator placed downstream of nutL. Proc Natl Acad Sci USA 1988; 85:2494-2498.

126. Devito J, Das A. Control of transcription processivity in phage lambda: Nus factors strengthen the termination-resistant state of RNA polymerase induced by N antiterminator. Proc Nat Acad Sci USA 1994; 91: 8660-8664.

127. Morgan EA. Insertions of Tn10 into an *E. coli* ribosomal RNA operon are incompletely polar. Cell 1980; 21:257-265.

128. Berg KL, Squires C, Squires CL. Ribosomal RNA operon anti-termination function of leader and spacer region boxB-boxA sequences and their conservation in diverse microorganisms. J Mol Biol 1989; 209:345-358.

129. Squires CL et al. Ribosomal RNA antitermination in vitro: Requirement for Nus factors and one or more unidentified cellular components. Proc Natl Acad Sci USA 1993; 90:970-974.

130. Robert J et al. The remarkable specificity of a new transcription termination factor suggests that the mechanisms of termination and antitermination are similar. Cell 1987; 51:483-492.

131. Hung SC, Gottesman ME. Phage HK022 nun protein arrests transcription on phage lambda DNA in vitro and competes with the phage lambda N antitermination protein. J Mol Biol 1995; 247:428-442.

132. Oberto J et al. Antitermination of early transcription in phage HK022: Absence of a phage-encoded antitermination factor. J Mol Biol 1993; 229:368-381.

133. Clerget M, Jin DJ, Weisberg RA. A zinc-binding region in the β' subunit of RNA polymerase is involved in antitermination of early transcription of phage HK022. J Mol Biol 1995; 248:768-780.

CODON CONTEXT, TRANSLATIONAL STEP-TIMES AND ATTENUATION

G. Wesley Hatfield

I. OVERVIEW

Charles Yanofsky and his co-workers have provided us with a detailed understanding of the attenuation mechanism for the regulation of the tryptophan operon of *Escherichia coli*. This understanding is based on the results of rigorous experimental documentation gathered over the course of the last 15 years. During this time, attenuation mechanisms have also been elucidated for several other operons required for the biosynthesis of amino acids and intermediary metabolites in bacteria. Most recently, Yanofsky and his co-workers have described a model for setting the basal level of transcriptional readthrough at the attenuator of the *trp* operon. This model, which is supported by a great deal of experimental evidence and has been reviewed elsewhere,[1-3] emphasizes the importance of the timing and synchronization of the movement of an RNA polymerase and a ribosome through the leader-attenuator region, i.e., the mechanistic importance of the relative transcription and translation elongation rates for attenuation. While much has been learned about the transit of an RNA polymerase molecule through the leader-attenuator region of the *trp* operon, little is known about the mechanisms that influence the movement of the ribosome. Indeed, little is known about the mechanisms that influence elongation rates and pausing during the translation of any messenger RNA.

In 1989, George Gutman and I described a strong codon pair utilization bias in bacteria, yeast and mammals and suggested that this bias might be related, perhaps in a predictable way, to the translation step-times of individual codon pairs.[4,5] In this article, I describe how we have used the basal level attenuation model for *trp* attenuation to test this suggestion. In Section II, I describe the attenuation mechanism of the *trp* operon and emphasize the temporal features of the mechanism that are proposed to set the basal level of transcriptional readthrough at the *trp* attenuator. In Section III, I review the results of our analyses

Regulation of Gene Expression in Escherichia coli, edited by E. C. C. Lin and A. Simon Lynch. © 1996 R.G. Landes Company.

of the codon pair utilization bias in *Escherichia coli* and discuss the evidence that this bias might be related to translational efficiencies associated with codon context. In Section IV, I describe our experimental evidence for a relationship between codon pair utilization pair bias and translational step-times. And, in Section V, I relate these results to the experimental data of others and discuss the features of codon context that might be most important for attenuation.

II. ATTENUATION

A. THE BASIC MECHANISM

The regulation of transcription termination at a site preceding the structural gene(s) of an operon is called attenuation. Transcription attenuation is now recognized as a common mechanism for the regulation of gene expression in bacteria and more general forms of attenuation have been described in eukaryotic cells and their viruses.[2,6,7] Attenuation has been demonstrated to regulate the expression of bacterial operons required for the biosynthesis of the amino acids tryptophan,[2,7-11] isoleucine and valine (the *ilvGMEDA* and *ilvBN* operons),[12-15] threonine,[16] leucine,[17] histidine[18,19] and phenylalanine.[20] The leader-attenuator regions of these operons share certain structural similarities. The leader region is the DNA sequence between the site of transcription initiation and the first structural gene of the operon. Each leader region encodes a short polypeptide coding sequence followed by a transcription termination site, the attenuator, near the 3' end of the leader sequence. The leader polypeptide coding sequence contains multiple codons for the amino acid(s) synthesized by the gene products of the operon. These are the regulatory codons. The role of this short polypeptide coding sequence is to monitor the cellular supply of the aminoacylated tRNA(s) for the regulatory codons. The leader polypeptide, itself, serves no other apparent functional role in the cell. If the aminoacylated tRNA(s) for the regulatory codons are in ample supply, then the leader polypeptide can be synthesized without interruption and transcription termination at the attenuator will be maximized. Thus, the structural genes of the operon will be expressed at a low, basal, level. If, on the other hand, the cells are starving for the amino acid(s) produced by the gene products of the operon then the aminoacylated tRNA(s) for the regulatory codons will be in short supply and transcription will proceed through the attenuator into the structural genes. Consequently, the catalysts for the biosynthesis of the limiting amino acid(s) will be produced and its supply will be replenished (reviewed in refs. 1-3).

The linkage of the translation and transcription functions necessary for attenuation is brought about by the alternative secondary structures that these leader RNAs can assume. The alternative, mutually exclusive, secondary structures of the 140 nucleotide long leader RNA of the *trp* operon is shown in Figure 4.1. Similar structures can be drawn for the leader RNAs of the other amino acid biosynthetic operons regulated by attenuation.[1,3] The 3:4 stem-loop structure in Figure 4.1A defines a Rho-independent transcription termination signal (a G + C rich stem-loop followed by a tract of uridines). This is the attenuator. The 1:2 stem-loop results in the formation of a transcriptional pause signal. The formation of this structure in the nascent leader RNA causes the transcribing RNA polymerase to pause after the synthesis of the stem 2 at nucleotide 92. Stem 2 of stem-loop 1:2 and stem 3 of stem-loop 3:4 can form yet a third alternative structure, the antiterminator, stem-loop 2:3 (Fig. 4.1B). The formation of

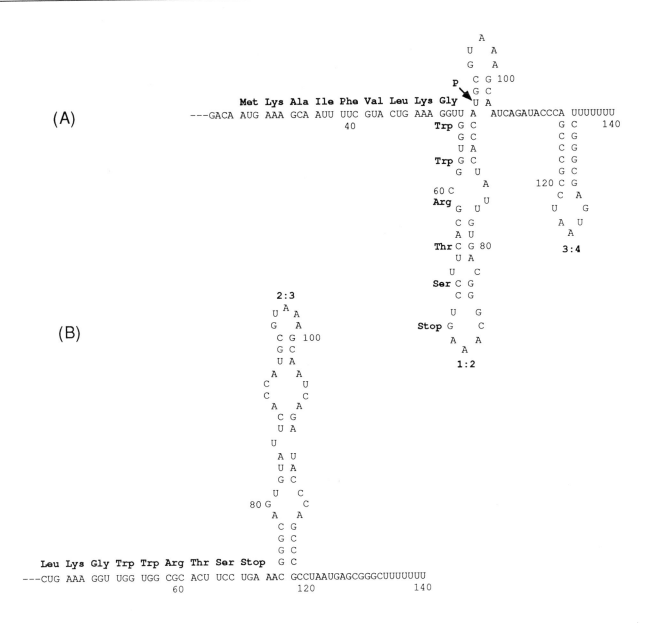

Fig. 4.1. Alternative secondary structures of the leader RNA transcript of the trp operon. (A) Pause structure, stem-loop 1:2; attenuator, stem-loop 3:4. (B) Antiterminator, stem-loop 2:3.

the antiterminator preempts the formation of the attenuator structure and effects deattenuation, transcription through the attenuator. The formation of these alternative structures (stem-loops 1:2 and 3:4 or stem-loop 2:3) is directed by the temporal positioning of a translating ribosome in the leader polypeptide coding region in relation to the transcribing RNA polymerase (reviewed in refs. 1-3).

The synchronization of ribosome and RNA polymerase movement through the leader-attenuator region is ensured by the presence of the transcriptional pause site. The paused RNA polymerase is released to resume transcription when the translating ribosome enters the stem 1 coding region and disrupts the stem-loop 1:2 pause structure. Once the translating ribosome releases the paused RNA polymerase, transcription and translation are synchronized. If the cellular levels of aminoacylated tRNAs are high, then translation through the leader polypeptide coding region will proceed unimpeded and translation and transcription will remain synchronized until the ribosome reaches the stop codon. The stop codon of the *trp* leader polypeptide coding region is in the loop region of the stem-loop 1:2 structure (Fig. 4.1). The presence of a ribosome at this site inhibits base pairings between stems 1 and 2 as well as 2 and 3 (Fig. 4.1B). The basal level of transcriptional readthrough at the attenuator is determined by the time it takes the ribosome to release the leader RNA template relative to the transcriptional speed of the RNA polymerase. If the ribosome were to dissociate from the leader RNA as soon as it reached the stop codon, before the synthesis of stem 3, then stem-loop 1:2 would be free to form followed by the synthesis, and the unchallenged formation, of the stem-loop 3:4 attenuator structure. This situation would result in a maximal level of transcription termination (superattenuation) at the attenuator. However, since the average ribosome release time is thought to be about 0.6 seconds[21] and the average transcription rate is about 50 nucleotides per second,[22] the released RNA polymerase will have synthesized stem 3 and be in stem 4 at the time the ribosome releases. In this case, the formation of the antiterminator structure, stem-loop 2:3 will compete with the formation of the alternative stem-loop 1:2 and 3:4 structures when the ribosome releases to set an average basal level of transcriptional readthrough at the attenuator. Thus, the basal level of attenuation in the *trp* operon is determined by the location of the transcribing RNA polymerase when the ribosome dissociates from the leader RNA template and by the relative probabilities for the formation of the competing RNA structures (reviewed in refs. 1,3).

Roesser and Yanofsky[23,24] have shown that mutations in *prfB*, encoding Release Factor 2 (RF2; UGA- and UAA-specific[25]), that delay ribosome release from the RNA template, increase transcription termination at the attenuator of the *trp* operon approximately 2-fold. These *prfB* mutations exert no effect on basal level expression in strains in which the naturally occurring *trp* leader polypeptide stop codon, UGA, is replaced with UAG. However, transcription termination at the attenuator is increased in these strains when they contain mutations in *prfA* which encodes Release Factor 1 (RF1; UAG- UAA-specific[25]). These experiments demonstrate the importance of the rate of ribosome release from the *trp* leader RNA for setting the basal level of transcription through the attenuator of the *trp* operon.

B. A Temporal Analysis of Basal level Attenuation in the Tryptophan Operon

To further illustrate the importance of the position of the RNA polymerase when the ribosome releases from the leader RNA for setting the basal level of attenuation in the *trp* operon, it is instructive to consider the details of the temporal events of transcription and translation through the leader-attenuator region of the *trp* operon. These events are displayed in Figure 4.2 in the order in which they are described in the following discussion. The attenuated leader RNA of the *trp* operon is 140 nucleotides long. If the average transcription rate through the *trp* leader is the same as the average transcription rate for *E. coli*, 50 bases per second,[22] and the frequency of translation initiation is one per second,[1,26] then a ribosome will initiate translation one second after the completion of the synthesis of the leader polypeptide translation initiation sequence. At this time, the RNA polymerase would be at base pair 77 and the ribosome would be centered on nucleotide 27 of the leader RNA (Fig. 4.2A). By the time the RNA polymerase reaches the transcriptional pause site (0.3 seconds later), the ribosome will be at nucleotide position 42, twelve nucleotides away from the stem 1 region of the stem-loop 1:2 pause structure (Fig. 4.2B). The

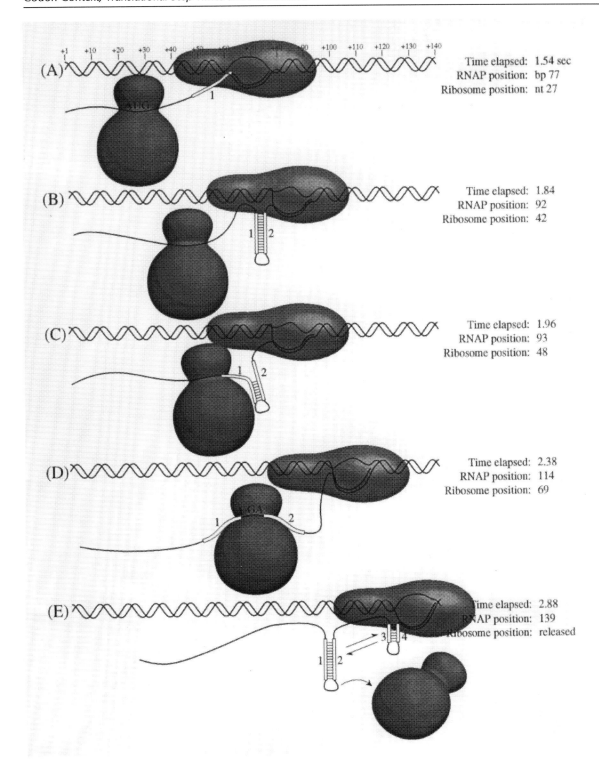

Time elapsed: 1.54 sec
RNAP position: bp 77
Ribosome position: nt 27

Time elapsed: 1.84
RNAP position: 92
Ribosome position: 42

Time elapsed: 1.96
RNAP position: 93
Ribosome position: 48

Time elapsed: 2.38
RNAP position: 114
Ribosome position: 69

Time elapsed: 2.88
RNAP position: 139
Ribosome position: released

RNA polymerase will remain at the pause site for about 0.1 seconds until the pause structure is disrupted by the translating ribosome (A ribosome masks about nine nucleotides on either side of the codon upon which it is centered.[27] For the purpose of this discussion, it is assumed that the ribosome must disrupt three base pairs of stem-loop 1:2 to release the paused RNA polymerase; Fig. 4.2C). At this point, the translating ribosome and the transcribing RNA polymerase are synchronized. If an ample supply of aminoacylated tryptophanyl-tRNA is present, translation and transcription will continue in synchrony until

Fig. 4.2. Temporal events during basal level transcription through the trp attenuator. The RNA polymerase (RNAP) and ribosome positions at the time elapsed since the initiation of transcription is given beside each panel. See text for discussion.

the translating ribosome arrives at the translation stop codon centered at nucleotide position 69 (Fig. 4.2D). At this time, the RNA polymerase will be at base pair position 114 in the region between the DNA sequences encoding stem 2 and stem 3. If, as described above, the ribosome were to release immediately upon its arrival at the stop codon in the middle of the stem-loop 1:2 structure, this structure would be free to form prior to the synthesis of stem 3. This would leave stem 3 free to base pair with stem 4 as these regions are synthesized and cause maximal termination (superattenuation) at the attenuator. However, since the average ribosome release time in *E. coli* is about 0.6 seconds,[21] the transcribing RNA polymerase will usually be near the end of the leader RNA when the ribosome releases (Fig. 4.2E). In this situation, competition for the formation of the alternative secondary structures will ensue. The more frequent formation of the antiterminator structure under this condition produces a higher basal level of transcription through the attenuator and a correspondingly higher basal level of expression of the structural genes of the operon.

Although the translational and transcriptional kinetic parameters assumed in this temporal analysis of the *trp* attenuation mechanism are based on *average* values obtained from the literature, they suffice to explain *trp* attenuation as we currently understand it, and they are in agreement with the values used by Suzuki et al[26] for the mathematical analysis of a stochastic model of *trp* attenuation including transcriptional pausing. In any case, this temporal analysis illustrates the point that the translation elongation rate of the leader polypeptide coding region of the *trp* attenuator (and presumably all translationally coupled attenuators) is of critical importance to the function of the attenuation mechanism. Changes in this rate should produce predictable effects on the basal level of transcription termination at the attenuator.

III. CODON CONTEXT AND TRANSLATIONAL EFFICIENCY

It should be clear from the above explanation of attenuation that "timing is everything." That is, the attenuation decision is influenced by the relative translation and transcription rates through the leader-attenuator region. Therefore, to understand the attenuation mechanism more thoroughly, we need to know how the transcription and translation rates through the leader-attenuator region are controlled. However, while we know quite a lot about transcriptional pausing in the *trp* leader,[2] we know relatively little about the factors that set the translational step-times through the *trp* leader polypeptide coding region.

Many reports have provided evidence that ribosomes translate mRNA with variable rates, and translational pauses which result in the in vivo accumulation of nascent polypeptide chains have been observed during the synthesis of several proteins.[28-35] Nevertheless, the factors that influence these variable translation rates have not been clarified. For some time, the conventional wisdom has been that infrequently used codons (for which the cognate tRNA is present at low concentration) are translated slowly, and that frequently used codons (for which the cognate tRNA is present at high concentration) are translated rapidly. However, a growing body of evidence is accumulating which supports the idea that translational elongation rates (translational step-times) are influenced by the compatibility of adjacent tRNA molecules on the surface of the translating ribosome, and that these rates are not related to codon usage. In this section, I present evidence that translational step-times are also related to a recently recognized, species-specific, highly biased, codon pair utilization pattern in *E. coli*, which might also be related to tRNA compatibilities. The evidence that codon pair utilization patterns are independent of codon usage and that they might be related to translational step-times has been reviewed elsewhere.[4]

If we accept that tRNA compatibilities play a major role in determining translational step-times, and that the regulation of translational step-times is an important aspect of protein synthesis, then a clear prediction emerges. The use of adjacent codons (codon pairs) that determine the positioning of tRNAs next to one another on the ribosome should be subject to a high degree of selection, and there should be a substantially nonrandom pattern of utilization of the 3721 (61^2) possible pairs of nonterminating codons in protein coding sequences. In this section, I review our evidence that codon pair utilization patterns are indeed highly biased, and that this bias is independent of biases in codon usage and dinucleotide and amino acid pair frequencies.

A computer analysis of the codon pair bias in *E. coli* was performed on a nonredundant collection of protein-coding regions consisting of 75,403 codon pairs in 237 sequences.[4,5] Codon usage frequencies for the 61 codons were determined for each sequence independently, and used to calculate the expected values for each of the 3721 codon *pairs* (EXP1). The use of codon frequencies in *each sequence*, as opposed to using global values, minimizes the contribution to the overall bias of differing codon usage between genes, which is known to be substantial.[36] A comparison of these expected codon pair usage values with the observed values yielded a set of chi square values which we refer to as CHISQ1. We calculated another set of expected values for each codon pair (EXP2) in such a manner as to remove that component of the codon pair bias associated with bias in *amino acid* nearest neighbors; this yielded a new set of chi square values (CHISQ2). Thus, the bias represented by CHISQ2 cannot be the consequence either of bias in codon usage per se (since the actual codon frequencies were used to calculate the expected values), or of bias in amino acid nearest neighbors. We applied an additional correction for the bias of neighboring nucleotides of adjacent codons (III-I dinucleotide frequencies), yielding a third set of chi square values corrected for both amino acid pair and III-I dinucleotide biases (CHISQ3). Dinucleotide bias is of minimal importance for *E. coli*, although it is a very significant factor for the mammalian databases.[4] Only CHISQ3 values are considered in this article; however, a computer database containing an alphabetical listing of all the codon pairs with observed (OBS1, OBS2 and OBS3), expected (EXP1, EXP2, and EXP3) and chi square (CHISQ1, CHISQ2 and CHISQ3) values for each pair based on its representation in *E. coli* protein coding sequences is available by anonymous ftp from the University of California, Irvine (ftp.uci.edu in subdirectory/mmg/codpair).

Several conclusions were drawn from our analyses of codon pair utilization in *E. coli* protein coding sequences:[4,5]

1. There is a high degree of bias in codon pair utilization in *E. coli*. Many codon pairs are observed many more times than expected (over-represented codon pairs) and many pairs are observed fewer times than expected (under-represented codon pairs). In fact, the sum of the CHISQ3 values of all codon pairs is more than 120 standard deviations removed from its expected value, and the manner in which these calculations were carried out ensures that this is not the simple consequence of nonrandomness in codon usage, amino acid nearest neighbors or dinucleotide frequencies.

2. This codon pair bias represents a *short-range* effect; analyzing codon pairs separated by two or three intervening codons removes more than 95% of the bias.

3. This codon pair bias is independent of the *abundance* of codon pairs. There is little correlation between codon pair usage and codon pair bias since abundant pairs can show high degrees of *either* over- *or* under-representation, as can rarely used pairs, and the correlation between the chi square and the abundance of codon pairs is poor (the *abundance* of a codon pair is simply defined as the number of times it is observed in the database; this, in turn, is related to the relative frequency of its constituent codons, since frequently used codons tend to be members of abundant pairs).

4. There is a high degree of *directionality* evident in this codon pair bias. There is very little correlation between the values of chi square for any given codon pair and its reverse counterpart (i.e., pair A-B vs. B-A).

5. Genes expressed at high versus low levels use very different proportions of highly over-represented versus under-represented codon pairs; specifically, those *genes expressed at high levels tend to avoid highly over-represented pairs* (in addition to the well-known avoidance of infrequently used codons). This suggests that (at least a portion of) codon pair bias is related to effects on the translation process.

Each of these features of codon pair utilization in *E. coli* are consistent with the hypothesis that codon pair bias is related to the translation process and correlated with the compatibilities of adjacent tRNA molecules on the surface of a translating ribosome.

IV. EFFECTS OF CODON PAIR BIAS ON TRANSLATIONAL STEP-TIMES

The results of our computer analyses show that the protein-coding sequences of prokaryotes (and yeast and mammals) exhibit a high degree of bias in codon pair usage.[4] Some codon pairs are used many more times than expected (based on the usage of the individual codons of the pair) and others many fewer times than expected. This observation, coupled with the evidence that tRNA-tRNA interactions on the surface of a translating ribosome and codon context influence translational elongation rates, suggested that there might exist a relationship between codon pair bias and the translational step-times of individual codon pairs. It occurred to us that the use of one codon next to another may have co-evolved with the abundance and structure of tRNA isoacceptors in order to allow control of the kinetics of translation of a growing polypeptide chain without simultaneously imposing constraints on amino acid sequence or protein structure. We, therefore, wished to determine if the translational step-times of over- and under-represented codon pairs might be demonstrably different. In order to do this, we needed an in vivo assay capable of measuring the relative translational step-times of selected codon pairs in a growing polypeptide. We have developed two such assays.

One assay takes advantage of the observation that ribosome pausing at a site near the beginning of an mRNA coding sequence can inhibit translation initiation by physically interfering with the attachment of a new ribosome to the message.[37] We have constructed a *trc::lacZ* translational fusion plasmid containing the *trc* promoter and the *lacZ* translation initiation sequences with a unique *Nco*I site at the ATG translation start codon and a *Bam*HI site preceding the ninth codon of the *lacZ* coding region of the *lac* operon. Codon pair substitutions near the beginning of the *lacZ* gene are created by the insertion of double-stranded synthetic DNA oligonucleotides into these unique restriction sites. These transcriptional fusion vectors are inserted into the chromosome of *E. coli* in single copy, and the effects of selected codon pairs on the level of *lacZ* translation are determined by measuring the steady state level of β-galactosidase activity in whole cells during steady state growth in a rich medium. To ensure that differences detected in β-galactosidase levels are due to translational events, and to ascertain any effects that the nucleotide changes might have on *lac* mRNA stability, the steady state levels of transacetylase activity (*lacA*) are measured along with the β-galactosidase activities.

The other assay is based on the observation that the transit time of a ribosome through the leader polypeptide coding sequence of the leader RNA of the *trp* operon of *E. coli* sets the basal level of transcriptional readthrough at the attenuator (see Section II). For this assay, we have constructed a *trp::lacZ* transcriptional fusion plasmid containing the *trpLep* promoter and leader-attenuator region described by Landick et al.[38,39] The leader polypeptide coding region of this construct contains unique *Pst*I and *Eco*RI sites that allow us to use double-stranded synthetic DNA oligonucleotides to replace the codon pair at positions 9 and 10 preceding the tandem Trp codons at positions 11 and 12 in the *trp* leader polypeptide coding sequence. It has been shown that nucleotide changes in these codons do not interfere with the secondary structures of *trp* leader RNA.[40,41] These transcriptional fusion vectors are inserted into the chromosome of *E. coli* in single copy, and the effects of selected codon pairs on the basal level of transcription through the attenuator into the *lacZ* gene is determined by measuring the steady state level of β-galactosidase activity in whole cells during steady state growth in a rich medium. Again, transacetylase activity is measured to detect any effects nucleotide changes in the *trp* leader RNA might have on the stability of the downstream *lac* mRNA. Ribosome pausing at codons 9 and 10 is expected to, at least partially, disrupt base pairings between stems 1 and 2 and cause deattenuation, increased transcription through the attenuator into the *lacZ* gene, and a higher β-galactosidase activity.

A. THE β-GALACTOSIDASE TRANSLATION INITIATION INHIBITION ASSAY

The specific effects of various codon pairs on the translation of β-galactosidase are shown in Table 4.1. For the β-galactosidase translation inhibition assay, the first eight codons of the leader polypeptide coding region of the *ilvGMEDA* leader-attenuator region were translationally fused to the ninth codon (CCC) of the *lacZ* coding sequence (Table 4.1, sequence A). The chi square values (CHISQ3) are given for each codon pair. Codon pairs that are used at their expected frequencies (i.e., exhibit no codon pair utilization bias) show a chi square value of 0.0. The more the usage of a codon pair deviates from its expected usage the higher the chi square value. For convenience, minus signs are used to identify chi square values of under-represented pairs. The CHISQ3 values for the codon pairs in the *E. coli* data base range from -52.5 for the most under-represented codon pair to 125.7 for the most over-represented codon pair. In this region of the *ilvGMEDA* leader polypeptide, as in other leader polypeptide coding sequences, mostly randomly used codon pairs are observed. That is, each of the codon pairs in the sequence possess chi square values very near 0.0. In sequence B of Table 4.1, the slightly under-represented, Ala-Leu, codon pair at sequence positions 3 and 4 of sequence A (GCC CUU; CHISQ3 = -5.4) is replaced with a more highly under-represented, Thr-Leu, codon pair (ACC CUG; CHISQ3 = -27.3). This change results in a 2-fold increase in the translatability of the *lacZ* coding sequence. The level of transacetylase activity in these cells is unaltered. In sequence C, only a single nucleotide of the highly under-represented, Thr-Leu, codon pair (ACC CUG; CHISQ3 = -27.3) of sequence B is changed to create a highly over-represented, Thr-Leu, codon pair (ACG CUG. CHISQ3 = 78.9). This single nucleotide change inhibits the translation of the *lacZ* message 10-fold. Again, the transacetylase activity is unchanged in these cells. Thus, this rather dramatic effect is at the level of translation and not at the level of transcription or message turnover. Also, the fact that the amino acid sequence of the β-galactosidase produced from sequences B and C are the same argues against the idea that the differences in β-galactosidase levels are due to altered enzyme activity or protein turnover. These results suggest, therefore, that the highly over-represented Thr-Leu codon pair (ACG CUG) is translated much more slowly than the highly under-represented Thr-Leu codon pair (ACC CUG).

We have previously determined that the codon pair bias in *E. coli* is directional.[4,5] That is, there is little correlation between the codon pair utilization bias of codon pair A-B and codon pair B-A. This directionality fits well with the directionality of protein synthesis and the idea that codon pair bias might be associated with the compatibility of adjacent tRNAs in the A- and P-sites of a translating ribosome. To determine if the translational efficiency of a given codon pair is

Table 4.1. Effect of codon pair bias on translation initiation of β-galactosidase mRNA

β - Galactosidase Coding Sequence and Chi Square Values of Codon Pairs		Specific Activity[1]	
		β - Galactosidase	Transacetylase
1 2 3 4 5 6 7 8 9 Met Thr Ala Leu Leu Arg Val Asp Pro			
(A) AUG ACA GCC CUU CUA CGA GUG GAU CCC CHISQ3: 0.1 -3.0 -5.4 -0.4 0.2 -0.6 -0.5 0.0		4120±990	288±39
Thr Leu (B) AUG ACA **A**CC C**U**G CUA CGA GUG GAU CCC CHISQ3: 0.1 -1.4 -27.3 0.1 0.2 -0.6 -0.5 0.0		9800±600	329±79
Thr Leu (C) AUG ACA AC**G** CUG CUA CGA GUG GAU CCC CHISQ3: 0.1 1.7 78.9 0.1 0.2 -0.6 -0.5 0.0		1030±120	250±55
Leu Thr (D) AUG ACA **CU**G **AC**C CUA CGA GUG GAU CCC CHISQ3: 0.1 -1.1 0.0 -0.3 0.2 -0.6 -0.5 0.0		3900±970	281±85
Ala Leu (E) AUG ACA **G**CG CUG AUA CGA GUG GAU CCC CHISQ3: 0.1 0.1 12.3 0.1 -1.0 -0.6 -0.5 0.0		2040±70	326±19

[1] nmol ONP/min/mg prot.

independent of its directionality, the orientation of the highly under-represented codon pair ACC CUG (CHISQ3 = -27.3) was reversed (Table 4.1, sequence D) to form the unbiased, randomly used, codon pair, CUG ACC. Even though five of the six nucleotides of codons 3 and 4 of sequences A and D are different, all of the codon pairs in both sequences exhibit CHISQ3 values near zero and the levels of β-galactosidase produced from each sequence are nearly equal. This result, and the observation that similar results are obtained when over- and under-represented codon pairs are placed at codon positions 6 and 7 instead of 3 and 4,[48] suggests that the nucleotide sequence in these regions of the mRNA does not influence its rate of translation initiation. The final sequence in Table 4.1, sequence E, contains a modestly over-represented codon pair GCG CUG (CHISQ3 = 12.3) and differs from sequence A by only one nucleotide. However, only one-half as much β-galactosidase is produced from the message containing this over-represented codon pair (sequence E) as is produced from the message containing random codon pairs (sequence A). Thus, these results again suggest that an over-represented codon pair is translated more slowly than an under-represented pair.

B. THE *TRP* ATTENUATION ASSAY

Although the results of the translation initiation inhibition assays suggested a correlation between codon pair utilization bias and translational step-times, we, nevertheless, wished to confirm these results in another way. To do this we developed the attenuation assay described above. Here, the coding sequence of the reporter gene is unaffected by the codon pair changes placed in the leader polypeptide coding region of the leader-attenuator region. With this assay, the β-galactosidase levels are a measure of transcriptional readthrough at the attenuator, and more rather than less β-galactosidase is expected if a slowly translated codon is placed in the leader.

The wild-type *trp* and *trpLep* leader polypetide coding sequences, along with the CHISQ3 values for the utilization bias of each codon pair, are shown in Figure 4.3. The wild-type *trp* leader polypeptide coding sequence is composed of randomly used codon pairs. Even the slightly over-represented Val-Leu codon pair (GUA CUG; CHISQ3 = 8.0) at positions 6 and 7 is used in a nearly random fashion since it is in only the sixth percentile of over-represented codon pairs (i.e., 94% of all over-represented codon pairs in *E. coli* are more over-represented than this pair). Landick et al[41] inserted a new CAG codon into the wild-type *trp* leader polypeptide coding sequence, between the seventh and eighth codons, to create the *Pst*I site of the *trpLep* leader. Interestingly, the resultant codon pair, CUG CAG, is *the* most under-represented codon pair in all *E. coli* protein coding sequences. On the other hand, these frequently used codons form a highly abundant codon pair. It is in the 87th percentile of codon pair abundance (i.e., only 13% of all codon pairs are used more often in *E. coli* protein coding sequences). The *Eco*RI site was placed after the stop codon of the polypeptide coding sequence in the loop of the 1:2 stem-loop of the *trpLep* leader by site-directed mutagenesis. Landick et al[41] have shown that the in vivo basal level of attenuation and the percent transcription readthrough at the attenuator in in vitro transcription assays are the same for the wild-type and *trpLep* leader-attenuator regions. In order to assess the effects of selected codon pairs on the basal level of transcription through the *trp* attenuator, we inserted the synthetic double-stranded DNA oligonucleotides shown in Table 4.2 into the unique *Pst*I and *Eco*RI sites of the *trpLep* leader sequence in the *trp::lacZ* tran-

trp Leader Polypeptide Coding Sequence

Met	Lys	Ala	Ile	Phe	Val	Leu	Lys	Gly	Trp	Trp	Arg	Thr	Ser	stop
AUG	AAA	GCA	AUU	UUC	GUA	CUG	AAA	GGU	UGG	UGG	CGC	ACU	UCC	UGA

CHISQ3:   0.0 -0.2 -2.2 0.0 -7.1 8.0 3.8 -0.8 -6.7 0.0 -4.7 -0.2 2.6

trpLep Leader Polypeptide Coding Sequence

Met	Lys	Ala	Ile	Phe	Val	Leu	Gln	Lys	Gly	Trp	Trp	Arg	Thr	Ser	stop
AUG	AAA	GCA	AUU	UUC	GUA	CUG	**CAG**	AAA	GGU	UGG	UGG	CGC	ACU	UCC	UGA

CHISQ3:   0.0 -0.2 -2.2 0.0 -7.1 8.0 -52.5 3.8 -0.8 -6.7 0.0 -4.7 -0.2 2.6

*Pst*1

Fig. 4.3. The trp *and* trpLep *leader polypeptide coding sequences. See text for discussion.*

scriptional fusion plasmids described above. In these new constructs, the randomly utilized *trpLep* Lys-Gly codon pair (AAA GGT; CHISQ3 = -0.8) at codon positions 9 and 10 is replaced by the under-represented Thr-Leu codon pair (ACC CTG; CHISQ3. = -27.3; Table 4.2, sequence B), or by the over-represented Thr-Leu codon pair (ACG CTG; CHISQ3 = 78.9; Table 4.2, sequence C). These transcriptional fusion plasmids were inserted, in single copy, into the chromosome of an *E. coli trpR, tna, ΔtrpE-A, polA* strain and the effects of selected codon pairs on the basal level of unrepressed transcription through the *trp* attenuator into the *lacZ* gene were determined by measuring the steady state level of β-galactosidase activity in whole cells during steady state growth in a rich medium.

Sequences B and C in Table 4.2 contain the same ACC CUG under-represented and ACG CUG over-represented codon pairs used in the translation initiation inhibition assay described above (Table 4.1). The data in Table 4.2 show that the basal level of transcription through the attenuator of the *trpLep* leader (Sequence A) and the *trpLep* 9,10, under-represented, leader (Sequence B) is the same. If translation through the leader containing this under-represented ACC CUG codon pair is faster, as demonstrated with the translation initiation inhibition assay, then why isn't superattenuation observed? That is, if the ribosome translates the leader faster and arrives at, and releases from, the stop codon earlier then why aren't the formations of stem-loops 1:2 and 3:4 favored as discussed in the temporal analysis of attenuation described above? This is probably due to the fact that since the leader polypeptide already contains randomly utilized, presumably rapidly translated, codon pairs, the addition of one more rapidly translated pair would not add significantly to the overall translation time of the leader polypeptide. For example, if each randomly utilized codon pair in the *trpLep* leader is translated with an average translation step-time of 0.06 seconds, then it will take 0.42 seconds for the ribosome to translate to the end of the leader sequence once it has released the paused RNA polymerase. If, however, the translation step-time of codon pair 9,10 is halved (as suggested by the translation initiation assay), then this ribosome transit time would be decreased by only 0.03 seconds. Perhaps, compared to an average ribosome release time of 0.6 seconds, this much of a variation in the arrival of the ribosome at the stop codon is not significant. We might also ask why the introduction of the highly under-represented codon pair CUG CAA at the *Pst*I site does not effect the basal level of transcription through the attenuator. It is probable that at the time the ribosome translates this codon pair it has not yet released the paused RNA polymerase. In this case, a more rapid transit through the early part of the leader polypeptide would not be expected to affect the timing of the attenuation mechanism.

When the same over-represented ACG CUG codon pair that was shown to inhibit the initiation of β-galactosidase translation (Table 4.1) is placed in the leader polypeptide coding region (Table 4.2, sequence C), the basal level of transcription through the attenuator is increased 2-fold. This result is consistent with a slow translational step-time (ribosome pausing) at this over-represented codon pair which results in a partial deattenuation of transcription through the attenuator of the *trp* operon. The observation that deattenuation rather than superattenuation is observed with this construct is significant. This means that the translational step-time of this over-represented codon pair is slower than the time it takes the released RNA polymerase to transcribe from the pause site to the end of the leader (about 0.9 seconds at an average transcription rate). If the translational step-time were much faster then it would only slow the arrival and release of the ribosome at the stop codon and cause superattenuation. The fact that

Table 4.2. Effect of codon pair bias on basal level attenuation in the trp operon

Double Stranded DNA Oligonucleotides Inserted into *pst*I and *Eco*R1 sites of *trpLep* Leader Region and Chi Square Values of Codon Pairs	Specific Activity[1] β-Galactosidase Transacetylase	

trpLep

Codon no.	7	8	9	10	11	12	13	14	15			
	Leu	Gln	Lys	Gly	Trp	Trp	Arg	Thr	Ser	stop		
		G	AAA	GGT	TGG	TGG	CGC	ACT	TCC	TG	1034±225	89±22
	AC	GTC	TTT	CCA	ACC	ACC	GCG	TGA	AGG	ACT TAA		
CHISQ3:	-52.5	- 2.0	-0.8	-6.7	0.0	-4.7	-0.2	2.6				
	*pst*I								*Eco*R1			

trpLep 9, 10 Under-represented

Codon no.	7	8	9	10	11	12	13	14	15			
	Leu	Gln	Thr	Leu	Trp	Trp	Arg	Thr	Ser	stop		
		G	ACC	CTG	TGG	TGG	CGC	ACT	TCC	TG	1397±298	93±22
	AC	GTC	TGG	GAC	ACC	ACC	GCG	TGA	AGG	ACT TAA		
CHISQ3:	-52.5	-0.2	-27.3	0.2	0.0	-4.7	-0.2	2.6				

trpLep 9, 10 Over-represented

Codon no.	7	8	9	10	11	12	13	14	15			
	Leu	Gln	Thr	Leu	Trp	Trp	Arg	Thr	Ser	stop		
		G	ACG	CTG	TGG	TGG	CGC	ACT	TCC	TG	2144±289	140±35
	AC	GTC	TGC	GAC	ACC	ACC	GCG	TGA	AGG	ACT TAA		
CHISQ3:	-52.5	-0.9	78.9	0.2	0.0	-4.7	-0.2	2.6				

[1] nmol ONP/min/mg prot.

only a partial deattenuation is observed might be because a ribosome stalled a codon 9 is expected to disrupt less of the stem-loop 1:2 structure than a ribosome stalled at the first Trp codon at position 11. This would allow partial base pairings in stem-loop 1:2 that can compete with base pairings in the antiterminator structure necessary for maximal deattenuation. This explanation can be tested by determining the amount of deattenuation that is effected by the placement of a stop codon at position 9.[41,48]

Although only a small number of codon pairs have been tested to date, the data from two different types of assays support the conclusion that over-represented codon pairs are translated more slowly than under-represented codon pairs. Also, the highly over-represented Thr-Leu codon pair, ACG CUG, that is translated slowly in both assay systems is composed of two frequently used codons. The Leu CUG codon is the most frequently used codon in *E. coli* protein coding sequences and the Thr ACG codon is the second most frequently used threonine codon. This demonstrates that even frequently used codons are translated slowly in certain contexts. The results of this experiment also confirm the importance the timing of ribosome movement through the leader polypeptide coding region for setting the basal level of transcription through the *trp* attenuator.

V. DISCUSSION

The evolutionary maintenance of an extreme bias in species-specific codon pair utilization patterns that are independent of amino acid pair biases, dinucleotide pair biases and codon usage, suggests an important biological role for codon context. It has long been suggested that codon context is an important parameter for determining translational efficiency, and we have provided preliminary evidence that the utilization bias of codon pairs is related to their translational efficiencies. The properties of the codon pair bias that we have described are compatible with the features of protein synthesis. For example, codon pair bias is limited to nearest neighbors; that is, codon pairs separated by two or three intervening codons are used as expected from their independent usage frequencies. This feature, along with the directionality of codon pair bias, is compatible with the idea that the bias is related to interactions between adjacent tRNA molecules in a growing polypeptide chain. Another feature of codon pair bias that links it to the translation process is the observation that genes expressed at high levels tend to avoid highly over-represented codon pairs. This observation coupled with our suggestion that highly over-represented codon pairs are translated slowly implies that the translational elongation rates of highly expressed genes are faster than those of genes expressed at much lower levels. While this is not a new suggestion, our explanation for the reasons for these differing rates is new. Previously, it has been suggested that genes expressed at high and low levels are translated at different rates because they, predominantly, contain frequently or infrequently used codons, respectively. Yet, our analyses of codon pair usage suggest that there is no correlation between over- and under-represented codon pairs and the usage frequency of the individual codons of the pair. I, therefore, suggest that codon usage in genes expressed at high and low levels is related *solely* to their substrate requirements. That is, codons in messages expressed at low levels can be serviced by aminoacylated-tRNA isoacceptor pools present at relatively low levels leaving the aminoacylated-tRNA isoacceptor pools present at higher levels free to service messages expressed at high levels.

Much evidence in support of the idea that there is no correlation between codon usage and translational efficiency has appeared in the literature during the last several years (for a review see ref. 4). Indeed, some of these data have been obtained by examining the effects of frequently and infrequently used codons in translationally coupled attenuation systems. For example, Bonekamp et al[42] used the translation rate sensitive *pyrE* attenuation system to show that the translation rates of individual codons are not related to tRNA isoacceptor abundance or codon usage. They replaced three codons in the leader-polypeptide coding sequence of the leader-attenuator region with sets of three, tandem, frequently or infrequently used codons. These experiments demonstrated that codon choice can influence translation rates and exert large effects (7- to 10-fold) on transcriptional readthrough at the *pyrE* attenuator. Interestingly, all of the codons used in their experiments form randomly used (presumably rapidly translated) codon pairs with one another. However, these tandem codons do, occasionally, form over-represented pairs with adjacent codons in the leader sequence and the presence of these over-represented pairs correlates well with the decreased rates of translation elongation observed by Bonekamp et al.[42] For example, their data indicate that the infrequently used isoleucine codon, AUU, can be translated as rapidly as the more frequently used AUA codon and more rapidly than the most frequently used isoleucine codon, AUC. In the context of the *pyrE* leader polypeptide coding sequence, only the AUC codon forms an over-represented (presumably slowly translated) codon pair with an adjacent codon. In another example, they compared the translational efficiency of the equally used histidine codons, CAC and CAU, which are serviced by a single tRNA isoacceptor. While it might be expected that these codons would be translated with equal efficiencies, the CAU codons were translated only about one-third as rapidly as the CAC codons. Again, this context effect correlates with the fact that only the CAC codon forms an over-represented codon pair with an adjacent codon in the *pyrE* leader. The same situation is observed with the cysteine UGU and UGC codons which are also serviced by a single tRNA isoacceptor. On the other hand, I cannot explain the observation that the tandem, frequently used, CUG leucine codons, which form randomly used codon pairs with adjacent codons in the *pyrE* leader polypeptide coding sequence, are translated at only about one-half of the rate of the isoleucine codons AUA and AUU which also form random pairs with their neighbors. Nevertheless, the infrequently used CUA leucine codon does produce an over-represented pair in this sequence and it is translated slower than the more frequently used CUG codon. Thus, it appears that strict correlations between codon pair utilization bias and translational efficiencies might be more closely correlated when synonymous codon changes are compared.

Others have examined the effects of codon changes in the leader polypeptide coding regions of amino acid biosynthetic operons on attenuation. Landick et al[41] changed the tandem *trp* codons in the *trpLep* leader polypeptide coding sequence (Fig. 4.3) from the randomly used Trp codon pair (UGG UGG; CHISQ3 = 0.0) to an Arg-Cys codon pair AGG UGC. These changes do not significantly change the thermodynamic stability of the *trpLep* leader RNA secondary structures or the in vitro RNA polymerase pause time.[41] Interestingly, this AGG-UGC codon pair does not appear in any of the 75,403 codon pairs in our database of *E. coli* protein coding sequences. Therefore, it is not possible to determine with statistical accuracy whether this is an over-represented or an under-represented codon pair. However, Landick et al[41] noted that this codon caused a translation-dependent 2-fold increase in the basal level of transcription through the *trp* attenuator. They suggested that the simplest explanation of this increased readthrough is slow translation of the rare AGG arginine codon. On the other hand, Robinson et al[43] have pointed out that the chloramphenicol transacetylase (CAT) gene, in which 25% of the codons are rare codons, can be successfully over-produced in *E. coli*; but, when four tandem AGG codons are placed in this sequence, the over-expression of the CAT gene is compromised. This is not observed when the frequently used Arg codons CGU are used in place of the rare codons. Robinson et al[43] concluded, therefore, that the minor arginyl-tRNA isoacceptor species can become rate limiting for protein synthesis but only under extreme circumstances. For example, when the rare codon is tandemly repeated in the coding sequence and the message is expressed at a high level. Thus, since we have shown that an over-represented codon pair can deattenuate the *trp* operon 2-fold (see Section IV.B), perhaps an equally plausible explanation for the affect of AGG UGC codon pair on the basal level of *trp* attenuation is codon context.

Another mutation in the *trpLep* leader isolated by Landick et al[41] contained, in addition to the AGG CGU codon pair, a change in the codon at position eight in the leader polypeptide coding sequence from CAG to CAA (Fig. 4.3). This single nucleotide change results in the transformation of a highly under-represented Leu-Gln codon pair (CUG CAG; CHISQ3 = -52.5) into a moderately over-represented Leu-Gln codon pair (CUG CAA; CHISQ3 = 21.6). This change caused a further deattenuation of *trp* operon expression (approx. 25%). This might be explained by ribosome pausing at this codon pair at positions 7 and 8 if this is close enough (nine base pairs away) to the 1:2 stem-loop structure to partially destabilize it and to release the paused RNA polymerase. A translational pause at this site might also function to enhance translation-dependent deattenuation from the AGG CGU codons in stem 1.

Codon replacements in the leader polypeptide coding sequences of other amino acid biosynthetic operons have also been examined.[44-47] In each of these cases, randomly used codon pairs were replaced with other randomly used pairs, and in no case were changes in the basal level of translation through the attenuators observed that could not be attributed to changes in leader RNA secondary structures important for attenuator function. In each case, however, the importance of the regulatory codons for deattenuation during end product amino acid limitation was documented. What is it about the context of these regulatory codons that make them extra sensitive to the in vivo levels of their cognate aminoacylated-tRNA isoacceptors? Two extraordinary features of leader polypeptide coding sequences are immediately apparent. Firstly, these sequences are predominantly composed of randomly utilized, presumably rapidly translated, codons. Secondly, the regulatory codons are invariably rare codons. Therefore, even a small slowing of the translation rate of a rare codon during amino acid limitation will be amplified compared to its effect on the overall elongation rate of a slowly translated sequence. Perhaps this is the major codon context effect operating in leader polypeptide coding sequences. Certainly the kinetic parameters of translation and transcription initiation frequencies and translation and transcription elongation rates are of crucial importance to the attenuation mechanism. It is, therefore, likely that slowly translated codon pairs have been selected against during the evolution of polypeptide coding sequences in leader-attenuator regions.

In this article, I have used some of our own data[48] and the data of others to support an apparent correlation between codon pair utilization bias and codon context effects. I suggest that this correlation is functionally related to the translation process through the interactions of tRNA molecules on the surface of translating ribosomes. The current data suggest that codon pairs that are observed in the coding sequences of *E. coli* proteins more times than is predicted by the usage of the individual codons of the pair (over-represented codon pairs) are translated slowly. It also appears that highly over-represented codon pairs are translated even more slowly than modestly over-represented pairs. As more data are obtained it is possible that a predictable pattern between translational step-times, codon pair usage and tRNA structure might emerge. As these data accumulate they can be used to analyze the importance of translational elongation rates and pausing in attenuation mechanisms as well as many other important biological problems such as translational pausing for in vivo protein folding. In the meantime, attenuation mechanisms will be exploited to obtain this basic information.

REFERENCES

1. Landick R, Yanofsky C. Transcription attenuation. In: Neidheidhardt FC, Ingraham JL, Low KB, Magasanik B, Schaechter M, Umbarger HE, eds. *Escherichia coli* and Salmonella typhimurium: Cellular and Molecular Biology. Washington, D.C.: American Society for Microbiology, 1987:1276-1301.

2. Landick R, Turnbough CL. Transcriptional attenuation. In: McKnight SL, Yamamoto K, ed. Transcriptional Regulation. Cold Spring Harbor, NY: Cold Spring Harbor Press, 1992:407-446.

3. Hatfield GW. A two ribosome model for attenuation. In: Ilan J, ed. Translational expression of gene expression 2. New York, NY: Plenum Publishing Corporation, 1993:1-22.

4. Hatfield GW, Gutman GA. Codon pair utilization bias in bacteria, yeast and mammals. In: Hatfield DL, Lee BJ, Pirtle RM, ed. Transfer RNA in Protein Synthesis. Boca Raton, LA: CRC Press, 1993:157-189.

5. Gutman GA, Hatfield GW. Nonrandom utilization of codon pairs in *Escherichia coli*. Proc Natl Acad Sci USA 1989; 86(10):3699-703.

6. Aloni Y, Hay N. Attenuation may regulate gene expression in animal viruses and cells. CRC Crit Rev Biochem 1985; 18(4):327-83.

7. Yanofsky C. Transcription attenuation. J Biol Chem 1988; 263(2):609-12.

8. Bertrand K, Korn LJ, Lee F, Yanofsky C. The attenuator of the tryptophan operon of *Escherichia coli*. Heterogeneous 3'-OH termini in vivo and deletion mapping of functions. J Mol Biol 1977; 117(1):227-47.

9. Bertrand K, Squires C, Yanofsky C. Transcription termination in vivo in the leader region of the tryptophan operon of *Escherichia coli*. J Mol Biol 1976; 103(2):319-37.

10. Bertrand K, Yanofsky C. Regulation of transcription termination in the leader region of the tryptophan operon of *Escherichia coli* involves tryptophan or its metabolic product. J Mol Biol 1976; 103(2):339-49.

11. Lee F, Squires CL, Squires C, Yanofsky C. Termination of transcription in vitro in the *Escherichia coli* tryptophan operon leader region. J Mol Biol 1976; 103(2):383-93.

12. Friden P, Newman T, Freundlich M. Nucleotide sequence of the ilvB promoter-regulatory region: a biosynthetic operon controlled by attenuation and cyclic AMP. Proc Natl Acad Sci USA 1982; 79(20):6156-60.

13. Hauser CA, Hatfield GW. Nucleotide sequence of the ilvB multivalent attenuator region of *Escherichia coli* K12. Nucleic Acids Res 1983; 11(1):127-39.

14. Lawther RP, Hatfield GW. Multivalent translational control of transcrip-

tion termination at attenuator of ilvGEDA operon of *Escherichia coli* K-12. Proc Natl Acad Sci USA 1980; 77(4):1862-6.

15. Nargang FE, Subrahmanyam CS, Umbarger HE. Nucleotide sequence of ilvGEDA operon attenuator region of *Escherichia coli*. Proc Natl Acad Sci USA 1980; 77(4):1823-7.

16. Gardner JF. Regulation of the threonine operon: tandem threonine and isoleucine codons in the control region and translational control of transcription termination. Proc Natl Acad Sci USA 1979; 76(4):1706-10.

17. Gemmill RM, Sessler SR, Calvo JM. leu operon of Salmonnella typhimurium is controlled by attenuation. Proc Natl Acad Sci USA 1979; 76:4941-4945.

18. Barnes WM. DNA sequence from the histidine control region: seven histidine codons in a row. Proc Natl Acad Sci USA 1978; 75(4281-4285).

19. DiNocera PP, Blasi F, DiLauro R, Frunzio R, Bruni CB. Nucleotide sequence of the attenuator region of the histidine operon of *Escherichia coli* K-12. Proc Natl Acad Sci USA 1978; 75:4276-4280.

20. Zurawski G, Brown K, Killingly D, Yanofsky C. Nucleotide sequence of the leader region of the phenylalanine operon of *Escherichia coli*. Proc Natl Acad Sci USA 1978; 75(9):4271-5.

21. Curran JF, Yarus M. Use of tRNA suppressors to probe regulation of *Escherichia coli* release factor 2. J Mol Biol 1988; 203(1):75-83.

22. Gausing K. Efficiency of protein and messenger RNA synthesis in bacteriophage T4-infected cells of *Escherichia coli*. J Mol Biol 1972; 71(3):529-45.

23. Roesser JR, Nakamura Y, Yanofsky C. Regulation of basal level expression of the tryptophan operon of *Escherichia coli*. J Biol Chem 1989; 264(21):12284-8.

24. Roesser JR, Yanofsky C. Ribosome release modulates basal level expression of the trp operon of *Escherichia coli*. J Biol Chem 1988; 263(28):14251-5.

25. Caskey CT. Peptide chain termination. Trends Biochem Sci 1980; 54:234-237.

26. Suzuki H, Kunisawa T, Otsuka J. Theoretical evaluation of transcriptional pausing effect on the attenuation of trp leader sequence. Biophys J 1986; 49:425-436.

27. Steitz JA. RNA-RNA interactions in ribosome translation initiation. In: Chambliss G, Craven GR, Davies J, Davis K, Kahan L, Nomura M, eds. Ribosomes: Structure Function and Genetics. Baltimore: University Park Press, 1980:479-495.

28. Protzel A, Morris AJ. Gel chromatographic analysis of nascent globin chains. Evidence of nonuniform size distribution. J Biol Chem 1974; 249:4594.

29. Lizardi PM, Mahdari V, Shields D, Candelas G. Discontinuous translation of silk fibroin in a reticulocyte cell-free system and in intact silk gland cells. Proc Natl Acad Sci USA 1979; 76:6211.

30. Candelas G, Candelas T, Ortiz A, Rodriguez O. Translational pauses during a spider fibroin synthesis. Biochem Biophys Res Commun 1983; 116:1033.

31. Varenne S, Knibiehler M, Cavard D, Morlon J, Lazdunski C. Variable rate of polypeptide elongation for colicins A, E2, and E3. J Mol Biol 1982; 159:57.

32. Chaney W, Morris A. Nonuniform size distribution of nascent peptides. The effect of messenger RNA structure upon the rate of translation. Arch Biochem Biophys 1979; 194:283.

33. Randall LL, Josefsson LG, Hardy SJS. Novel intermediates in the synthe-

sis of maltose-binding protein in *Escherichia coli*. Eur J Biochem 1980; 107:375.

34. Abraham AK, Pihl A. Variable rate of polypeptide chain elongation in vitro. Eur J Biochem 1980; 106:257.

35. Wolin SL, Walter P. Ribosome pausing and stacking during translation of a eukaryotic mRNA. Embo J 1988; 7(11):3559-69.

36. Gouy M, Gautier C. Codon usage in bacteria: correlation with gene expressivity. Nucleic Acids Res 1982; 10:7055-61.

37. Liljenstrom H, von Heinje G. Translation rate modification by preferential codon usage. J Theor Biol 1987; 124:43-8.

38. Kuroda MI, Yanofsky C. Evidence for the transcript secondary structures predicted to regulate transcription attenuation in the trp operon. J Biol Chem 1984; 259(20):12838-43.

39. Oxender DL, Zurawski G, Yanofsky C. Attenuation in the *Escherichia coli* tryptophan operon: role of RNA secondary structure involving the tryptophan codon region. Proc Natl Acad Sci USA 1979; 76(11):5524-8.

40. Kolter R, Yanofsky C. Genetic analysis of the tryptophan operon regulatory region using site-directed mutagenesis. J Mol Biol 1984; 175(3): 299-312.

41. Landick R, Yanofsky C, Choo K, Phung L. Replacement of the *Escherichia coli* trp operon attenuation control codons alters operon expression. J Mol Biol 1990; 216(1):25-37.

42. Bonekamp F, Dalbege H, Christensen T, Jensen KF. Translation rates of individual codons are not correlated with tRNA abundances or with frequencies of utilization in *Escherichia coli*. J Bact 1989; 171(11):5812-5816.

43. Robinson M, Lilley R, Little S et al. Codon usage can affect efficiency of translation of genes in *Escherichia coli*. Nucleic Acids Res 1984; 12:6663.

44. Lynn SP, Burton WS, Donohue TJ, Gould RM, Gumport RI, Gardner JF. Specificity of the attenuation response of the threonine operon of *Escherichia coli* is determined by the threonine and isoleucine codons in the leader transcript. J Mol Biol 1987; 194(1):59-69.

45. Chen JW, Bennett DC, Umbarger HE. Specificity of attenuation control in the ilvGMEDA operon of *Escherichia coli* K-12 [published erratum appears in J Bacteriol 1991 May; 173(10):3269]. J Bacteriol 1991; 173(7):2328-40.

46. Chen JW, Harms E, Umbarger HE. Mutations replacing the leucine codons or altering the length of the amino acid-coding portion of the ilvGMEDA leader region of *Escherichia coli*. J Bacteriol 1991; 173(7): 2341-53.

47. Carter PW, Weiss DL, Weith HL, Calvo JM. Mutations that convert the four leucine codons of the Salmonella typhimurium leu leader to four threonine codons. J Bacteriol 1985; 162(3):943-9.

48. Irwin B, Heck JD, Hatfield GW. Codon pair utilization biases influence translational elongation step times. J Biol Chem 1995; 270(39):22801-6.

CHAPTER 5

CONTROL BY ANTISENSE RNA

Brian N. Zeiler and Robert W. Simons

INTRODUCTION

Antisense RNAs are small, diffusible, untranslated RNAs that pair to complementary regions on specific target RNAs, altering the expression or function of those RNAs post-transcriptionally. Antisense RNA control is well documented in bacteria, especially in their accessory elements—plasmids, bacteriophages and transposable elements (see Table 5.1). In all of these cases, control is negative. However, mechanisms for positive control are quite plausible. The biological processes inhibited vary widely, as do the mechanisms by which antisense RNAs inhibit those processes. Some antisense RNAs bind directly at their sites of action, while others bind at a distance, altering target RNA structure at the actual site of action. Accessory proteins are involved in a few instances. In most cases, the antisense and target RNAs are transcribed from opposite strands of the same DNA segment and thus contain regions of complete complementarity. In a growing number of cases, however, the antisense and target RNAs are expressed from unlinked genes and complementarity is substantial but incomplete. Antisense RNAs usually contain one or more stem-loop structures. The loop domains are often found to be important determinants of specificity for antisense/target RNA pairing. The stem domains frequently determine metabolic stability. Target RNA structure can be exceptionally complicated and is frequently (if not always) critical for efficient pairing. This chapter will describe a limited number of biological cases where antisense RNA control has been documented, emphasizing the underlying principles of antisense RNA structure, target RNA pairing, and modes of action. The reader is referred to other recent reviews[1-6] and the primary literature for further detailed information.

ANTISENSE RNAs CONTROL DIVERSE BIOLOGICAL FUNCTIONS

Antisense RNA control has been documented or strongly implicated in gram-positive and -negative eubacteria, archaebacteria, and a variety of eukaryotic cells.[4,6] Antisense RNAs control a diverse array of biological functions including plasmid replication, cell division, transposition, stress response, cell death, and bacteriophage development (Table 5.1). Given the diversity of biological functions regulated, it is unlikely that antisense control is limited to any particular organism or

Regulation of Gene Expression in Escherichia coli, edited by E. C. C. Lin and A. Simon Lynch. © 1996 R.G. Landes Company.

genetic element. Indeed, it is hard to imagine a gene whose expression could not be controlled by antisense RNA. In this section, a limited number of examples of biological control by antisense RNA are described, in an attempt to document its wide range of action.

PLASMID BIOLOGY

Antisense control was first discovered in plasmid ColE1[1,2] and subsequently found to inhibit a variety of other plasmid functions (see Table 5.1). In all of these cases, the antisense and target RNAs are transcribed from opposite DNA strands of the same region, by opposing promoters.

Plasmid replication

The replication frequency of bacterial plasmids is carefully regulated, so that plasmid copy numbers remain reasonably stable within bacterial cells growing under different conditions. Because genes and/or sites required for regulating replication are located on the plasmids themselves, they control their own copy numbers in a gene-dosage dependent manner: as the plasmid copy number rises and falls, so does the concentration of the regulator, adjusting the copy number accordingly.[7] In many plasmids, antisense RNAs are the principal elements controlling copy number. Control usually targets some rate-limiting function required for the initiation of replication. For example, in plasmid ColE1, a ≈ 108 nt RNA, RNA I, binds to and inhibits the required processing of a preprimer for DNA synthesis. In all other known cases, the antisense RNA inhibits synthesis of an essential replication initiation protein. This can occur by various means, as described below. Antisense inhibition of plasmid replication is well adapted to its task: these RNA regulators are expressed and turned over at high rates, so their levels can respond quickly to rapid changes in plasmid copy number.[4] Antisense RNA/target RNA interactions are usually very specific, and this feature determines the compatibility relationships that exist between groups of related plasmids, as explained below.[8] The fact that antisense RNA control regulates plasmid replication in very diverse ways suggests that it is generally an efficient and effective means to achieve such control.

Plasmid conjugation

Conjugative transfer of F-like plasmids from one cell to another requires expression of more than 30 *tra* genes, which in turn require expression of a positive transcription factor, TraJ.[9,10] The *traJ* gene is normally inhibited by an antisense RNA, FinP.[11,12] Occasionally, however, *traJ* is expressed and transfer occurs. Once in the recipient cell, FinP inhibition is slow to manifest

Table 5.1. Partial list of documented or proposed cases of antisense RNA control in prokaryotes[1]

Biological system	Biological function controlled	Target RNA	Antisense RNA	Level of control
Plasmids				
ColE1	replication	RNA II	RNA I	RNA processing
R1	replication	*repA* mRNA	CopA RNA	translation
R1	conjugation	*traJ* mRNA	FinP RNA	translation
R1	host killing	*hok* mRNA	Sok RNA	translation
pT181	replication	*repC* mRNA	ctRNA	transcription termination
Mobile genetic elements				
IS*10*	transposition	*tnp* mRNA	RNA-OUT	translation
Bacteriophages				
λ	development	cII mRNA	OOP RNA	mRNA stability
λ	late gene exp'n	*Q* mRNA	PαQ RNA	translation
P22	antirepression	*ant* mRNA	Sar RNA	translation
P22	exclusion	*sieB* mRNA	Sas RNA	translation
Bacterial genes				
E. coli	stress response	*ompF* mRNA	MicF RNA	translation
E. coli	cell division	*ftsZ* mRNA	DicF RNA	translation/decay
E. coli	cell division	*ftsZ* mRNA	StfZ RNA	translation

[1]For additional examples, the reader should see other recent reviews.[1-6]

itself once again, and *traJ* expression continues for some time so that additional rounds of conjugation to other cells ensue. After several hours, however, effective FinP levels rise, restoring inhibition of *traJ* gene expression and halting further plasmid transmission. In this way, F-like plasmids are able to spread rapidly throughout a new population of cells, and then to cease any further conjugation. Both FinP stability and FinP RNA/*traJ* mRNA pairing are increased by the plasmid-encoded FinO protein, and it may be that control of FinO level or activity provides the key to this interesting temporal control of *traJ* expression.[13,14]

Plasmid-mediated post-segregational killing

One of the many ways that plasmids ensure stable maintenance within a bacterial population is to use a post-segregational host-killing system.[15] A remarkable feature of these systems is their memory: they kill only *after* the plasmid is lost at cell division. The killer system of plasmid R1 encodes the Hok (host killing) protein, whose expression renders the cell membrane highly permeable, resulting in cell death.[16,17] A small antisense RNA, Sok (suppressor of killing), prevents *hok* gene expression in cells containing the plasmid.[18,19] However, daughter cells that fail to inherit the plasmid nevertheless acquire pre-existing *hok* mRNA and the short-lived Sok RNA from the mother cell's cytoplasm. In a complicated and fascinating pathway of *hok* RNA processing and decay (see below), the Hok protein is translated only in these plasmid-less cells, resulting in their death.[20,21]

BACTERIOPHAGES

Antisense control is involved in several different aspects of the developmental pathways of temperate bacteriophages, where it is secondary to other controlling elements. Like several of the plasmid cases, these systems can be complicated, revealing the complexities of post-transcriptional control.

λ OOP RNA and *c*II expression

The λ CII protein activates the P_{RE} and Pi promoters, which transcribe the CI repressor and integrase genes, respectively, thereby participating in the choice between lytic and lysogenic pathways.[22] *c*II expression is regulated, in part, by the ≈77 nt OOP antisense RNA, whose biological function has long been debated. OOP RNA is complementary to the major rightward transcript across a region including the 3'-terminal ≈55 nt of the *c*II gene, and inhibits *c*II expression by initiating the degradation of that region of the mRNA.[23-26] Phages lacking OOP RNA have normal growth and lysogeny. However, OOP expression is weakly repressed by LexA, the principal repressor of the SOS regulon, and *oop⁻* phages exhibit 3-fold more *c*II mRNA and 2-fold lower phage burst size following UV induction. This observation suggests that OOP plays a small but significant role in adjusting CII levels during induction, thereby promoting lysis.

λ PαQ RNA and late gene expression

The CII protein also appears to delay late gene expression, so that lysogeny can occur without high levels of lysis protein expression.[22] This is probably mediated, in part, by the ≈220 nt PαQ RNA, which inhibits the synthesis of Q protein, the activator of late gene expression.[27] PαQ RNA, expressed from a third CII-dependent promoter, located within and antisense to the Q gene, probably inhibits Q gene translation. The biological role of PαQ RNA is revealed by the observation that PαQ⁻ phages show a substantially decreased frequency of lysogeny.

P22 Sar RNA and antirepression

The P22 antirepressor (Ant) inhibits DNA binding by the principal repressors of many different lambdoid phages, including the C2 repressor of P22 itself. Early in P22 infection, Ant is expressed at low levels. This may be advantageous: any heterologous, resident lambdoid prophages present in the infected cell would be induced at that time, increasing the opportunity for recombination between two phages, resulting in new phage types with possible adaptive advantage. However, Ant expression must be carefully controlled: too much expression early or late in infection will block, respectively, establishment and maintenance of lysogeny. Ant expression is carefully controlled at several levels, including antisense control by the Sar RNA, which binds to the *ant* ribosome binding site (RBS) and probably prevents its translation.[28-30] The observation that P22 *sar⁻* phages form clear plaques indicates the biological significance of the Sar RNA.

P22 Sas RNA and phage exclusion

One of the very few genes expressed from the P22 prophage is *sieB*, which prevents the lytic growth of certain superinfecting phages. Superinfecting P22 phages are, themselves, immune to this exclusion. Interestingly, the *sieB* locus encodes two functions, SieB, which is the exclusion factor per se, and Esc, which provides P22 its means of escape from SieB exclusion. Esc, which is in essence a truncated version of the SieB protein, arises by translation initiation within the *sieB* coding sequence.[31,32] It is the relative level of these two proteins (SieB and Esc) that determines exclusion versus escape (Esc probably inhibits SieB function through subunit poisoning). The P22 Sas RNA appears to modulate these relative levels as follows: In the P22 prophage, SieB and Esc are expressed from the P_{SIEB} promoter, such that SieB is in excess and exclusion is manifest. In the prophage, the opposing P_L promoter is repressed. However, in the superinfecting P22 phage, both the P_{SIEB} and P_L promoters are active, and the ≈105 nt Sas RNA, most likely processed from the P_L transcript, is expressed. Sas RNA is complementary to the *sieB* RBS and presumably inhibits its translation, but not that of the downstream *esc* gene. This situation favors translation of Esc, rendering P22 immune to exclusion. Consistent with this model, P22 phages defective in this translational switch commit SieB-mediated suicide.

INHIBITION OF TRANSPOSITION

IS*10* is a very compact mobile genetic element that is responsible for transposition of the well-characterized tetracycline-resistance transposon, Tn*10*. IS*10* encodes a transposase function, Tnp, which acts at the ends of IS*10* or Tn*10* to bring about transposition and related events.[33] Tnp expression is carefully regulated at a number of levels, including antisense control by RNA-OUT, which is complementary to the *tnp* RBS and inhibits ribosome binding.[34-36] The biological role of IS*10* antisense RNA control is probably to limit the accumulation of this element in the cell, thereby lessening the chance for deleterious transposition events. Indeed, a single IS*10* copy is under very little antisense control, but inhibition increases as copy numbers rise, leading to a proportionate decrease in Tnp expression and transposition.[34] This "counting" effect is made efficient by the preferential cis-action of the Tnp protein.[33] Importantly, RNA-OUT is expressed at only moderate levels but has unusually high stability,[37] consistent with its need to respond only slowly to gradually changing IS*10* copy numbers.

BACTERIAL GENE EXPRESSION

Since the first discovery of antisense control of bacterial genes,[38] several additional cases have been described or proposed, and it is now generally accepted that many bacterial genes may be under such control. Unlike the plasmid, phage and transposon cases, many of these bacterial genes are inhibited by antisense RNAs expressed from unlinked genes, and the antisense and target RNAs are only partially complementary to one another.

Response to environmental stress

The OmpF protein of *E. coli* is a major outer membrane porin whose expression is maximal at low temperature and low osmolarity.[39] OmpF expression is activated by the *ompR-envZ* two-component regulatory system, and inhibited post-transcriptionally by the ≈93 nt MicF RNA.[38,40] The *micF* and *ompF* genes are unlinked, and MicF RNA is only ≈70% complementary to the *ompF* RBS. Control almost certainly occurs at the level of translation initiation. A 70 kDa *E. coli* protein binds MicF RNA specifically, but its role in *ompF* regulation is unknown.[41,42] Initial evidence, based on experiments with multicopy *micF* genes, led to the proposal that MicF plays a major role in OmpF osmoregulation,[38] but later work showed that deletion of the chromosomal *micF* gene had only a minor effect on osmoregulation, casting doubt on this proposal.[43,44] However, subsequent studies revealed that MicF clearly inhibits *ompF* expression in response to elevated temperature,[45,46] exposure to salicylate[47] and redox stress,[48] and to be at least partially responsible for the decreased OmpF levels seen in *tolC* (toluene resistant) mutants,[49] *marC* (multiple antibiotic resistance) mutants,[49,50] and in cells resistant to activated macrophages following exposure to nitric oxide.[51] Thus, MicF repression of OmpF appears to be involved in the cell's defense against certain xenobiotic molecules.[48]

Cell division

The control of *E. coli* cell division is ill-defined, but may be initiated by the FtsZ protein. Indeed, FtsZ expression is inhibited by several cell-division inhibitors: the SulA (SfiA) protein expressed during SOS induction,[52] the MinC and MinD proteins involved in placement of the division septum,[53] and the DicB protein encoded in a defective prophage.[54,55] The *dicB* operon specifies yet another inhibitor of *FtsZ* expression, the 53 nt DicF RNA which, like MicF, is unlinked to and only partially complementary to is target, the *ftsZ* RBS.[56-58] Interestingly, antisense transcription of the *ftsZ* gene itself produces the antisense StfZ RNA, which is also complementary to the *ftsZ* RBS. StfZ RNA is probably responsible for the inhibition of cell division observed with multicopies of this region.[59] The biological significance of the DicF and StfZ RNAs remains unclear.

ANTISENSE RNAs CONTROL GENE EXPRESSION AT MANY DIFFERENT POST-TRANSCRIPTIONAL LEVELS

Several antisense systems have been studied in sufficient detail to discern the actual mechanisms of control involved. These various modes of action include premature transcription termination, altered RNA processing, facilitated RNA decay and inhibition of translation. This diversity reflects both the complexities of post-transcriptional expression and the unique ways in which antisense RNAs control these events.

PREMATURE TRANSCRIPTION TERMINATION

Replication of the staphylococcal plasmid pT181 requires the RepC function,[60] whose expression is controlled by an antisense RNA, ctRNA.[61] In a mechanism reminiscent of transcriptional attenuation of the *E. coli* tryptophan operon, ctRNA binds to its complementary region on the *repC* mRNA leader and induces premature transcription termination. This occurs through an alteration in *repC* mRNA conformation, as shown in Figure 5.1. The *repC* leader is thought to form either of two mutually exclusive stem-loop structures, I-III or III-IV. The III-IV structure is a Rho-independent transcriptional terminator; structure I-III has no apparent effect on transcription. In the absence of antisense control, structure I-III forms in preference to III-IV, and no termination occurs. However, when ctRNA binds, it is proposed to prevent I-III formation, in turn allowing structure III-IV to form and transcription termination to occur. Such alteration of RNA conformation is a recurrent theme in many other antisense control mechanisms.

RNA PROCESSING

Plasmid ColE1 replication has been analyzed in great detail and represents the best understood case of antisense control (reviewed in refs. 1, 2). Replication initiates with the transcription of a preprimer RNA, RNA II, which forms an unusually persistent hybrid with the DNA template strand (Fig. 5.2).[62] This hybrid is then cleaved by ribonuclease H at a specific site in RNA II $\approx$550 nt from its 5' end, creating the 3' end where DNA synthesis begins (the origin).[63] In order for this persistent RNA/DNA hybrid to form, RNA II must fold into a complicated set of secondary and tertiary structures, which arise during the process of RNA II synthesis.[64-66] The first $\approx$110 nt of RNA II fold into a structure comprising three stem-loop elements. As transcription continues, stem-loop III of RNA II converts to a more stable configuration, structure IV. Once structure IV forms, sequences from +200 to the origin are committed to fold into the conformation that will enable the preprimer to base-pair to the template DNA at certain critical positions, resulting in the stable hybrid and enabling proper processing at the origin.

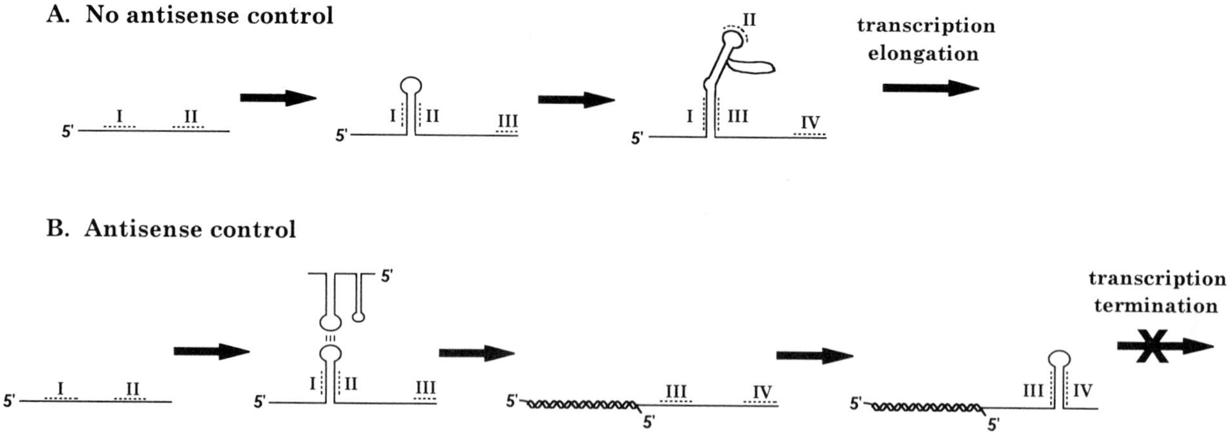

Fig. 5.1. Antisense RNA induces premature transcription termination in plasmid pT181. (A) In the absence of control, the repC leader folds into structure I:III (the structure of the intervening 72 nt region has not been defined), permitting transcription elongation. (B) Antisense RNA (ctRNA) binding to the leader prevents formation of structure I:III, allowing formation of the alternative III:IV structure, which prematurely terminates transcription.

The ≈108 nt antisense RNA, RNA I, is expressed from the same region and is complementary to the 5' end of RNA II. When RNA I binds to RNA II, it triggers a conformational change in the preprimer that prevents persistent hybrid formation, thus inhibiting processing and replication initiation (see Fig. 5.2).[64,65] However, the timing of these events is crucial. If RNA I/RNA II pairing occurs after the synthesis of about the first 150 nt of RNA II, RNA I binding can still occur but RNA II is already committed to stable hybrid formation and subsequent initiation of DNA synthesis. Thus, the regulation of ColE1 replication represents a far more complex array of RNA structures and conformational changes than is apparently the case in plasmid pT181. Nevertheless, the underlying principle of modulation of target RNA structure remains the same in both instances.

A plasmid-encoded protein, Rom, binds to and stabilizes an intermediate in the RNA I-RNA II pairing process, enhancing control.[67-70] Several *E. coli* functions also modulate ColE1 antisense control. Ribonuclease E cleaves RNA I near its 5' end, initiating rapid RNA I decay.[71] The PcnB protein polyadenylates the 3' end of RNA I, targeting it for rapid decay by an unknown mechanism.[72,73] Both of these effects increase ColE1 plasmid copy number, but it is not clear if these activities are exploited to regulate copy number under different conditions.

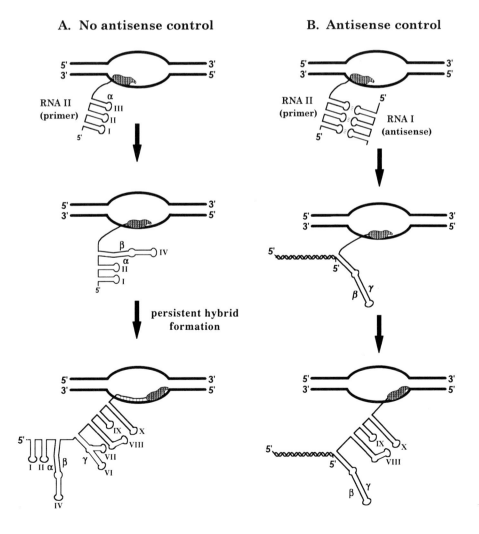

A. No antisense control

B. Antisense control

Fig. 5.2. Antisense RNA inhibits primer maturation in plasmid ColE1. (A) In the absence of control, the primer RNA (RNA II) folds into a specific structure that engenders persistent RNA/DNA hybrid formation, in turn allowing RNA II processing by RNaseH and the inception of DNA synthesis (see text for details). (B) Antisense RNA (RNA I) binding to RNA II prevents proper RNA II folding and subsequent persistent hybrid formation.

FACILITATED RNA DECAY

The OOP RNA of bacteriophage λ pairs to the major rightward transcript across a region including the 3' end of the *cII* open reading frame (Fig. 5.3). This forms a double-stranded RNA species that is rapidly cleaved by ribonuclease III at specific sites in the *cII* mRNA, initiating its rapid decay.[25] Interestingly, this facilitated decay does not extend to the region of the mRNA encoding the *O* gene. Importantly, the effects of OOP RNA on *cII* expression are dependent on an intact *rnc* gene, which encodes ribonuclease III. The double-stranded RNA species that arise in the IS*10* and plasmid R1 cases are also cleaved and destabilized by ribonuclease III, but such cleavage is not necessary for antisense inhibition in these cases, because it is preceded by inhibition of translation.[35] These comparisons point out the need to carefully distinguish primary and secondary effects of facilitated mRNA decay.

INHIBITION OF TRANSLATION

The far most common mode by which antisense RNAs repress gene expression is the inhibition of translation. However, the mechanisms involved range from direct blockage of ribosome binding by antisense RNAs complementary to the target RBS, to complex alterations of RNA secondary structure by antisense RNA binding at a distance, and to effects mediated through translational coupling between adjacent genes. The simplest case is IS*10* (Fig. 5.4), where the antisense RNA, RNA-OUT, is complementary to the RBS of the *tnp* mRNA, RNA-IN. In vitro studies show that RNA-OUT/RNA-IN pairing directly blocks ribosome binding without otherwise perturbing target RNA structure.[36] This simple mechanism is likely to occur in many if not all cases where the antisense RNA pairs directly to the target gene RBS.

The mechanism in plasmid R1 is more complex (Fig. 5.5). CopA antisense RNA binds to a region of the *repA* mRNA termed CopT (for Cop target), inhibiting RepA synthesis post-transcriptionally.[74,75] However, because the *repA* RBS is ≈80 nt downstream of CopT, several indirect mechanisms were proposed for its effects, including transcriptional attenuation, altered mRNA conformation, and facilitated decay. However, the most likely mechanism became clear with the discovery of another, small open reading frame, *tap*, which lies immediately downstream of the CopT site.[76] Importantly, *tap* translation is inhibited by CopA/CopT binding. Furthermore, the 3' end of *tap* overlaps the *repA* RBS, and *repA* translation is coupled to that of *tap* (i.e., *repA* is translated only when *tap* is translated). Thus, *repA* translation is regulated indirectly by CopA at the level of initiation of *tap* translation. Even though the CopA RNA does not actually overlap the *tap* RBS, CopA/CopT pairing prevents ribosome binding at the *tap* RBS in vitro (C. Malmgren, P. Romby and E.G.H. Wagner, unpublished observations).

Fig. 5.3. Antisense RNA initiates rapid decay of the λ cII mRNA. The OOP antisense RNA is complementary to the 3' terminus of the cII mRNA, and the OOP/cII mRNA duplex is cleaved by RNaseIII, initiating its rapid decay.

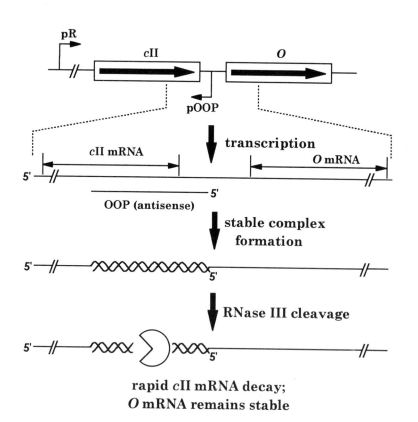

**rapid *cII* mRNA decay;
O mRNA remains stable**

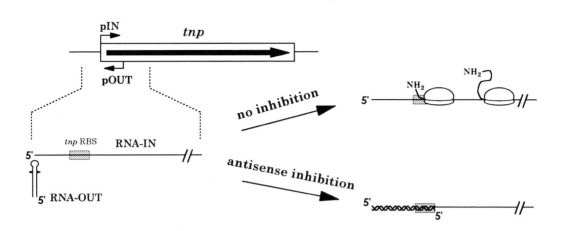

Fig. 5.4. Antisense RNA directly inhibits IS10 transposase translation. In a simple mechanism, the antisense RNA (RNA-OUT) binds to the transposase (tnp) mRNA across a region including the ribosome binding site, thereby directly inhibiting ribosome binding.

A. No antisense control

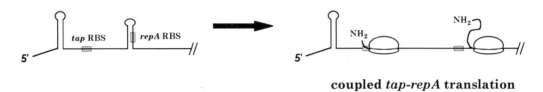

coupled *tap-repA* translation

B. Antisense control

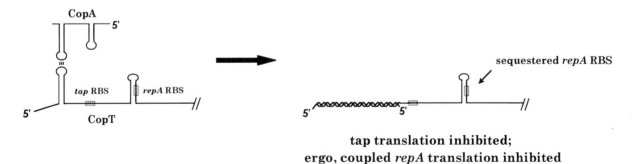

tap translation inhibited;
ergo, coupled *repA* translation inhibited

Sok RNA inhibition of *hok* translation has some parallels to the CopA/CopT case, but is even more complicated[20,21] (Fig. 5.6). In addition to *hok* and *sok*, this locus encodes a third gene, *mok* (modulator of killer), which completely overlaps the *hok* gene but in a different reading frame. Furthermore, *hok* gene expression is coupled to that of *mok* (presumably by translational coupling), and Sok RNA inhibits *hok* expression indirectly by binding to the *mok* RBS and blocking translation. Other interesting aspects of the *hok* mRNA further complicate this regulatory circuit. The *hok* mRNA has two principal forms, the full length species and a processed version lacking the 3'-terminal ≈70 nt. The full-length species is not efficiently translated whereas the truncated mRNA is. This results from different structure in the two mRNAs. The 3'-end of the full-length species is partially complementary to, and pairs with, the *mok* RBS at the 5' end. This forms a so-called *fbi* (fold-back-inhibition) structure, which blocks ribosome binding. The truncated mRNA lacks this structure. Therefore, the long *hok* mRNA is

Fig. 5.5. Antisense RNA indirectly inhibits translation of the repA gene in plasmid R1. (A) In the absence of control, translation of the small tap gene enables coupled translation of the downstream repA gene. (B) Antisense RNA (CopA RNA) directly inhibits tp translation, indirectly inhibiting coupled translation of the repA gene.

A. Plasmid containing cells
(no *hok* expression)

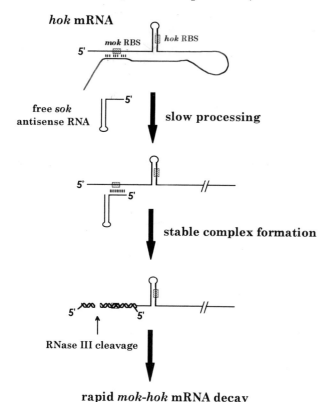

B. Plasmid free cells
(*hok* expression)

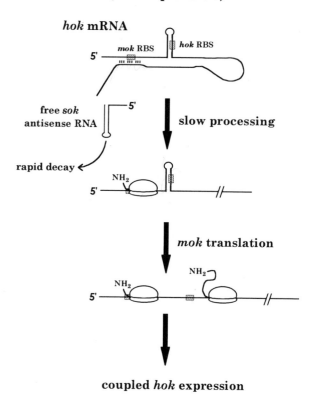

Fig. 5.6. Antisense RNA inhibits plasmid R1 hok gene translation in a complicated mechanism. (A) In plasmid-containing cells, the mok/hok mRNA folds back onto itself, preventing translation of the mok gene. This in turn prevents hok gene expression, whose translation is coupled to that of the upstream mok gene. When the 3' end of this reservoir of hok mRNA is slowly processed away, the antisense RNA (sok) inhibits its mok translation and, thereby, hok translation. (B) In cells that have lost the R1 plasmid, the short-lived sok antisense RNA has been degraded by the time mRNA processing occurs, allowing translation of mok and hok (see text for details).

translationally inert but serves as a reservoir for slow processing to the shorter, translationally active species. Importantly, the *fbi* structure also prevents Sok RNA binding. Thus, while the Sok RNA inhibits *hok* expression from truncated mRNAs, it does not affect the reservoir of those mRNAs. These complicated features of the *mok/hok/sok* system conspire to bring about post-segregational killing, as follows: In cells containing the plasmid, the *hok* mRNA reservoir is slowly converted to active mRNA, which is then rapidly inhibited by Sok RNA, ensuring that Hok is not expressed. However, when the plasmid is lost, the unstable Sok RNA is quickly degraded, allowing the more stable processed Hok mRNA to accumulate and be translated. This remarkable mechanism explains the "memory" of the host-killing system and illustrates the potential complexities of post-transcriptional control.

ANTISENSE RNAs PAIR TO THEIR TARGET RNAs BY DEFINED MECHANISMS

Genetic and biochemical studies with three unrelated antisense systems (ColE1, R1 and IS*10*) reveal how antisense RNAs pair efficiently and specifically to their target RNAs. While the pathways differ in detail, they share several underlying principles, which may dictate how all antisense RNAs pair to their targets.

RNA I/RNA II PAIRING IN COLE1

In plasmid ColE1, RNA I/RNA II pairing occurs in several, sequential steps (Fig. 5.7). In the first, which is rate-limiting, the RNAs

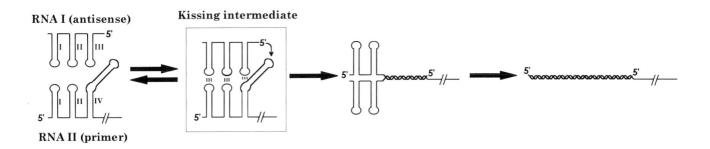

RNA I (antisense)   **Kissing intermediate**

RNA II (primer)

form reversible base pairs between the loops of RNA I and complementary regions in RNA II: loops I and II of RNA I interact with the corresponding loops of RNA II, and loop III of RNA I interacts with either loop III or an elbow of structure IV in RNA II.[77] The initial loop-loop interactions in this "kissing" intermediate enable the RNAs to associate specifically with one another, but topological constraints prevent kissing from leading directly to further base-pairing without disruption of the stems (the RNA strands must twist around each other in order for double-stranded RNA to form, and this cannot occur between loops that are effectively closed by stems). In the second step, base-pairing between the free 5' end of RNA I and its complementary region in RNA II initiates the formation of a more stable complex, as the kissing interactions give way.[77-79] Finally, pairing propagates from the 5' end of RNA I to the full extent of complementarity (no topological constraint operates here, as the 5' end of RNA I is free to twist around RNA II to form the double helix). Genetic evidence strongly supports this pairing model (reviewed in ref. 5). Mutations in the loops alter primarily the specificity of pairing, such that a mutant RNA I molecule will pair efficiently to its cognate (homologous) RNA II molecule, but not to a noncognate (heterologous) RNA II species. This is consistent with initiation of pairing by loop/loop kissing. Deletion of the 5' end of RNA I prevents stable complex formation altogether, consistent with its involvement in an obligatory intermediate. In general, mutations that increase or decrease the rate of RNA I/RNA II pairing have reciprocal effects on plasmid copy number. The plasmid-encoded Rom protein facilitates control by increasing the rate of stable complex formation, without a discernible effect on subsequent steps in the propagation of pairing.[69,80] Specifically, Rom decreases the equilibrium dissociation constant for the unstable kissing complex (boxed in Fig. 5.7), to which it specifically binds.[67-70] In this way, Rom increases the effectiveness of antisense control, reducing plasmid copy number a few fold.

Antisense control also determines the compatibility properties of ColE1.[81] Two distinguishable ColE1-type plasmids will not stably coexist in the same cell if they have identical sequence specificities in their RNA I/RNA II interactions (these specificity effects map to the loops, as described above). This effect is termed incompatibility and results from the ability of the trans-acting antisense RNAs from either plasmid to inhibit the cis-acting RNA II primer of the other. This eventually leads to the loss of one or the other plasmid type, manifesting incompatibility. On the other hand, when the RNA I/RNA II regions of the two plasmids have different specificities (i.e., different sequences in one more loops), heterologous inhibition does not occur and the plasmids can stably coexist (compatibility).

Fig. 5.7. The process of RNA I/RNA II pairing in ColE1 involves several discrete steps. Pairing initiates with loop-loop interactions between the two RNAs. This reversible "kissing" complex (boxed) is stabilized by the Rom protein. Subsequent interactions between the free 5' end of RNA I and a loop in RNA II lead to stable complex formation (see text for details).

CopA/CopT Pairing in Plasmid R1

The process of CopA/CopT pairing is very similar to the ColE1 case, consisting of at least two distinguishable steps and confronting the same topological constraints (Fig. 5.8). Reversible kissing occurs through loop-loop contacts between these RNAs.[82] The second step, which leads rapidly to complete duplex formation, requires the single-stranded regions between two stem-loops of CopA and the corresponding single-stranded region in CopT.[83] Genetic and kinetic analyses suggest that, like the ColE1 case, formation of the early loop/loop intermediate is the rate-limiting step in the pathway leading to the stable complex. As in ColE1, mutations in loop II define the compatibility properties of R1.

RNA-OUT/RNA-IN Pairing in IS*10*

RNA-OUT consists of a single stem-loop structure (Fig. 5.9). Complementarity to the target, RNA-IN, begins at the top of the RNA-OUT loop and extends down one side of the stem. In vitro, fully duplexed RNA-OUT/RNA-IN complexes appear within 30 seconds of the initiation of pairing, but no intermediates have been detected (B. Zeiler, unpublished observations).[84,85] However, genetic studies suggest that pairing begins with the formation of 2-3 base pairs between the 5' end of RNA-IN and the RNA-OUT loop. Mutations in this region alter the specificity of pairing.[37,84,85] The relatively loose structure of the RNA-OUT loop region is important for pairing: mutations that stabilize the stem in that region abolish pairing altogether. This and related observations suggest that once the initial interactions occur, the 5' end of RNA-IN passes through the RNA-OUT loop as the RNAs form double helix. Thus, RNA-OUT/RNA-IN pairing differs in a fundamental way from the multistep process described for the ColE1 and R1 cases: RNA-OUT/RNA-IN pairing initiates with a loop-linear interaction which then proceeds immediately to stable complex formation without topological constraint. Interestingly, a wholly different pathway, initiated by linear-linear contacts that presumably lead directly to stable complex formation, has been proposed for *hok/sok* pairing (Fig. 5.6).[21] These and other important considerations of RNA/RNA pairing pathways are discussed in greater detail elsewhere (B. Zeiler and R. Simons, submitted to *Molec. Microbiol.*).

OVERVIEW

Antisense RNA control of gene expression in prokaryotic cells and their accessory elements is now well documented, illustrating a wide variety of control mechanisms and biological effects, often quite complex. It is certain that additional examples will emerge in the future, very likely illustrating new and unique mechanisms by which these

Fig. 5.8. The process of CopA/CopT pairing in plasmid R1. CopA/CopT RNA pairing is essentially identical to that of ColE1 (see text for details).

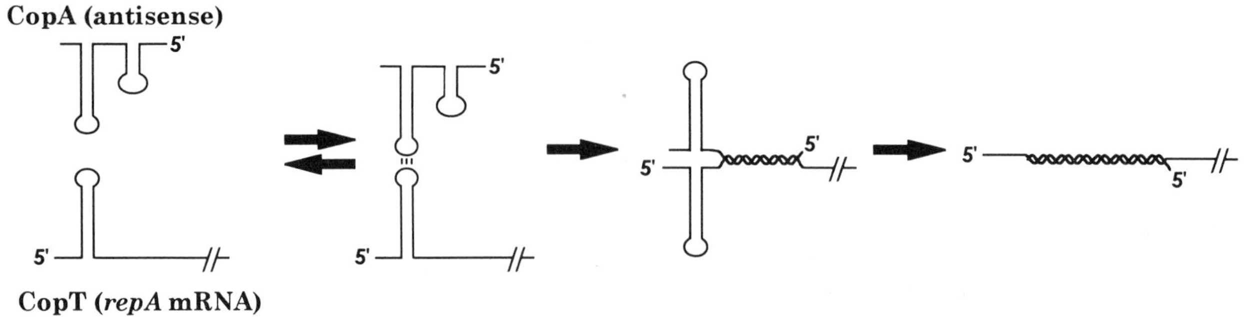

CopA (antisense)

CopT (*repA* mRNA)

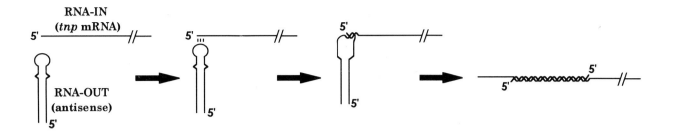

novel regulatory molecules exert their effects. Ongoing studies on RNA pairing mechanisms should continue to advance our understanding of the RNA folding problem. Proteins or other factors that modulate such interactions are of particular interest. A question that remains to be fully addressed is how antisense RNA systems evolve, and what special advantages such systems might have over regulatory proteins.

Fig. 5.9. The process of RNA-OUT/RNA-IN pairing in IS10. Unlike plasmids ColE1 and R1, RNA pairing in IS10 initiates by interactions between the linear 5' end of the target RNA and a loop in the antisense RNA, followed immediately by stable complex formation.

REFERENCES

1. Tomizawa J. Evolution of functional structures of RNA. In: R F Gasteland and J F Atkins, eds. The RNA World. Cold Spring Harbor: Cold Spring Harbor Laboratory Press, 1993:419-445.

2. Eguchi Y, Itoh T, Tomizawa J. Antisense RNA. Annu Rev Biochem 1991; 60:631-652.

3. Simons RW. Naturally occurring antisense RNA control—a brief review. Gene 1988; 72:35-44.

4. Wagner EGH, Simons RW. Antisense RNA control in bacteria, phages, and plasmids. Annu Rev Microbiol 1994; 48:713-742.

5. Simons RW, Kleckner N. Biological regulation by antisense RNA in prokaryotes. Annu Rev Genet 1988; 22:567-600.

6. Simons RW. The control of prokaryotic and eukaryotic gene expression by naturally occurring antisense RNA. In: Crooke ST, Lebleu B, eds. Antisense Research and Applications. Boca Raton: CRC Press, 1993: 97-123.

7. Nordström K. Control of plasmid replication: theoretical considerations and practical solutions. In: Helinski DR, Cohen SN, Clewell DB, Jackson DA, Hollaender A, eds. Plasmids in Bacteria. New York: Plenum, 1985:189-214.

8. Novick RP, Hoppensteadt FC. On plasmid imcompatibility. Plasmid 1978; 1:421-434.

9. Ippen-Ihler K, Minkley EG. The conjugation system of F, the feritility system of *Escherichia coli*. Annu Rev Genet 1986; 20:593-624.

10. Willets NS, Skurray R. Structure and function of the F factor and mechanism of conjugation in *Escherichia coli* and *Salmonella typhimurium*. In: Neidhardt FC, ed. Cellular and Molecular Biology. Washington, D.C. American Society of Microbiology, 1987:1110-1133.

11. Dempsey WB. Transcript analysis of the plasmid R100 *traJ* and *finP* genes. Molec Gen Genet 1987; 209:533-544.

12. Finlay BB, Frost LS, Paranchych W, Willetts NS. Nucleotide sequences of five IncF plasmid *finP* alleles. J Bacteriol 1986; 167:754-757.

13. van Bieson T, Soderbom, F, Wagner, EG, Frost LS. Structural and functional analyses of the FinP antisense RNA regulatory system of the F conjugative plasmid. Molec Microbiol 1993; 10:35-43.

14. van Biesen T, Frost LS. The FinO protein of IncF plasmids binds FinP antisense RNA and its target, *traJ* mRNA, and promotes duplex formation. Molec Microbiol 1994; 14:427-436.

15. Gerdes K, Poulsen LK, Thisted T, Nielsen AK, Martinussen J, Andreasen PH. The *hok* killer gene family in gram-negative bacteria. The New Biologist 1990; 2:946-956.

16. Gerdes K, Bech FW, Jorgensen ST, Lobner-Olesen A, Rasmussen PB, Atlung T, Boe L, Karlstrom O, Molin S, Von Meyenburg K. Mechanism of postsegregational killing by the *hok* gene product of the *parB* system of plasmid R1 and its homology with the *relF* gene product of the *E. coli relB* operon. EMBO J 1986; 5:2023-2029.

17. Gerdes K, Rasmussen PB, Molin S. Unique type of plasmid maintenance function: Postsegregational killing of plasmid-free cells. Proc Natl Acad Sci USA 1986; 83:3116-3120.

18. Gerdes K, Helin K, Christensen OW, Lobner-Olesen A. Translational control and differential RNA decay are key elements regulating postsegregational expression of the killer protein encoded by the *par B* locus of plasmid R1. J Molec Biol 1988; 203:119-129.

19. Nielsen AK, Thorsted P, Thisted T, Wagner EGH, Gerdes K. The rifampicin-inducible genes *srnB* from F and *pnd* from R483 are regulated by antisense RNAs and mediate plasmid maintenance by killing of plasmid-free segregants. Molec Microbiol 1991; 5:1961-1973.

20. Thisted T, Nielsen AK, Gerdes K. Mechanism of post-segregational killing: translation of Hok, SrnB and Pnd mRNAs of plasmids R1, F and R483 is activated by 3'-end processing. EMBO J 1994; 13:1950-1959.

21. Thisted T, Sorensen NS, Wagner EGH, Gerdes K. Mechanism of postsegregational killing: Sok antisense RNA interacts with Hok mRNA via its 5'-end singe-stranded leader and competes with the 3'-end of Hok mRNA for binding to the *mok* translational initiation region. EMBO J 1994; 13:1960-1968.

22. Echols H. Bacteriophage λ development: temporal switches and the choice of lysis or lysogeny. Trends in Genetics 1986; 2:26-30.

23. Krinke L, Wulff DL. OOP RNA, produced from multicopy plasmids, inhibits λ *c*II gene expression through an RNase III-dependent mechanism. Genes & Development 1987; 1:1005-1013.

24. Krinke L, Mahoney M, Wulff DL. The role of the OOP antisense RNA in coliphage λ development. Molec Microbiol 1991; 5:1265-1272.

25. Krinke L, Wulff DL. RNase III-dependent hydrolysis of λ *c*II-*O* gene mRNA mediated by λ OOP antisense RNA. Genes & Development 1990; 4:2223-2233.

26. Takayama KM, Houba-Herin N, Inouye M. Overproduction of an antisense RNA containing the *oop* RNA sequence of bacteriophage λ induces clear plaque formation. Molec Gen Genet 1987; 210:184-186.

27. Hoopes BC, McClure WR. A cII-dependent promoter is located within the Q gene of bacteriophage λ. Proc Natl Acad Sci USA 1985; 82:3134-3138.

28. Liao S, Wu T, Chiang CH, Susskind MM, McClure WR. Control of gene expression in bacteriophage p22 by a small antisense RNA. I. Characterization in vitro of the P_{sar} promoter and the *sar* RNA transcript. Genes & Development 1987; 1:197-203.

29. Wu T, Liao S, McClure WR, Susskind MM. Control of gene expression in bacteriophage p22 by a small antisense RNA. II. Characterization of mutants defective in repression. Genes & Development 1987; 1:204-212.

30. Jacques J, Susskind MM. Use of electrophoretic mobility to determine the secondary structure of a small antisense RNA. Nucleic Acids Res 1991; 19:2971-2977.

31. Ranade K, Poteete AR. Superinfection exclusion (*sieB*) genes of bacteriophages P22 and λ. J Bacteriol 1993; 175:4712-4718.

32. Ranade K, Poteete AR. A switch in translation mediated by an antisense RNA. Genes and Development 1993; 7:1498-1507.

33. Kleckner N. Transposon Tn*10*. In: Berg D, Howe M, eds. Mobile DNA. Washington, DC: American Society of Microbiology, 1989:227-268.

34. Simons RW, Kleckner N. Translational control of IS10 transposition. Cell 1983; 34:673-691.

35. Case CC, Simons EL, Simons RW. The IS*10* transposase mRNA is destabilized during antisense RNA control. EMBO J 1990; 9:1259-1266.

36. Ma C, Simons RW. The IS10 antisense RNA blocks ribosome binding at the transposase translation initiation site. EMBO J 1990; 9:1267-1274.

37. Case CC, Roels SM, Jensen PD, Lee J, Kleckner N, Simons RW. The unusual stability of the IS10 anti-sense RNA is critical for its function and is determined by the structure of its stem-domain. EMBO J 1989; 8:4297-4305.

38. Mizuno T, Chou M, Inouye M. A unique mechanism regulating gene expression: Translational inhibition by a complementary RNA transcript (micRNA). Proc Natl Acad Sci USA 1984; 81:1966-1970.

39. Czonka LN. Physiological and genetic responses of bacteria to osmotic stress. Micorbiol Rev 1991; 53:121-147.

40. Andersen J, Delihas N, Ikenaka K, Green PJ, Pines O, Ilercil O, Inouye M. The isolation and characterization of RNA coded by the *micF* gene in *Escherichia coli*. Nucleic Acids Res 1987; 15:2089-2101.

41. Esterling L, Delihas N. The regulatory RNA gene *micF* is present in several species of gram-negative bacteria and is phylogenetically conserved. Molec Microbiol 1994; 12:639-646.

42. Andersen J, Delihas N. *micF* RNA binds to the 5' end of *ompF* mRNA and to a protein from *Escherichia coli*. Biochemistry 1990; 29:9249-9256.

43. Aiba H, Matsuyama S, Mizuno T, Mizushima S. Function of *micF* as an antisense RNA is osmoregulatory expression of the *ompF* gene in *Escherichia coli*. J Bacteriol 1987; 169:3007-3012.

44. Matsuyama S, Mizushima S. Construction and characterization of a deletion mutant lacking *micF*, a proposed regulatory gene for OmpF synthesis in Escherichia coli. J Bacteriol 1985; 162:1196-1202.

45. Andersen J, Forst SA, Inouye M, Delihas N. The function of *mic F* RNA. J Biol Chem 1989; 264:17961-17970.

46. Coyer J, Andersen J, Forst SA, Inouye M, Delihas N. *micF* RNA in *ompB* mutants of *Escherichia coli*. Different pathways regulate *micF* RNA levels in response to osmolarity and temperature change. J Bacteriol 1990; 172:4143-4150.

47. Rosner JL, Chai T, Foulds J. Regulation of OmpF porin expression by salicylate in *Escherichia coli*. J Bacteriol 1991; 173:5631-5638.

48. Chou JH, Greenberg JT, Demple B. Post-transcriptional repression of *Escherichia coli* OmpF protein in response to redox stress: Positive control of the *micF* antisense RNA by the *soxRS* locus. J Bacteriol 1993; 175:1026-1031.

49. Misra R, Reeves PR. Role of *micF* in the *tolC*-mediated regulation of Ompf, a major outer membrane protein of *Escherichia coli* K-12. J Bacteriol 1987; 169:4722-4730.

50. Cohen SP, McMurray LM, Levy SB. *marA* locus causes decreased expression of OmpF porin in multiple-antibiotic-resistant (Mar) mutants of *Escherichia coli*. J Bacteriol 1988; 170:5416-5422.

51. Nunoshiba T, DeRojas-Walker T, Wishnok JS, Tannenbaum SR, Demple B. Activation by nitric oxide of an oxidative-stress response that defends *Escherichia coli* against activated macrophages. Proc Natl Acad Sci USA 1993; 90:9993-9997.

52. Huisman OR, D' Ari R, Gottesman S. Cell division control in *Escherichia coli*: specific induction of the SOS function SfiA protein is sufficient to block division. Proc Natl Acad Sci USA 1984; 81:4490-4494.

53. de Boer PAJ, Crossley RE, Rothfield LI. A division inhibitor and a topological specificity factor coded by the minicell locus determine the proper placement of the division site in Escherichia coli. Cell 1989; 56:641-649.

54. de Boer PAJ, Crossley RE, Rothfield LI. Central role of the *Escherichia coli minC* gene product in two different division-inhibition systems. Proc Natl Acad Sci USA 1990; 87:1129-1133.

55. Labie C, Bouché F, Bouché J. Minicell-forming mutants of *Escherichia coli*: suppression of both DicB- and MiniD-dependent division inhibition by inactivation of the MinC gene product. J Bacteriol 1990; 172: 5852-5855.

56. Bouché F, Bouché J. Genetic evidence that DicF, a second division inhibitor encoded by the *Escherichia coli dicB* operon, is probably RNA. Molec Microbiol 1989; 3:991-994.

57. Faubladier M, Cam K, Bouché J. *Escherichia coli* cell division inhibitor DicF-RNA of the *dicB* operon: Evidence for its generation in vivo by transcription termination and by RNase III and RNase E-dependent processing. J Molec Biol 1990; 212:461-471.

58. Tétart F, Bouché J. Regulation of the expression of the cell-cycle gene *ftsZ* by DicF antisense RNA. Division does not require a fixed number of FtsZ molecules. Molec Microbiol 1992; 6:615-620.

59. Dewar SJ, Donachie WD. Antisense transcription of the *ftsZ-ftsA* gene junction inhibits cell division in *Escherichia coli*. J Bacteriol 1993; 175:7097-7101.

60. Novick RP. Staphylococcal plasmids and their replication. Annu Rev Microbiol 1989; 43:537-565.

61. Novick RP, Iordanescu S, Projan SJ, Kornblum J, Edelman I. pT181 plasmid replication is regulated by a countertranscript-driven transcriptional attenuator. Cell 1989; 59:395-404.

62. Masukata H, Tomizawa J. A mechanism of formation of a persistent hybrid between elongating RNA and template DNA. Cell 1990; 62:331-338.

63. Itoh T, Tomizawa J. Formation of an RNA primer for initiation of replication of ColE1 DNA by ribonuclease H. Proc Natl Acad Sci USA 1980; 77:2450-2454.

64. Masukata H, Tomizawa J. Control of primer formation for ColE1 plasmid replication: Conformational change of the primer transcript. Cell 1986; 44:125-136.

65. Tomizawa J. Control of ColE1 plasmid replication: Binding of RNA I to RNA II and inhibition of primer formation. Cell 1986; 47:89-97.

66. Wong EM, Polisky B. Alternative conformations of the ColE1 replication primer modulate its interaction with RNA I. Cell 1985; 42:959-966.

67. Eguchi Y, Tomizawa J. Complex formed by complementary RNA stem-loops and its stabilization by a protein: Function of ColE1 rom protein. Cell 1990; 60:199-209.

68. Eguchi Y, Tomizawa J. Complexes formed by complementary RNA stem-loops: their formation, structures and interaction with ColE1 Rom protein. J Molec Biol 1991; 220:831-842.

69. Tomizawa J. Control of ColE1 plasmid replication: initial interaction of RNA I and the primer transcript is reversible. Cell 1985; 40:527-535.

70. Tomizawa J. Control of ColE1 plasmid replication—interaction of Rom protein with an unstable complex formed by RNA I and RNA II. J Molec Biol 1990; 212:695-708.

71. Lin-Chao S, Cohen SN. The rate of processing and degradation of antisense RNAI regulates the replication of ColE1-Type plasmids in vivo. Cell 1991; 65:1233-1242.

72. Xu F, Lin-Chao S, Cohen SN. The *Escherichia coli pcnB* gene promotes adenylylation of antisense RNAI of ColE1-type plasmids in vivo and deg-

radation of RNAI decay intermediates. Proc Natl Acad Sci USA 1993; 90:6756-6760.

73. He L, Söderbom F, Wagner EGH, Binnie U, Binns N, Masters M. PcnB is required for the rapid degradation of RNAI, the antisense RNA that controls the copy number of ColE1-related plasmids. Molec Microbiol 1993; 9:1131-1142.

74. Light J, Molin S. Post-transcriptional control of expression of the *repA* gene of plasmid R1 mediated by a small RNA molecule. EMBO J 1983; 2:93-98.

75. Womble DD, Dong X, Wu RP, Luckow VA, Martinez AF, Rownd RH. IncFII plasmid incompatibility product and its target are both RNA transcripts. J Bacteriol 1984; 160:28-35.

76. Blomberg P, Nordstrom K, Wagner EGH. Replication control of plasmid R1: RepA synthesis is regulated by CopA RNA through inhibition of leader peptide translation. EMBO J 1992; 11:2675-2683.

77. Tomizawa J. Control of ColE1 plasmid replication: the process of binding of RNA I to the primer transcript. Cell 1984; 38:861-870.

78. Tomizawa J. control of Cole1 plasmid replication—intermediates in the binding of RNA I and RNA II. J Molec Biol 1990; 212:683-694.

79. Tamm J, Polisky B. Characterization of the ColE1 primer-RNA1 complex: Analysis of a domain of ColE1 RNA1 necessary for its interaction with primer RNA. Proc Natl Acad Sci USA 1985; 82:2257-2261.

80. Tomizawa J, Som T. Control of ColE1 plasmid replication: Enhancement of binding of RNA I to the primer transcript by the Rom protein. Cell 1984; 28:871-878.

81. Tomizawa J, Itoh T. Plasmid ColE1 incompatibility determined by interaction of RNA I with primer transcript. Proc Natl Acad Sci USA 1981; 78:6096-6100.

82. Persson C, Wagner EGH, Nordstrom K. Control of replication of plasmid R1: formation of an initial transient complex is rate-limiting for antisense RNA—target RNA pairing. EMBO J 1990; 9:3777-3785.

83. Persson C, Wagner EGH, Nordstrom K. Control of replication of plasmid R1: structures and sequences of the antisense RNA, CopA, required for its binding to the target RNA, CopT. EMBO J 1990; 9:3767-3775.

84. Kittle JD, Simons RW, Lee J, Kleckner N. Insertion sequence IS*10* antisense pairing initiates by an interaction between the 5' end of the target RNA and a loop in the anti-sense RNA. J Molec Biol 1989; 561-572.

85. Kittle JD. Regulation of Tn*10*: RNA-IN, RNA-OUT and their antisense pairing reaction. Ph.D. Dissertation, Harvard University, 1988.

CHAPTER 6

TRANSLATIONAL CONTROL OF GENE EXPRESSION IN *E. COLI* AND BACTERIOPHAGE

Mathias Springer

I. INTRODUCTION

Gene expression can be regulated in response to very different stimuli. These may be external, such as the cellular growth medium, or internal, in response to a specific need at a given stage of the cell cycle or development. In many cases, the regulation is transcriptional. The discovery that translation is a level of gene expression at which regulation can take place goes back to the observation that transcription is not necessary for early phases of embryonic development in many organisms.[1] Gene expression following fertilization relies completely on maternal mRNAs which are translationally repressed until then.[2] Translational control in prokaryotes was discovered first with RNA bacteriophages where translation is the sole possible level of regulation,[3,4] later in DNA phages and finally in *E. coli*. Even if much remains to be understood about translational regulation in this bacterium, very few genes from other prokaryotes or from eukaryotes have been studied in the same kind of molecular detail as those of coliphages and *E. coli* itself.

Translational control can be effected by either proteins or RNAs, namely, antisense RNAs. Regulation by antisense RNAs is the subject of another chapter of this book. In the present chapter, we shall concentrate on protein mediated translational control. Several recent excellent reviews have covered the field of translational control in RNA phages,[5] bacteriophage T4,[6] bacteriophage λ[7] and *E. coli*.[8] Some other, more general reviews, have also appeared.[9,10]

Protein mediated translational regulation is very often negative, i.e., the protein represses the translation of its target gene. The only prokaryotic exceptions are found in phages. The expression of several early bacteriophage T7 genes is activated by cleavages by the host endonuclease, RNase III.[11] The bacteriophage Mu *mom* mRNA translation is positively controlled by phage Com protein.[12,13] In bacteriophage λ,

Regulation of Gene Expression in Escherichia coli, edited by E. C. C. Lin and A. Simon Lynch. © 1996 R.G. Landes Company.

three positive translational controls have been shown to regulate the lysogenic pathway: cIII and N mRNAs are positively regulated by the host RNase III and the translation of the cII mRNA has been shown to be stimulated by the integration host factor (IHF).[7]

The most common form of translational control is negative feedback regulation, i.e., the product of the gene inhibits the translation of its own mRNA. Negative feedback was discovered in bacteriophage T4 gene 32, but is very common in *E. coli* where the expression of many genes encoding components of the translational machinery, mainly ribosomal proteins[8] but also a translation initiation factor[14] and at least one aminoacyl-tRNA synthetase,[15] is regulated in such a way. The expression of quite a few RNases[16-18] is also feedback regulated at the post-transcriptional level.

Ribosomal proteins are often expressed from very long operons. In this case, it is almost invariably the translation of the first cistron that is controlled by one of the products of the operon. The negative regulation of the first cistron is then transmitted to the promoter distal ones by translational coupling.[19] Translational coupling in ribosomal protein operons could be caused by the formation of secondary structures between the ribosomal binding site (RBS) of one cistron and sequences in the translated region of the upstream cistron which sequester the RBS. The formation of such secondary structures inhibits translation initiation of the downstream cistron unless the upstream gene is translated. If the translation of the upstream cistron is repressed, the lack of translating ribosomes permits the sequestration of the next RBS and so on, thus preventing the translation of all the downstream genes of the operon. Equimolar synthesis of the products of the different cistrons could be explained if the only ribosomes used to initiate translation are those which terminate translation upstream. This "continuous translation" scheme could be facilitated by the fact that, in most cases, the intercistronic regions in ribosomal protein operons are between 5 and 20 nucleotides long, i.e., completely covered by the terminating ribosome. Although the role of mRNA secondary structure in translational coupling was proposed a long time ago, it is only recently that such an inhibitory structure was in fact shown to exist in a ribosomal protein operon.[20] Coupling might not always be tight enough for translation inhibition to propagate through very long operons. In the case of the longest ribosomal protein operon, S10, which consists of 11 genes, autogenous control operates at the level of translation and transcription. The purpose of the transcriptional regulation might be to compensate for incomplete translational coupling.[8]

In the case of RNA bacteriophage MS2, the expression of the lysis gene is translationally coupled to that of the coat protein. The coupling is caused by a secondary structure that forms within the upstream coat protein cistron which masks the RBS of the overlapping lysis gene.[21] However, in this case, the translation of the coat protein is not enough to activate the translation of the lysis cistron. Activation requires translation termination at the nearby coat termination codon.[22] There is good evidence that the activation is caused by ribosomes scanning around the coat protein termination codon.[23] Translational coupling in RNA phages and the role of RNA secondary structure in this phenomenon have been reviewed in detail.[24] It is not known whether it is the translation, or the termination event at the end of the upstream cistrons that is responsible for unmasking the downstream RBS in the case of the ribosomal protein operons.

There are two known exceptions to the rule that translational regulation acts directly on the first cistron among the ribosomal protein operons. In the *spc* operon, the operator is located at the start of the third gene of the operon and negative feedback directly affects the translation of this cistron.[25] The expression of the two first genes of the operon is controlled by this same internal operator, but by a mechanism termed retroregulation (named so because the *cis*-acting site is located *after* the target of the control), which apparently involves mRNA degradation rather than a direct translational block.[26] In the case of the *str* operon, the repressor binding site is located between the first and the second genes of the operon. Again, the expression of the first cistron appears to be regulated by the site that regulates the expression of the second cistron, possibly also by retroregulation. Interestingly, in the case of the *str* operon, binding of the repressor (the ribosomal protein S7) does not cause translational repression in a classical way.[27,28] It has been shown that the binding of S7 to the target mRNA site does not interfere with 30S binding to the RBS of the second cistron. In fact, the majority of the expression of the second cistron is translationally coupled to that of the first, and ribosomal protein S7 seems to inhibit this coupling that is essential for the expression of the second cistron.

Most of the translational repressors recognize only one operator even if, due to the polycistronic nature of many operons, they control the expression of several genes. The T4 RegA protein is exceptional in the sense that it binds to several operators controlling the expression of dispersed genes.[6] Another translational repressor known to act on several different operators is the phage M13/f1 gene V protein. It binds to an operator located in the leader of the gene II mRNA and not only represses its own translation,[29] but also that of gene X, I, III apparently with different efficiencies.[30] In contrast to the "classical" translational regulators (that affect the translation initiation by binding to an operator), RNases (which affect translation initiation mostly by mRNA cleavage) are, in general, able to modulate the expression of several genes.

An important characteristic of translational regulators is that they are often involved in cellular processes other than regulation. The possible exceptions are the phage Mu Com protein, which seems to be only involved in positive regulation of the translation of the *mom* gene, and the phage T4 RegA protein, whose only demonstrated function is the regulation of a dozen T4 genes at the translational level. The second function of translational repressors is, almost invariably, related to nucleic acids binding. The SecA protein, which negatively regulates the translation of its own mRNA,[31] is the only translational repressor with a function not obviously involved in DNA or RNA binding. The SecA ATPase is suspected of catalyzing the rate limiting step in protein secretion and is essential for preprotein binding to the inner membrane and translocation.[32] SecA synthesis is translationally derepressed when the protein secretion pathway is blocked.

After this brief introduction to the subject, the main characteristics of translation initiation, the step at which most controls work, will be briefly described. Then follows a summary of some general properties of the main effectors of most controls, the operators and the repressors. The aim is to avoid making a catalogue of existing systems, but rather to try describing some of their important properties with a few examples. I then describe some of the most representative mechanisms of translational regulation. I further describe how translation can be controlled by mRNA maturation and discuss the relationship between control and mRNA degradation. Finally, I relate these control mechanisms to cellular physiology.

II. TRANSLATION INITIATION

Translational control generally acts at the level of initiation, the step which is usually considered to be rate limiting in translation. This is certainly true in many cases, since ribosomal binding site manipulations are known to cause very strong variations in expression of many different genes.[33,34] In the case of bacteriophage λ, 500-fold differences in expression between distinct late genes can be explained by differential translation initiation.[35] In many cases, a direct selection of mutants for increased expression has yielded mutations in the translation initiation site.[36,37] However, in exceptional cases, which will not be reviewed here, translation has also been shown to be controlled at either the elongation,[38] or the termination step.[39,40]

Translation initiation has been reviewed several times in recent years[41,42] and we shall only give a very brief description of the main components involved. In addition to the ribosome and mRNA, initiation requires a specific tRNA (tRNA$_f^{Met}$), at least three initiation factors, and the hydrolysis of a GTP molecule. The three characterized initiation factors, IF1, IF2 and IF3, bind to the 30S ribosomal subunit (IF1 and IF2 binding is dependent on that of IF3). They all (IF1 only in combination with IF2 and IF3) stimulate the rate of ternary complex (30S-mRNA-tRNA$_f^{Met}$) formation. IF3 also inhibits association between the two ribosomal subunits and favors dissociation of ternary complexes with noninitiator tRNAs. IF2 is a G-binding protein that recognizes, and probably positions, fMet-tRNA$_f^{Met}$ in the ternary complex. A kinetic scheme, similar to that extensively applied to transcription,[43] has been proposed for translation initiation.[42] The 30S subunit, to which initiation factors are fixed, binds each of the ligands (mRNA and fMet-RNA$_f^{Met}$) in a random order to form a ternary preinitiation complex with the ligands not interacting in a productive way. A rate-limiting first-order conformational change permits the formation of the active initiation complex. Upon binding of the 50S subunit, IF3 and IF1 leave the active initiation complex and the 70S initiation complex forms. Finally, IF2 promotes positioning of the fMet-RNA$_f^{Met}$ in the P site and is ejected from the complex; GTP is hydrolyzed and translation elongation starts.

Translation initiates on a specific fragment of the mRNA that is characterized by the presence of an initiator codon (usually AUG but sometimes GUG or UUG, and other codons exceptionally) located 5 to 9 nucleotides downstream of the 3-9 nucleotide Shine-Dalgarno (SD) sequence, which has been shown to base pair with the 3' end of the 16S RNA during initiation.[44] Messenger RNAs sequences other than the SD sequence are suspected of being capable of pairing with the 16S RNA.[45,46] Statistical analysis of translation initiation sites shows a nonrandom distribution of nucleotides between the position -20 to +14 (if the first nucleotide of the initiation codon is +1). This corresponds almost exactly to the region of mRNA protected from RNase degradation by ribosomes and to the end of the reverse transcriptase extension from primers annealed to the mRNA downstream of the RBS.[41]

The secondary structure generally hinders translation initiation. Expression of the coat protein of RNA phage MS2 can be quantitatively correlated to the stability of a hairpin structure containing the RBS.[33] Some of the characteristics of initiation sites are interdependent. For instance, it has recently been shown that a strong SD interaction can compensate for a structured initiation region.[47]

III. TRANSLATIONAL OPERATORS

1. GENERAL PROPERTIES OF TRANSLATIONAL OPERATORS

An operator is generally defined as the site of the *cis*-acting mutations that affect control. This definition is independent of the mechanism and is also generally accepted when regulation occurs at the translational level. In the following paragraphs, I shall restrict myself to the description of translational operators recognized by repressors that inhibit translation initiation by binding and not by mRNA cleavage. Translational regulation by mRNA cleavage will be described separately.

The operators governing translational regulation are located within the transcribed region and, very often, close to the translation initiation site of the controlled gene. In the case of polycistronic operons, the operators are often, but not always, found upstream of the translation initiation site of the first cistron as described above.

There is absolutely no rule as far as the size of translational operators is concerned. In the case of *infC*, the structural gene for translation initiation factor IF3, whose expression is negatively autoregulated at the translational level, the operator is a trinucleotide (the initiation codon of the gene). Also, among the smaller operators is that of the bacteriophage T4 gene 44, which is about 15 nucleotides long, and whose expression is repressed by the RegA protein.[6] However, much larger operators of about 200 to 400 nucleotides have been reported in the case of the L10-L12[48] or IF3-L35-L20 ribosomal protein operons[49] and the *S. typhimurium cob* operon.[50,51]

As regards the secondary structure of translational operators, again no general rule can be drawn from the literature. The T4 gene 32 operator extends over about 100 nucleotides which are mainly unstructured (Fig. 6.1),[52] although a pseudoknot at the 5' end of the operator is used as a nucleation point for cooperative gp32 binding, and was recently shown to be essential for regulation in vitro.[53] The bacteriophage T4 RegA protein also seems to recognize operators characterized by low secondary structure.[6] In most cases, however, translational operators are highly structured. The best studied of all the translational operators, in terms of repressor binding, is the replicase gene of RNA coliphage R17. Its expression is repressed by the coat protein and its operator consists of a 21 nucleotide long stem and loop structure with a single nucleotide bulge in the stem (Fig. 6.1).[54] Most of the translational operators consist of one or several stem and loop structures (containing bulges of various sizes) separated by single stranded regions (Fig. 6.1). Pseudoknots or pseudoknot-like long range interactions can be found in many RNAs and are associated with several biological phenomena.[55] It is thus not surprising that, besides T4 gene 32, at least two other operators contain pseudoknots. In the case of the α operon[56] and of the gene coding ribosomal protein S15,[57] such pseudoknots are well characterized and, most importantly, are essential for control.

The operator, by definition, includes the binding site of the regulatory protein. Mutations that affect this binding will affect control, i.e., give an operator-constitutive phenotype in the case of negative control. In addition to the binding site for the regulatory protein, an operator may also include the RBS. This is obviously the case when the binding sites for the regulator overlap the RBS, as occurs in the bacteriophage T4 RegA sensitive operators. In cases where the operator and the RBS are close but not overlapping, the RBS can also be

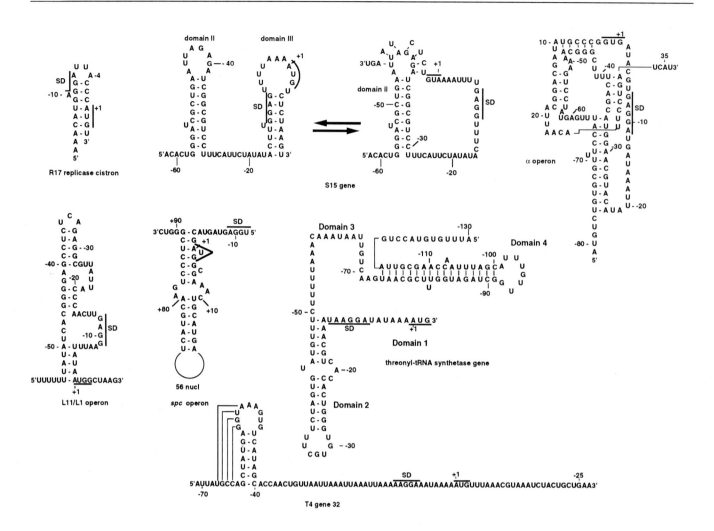

Fig. 6.1. Secondary structure of selected translational operators. The name of the genes is given under the structures. Unless specified, the genes are from E. coli. The initiation codon and the Shine-Dalgarno (SD) sequence are under- or overlined. The indicated 5' ends do not necessarily correspond to the true 5' ends of the transcripts. The A of the AUG is +1 in all cases. See text for references.

considered part of the operator if the regulatory protein and the ribosome compete for mRNA binding. In this case, it is possible to isolate RBS mutants that do not affect the binding of the regulatory protein but nevertheless affect regulation because ribosome binding is unable to compete normally with the binding of the repressor. Such RBS mutations have been isolated in the case of the gene for *E. coli* threonyl-tRNA synthetase, which is under negative feedback control.[58] In addition to the RBS and the regulatory protein binding site, an operator might also include regions of the mRNA that are essential for the RBS to "feel" the binding of the regulatory protein. The need for such a function is obvious in cases where the binding site of the regulatory protein is far away from the RBS. In the L7-L10/L12 operon, the repressor binding site is located 140 nucleotides 5' to the translation initiation codon.[59] Repression was proposed to be due to the masking of the RBS by a conformational change of the leader mRNA following repressor binding.[60] Obviously in such cases, mutations that affect the folding of the inhibitory structure will have an effect on regulation, i.e., define regions of the operator. This type of mutation has also been isolated in a case where the regulatory protein binding site physically overlaps the RBS, namely in the operator of the α operon, whose expression is negatively regulated by ribosomal protein, S4 (see below).

2. TOOLS TO STUDY OPERATORS

Systems where the expression of several operons is controlled by the same regulatory protein have often been very useful in recognizing the *cis*-acting regulatory regions of diverse regulons, especially in the case of transcriptional regulation. A simple search for common sequences in the promoter regions has frequently shown the precise location of the operator. Regions of nucleotide conservation, in general, define the regulatory protein binding site. The situation is different with translational control, where regulatory proteins often only recognize a target in a single gene. It is, however, possible to rely on sequence comparisons from phylogenetically related organisms in searching for nucleotides essential for the recognition process. In this case, sequence comparisons are meaningful only if they are experimentally shown to be recognized by the same regulatory protein.

The main tool that has been used to try to define the nucleotides within an operator region that are essential for regulation has been localized mutagenesis and subsequent analysis of the in vivo or in vitro effects of these nucleotide changes. However, because of the generally large size of translational operators (often more than 50 nucleotides) saturation mutagenesis has rarely been performed on a whole operator. A difficulty with the brute force mutagenic approach is related to the possible global structural changes that any point mutation may cause. Thus, the interpretation of the effect of any nucleotide change is possible only if the structure of the mutated RNA has been checked in solution with adequate chemical or enzymatic probes.

A powerful tool, SELEX (Systematic Evolution of Ligands by Exponential Enrichment), has recently been used to study RNA-protein interaction. It consists of the selection of a particular sequence from a random pool of nucleic acid sequences (prepared by direct DNA synthesis or, in the case of RNA, by subsequent transcription) using its binding specificity to partition reiteratively. This method has been used in the case of the gene for bacteriophage T4 DNA polymerase (gene 43), whose expression is negatively autoregulated at the translational level.[61] The gene 43 operator is about 35 nucleotides long and folds into a stem and loop structure (Fig. 6.1) which binds the polymerase. The 8 nucleotides of the loop have been randomized and two RNA sequences were selected after four rounds of enrichment (optimal binding to the immobilized T4 DNA polymerase) and amplification (cDNA synthesis and PCR amplification). One of the sequences corresponded to the wild-type loop of the operator and the other to a quadruple mutant, which was shown to bind the T4 DNA polymerase with the same affinity as the wild-type operator. The new operator, if placed in T4, shows normal control by gene 43 protein in vivo.[6] The latter result was by no means obvious, since selection was driven by the affinity between the repressor and the operator only. The whole process of regulation involves, in addition to repressor binding, a more complicated process (see above) not necessarily addressed in a simple selection procedure.

3. OPERATOR RECOGNITION SPECIFICITY: THE EXAMPLE OF R17 COAT PROTEIN AND THE REPLICASE CISTRON

In no case was the operator recognition problem studied as thoroughly as for the RNA coliphage R17 replicase cistron, whose translation is negatively regulated by the coat protein.[5] A 21 nucleotide long RNA fragment (Fig. 6.1) was shown to bind the coat protein with an affinity similar to that of the whole R17 RNA. Coat protein binding to more than 100 mutants of this 21 nucleotide long molecule shows that: (1) At some positions such as A(-17) and A(-16), nucleotide changes have no effect. However, deletions in these positions affect binding, indicating that the interaction with the coat protein involves the sugar-phosphate backbone rather than a specific functional group. (2) At the bulged -10 position, a purine is essential. Although the role of this bulged purine has been investigated in great detail, it is not known if it makes direct contacts with the repressor or if it is essential for the general shape of the helical part of the operator. (3) In the loop (position -7 to -4), a purine is required at position -7, the U at -6 can be changed to any nucleotide, a pyrimidine is required at position -5 and A is absolutely required at position -4. This indicates that three positions in the loop (-7, -5 and -4), which have either partial or total nucleotide specificity, make contact with the protein or may be part of a more complex folding that is disrupted when nucleotides are changed. However, there is quite strong evidence that the nucleotide at position 5 makes direct contacts with the protein (see below). (4) Many changes made in the stem indicate that the coat protein does not interact with the helical part of the operator although base pairing is essential. Thus, either the coat protein makes contacts with the sugar-phosphate backbone or the base pairing is essential to maintain the loop in a conformation that permits its essential residues to be recognized by the coat protein. Construction of variants with as many as ten mutations (in a 21 nucleotide long molecule), keeping the general structure and the two essential As of the loop intact, was shown to be recognized almost normally by the coat protein. This indicates that the specific nucleotides of the loop are not only necessary but, within a specific tertiary structure, sufficient for recognition.

Some of the recognition properties of the R17 replicase operator can certainly be generalized; for instance, the precise tertiary structure requirements which indicate that extensive contacts occur between the RNA and the protein. These contacts may be with specific nucleotides or the sugar-phosphate backbone. The latter interactions may be predominant if the tertiary structure presents the corresponding atoms in precise manner to interact with the protein. This is quite different from most DNA-protein interactions between transcriptional regulatory proteins and their respective operators.

The double stranded regions are not specifically recognized in the R17 coat protein/replicase operator interaction. However, this is unlikely to be the case for all translational repressors. The recognition of tRNA by aminoacyl-tRNA synthetases was shown to rely on interactions with double stranded regions of the acceptor arm of the tRNA.[62] X-ray structures of co-crystals between tRNA^{Glu} and tRNA^{Asp} with their synthetases show that the acceptor arm is specifically recognized by interactions with the minor groove in the first case, and with the extremity of the major groove in the second. Similar interactions probably exist in the case of operator-repressor interactions in translational control. An indication that this may be true, comes from mutations in double stranded regions that are not completely compensated when double stranded structure is restored.[25] In the case of the operator of the α operon, C(-43) and G(+4) changes to G(-43) and C(+4) (Fig. 6.1) abolish S4 binding either singly or in combination.[63] Although this can be interpreted as the formation of a G(+4) and C(-43) base pair that is recognized specifically, alternative explanations are still possible. In the case of the operator for ribosomal protein S15, recent experiments seem to indicate that the U(-49)-G(-36) base pair in the lower stem of the pseudoknot (Fig. 6.1) is essential for control.[64] Many changes in this U-G base pair,[64] including its replacement by C-G or U-A, abolish control (L. Bénard and C. Portier, unpublished experiments). This situation may be somewhat reminiscent of the specificity of tRNA^{Ala} recognition which also relies on a G-U base pair in the acceptor stem.[62,65]

4. The Importance of Cooperativity in Operator Recognition

Cooperative binding of the repressor to its operator has been shown to be essential in explaining the regulation of bacteriophage T4 gene 32, and of several other T4 genes that are under RegA translational control. The gene 32 operator and sites recognized by the RegA protein have in common a rather low occurrence of secondary structure, and bind their repressors with a relatively low affinity, even among translational repressors (Table 6.1). However, the gene 32 product represses its own mRNA about 3- to 4-fold more efficiently than other T4 mRNAs.[66] It has been proposed that specificity comes from the cooperative binding of the gene 32 protein to its operator.[67] Preferential binding of gene 32 protein to its own RNA is also achieved, as outlined above, by a specific interaction with the pseudoknot at the 5' end of the operator, that seems to serve as a nucleation site for gene

Table 6.1. Repressor-operator binding affinities and other relevant values

Repressor/Nucl. Acid	Ka (M^{-1})	References
Phages		
gp43/ gene 43 operator	6.7 x 10^8	cited in Miller, Karam & Spicer[6]
gp32/T4 mRNA (average base compos.)	4 x 10^6	cited in Miller, Karam & Spicer[6]
gp32/gene 32 mRNA	1.5 x 10^7	cited in Miller, Karam & Spicer[6]
gp32/T4 ssDNA	10^8	cited in Miller, Karam & Spicer[6]
RegA/gene 44 operator	10^7	Webster & Spicer[103]
R17 coat protein/R17 21-mer	6 x 10^8	Witherell, Gott & Uhlenbeck[5]
R17 coat protein/unspecific RNA	< 10^2	cited in Witherell, Gott & Uhlenbeck[5]
E. coli		
S4/*alpha* operon operator	1.2 x 10^7	Tang & Draper[63]
S4/16S rRNA	1.3 x 10^7	Vartikar & Draper[182]
S4/unspecific (tRNA)	about 10^5	Deckman, Draper & Thomas[119]
S8/*spc* operon operator	5 x 10^6	Gregory et al[71]
S8/16S rRNA	5 x 10^7	Mougel, Ehresmann & Ehresmann[114]
S15/*rpsO* operator	2 x 10^7	Bénard et al[64]
S15/16S rRNA	0.9 x 10^7	Schwartzbauer & Craven[183]
ThrRS/ *thrS* operator	10^8	Brunel et al[126]
ThrRS/tRNA^{Thr} (1/Km)	2 x 10^7	Brunel et al[126]
SecA/*secA* operator	7.5 x 10^6	Dolan & Oliver[184]
30S/mRNA		
30S/002mRNA	1.2 x 10^7	Calogero et al[185]
30S/T4 gene 32 mRNA	2 x 10^7	cited in Draper[120]
Other relevant constants		
X. laevis TFIIIA/5S	1.4 x 10^9	cited in Dolan & Oliver[184]
IF3/30S	4.3 x 10^7	Gualerzi & Pon[42]
Concentrations		
Free 30S: 1-10 µM (10-15% of total)		cited in Draper[120]
Free r-proteins: 1-2 µM		cited in Draper[120]
Gene 32 protein: 3 µM		cited in McPheeters, Stormo & Gold[52]

32 binding.[52] Together, the data suggest that the specificity of the control relies on the pseudoknot structure to which gene 32 binds first, but which is too far away from the RBS to repress translation. As soon as the concentration of gene 32 protein rises over a certain threshold, translation is repressed by the cooperative binding of gene 32 protein to the rest of the operator.

The T4 RegA protein represses the expression of a whole regulon consisting of the gene 45-44-62-regA cluster and other scattered genes.[6] The RegA sensitive operators are always in the immediate vicinity of the initiator AUG and are generally unstructured. The sensitivity of the regulated genes to RegA repression varies over an approximate 20-fold range and the regions of the mRNA protected by the protein vary from 12 to 48 nucleotides in length. Thus, although the sequences of 13 RegA sensitive sites are known, a meaningful consensus is difficult to propose and the nucleotides that are specifically recognized by the RegA protein are still unknown. RegA protein was shown to bind $d(T)_{10}$, which is long enough to carry one RegA protein binding site, with a lower affinity than polyd(T), suggesting cooperative binding to some of its natural operators, possibly those which have been shown to be protected by the protein over relatively long distances. Although the nature of the specific interactions is not known, it seems that RegA mediated regulation is caused, depending on the sites, by a combination of site specific and cooperative binding.

5. MIMICRY BETWEEN OPERATORS AND PRIMARY BINDING SITES

As outlined above, the translational repressors are often, in addition to their regulatory role, functionally implicated in binding either RNA or DNA. This quite systematic double commitment, regulation and an involvement in another cellular activity, is unique to translational regulation. Gene 32 protein is involved in DNA replication, recombination and repair.[68] Its primary role is to bind to single stranded DNA in a sequence independent and cooperative way, to stabilize the separated strands and to permit the movement of the replisome. Gene 32 protein binds to ssDNA with high affinity and to global T4 mRNA with a much lower affinity (Table 6.1). The fact that gene 32 protein binds to RNA with a lower affinity than to DNA, led to the model in that gp32, once synthesized, preferentially binds to single stranded DNA and, when single stranded DNA is saturated, binds its own mRNA.[67] Thus, regulation relies on the competition between gene 32 protein binding to ssDNA and to its own mRNA. This competition between single stranded DNA and gene 32 mRNA binding causes translational repression to be effective only at a specific period of phage development. Both interactions rely on cooperativity, although operator recognition also relies on the specific recognition of the pseudoknot at low gene 32 protein concentrations. Both the competition between function and regulation and the similar fashion in which the regulatory protein recognizes its competing nucleic acid target sites (cooperatively, in the present case) are probably quite general features of translational control.

For the majority of the translational repressors, the primary function is to bind RNA. In ribosomal protein operons, the effector of feedback regulation is, in most cases, one of the operon-products, which binds directly to rRNA. Based on the apparent similarities between some of the ribosomal protein translational operators and their binding sites on rRNA, Nomura and co-workers proposed that recognition of both RNA ligands would be related.[69] The small size of the ribosomal proteins, which would unlikely be able to accommodate two separate

RNA binding domains, made this hypothesis particularly attractive. More importantly, this hypothesis led to the model, similar to that proposed for T4 gene 32, in that a regulatory ribosomal protein binds primarily to its site on rRNA and participates in ribosome assembly, and, once rRNA sites are saturated, the protein starts to bind to its operator to inhibit translation. This competition between function and control, in addition to being a general feature of translational regulation, is also the basis of a general model that relates the synthesis of the components of the translational machinery to the physiological state of the cell (described later).

Nomura's hypothesis was suggested prior to thorough studies of how a control ribosomal protein recognizes both its operator and its rRNA binding site. At the present time, mimicry between rRNA and mRNA recognition seems to be true only in a limited number of cases. In the L11-L1 operon, a clear similarity between the operator and the RNA binding site of L1 was shown to be significant because changes in corresponding nucleotides in the two sites were shown to have analogous effects.[70] The mutations in the operator caused a regulation deficiency whereas the changes in the 23S rRNA binding site of L1 were shown to interfere with its ability to titrate L1 in vivo. It should be noted that, although the mimicry is quite convincing, L1 binding to the operator has never been reported. Another case of mimicry is that between the operator of *spc* operon and the 16S rRNA binding site of S8.[25,71] Using S8 binding to in vitro synthesized RNA fragments of various sizes, the rRNA site of S8 was reduced to a stem and loop structure of about 30 nucleotides with an asymmetrical bulge.[72,73] Similar studies reduced the operator size to about 80 nucleotides without a major change in the affinity for S8 binding.[72] The secondary structure of the operator differs mainly from that of the rRNA site by the presence of two bulged bases present on the mRNA. The operator fragment has about 4 times lower affinity for S8 compared to the rRNA site. Insertions in the rRNA binding site that create bulged bases at positions equivalent to those of the mRNA cause a decrease in affinity for S8 to a level equivalent to that of the mRNA. Conversely, deletions of the bulged bases in the operator (between +1 and +10 in Fig. 6.1) increase the affinity for S8 to levels equivalent to that of the rRNA binding site.[72]

The most convincing case of mimicry is that between the operator of the threonyl-tRNA synthetase gene (*thrS*) and the natural substrates of the enzyme, the tRNA^{Thr} isoacceptors. The expression of the threonyl-tRNA synthetase gene is under negative translational feedback control.[74] The operator is about 120 nucleotides long (Fig. 6.1) and consists of four structural domains.[75] The synthetase has been shown to bind two of the four domains and inhibits ribosome binding.[76] The loop of domain 2 (Fig. 6.1) is 7 nucleotides long, as is the anticodon loop of the tRNAs. Both the operator and tRNA^{Thr} isoacceptors carry a GU at equivalent positions in the loop. In the tRNA, the GU comprises the second and third bases of the anticodon that are conserved in the four isoacceptors. These two anticodon bases were shown to be essential for aminoacylation,[77] whereas the first, which differs in different isoacceptors, is not recognized specifically. A thorough mutational analysis of the three bases of the loop in domain 2, equivalent to those of the anticodon in tRNA^{Thr}, shows that the first base can be changed without major effect on control, whereas any change in the second or third base abolishes control (J. Caillet et al, unpublished). This recognition pattern of the mRNA is very reminiscent of that of the tRNA. Moreover, an in vitro synthesized operator was shown to be a competitive inhibitor in the aminoacylation reaction which indicates that both ligands are recognized by the same, or overlapping site,s of the synthetase.[78] The proof that the resemblance between the anticodon-like arm of domain 2 and the true anticodon arm of tRNA^{Thr} reflects an authentic functional similarity came from experiments based on tRNA identity rules. By definition, the identity elements are the nucleotides that mediate the correct recognition of the tRNA isoacceptor set by its cognate synthetase.[62] The postulate was that, if the *thrS* leader mRNA carries a true tRNA^{Thr}-like structure, it should obey tRNA identity rules so that its identity could be changed from tRNA^{Thr} to that of another tRNA, causing control of the *thrS* gene to be dependent upon another synthetase. The anticodon of tRNA^{Met} (CAU) is known to be a major identity element for methionyl-tRNA synthetase recognition, since its transfer to tRNA^{Val}, tRNA^{Phe} and tRNA$_2^{Ile}$ leads to efficient misacylation by methionyl-tRNA synthetase.[79] Based on similar experiments, the NGU anticodon of tRNA^{Thr} isoacceptors was found to be a major identity element.[80,81] The simple replacement of CGU, the anticodon-like sequence of the *thrS* operator by CAU, was sufficient to abolish control by threonyl-tRNA synthetase and establish control by methionyl-tRNA synthetase in vivo.[82] This identity switch was shown to be due to methionyl-tRNA synthetase recognition of the switched *thrS* operator in vitro.[78]

For most of the other genes controlled by translational feedback, and for which the autogenous repressors are functionally involved in RNA recognition, no meaningful mimicry between function and regulation has been characterized at the present time. This does not mean that common recognition features between the two processes do not exist, but that these may be difficult to detect without knowing 3D structures of the RNA operators.

6. CONFORMATIONAL CHANGES IN OPERATORS

RNA conformational changes are essential in many biological phenomena, ranging from replication (in the case of many plasmids) to control of gene expression at both the transcriptional (attenuation in the *trp* operon, for instance) and the translational level. Some RNA conformational changes induced by protein binding have been characterized down to the atomic level. Work on aminoacyl-tRNA synthetases bound to their cognate tRNAs has shown that specificity partly relies on the ability of the RNA to change its conformation when bound to its synthetase. For instance, in the case of tRNA^{Gln}, the last base-pair of the acceptor arm is disrupted in the complex with the synthetase. In the case of tRNA^{Asp}, the interaction with the synthetase produces a very large conformational change in the anticodon loop.[83]

Conformational changes were shown to be essential for translational regulation in the case of the phage lambda cIII gene[84] and the bacteriophage Mu *com* gene.[12,13] In *E. coli*, mRNA conformational changes, which pertain to translational regulation, were first postulated in the L10-L7/L12 operon, whose expression is negatively regulated at the translational level by the cellular concentration of L10.[48,60] These changes were postulated to explain several peculiarities of the operon, such as the exceptionally long distance (140 nucleotides) between the L10 binding site and the first cistron of the operon, but also the properties of some *cis*-acting mutations that abolish control. Although the model fits the data, and a part of the structure of the operator was experimentally studied,[85] the postulated conformational change remains to be proven in this case.

In *E. coli*, the most detailed studies of conformational changes have been with the α operon and the ribosomal S15 gene. The α operon operator, which is recognized by the S4 ribosomal protein, consists of a very compact multi-pseudoknotted structure (Fig. 6.1) which has been proposed on the basis of both nuclease mapping experiments and the S4 binding properties of an extensive set of mutant operators.[56,86] Conformational changes of the RNA, originally suspected from structural probing data, were confirmed by the existence of mutations in the operator that abolish control but do not affect S4 binding.[63] The latter finding indicates that repressor binding per se is not enough for regulation, and that an mRNA conformational change is necessary for repression. In other words, the operator consists of two functionally distinct (although physically overlapping) sites, the RBS and the repressor binding site, which are allosterically coupled. The existence of two conformations for the α operon mRNA has been confirmed by recent experiments.[87] An RNA fragment containing the operator was shown to melt as a single transition between two conformations that were proposed to correspond to two functionally different states (one active for translation and the other not). These two conformations were also observed, in a very complete set of experiments, by reverse transcriptase primer extension (toeprint). In the inactive state, the structure of the operator stimulates 30S binding (binary complex formation), but inhibits the rate of isomerization to the active conformation. The inactive to active transition is necessary for the formation of the ternary complex in the presence of initiator tRNA. In other words, the mRNA conformation can affect the kinetics of productive initiation complex formation. Interestingly, further experiments show that the two conformations differ in their S4 binding properties, and that S4 traps the mRNA in a conformation capable of binding the 30S subunit but unable to switch to the active conformation which permits ternary complex formation.[88] Unfortunately, the change in structure, which may be quite subtle, between the active and inactive conformation is not known.

The expression of gene for the ribosomal protein S15 is also negatively autoregulated at the translational level.[89] The leader mRNA of the S15 gene is organized into three domains (two of them shown in Fig. 6.1): domain I and II consist of two stem and loop structures and domain III can adopt either a stem and loop structure, or fold back on domain II (as shown in Fig. 6.1) to form a pseudoknotted structure.[57] The conformational change has been inferred on the basis of a very complete set of structure-mapping experiments with both wild-type and mutant operators. In particular, a mutation that changes a C at position -15 to G (Fig. 6.1) and destroys a C-G pair in domain III, was shown to eliminate the conformational heterogeneity and stabilize the pseudoknotted structure. Mutations that strongly destabilize the pseudoknot abolished control, which clearly shows that the formation of the pseudoknot is essential for control. On the other hand, the C to G mutation at -15, which destabilizes the stem-loop conformation of domain III and stabilizes the pseudoknot, rather increases control, indicating that the conformational change per se is not essential for the regulation.

These examples demonstrate the existence of quite significant conformational changes in translational operators which, in most cases, seem essential for regulation. In addition to these major changes, other, more subtle changes probably occur during the binding of the regulatory protein to its target operator. Although important in some cases, conformational change of the mRNA is most probably not systematically required for regulation. For instance, when the R17 coat protein binds to its operator, the hairpin probably remains intact.[90]

IV. TRANSLATIONAL REPRESSORS

1. GENERAL PROPERTIES OF TRANSLATIONAL REPRESSORS

Although many translational operators have been characterized in a quite detailed way in *E. coli*, studies of their corresponding regulatory proteins lag, in general, far behind. The situation with phages is fortunately different since precise functional and/or structural information is available for some repressors. I shall restrict the description of repressors to those that inhibit translation by RNA binding, as opposed to those that act by direct mRNA cleavage (see later).

Translational repressors do not seem to belong to a class of proteins with a particular quaternary structure. For instance, most of the regulatory ribosomal proteins are believed to act as monomers, with the possible exception of L10 which forms a pentameric complex made up of one L10 and four L7/L12 subunits, where L7 and L12 are identical except for the N-terminal acetylation of L7. The complex seems to be more efficient in repressing the translation of the ribosomal protein L10-L7/L12 operon than the isolated L10 protein.[59] Repressors other than ribosomal proteins are often active as dimers, as in the case of both the RNA phage R17 coat protein and threonyl-tRNA synthetase of *E. coli* (see below). On the other hand, the TRAP protein in *B. subtilis* seems to consist of 11 or 12 identical subunits of 8 kDa, each subunit recognizing one GAG or UAG repeat, with several protein-RNA interactions being required for stable association (see below). Binding of translational repressors can depend either completely on specific protein-RNA interactions, or on both specific interactions with the RNA coupled with cooperative protein-protein interactions.

2. THE R17/MS2 COAT PROTEIN: A REPRESSOR CASE STUDY

The X-ray structure of the MS2 virus has been determined to 3.3 Å resolution and shows that the protein shell is composed of 60 triangular units made out of three coat protein monomers arranged with a 3-fold symmetry.[91] This icosahedral structure gives two types of coat protein dimer arrangements. Although the RNA component was not visible, the general structure of the coat protein was shown to consist of five β-strands and two α-helices. The residues involved in various protein-protein contacts essential for capsid formation were also characterized from the 3D structure of the virus. R17 coat protein is 129 amino acids long and is identical to that of phage MS2. R17 coat protein binds to the replicase operator fragment as a dimer.[92] The two types of dimers observed in the crystal structure are presumed to resemble the dimer found in solution. At high concentrations, the coat protein dimer/operator complex polymerizes into phage-like particles. Many of the amino acids that face the interior of the virion, where the RNA is located, are hydrophilic and were suspected to form the RNA binding domain. The amino acids responsible for RNA-protein interactions were probed by genetic selection. First, mutations in the MS2 coat protein which compensated for operator mutations were selected.[93] All of the mutant coat proteins isolated were shown to have a super-repressor phenotype, i.e., to repress replicase expression from a wild-type operator more than the wild-type coat protein. Most of the mutants identified were defective in virus-like particle formation and caused a decrease in the concentration of coat protein multimers (> 2). Since the dimeric form is favored over the multimeric form for repression,[94] it was proposed that the super-repressor phenotype was caused by an increase in the dimer concentration due to the defect in the assembly of the virus-like particles. Genetic selections supplemented with biochemical screens which eliminated mutants with assembly defects and intrinsic instability, yielded mutants which were shown to be mainly defective in binding the operator.[95] The mutations caused amino acid changes mainly in 3 β-sheets located on the inner surface of the virion, as might be expected, and which may be part of the RNA binding site. One of the β-sheets, where the changes were found, is also known to be involved in making hydrogen bonds with the equivalent β-sheet of the other subunit in the dimer. Beta-sheet mediated nucleic acid recognition is known to occur in several bacterial regulatory and histone-like proteins, such as HU.[96]

As outlined above, the R17 replicase operator requires a pyrimidine at position -5 (Fig. 6.1). If the wild-type U is changed to a C, the operator binds to the coat protein with a more than 50-fold increased affinity. Among all of the *cis*-acting operator mutants tested, this is the only case where affinity was increased. It has been proposed that the nucleotide in the -5 position makes a transient covalent bond with Cys46 of the coat protein and that the U to C change facilitates the reaction.[5] Whether this transient covalent bond forms or doesn't form is still an open question, but there is strong evidence that the -5 nucleotide of the operator is in close proximity to the Cys46 residue. Although Cys46 can be changed to several other amino acids without any effect on control, with the wild-type operator, a coat protein Cys46 to Ala46 mutant was shown to bind to the C(-5) operator with the same affinity as the wild-type operator. Thus, this mutant coat protein does not show the increased affinity that wild-type coat protein shows for that specific operator mutant. In addition, the Cys46 to Ala46 coat protein mutant showed only a weak preference for pyrimidines at the -5 position in the operator. In conclusion, the Cys46 to Ala46 mutant shows effects that are specific to operator changes at the -5 position. Since the lack of specificity of this coat protein mutant concerns only that position, Cys46 is probably in direct contact with the operator at -5.

The fact that the active form of the coat protein is a dimer, raises questions about the recognition of the asymmetric operator. One possible solution is that each monomer carries half the recognition site. In another model, each monomer has its own binding site but the binding of the operator to the first monomer occludes the site of the second.

3. Specific Motifs in Translational Repressors

Bacteriophage T4 gene 32 product (gp32) is a protein of 301 amino acids which consists of three domains: (1) The N-terminal domain (1-20) is involved in gp32-gp32 interactions and is essential for the cooperativity with which the protein binds to single stranded DNA and RNA. (2) The central domain (21-253) is responsible for the interaction of the protein with nucleic acids. (3) The C-terminal end (254-301) binds to other T4 replication proteins. Chemical modification, H1 NMR studies, and site-directed mutagenesis have shown that tyrosine residues of the central domain are involved in nucleic acid binding.[97] This central domain also carries a zinc domain of the Cys-Cys-His-Cys type.[98] The protein has been shown to contain Zn(II) and mutants in the central domain have been shown to revert to a wild-type phenotype in the presence of $ZnSO_4$.[99] Removal of Zn(II) from the protein results in a protein that is still able to bind to single stranded RNA but is unable to repress the translation of its own mRNA in vitro. As discussed above, the gene 32 operator consists of an unstructured region preceded by a pseudoknot, which serves as a nucleation center for cooperative gp32 binding. Thus, the properties of gp32 devoid of Zn(II) lead to the hypothesis that the Zn motif is essential for the recognition of the pseudoknot.[100] This situation is reminiscent of that of retroviral nucleocapsid proteins (NCPs) which carry a Zn(II) domain of the same type and are known to be essential for NCP binding to a structured region of HIV-1 RNA.[101] Finally, an RNA recognition motif (RRM), responsible for interaction with nucleic acids, has been recently proposed to exist in the central domain of the T4 gene 32 protein.[102]

The bacteriophage T4 RegA protein is 122 amino acids long and binds to both specific and nonspecific nucleic acids.[103] RegA protein binding to nucleic acids shows the following properties: (1) both sugar and base specificity, with a strong preference for ribo- over deoxyribo- nucleotides; (2) a binding site of about nine nucleotides (when bound to poly-rU); (3) low cooperativity; and (4) a 100-fold higher affinity for target RNA than for nontarget RNA. Photochemical cross-linking studies with oligo-dT identified Phe106, and to a lesser extent, Cys36 as sites involved in nucleic acid recognition.[104] A C-terminal peptide (from 95 to 122) was shown to be able to cross-link and to bind to nucleic acids (but without the specificity of the complete protein). Both phylogenetic comparisons and mutational approaches identified an N-terminal region from Val15 to Ala25 and an Arg-rich region from Arg70 to Ser73 as essential for translational repression.[105] The muta- tional studies also identified the C-terminal region as essential for re- pression. Interestingly, the Val15 to Ala25 region shows similarity to the helix-turn-helix motif of many DNA binding regulatory proteins. Two mutations in the site corresponding to the helix-turn-helix motif of the RegA protein cause a change in the specificity of recognition of the different regulated genes. Such a motif has also been found in the TRAP protein, the product of the *mtrB* gene of *B. subtilis*, which is responsible for the transcriptional attenuation of the *trp* operon in response to changes in the cellular concentration of tryptophan.[106] The TRAP protein has been proposed to bind to a segment of the nascent transcript that is part of an antiterminator structure. This binding prevents the formation of the antiterminator and thus allows the formation of the overlapping terminator. Recent experiments have shown that the TRAP protein binds to GAG and UAG repeats in both the *trp* leader[107] and in the ribosomal binding site of the *trpG* gene, whose expression is know to be negatively regulated by L-Tryptophan and *mtrB*. Taken together, these results are a strong indication that TRAP regulates the expression of *trpG* at the translational level. Interestingly, three transdominant negative mutations have been found at the site of the putative helix-turn-helix motif of TRAP. This is exactly the kind of phenotype that is associated with mutations in the equivalent motif in DNA binding regulatory proteins.[108]

Although more work is essential to prove that the helix-turn-helix motif in translational repressors does in fact correspond to the struc- ture that has been shown to exist in transcriptional regulators, the phenotypes associated with mutations in these motifs seem to indicate that they are involved in RNA recognition. Another RNA binding protein, the *bglG* gene product, which controls antitermination in the *E. coli bgl* operon by binding to a specific secondary structure and preventing the formation of a terminator,[109] also contains a putative helix-turn-helix motif.[105] This motif is located in the N-terminal end of the protein, which has been implicated in RNA binding.[109] Interestingly, the C-terminal end of the L7/L12 protein, whose 3-D structure is known at 1.3 Å resolution, carries a helix-turn-helix motif.[110] However, since the L10-L7/L12 complex is thought to interact with RNA through the L10 subunit, the meaning of this observation is still obscure. The fact that this motif may have a role in RNA binding raises the question of how a motif, known to bind to duplex B-form DNA, manages to recognize RNA which, even when double stranded, is in the A-form. The major deep groove of duplex RNA is too narrow to accommodate an α-helix, although the very shallow minor groove might be capable of doing so. An interesting alternative has recently been proposed in the case of T4 gene 44 whose expression is translationally regulated by RegA. NMR data indicate that the gene 44 operator is single stranded and that the phosphodiester torsion angles and the ribose 2' puckering resemble that of a B-form DNA duplex.[111]

4. *E. COLI* REPRESSORS

Structural and/or functional information about bacterial translational repressors, as distinct from phage, is scarce. One rare exception is the ribosomal protein S8 which regulates the expression of the *E. coli spc* operon. Overproduction of S8 alone is detrimental to cell growth, most probably because it causes translation repression of L14 and, through translational coupling, a decrease of the expression of the other essential cistrons of the operon. This detrimental effect has been used to select for mutants in S8 that do not cause growth inhibition.[112] In this way 39 mutations were isolated at 34 different sites within the S8 protein. 19 of the mutant proteins were shown to have a stability similar to wild type. The mutated sites are scattered throughout the 129 residue-long protein. The amino acid changes affect residues that are often phylogenetically conserved. Although the selection is thought to yield S8 mutant proteins unable to bind *spc* operon mRNA normally, their RNA binding properties have not yet been investigated. S8 has probably a domain involved in protein-protein interaction, since Lys93 can be cross-linked to S5.[113] In addition, Lys30, Lys68 and Lys86 of S8 are accessible for chemical reaction in 30S subunits and are thus probably on the surface of the ribosome.[113] Finally, Cys126 modification by several chemicals impairs the interaction with 16S rRNA.[114] A change of this Cys residue to Ser impairs, not only the binding of S8 to rRNA, but also affects its thermal stability.[115] If the same Cys residue is changed to Ala, no effect can be observed. These results indicate that, although the Cys residue is probably not in a direct contact with the RNA, it is part of the RNA binding domain of S8.

In the case of *E. coli* threonyl-tRNA synthetase, which binds to its mRNA at a site that contains an anticodon domain-like structure (above), two different selections for *trans*-acting mutants that affect control were carried out.[80] In the first, mutations in the synthetase which could compensate for operator constitutive mutations and re-establish control were screened. In a second screen, mutations in the synthetase that were defective in control were isolated. The compensatory mutants turned out to behave as super-repressors, i.e., they repressed expression from a wild-type threonyl-tRNA synthetase operator much more than the wild-type synthetase. Two independently isolated super-repressors had the same double GluGlu to HisHis change at contiguous residues in the central part of the protein. Compensation occurred in a partially allele-specific way: some, but not all, of the operator constitutive mutations could be compensated by the mutant synthetase. Interestingly, the mutated synthetase was shown to have an increased affinity for tRNA^{Thr}, whereas the K_m for the two other substrates of the enzyme, ATP and Thr, are unchanged. Recent experiments (B. Beltchev, unpublished results) indicate that the purified mutant enzyme also shows a strongly increased affinity for the operator. The double change occurs in a region close to motif 1, a sequence that is common to all type 2 synthetases and which may be involved in dimerization. Analysis of the quaternary structure of the synthetase shows clearly that this mutant enzyme has a dimerization defect. Among the mutants defective in control, three were studied in detail. These synthetases were shown to be as stable as the wild-type enzyme and to have a decreased affinity for tRNA^{Thr}; the K$_m$ for the two other substrates of the enzyme, ATP and Thr, were again unchanged. The mutants had alterations in the C-terminal end of the protein, in a region suspected to be the anticodon binding domain of a subgroup (2a) of the type 2 synthetases, to which threonyl-tRNA synthetases belongs.[116] The fact that in both types of mutants the affinity for the operator and for the tRNA^{Thr} vary in coordinate fashion is another good indication that, in the case of threonyl-tRNA synthetase, the operator is recognized in a similar fashion to the natural substrate of the enzyme (see above). Potential Cys-His Zn(II) binding sites exist in many aminoacyl-tRNA synthetases, and Zn has been found in several of the enzymes that have been investigated (reviewed in Miller[117]). Threonyl-tRNA synthetase from *E. coli* and other organisms also carries a Cys-His Zn(II) motif. Chemical modification of Cys residues of threonyl-tRNA synthetase was shown to cause a release of one Zn^{2+} ion per subunit and abolish aminoacylation.[118]

V. MECHANISMS OF CONTROL

1. GENERAL REMARKS

Gene expression can be regulated at the level of translation in many different ways. Control is often mediated by what can be called the "classical" translational equivalent of the Jacob-Monod model, where the binding of a regulatory protein to a translational operator (which is, in this case, located on the mRNA) affects translation initiation. However, several completely different mechanisms of translational control have been characterized, including an example were the repressor binds to the ribosome, and possibly, another example involving translation initiation factor modification. Mechanisms where regulation is a direct consequence of mRNA cleavage are described separately in part VI. We first describe "classical" regulation which can operate at different steps of translation initiation; for instance, at early steps, such as the formation of the preinitiation complex, by decreasing the affinity with which the 30S ribosome binds to the RBS. Repressors have also been shown to inhibit subsequent steps of initiation by acting, for example, at the rate limiting step, which seems to be a conformational rearrangement of the ternary preinitiation complex.

Translational repressors bind to their targets with a much lower affinity than the transcriptional repressors (Table 6.1). This is specially true for the phage T4 gene 32 for which specificity of control is explained by cooperative binding of the repressor to the operator (see above). Relatively low affinity constants for the repressor-operator interaction seem also to be the rule in *E. coli*. The affinity constants for unspecific binding are often comparatively high (about 10^5 M^{-1} in the case of S4, the regulator of the α operon[119]). The affinity between messengers and ribosomes is in general around 10^7 M^{-1}, i.e., about the same order of magnitude as that of the translational repressors for their operators. These numbers and the fact that the supposed concentration of free ribosomes (about 15% of total, i.e., between 1-10 μM depending on growth rate) is much higher than the concentration of most repressors raise some questions about how translational regulation really works in *E. coli.*

In the case of ribosomal proteins, simple thermodynamic calculations with the classical regulation model, in which the repressor competes directly with the ribosome for mRNA binding, indicate that either much higher repressor-operator affinities (or specificities) or higher repressor concentrations than those measured are needed to obtain the in vivo level of repression.[120] The same theoretical considerations indicate that the competition (also called displacement) model makes more sense if mRNA degradation is taken into account.[120] A completely different model, called the entrapment model, seems to be compatible with lower repressor concentration and/or lower affinity for the target operator. In this model the repressor binds to both mRNA and to the 30S-mRNA complex with the same affinity, but slows to zero the isomerization rate of the ternary preinitiation complex to the initiation complex.

2. THE REPRESSOR AND RIBOSOME COMPETE FOR mRNA BINDING

An experimental test that has proven to be extremely useful to differentiate between competition and entrapment is the primer extension inhibition (or toeprint) technique.[121] The initiation complex (consisting of 30S subunit, initiator tRNA and mRNA) stops reverse transcriptase extension from a primer hybridized downstream of the translation initiation site. The strength of the reverse transcriptase stop caused by the ribosome, versus that of the full length reverse transcript (to the 5' end of the mRNA), is a quantitative measure of the formation of the initiation complex.

Once formed, the ternary complex is essentially irreversible in the absence of initiation factors.[87] Thus, under these conditions, the toeprint is a measure of 30S-mRNA association rate. The same technique, with a few modifications, permits the detection of binary complexes between mRNA and the 30S subunit in the absence of initiator tRNA.[122]

If competition between repressor and ribosome binding to the mRNA is responsible for translation regulation, one would expect that, in the presence of sufficiently high repressor concentrations, the ribosome will not bind to the mRNA and the toeprint will disappear. This should be true for both ternary and binary complex formation. In the case of the entrapment model on the contrary, one does not expect the toeprint to disappear with increasing repressor concentrations, since in this case, both repressor and ribosome can bind the same mRNA.

Bacteriophage T4 gene 32 mRNA, gives a toeprint at the classical +16 position.[121] Preincubation in the presence of gene 32 protein inhibits the appearance of the toeprint. Gene 32 protein itself does not inhibit reverse transcription extension although it is known to bind cooperatively the same region of the mRNA. The disappearance of the toeprint is a good indication that gene 32 slows the association of the mRNA with the ribosome. This result also suggests that, on the gene 32 mRNA, the simultaneous presence of gene 32 protein and ribosomes is impossible, and is thus in favor of the competition model. Similar results were obtained with the T4 RegA protein which caused the disappearance of the 30S-initiator tRNA toeprint on the rIIB mRNA, one of the T4 mRNAs subject to RegA mediated repression.[123] The disappearance of the toeprint is specific since mutations in the RBS of rIIB that are known to abolish RegA mediated control also abolish the disappearance of the toeprint in the presence of the RegA protein.[124]

In *E. coli*, threonyl-tRNA synthetase mRNA also gives a toeprint at position +16.[76] The reverse transcriptase stop disappears when threonyl-tRNA synthetase is added before ternary complex formation. The disappearance of the toeprint is not observed with mutants operators which were shown to abolish control in vivo. Also, the toeprint was shown to reappear under conditions where tRNA^{Thr} (but not another control tRNA) was added to the incubation mixture, indicating that this specific tRNA (which binds the synthetase) can compete out the inhibitory effect of the synthetase. Moreover, binary complex formation between the mRNA and 30S ribosomes was also shown to be inhibited by addition of threonyl-tRNA synthetase (P. Romby, personal communication). Finally, increasing concentrations of threonyl-tRNA synthetase inhibits the binding of initiator tRNA to the preformed binary 30S-mRNA complex, also indicating that synthetase binding to the mRNA is incompatible with ternary complex formation.[125] In vivo, mutations in the RBS are expected to affect regulation due to the competition between ribosomes and the synthetase for mRNA binding. If a RBS change causes an increase in affinity for the ribosomes, they should compete more efficiently with the repressor and decrease control. On the other hand, an RBS change that decreases the affinity for the ribosomes should cause an increase of control. This type of competition can be presumed to exist if RBS mutations have this particular phenotype only when the mutations are shown to be "pure" RBS mutations, i.e., to affect only ribosome and not repressor binding. In the case of threonyl-tRNA synthetase, a mutation that increased the length of the Shine-Dalgarno sequence from 6 to 9 nucleotides was shown to increase expression and to abolish control.[58] On the other hand, a mutation that decreased the length of the Shine-Dalgarno was shown to decrease expression and to cause a slight increase in control. Both mutations were shown not to affect the affinity with which the synthetase binds to the operator, but only to modify the intensity of the toeprint, indicating a change in capacity to bind the ribosome.[126]

Taken together, the toeprint data strongly indicate that, in the above examples, translational regulation is due to the binding of the repressor to the mRNA and that this binding somehow competes with 30S binding. In both the gene 32 and the RegA cases, the repressor binding site overlaps with the RBS which suggests that the repressor acts by simple steric hindrance. In the case of threonyl-tRNA synthetase, the Shine-Dalgarno sequence is also close enough to the repressor binding site for repression to be caused by steric hindrance.

As outlined above, the efficiency of the control by simple competition might be increased if mRNA degradation is taken into account. Other features of translational operators might also increase the efficiency of this type of regulation. For instance, thermodynamic considerations indicate that control by competition is more efficient when the repressor binds to a double stranded target (in which the RBS is sequestered) than to a single stranded region that is free to bind 30S ribosomal subunits.[127] The reason is that, in the first case, the majority of the RNA will be in a folded state ready to bind the repressor and not the ribosome, whereas in the second case, both the ribosome and the repressor are able to bind the RNA at any moment. Obviously, the stronger the structure, the more repressor binding occurs relative to ribosomes, i.e., the more effective is the repression. However, if the structure is too stable, it will only bind the repressor and never the ribosome, and constitutively prevent translation. In other words, the most favorable situation is when the structure is stable but sufficiently weak to yield maximal expression in the absence of repressor. Such predictions seem to be fulfilled in the case of *E. coli* threonyl-tRNA synthetase, where mutations were recently isolated that sequester the RBS (C. Brunel, P. Romby and C. Sacerdot, unpublished results). When the sequestering structure is very weak or absent (wild-type situation), control is normal; when sequestration is stronger, control increases at the expense of expression; and when the sequestering structure is very stable, no translation occurs.

3. THE REPRESSOR TRAPS THE INITIATION COMPLEX IN A NONPRODUCTIVE CONFORMATION

Suggested on largely theoretical grounds as a possible model for translational regulation,[120] the entrapment model has recently been shown to explain regulation of the α operon and of the gene for ribosomal protein S15.

In the case of the α operon, a complete set of toeprint experiments was performed.[87,88] When the α operon mRNA is incubated with both S4 and ribosomes, a toeprint can be observed, indicating that S4 and the 30S subunit are able to bind together on the mRNA. As discussed above (section III), the α operon mRNA can fold in two different conformations which both bind the ribosome, but which differ in their capacity to form an active ternary complex in the presence of initiator tRNA. High temperature shifts the equilibrium to the active form. At high temperature, the mRNA gives a classical toeprint (corresponding to an active ternary complex) at +16, when incubated with 30S in the presence of initiator tRNA, but without S4. On the contrary, at low temperature, the ternary complex cannot form, whereas the binary complex can do so in the presence of S4. These results and other kinetic experiments indicate that the mRNA exists in either of two conformations, one preferentially binds S4 and the other, the active one, binds initiator tRNA in the presence of 30S subunits. It seems that S4 binds to the mRNA with the ribosome and traps the mRNA in a conformation that precludes active ternary complex formation. Unfortunately, the differences between the active and the inactive conformation are not known, although it is reasonable to speculate that in the active form, the GUG initiation codon is free to pair with the initiator tRNA.

The mRNA of the gene encoding the ribosomal protein S15 also folds in two different conformations (see section III), one of which contains a pseudoknot (Fig. 6.1).[57] When the mRNA is incubated in the presence of 30S subunits alone, a toeprint is found at +10, at the start of the pseudoknot, suggesting that 30S subunits stabilize the pseudoknot.[125] When the mRNA is incubated with S15 alone, weak but distinct toeprints appear at several positions around +10 suggesting, as was the case in the α operon, that repressor binding also stabilizes the pseudoknot. When the mRNA is incubated with the 30S and initiator tRNA, a classical toeprint at position +17 indicates the formation of an active ternary complex. In the presence of S15, the toeprint appears at +10 and not at +17. This indicates that in the presence of S15, a toeprint at a position identical to that of the binary complex (mRNA and ribosomes) is seen. The complex formed in the presence of all components and S15 was shown to contain the 30S subunit, S15, and initiator tRNA.[125] Therefore, it seems that S15 traps the ribosomes and initiator tRNA in a form different from the active ternary complex, i.e., in some unproductive ternary complex. The main difference between the α operon and S15 resides in the fact that the equilibrium between the active and inactive conformation is an essential component of the control in the first case only, since the role of S4 is to trap the mRNA (with the ribosome) in the inactive conformation. The two systems also differ by virtue of the fact that, in the case of the α operon, a substantial number of mutations that affect the structure of the pseudoknot and abolish control, do not affect the affinity with which S4 binds to the mRNA.[63] This seems to indicate that repression requires not only repressor binding but a further conformational change of the mRNA. In the case of the operator of the S15 gene, all mutations screened which abolished control also affected S15 binding to the mRNA. This difference could mean that, in the latter case, trapping does not require a conformational change of the mRNA but is, instead, due to specific contacts between S15 and the ribosome.

4. THE REPRESSOR BINDS TO THE RIBOSOME AND NOT TO THE TRANSLATIONALLY REGULATED mRNA

Translation initiation factor IF3, among other activities (see Section II), kinetically favors the formation of the ternary complex. The gene for IF3 has been sequenced in several bacterial species, including *E. coli*, and all start with exceedingly rare initiation codons. These are mostly AUU codons[128-131] but AUC has been found in the *dsg* gene which encodes a protein similar to IF3 in *M. xanthus*.[132] In addition, the expression of translation initiation factor IF3 is negatively autoregulated at the translational level, and a change of the AUU initiation codon to AUG was shown to abolish control.[14,133] This clearly shows that the abnormal AUU codon is necessary for regulation. Recent experiments indicate that if the normal AUG initiation codon of several genes is changed to AUU, these genes become regulated by IF3 levels in the cell (J. Sussman, J. Chang and R. Simons, personal communication; K. Engst, C. Sacerdot and M. Springer, unpublished experiments). This indicates that the AUU initiation codon is not only necessary but may even be sufficient for regulation. If the classical regulation model by which the repressor binds to its own mRNA and affects ribosome attachment is valid in this case, IF3 should recognize and bind to an AUU sequence when placed in the context of an RBS. However, regulation by IF3 levels can be obtained with other abnormal initiation besides AUU. These can vary at either of the three positions of the initiation codon. Thus, it is not AUU which seems to be necessary and sufficient for regulation by IF3, but the presence of an abnormal initiation codon in the RBS. Since IF3 regulation is found with initiation codons which differ from AUU at every position, there is no specific nucleotide that is responsible for control, and thus it seems unlikely that IF3 binds to its mRNA. Since IF3 is known to bind to the 30S subunit, it is reasonable to propose that IF3 regulates its expression in vivo by binding the ribosome rather than its mRNA. Once on the ribosome, the factor is able, by a mechanism that remains to be identified, to differentiate between normal and abnormal initiation codons.

This is exactly the role for IF3 proposed on the basis of earlier in vitro experiments that showed that IF3 stimulates translation initiation with the usual initiation codon AUG and inhibits initiation with some other codons,[134] in particular AUU.[135] Experiments using the toeprint technique have shown that IF3 selectively favors the binding of initiator tRNA versus elongator tRNA to the 30S subunit at the

RBS.[136] This selectivity is only found with AUG, GUG and UUG initiation codons but not with AUU. These experiments show quite clearly that IF3 is able to "inspect" the initiation codon and the anticodon domain of the initiator tRNA. It thus seems that, in this case, regulation is explained by the activity of the factor. When the 30S subunit attaches to an mRNA with a standard initiation codon, IF3 somehow accelerates the rate at which the preternary complex changes to a productive ternary complex. If the 30S attaches to its own mRNA with its AUU initiation codon, IF3 rather accelerates dissociation of the preternary complex. Recent in vitro experiments have shown that IF2 dependence is increased when translation is initiated with AUU instead of AUG.[135] The same set of experiments indicate that IF3-stimulated initiation at AUG and IF3-stimulated dissociation at AUU codons give superimposable curves that saturate at an IF3/30S ratio of 1. Both the stimulation and the dissociation are reduced with rRNA mutants which were shown to be defective in IF3 binding. This indicates that stimulation of AUG-dependent initiation and dissociation of initiation complexes made with AUU are due to IF3 binding to the same site on the ribosome.

This type of regulation, where the repressor binds to the ribosome, might not be restricted to IF3. The gene for ribosomal protein S20, whose expression is under negative feedback control, starts with an UUG initiation codon that was shown to be essential for regulation.[137] The existence of such a control and the fact that it was not possible to demonstrate binding of S20 to its own mRNA[138] have been interpreted as an indication that S20 performs its regulatory task bound to the 30S subunit. The gene for ribosomal protein S1 is also under negative feedback control in vivo.[139] In vitro, addition of purified protein S1 to a coupled transcription-translation system causes a specific reduction in the synthesis of S1.[140] Addition of various subdomains of S1 showed that the N-terminal fragment, which possesses the ribosome binding domain, was active in repression. In contrast, fragments from the C-terminal region, containing the nucleic acid binding domain, had no repression activity. Induction of the N-terminal domain in vivo caused a reduction in the synthesis of the chromosomally encoded protein S1, thus confirming that the N-terminal part of the protein is responsible for the feedback. Since the domain of the protein responsible for the binding to the ribosome is sufficient for control, it is possible that feedback regulation of S1 occurs, as in the case of IF3, while the protein is bound to the ribosome and not to its own mRNA. The underlying hypothesis, in these cases, is that, in the absence of these proteins on the ribosome, their respective mRNA are translated more efficiently.

5. Initiation Factor Modification and Control

Many forms of global regulation occur by initiation factor modification (mainly phosphorylation) in eukaryotic cells.[141] In prokaryotes, T7 gene 0.7 encodes a serine/threonine-specific, cAMP-independent protein kinase which specifically stimulates late gene expression under sub-optimal growth conditions. The gene is required for efficient development in cells containing the ColIb factor. Under these conditions, phage late protein synthesis seems to be specifically inhibited in the absence of gene 0.7 product. The RNA polymerase β' subunit and RNase III were known for quite a while to be phosphorylated by the T7 kinase, but more recent experiments show that the kinase phosphorylates more than 90 *E. coli* proteins.[142,143] Strongly phosphorylated proteins include the initiation factors (IF1, IF2 and IF3), the ribosomal proteins S1 and S6 and the elongation factor G. Migration shift analysis on 2-D gels shows that more than 90% of the cell initiation factors are phosphorylated under these experimental conditions (R. S. Provost and J. W. B. Hershey, personal communication). In the same study, parallel purification of phosphorylated and nonphosphorylated IF2 and IF3 shows that phosphorylation causes a 2- to 3-fold decrease of the activity of the factors in an fMet-tRNA-ribosome binding assay. Activity is restored upon treatment with potato acid phosphatase, which shows that the decreased activity is really due to phosphorylation. It remains to be shown that, as suspected from the in vivo studies, translation of T7 late mRNA is specifically enhanced by phosphorylation.

VI. TRANSLATIONAL CONTROL AND mRNA PROCESSING AND/OR DEGRADATION

1. TRANSLATIONAL CONTROL AND mRNA PROCESSING

In all the models described in the previous section, control is thought to occur primarily via the binding of a regulatory protein to its target operator. As discussed later in this section, this primary binding event often has an important effect on the degradation rates of the mRNA being regulated. However, in other cases, mRNA processing seems to be the primary source of regulation. Such examples have been shown to occur in both phage and bacteria.

Phage T7 early transcripts are cleaved at five different sites by RNase III.[144] Gene 0.3 expression is enhanced by RNase III processing although both processed and unprocessed transcripts are very stable.[11] The stem and loop structure recognized and cleaved by RNase III is adjacent to the Shine-Dalgarno sequence of the 0.3 cistron. This structure, because of its position next to the Shine-Dalgarno sequence, may impair ribosome binding. Both in vivo and in vitro data indicate that the activation occurs via enhanced translation of the mature mRNA. It thus seems plausible that the mechanism of activation is the removal of an inhibitory sequence of the mRNA.

Gene 1.1 and 1.2 expression is under *negative* control by RNase III.[145] The inhibitory sequence seems to be located in the RNase III site at the 3' end of the transcript carrying both the 1.1 and 1.2 cistrons. There is quite convincing evidence that a 29 nucleotide sequence near the 3' end of the transcript can pair with the RBS of cistron 1.1 and inhibit translation. The effect is then transmitted to the 1.2 cistron by translational coupling. The negative effect of this 29 nucleotide long stretch is only observed with the processed transcript, since in the unprocessed mRNA, this stretch is base-paired to form the stem of the RNase III recognition site and is thus not free to pair with the RBS of gene 1.1.

A situation similar to that of T7 gene 0.3 occurs in bacteriophage lambda where gene N expression is also increased by a processing event that occurs just upstream of the Shine-Dalgarno sequence. The RNase III processing does not seem to affect mRNA stability (reviewed in Court[16]). Again, activation seems to be due to the elimination of an RNase III site, which in its unprocessed form, inhibits the translation initiation of the N mRNA.

Interestingly, RNase III seems to be able to activate gene expression in the absence of processing. Two such cases have been described in bacteriophage lambda. Firstly, in the cIII gene, a stem and loop structure that resembles an RNase III site straddles the RBS.[84,146] The activation by RNase III does not seem to be explained by maturation, since the enzyme cleaves that site only very weakly in vitro. The cIII mRNA appears to exist in two alternative conformations. Structural probing in vitro and the properties of a set of mutants indicate that one conformation is inactive, and the other active for translation. Activation by RNase III seems to be due to the stabilization of the mRNA in a conformation that permits translation initiation. The second example of RNaseIII activation without apparent processing occurs in the lambda *int* gene, which, when expressed from the p$_L$ promoter, is under negative control by RNase III.[147] This regulation is due to an increase in mRNA degradation that is initiated by an RNase III processing event just downstream of the *int* gene. On the contrary, when expressed from the p$_I$ promoter, *int* expression is activated by RNase III. When expressed from the latter promoter, the 3' end of the transcript does not contain the whole RNase III recognition site present in the p$_L$ transcript. It has been proposed that the binding of RNase III to this uncomplete recognition site causes stabilization of the message and hence, activation of expression. However, other alternative explanations are possible.

The expression of about 10% of the *E. coli* proteins is affected in mutant strains (*rnc*) deficient in RNase III,[148] which indicates that RNase III may have a quite global regulatory role. The mechanism of regulation of gene expression by this enzyme is understood in the case of the *rnc* gene encoding RNase III itself, and that of the *metY-infB-nusA* and *rpsO-pnp* operons. The *rnc* gene is the first cistron of an operon that also carries the genes for *era* and *recO*. The biological role of *era* is unknown, and *recO* belongs to the RecF recombination and repair pathway. RNase III negatively regulates the expression of its own gene[149] apparently by a processing event 31 nucleotides upstream of the AUG of *rnc*. This processing was shown to destabilize the mRNA downstream of the cleavage site, which is suspected to be directly responsible for decreased expression. The processing is thought to remove elements essential to mRNA stability that are located upstream of the cleavage site.

The situation is somewhat similar in the *metY-infB-nusA* and *rpsO-pnp* operons where RNase III was shown to cause destabilization of the mRNA downstream of the maturation site. In the case of the *metY-infB-nusA* operon, the destabilization caused by the cleavage is not followed by decreased expression of the downstream genes.[150] However, decreased expression downstream of the cut was observed with the *rpsO-pnp* operon.[151] In the latter case, the RNase III cut occurs between the *rpsO* and the *pnp* genes, encoding ribosomal protein S15 and the 3' to 5' exonuclease, polynucleotide phosphorylase (PNPase), respectively. The expression of the *pnp* gene is under strong negative feedback control at the translational level.[17] Surprisingly for a 3' to 5' exonuclease, this regulation depends on *cis*-acting sites that are located 5', and not 3', from the *pnp* gene. Moreover, the feedback is lost in an *rnc* strain, or if the RNase III site has been removed by mutation; thus, the regulation is dependent on RNase III processing. These results raise the possibility that PNPase acts as a classical translational repressor by binding to its processed (but not to its precursor) leader mRNA to decrease translation. In this case, the destabilization of the *pnp* message in an *rnc*$^+$ strain would be the consequence rather than the cause of its bad translation when PNPase is in excess. Recent experiments indicate that *pnp* mRNA destabilization occurs only in the presence of both RNase III and PNPase. In other words, the removal of secondary structures supposed to stabilize *pnp* mRNA by the RNase III cut is not sufficient to promote degradation of the downstream mRNA, the presence of PNPase is also required. No mutation in the processed leader completely abolishes negative feedback, which argues somewhat against classical translational regulation by simple binding and subsequent ribosome inhibition. It is quite possible that PNPase relies on another protein for the specific recognition of its leader. Such a protein may be RNase III itself.

RNase III and PNPase are not the only nucleases able to regulate gene expression at the translational level. RegB, a bacteriophage T4 encoded endonuclease, was shown to cleave the GGAG tetranucleotide, which is often found in Shine-Dalgarno sequences.[152] As expected, cleavage of a Shine-Dalgarno sequence has been shown to be the cause of functional inactivation of several T4 early mRNAs. It has also been shown that the RegB nuclease cleaves GGAG sequences either within coding sequences, or in intergenic regions of T4 early mRNAs. In the latter case, it was shown that maturation destabilizes T4 transcripts. In the absence of RegB, bulk T4 early mRNA half-life rises from 9 to 32 minutes. Interestingly, RegB does not affect either middle or late T4 transcripts. Middle transcripts are almost devoid of GGAG sequences and thus escape RegB cleavage. Late transcripts do contain GGAG sequences in about the same proportion as early genes. In this case, escape from RegB mediated processing may be due to the instability of the endonuclease, whose expression is shut off early in the cycle. Another possibility is that the RegB nuclease works in conjunction with another factor that is modified in the late period. Interestingly, purification of the RegB protein indicates that its activity can be strongly enhanced by the ribosomal protein S1 which may thus be a candidate factor for modification in the late period (reviewed in Sanson[153]).

2. REGULATION AT THE LEVEL OF TRANSLATION AND THE ROLE OF mRNA DEGRADATION

In phage, mRNA translatability and stability do not seem to be coupled in a strict way. For instance, the T4 RegB endonuclease cuts the Shine-Dalgarno sequence of the *motA* mRNA, encoding the T4 middle gene transcriptional activator, and causes functional inactivation with only a minor effect on stability: the processed *motA* mRNA is only destabilized by 40% after cleavage.[154] A minor effect on mRNA degradation is also observed in the case of T4 gene 32: the half-life of this mRNA drops only from 30 to 15 minutes under conditions of very strong repression.[155] Bacteriophage M13 gene V translationally represses the expression of its own gene, as described above, and that of genes I, II, III and X to various extents.[30] Interestingly, the steady state cellular concentration of gene I and X mRNA seems to be increased under repressed conditions in a simplified in vivo system where gene I and X expression is measured using fusions to *lacZ* cloned on plasmids, and where gene V product is synthesized from another compatible plasmid *in trans*.

E. coli genes, on the other hand, often seem to have coupled translation and mRNA stability. Detailed studies with hybrid *lacZ* genes clearly show that an increase of translation causes mRNA stabilization.[156,157] Translation is certainly not the only element involved in mRNA stability, since it is possible to modulate mRNA half-life without changing its translatability by modifying its secondary structure at both the 5'[158] and 3' ends (reviewed in Ehretsmann[159]). As far as translationally regulated genes are concerned, the steady-state mRNA level seems to decrease under repressed conditions in several ribosomal protein genes[160,161] and in the threonyl-tRNA synthetase gene (M. Comer and M. Springer, unpublished results). This means that, in those cases, the steady-state mRNA level parallels the quantity of protein. This raises the general question about what comes first in vivo: Is the decrease in protein synthesis under repressed conditions a result of a true translational block, which causes subsequent mRNA degradation, or the consequence of increased mRNA degradation, which causes a subsequent translational decrease? In other words, are "classical" translational repressors real translational inhibitors, or simply proteins that enhance the recognition of the controlled mRNA by nucleases such as RNase E, that are generally involved in mRNA degradation? In vitro data clearly point to the fact that, with the exception of RNases themselves, translational repressors act primarily at the level of translation initiation. In the case of the L1-L11 operon[160] and the S20 gene,[162] steady-state mRNA levels were shown to be coupled to translational activity in vivo; in these cases, *cis*-acting mutations which decrease translation independently of regulation have been shown to increase mRNA turnover. It therefore makes sense to suppose that translational repressors, although acting *in trans*, cause translational blocks in a way similar to the *cis*-acting mutations, resulting in an increase of mRNA turnover. Taken together, the in vitro and in vivo data point to the fact that translational repression comes first and mRNA turnover is a result. However, it is still possible that mRNA degradation participates in control more than suspected. As stated above, simple thermodynamic considerations indicate that, in the case of ribosomal proteins, the high free ribosome concentration, the relatively low regulatory ribosomal protein concentration, and their low affinity for their mRNA targets makes it difficult to understand how control really works. It was proposed that mRNA degradation might substantially amplify the effect of repressors on the translational rate of their respective mRNAs.[120] Thus, the real role of mRNA turnover in translational repression still remains to be investigated. At the present time, there is no general answer to that question, since there does not seem to be a general pathway of mRNA degradation. The rate limiting step is most probably an endonuclease cleavage (often performed by RNase E), followed by exonuclease chewing by PNPase or RNase II (reviewed in Ehretsmann[159]). Although there is good evidence that mRNA degradation often starts with an endonucleolytic cleavage in the 5' end of mRNAs,[163] there are other cases where this cleavage occurs at the 3' end of the mRNA. For example, in the *rpsO* gene which encodes ribosomal protein S15 and which is under translational negative feedback control, the limiting step for degradation is an endonucleolytic cleavage by RNase E which removes a stabilizing secondary structure at the 3' end of the gene.[164]

A more direct role of mRNA degradation in control has been proposed in both the *spc* and *str* operons. In the *spc* operon, the operator is located between the second and third cistron.[25] The repressor (the ribosomal protein S8, the product of the fifth gene) binds on the *spc* mRNA at the operator site and directly represses the translation of the third cistron. However, the expression of the first two genes is also negatively controlled by the cellular level of S8. The same operator seems to be responsible for the regulation of both upstream and downstream genes.[26] This retroregulation seems to be achieved by mRNA degradation by PNPase and/or RNase II under conditions where S8 binds to the operator between the second and third cistron. Interestingly, the same retroregulation is observed when the translation of the third cistron is lowered by a mutation acting *in cis*. It is therefore possible that *spc* retroregulation uses the general degradation pathway; the low translation of the third cistron unmasks an endonucleolytic site (most probably an RNase E site) which, if cleaved, permits the 3' to 5' exonucleases to degrade the mRNA of the upstream cistrons. In the *str* operon, the operator is located between the first and second cistrons and the repressor (the ribosomal protein S7, product of the second cistron) directly represses the translation of its own cistron. The expression of the first cistron is also under S7 negative feedback control and the site responsible for the feedback is again the translational operator. Therefore, it seems quite probable that retroregulation controls this operon in a similar fashion to the *spc* operon.

VII. THE ROLE OF TRANSLATIONAL CONTROL IN GROWTH RATE REGULATION

As already mentioned, most of the true translational regulators are involved in other cellular activities besides their role in control of gene expression. These other activities often provide the physiological link between a specific control mechanism and a developmental program in phage, or more global controls such as growth rate dependent regulation in bacteria.

As discussed above, the T4 gene 32 product first binds to single stranded DNA and once these primary binding sites are saturated, to its own mRNA. This basic scheme of competition between function and regulation extends to many other examples of translational control and, in particular, to the growth rate dependent control of ribosomal proteins. Many components of the translational machinery are synthesized in higher amounts at high growth rate. For example, the number of ribosomes per cell varies with growth rate from about 7,000 to about 70,000.[165] The synthesis of rRNA is itself subject to direct growth rate control although the mechanism of this control is only partially understood. For many years, "magic spot" (guanosine 3'-diphosphate 5'-diphosphate, or ppGpp) was thought to be responsible for both stringent and growth rate control. However, experiments have clearly shown that growth rate control (but not stringent control) can be achieved in the absence of ppGpp.[166] The rRNA promoters are activated both in vivo and in vitro by the product of the *fis* gene,[167] whose expression is sensitive to nutritional upshift[168] and could be responsible for growth rate control of rRNA synthesis. Since rRNA synthesis in strains deleted for the *fis* gene are still under growth rate control, the Fis protein is not the unique mediator of this regulation. At the present time, it seems reasonable to suppose that growth rate regulation is achieved through several regulatory devices.

As regards the synthesis of ribosomal proteins, an attractive model based on the same kind of competition as in the T4 gene 32 case, may explain growth rate control.[169,170] As already mentioned, each ribosomal protein operon contains a cistron encoding a ribosomal control protein that can bind either to its specific site on the rRNA or to a site on its own mRNA and inhibit the translation of the operon. Under conditions of rRNA excess, the control protein will bind preferentially to its site on the rRNA and be assembled into the ribosome. Under conditions where rRNA concentrations are limiting, the primary ligand is not available and the ribosomal control protein will bind to its mRNA and decrease its translation. Thus, through the competition between rRNA and mRNA for the binding of the regulatory ribosomal proteins, ribosomal protein synthesis rates will parallel rRNA synthesis rates. Since regulatory ribosomal proteins generally bind rRNA directly, an increase of rRNA synthesis will derepress all the ribosomal protein operons that are under translational feedback control. This mechanism for growth rate control of ribosomal proteins means that rRNA is the true effector of the control and titrates the translational repressors (the regulatory ribosomal proteins). This titration derepresses the ribosomal protein operons and implies that the negative feedback is essential for growth rate control. Such a model seems to be valid in the case of the L11–L1 operon[169] but may not be true in other cases.[171]

A similar model has been proposed to link RNase III synthesis to growth rate.[16] At high growth rate, a large portion of the RNase III molecules are thought to be titrated out by precursor rRNAs, which constitute about 70% of total RNA. Thus, under shift-up conditions, where rRNA synthesis increases, free RNase III will decrease. This will cause an increase in the expression of a certain number of key genes that are negatively controlled by the enzyme: its own gene (*rnc*) and possibly, also that of nucleotide phosphorylase (*pnp*). The increase in nuclease synthesis might lead to a transient increase in mRNA turnover after shift-up. Under steady-state conditions, RNase III synthesis might also be growth rate-regulated but its actual free concentration will depend on the number and properties of the different available substrates, which are not always processed with the same efficiency. This may lead to growth rate-dependent differential processing at certain sites. In fact, growth rate-dependent mRNA stability has been observed in some mRNAs in particular that of *ompA*, one of the most stable mRNAs of the cell.[172] This exceptional stability is due, at least in part, to its 5' untranslated region which, if cleaved, decreases stability without affecting translational yield.[173] Replacement of the 5' end of an unstable mRNA with that of *ompA* causes stabilization and growth-rate dependent stability.

The secondary structure, per se, may also have an interesting regulatory role in growth rate-dependent control since RBS mutations that modify mRNA structure can cause differential effects in response to alterations in cellular growth conditions.[174] For instance, certain RBS mutations can cause expression to increase with growth rate while others have the opposite effect. The molecular interpretation of these results is difficult since many elements of the transcription and translation machinery are changed with bacterial growth rate, nevertheless the important conclusion remains: slight changes in a translation initiation site can completely alter the growth rate response. This kind of mRNA secondary structure dependent effect might be essential in order to explain why many elements of the translation machinery, although not under negative feedback, are growth rate regulated.

VIII. CONCLUSIONS AND PERSPECTIVES

In this section, some of the important points concerning operator-regulatory protein recognition and regulatory mechanisms are underlined. I also comment on other possible, but yet unreported, regulatory mechanisms. Finally, some possible advantages of translational over transcriptional control of gene expression are considered.

Based mainly on structural and biochemical studies of complexes between aminoacyl-tRNA synthetases and their cognate tRNAs, the source of sequence specificity in RNA-protein interactions seems to come from: (1) interactions between bases of the RNA in single stranded stretches, loops, or bulges, and protein domains that are either complementary in shape, or capable of specific hydrogen bonds; (2) interactions between the minor groove of double stranded regions, and possibly the major grooves at the end of helices, with specific residues of the protein; (3) conformational changes in the RNA that permit specific sequences to bind to the protein with a higher affinity than other sequences.[175] There is no reason to believe that translational repressors (or activators) bind to their target operators with different specificity rules. Although 3D structural data have not yet been obtained, the existence of many operator mutants in characterized single-stranded regions which affect repressor binding is a good indication for specific interactions with single-stranded regions. Most operators contain double-stranded regions, but sequence-specific interactions with such regions still have to be firmly established although there is good evidence for such an interaction in the case of the *E. coli* ribosomal protein S15 operator. Conformational changes in operators have been shown to be essential for regulation in only a limited number of cases. However, future studies will most probably tell us that they are more general than suspected. These conformational changes will likely turn out to be either small adaptations to the bound regulatory protein, or major structural changes spanning over several hundred nucleotides. In operators where long distance interactions seem to explain regulation, RNA folding kinetics may play an essential role. With operators that are about 200 nucleotides long, transcription takes a few seconds to be completed which must be enough for the operator to fold and form a 3D structure that is recognized by the regulatory protein and/or the ribosome. Because the final conformation has to form within a restricted time for regulation to work, the number of folding intermediates must be limited, i.e., a folding pathway must exist. Such pathways still remain to be investigated.

Proteins seem to use very different strategies to recognize RNA in a sequence-specific way. Aminoacyl-tRNA synthetases recognize their cognate tRNAs through multiple contacts, whereas other proteins do so by using specific binding motifs. Examples of such motifs are the arginine-rich segments of the HIV Tat protein and the RNA recognition motif (RRM) shared by many RNA binding proteins. Translational regulators could recognize their targets by a combination of these strategies. Although 3D structures of translational regulators with their targets are not yet available, there is good evidence that specific protein motifs are involved in recognition. This is well documented in T4 gene 32 protein which carries a Zn domain most probably involved in specific RNA binding. Strangely enough, several translational regulators seem to carry a helix-turn-helix domain which seems to be involved in RNA recognition. The role of this motif in DNA recognition by transcriptional regulators is extremely well documented; its real role in RNA recognition remains to be investigated. Protein-protein contacts are also essential for operator recognition in a certain number of cases. This is true when the repressors bind cooperatively to their target operators and when the repressors are active in a multimeric form.

Translational regulation in *E. coli* and its phages can be the result of the binding of a regulatory protein to its target operator or mRNA processing by a nuclease. Processing of mRNA has been reported to either activate translation by eliminating inhibitory structures, or inhibit translation by cleaving mRNAs in a region essential for translation, such as the Shine-Dalgarno sequence. In phage, mRNA processing is not always followed by an increase of mRNA degradation, whereas the contrary seems to be true in *E. coli*. The distinction between the two mechanisms (processing or simple binding) may not always be very clear cut, since RNase III seems to act as an activator in the absence of cleavage, and polynucleotide phosphorylase, which is a 3' to 5' exonuclease, has been proposed to bind to the 5' leader of its own mRNA and act as a "classical" translational repressor. As discussed above (see also Table 6.1), the affinity of these "classical" translational regulators for their targets is low when compared to most transcriptional regulators. However, the situation completely differs from transcriptional control since most translational regulators are present in very high concentrations in the cell (such as ribosomal proteins). Nevertheless, the concentration of free repressors is generally much lower since they are more often bound to DNAs or RNAs with equal or even higher affinities than to their target operators. This competition between primary binding sites and target operators is the basis for this type of control mechanism and takes into account the general physiological state of the cell (such as growth rate) for bacteria, or the specific step of the developmental program in phages. Although the concentrations of free repressor and free 30S ribosomal subunits are often a subject of debate, it seems that simple competitive inhibition of ribosome binding at the RBS by repressor binding at the overlapping or adjacent operator is not always sufficient to explain control. Thus, alternative schemes have been proposed and confirmed by experiments, whereby the ribosome and repressor bind together on the mRNA in a nonproductive way. The latter scheme is equivalent to noncompetitive inhibition of translation. The efficiency of simple competitive inhibition may be increased if mRNA degradation is taken into account.

This has still to be conclusively confirmed by experimental data and may mean that mRNA degradation is more important for repression efficiency than suspected. In addition to mRNA degradation, mRNA conformational changes may increase control in the competitive inhibition scheme. In conclusion, it seems that the binding of a regulatory protein to an operator can act on translation by a variety of ways: the regulatory protein may cover the RBS and directly inhibit ribosome binding; it may also cause sequestration of the RBS into a secondary structure that will restrain ribosome attachment, and it can also cause the ribosome to be trapped on the mRNA in a conformation incompatible with translation initiation. How trapping occurs and what kind of conformation the trapped initiation complex corresponds to are still unknown. In addition to processing or binding of a regulatory protein, alternative schemes of regulation acting on mRNA level could be envisaged: mRNA modification has been reported in eukaryotes not only at the 5' end (capping) but also at internal A residues.[176] Although no such modifications have yet been reported for prokaryotes, one might envisage that nucleotide modification of an RBS or an operator could modify the interaction with the ribosome and/or a regulatory protein. Modification of tRNA is extremely sensitive to environmental changes.[177] An equivalent sensibility of mRNA modification could lead to translational control in response to growth medium changes. RNA is known to be able to bind small molecules in a specific way.[178] Thus, it may be possible for a folded leader mRNA to bind small metabolites. The binding of metabolites could change the structure of the mRNA and cause regulation in response to their cellular concentration, i.e., possibly to their presence in the growth medium. Strangely enough, in the *S. typhimurium cob* operon, whose expression is negatively regulated at the translational level by the vitamin B12 concentration, powerful genetic screens and direct searches for proteins binding the mRNA have failed to identify a regulatory gene or protein.[50] Although negative evidence does not mean that such a protein does not exist, or that B12 or one of its derivatives is the direct effector of the control, it is interesting to note that an RNA that specifically binds cyano-B12 has been isolated by in vitro selection.[179]

Finally, translational regulation may also be the result of events acting on other molecules than the mRNA. A reported example is IF3, which binds to the ribosome and most probably not to its mRNA to regulate its translation. IF3 is special in the sense that the regulation mechanism seems to be based on its function, which is "inspection" of the codon-anticodon interactions at the initiation event. It would be interesting to know whether or not binding of a regulatory protein to ribosomes could also cause regulation in a less particular case. Another possible control mechanism is initiation factor modification as it occurs in eukaryotes. Although initiation factor modification by the T7 protein kinase has been reported, whether the phosphorylation affects regulation has still to be investigated.

What are the possible advantages of regulating gene expression at the translational rather than the transcriptional level? It is obvious that translational control acts at a later step and has the supplementary energy cost of mRNA synthesis. In some cases, regulation simply cannot occur at transcription: regulation in RNA phage obviously has to take place at a post-transcriptional step. In DNA phages, specially in T7, but often in T4, mRNAs are quite stable. Under those conditions, regulation at the post-transcriptional level has several advantages. Shut-off of expression is faster since it is not delayed by long mRNA half-lives. Derepression of stable mRNAs is also faster, since there is no delay required for mRNA synthesis: the translational repressor leaves its mRNA, allowing immediate translation.

Obviously, another advantage of translational control is the ability to differentially regulate genes that are expressed from the same transcription unit. This permits phage to specifically regulate the expression of one out of a set of transcriptionally active genes at one specific moment of the developmental program. Such differential regulation also exists in bacteria in several ribosomal protein operons. The gene for the α subunit of RNA polymerase is the third cistron of the α operon which also encodes four ribosomal proteins. All the ribosomal protein cistrons are under the control of S4 including the fourth cistron; however, the third cistron, encoding the α subunit, escapes that regulation.[180] A similar situation occurs in the *spc* operon which is made of 12 cistrons where the 11th, encoding the SecY protein, is not under the control of S8 (reviewed in Zengel[8]). As far as the speed of control is concerned, the *raison d'être* of translational control in *E. coli* is somewhat less obvious than in phage since mRNA half-life is, in general, quite short, and repressed mRNAs are degraded. Transcriptional shut-off will always be delayed by mRNA half-life, i.e., about a few minutes. Can such a time-lag be a problem for the cell? The answer is most probably yes, and specially so in translation. As an example, the turnover of tRNA which has an average value of 4 s⁻¹, i.e., each tRNA isoacceptor family is charged by its synthetase and uncharged by translation about four times per second.[181] This means that any amino acid shortage will be "felt" by the cell in less than one second. Since, the cell is mainly engaged in the synthesis of the elements of the translation machinery at high growth rate, fast regulatory solutions may be essential to the survival of the cell. Besides, occurring at the translational level, another cause of speed in the regulatory response is the direct relationship between the effector and the control. The effector, which in the case of ribosomal proteins can be considered to be rRNA, is the first molecule whose synthesis is shut down in the case of nutritional shift down. In Nomura's model, the lack of free rRNA will cause the regulatory ribosomal proteins to bind to their target mRNAs and repress translation. The speed at which regulation will work is only limited by the time it takes the control ribosomal proteins to find their target operators.

In conclusion, translational control may have several advantages over transcriptional control, and in particular, the speed of regulation. In addition, translational control can be accomplished by a variety of mechanisms acting on different components of the initiation machinery. Speed and mechanistic variety may be important features to achieve efficient regulation of genes which are often involved in central housekeeping tasks. Future studies will probably reveal that this mechanistic variety is even wider than suspected at the present time.

ACKNOWLEDGMENTS

I thank Ciaran Condon, Marc Dreyfus and Claude Portier for helpful comments on the manuscript. I appreciate the help of many colleagues who provided reprints or preprints. The work in the laboratory was supported by grants from the C.N.R.S. (URA1139) and from the Ministère de la Recherche et de la Technologie.

REFERENCES

1. Gross PR, Malkin LI, Moyer WA. Templates for the first proteins of embryonic development. Proc Natl Acad Sci USA 1964; 51:407.
2. Newport J, Kirschner M. A major developmental transition in early *Xenopus* embryos: I. characterization and timing of cellular changes at the midblastula stage. Cell 1982; 30:675-686.
3. Lodish HF, Cooper S, Zinder N. Host-dependent mutants of the bacteriophage f2 IV. On the biosynthesis of the viral RNA polymerase. Virology 1964; 24:60-70.
4. Lodish H, Zinder N. Mutants of the bacteriophage f2 VIII. Control mechanisms for phage-specific synthesis. J Molec Biol 1966; 19:333-348.
5. Witherell GW, Gott JM, Uhlenbeck OC. Specific interaction between RNA phage coat proteins and RNA. Prog Nucl Acid Res Molec Biol 1991; 40:185-220.
6. Miller ES, Karam JD, Spicer E. Control of translation initiation: mRNA structure and protein repressors. In: Karam JD, ed. The molecular biology of bacteriophage T4. Washington DC: American Society for Microbiology, 1994:193-205.
7. Oppenheim A, Kornitzer D, Altuvia S et al. Post-transcriptional control of the lysogenic pathway in bacteriophage λ. Prog Nucleic Acid Res Molec Biol 1993; 46:37-49.
8. Zengel JM, Lindahl L. Diverse mechanisms for regulating ribosomal protein synthesis in *E. coli*. Prog Nucleic Acid Res Molec Biol 1994; 47:331-369.
9. McCarthy JEG, Gualerzi C. Translational control of prokaryotic gene expression. Trends in Genet 1990; 6:78-85.
10. Lindahl L, Hinnebush A. Diversity of mechanisms in the regulation of translation in prokaryotes and lower eukaryotes. Current Opinion in Genet and Devel 1992; 2:720-726.
11. Dunn JJ, Studier FW. Effect of RNAse III cleavage on translation of bacteriophage T7 mRNAs. J Molec Biol 1975; 99:487-499.
12. Wulczyn FG, Kahmann R. Translational stimulation: RNA sequence and structure requirements for binding of Com protein. Cell 1991; 65:259-269.
13. Hattman S, Newman L, Murthy HMK et al. Com, the phage Mu *mom* translational activator, is a zinc-binding protein that binds specifically to its cognate mRNA. Proc Natl Acad Sci USA 1991; 88:10027-10031.
14. Butler JS, Springer M, Dondon J et al. *Escherichia coli* protein synthesis initiation factor IF3 controls its own gene expression at the translational level in vivo. J Mol Biol 1986; 192:767-780.

15. Springer M, Graffe M, Butler JS et al. Genetic definition of the translational operator of the threonine tRNA ligase gene in *Escherichia coli*. Proc Natl Acad Sci USA 1986; 83:4384-4388.

16. Court D. RNA processing and degradation by RNAse III. In: Belasco J, Brawerman G, eds. Control of mRNA stability. San Diego, CA: Academic Press, 1993:71-116.

17. Robert-Lemeur M, Portier C. *E. coli* polynucleotide phosphorylase expression is autoregulated through an RNAse III-dependent mechanism. EMBO J 1992; 11:2633-2641.

18. Jain C, Belasco JG. RNase E autoregulates its synthesis by controlling the degradation rate of its own mRNA in *E. coli*. Unusual sensistivity of the *rne* transcript to RNase E activity. Genes & Devel 1995; 9:84-96.

19. Nomura M, Gourse R, Baughman G. Regulation of the synthesis of ribosomes and ribosomal components. Ann Rev Biochem 1984; 53:73-117.

20. Lesage P, Chiaruttini C, Dondon J et al. Messenger RNA secondary structure and translational coupling in the *E. coli* operon encoding translation initiation factor IF3 and the ribosomal proteins, L35 and L20. J Molec Biol 1992; 228:366-386.

21. Schmidt BF, Berkhout B, Overbeek GP et al. Determination of the RNA secondary structure that regulates lysis gene expression in bacteriophage MS2. J Molec Biol 1987; 195:505-516.

22. Berkhout B, Schmidt BF, van Strien A et al. Lysis gene of bacteriophage MS2 is activated by translation termination at the overlapping coat gene. J Molec Biol 1987; 195:517-524.

23. Adhin MR, van Duin J. Scanning model for translation reinitiation in Eubacteria. J Molec Biol 1990; 213:811-818.

24. de Smit MH, van Duin J. Control of prokaryotic translation initiation by mRNA secondary structure. Prog Nucl Acid Res and Molec Biol 1990; 38:1-35.

25. Cerretti DP, Mattheakis LC, Kearney KR et al. Translational regulation of the *spc* operon in *E. coli*. Identification and structural analysis of the target site for S8 repressor protein. J Molec Biol 1988; 204:309-329.

26. Mattheakis L, Vu L, Sor F et al. Retroregulation of the synthesis of ribosomal proteins L14 and L24 by feedback repressor S8 in *E. coli*. Proc natl Acad Sci USA 1989; 86:448-452.

27. Saito K, Nomura M. Post-transcriptional regulation of the *str* operon in *E. coli*. Strucural and mutational analysis of the target site for translational repressor S7. J Molec Biol 1994; 235:125-139.

28. Saito K, Mattheakis LC, Nomura M. Post-transcriptional regulation of the *str* operon in *E. coli*. Ribosomal protein S7 inhibits coupled translation of S7 but not its independent translation. J Molec Biol 1994; 235:111-124.

29. Michel B, Zinder N. Translation repression in bacteriophage f1: characterisation of the gene V protein target on the gene II mRNA. Proc Natl Acad Sci USA 1989; 86:4002-4006.

30. Zaman G, A. S, Kaan A et al. Regulation of expression of the genome of bacteriophage M13. Gene V protein regulated translation of the mRNAs encoded by genes I, III, V and X. Bioch Bioph Acta 1991; 1089:183-192.

31. Schmidt MG, Dolan KM, Oliver DB. Regulation of *Escherichia coli secA* messenger RNA translation by a secretion-responsive element. J Bacteriol 1991; 173:6605-6611.

32. Oliver DB. SecA Protein—Autoregulated ATPase Catalysing Preprotein Insertion and Translocation Across the Escherichia-Coli Inner Membrane. Mol Microbiol 1993; 7:159-165.

33. de Smit MH, van Duin J. Secondary structure of the ribosome binding site determines translation efficiency: a quantitative analysis. Proc Natl Acad Sci USA 1990; 87:7668-7672.

34. Ringquist S, Shinedling S, Barrick D et al. Translation initiation in *Escherichia coli*—sequences within the ribosome-binding site. Mol Microbiol 1992; 6:1219-1229.

35. Ray PN, Pearson ML. Functional inactivation of bacteriophage λ morphogenic gene mRNA. Nature 1975; 253:647-650.

36. Hall MH, Gabay J, Débarbouillé M et al. A role for mRNA secondary structure in the control of translation initiation. Nature 1982; 295: 616-618.

37. Chapon C. Expression of *malT*, the regulator gene of the maltose regulon in *E. coli*, is limited both at transcription and translation. EMBO J 1982; 1:369-374.

38. Yu YT, Snyder L. Translation elongation factor Tu cleaved by a phage-exclusion system. Proc Natl Acad Sci USA 1994; 91:802-806.

39. Craigen WJ, Cook RG, Tate WP et al. Bacterial peptide chain release factors: conserved primary structure and possible frameshift regulation of release factor 2. Proc Natl Acad Sci, USA 1985; 82:3616-3620.

40. Curran JF, Yarus M. Use of tRNA suppressors to probe regulation of *E. coli* release factor 2. J Molec Biol 1988; 203:75-83.

41. Gold L. Post-transcriptional regulatory mechanisms in *Escherichia coli*. Ann Rev Biochem 1988; 57:199-233.

42. Gualerzi CO, Pon CL. Initiation of mRNA translation in prokaryotes. Biochemistry 1990; 29:5882-5888.

43. Hoopes BC, McClure WR. Strategies in regulation of transcription initiation. In: Neidhardt FC, ed. *Escherichia Coli* and *Salmonella typhimurium*. Washington, DC: American Society for Microbiology, 1987:1231-1240.

44. Hui A, deBoer HA. Specialized ribosome system: preferential translation of a single mRNA species by a subpopulation of mutated ribosomes in *Escherichia coli*. Proc Natl Acad Sci USA 1987; 84:4762-4766.

45. Thanaraj TA, Pandit MW. An additional ribosome-binding site on mRNA of highly expressed genes and a bifunctional site on the colicin fragment of 16S rRNA from *E. coli*: important determinants of the efficiency of translation initiation. Nucleic Acids Res 1989; 17:2973-2985.

46. Sprengart ML, Fatscher HP, Fuchs E. The initiation of translation in *E. coli*: apparent base pairing between the 16S rRNA and downstream sequences of the mRNA. Nucl Acids Res 1990; 18:1719-1723.

47. de Smit MH, van Duin J. Translation initiation on structured messengers. Another role for the Shine-Dalgarno interaction. J Molec Biol 1994; 235:173-184.

48. Friesen JD, Tropak M, An G. Mutations in the *rplJ* leader of *Escherichia coli* that abolish feedback regulation. Cell 1983; 32:361-369.

49. Lesage P, Truong HN, Graffe M et al. Translated translational operator in *Escherichia coli*: autoregulation in the *infC-rpmI-rplT* operon. J Molec Biol 1990; 213:465-475.

50. Richter-Dahlfors AA, Ravnum S, Andersson DI. Vitamin B12 represion of the *cob* operon in *S. typhimurium*: translational control of the cbiA gene. Molec Microbiol 1994; 13:541-553.

51. Richter-Dahlfors AA, Andersson DI. Cobalamine (vitamin B12) repression of the Cob operon in *S. typhimurium* requires sequences within the leader and the first translated open reading frame. Molec Microbiol 1992; 6:743-749.

52. McPheeters DS, Stormo GD, Gold L. Autogenous regulatory site on the bacteriophage T4 gene 32 messenger RNA. J Molec Biol 1988; 201:517-535.

53. Shamoo Y, M. KK, Konigsberg WH et al. Translation repression by the bacteriophage T4 gene 32 protein involves specific recognition of an RNA pseudoknot structure. J Molec Biol 1993; 232:89-104.

54. Romaniuk PJ, Lowary P, Wu HN et al. RNA binding site of R17 coat protein. Biochemistry 1987; 26:1563-1568.

55. ten Dam E, Pleij K, Draper D. Structural and functional aspects of RNA pseudoknots. Biochemistry 1992; 31:11665-11676.

56. Tang CK, Draper DE. Unusual mRNA pseudoknot structure is recognized by a translational repressor. Cell 1989; 57:531-536.

57. Philippe C, Portier C, Mougel M et al. Target site of *E. coli* ribosomal protein S15 on its mRNA. Conformation and interaction with the protein. J Molec Biol 1990; 211:415-426.

58. Brunel C, Caillet J, Lesage P et al. The domains of the *E.coli* threonyl-tRNA synthetase translational operator and their relation to threonine tRNA isoacceptors. J Molec Biol 1992; 227:621-634.

59. Johnsen M, Christensen T, Dennis PP et al. Autogenous control: ribosomal protein L10-L12 complex binds to the leader sequence of its mRNA. EMBO J 1982; 8:999-1004.

60. Christensen T, Johnsen M, Fiil NP et al. RNA secondary structure and translation inhibition: analysis of mutants in the *rplJ* leader. EMBO J 1984; 3:1609-1612.

61. Tuerk C, Gold L. Systematic evolution of ligands by exponential enrichment: RNA ligands to bacteriophage T4 DNA polymerase. Science 1990; 249:505-510.

62. McClain WH. Rules that govern tRNA identity in protein synthesis. J Molec Biol 1993; 234:257-280.

63. Tang CK, Draper DE. Evidence for allosteric coupling between the ribosome and repressor binding sites of a translationally regulated mRNA. Biochemistry 1990; 29:4434-4439.

64. Bénard L, Philippe C, Dondon L et al. Mutational analysis of the pseudoknot structure of the S15 translational operator from *E. coli*. Molec Microbiol 1994; 14:31-40.

65. Hou YM, Schimmel P. A simple structural feature is a major determinant of the identity of a transfer RNA. Nature 1988; 333:140-145.

66. Lemaire G, Gold L, Yarus M. Autogenous translational repression of bacteriophage T4 gene 32 expression in vitro. J Molec biol 1978; 126:73-90.

67. von Hippel PH, Kowalczykowski SC, Lonberg N et al. Autoregulation of gene 32 expression. Quantitative evaluation of the expression and function of the bacteriophage T4 gene 32 (single-stranded DNA binding) protein system. J Molec Biol 1982; 162:795-818.

68. Alberts BM, Frey L. T4 bacteriophage gene 32: a structural protein in the replication and recombination of DNA. Nature 1970; 227:1313-1318.

69. Nomura M, Yates JL, Dean D et al. Feedback regulation of ribosomal protein gene expression in *E. coli*: structural homology of ribosomal RNA and ribosomal protein mRNA. Proc Natl Acad Sci USA 1980; 77: 7084-7088.

70. Said B, Cole JR, Nomura M. Mutational analysis of the L1 binding site of 23S rRNA in *Escherichia coli*. Nucleic Acids Res 1988; 22:10529-10545.

71. Gregory RJ, Cahill PBF, Thurlow DL et al. Interaction of *E. coli* ribosomal protein S8 with its binding sites in ribosomal RNA and messenger RNA. J Molec Biol 1988; 204:295-307.

72. Wu H, Jiang L, Zimmermann RA. The binding site for ribosomal protein S8 in 16S rRNA and *spc* mRNA from *E. coli*: minimum structural requirements and the effects of single bulged bases on S8-RNA interaction. Nucl Acids Res 1994; 22:1687-1695.

73. Mougel M, Allmang C, Eyermann F et al. Minimal 16S rRNA binding site and role of conserved nucleotides in *E. coli* ribosomal protein S8 recognition. Eur J Biochem 1993; 215:787-792.

74. Springer M, Plumbridge JA, Butler JS et al. Autogenous control of *Escherichia coli* threonyl-tRNA synthetase expression in vivo. J Molec Biol 1985; 185:93-104.

75. Moine H, Romby P, Springer M et al. Messenger RNA structure and gene regulation at the translational level in *Esherichia coli*: the case of threonine:tRNA^Thr ligase. Proc Natl Acad Sci USA 1988; 85:7892-7896.

76. Moine H, Romby P, Springer M et al. *E. coli* threonyl-tRNA synthetase and tRNA^Thr modulate the binding of the ribosome to the translation initiation site of the *thrS* mRNA. J Molec Biol 1990; 216:299-310.

77. Hasegawa T, Miyano M, Himeno H et al. Identity determinants of *E. coli* threonine tRNA. Biochem Biophys Res Commun 1992; 184:478-484.

78. Romby P, Brunel C, Caillet J et al. Molecular mimicry in translational control of *E. coli* threonyl-tRNA synthetase gene. Competitive inhibition in tRNA aminoacylation and operator-repressor recognition switch using tRNA identity rules. Nucleic Acids Res 1992; 20:5633-5640.

79. Schulman LH. Recognition of transfer RNAs by aminoacyl-transfer RNA synthetases. Progress in Nucleic Acid Res and Molecular Biology 1991; 41:23-87.

80. Springer M, Graffe M, Dondon J et al. tRNA-like structures and gene regulation at the translational level: a case of molecular mimicry in *E. coli*. EMBO J 1989; 8:2417-2424.

81. Schulman LH, Pelka H. An anticodon change switches the identity of *E.coli* tRNA^Met_m from methionine to threonine. Nucleic Acids Res 1990; 18:285-289.

82. Graffe M, Dondon J, Caillet J et al. The specificity of translational control switched using tRNA identity rules. Science 1992; 225:994-996.

83. Cavarelli J, Rees B, Thierry JC et al. Yeast aspartyl-tRNA synthetase: a structural view of the aminoacylation reaction. Biochimie 1993; 75:1117-1123.

84. Altuvia S, Kornitzer D, Teff D et al. Alternative mRNA structures of the cIII gene of bacteriophage lambda determine the rate of its translation initiation. J Molec Biol 1989; 210:265-280.

85. Climie SC, Friesen JD. Feedback regulation of the *rplJL-rpoBC* ribosomal protein operon of *Escherichia coli* requires a region of mRNA secondary structure. J Molec Biol 1987; 198:371-381.

86. Deckman IC, Draper DE. S4-α mRNA translation regulation complex. II. Secondary structures of the RNA regulatory site in the presence and absence of S4. J Molec Biol 1987; 196:323-332.

87. Spedding G, Gluick TC, Draper D. Ribosome initiation complex formation with the pseudoknotted α operon mRNA. J Molec Biol 1993; 229:609-622.

88. Spedding G, Draper D. Allosteric mechanism for translational repression in the E. coli α operon. Proc Natl Acad Sci USA 1993; 90:4399-4403.

89. Portier C, Dondon L, Grunberg-Manago M. Translational autocontrol of the *E.coli* ribosomal protein S15. J Molec Biol 1990; 211:407-414.

90. Gralla JA, Steitz JA, Crothers DM. Direct physical evidence for secondary structure in an isolated fragment of R17 bacteriophage mRNA. Nature 1974; 248:204-208.

91. Valegard K, Liljas L, Fridborg K et al. The three-dimensional structure of the bacterial virus MS2. Nature 1990; 345:36-41.

92. Beckett D, Uhlenbeck OC. Ribonucleoprotein complexes of R17 coat protein and a translational operator analog. J Molec Biol 1988; 204:927-938.

93. Peabody DS, Ely KR. Control of translation repression by protein-protein interaction. Nucl Acids Res 1992; 20:1649-1655.

94. Carey J, Cameron V, de Haseth PL et al. Sequence-specific interaction of R17 coat protein with its ribonucleic binding site. Biochemistry 1983; 22:2601-2610.

95. Peabody DS. The RNA binding site of bacteriphage MS2 coat protein. EMBO J 1993; 12:595-600.

96. Phillips SEV. Specific β-sheet interactions. Curr Opin Str Biol 1991; 1:89-98.

97. Shamoo Y, Ghosaini LR, Keating KM et al. Site-specific mutagenesis of T4 gene 32: the role of tyrosine residues in protein-nucleic acid intaractions. Biochemistry 1989; 28:7409-7414.

98. Berg J. Zinc fingers and other metal binding domains. J Biol Chem 1990; 265:6513-6616.

99. Gauss P, Boltrek Krassa K, McPheeters DS et al. Zinc(II) and the single-stranded DNA binding protein of bacteriophage T4. Proc Natl Acad Sci USA 1987; 84:8515-8519.

100. Shamoo Y, Webster KR, Williams KR et al. A retrovirus-like Zinc domain is essential for translational repression of bacteriophage T4 gene 32. J Biol Chem 1991; 266:7967-7970.

101. Darlix JL, Gabus C, Nugeyre MT et al. *Cis* elements and *trans* acting factors involved in the RNA dimerisation of human immunodeficiency virus HIV-1. J Mol Biol 1990; 216:689-699.

102. Kim YJ, Baker BS. Isolation of RRM-type RNA-binding protein genes and the analysis of their relatedness by using a numerical approach. Molec Cell Biol 1993; 13:174-183.

103. Webster KR, Spicer EK. Characterization of bacteriophage T4 regA protein-nucleic acid interactions. J Biol Chem 1990; 265:19007-19014.

104. Webster KR, Keill S, Konigsberg W et al. Identification of aminoacid residues at the interface of a bacteriophage T4 regA protein-nucleic acid complex. J Biol Chem 1992; 267:26097-26103.

105. Joswik CE, Miller ES. Regions of bacteriophage T4 and RB69 RegA translational repressor proteins that determine RNA binding specificity. Proc Natl acad Sci USA 1992; 89:5053-5057.

106. Babitzke P, Yanofsky C. Reconstitution of *Bacillus subtilis trp* attenuation in vitro with TRAP, the *trp* RNA-binding attenuation protein. Proc Natl Acad Sci USA 1993; 90:133-137.

107. Babitzke P, Stults JT, Shire SJ et al. TRAP, the trp RNA binding attenuation protein of *B. subtilis*, is a multisubunit complex that appears to recognize G/UAG repeats in the *trpEDCFBA* and *trpG* transcripts. J Biol Chem 1994; 269:16597-16604.

108. Schleif B. DNA binding by proteins. Science 1988; 241:1182-1187.

109. Houman F, Diaz-Torres M, Wright A. Transcriptional antitermination in the *bgl* operon of E. coli is modulated by a specific RNA binding protein. Cell 1990; 62:1153-1163.

110. Rice PA, Steitz TA. Ribosomal protein L7/L12 has a helix-turn-helix motif similar to that found in DNA binding regulatory proteins. Nucl Acids Res 1989; 17:3757-3762.

111. Szewczak AA, Webster KR, Spicer EK et al. An NMR characterization of the RegA protein binding site of bacteriophage gene 44 mRNA. J Biol Chem 1991; 266:17832-17837.

112. Wower I, Kowaleski MP, Sears LE et al. Mutagenesis of ribosomal protein-S8 from *Escherichia-Coli*—Defects in Regulation of the *spc* Operon. J Bacteriol 1992; 174:1213-1221.

113. Allen G, Capasso R, Gualerzi C. Identification of the aminoacid residues of protein S5 and S8 adjacent to each other in the 30S ribosomal subunit of *E. coli*. J Biol Chem 1979; 254:9800-9806.

114. Mougel M, Ehresmann B, Ehresmann C. Binding of *E. coli* ribosomal protein S8 to 16S rRNA: kinetic and thermodynamic characterization. Biochemistry 1986; 25:2756-2765.

115. Wu H, Wower I, Zimmermann RA. Mutagenesis of ribosomal protein S8 from *E. coli*: expression, stability and RNA-binding properties of S8 mutants. Biochemistry 1993; 32:4761-4768.

116. Cusack S. Sequence, strucure and evolutionary relationships between class 2 aminoacyl-tRNA synthetases: an update. Biochimie 1993; 75:1077-1081.

117. Miller TW, Schimmel P. A metal-binding motif implicated in RNA recognition by an aminoacyl-tRNA synthetase and by a retroviral gene product. Molec Microbio 1992; 6:1259-1262.

118. Nureki O, Kohno T, Sakamoto K et al. Chemical modification and mutagenesis studies on Zinc binding of aminoacyl-tRNA synthetases. J Biol Chem 1993; 268:15368-15373.

119. Deckman IC, Draper DE, Thomas MS. S4-α mRNA translation regulation complex. I. Thermodynamics of formation. J Molec Biol 1987; 196:313-322.

120. Draper DE. Translational regulation of ribosomal proteins in *E.coli*. In: J. Ilan, ed. Translational regulation of gene expression. New York: Plenum publishing corporation, 1987:1-25.

121. Hartz D, McPheeters DS, Gold L. Selection of the initiator tRNA by *E.coli* initiation factors. Genes & Develop 1989; 3:1899-1912.

122. Hartz D, McPheeters DS, Green L et al. Detection of *Escherichia coli* ribosome binding at translation initiation sites in the absence of tRNA. J Molec Biol 1991; 218: 99-105.

123. Winter RB, Morrissey L, Gauss P et al. Bacteriophage T4 regA protein binds to mRNAs and prevents translation initiation. Proc Natl Acad Sci USA 1987; 84:7822-7826.

124. Unnitham S, Green L, Morrissey L et al. Binding of the bacteriophage T4 RegA protein to mRNA targets: an AUG is required. Nucl Acids Res 1990; 18:7083-7092.

125. Philippe C, Eyermann F, Bénard L et al. Ribosomal protein S15 from *E. coli* modulates its own translation by trapping the ribosome on the mRNA initiation loading site. Proc Natl Acad Sci USA 1993; 90: 4394-4398.

126. Brunel C, Romby P, Moine H et al. Translational regulation of the *E. coli* threonyl-tRNA synthetase gene: structural and functional importance of the *thrS* operator domains. Biochimie 1993; 75:1167-1179.

127. de Smit MH. Regulation of translation by mRNA structure. 1994. Ph.D. thesis. University of Leiden.

128. Sacerdot C, Fayat G, Dessen P et al. Sequence of a 1.26-kb DNA fragment containing the structural gene for initiation factor IF3: presence of an AUU initiator codon. EMBO J 1982; 1:311-315.

129. Pon C, Brombach M, Thamm S et al. Cloning and characterization of a gene cluster from *B. stearothermophilus* comprising *infC, rpmI, rplT*. Mol Gen Genet 1989; 218:355-357.

130. Liveris D, Schwartz JJ, Geertman R et al. Molecular cloning and sequencing of *infC*, the gene encoding translation initiation factor-IF3, from four enterobacterial species. FEMS Microbiol Lett 1993; 112:211-216.

131. Hu WS, Wang RYH, Shih JWK et al. Identification of a putative *infC-rpmI-rplT* operon flanked by long inverted repeats in *Mycoplasma fermentans* (Incognitus Strain). Gene 1993; 127:79-85.

132. Cheng YL, Kalman LV, Kaiser D. The *dsg* gene of Myxococcus xanthus encodes a protein similar to translation initiation factor IF3. J Bacteriol 1994; 176:1427-1433.

133. Butler JS, Springer M, Grunberg-Manago M. AUU to AUG mutation in the initiator codon of the translation initiation factor IF3 abolishes trans-

lational autocontrol of its own gene (*infC*) in vivo. Proc Natl Acad Sci USA 1987; 84:4022-4025.

134. Berkhout B, van der Laken CJ, van Knippenberg PH. Formyl-methionyl-tRNA binding to 30S ribosomes programmed with homopolynucleotides and the effect of translational initiation factor 3. Biochim Biophys Acta 1986; 866:144-153.

135. Lateana A, Pon CL, Gualerzi CO. Translation of mRNAs with degenerate initiation triplet AUU displays high initiation factor 2 dependence and is subject to initiation factor 3 repression. Proc Natl Acad Sci USA 1993; 90:4161-4165.

136. Hartz D, Binkley J, Hollinsworth T et al. Domains of the initiator tRNA and initiation codon crucial for initiator tRNA selection by *E.coli* IF3. Genes and Develop 1990; 4:1790-1800.

137. Parsons GD, Donly BC, Mackie GA. Mutations in the leader sequence and initiation codon of the gene for ribosomal protein S20 (*rpsT*) affect both translational efficiency and autoregulation. J Bacteriol 1988; 170:2485-2492.

138. Donly BC, Mackie GA. Affinities of ribosomal protein S20 and C-terminal deletion mutants for 16S rRNA and S20 mRNA. Nucleic Acid Res 1988; 16:997-1010.

139. Rasmussen MD, Sorensen MA, Pedersen S. Isolation and characterization of mutants with impaired regulation of *rpsA*, the gene encoding ribosomal protein S1 of *Escherichia coli*. Molec Gene Genet 1993; 240:23-28.

140. Skouv J, Schnier J, Rasmussen MD et al. Ribosomal protein S1 of *Escherichia coli* is the effector for the regulation of its own synthesis. J Biol Chem 1990; 265:17044-170049.

141. Hershey JWB. Translational control in mammalian cells. Annu Rev Biochem 1991; 60:717-755.

142. Robertson ES, Nicholson AW. Phosphorylation of *E. coli* translation initiation factors by the bacteriophage T7 protein kinase. Biochemistry 1992; 31:4822-4827.

143. Robertson ES, Nicholson AW. Phosphorylation of elongation factor G and ribosomal protein S6 in bacteriophage T7-infected *E. coli*. Molec Microbiol 1994; 11:1045-1057.

144. Dunn JJ, Studier FW. T7 early RNAs and *E. coli* ribosomal RNAs are cut from large precursor RNAs in vivo by ribonuclease III. Proc Natl Acad Sci USA 1973; 70:3296-3300.

145. Saito H, Richardson CC. Processing of mRNA by RNAse III regulates expression of gene 1.2 of bacteriophage T7. Cell 1981; 27:533-542.

146. Altuvia S, Kornitzer D, Kobi S et al. Functional and structural elements of the mRNA of the cIII gene of bacteriophage lambda. J Molec Biol 1991; 218:723-733.

147. Guarneros G. Retroregulation of bacteriophage λ int gene expression. In: Clarke A, Compas RW, Cooper M, Eisen H, Goebel W, Koprowski H, Melchers F, Olldstone M, Vogt PK, Wagner H, Wilson I, eds. Current topics in Microbiology and Immunology. Berlin: Springer-Velag, 1988:1-19.

148. Gitelman DR, Apirion D. The synthesis of some proteins is affected in RNA processing mutants of *E. coli*. Biochem and Biophys Res Com 1980; 96:1063-1070.

149. Bardwell JCA, Régnier P, Chen S-M et al. Autoregulation of RNase III operon by mRNA processing. EMBO J 1989; 8:3401-3407.

150. Régnier P, Grunberg-Manago M. Cleavage by RNase III in the transcripts of the *metY-nusA-infC* operon of *E. coli* releases the tRNA and initiates the decay of downstream mRNA. J Molec Biol 1989; 210:293-302.

151. Portier C, Dondon L, Grunberg-Manago M et al. The first step in the functional inactivation of the *E. coli* polynucleotide phosphorylase mes-

senger is a ribonuclease III processing at the 5' end. EMBO J 1987; 6:2165-2170.

152. Uzan M, Favre R, Brody NE. A nuclease that cuts specifically in the ribosome binding site of some T4 mRNAs. Proc Natl Acad Sci USA 1988; 85:8895-8899.

153. Sanson B, Uzan M. Post-transcriptional controls in bacteriophage T4: roles of the sequence-specific endonuclease RegB. FEMS Microbiol Rev 1994; In press:

154. Sanson B, Uzan M. Dual role of the sequence-specific bacteriophage T4 endoribonuclease RegB: mRNA inactivation and mRNA destabilisation. J Molec Biol 1993; 233:429-446.

155. Russel M, Gold L, Morisett H et al. Translational, autogenous regulation of gene 32 expression during bacteriophage T4 infection. J Biol Chem 1976; 251:7263-7270.

156. Yarchuk O, Jacques N, Guillerez J et al. Interdependence of translation, transcription and mRNA degradation in the *lacZ* gene. J Molec Biol 1992; 226:581-596.

157. McCormick JR, Zengel JM, Lindahl L. Correlation of translation efficiency with the decay of *lacZ* mRNA in *E. coli*. J Molec Biol 1994; 239:608-622.

158. Emory SA, Bouvet P, Belasco JG. A 5' terminal stem-loop structure can stabilise mRNA in *E. coli*. Genes & Development 1992; 6:135-148.

159. Ehretsmann CP, Carpoussis AJ, Krisch HM. mRNA degradation in procaryotes. FASEB J 1992; 6:3186-3192.

160. Cole JR, Nomura M. Changes in the half-life of ribosomal protein messenger RNA caused by translational repression. J Molec Biol 1986; 188:383-392.

161. Singer P, Nomura M. Stability of ribosomal protein mRNA and translational feedback regulation in *E. coli*. Molec Gene Genet 1985; 199:543-546.

162. Rapaport LR, Mackie GA. Influence of translational efficiency on the stability of the mRNA for ribosomal protein S20 in *E. coli*. J Bact 1994; 176:992-998.

163. Belasco J, Higgins C. Mechanisms of mRNA decay in bacteria: a perspective. Gene 1988; 72:15-23.

164. Régnier P, Hajnsdorf E. Decay of mRNA encoding ribosomal protein S15 of *E. coli* is initiated by an RNase E-dependent endonucleolytic cleavage that removes the 3' stabilising stem and loop structure. J Molec Biol 1994; 217:283-292.

165. Bremer H, Dennis PP. Modulation of chemical composition and other parameters of the cell by growth rate. In: Neidhardt, ed. *Escherichia Coli* and *Salmonella Typhimurium*. Cellular and Molecular Biology. Washington, DC: American Society for Microbiology, 1987:1527-1542.

166. Gaal T, Gourse RL. Guanosine 3'-diphosphate 5'-diphosphate is not required for growth rate-dependent control of rRNA synthesis in *Escherichia coli*. Proc Natl Acad Sci USA 1990; 87:5533-5537.

167. Ross W, Thompson JF, Newlands JT et al. *E. coli* Fis protein activates ribosomal RNA transcription in vitro and in vivo. EMBO J 1990; 9:3733-3742.

168. Ninnemann O, Koch C, Kahmann R. The *E. coli fis* promoter is subject to stringent control and autoregulation. EMBO J 1992; 11:1075-1083.

169. Cole JR, Nomura M. Translation regulation is responsible for growth-rate-dependent and stringent control of the synthesis of ribosomal proteins L11 and L1 in *E. coli*. Proc Natl Acad Sci USA 1986; 83:4129-4133.

170. Lindahl L, Zengel JM. Ribosomal genes in *Escherichia coli*. Ann Rev Genet 1986; 20:297-326.

171. Lindahl L, Zengel J. Autogenous control is not sufficient to ensure steady-state growth rate-dependent regulation of the S10 ribosomal operon of *Escherichia coli.* J Bacteriol 1990; 172:305-309.

172. Nilsson G, Belasco JG, Cohen SN et al. Growth rate dependent regulation of mRNA stability in *E. coli.* Nature 1984; 312:75-77.

173. Emory SA, Belasco JG. The *ompA* untranslated RNA segment functions in *E. coli* as growth rate mRNA stabiliser whose activity is unrelated to translation efficiency. J Bact 1990; 172:4472-4481.

174. Jacques N, Guillerez J, Dreyfus M. Culture conditions differentially affect the translation of individual *Escherichia coli* mRNAs. J Molec Biol 1992; 226:597-608.

175. Steitz TA. Similarities and differences between RNA and DNA recognition by proteins. In: Gesteland RF, Atkins JF, eds. The RNA world. Cold Spring Harbor, NY: Cold Spring Harbor Laboratory Press, 1993:219-237.

176. Narayan P, Rottman F. Methylation of mRNA. In: Meister A, ed. Advances in enzymology. New York: John Wiley & Sons, 1992:255-285.

177. Björk GR. Biosynthesis and function of modified nucleosides. In: Söll D, Rajbhandary UL, eds. tRNA. Washington, DC: American Society for Microbiology, 1995:165-205.

178. Szostack JW, Ellington AD. In vitro selection of functional RNA sequences. In: Gesteland RF, Atkins JF, eds. The RNA world. Cold Spring Harbor, NY: Cold Spring Harbor Laboratory Press, 1993:511-533.

179. Lorsch JR, Szostack JW. In vitro selection of RNA haptamers specific for cyanocobalamin. Biochemistry 1994; 33:973-982.

180. Thomas MS, Bedwell DM, Nomura M. Regulation of α operon gene expression in *E. coli.* A novel form of translational coupling. J Molec Biol 1987; 196:333-345.

181. Jakubowski H, Goldman E. Quantities of individual aminoacyl-tRNA families and their turnover in *E. coli.* J Bacteriol 1984; 158:769-776.

182. Vartikar JV, Draper DE. S4-16S ribosomal RNA complex. Binding constant measurement and specific recognition of a 460-nucleotide region. J Molec Biol 1989; 209:221-234.

183. Schwartzbauer J, Craven GR. Apparent association constants for *E. coli* ribosomal proteins S4, S7, S8, S15, S17 and S20. Nucleic Acids Res 1981; 9:2223-2237.

184. Dolan K, Oliver DB. Characterisation of *E. coli* SecA protein binding site on its mRNA involved in autoregulation. J Biol Chem 1991; 266:23329-23333.

185. Calogero RA, Pon CL, Canonaco MA et al. Selection of the mRNA translation initiation region by *E. coli* ribosomes. Proc Natl Acad Sci USA 1988; 85:6427-6431.

EFFECTS OF DNA SUPERCOILING ON GENE EXPRESSION

James C. Wang and A. Simon Lynch

SYNOPSIS

It is well known that open complex formation between promoters and RNA polymerase is thermodynamically favored by negative supercoiling of the DNA template. The effects of template supercoiling on the kinetics of transcription are, however, much more complex even in simple cases involving no regulatory factors; *a priori* predictions of such effects are at best tenuous. In this chapter, we focus on insights gained from experimental data accumulated in the past two decades on how template supercoiling and transcription affect each other. We begin with a review of the historical link between DNA supercoiling and transcription. This introduction is followed by a brief account of how gene expression is affected upon decreasing the cellular level of gyrase, a DNA topoisomerase that negatively supercoils DNA, or DNA topoisomerase I, an enzyme that specifically relaxes negatively supercoiled DNA. Mechanistic considerations are then presented for several cases of increasing complexity: from the simplest case in which the rate of transcription is determined by that of open complex formation between RNA polymerase and promoter, to cases involving regulatory and auxiliary DNA binding proteins, and finally to cases in which the rate of transcription is determined by a step that occurs after open complex formation. The twin-supercoiled-domain model of transcription is then outlined to emphasize the interplay between transcription and template topology. Several predicts of the model, including the division of labor between gyrase and DNA topoisomerase I during transcription, the existence of localized supercoiled domains, and the interdependence of adjacent transcription units, are described; the available evidence in support of these predictions is summarized. The possibility of heteroduplex formation between nascent RNA and the template strand of DNA is also discussed. The coverage of the literature on these topics is representative rather than comprehensive. Our emphasis is to outline a conceptual framework, rather than to provide a detailed review of available data. As the present volume specifically

Regulation of Gene Expression in Escherichia coli, edited by E. C. C. Lin and A. Simon Lynch. © 1996 R.G. Landes Company.

deals with the bacterium *Escherichia coli*, we have also omitted discussions of studies of other organisms, with the exception of a few studies of organisms that are most closely related to *E. coli*.

INTRODUCTION

Studies implicating an influence of DNA supercoiling on transcription date back some three decades. It was observed in 1965 that the two replicative forms RF I and RF II of phage φX174 differ in their efficiency as templates for transcription by *Escherichia coli* RNA polymerase in vitro.[1] Following the discovery of supercoiled DNA in the same year,[2] it became known that RF I, the better template, is negatively supercoiled. Furthermore, the higher template efficiency of RF I appeared to be a result of its negatively supercoiled conformation, rather than the absence of the nick present in RF II.[3]

Around the same time, the accumulation of biochemical evidence provided strong support of the notion that initiation of transcription from a promoter by RNA polymerase is preceded by the formation of an open complex, in which base-pairing within a small region of the promoter is disrupted.[4,5] Because the supercoiling of a DNA is closely tied to its helical structure, the discovery of supercoiled DNA also provided a sensitive tool for probing the unwinding of the DNA double helix. It became known in the early seventies that in the open complex as well as the elongation complex of *E. coli* RNA polymerase, the DNA helix is unwound significantly.[6] These earlier findings were confirmed and refined,[7-10] and spectroscopic and chemical reactivity measurements indicated that DNA unwinding in an open complex is mostly the result of disrupting base stacking and pairing.[11-15]

It was a prediction of the thermodynamics of DNA supercoiling[16] that the unpairing of even a small stretch of DNA would be greatly favored if the DNA is in a negatively supercoiled form.[17-19] In agreement with the finding based on transcription of the two replicative forms of φX174, studies using templates comprised of DNA rings supercoiled to different extents soon demonstrated that transcription by purified *E. coli* RNA polymerase is strongly affected by template supercoiling.[20-24]

While the in vitro experiments described above were lending support to the idea that DNA supercoiling affects transcription, tests of such effects inside bacterial cells were also being carried out at the same time. Intercalating agents were used in a few early attempts to perturb supercoiling of intracellular DNA.[25,26] However, it was the discovery of *E. coli* DNA gyrase, the product of the *gyrA* and *gyrB* genes that negatively supercoils DNA in vitro,[27] and the identification of antibiotics that specifically target the enzyme,[28] that provided a much needed tool in probing the dependence of transcription on the supercoiling of intracellular DNA. Similarly, identification of the *topA* gene encoding bacterial DNA topoisomerase I, an enzyme that relaxes specifically negatively supercoiled DNA,[29] and the identification of mutants deficient in the enzyme,[30,31] provided another handle in the study of the physiological effects of DNA supercoiling.

Historically, mutations in *topA*, *gyrA* and *gyrB* were known to affect gene expression before the identification of the products of these genes. In the early sixties, genetic studies of *Salmonella typhimurium* led to the identification of an extragenic suppressor of a mutation *leu500* in the leucine biosynthesis operon; *leu500* was shown to confer leucine auxotrophy by reducing the level of transcription of the *leuABCD* operon, and the suppressor was found to elevate the level of transcription of the operon in a *leu500* background.[32] Nearly two decades later, the suppressor gene was found to be identical to *topA*, the structural gene of DNA topoisomerase I,[30,31,33] and *leu500* was shown to be a transition mutation in the -10 region of the *leu* promoter.[34] In a similar development, mutations in loci termed *hisU* and *hisW*, which were known to affect expression of the histidine biosynthesis operon as well as a number of other transcription units in *S. typhimurium*,[35] turned out to map respectively in *gyrB* and *gyrA*, the structural genes of the gyrase subunits.[36,37] In combination, these findings provided the strongest evidence that supercoiling of intracellular DNA may affect transcription.

THE DEPENDENCE OF TRANSCRIPTION ON THE CELLULAR LEVEL OF DNA GYRASE AND DNA TOPOISOMERASE I

Several systematic studies were carried out to assess the dependence of gene expression on intracellular levels of DNA topoisomerases.[26,38-43] Upon the addition of coumermycin, an inhibitor of the DNA-dependent ATPase activity of gyrase, the levels of transcription from 40-70% of promoters on random *E. coli* DNA fragments cloned into a multicopy plasmid were found to exhibit upward or downward shifts.[40] From studies with Mu*dlac* fusions randomly inserted into the *S. typhimurium* chromosome, transcription from a similar portion of promoters was found to increase or decrease upon exposure of the cells to the same gyrase inhibitor.[41]

More recently, it was shown that for a large number of proteins the relative abundance is altered upon reducing the cellular level of *E. coli* DNA topoisomerase I through the introduction of a nonlethal mutation *topA10*, or upon lowering gyrase activity through the introduction of a nonlethal mutation *gyrB226*.[44] For most of the proteins, changes in their relative abundance due to the presence of the *topA10* or *gyrB226* mutation were observed to be less than 2-fold. The increase or decrease in the relative abundance does not appear to be, however, a simple function of the extent of supercoiling of intracellular DNA. The relative abundance of a significant fraction of all proteins examined is increased by either the *topA10* mutation, which is supposed to increase the degree of negative supercoiling, or the *gyrB226* mutation, which is supposed to decrease the degree of negative supercoiling; similarly, several other proteins showed a reduction of their relative abundance by either mutation. Because of the pleiotropic effects of mutations in the topoisomerase genes, including effects on tRNA synthesis,[45,46] interpretation of the protein synthesis data is not straightforward.

A comparison of mRNA levels in about 400 segments of the entire *E. coli* genome was reported recently.[47] In general, only minor differences are noticeable between mRNA levels in a *topA10* strain and those in an isogenic *topA+* control; the largest change being a 2-fold drop in mRNA synthesis in the *topA10* strain from a region located at 43.8 minutes of the *E. coli* genome. Because groups of mRNAs from large chromosomal segments, with an average size of about 11 kb, rather than mRNAs from individual genes were examined in these experiments, upward and downward shifts in different transcription units within the same chromosomal region would tend to reduce any *topA*-dependent difference in mRNA synthesis from the region as a whole.

Plausible effects of template supercoiling on the expression of individual genes in vivo have been tabulated in recent reviews,[43,48] and will not be discussed here. The cases of particular interest are the ways the DNA topoisomerase genes *topA*, *gyrA* and *gyrB* themselves respond to changes in the cellular level of DNA topoisomerase I or gyrase.

The administration of coumermycin to *E. coli* cells, or a shift of the growth temperature to inactivate a thermal sensitive *gyrA* or *gyrB* mutant of *E. coli*, was found to increase the rate of synthesis of GyrA and GyrB protein.[38] Inhibition of gyrase was also found to reduce *topA* expression.[49-50] Thus there appears to be a homeostatic regulation of supercoiling of intracellular DNA: gyrase expression seems to increase with a decrease in DNA negative supercoiling, and the opposite seems to occur for DNA topoisomerase I expression. The possibility of an interplay between the two topoisomerases, modulated by the level of supercoiling of intracellular DNA, is also supported by the finding that certain mutations in *gyrA* or *gyrB* can compensate for an otherwise lethal mutation in *topA*.[51,52] It is apparent, however, that some of the *topA*-compensatory mutations do not map in any of the known DNA topoisomerase genes.[53,54] A cautionary note is also needed in inferring that dependence of the expression of a particular gene on intracellular topoisomerase levels is a reflection of a dependence on template topology. Because a reduction or increase in the cellular level of DNA gyrase or DNA topoisomerase I can affect the expression of a large number of genes as well as other cellular processes, such a change may affect the expression of a particular gene by ways that are not directly related to the topological state of that gene.

MECHANISTIC CONSIDERATIONS

General aspects. The kinetics of transcription is rather complex even in the absence of regulatory factors. As initially proposed by the groups of Zillig and Chamberlin,[4,5] and expended by others (reviewed in refs. 55-58), the initiation of a transcript involves a succession of steps: the binding of an RNA polymerase holoenzyme R to a promoter P to form a closed complex RP_c, the isomerization of the closed complex to an open complex RP_o in which a short stretch of DNA is unpaired, and the clearance of the polymerase from the open complex concomitant with the initiation of RNA synthesis. The rate by which a polymerase holoenzyme successfully synthesizes RNA chains from a particular promoter can be expressed by

$$\text{Rate} = \theta \, k_2 \phi \qquad\qquad (1)$$

where θ is the occupancy of the promoter, k_2 is the rate of isomerization of RP_c to RP_o, and ϕ is a variable dependent on the rate of promoter clearance and the fraction of initiation events that leads to the synthesis of full length transcripts.[56,58] The occupancy θ of a promoter is related to the equilibrium constant K_b for closed complex formation by

$$\theta = K_B[R]/\{1+K_B[R]\} \qquad\qquad (2)$$

where [R] is the free polymerase concentration. The promoter is fully occupied ($\theta = 1$) if $K_B[R] \gg 1$, and is mostly unoccupied (θ close to 0) if $K_B[R] \ll 1$; half occupancy ($\theta = 0.5$) occurs when the free holoenzyme concentration is $1/K_B$, which usually falls within the range of 1 nM to 1 μM depending on the particular promoter. Because DNA supercoiling may affect each of the parameters θ, k_2 and ϕ, some of which are themselves composite parameters lumped together for convenience or for lack of sufficient quantitative information, predicting the effect of template supercoiling on the rate of transcription from a particular promoter is difficult. Several examples are discussed below.

EFFECTS OF DNA SUPERCOILING ON OPEN COMPLEX FORMATION BETWEEN RNA POLYMERASE AND DNA

For a significant fraction of promoters, template supercoiling primarily affects transcription through its effect on the rate of open complex formation. Whereas thermodynamically DNA negative supercoiling is expected to favor open complex formation, its effect on the rate of open complex formation is generally unpredictable. It is useful to distinguish two limiting cases: when promoter occupancy is high, template supercoiling affects the rate of open complex formation through its effect on k_2, the rate of isomerization; when promoter occupancy is low, template supercoiling affects the rate of open complex formation through its effect on the product of K_B and k_2 (see Eqs. 1 and 2). Experimental findings for several specific examples are summarized below.

For the RNA I promoter of pBR322, k_2 is increased by two orders of magnitude when σ, the specific linking difference or superhelical density, is changed from 0 to about -0.06.[59] In contrast, k_2 for the *lacUV5* promoter is significantly decreased over the same range of template superhelicity.[10,60] Whereas negative supercoiling does not affect K_B of the pBR322 RNA I promoter significantly, it increases greatly that of the *lacUV5* promoter and decreases those of several other promoters (reviewed in ref. 56).

The strong dependence of K_B on negative superhelicity exhibited by some of the promoters seems surprising. In principle, because K_B is an equilibrium rather than a rate constant, the effects of supercoiling on K_B should be readily predictable: if the formation of RP_c is associated with a change in the twist (ΔTw) and a change in the writhe (ΔWr) of the promoter segment, then

$$\Delta \ln K_B = 210\sigma\Delta Lk° \qquad (3)$$

where $\Delta \ln K_B$ is the difference between the value of $\ln K_B$ when the promoter is in a supercoiled DNA with a specific linking difference σ, and the value of $\ln K_B$ when the same promoter is in a relaxed DNA, and $\Delta Lk°$ is the sum of ΔTw and ΔWr.[17-19,61] Thus $\ln K_B$ is expected to be linearly dependent on σ, with a slope equal to $210\Delta Lk°$. For the *lacUV5* promoter, Amouyal and Buc[10] estimated through the use of Equation 3 that the experimentally observed dependence of K_B on σ corresponded to a $\Delta Lk°$ of -1.25. In other words, the formation of the closed complex appears to be associated with a substantial distortion of the promoter DNA: the changes in the twist and writhe of the promoter segment amount to a sum of -1.25 turns. The magnitude of this $\Delta Lk°$ is rather large when compared with the corresponding $\Delta Lk°$ for open complex formation, which has been measured to be -1.7 turns for *lacUV5* promoter from DNA linking number distributions in the presence and absence of RNA polymerase.[9,10]

Furthermore, because K_B depends on σ in different ways for different promoters, it would appear that topological unwinding of the DNA occurs to different degrees in different promoter-polymerase closed complexes. The interpretation of the dependence of K_B on σ is further complicated by observations that it is not monotonic in the case of the wild-type *ada* promoter, or a transversion mutant of it in which a stretch of seven As, centered around position -60 of the promoter, is changed to A_3TA_3.[62] In vitro measurements with the wild-type and mutant *ada* promoters showed that although the mutation has little effect on k_2, it affects K_B significantly. Under conditions in which the rate of open complex formation is dependent on $K_B \cdot k_2$, the supercoiling responsiveness of the wild-type promoter and that of the mutant promoter are very different: for the wild-type promoter the product $K_B \cdot k_2$ is insensitive to σ, but for the mutant promoter it exhibits a distinct maximum around a value of σ of -0.05.[62]

Several factors are likely to be relevant in understanding the complexity of the dependence of K_B on template superhelicity. First, in formulating the rate expression given in Equation 1, the closed complex represents a kinetic intermediate in the path to open complex formation. It is an oversimplification assuming that only one intermediate is involved, and multiple intermediates are likely to be the rule rather than the exception.[63-67] For different promoters in templates supercoiled to different degrees, the closed complex specified by the single intermediate kinetic formulation may not represent the same complex among multiple kinetic intermediates,[68,69] and protein-DNA contacts in the complex may change as the superhelicity of the DNA is altered. Second, the thermodynamic formulation leading to Equation 3 takes no explicit consideration of DNA bending by a polymerase, and the free energy of supercoiling is expressed in terms of twist and writhe of the DNA. Therefore if a macromolecule causes a smooth planar bend of the DNA so that neither the twist nor the writhe of the DNA is affected, Equation 3 would predict that supercoiling has no effect on the binding of this molecule. It is plausible, however, that the presence of hairpin bends at the tips of supercoiled DNA loops may favor the binding of molecules, such as RNA polymerases, that tend to bend the DNA double helix.[70] Third, it is also plausible that in some of the experimental measurements, an apparent decrease in promoter activity might be due to competition for RNA polymerase by other sites that are generated in the template as its negative superhelicity is increased, as noted in the transcription of phage λ early genes in vitro.[20] Such competing sites may or may not serve as promoters for RNA synthesis, but their binding of the polymerase would cause an apparent reduction of synthesis at the promoter under investigation.

In addition to providing mechanistic insight, measurements of the dependence of K_B and k_2 on DNA supercoiling in vitro may also provide clues on the transcription process in vivo. In the case of the *lacUV5* promoter, in vitro results show that $K_B \cdot k_2$ is greatly increased, while k_2 itself is decreased, when the DNA becomes more negatively supercoiled.[10,60] In vivo, inhibition of DNA gyrase appears to increase expression from *lacUV5*.[26] Together, these results suggest that *lacUV5* is probably fully occupied in vivo, and the rate limiting step is the isomerization step converting the closed complex to the open complex, or a step after open complex formation.[56] This conjecture is reasonable in view of the relatively high K_B value for this promoter.

MUTATIONS THAT AFFECT THE RESPONSIVENESS OF A PROMOTER TO TEMPLATE SUPERCOILING

Several recognition elements within a promoter are of particular importance in its interactions with RNA polymerase. The two hexamer sequences in the -10 and -35 region are well-known, and more recently a region around -50 has been implicated to interact with the carboxyl terminus of the α subunit of RNA polymerase (see for example, ref. 71). In addition, in several cases the transcribed region from +1 to +20 has been shown to influence promoter clearance but not open complex formation.[72] For promoters such as RNA I and *lacUV5*, the upstream element around -50 has little effect on their function, but for promoters such as *ada* and *hisR* the upstream region has a strong effect (see below).

Because even in simple cases the rate of open complex formation may depend on both K_B and k_2, it is not surprising that the supercoiling responsiveness of a particular promoter is determined by the promoter as a whole rather than by any one of the three elements, or the spacing between one element and another. Examination of several mutants of the *lac* p^s promoter, one with a T to A change at position -8 to give the *lacUV5* promoter, one with a C to A change at position -32, and a third with the deletion of a base pair at position -18, showed that the supercoiling response is not just mediated by the region of base unpairing: each mutation alters the extent of stimulation of transcription by negative supercoiling of the template, as well as the degree of negative superhelicity corresponding to maximal stimulation.[73] Direct chemical probing of open complex formation on linear DNA templates also showed that although the *lacUV5* and tac promoters have identical sequences in the unpaired region, open complex formation for the tac promoter occurs at a temperature 3°C lower than for the *lacUV5* promoter.[14] In the case of the supercoiling sensitive *S. typhimurium hisR* gene, a single C to T change at position -7 relieves the response of the promoter to supercoiling, and suppresses the His-constitutive phenotype of the *hisU* (*gyrB*) mutation mentioned earlier.[45] Interestingly, the same mutation was identified independently as the suppressor of a *hisR* promoter down-mutation due to the deletion of three base pairs centered around position -70, a deletion that has been shown to alter DNA bending at that locus.[74]

The supercoiling responsiveness of the gyrase gene promoters themselves has been analyzed in some detail (reviewed in ref. 43), and several findings are particularly significant. First, in an in vitro transcription-translation system, relaxed or linear DNA was shown to be a better template,[38,75,76] as anticipated from in vivo results that inhibition of gyrase stimulates *gyrA* and *gyrB* expression. Second, stimulation of *gyrA* expression in extracts by the addition of novobiocin, an analog of coumermycin and an inhibitor of the DNA-dependent ATPase activity of gyrase, was found to depend on the concentration of DNA in the extract, suggesting the involvement of a titratable DNA-binding factor in this supercoiling response.[76] The binding of such a factor is presumably not specific to the *gyrA* promoter, however, as DNA without *gyrA* sequences appears to be equally effective in titrating the factor. The effects of supercoiling on promoters that are dependent on regulatory factors will be discussed in the section below. Third, only relatively short sequences of the gyrase promoters are necessary for the observed stimulation of transcription by inactivation of gyrase, a minimal sequence of 21 bp containing the -10 consensus hexamer, the transcription start site, and the first few transcribed nucleotides appears to be sufficient.[39,40] It is presently unknown, however, whether the rate of transcription from the gyrase promoters is determined by open complex formation or a later step; this point will be discussed further in the sections below.

EFFECTS OF REGULATORY PROTEINS

In cases where the initiation of transcription at a promoter involves regulatory proteins and/or auxiliary DNA binding proteins such as Fis, IHF and H-NS, template supercoiling may also affect gene expression through its effects on the binding of a regulatory or auxiliary protein to DNA, and through its effects on interactions between such proteins and RNA polymerase. One of the best characterized *E. coli* regulatory proteins is the catabolite activator protein (CAP), also known as the cAMP receptor protein (CRP), which can act either as a transcriptional activator or repressor depending on the location of its binding site in a particular promoter. Extensive studies of CAP in the *lac-* and *gal*-type of promoters, in which the CAP binding site is centered at -61 and -41 respectively, showed that activation of such a promoter can occur by an increase in K_B, k_2, or both (reviewed in ref. 69). An important structural change underpinning CAP-mediated activation of transcription appears to be the bending of DNA. A number of studies have shown that the insertion of bent sequences upstream of the polymerase binding sites in promoters that are normally activated by CAP can enhance k_2 in the absence of CAP.[68,77,78] Many other regulatory and auxiliary DNA-binding proteins are also known to effect DNA bending or to bind preferential to curved DNA (for a comprehensive recent review on plausible roles of DNA bending in prokaryotic gene expression, see ref. 79). In the case of the phage λ P_L promoter, in vitro transcription on linear and supercoiled plasmids showed that negative supercoiling stimulates k_2 by 16-fold but has only a minor effect on K_B; the binding of IHF to a site centered around -86 increases K_B by about 3-fold with either linear or supercoiled template.[80]

For the enhancer-dependent *glnHp2* promoter of the *E. coli glnHPQ* operon, isomerization from closed to open complex is promoted by IHF bound to sites located between the enhancer and the bound RNA polymerase containing σ^{54}; stimulation of transcription by IHF requires that the protein be on the same side of the DNA double helix as the enhancer-bound activator and RNA polymerase.[81,82] When contacts between two DNA bound proteins are responsible for the formation of a DNA loop in between, the spacing between the proteins along the DNA and their relative angular orientations are directly coupled because of the helical geometry of DNA.[83] Supercoiling affects the twist and writhe of DNA segments in space as well as the binding of the proteins involved to DNA, and thus DNA loop formation is dependent on supercoiling (reviewed in refs. 84-86; see also refs. 87-92).

EFFECTS OF DNA SUPERCOILING ON PROMOTER CLEARANCE, POLYMERASE PAUSING AND THE TERMINATION OF TRANSCRIPTION

Although open complex formation is a critical step in the transcription process, it is not always the rate-determining step in vivo or under a given set of experimental conditions in vitro. Some of the steps that follow open complex formation, such as initiation of RNA synthesis, release of the sigma factor of RNA polymerase holoenzyme, the binding of auxiliary factors to the σ-free core enzyme in the elongation complex, stalling of an elongating polymerase at pause sites and the premature termination of RNA synthesis, may also limit the rate of synthesis of full length RNA chains. There is a general paucity of data on the effects of template supercoiling on transcription beyond the stage of open complex formation, and this lack of information is the primary reason for lumping all steps following open complex formation into a single parameter (φ) in Equation 1. For the *lacUV5* promoter, it was reported that under typical in vitro transcription conditions, open complex formation is probably not the rate-determining step.[93] Furthermore, for several mutant *lac* promoters there appears to be an inverse correlation between the rate of productive initiation and the rate of open complex formation.[94] These observations led to the suggestion that in terms of binding energies, contacts that are made in the open complex may compete with conformational changes or movement of the polymerase upon initiation of RNA synthesis.[65,66] In the case of CAP-dependent transcription of the *malT* gene, measurements on the rate of synthesis of short oligonucleotides and the rate of synthesis of long transcripts led to the conclusion that the cAMP·CAP complex activates *malT* transcription by facilitating the clearance of the polymerase from the initiation complex.[95] In their analysis of the dependence of GyrA and GyrB synthesis on DNA supercoiling, Menzel and Gellert[39,43] also suggested that promoter clearance might be the rate-limiting step.

In vitro studies of the effects of template supercoiling on RNA polymerase pausing in the *E. coli* ribosomal RNA leader region indicate that pausing at specific sites is more prominent in negatively supercoiled templates either in the absence or presence of the transcription elongation factor NusA.[96] The authors suggested that at the pausing sites supercoiling induced minor structural distortions in the DNA template may affect optimal interactions in the elongation complex; because the rate of rRNA synthesis in cells growing in rich media is limited by the rate of elongation, the effects of supercoiling on pausing may be physiologically significant in the regulation of rRNA synthesis.

Whereas negative supercoiling of the template accentuates pausing at specific sites, the average step time for the lengthening of the RNA chain appears to be shorter with negatively supercoiled than with relaxed template, which results in a faster overall elongation rate with the former template.[96] It is not established, however, whether the average step time may lengthen, in the case of highly supercoiled templates, when the length of a transcript is very long. The tight coiling of a pair of DNA helices in an interwound supercoiled DNA loop might interfere with the circling of a long nascent transcript around the temple DNA, and may thus slow down elongation.[21]

In a highly supercoiled DNA, major structural changes may occur at particular nucleotide sequences, and these changes may in turn affect transcription elongation. In vitro, the formation of a left-handed Z-helical form of a stretch of alternating CG base pairs in a negatively supercoiled DNA template has been shown to block completely the passage of a transcribing *E. coli* RNA polymerase,[97,98] whereas the same enzyme can move through a Z-form stretch comprised of alternating TG pairs.[97] For phage T7 RNA polymerase, only a small fraction of the elongating complexes appear to abort at the B-Z junctions formed by alternating CG base pairs.[99] The effects of supercoiling induced cruciform formation on transcription have also been examined. For a 68-bp perfect palindrome in a negatively supercoiled DNA, passage of a transcribing polymerase through it does not trigger cruciform formation under conditions that are thermodynamically in favor of cruciform formation but the extrusion of this structure is forbidden kinetically (A. Courey and J.C.W., cited in ref. 97). Interestingly, in the case of a 48 bp palindrome in a negative supercoiled DNA, a preformed cruciform does not appear to block transcription elongation; rather, the passage of the transcribing polymerase converts the cruciform back to the palindromic noncruciform state.[100] In the latter study, the noncruciform state is presumably the structure favored thermodynamically in a DNA with a specific linking difference of -0.05, that of the template used, under the conditions of transcription.

Supercoiling induced changes in the DNA template provide a plausible structural basis of direct effects of supercoiling on elongation of transcription; because such structural changes could affect the binding of proteins and or other molecules, indirect effects are equally possible. Few in vivo examples are available, however, and the physiological significance of supercoiling induced structural changes on transcription remains uncertain. In the case of the histidine biosynthesis operon, it is known that reducing the cellular level of gyrase leads to de-attenuation of the operon. This effect, however, does not appear to involve a structural transition in the template; rather, it appears that expression of a cluster of tRNA^{His} genes (the *hisR* locus) is reduced upon decreasing the cellular level of gyrase, and that this reduction in tRNA^{His} in turn affects *his* attenuation.[45]

A case that is likely to be physiologically significant is the effect of template supercoiling on the termination of transcripts from the *gyrA* promoter at the phage λ T$_{oop}$ terminator.[101] In vivo, transcripts originating from the *gyrA* promoter are less likely to terminate at this terminator (20% read-through) than those from the *galOP* or *topA* promoter (a few percent read-through). Relaxation of the template by the addition of coumermycin to inactivate gyrase decreases further *gyrA*-directed transcripts at λ T$_{oop}$ (60% read-through), but has little effect on termination of *galOP*- and *topA*-directed transcripts at T$_{oop}$. As described earlier, a minimal sequence of 21 bp, including the -10 consensus hexamer, the transcription start site and the first few transcribed nucleotides, is sufficient for the stimulation of *gyrA* transcription upon inactivation of DNA gyrase.[39,40] It is thus significant that sequences at the beginning of the transcribed regions, from +1 to about +25, have been found to affect termination at a number of rho-independent terminators.[102,103] As mentioned before, the sequence of the same transcribed region also affects promoter clearance.[72] How the early transcribed sequences affect termination and how supercoiling influences such a process are thought-provoking and attest to the intricate interplay among the RNA polymerase, the DNA template, the nascent RNA, and the ribonucleoside triphosphates. The structural features of the transcription apparatus at pausing and termination sites and how template supercoiling affects these features are of much interest. Based on linking number measurements of plasmids in various *rho* mutants, it has been suggested that DNA unwinding in the transcription apparatus may be affected by the Rho protein, especially at the Rho-mediated termination sites.[104]

SUPERCOILING OF THE DNA TEMPLATE BY TRANSCRIPTION

The twin-domain model of transcriptional supercoiling of DNA,[105] and the cumulation of evidence in support of the model, have been discussed in a number of recent reviews.[106-109] Briefly, the model postulates that under certain conditions the rotation of a transcribing polymerase around its template may be hindered, and the DNA may therefore be forced to rotate as the enzyme tracks along it. Several different ways of hindering the rotation of a polymerase around the DNA double helix have been discussed. The polymerase may be directly anchored on the template, for example, through contacts with a template bound regulatory protein.[105,110,111] That direct anchoring of the polymerase can lead to DNA supercoiling, even when the DNA is in a linear form, was demonstrated by the construction and characterization of a chimeric protein in which the DNA binding domain of yeast GAL4 protein is fused to phage T7 RNA polymerase.[112] The polymerase may also anchor to the template through the formation of triplex or heteroduplex structures between the nascent RNA and the DNA template,[105,113] or by anchoring to the inner cell membrane if an integral membrane protein or a protein for export is synthesized cotranscriptionally.[105,114-117]

In relation to the membrane-anchoring possibility, it is interesting that a suppressor of gyrase deficiencies has been identified as a deletion of three of the four tandemly repeated copies of *argV*, which encodes the major arginine isoacceptor tRNA.[46] The authors suggested that the slowdown of protein synthesis resulting from this deletion might reduce the cellular demand for gyrase, a large fraction of which would otherwise be engaged in the relaxation of positive supercoils resulting from cotranscriptional anchoring of mRNAs encoding membrane proteins.

Any mechanism that forces the DNA to rotate relative to the transcribing polymerase would generate positive supercoils ahead of the enzyme and negative supercoils in its wake. The superhelicity of these oppositely supercoiled domains is in turn modulated by two processes: the enzymatic actions of the DNA topoisomerases, and diffusional pathways through which opposing supercoils can neutralize each other. Several specific aspects of the twin-supercoiled-domain model of transcription are discussed below.

ASYMMETRIC DISTRIBUTION OF DNA GYRASE AND DNA TOPOISOMERASE I ALONG A TRANSCRIPTIONAL UNIT IN BACTERIA

Because the binding of gyrase to DNA induces a positive writhe in the DNA segment,[118,119] this enzyme is expected to bind preferentially to positively supercoiled DNA, and least favorably to negatively supercoiled DNA. Experimentally, it has been observed that relaxed DNA binds gyrase more strongly than negatively supercoiled DNA.[120] The twin-supercoiled-domain model of transcription would therefore predict that gyrase would be located preferentially to the downstream side of a transcript, where it effectively removes the positive supercoils; this prediction is supported by available data.[121,122] Bacterial DNA topoisomerase I, on the other hand, is known to act preferentially on negatively supercoiled DNA;[29] the enzyme is therefore expected to be located preferentially to the upstream side of a transcript, where negative supercoils are generated.[105] From the perspective of the twin-supercoiled-domain model of transcription, the two enzymes with apparently opposing activities in vitro are in fact acting in concert in vivo to prevent excessive supercoiling of DNA during transcription, or by other processes involving the tracking of a macromolecular assembly along DNA.

LOCALIZED SUPERCOILED REGIONS

According to the twin-supercoiled-domain model, the degree of DNA supercoiling in a particular region is not only dependent on the cellular levels of the DNA topoisomerases and their accessibility to this region, but also on the levels of transcription, the relative orientations of the transcriptional units, and the presence of other barriers to rotating the DNA template.[105] Locally, even the sign of DNA supercoiling may alter in a spatially and temporally dependent way, and thus the average linking number of a plasmid is a rather poor indicator of the degree of supercoiling at a particular location. To probe the local degree of supercoiling, it is necessary to utilize supercoiling-dependent DNA structural transitions such as the flipping of an alternating CG-sequence from the right-handed B-helical form to the left-handed Z-helical form,[123-126] the extrusion of a pair of hairpins from an alternating AT-segment,[127-129] or the formation of triple-stranded structures.[130] Using CG tracts of various lengths as supercoiling-sensitive probes, Rahmouni and Wells showed that under steady-state transcription the region upstream of the promoter experiences an increase in negative supercoiling, whereas the opposite is found downstream of the promoter.[126]

It should be pointed out that the sizes of intracellular DNA domains that are supercoiled to different extents, as predicted by the twin-domain model, are of the order of the sizes of transcription units. These domains should not be confused with the much larger "topological" domains in nucleoids.[131-134] No significant difference in the average extents of supercoiling of the larger chromosomal domains was detected in two recent studies.[135,136]

Interdependence of Adjacent Transcription Units

The prediction of the twin-domain model that the local degree of supercoiling is dependent on transcription raised the interesting possibility that the expression of one gene may affect the expression of an adjacent gene through transcriptional supercoiling;[106,137] effects of neighboring genes encoding proteins for cotranscriptional membrane localization may be particularly significant in gyrase or DNA topoisomerase I mutants. This interdependence can be generalized to couplings between any DNA supercoiling-sensitive process and a tracking process which may generate supercoiled domains; plausible effects between replication and transcription, and between transcription and site-specific recombination, are some of the examples.[138-140]

Interdependence between adjacent transcriptional units has been examined in great detail in the case of the activation of the *leu500* promoter by mutations in *topA*. Whereas suppression of the chromosomally located defective *leu500* promoter in a *topA* strain is readily demonstrated,[32] this suppression was not observed when the gene was transplanted from its chromosomal context to a plasmid.[141] Activation of a plasmid-borne *leu500* promoter occurs, however, when a *tetA* transcription unit encoding a membrane protein is placed within a short distance of it, such that the two transcription units are transcribed divergently.[142-145] Furthermore, in agreement with the observation that the strong effect of *tetA* expression on plasmid supercoiling requires translation of the tet message and insertion of the nascent polypeptide into membrane,[114,116,117] synthesis and membrane insertion of TetA protein were also found to be important for the activation of the *leu500* promoter.[143-145]

Transcriptional Supercoiling and R-Loop Formation

Heteroduplex or R-loop formation between nascent RNA and the transcribed strand of the DNA template was first suggested in 1965, and the 3' portion of the RNA was implicated in this process.[1] Negative supercoiling of the DNA template is necessary for the formation and stabilization of the RNA-DNA heteroduplex,[3,21,146] but mechanistically little is known about this process. In vitro experiments with phage PM2 DNA, which is highly negatively supercoiled with a specific linking difference of about -0.11,[147] showed that R-loop formation is undetectable unless the RNA polymerase is denatured.[23] R-loop formation without denaturing the transcribing polymerase was clearly demonstrated, however, in a series of elegant experiments on the initiation of DNA replication in colicin E1.[148-152] Here the synthesis of the primer RNA is initiated 555 nucleotides upstream of the start site of DNA synthesis, and RNA-DNA hybrid formation begins 10-20 nucleotides upstream of the start site. A cut is then introduced by

RNase H within a cluster of three As around the start site, and the 3' hydroxyl group generated by this cleavage allows the initiation of leading-strand DNA synthesis by DNA polymerase I. The structure of the primer RNA is essential for the formation of the RNA-DNA hybrid. In the presence of DNA gyrase and absence of RNase H in vitro, the length of the RNA-DNA heteroduplex can be extended to thousands of base pairs by the transcribing polymerase.[153] The formation of such a long heteroduplex is likely to be responsible for the initiation of the replication of ColE1-type plasmids in mutant *E. coli* cells lacking RNase H (reviewed in ref. 154). Similarly, in RNase H-deficient *E. coli* strains which additionally lack either DnaA protein, or the *oriC* origin of replication normally used in chromosomal replication, initiation of DNA replication is most likely dependent on R-loop formation at secondary origins termed *oriK* (for a recent review, see ref. 155). The formation of R-loops in vitro was also implicated by Drolet et al when transcription by *E. coli* RNA polymerase was carried out in vitro in the presence of DNA gyrase.[156]

The results cited above, especially those on the initiation of ColE1 DNA replication, indicate that in general R-loop formation is dependent not only on the topology of the DNA template, but also on the structure of the transcript and interactions among the transcript, the template and the transcribing polymerase. The topology of R-loop formation between a nascent transcript and the DNA template strand is of much interest. R-loop formation between a negatively supercoiled DNA and an RNA attached to it is thermodynamically favorable.[21] For a 10 kb DNA with a specific linking difference of -0.11, a stable heteroduplex close to 1 kb in length would be expected. Therefore the absence of long stretches of RNA-DNA hybrid in PM2 DNA undergoing active transcription by *E. coli* RNA polymerase provides strong evidence that R-loop formation is normally kinetically forbidden.[23] The transcribing polymerase is presumably involved in preventing RNA-DNA hybrid formation, as R-loop formation occurs readily upon denaturation of the polymerase.[23]

The formation of a plectonemic RNA-DNA duplex between an RNA and a double-stranded DNA ring can be viewed as a two-step process: the initial formation of a few RNA-DNA base pairs near one end of the RNA is followed by the rotation of that end around the DNA strand complementary to the RNA. Because hybridization between DNA or RNA fragments and negatively supercoiled DNA from natural sources is undetectable under physiological conditions, the high efficiency of R-loop formation upon denaturation of a transcribing polymerase suggests that the few RNA-DNA base pairs in a transcription complex are sufficient to initiate RNA-DNA heteroduplex formation, and that denaturation of the polymerase serves the purpose of freeing the 3' end of the transcript to permit the formation of a plectonemic joint between the RNA and one strand of duplex DNA.[21,23,146] Such an interpretation leaves a dilemma, however, in regard to the high efficiency of R-loop formation at particular sites without denaturation of the transcribing polymerase, such as at the origin of replication of ColE1.

CONCLUDING REMARKS

The supercoiling of a DNA affects many of its biological transactions. During various steps of transcription, complex and intricate interactions take place among the participants of the process, including the polymerase and various factors interacting with it, regulatory proteins and auxiliary DNA-binding proteins, the nascent RNA and the

DNA; in this process, the DNA should be viewed as an active participant rather than a passive template along which all actions occur. From our understanding of how DNA supercoiling affects its structure and its interactions with other molecules, it is not surprising that supercoiling affects many steps of transcription. It is also clear that the helical geometry of DNA requires a rotational movement of the transcribing polymerase relative to its template, and that this requirement may under certain conditions drive the transient formation of oppositely supercoiled domains in the template. Whereas a rough sketch is now possible on how DNA supercoiling and transcription are interrelated, there is a general paucity of data at a more detailed level. The lack of knowledge about the nucleoprotein organization and higher order structures of the chromosome inside *E. coli* often casts a shadow on conclusions drawn from in vitro studies. Being pleiotropic in nature, the effects of supercoiling in vivo should be interpreted with caution, as causality is often difficult to establish. Conclusions based on the average degree of supercoiling of a plasmid or intracellular DNA may also be misleading, and more direct measurements of the local degrees of supercoiling, using properly placed supercoiling-sensitive structural probes, should be helpful. In spite of the lack of definitive answers to many questions on supercoiling and transcription, it seems clear that the DNA topoisomerases have evolved to control and modulate the extents of supercoiling, otherwise the delicate balance among various cellular processes may be severely perturbed, with disastrous consequences for the cell.

ACKNOWLEDGMENTS

Other than the unpublished data referred to in the text, only material published prior to September 1994 was included. We thank all authors who kindly provided reprints and preprints. Work on DNA topology and its biological consequences in this laboratory in the past three decades was mainly supported by grants from the National Institutes of Health and the National Science Foundation.

REFERENCES

1. Hayashi M. A DNA-RNA complex as an intermediate of in vitro genetic transcription. Proc Natl Acad Sci USA 1965 54:1736-1743.
2. Vinograd J, Lebowitz J, R. Radloff et al. The twisted circular form of polyoma viral DNA. Proc Natl Acad Sci USA 1965; 53:1104-1111.
3. Hayashi Y, Hayashi M. Template activities of the φX-174 replicative allomorphic deoxyribonucleic acids. Biochemistry 1971;10:4212-4218.
4. Walter G, Zillig W, Palm P et al. Initiation of DNA-dependent RNA synthesis and the effect of heparin on RNA polymerase. Eur J Biochem 1967; 3:194-201.
5. Chamberlin MJ. The selectivity of transcription. Annu Rev Chem 1974; 43:721-745.
6. Saucier J-M, Wang JC. Angular alteration of the DNA helix by *E. coli* RNA polymerase. Nature New Biology 1972; 239:167-170.
7. Wang JC, Jacobsen JH, Saucier J-M. Physicochemical studies on interactions between DNA and RNA polymerase. Unwinding of the DNA helix by Escherichia coli RNA polymerase. Nucl Acids Res. 1977; 4:1225-1241.
8. Mirkin SM, Bogdanova ES, Gorlenko ZM et al. DNA supercoiling and transcription in *Escherichia coli*: Influence of RNA polymerase mutations. Molec Gen Genet 1979; 177:169-175.
9. Gamper HB, Hearst JE. A topological model of transcription based on unwinding angle analysis of *E. coli* RNA polymerase binary, initiation and ternary complexes. Cell 1982; 29:81-90.

10. Amouyal M, Buc H. Topological unwinding of strong and weak promoters by RNA polymerase. J Mol Biol 1987; 195:795-808.

11. Hsieh T-S, Wang JC. Physicochemical studies on interactions between DNA and RNA polymerase. Ultraviolet absorption measurements. Nucl Acids Res 1978 5:3337-45.

12. Melnikova A, Beakealashvilli R, Mirzabekov AD. A study of unwinding of DNA and shielding of the DNA grooves by RNA polymerase by using methylation with dimethylsulfate. Eur J Biochem 1978; 84:301-9.

13. Siebenlist U. RNA polymerase unwinds an 11-base pair segment of a phage T7 promoter. Nature 1979; 279:651-2.

14. Kirkegaard K, Buc H, Spassky A et al. Mapping of single-stranded regions in duplex DNA at the sequence level: Single-stranded specific cytosine methylation in RNA polymerase-promoter complexes. Proc Natl Acad Sci USA 1983; 80:2544-48.

15. Sasse-Dwight S, Gralla JD. KMnO$_4$ as a probe for *lac* promoter DNA melting and mechanism in vivo. J Biol Chem 1989; 264:8074-81.

16. Bauer W, Vinograd J. Interaction of closed circular DNA with intercalative dyes. II. The free energy of superhelix formation in SV40 DNA. J Mol Biol 1970; 47:419-35.

17. Davidson N. Effect of DNA length on the free energy of binding of an unwinding ligand to a superhelical DNA. J Mol Biol 1972; 66:307-9.

18. Wang JC, Barkley MD, Bourgeois S. Measurements of unwinding of lac operator by repressor. Nature 1974; 251:247.

19. Hsieh T-S, Wang JC. Thermodynamic properties of superhelical DNAs. Biochemistry 1975; 14:527-35.

20. Botchan P, Wang JC, Echols H. Effect of circularity and superhelicity on transcription from bacteriophage λ DNA. Proc Natl Acad Sci USA 1973; 70:3077-81.

21. Wang JC. Interactions between twisted DNAs and enzymes: The effects of superhelical turns. J Mol Biol 1974; 87:797-816.

22. Richardson JP. Effects of supercoiling on transcription from bacteriophage PM2 deoxyribonucleic acid. Biochemistry 1974; 13:3164-9.

23. Richardson JP. Initiation of transcription by *Escherichia coli* RNA polymerase from supercoiled and nonsupercoiled bacteriophage PM2 DNA. J Mol Biol 1975; 91:477-87.

24. Botchan P, An electron microscopic comparison of transcription on linear and superhelical DNA. J Mol Biol 1976; 105:161-76.

25. Sankaran L, Pogell BM. Differential inhibition of catabolite-sensitive enzyme induction by intercalating dyes. Nature New Biol 1973; 245:257-60.

26. Sanzey B. Modulation of gene expression by drugs affecting deoxyribonucleic acid gyrase. J Bacteriol 1979; 138:40-47.

27. Gellert M, Mizuuchi K,O'Dea MH et al. DNA gyrase: An enzyme that introduces superhelical turns into DNA. Proc Natl Acad Sci USA 1976; 73:3872-6.

28. Gellert M, O'Dea MH, Itoh T et al. Novobiocin and coumermycin inhibit DNA supercoiling catalyzed by DNA gyrase. Proc Natl Acad Sci USA 1976b; 73:4474-8.

29. Wang JC. Interaction between DNA and an *Escherichia coli* protein ω. J Mol Biol 1971; 55:523-33.

30. Sternglanz R, DiNardo S, Voelkel KA et al. Mutations in the gene coding for *Escherichia coli* DNA topoisomerase I affect transcription and transposition. Proc Natl Acad Sci USA 1981; 78:2747-51.

31. Trucksis M, Depew RW. Identification and localization of a gene that specifies production of *Escherichia coli* DNA topoisomerase I. Proc Natl Acad Sci USA 1981; 78:2164-8.

32. Mukai FH, Margolin P. Analysis of unlinked suppressors of an 0° mutation in Salmonella. Proc. Natl. Acad. Sci. USA 50:140-148.

33. Margolin P, Zumstein L, Sternglanz R et al. The *Escherichia coli supX* locus is *topA*, the structural gene for DNA topoisomerase I. Proc Natl Acad Sci USA 1985; 82:5437-41.

34. Gemmill RM, Tripp, Friedman SB et al. Promoter mutation causing catabolite repression of the *Salmonella typhimurium* leucine operon. J Bacteriol 1984; 158:948-53.

35. Roth JR, Anton DN, Hartman PE. Histidine regulatory mutants in *S. typhimurium*: I. Isolation and general properties. J Mol Biol 1966; 22:305-23.

36. Rudd KE, Menzel R. *his* operons of *Escherichia coli* and *Salmonella typhimurium* are regulated by DNA supercoiling. Proc Natl Acad Sci USA 1987; 84:517-21.

37. Toone MW, Rudd KE, Friessen JD. Mutations causing aminotriazole resistance and temperature sensitivity reside in *gyrB*, which encodes the B subunit of DNA gyrase. J Bacteriol 1992l 174:5479-81.

38. Menzel R, Gellert M. Regulation of the genes for *E. coli* DNA gyrase: Homeostatic control of DNA supercoiling. Cell 1983; 34:105-13.

39. Menzel R, Gellert M. Modulation of transcription by DNA supercoiling: A deletion analysis of the *Escherichia coli gyrA* and *gyrB* promoters. Proc Natl Acad Sci USA 1987; 84:4185-9.

40. Menzel R, Gellert M. Fusions of *Escherichia coli gyrA* and *gyrB* control regions to the glactokinase gene are inducible by coumermycin treatment. J Bacteriol 1987b; 169:1272-8.

41. Jovanovich SB, Lebowitz J. Estimation of the effect of coumermycin A$_1$ on *Salmonella typhimurium* promoters by using random operon fusions. J Bacteriol 1987; 169:4431-5.

42. Pruss GJ, Drlica K. DNA supercoiling and transcription. Cell 1989; 56:521-3.

43. Menzel R, Gellert M. The biochemistry and biology of DNA gyrase. In: Liu L, ed. DNA Topoisomerases and Their Applications in Pharmacology. Academic Press, Orlando, FL In press.

44. Steck TR, Franco RJ, Wang J-Y et al. Topoisomerase mutations affect the relative abundance of many *Escherichia coli* proteins. Mol Microbiol 1993; 10:473-81.

45. Figueroa N, Wills N, Bossi L. Common sequence determinants of the response of a prokaryotic promoter to DNA bending and supercoiling. EMBO J 1991; 10:941-9.

46. Blanc-Potard A-B, Bossi L. Phenotypic suppression of DNA gyrase deficiencies by a deletion lowering the gene dosage of a major tRNA in *Salmonella typhimurium*. J Bacteriol 1994; 176:2216-26.

47. Chuang S-E, Daniels DL, Blattner FR. Global regulation of gene expression in *Escherichia coli*. J Bacteriol 1993; 175:2026-36.

48. Drlica K. Control of bacterial DNA supercoiling. Mol Microbiol 1992; 6:425-33.

49. Tse-Dinh Y-C. Regulation of the *Escherichia coli* DNA topoisomerase I gene by DNA supercoiling. Nucleic Acids Res 1985; 13:4751-63.

50. Tse-Dinh Y-C, Beran-Steed RK. Multiple promoters for transcription of *E. coli* topoisomerase I gene and their regulation by DNA supercoiling. J Mol Biol 1982; 202:735-42.

51. Pruss GJ, Manes SH, Drlica K. *Escherichia coli* DNA topoisomerase I mutants: Increased supercoiling is corrected by mutations near gyrase genes. Cell 1982; 31:35-42.

52. DiNardo S, Voelkel KA, Sternglanz R et al. *Escherichia coli* DNA topoisomerase I mutants have compensatory mutations in DNA gyrase genes. Cell 1982; 31:43-51.

53. Raji A, Zabel DJ, Laufer CS et al. Genetic analysis of mutations that compensate for loss of *Escherichia coli* DNA topoisomerase I. J Bacteriol 1985; 162:1173-79.

54. Richardson SMH, Higgins CF, Lilley DMJ. The genetic control of DNA supercoiling in *Salmonella typhimurium*. EMBO J 1984; 3:1745-52.

55. von Hippel PH, Bear DG, Morgan WD et al. Protein-nucleic acid interactions in transcription: A molecular analysis. Annu Rev Biochem 1984; 53:389-446.

56. McClure WR. Mechanism and control of transcription initiation in prokaryotes. Annu Rev Biochem 1985; 54:171-204.

57. Record Jr MT, Ha J-H, Fisher MA. Use of equilibrium and kinetic measurements to determine the thermodynamic origins of stability and specificity and mechanism of formation of site specific complexes between proteins and helical DNA. Methods Enzymol 1991; 208:291-343.

58. von Hippel PH, Yager TD, Gill SC. Quantitative aspects of the transcription cycle in *Escherichia coli*. In: McKnight S, Yamamoto K, eds. Transcriptional Regulation. Cold Spring Harbor NY: Cold Spring Harbor Laboratory Press, 1992:179-201.

59. Wood DC, Lebowitz J. Effect of supercoiling on the abortive initiation kinetics of the RNA I promoter of *colE1* plasmid DNA. J Biol Chem 1984; 259:11184-87.

60. Malan TP, Kolb A, Buc H et al. Mechanism of CRP-cAMP activation of *lac* operon transcription initiation activation of the P1 promoter. J Mol Biol 1984; 180:881-909.

61. Buc H, Amouyal M. Superhelix density as an intensive thermodynamic variable. Eckstein F, Lilley DMJ eds. Nucleic Acids and Molecular Biology. vol. 6. New York: Springer-Verlag, 1992:23-54.

62. Bertrand-Burggraf E, Dunand J, Fuchs RPP et al. Kinetic studies of the modulation of *ada* promoter activity by upstream elements. EMBO J 1990; 9:2265-71.

63. Buc H, McClure WR. Kinetics of open complex formation between *Escherichia coli* RNA polymerase and the *lac* UV5 promoter. Evidence for a sequential mechanism involving three steps. Biochemistry 1985; 24:2712-23.

64. Straney DC, Crothers DM. Intermediates in transcription initiation from the *Escherichia coli lac* UV5 promoter. Cell 1985; 43:449-54.

65. Straney DC, Crothers DM. Comparison of the open complexes formed by RNA polymerase at the *E. coli lac* UV5 promoter. J Mol Biol 1987; 193:279-92.

66. Straney DC, Crothers DM. A stressed intermediate in the formation of stably initiated RNA chains at the *Escherichia coli* lac UV5 promoter. J Mol Biol 1987 193:267-278.

67. Schickor P, Metzger W, Werel W et al. Topography of intermediates in transcription initiation of *E. coli*. EMBO J 1990; 9:2215-20.

68. Lavigne M, Herbert M, Kolb A et al. Upstream curved sequences influence the initiation of transcription at the *Escherichia coli* galactose operon. J Mol Biol 1992; 224:293-306.

69. Crothers DM, Steitz TA. Transcriptional activation by *Escherichia coli* CAP protein. McKnight SL, Yamamoto KR eds. Transcription Regulation. Cold Spring Harbor NY: Cold Spring Harbor Laboratory Press, 1992:501-34.

70. Heggeler-Bordellier B, Wahli W, Adrian M et al. The apical localization of transcribing RNA polymerases on supercoiled DNA prevents its rotation around the template. EMBO J 1992; 11:667-72.

71. Ross W, Gosink KK, Salomon J et al. A third recognition element in bacterial promoters: DNA binding by the α subunit of RNA polymerase. Science 1993; 262:1407-13.

72. Kammerer W, Deuschle U, Gentz R et al. Functional dissection of *Escherichia coli* promoters: Information in the transcribed region is involved in late steps of the overall process. EMBO J 1986; 5:2995-3000.

73. Borowiec JA, Gralla JD. Supercoiling response of the *lac* p^s promoter in vitro. J Mol Biol 1985; 184:587-98.

74. Bossi L, Smith DM. Conformational change in the DNA associated with an unusual promoter mutation in a tRNA operon of *Salmonella*. Cell 1984; 39:643-52.

75. Yang H-L, Heller K, Gellert M et al. Differential sensitivity of gene expression in vitro to inhibitors of DNA gyrase. Proc Natl Acad Sci USA 1979; 76:3304-8.

76. Carty M, Menzel R. Inhibition of DNA gyrase activity in an in vitro transcription-translation system stimulates *gyrA* expression in a DNA concentration dependent manner. J Mol Biol 1990; 214:397-406.

77. Bracco L, Kotlarz D, Kolb A et al. Synthetic curved DNA sequences can act as transcriptional activators in *Escherichia coli*. EMBO J 1989; 8:4289-96.

78. Gartenberg MR, Crothers DM. Synthetic DNA bending sequences increase the rate of in vitro transcription initiation at the *Escherichia coli lac* promoter. J Mol Biol 1991; 219:217-30.

79. Pérez-Martín J, Rojo F, deLorenzo V. Promoters responsive to DNA bending: A common theme in prokaryotic gene expression. Microbiol Rev 1994; 58:268-90.

80. Giladi H, Koby S, Gottesman ME et al. Supercoiling, integration host factor, and a dual promoter system, participate in the control of the bacteriophage λ pL promoter. J Mol Biol 1992; 224:937-48.

81. Claverie-Martin F, Magasanik B. Role of integration host factor in the regulation of the *glnHp2* promoter of *Escherichia coli*. Proc Natl Acad Sci USA 1991; 88:1631-35.

82. Claverie-Martin F, Magasanik B. Positive and negative effects of DNA bending on activation of transcription from a distant site. J Mol Biol 1992; 227:996-1008.

83. Hochschild A, Ptashne M. Cooperative binding of λ repressors to sites separated by integral turns of the DNA helix. Cell 1986; 44:681-7.

84. Wang JC, Giaever G. Action at a distance along a DNA. Science 1988; 240:300-4.

85. Hochschild A. Protein-protein interactions and DNA loop formation. Cozzarelli NR, Wang JC eds. DNA topology and its biological effects. Cold Spring Harbor, Cold Spring Harbor Laboratory Press, 1990:107-38.

86. Matthews KS. DNA looping. Microbiol Rev 1992; 56:123-6.

87. Kramer H, Amouyal M, Nordheim A et al. DNA supercoiling changes the spacing requirement of two *lac* operators for DNA loop formation with *lac* repressor. EMBO J 1988; 7:547-56.

88. Kramer H, Niemoller M, Amouyal M. *lac* repressor forms loops with linear DNA carrying two suitably spaced *lac* operators. EMBO J 1987; 6:1481-91.

89. Lobell R, Schlief RF. DNA looping and unlooping by AraC protein. Science 1990; 250:528-32.

90. Whitehall S, Austin S, Dixon R. DNA supercoiling response of the σ^54-dependent *Klebsiella pneumoniae nifL* promoter in vitro. J Mol Biol 1992; 225:591-607.

91. Whitehall S, Austin S, Dixon R. The function of the upstream region of the σ^54-dependent *Klebsiella pneumoniae nifL* promoter is sensitive to DNA supercoiling. Mol Microbiol 1993; 9:1107-17.

92. Law SM, Bellomy GR, Schlax PJ et al. In vivo thermodynamic analysis of repression with and without looping in *lac* constructs. J Mol Biol 1993; 230:161-73.

93. Stefano JE, Gralla JD. Kinetic investigation of the mechanism of RNA polymerase binding to mutant *lac* promoters. J Biol Chem 1979; 255:10423-30.

94. Carpousis AJ, Stefano JE, Gralla JD. 5' Nucleotide heterogeneity and altered initiation of transcription at mutant *lac* promoters. J Mol Biol 1982; 157:619-33.

95. Menedez M, Kolb A, Buc H. A new target for CRP action at the *malT* promoter. EMBO J. 1987; 6:4227-34.

96. Krohn M, Pardon B, Wagner R. Effects of template topology on RNA polymerase pausing during in vitro transcription of the *Escherichia coli rrnB* leader region. Mol Microbiol 1992; 6:581-9.

97. Peck LJ, Wang JC. Transcriptional block caused by a negative supercoiling induced structural change in an alternating CG sequence. Cell 1985; 40:129-37.

98. Brahms JG, Dargouge O, Brahms S et al. Activation of transcription by DNA supercoiling. J Mol Biol 1985; 181:455-65.

99. Dröge P, Pohl FM. The influence of an alternate template conformation on elongating phage T7 RNA polymerase. Nucleic Acids Res 1991; 19:5301-6.

100. Morales NM, Cobourn SD, Müller UR. Effect of in vitro transcription on cruciform stability. Nucleic Acids Res 1990; 18:2777-82.

101. Carty M, Menzel R. The unexpected antitermination of *gyrA*-directed transcripts is enhanced by DNA relaxation. Proc Natl Acad Sci USA 1989; 86:8882-6.

102. Telesnitsky APW, Chamberlin MJ. Sequences linked to prokaryotic promoters can affect the efficiency of downstream termination sites. J Mol Biol 1989; 205:315-30.

103. Goliger JA, Yang X, Guo H-C et al. Early transcribed sequences affect termination efficiency of *Escherichia coli* RNA polymerase. J Mol Biol 1989; 205:331-41.

104. Arnold GF, Tessman I. Regulation of DNA superhelicity by *rpoB* mutations that suppress defective Rho-mediated transcription termination in *Escherichia coli*. J Bacteriol 1988; 170:4266-71.

105. Liu LF, Wang JC. Supercoiling of the DNA template during transcription. Proc Natl Acad Sci USA 1987; 84:7024-7.

106. Wang JC. Template topology and transcription. McKnight S, Yamamoto K, eds. Transcriptional Regulation. Cold Spring Harbor NY: Cold Spring Harbor Laboratory Press, 1992:1253-69.

107. Wang JC, Lynch AS. Transcription and DNA supercoiling. Current Op in Gen and Dev 1993; 3:764-8.

108. Dröge P. Protein tracking-induced supercoiling of DNA: A tool to regulate DNA transactions in vivo? Bioessays 1993;16:91-9.

109. Cook DN, Ma D, Hearst JE. Nucleic Acids and Molecular Biology. In: Eckstein F, Lilley DMJ, eds. New York: Springer-Verlag, In press.

110. Wang JC. DNA topoisomerases. Jerusalem Symposium. Ann Rev Biochem 1985; 54:665-97.

111. Wang JC. Recent Studies of DNA Topoisomerases. Harvey Lecture. Biochim Biophys Acta 1987; 909, 1-9.

112. Ostrander EO, Benedetti P, Wang JC. Template supercoiling by a chimera of yeast *GAL4* protein-phage T7 RNA polymerase chimera. Science 1990; 249:1261-65.

113. Reaban ME, Griffith JA. Induction of RNA-stabilized DNA conformers by transcription of an immunoglobulin switch region. Nature 1990; 348:342-4.

114. Lodge JK, Kazic T, Berg DE. Formation of supercoiling domains in plasmid pBR322. J Bacteriol 1989; 171:2181-87.

115. Cook DN, Ma D, Pon NG et al. Dynamics of DNA supercoiling by transcription in *Escherichia coli*. Proc Natl Acad Sci USA 1992; 89:10603-7.

116. Lynch AS, Wang JC. Anchoring of DNA to the bacterial cytoplasmic membrane through cotranscriptional synthesis of polypeptides encoding membrane proteins or proteins for export: A mechanism of plasmid hypernegative supercoiling in mutants deficient in DNA topoisomerase I. J Bacteriol 1993; 175:1645-55.

117. Ma D, Cook DN, Pon NG et al. Efficient anchoring of RNA polymerase in *Escherichia coli* during coupled transcription-translation of genes encoding integral inner membrane polypeptides. J Biol Chem 1994; 269:15362-70.

118. Liu LF, Wang JC. *Micrococcus luteus* DNA gyrase: Active components and a model for its supercoiling of DNA. Proc Natl Acad Sci USA 1978; 75:2098-2102.

119. Liu LF, Wang JC. DNA-DNA gyrase complex: the wrapping of the DNA duplex outside the enzyme. Cell 1978;15:979-84.

120. Higgins NP, Cozzarelli NR. The binding of gyrase to DNA: Analysis by retention by nitrocellulose filters. Nucleic Acids Res 1982; 10:6833-47.

121. Koo H-S, Wu H-Y, Liu LF. Effects of transcription and translation on gyrase-mediated DNA cleavage in *Escherichia coli*. J Biol Chem 1990; 265:12300-5.

122. Condemine G, Smith CL. Transcription regulates oxolinic acid-induced DNA gyrase cleavage at specific sites on the *E. coli* chromosome. Nucleic Acids Res 1990; 18:7389-96.

123. Rahmouni AR, Wells RD. Stabilization of Z DNA in vivo by localized supercoiling. Science 1989; 246:358-63.

124. Kochel TJ, Sinden RR. Hyperreactivity of *B-Z* junctions to 4,5',8-trimethylpsoralen photobinding assayed by an exonuclease III/photoreversal mapping procedure. J Mol Biol 1989; 205:91-102.

125. Jaworski, A., N.P. Higgins NP, Wells RD et al. Topoisomerase mutants and physiological conditions control supercoiling and Z-DNA formation in vivo. J Biol Chem 1991; 266:2576-81.

126. Rahmouni AR, Wells RD. Direct evidence for the effect of transcription on local DNA supercoiling in vivo. J Mol Biol 1992; 223:131-44.

127. McClellan JA, Boublíkova P, Palecek E et al. Superhelical torsion in cellular DNA responds to directly to environmental and genetic factors. Proc Natl Acad Sci USA 1990; 87:8373-7.

128. Dayn A, Malkhosyan S, Mirkin SM. Transcriptionally driven cruciform formation in vivo. Nucleic Acids Res 1992; 20:5991-7.

129. Bowater R, Chen D, Lilley DMJ. Elevated unconstrained supercoiling of plasmid DNA generated by transcription and translation of the tetracycline resistance gene in eubacteria. Biochemistry 1994; In press.

130. Kohwi Y, Malkhosyan SR, Kohi-Shigematsu T. Intramolecular dG·dG·dC triplex detected in *Escherichia coli* cells. J Mol Biol 1992; 223:817-22.

131. Worcel A, Burgi E. On the structure of the folded chromosome of *Escherichia coli*. J Mol Biol 1972; 71:127-47.

132. Pettijohn DE, Hecht R. RNA molecules bound to the folded bacterial genome stabilize DNA folds and segregate domains of supercoiling. Cold Spring Harbor Symp Quantitative Biol 1974; 38:31-42.

133. Pettijohn DE, Pfenninger O. Supercoils in prokaryotic DNA restrained in vivo. Proc Natl Acad Sci USA 1980; 77:1331-5.

134. Sinden RR, Pettijohn DE. Chromosomes in living *Escherichia coli* cells are segregated into domains of supercoiling. Proc Natl Acad Sci USA 1981; 78:224-8.

135. Miller WG, Simons RW. Chromosomal supercoiling in *Escherichia coli*. Mol Microbiol 1993; 10:675-84.

136. Pavitt GD, Higgins CF. Chromosomal domains of supercoiling in *Salmonella typhimurium*. Mol Microbiol 1993; 10:685-96.

137. Lilley DMJ, Higgins CF. Local DNA topology and gene expression: The case of the *leu-500* promoter. Mol Microbiol 1993; 5:779-83.

138. Asai T, Chen C-P, Nagata T et al. Transcription in vivo within the replication origin of the *Escherichia coli* chromosome: A mechanism for activating initiation of replication. Mol Gen Genet 1992; 231:169-78.

139. Dröge P, Transcription-driven site-specific DNA recombination in vitro. Proc Natl Acad Sci USA 1993; 90:2759-63.

140. Kohwi Y, Panchenko Y. Transcription-dependent recombination induced by triple-helix formation. Genes & Devel 1993; 7:1766-78.

141. Richardson SM, Higgins CF, Lilley DM. DNA supercoiling and the *leu500* mutation of *Salmonella typhimurium*. EMBO J 1988; 7:1863-9.

142. Chen D, Bowater R, Dorman CJ et al. Activity of a plasmid-borne *leu-500* promoter depends on the transcription and translation of an adjacent gene. Proc Natl Acad Sci USA 1992; 89:8784-8.

143. Chen D, Bowater RP, Lilley DMJ. Activation of the *leu-500* promoter: A topological domain generated by divergent transcription in a plasmid. Biochem 1993; 32:13162-70.

144. Chen D, Bowater R, Lilley DMJ. Topological promoter coupling in *Escherichia coli*: Δ*topA*-dependent activation of the *leu-500* promoter on a plasmid. J Bacteriol 1994; 176:3757-64.

145. Tan J, Shu L, Wu H-Y. Activation of the *leu-500* promoter by adjacent transcription. J Bacteriol 1994; 176:1077-86.

146. Liu LF, Wang JC. In vitro DNA synthesis of primed covalently closed double-stranded templates. I. Studies with *Escherichia coli* DNA polymerase I. In: Goulian M, Hanawalt P, eds. DNA Synthesis and Its Regulation. Menlo Park: Benjamin Inc., 1975:38-63.

147. Wang JC. The Degree of unwinding of the DNA helix by ethidium. I. Titration of twisted PM2 DNA molecules in alkaline cesium chloride density gradients. J Mol Biol 1974; 89:783-801.

148. Itoh T, Tomizawa J. Formation of an RNA primer for initiation of replication of ColE1 DNA by ribonuclease H. Proc Natl Acad Sci USA 1980; 77:2450-4.

149. Selzer G, Tomizawa JI. Specific cleavage of the p15A primer precursor by ribonuclease H at the origin of DNA replication. Proc Natl Acad Sci USA 1982;79:7082-6.

150. Masukata H, Tomizawa J. Effects of point mutations on formation and structure of RNA primer for ColE1 replication. Cell 1984; 36:513-22.

151. Masukata H, Tomizawa J. Control of primer formation of ColE1 plasmid replication: Conformational change of the primer transcript. Cell 1986; 44:125-36.

152. Masukata H, Dasgupta S, Tomizawa J. Transcriptional activation of ColE1 DNA synthesis by displacement of the nontranscribed strand. Cell 1987; 51:1123-30.

153. Parada CA, Marians KJ. Mechanism of DNA A protein-dependent pBR322 DNA replication. J Biol Chem 1991; 66:18895-906.

154. Marians KJ. Prokaryotic DNA replication. Annu Rev Biochem 192; 61:673-719.

155. Asai T, Kogoma T. Minireview: D-loops and R-loops: Alternative mechanisms for the initiation of chromosome replication in *Escherichia coli*. J Bacteriol 1994; 176:1807-12.

156. Drolet M, Bi X, Liu LF. Hypernegative supercoiling of the DNA template during transcription elongation in vitro. J Biol Chem 1994; 269:2068-74.

THE HU AND IHF PROTEINS: ACCESSORY FACTORS FOR COMPLEX PROTEIN-DNA ASSEMBLIES

Howard A. Nash

I. PERSPECTIVE

The apparatus required for the regulation of gene expression varies widely in complexity. At one extreme, regulation can be achieved by one or two proteins acting at a DNA locus that encompasses 30-50 bp. However, such a simple picture is not usually (if ever) the whole story. Much more common are complex systems that typically involve several proteins acting over much larger segments of DNA. At these extended loci, one finds that function depends not only on a principal actor and a regulator (e.g., RNA polymerase and a transcriptional activator) but also on additional components. This chapter focuses on two of the best studied of such accessory factors: HU and IHF. We review the structure of these proteins, their influence on DNA conformation and topology, the regulation of their expression, their function in several well-studied systems, and their roles in homeostasis of *E. coli*. In order to focus on these questions, the present work has not presented all that is known about these proteins. More information can be found in several other surveys.[1-6] For reviews of other accessory factors of *E. coli*—e.g., Fis, H-NS, Crp and Lrp—the reader is directed to chapters 12 and 20 of this volume and to other sources.[7-11]

Our current understanding of the physiological roles for HU and IHF proteins and the mechanism(s) by which they achieve these roles is rather patchy. This unevenness reflects in part the history of interest in these proteins. Both HU and IHF were discovered by their ability to influence macromolecular metabolism involving the DNA of bacteriophage lambda—HU as an activity found in extracts of uninfected *E. coli* that stimulated in vitro transcription of lambda DNA[12] and IHF as an activity found in similar extracts that was essential for in vitro lambda integrative recombination.[13] Bacteriophage Mu also played an important role in the development of our understanding of these

Regulation of Gene Expression in Escherichia coli, edited by E. C. C. Lin and A. Simon Lynch. © 1996 R.G. Landes Company.

proteins. Lytic growth of this phage is severely depressed in IHF mutants[14] and biochemical studies showed that the transposition process that primes Mu replication has a strong dependence on HU.[15] Thus, the importance of HU and IHF for the life cycle of parasites of *E. coli* was appreciated in some detail before roles for these proteins in the host organism itself were investigated. The resulting differential distribution in our knowledge is coming more into balance but many of the examples cited in this review will of necessity come from experiments on phages, plasmids and transposons. Similarly, because of the dramatic and specific effects of HU and IHF on recombination reactions, much of the early work focused on their action in these systems. It is only more recently that attention has been turned to the effect of these proteins on gene expression. Accordingly, our fund of knowledge in these areas is smaller but the principles established from study of HU and IHF in other systems seem to be applicable to their use in gene expression.

II. STRUCTURE

Purification of HU and IHF from *E. coli* revealed that each is a protein of molecular weight about 20,000 that contains two nonidentical subunits. The polypeptide sequence of these proteins, obtained from the purified subunits of HU and from the gene sequence of IHF, showed that all four subunits are homologous. They can be aligned without major gaps and have more than 45 percent identical or similar residues within that alignment (Fig. 8.1). No other members of the HU/IHF family have been found in *E. coli* but protein and gene sequencing have identified homologs in virtually every branch of the eubacterial kingdom, a few archaebacteria, a bacteriophage[16] and even one animal virus.[17] In many of these organisms it appears that HU is a dimer of identical subunits.[1] Indeed, in *E. coli* each of the two HU subunits readily forms homodimers[18] and these provide substantial HU function.[19,20] In contrast, individual IHF subunits of *E. coli* function poorly, in large part because they form homodimers quite weakly.[21,22] Dimer formation in the HU/IHF family does not appear to be promiscuous. For example, there is no evidence for dimers involving one IHF subunit and one HU subunit nor have mixed dimers involving subunits from different bacterial species been reported.

Our understanding of the molecular structure of this family of proteins is based almost exclusively on the X-ray crystallographic analysis of the homodimeric HU protein of *B. stearothermophilus*.[23,24] This analysis revealed that HU employs a novel fold (Fig. 8.2A). The amino-terminal half of the protein is dominated by two α helixes connected by a turn; the carboxy-terminal residues also form an α helix. The remainder of the protein is dominated by a three-stranded β sheet structure which includes a centrally located β ribbon extension. The existence of these structural features has been largely confirmed by NMR spectroscopy.[29] The dimer interface is also novel; it is largely dependent upon interactions between eight hydrophobic residues that are scattered through out the primary sequence (Fig. 8.2A). The alignment of members of the HU/IHF family (Fig. 8.1) reveals virtually perfect conservation of the interfacial hydrophobic residues and excellent conservation of residues involved in tight turns. The preservation of sequences that contribute to important structural elements confirms the homology between family members and strongly predicts that each of them will have a fold quite similar to that of HU of *B. stearothermophilus*.

```
1  Ihfα-ec  MALT    KAEMSEYLFDK - LGLSKRD   AKELVELFFEEIRRALE NG EQVKL SGF GNFD LRDKNQRPGR NPKT GEDIPITARRVV TFRPGQ KLKSRV ENASPKDE*
2  Ihfβ-ec  MT      KSELIERLATQQ  SHIPAKT   VEDAVKEMLEHMASTLA QG ERIEI RGF GSFS LHYRAPRTGR NPKT GDKVELEGKYVP HFKPGK ELRDRA NIYG*
3  Hu1-ec   MN      KSQLIDKIAAG - ADISKAA   AGRALDAIIASVTESLK EG DDVAL VGF GTFA VKERAARTGR NPQT GKEITIAAAKVP SFRAGK ALKDAV N*
4  Hu2-ec   MN      KTQLIDVIAEK - AELSKTQ   AKAALESTLAAITESLK EG DAVQL VGF GTFK VNHRAERTGR NPQT GKEIKIAAANVP AFVSGK ALKDAV K*
5  Ihfα-rc  MSEKTLT  RMDLSEAVFRE - VGLSRNE   SAQLVETVLQHMSDALV RG ETVKI SSF GTFS VRDKTSRMGR NPKT GEEVPISPRRVL SFRPSH LMKDRV AERNAK*
6  Ihfβ-rc  MI      RSELIAKIAEEN  PHLFQRD   VEKIVNTIFEEIIEAMA RG DRVEL RGF GAFS VKKRDARTGR NPRT GTSVAVDEKHVP FFKTGK LLRDRL NGGEE*
7  Hu-ta    MVG     ISELSKEVAKK - ANTTQKV   ARTVIKSFLDEIVSEAN GG QKINL AGF GTFE RRTQGPRKAR NPQT KKVIEVPSKKKF VFRASS KIKYQQ *
8  Hu-rm    MN      KNELVAAVADK - AGLSKAD   ASSAVDAVFETIQGELK NG GDIRL VGF GNFS VSRREASKGR NPST GAEVDIPARNVP KFTAGK FLKDAV N*
9  Hu-bs    MN      KTELINAVAET - SGLSKKD   ATKAVDAVFDSITEALR KG DKVQL IGF GNFE VRERAARKGR NPQT GEEMEIPASKVP AFKPGK ALKDAV K*

                    α1              t          α2            β1   β2  t   outgoing  returning t   β3     α3
```

Fig. 8.1. Amino acid sequences of the HU/IHF family. Each protomer is indicated by an abbreviated name followed by a code identifying the host organism: ec for E. coli; rc for R. capsulatum; ta for T. acidophilum; rm for R. melitoti; bs for B. stearothermophilus. Except for the β subunit of the IHF from rc,[179] amino acid sequences are taken from the GenBank and NBRF PIR databases. The sequences of many other family members are known; those presented here are chosen as examples of the diversity and ubiquity of the family. The symbols underneath line 9 indicate units of secondary structure determined from X-ray crystallography of the HU protein of B. stearothermophilus.[23,24] The eight hydrophobic residues that are shown in boldface are the most important contributors to the hydrophobic core of the HU-bs dimer.[23]

III. INTERACTION WITH NUCLEIC ACIDS

A. DNA Binding

Both HU and IHF bind to nucleic acids and this binding requires neither cofactors nor additional proteins. Early studies on HU protein used techniques as varied as electron microscopy, nuclease protection, affinity chromatography and nitrocellulose filtration (reviewed in ref. 30). These studies indicated that HU can interact with RNA and single-stranded or double-stranded DNA in a way that is not strongly sequence-dependent. There may be some specificity in the interaction with duplex DNA since HU can induce enhancements of reactivity to DNAase in some segments of DNA in preference to others (for examples see refs. 31,32). However, at high protein:DNA ratios contacts can occur at almost every base pair of a long duplex.[33] Recently, the nonspecific mode of HU binding to DNA has been assessed more

A

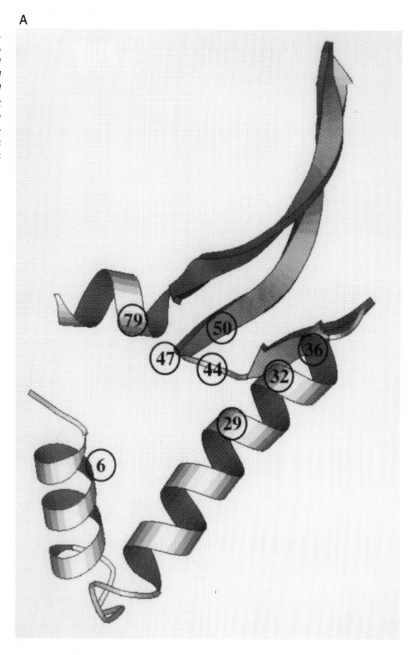

Fig. 8.2A. The molecular structure of HU protein, this figure displays the peptide backbone and α carbon atoms of one subunit of the HU protein of B. stearothermophilus, *as determined by X-ray crystallography (coordinates provided by Dr. S. White, Duke University). The eight hydrophobic residues that contribute to the dimeric interface are marked with circles. Because the tip of the "arm" of each subunit is disordered in the crystal,[23,24] no position is indicated for residues 59-69.*

quantitatively using electrophoretic mobility shift assays (EMSA).[34] At 200 mM salt, a weak interaction (K_d~13 µM) with a site that covers ~9 bp has been observed; this interaction is characterized by a low-level cooperativity and a modest enhancement in affinity at lower ionic strength. A few studies have compared the DNA binding behavior of homodimers of the different subunits of *E. coli* HU with that of the predominant heterodimer; only modest differences have been observed.[35-37]

In contrast to the weak and nonspecific binding to duplex DNA that is displayed by HU protein, IHF is characterized by tight binding to unique sites. These sites have been identified principally by protection assays and EMSAs. Protection from attack by hydroxyl radicals provides the most precise estimate of the size of an IHF site, typically 28-30 bp.[25,38,39] Both assays indicate that IHF binds its specific sites with an affinity of 1-20 nM.[40,41] Recent studies[42-44] have confirmed these estimates and have also determined that the specificity ratio of IHF for its natural targets vs. nonspecific nucleic acids is roughly 1,000 to 10,000. In the few cases where it has been examined, cooperativity between IHF sites is undetectable or weak.[40,45]

From phages, plasmids and transposons a large number of specific IHF binding sites have been identified; a smaller number have been identified in regulatory regions of *E. coli* operons. Almost all of these sites can be readily aligned.[46] As might be expected, in this alignment nonrandom DNA sequence is found to be coextensive with the IHF footprint. However, with the exception of 9 residues, sequence conservation is modest (Fig. 8.3A). This indicates that, although IHF clearly

Fig. 8.2B. The molecular model for the IHF-DNA complex, this figure displays a model[25] in which the dimer of IHF is assumed to have the identical fold to that of HU. A 28 bp stretch of DNA, encompassing almost all of a typical IHF site, is draped around this dimer. The draping is symmetrical, to accommodate a roughly symmetrical hydroxyl radical footprint.[25] However, the figure (redrawn, with modification, from 27) shows that the most conserved DNA sequence elements contact the arm of one subunit and the flank of another. Genetic[26] and biochemical[28] experiments indicate that these are from the α and β subunits of IHF, respectively.

B

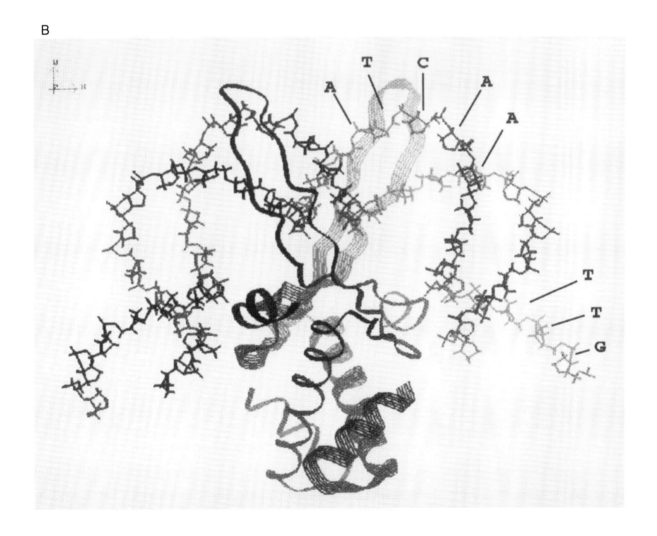

A

A	15	7	12	16	13	6	6	3	11	14	9	8	13	14	20	21	5	0	27	21	5	15	19	6	2	0	20
T	6	14	7	7	9	12	13	13	5	7	14	13	10	8	7	5	22	0	0	0	11	5	2	2	24	26	0
C	2	1	6	1	3	8	5	9	7	5	1	4	2	4	0	0	0	27	0	3	9	4	3	6	0	1	0
G	4	5	2	3	2	1	3	2	4	1	3	2	2	1	0	1	0	0	0	3	2	3	3	13	1	0	7
															A	A	T	C	A	A					T	T	A

B

	A	C	A	A	A	A		C		T	G	C	A	A	A	T	T	C	A	A	T	A		A	T	T	G
A	26	2	26	27	21	25	3	0	9	0	1	3	26	26	28	2	0	0	28	28	0	27	8	27	0	0	0
T	2	2	0	1	1	2	14	6	5	24	1	1	1	0	0	26	28	0	0	0	25	0	16	0	28	28	0
C	0	24	2	0	0	0	5	22	3	3	1	24	1	0	0	0	0	28	0	0	0	3	0	0	0	0	0
G	0	0	0	0	6	1	6	0	11	1	25	0	0	2	0	0	0	0	0	0	3	1	1	1	0	0	28

Fig. 8.3. Nucleotide conservation in IHF binding sites. Panel A presents a compilation of nucleotides found in a set of 27 aligned IHF sites.[46] A consensus sequence, listing the identity of each nucleotide found in at least 20 of the 27 aligned sequences is given below the alignment. Panel B presents a compilation of nucleotides found in a set of 28 IHF sites that are flanked by PU (REP) elements.[48] The sequences are aligned as described[48] except that no gaps are permitted. A consensus sequence, listing the identity of nucleotides found in at least 20 of the 28 aligned sequences, is given above the compilation.

has preferences at almost 30 positions, the protein can tolerate a variety of sequences in its binding site. Even the nine most conserved residues are somewhat subject to variation, further indicating flexibility in the target sequence. It should be noted that these nine residues are clustered in one half of the region (Fig. 8.3A), suggesting some asymmetry in the recognition process.[21] Furthermore, the consensus sequence derived from these residues shows little or no inverted repeat character.

A special set of more than 50 IHF sites has recently been identified in association with a subset of REP (also called PU) elements, a family of several hundred quasi-palindromic sequences that are usually located in intergenic regions. The targets for IHF that are situated between REP elements, the so-called RIB or RIP class of IHF sites,[47,48] bind the protein about as tightly as a typical site and contain its most conserved features. However, RIB/RIP sites are much more homogeneous in DNA sequence at positions that are weakly conserved in the set of typical sites (Fig. 8.3B). Since the function of the repeated elements remains speculative,[49] we can only guess whether the conservation in the IHF sites associated with them reflects selection for a particular function of these elements or is the unselected consequence of gene conversion (which might be expected to occur preferentially in a long block of repeated sequence). We also do not know how many of the RIB/RIP sites account for the 70 anonymous binding sites for IHP that were identified by a novel two-dimensional EMSA.[50]

B. Deformation of DNA

HU binding perturbs the overall structure of duplex DNA. When small plasmids are incubated with the protein and then relaxed by a topoisomerase, modest levels of negative supercoiling remain upon deproteinization.[51,33] There is no evidence to suggest that this effect is due to protein-induced melting of regions of the double helix, or even a uniform decrease in the twist of DNA. Instead, all the available evidence suggests that HU changes the shape of the double helix so that, instead of being roughly linear, its axis follows a path that is contorted. This can also be inferred from the appearance of HU-DNA complexes in the electron microscope.[51] However, the most convincing experiments follow from the effects of HU on the circularization of small linear fragments. The inherent stiffness of duplex DNA mitigates against the circularization of fragments of less than 500 bp in length[52] but, in the presence of HU, circles of 60-100 bp are readily obtained.[53,54]

Like HU, IHF also perturbs the shape of duplex DNA. However in this case, the specificity of IHF binding makes available a variety of shape-sensitive EMSA techniques such as cyclic permutation[55] and phasing[56] that are difficult to apply in the case of HU. These techniques, as well as electron microscopy, lead to the conclusion that IHF introduces a very substantial distortion of the double helix axis.[57-60] Although there is no definitive basis for quantitating the parameters of the distortion, studies employing most methods agree that, if IHF were to induce a hinge-like bend, the bend angle would be 140° or greater; this would convert the shape of DNA from something approximating a straight line to something resembling a hairpin. Interestingly, in contrast to HU, IHF has little or no effect on the overall topology of DNA. This is deduced from the very similar affinity of IHF for sites on supercoiled and nonsupercoiled templates (B.D. Lavoie and G. Chaconas, personal communication).[43] Thus, like the eukaryotic TBP protein,[61,62] IHF probably achieves its distortion of DNA by altering writhe and twist in ways that largely compensate for each other.

Agents that deform the helical axis should bind preferentially to DNA that is prebent.[63] For the proteins reviewed here, this expectation appears to be borne out. In the case of IHF, circularization of short fragments containing IHF sites significantly increases their affinity for the protein.[64,65] For HU, the DNA distortions associated with cruciform DNA create a highly preferred DNA binding site,[66] one whose affinity for HU is increased by a thousand fold.[34] Simple theory also predicts that IHF and HU will prefer to bind to DNA whose flexibility is greater than average. This may be the basis for the preponderance of A-T base pairs within typical IHF sites (Fig. 8.3A). It may also be the basis for the specific binding of HU to DNA that contains single-strand nicks or gaps.[67]

C. PROTEIN-DNA CONTACTS

The site-specific nature of IHF binding has enabled a variety of studies that map the contact surface on the DNA and, to a lesser extent, the contact surface on the protein. Within the 30 bp region that is footprinted by IHF, the pattern of protection against attack by DMS and hydroxyl radicals is easily accommodated if one assumes that the principal interactions with IHF involve the minor groove of the DNA target.[68,25] (However, this inference is questioned by recent experiments that assess the strength of IHF binding when normal bases are replaced by analogs.[44]) The detailed footprint, the finding that a single dimer of IHF suffices to protect a 30 bp target[25] and the assumption that the fold of IHF is nearly identical to that of HU provide the ingredients for a model of the protein-DNA interaction (Fig. 8.2B). In this model each monomer of the dimer contributes two DNA binding surfaces—a β ribbon "arm" and a segment from the side or "flank" of the subunit both of which contact the minor groove of the target. Genetic studies that identify residues of the protein likely to be in contact with the DNA[26,27,42,69] are consistent with the model, as are photo-crosslinking experiments that probe more directly for positions of close contact.[28] Of course, the model is both speculative and approximate; for example, it provides no insight into the way the protein recognizes a specific DNA sequence. Co-crystals of IHF and one of its target DNA sites have been prepared (S. Yang and P. Rice, unpublished observations); the solution of the molecular structure contained in these crystals is eagerly awaited.

HU cannot by itself be localized on duplex DNA and there is thus no detailed footprint for this protein bound to naked DNA. However, HU can be localized to a specific site when cooperating with other proteins (see below). In one such case, coupling of a chemical nuclease to specific residues of HU produces a footprint that is remarkably consistent with the one observed for IHF (B. D. Lavoie and G. Chaconas, personal communication).[70] This observation strongly suggests that the two proteins have very similar modes of binding to DNA.

IV. CONTROL OF INTRACELLULAR CONCENTRATION AND ACTIVITY

The capacity of *E. coli* to respond to an altered environment is impressive. Coordinated responses of multi-gene systems to variations in growth rate, limitation of nitrogen or carbon sources, temperature extremes, DNA damage, and osmotic pressure have been well documented.[71] The environmental state can have large effects on some of the accessory factors that participate in complex protein-DNA assemblies. For example: (a) changes in the carbon source produce large alterations in the concentration of the cAMP cofactor that promotes sequence-specific binding by CRP protein;[72] and (b) dilution of stationary phase cells into fresh medium raises the level of messenger encoding Fis protein by a factor of 1000.[73,74] In contrast there is little evidence for such dramatic regulation in the concentration or activity of HU and IHF.

A. REGULATION OF HU

There is a virtual absence of reports of regulation of the abundance of HU protein in different growth conditions. We presume that the reported content of 2 to 5 ng of HU per µg of total cell protein[35,75] represents an average that varies little, if at all, in the life of *E. coli*. The constancy of the amount of HU per cell may in part reflect the operation of an autoregulatory circuit in which excess accumulation of HU depresses transcription from the promoters for its synthesis.[76] In addition, there is no evidence for a cofactor that specifically influences the binding of HU to DNA and thus no known opportunity for regulation of its activity.

B. REGULATION OF IHF

IHF increases in abundance upon entry into stationary phase. As measured either by DNA binding activity of crude extracts[77] or immunoactivity in whole cell extracts,[75] both IHF subunits increase about 5-fold above the level of 0.5 to 1.0 ng of IHF per µg of cell protein that is found during exponential growth.[75] Since the subunits of IHF appear to be stable under standard conditions (for an exception, see ref. 78), it is probable that the increased abundance in stationary phase largely reflects increased synthesis of IHF. Indeed, increased transcription is found for both genes as cells enter stationary phase.[79] The genes for both IHF subunits lie just downstream from genes involved in protein synthesis;[80,81] readthrough from the promoters of those genes could make regulation of IHF quite intricate. Other evidence of cellular regulation of IHF includes autoregulation, inferred from an increased abundance of both subunits in an IHF mutant strain.[75,82] The phenomenon probably reflects IHF binding to the promoters of genes encoding its subunits[79] and could contribute to depressed IHF levels during outgrowth from stationary phase.[75] Although an early report claimed a 3- to 4-fold SOS induction of IHF,[82] subsequent studies found much smaller induction ratios upon inactivation of LexA protein;[80,83] no binding sites for this repressor are apparent in the promoter regions for the IHF subunits.

Taken together, the data cited above indicate that regulation of IHF levels occurs but is not dramatic. However, environmental influences may be critical in some contexts. For example, the segment of DNA that controls the expression of the *him*A gene of *S. typhimurium* (encoding the α subunit of IHF) is included in a short list of genetic elements whose expression is differentially regulated upon growth in animal cells.[84] Given the high concentrations of IHF found under standard growth conditions (see below), this is a surprising result and one that needs further investigation. For example, it would be of interest to know if the *hip* gene (encoding the β subunit of IHF) were similarly regulated. If so, growth in animal cells would be identified as a condition requiring extremely high levels of IHF; if not, the disproportionate synthesis of IHF subunits might indicate a specific role for a homodimer.

C. CELLULAR CONCENTRATION OF HU AND IHF

If one takes the usual values for the protein content and volume of a cell,[85] one can convert the measured abundance of HU and IHF into cellular concentrations. For HU the average value of 2.5 ng per µg of cell protein[75] implies there are about 12,000 molecules (dimers) of HU per cell, which corresponds to an intracellular concentration of about 20 µM. This value is above the reported dissociation constant of HU for a random sequence of DNA.[34] Thus, if intracellular conditions are comparable to those used to determine the in vitro binding constant, much of the intracellular HU should be bound to chromosomal DNA. For IHF, the log phase abundance of 0.75 ng of IHF per µg of cell protein implies that there are about 3,500 molecules (dimers) of IHF per cell, yielding an intracellular concentration of about 6 µM. This value is also close to the dissociation constant of IHF for nonspecific DNA,[42-44] but is more than 1,000-fold higher than the dissociation constant of IHF for one of its typical sites. Because the number of specific binding sites for IHF in *E. coli* is probably less than 200, the distribution of IHF inside the cell will be largely governed by the density and availability of nonspecific sites, just as described twenty years ago for Lac repressor.[86-88] However, as discussed in a later section, little is known about either the availability of chromosomal nonspecific DNA binding sites for HU and IHF or the contribution of cellular RNA to nonspecific binding of these proteins. Despite these theoretical limitations, a rough evaluation of the intracellular distribution of IHF can be made. Weisberg and colleagues showed that several biological systems respond to graded changes in the total intracellular content of IHF, suggesting that the protein's effective concentration is limiting.[75] More recently, an in vivo footprinting study analyzed the occupancy of a series of IHF binding sites whose in vitro affinity had been measured; the data suggest that the effective intracellular concentration of IHF is only about 40 nM.[43] This value implies that more than 95% of the intracellular IHF is indeed sequestered, presumably by binding to nonspecific targets.

V. PARTICIPATION OF IHF AND HU IN WELL-CHARACTERIZED BIOCHEMICAL PROCESSES

A. IHF PROTEIN

Studies of lambda site-specific recombination introduced the concept that IHF acts as an architectural element in protein-DNA assemblies, that is, an agent whose principal role is to expedite the construction of a compact nucleoprotein array. This notion now dominates much, but not all, of our current view about the way IHF acts as an accessory factor. According to this view, because duplex DNA is relatively stiff, it is difficult to connect protein binding sites that are separated by 30-300 bp.[52] IHF overcomes this difficulty by introducing severe deformation of the duplex at specific places, thereby shortening the distance between the sites that need to be brought together to form the desired structure. The image of a segment of DNA bent by a protein to resemble the letter Vee so as to facilitate contact between binding sites for other proteins that are located at the tips of the Vee has been nicknamed the "compass model" for architectural elements.[89]

1. Lambda integration and excision

Biochemical studies showed that the catalytic protein in lambda site-specific recombination is the phage-encoded Int protein. IHF must therefore function as an accessory component in these reactions. When the specific nature of IHF binding was discovered,[68] it seemed particularly telling that IHF not only binds to specific places within attachment sites (the regions of DNA at which integration and excision take place) but that these IHF binding sites are interspersed with those for Int protein. This arrangement, coupled with experiments showing that the lambda attachment site assumes a complicated higher-order structure during recombination, led to the speculation that IHF might be involved in the folding of the attachment site into an active form.[90] With the demonstration that IHF binding to its targets induces a severe distortion, this idea became more focused: IHF might function at the attachment sites as an architectural element whose principal role is to bend DNA so as to permit assembly of Int into a higher-order structure.[91]

To test one aspect of this hypothesis, Steven Goodman[92,93] engineered "bend swap" experiments in which one or more IHF binding sites were replaced by unrelated modules that introduced DNA distortion. The success of these experiments provide compelling evidence that deformation of DNA is a principal function for IHF in lambda site-specific recombination. However, one must note that, because no bend swap chimera functions with perfect efficiency, secondary roles for IHF cannot be ruled out.

Work by Landy and colleagues has provided a concrete example of how the architectural function of IHF contributes to formation of a higher-order structure.[94] They studied *att*L, an attachment site involved in lambda excision, and showed that IHF enhances the capacity of Int to bind to one part of the recombination locus, the core. The enhancement depends not only on a specific IHF binding site adjacent to the core but on a second part of *att*L, the arm. Since core and arm regions are recognized by different domains of Int protein, they proposed that bending by IHF promotes the bridging of core and arm sites by a single protomer of Int. Subsequent work[95] indicated that, of the two core binding sites and three arm binding sites present in *att*L, one particular pair was involved in such bridging (Fig. 8.4A).

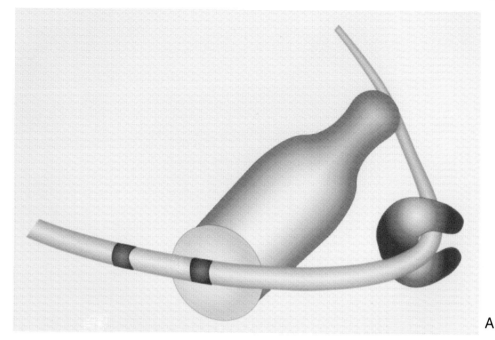

Fig. 8.4. IHF contributions to the architecture of attachment sites. In each diagram, the DNA of an attachment site is represented by a gray cylinder; a pair of black bands indicates a portion of this DNA (the core) at which strand exchange occurs. Int protein is drawn in the shape of a bowling pin; the tip of the pin carries the domain responsible for binding to the arm of attachment site DNA and the bottom of the pin carries the catalytic domain. IHF is drawn in the shape of a crescent that encircles and kinks the DNA.

A

Fig. 8.4A shows the critical elements of an architectural proposal for attL.[95] IHF-induced distortion at the lone IHF binding site of attL assists Int that is bound to one arm site of this DNA in reaching one target in the core. The other target is proposed[180] to be bound by Int bound to a partner attachment site that has been deformed by IHF and Xis into a complementary shape (not shown).

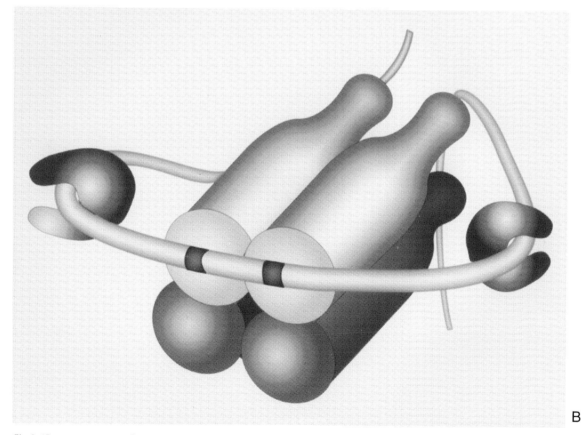

B

Fig. 8.4B presents a speculative proposal for the architecture of attP. The DNA of attP is wrapped into a coiled structure. Negative supercoiling and three specifically bound IHF dimers (one of which is hidden from the reader's view) contribute to this wrapping. In the final structure, the catalytic domains of four Int proteins are precisely positioned; two are bound to the core of attP and two are available to interact with the core of attB.[181] In contrast to the case of attL, there is no strong evidence to indicate the way in which Int protomers bound to the core connect to binding sites in the arms of attP; the arrangement shown is plausible but untested.

In contrast to the relatively simple case of *att*L, a more complex picture is needed to explain IHF action at *att*P, the phage attachment site that directs viral integration. This locus spans ~250 bp, must be present on a supercoiled circle in order to be active in recombination and has three essential IHF sites (reviewed in refs. 96, 97). Understanding how the multiple bends introduced by these IHF sites contribute to the functional architecture of *att*P will be a formidable task. We cannot even be certain how the three IHF sites will combine to shape *att*P in the absence of Int. This is because the available electrophoretic and electron microscopic methods provide no estimate of any rotation of helical segments that may be associated with a hinge-like bend, a parameter that is essential for calculating the overall path induced by multiple bends.[98] Nevertheless, the success of the bend swap experiment with an IHF site within *att*P[92] encourages the view that the deformations introduced by each of the three IHF sites, taken together with the compaction introduced by supercoiling, combine to yield a structure in which the arms of *att*P are tightly folded so as to contact and thereby stabilize a cluster of Int protomers. A speculative cartoon of this architecture is shown in Figure 8.4B.

2. Transcriptional activation at σ[54]-dependent promoters

The *rpo*N (also called *ntr*A) gene encodes a polypeptide of M_r ~54,000 that functions as an alternative σ factor. The form of RNA polymerase that contains σ[54] recognizes and binds to a distinct set of promoters. Typically, conversion of such σ[54] holoenzyme-promoter complexes from the closed, inactive state to the open, active state depends on contact with an activator protein whose binding site lies far upstream or downstream from the promoter (reviewed in ref. 99). Kustu and colleagues pointed out that in one set of such promoters, those regulated by activators encoded by the *nif*A gene, an IHF site is positioned between the promoter and activator elements.[100] They showed that IHF is important for activation of one of these promoters, p*nif*H, and proposed that bending by IHF assists contact between the activator and RNA polymerase, i.e., similarly to the way it assists λ recombination. A similar arrangement of promoter, activator and IHF binding sites has been found for σ[54]-dependent systems in many eubacteria. For example, activation of the *E. coli* *gln*Hp2 promoter by nitrogen regulator I (the phosphorylated form of the *ntr*C gene product) strongly depends on IHF protein binding to a site that lies between the two elements.[101] At the *nif*H promoter (and one other σ[54] promoter with a similar arrangement of activator and IHF sites), successful bend swaps indicate that for these systems, the principal role of IHF is to deform DNA.[102,103]

3. Repression and replication enhancement of bacteriophage Mu

The region of DNA that is devoted to the control of early events in the bacteriophage Mu life cycle contains a single IHF binding site.[104,105] This site is flanked by sequence determinants that direct the binding of the phage repressor. IHF enhances the stability of repressor bound to these sites,[106,107] presumably by acting as an architectural element that permits bridging contacts between repressor protomers. This arrangement should improve repression under conditions in which repressor protein is limiting. Remarkably, IHF binding to this site also appears to be important for the lytic development of Mu. The same segments of DNA that direct the binding of Mu repressor appear to also direct the binding of Mu transposase.[108,109] Transposase bound to this region acts as an enhancer of replicative transposition and IHF is needed for enhancer action, presumably by assisting the assembly of multiple protomers of transposase.[110] It thus appears that, within the same segment of DNA, IHF can assist the formation of two distinct higher-order structures that respectively serve the lytic and lysogenic outcomes of Mu infection. That is to say, IHF is not itself involved in the regulation of the critical epigenetic switch between these states; it simply enables both options.

It is important to note that IHF stimulation of enhancer function in vitro is only observed on DNA molecules that have modest superhelicity, typically in the range believed to characterize intracellular conditions.[111] At very low superhelical densities, the reaction fails except under special conditions,[110] and at superhelical densities higher than physiologic the Mu enhancer works well without IHF. The dispensability of IHF at high levels of supercoiling can be rationalized according to the following considerations. In the enhancer, the binding sites for Mu transposase are separated by about 100 bp; IHF presumably serves to juxtapose protomers of transposase that are bound to these sites. As the degree of supercoiling is raised, DNA tends to increasingly fold on itself, making a tighter superhelix.[112] At a certain superhelicity, this tendency will balance the energetic cost of deforming DNA and permit juxtaposition of the two closely spaced transposase binding sites, either without any accessory factor or with the weaker-binding HU in place of IHF (see above). A similar argument could account for the dispensability of IHF at the *gln*Hp2 promoter on supercoiled vs. linear templates.[101] However, it should be recalled that IHF is needed for *att*P function even on highly supercoiled DNA,[90,96] so that dispensability of IHF at high superhelical densities is not universal.

4. Replication of *ori*C-dependent minichromosomes

The *E. coli* chromosome contains a single locus, *ori*C, that is normally used as the replication initiator. Analysis of *ori*C has been most intensively studied with plasmid minichromosomes that contain 1 to 2 kb of sequence from the *ori*C region (reviewed in ref. 113). Within the 300 bp segment of DNA from this region that is essential for replication of minichromosomes lies a single IHF binding site.[114,115] The IHF consensus elements found at this site are conserved in the minimal origins from many other bacteria, suggesting that IHF binding to *ori*C is universal. Mutation of the IHF target sequence in the *ori*C of *E. coli* damages the origin, as judged by failure of transformation with the altered minichromosome.[116] Further support for the importance of this element comes from the observations that IHF⁻ strains have partial defects in transformation by *ori*C plasmids, e.g., when using minichromosomes that have alterations in segments outside the minimal origin[117] or when transforming into strains with a defective DNA polymerase I.[115]

How does IHF binding to *ori*C influence the origin? The critical event in the early stages of replication initiation appears to be the separation of complementary strands in one part of the minimal origin, a segment containing three copies of a 13 bp sequence.[113] The melting of *ori*C DNA depends on the DnaA initiator protein, which binds tightly to four repeated sequences (*dna*A boxes) that are scattered from 20-200 basepairs from the edge of the 13 bp repeats. Even when bound to a highly supercoiled minichromosome plasmid, highly purified DnaA cannot by itself achieve melting of the 13-mer region but requires the participation of IHF.[118] In exponentially growing cells, IHF binding to *ori*C of minichromosomes is a cyclic phenomenon, occurring only at the time of initiation.[119] Because there is no evidence of cell cycle regulation of the abundance or activity of IHF, the protein may be cooperating with a cyclically-regulated factor, perhaps DnaA protein, in binding to *ori*C. This scenario would explain the observation[120] that an IHF mutation exacerbates the growth defect of a conditional *dna*A mutation. Taken together, the existing experiments make it seem plausible that the distortion introduced by IHF promotes contacts between protomers of DnaA bound to the four *dna*A boxes or assists the recruitment of additional copies of this protein to the origin; melting of the origin would follow the construction of this new architecture.

We are only beginning to relate studies of minichromosomes to the function of *ori*C in its normal genomic context. The observation that *E. coli* replicates from *ori*C in strains that are defective for IHF (and HU; see below),[117] suggests that additional elements that are not found in minichromosomes provide redundancy to origin function. However, IHF⁻ strains replicate asynchronously[121] indicating that the cell's capacity to transmit physiological signals (such as cell mass) to the initiation apparatus is compromised by the loss of IHF.

5. Transposition of *IS*10 and *Tn*10

The insertion element IS10 is capable of moving from one genetic location to another by cleavage and ligation events involving the termini of the element. In transposon *Tn*10, a pair of *IS*10 elements flank a tetracycline resistance gene, an arrangement that permits a variety of genetic outcomes depending on which pair of ends from the two *IS*10 elements participate in transposition (reviewed in ref. 122). One of the termini of *IS*10, the outside end, comprises two elements: (a) a binding site for a transposase that catalyzes the cleavage and joining reactions; and (b) a binding site for IHF protein. In vitro studies with partially purified transposase showed that IHF can influence the reactions involving the outside ends of *Tn*10.[123] The recent preparation of highly active, highly purified transposase[124] has enabled a closer look at reactions that involve the outside ends of *IS*10. Interestingly, two distinct effects of IHF have been found.[122] First, IHF can stimulate the rate and extent of the overall transposition reaction, especially under suboptimal conditions such as low degrees of supercoiling of the DNA carrying the *IS*10 ends. Much of this IHF effect is probably manifested at the earliest stage of the reaction, assembly of the transposase and DNA into a complex in which two outside ends are paired. Second, IHF can alter the distribution of the products of transposition, favoring those that arise from a particular kind of intramolecular rearrangement. Such "channeling" of the transposition pathway probably reflects the influence of IHF on the degree of compaction with which the outside ends are held by the transposase. Evidence for both stimulatory and channeling effects can be seen in in vivo studies of *Tn*10 transposition.[125] Here, however, the magnitude of the two effects depends upon several additional features such as length of the transposon and whether it is borne on a plasmid or the *E. coli* chromosome. This added complexity may provide useful clues to ways in which the organization of DNA inside cells differs from that observed in vitro.

6. Direct repression and activation of transcription

Although the compass model[89] is valid for many of the systems that use IHF, this postulated mode of action is probably not the universal explanation for IHF action. In addition to the common pattern of IHF sandwiched between other parts of a multiprotein array, it sometimes appears that IHF sites are adjacent to or overlap with a binding site for only one other protein. This arrangement is epitomized by several cases in which IHF and purified RNA polymerase are the only protein components of a transcription system (reviewed in ref. 126). The compass model would suggest that IHF bending influences transcription in these cases by bringing RNA polymerase into contact with upstream DNA sequences. This possibility might also explain the effects seen when sequence-directed bends are juxtaposed to promoters (reviewed in ref. 182). On the other hand, it is just as easy to imagine other possibilities for IHF action. One involves steric interference between IHF and RNA polymerase. For example, at the ilvP$_G$1 promoter of *E. coli*, IHF binding to a site that closely abuts the promoter appears to prevent RNA polymerase from binding; IHF repression of ilvP$_G$1 probably reflects this interference.[127,128] A similar arrangement is found at the promoter for Tn10 transposase and steric blockage could well explain the observed IHF repression.[75,129] Another possible way IHF could control promoters is by direct contact between the protein and RNA polymerase. This possibility has been invoked to explain the observation that IHF activates the bacteriophage lambda P$_L$ promoter. Evidence in support of the hypothesis includes: (1) the effect of IHF is to increase the affinity with which RNA polymerase binds to the promoter to make a closed complex;[130] (2) the region of DNA between the IHF site and the promoter appears to be distorted (as if contacts between the two proteins had looped the intervening DNA);[131] and (3) stimulation of the promoter by IHF requires the same domain of the α subunit of RNA polymerase that responds (in other promoters) to direct contact by the CRP activator.[132] The domain of IHF involved in the putative contact with RNA polymerase has not yet been identified nor has the physiological significance of the activation. Elucidation of this case may also shed light on the way in which IHF binding to the early control region of bacteriophage Mu (see above) activates in vitro transcription of the early promoter.[104,133,134]

B. HU PROTEIN

1. Transposition of bacteriophage Mu

Efficient in vitro transposition of bacteriophage Mu requires HU protein.[15] This requirement is seen even when highly supercoiled substrates are used and thus is distinct from the requirement for IHF action at the enhancer (see above). The essential role for HU lies at an early stage in the reaction, one that leads to the formation of a stable synaptic complex (SSC, or type 0 transpososome). The formation of the SSC is a complex process in which multiple copies of Mu transposase cooperate to assemble a tetramer that juxtaposes the ends of Mu; these "donor ends" will become attached to the target DNA during later stages of transposition. Under physiological conditions, tetramer assembly requires: (1) a pair of donor ends oriented as an inverted repeat on a supercoiled DNA; (2) transposase protomers competent for binding to these ends; (3) the recombinational enhancer, described above, that binds a domain of the transposase protein that is distinct from that which binds to the ends; and (4) HU (see ref. 135 for a review). Since HU can be removed from the resulting SSC without detriment to later steps in the transposition pathway (ref. 136 and references therein), it is clear that the principal role for this protein is in the formation of a higher order structure. However, our understanding of the process is only partial. The relatively small amount of HU required for the process[70,137] argues against the possibility that HU simply coats the DNA in order to produce general condensation of the supercoiled donor and/or take up some of its superhelical tension. Moreover, electron microscopic analysis of another transposition intermediate (the cleaved donor complex or type I transpososome) indicates that all detectable HU is colocalized together with Mu transposase at the donor ends.[138] Fine mapping of HU within this structure has been achieved by an elegant experiment in which HU was converted by chemical modification into a nuclease.[70] This technique indicates that one dimer of HU is sandwiched between two binding sites for Mu transposase that are located at one particular Mu end. This is so reminiscent of the arrangement of IHF in structures like that of the bacteriophage lambda attachment sites or the *nif*A-activated promoters (see above) that it is attractive to imagine HU is also playing an architectural role at the ends of Mu. However, the situation may be more complex since at least one additional dimer of HU is contained in the type 1 transpososome whose location is not revealed by the chemical nuclease strategy.[139]

2. Site-specific inversion of the flagellin promoter

Two proteins are required as accessories for the recombination system responsible for the classical phenomenon of phase variation in *S. typhimurium*. These proteins, found in both *E. coli* as well as *S. typhimurium*, work together with a recombinase, Hin, that promotes recombination between inversely repeated loci, the *hix* sites. *hix* sites flank the flagellin promoter and the resulting rearrangement modulates the expression of a *Salmonella* surface antigen (see ref. 140 for a review). One of the accessory proteins for this recombination, Fis, binds to a pair of inverted repeats that normally lie about 100 bp from one *hix* site. Fis stimulates inversion even when the cluster of Fis binding sites is artificially moved by several kbp. A variety of biochemical, electron microscopic, and topological studies indicate that Fis bound to this "recombinational enhancer" cooperates with Hin protein bound to the two *hix* sites to form a complex structure, the invertasome. In this assembly, the three segments of DNA bound by Hin and Fis are interwoven.

The second accessory protein needed in this system is HU. However, recombination becomes largely independent of HU when the enhancer is moved from its natural position so as to be quite distant from either *hix* site.[141] Since loops of naked DNA form more easily as length increases,[52] this observation has been taken to imply that HU functions by assisting the looping of DNA between the enhancer and the nearby *hix* site (Fig. 8.5). In favor of this view is the fact that only 15-20 HU dimers are needed per molecule of DNA,[143] an amount sufficient to coat the 100bp loop but not the entire plasmid. The finding that extracts of HU⁻ cells fail to stimulate Hin-promoted recombination[143] suggests that no other bacterial protein can assist looping in this system. Nonetheless, not all workers have found that Hin-promoted recombination is defective in HU⁻ deficient strains.[144,145] A possible explanation for the varying results is that different physiological conditions make it easier or harder for DNA to achieve looping in the absence of HU protein. This scenario might also explain why an in vitro study failed to detect an effect of HU in the closely related Gin-promoted recombination system[146] while in vivo studies show that this process is very depressed in HU⁻ strains.[145]

The strong dependence of in vitro Hin-promoted recombination on HU protein enabled a search by Johnson and colleagues for

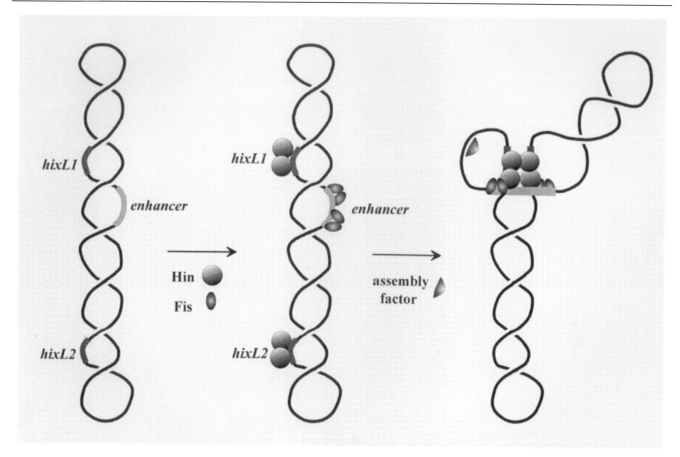

stimulatory factors from other organisms that could assist invertasome formation. From HeLa cell nuclear extracts, the principal activity proved to be the HMG1 and HMG2 proteins.[54] This surprising result, confirmed by the effectiveness of calf thymus and yeast homologs,[147] greatly strengthens the assertion that deformation of DNA is the principal role of HU in this system. This is because, although the HMG proteins are unrelated to HU in primary, secondary or tertiary structure, they resemble HU in their effect on DNA conformation (reviewed in ref. 148). Like HU, HMG1 and HMG2 can increase the efficiency of circularization of short linear fragments of DNA. And both HU and the HMG proteins bind preferentially to prebent DNA, as evidenced by their strong affinity for cruciforms.[34,148] The facile replacement of HU in the invertasome by eukaryotic DNA deformation proteins that are unrelated to it not only indicates the HU is an architectural element, but it provides strong evidence against an important role for contact between HU and Fis or Hin. The identical strategy has been used to rule out protein-protein interactions as essential for HU action in several other systems;[139,149] these "protein swap" experiments provide a perfect complement to the "bend swap" experiments (see above) used to probe IHF action.

C. INTERCHANGEABILITY OF HU AND IHF

Because IHF binding is characterized by high affinity and specificity while HU binding is typically low affinity and nonspecific, it might be imagined that successful replacement of IHF by HU would be rare. This is not the case. For example, assembly of a functional *att*L site can be accomplished by Int either with IHF or HU.[93] Remarkably, when tested under conditions where the concentration of

Fig. 8.5. HU contribution to the formation of the invertasome. A supercoiled circle of duplex DNA is cartooned as a continuous line. The positions on this circle of two hix sites and a recombinational enhancer are indicated. Binding of Hin protein to the hix sites and Fis protein to the enhancer is insufficient to make an active recombination machine, presumably because the segment of DNA between the enhancer and one Hix site is too short to permit their mutual interaction. DNA deformation introduced by HU protein binding to this segment obviates the inflexibility and permits assembly of an invertasome. Reprinted with permission from Paull TT et al, Biochimie 1994; 76:992-1004.

accessory protein reflects its affinity in the resulting complex, HU is needed at only 5-fold higher concentration than IHF[149] despite the fact that the intrinsic affinity of HU for random DNA is 1,000-fold lower than the affinity of IHF for the site in *att*L.[34,42-44] Moreover, the ability of IHF to assist in the formation of an *att*L intasome is not greatly impaired when either the IHF binding site or the protein itself is altered so as to eliminate specific binding.[149] Under all these conditions, the *att*L intasome has nearly the same electrophoretic mobility as that made when IHF protein is used at its normal target. Presumably, in each case the intasome structure resembles that shown in Fig. 8.4A. In essence, nonspecific architectural elements like HU or mutant IHF have gained site-specificity by cooperating with Int to build a specific structure.[149] This scenario provides an explanation for the observation that only a few molecules of HU suffice to build an invertasome or transpososome.[143,137] Rather than being distributed nonspecifically over the entire chromosome, the HU molecules are preferentially trapped in the higher-order structure.

Just as with *att*L, HU successfully replaces IHF in several other in vitro systems, e.g., *ori*C minichromosomes[118] and transposon *Tn*10.[123] As might be expected from its capacity to bind tightly to its targets in these loci, in general, it appears that IHF is somewhat more effective than HU, i.e., works at lower concentration and produces a higher yield of the active structure. However, there are a few cases, mostly studied in phages and plasmids, where specific IHF action is demanded for function. For example, IHF was discovered because in vitro lambda integrative recombination is absolutely dependent on the protein: HU is an ineffective replacement for IHF at *att*P.[150,93] Presumably the complexity and/or the energetics of the *att*P intasome (Fig. 8.4B) can only be satisfied by a high affinity architectural element that introduces multiple bends at specific places. A similar scenario may explain the requirement for IHF at the origin of replication of plasmid pSC101.[151,152]

Given the fact that IHF can bind to DNA nonspecifically,[149] one might expect IHF to functionally replace HU protein. This has indeed been observed in some systems,[137] but in others, like Hin-promoted recombination,[143] such replacements are ineffective. This suggests that the nonspecific mode of IHF binding is not identical to that of HU, a proposition supported by the details of experiments in which IHF is forced to bind nonspecifically to the lambda attachment site.[149] Thus, just as the effectiveness of HU replacing IHF depends upon detailed anatomy and energetics, the use of IHF as a nonspecific replacement for HU varies from case to case.

VI. UNFINISHED BUSINESS

The previous sections of this chapter have presented a wealth of detail about the way HU and IHF interact with DNA and how such interactions contribute to the formation of complex DNA-protein assemblies. Despite this knowledge, some very fundamental questions about the biological role of HU and IHF remain unanswered. In this section we discuss some of these remaining mysteries in hopes of stimulating additional research.

A. PHENOTYPE OF IHF-DEFICIENT STRAINS

Although IHF mutants are viable, closer scrutiny reveals that they exhibit a variety of subtle changes in bacterial physiology. In addition to the specific cases of altered promoter usage described above, there are many other changes in gene expression.[5,126] For example, when bacterial proteins are displayed on a two-dimensional gel, IHF+ and

IHF⁻ strains differ qualitatively and/or quantitatively at dozens of individual protein spots.[5] Some of these differences certainly represent effects on promoter utilization of the kind described in the above sections. Other differences may reflect post-transcriptional action of IHF, a possibility that has been demonstrated in the case of the bacteriophage λ cII protein but one that operates by an as yet unknown mechanism.[153,154]

Given the diversity of IHF effects on *E. coli*, one wonders whether there is a coherence to the pleiotropy. In other words, is there a principal role for IHF in the economy of the cell, something akin to the role of CRP protein in the global carbon energy modulon?[72] If so, this role might account for the continued existence of IHF in the evolutionary history of bacteria. Accordingly, many of the well-analyzed systems that use IHF would be viewed merely as scavengers that exploit a handy architectural tool, one whose presence is insured by the dependence of the cell for IHF in its most essential job.

In searching for a "master role" for IHF, one tends to focus on systems in which removal of IHF causes a qualitative change. One possibility is the requirement of IHF for synchronous firing of replication origins.[121] However, it is hard to devise methods for assessing the contribution of such synchrony to the fitness of a cell. Regulation of gene expression is another attractive possibility for the essential focus of IHF action. However, many of the reported effects of mutations in IHF subunits or IHF sites are modest—e.g., changes of 2- to 5-fold in gene expression.[5] Moreover, in at least one case removal of IHF causes opposite effects when the test system is plasmid-borne vs. chromosomal.[125] Despite these caveats, three cases where IHF is responsible for large effects are prominent: (a) the induction (described above) by the *nif*A regulator of the *nif*HDK operon, encoding the subunits of dinitrogenase and dinitrogenase reductase;[103,100] (b) the induction by the *nar*L regulator of the genes encoding nitrate reductase;[155,156] and (c) the induction by the *tdc*R regulator of the gene encoding threonine dehydratase.[157] In each case the capacity of the regulator to produce a marked increase in enzyme levels is virtually abolished by mutations in the IHF subunits and/or IHF sites that lie upstream of the relevant promoters. Remarkably, in the host organism that normally contains the *nif*A, *nar*L, or *tdc*R regulator, induction takes place only under anaerobic or microaerobic conditions and leads to the synthesis of ammonia from alternate sources (nitrogen, nitrate and threonine, respectively). This suggests that IHF may be essential for the diverse cellular responses to limitation in a preferred source of nitrogen, especially during simultaneous oxygen deprivation. Of course, it is not possible to be confident that assisting gene expression under such dual limitation is the raison d'etre for IHF. However, it should be noted that in all three cases, inappropriate gene expression by cross-talk from homologous regulators would be wasteful and possibly harmful. In this context, the dual capacity of IHF to favor the response of a promoter to one regulator whose position is optimally located[158] and to disfavor the action of extraneous activators[159] may be particularly critical. It may also be the case that the abundant expression of IHF under aerobic conditions (see above) serves to insure that an adequate supply of the protein is available for a prompt response to the dual challenge of oxygen and ammonia deprivation.

B. Phenotype of HU-deficient Strains

Strains that are simultaneously defective for both subunits of the *E. coli* HU protein are viable, but a number of observations indicate that the mutant cells are grossly abnormal. For example, compared to isogenic wild-type controls, HU⁻ cultures display a long lag phase upon outgrowth from saturated cultures and long doubling times during exponential growth, especially at temperatures below 30°.[19] Moreover, colonies tend to be small in size and to contain anucleate and/or filamentous cells.[19,20] Some of these phenotypes are genetically unstable,[20] indicating the rapid accumulation of suppressor mutations. Despite these unsubtle alterations in the physiology of cells missing HU protein, no specific and dramatic defects in macromolecular metabolism have been reported. Chromosomal replication initiates at *ori*C in HU⁻ strains.[160] Moreover, evidence that HU regulates promoters other than its own[76] is negative.[20,76] There appear to be no studies of the in vivo role of HU on post-transcriptional events. This is surprising because HU has a significant capacity to interact with RNA in vitro (reviewed in ref. 30). Indeed, in addition to its discovery as a factor influencing in vitro transcription,[12] the protein was discovered independently as a nonstoichometric component of ribosomes and as a tRNA binding protein.[30] Measurement of messenger lifetime and/or in vivo translation efficiency might be useful in understanding the altered cellular growth behavior of HU mutants.

Although it would be satisfying to identify a specific defect responsible for the phenotypes of HU mutants, it is also possible that the observed behavior of these mutants is the sum of many small effects. In this regard, it is important to consider the suggested role of HU as an aid to protein recognition in situations that require DNA flexibility.[161] Specifically, HU has been reported to lower the concentration of Lac repressor, CRP protein, and IHF protein required to bind to their canonical sites.[161,162] If this phenomenon indeed reflects a general property of HU, such as its capacity to distort DNA, the protein could potentially be involved in hundreds of DNA transactions. On the other hand, it is hard to judge the physiological relevance of the published findings since the in vivo behavior of the systems in question have not been compared in HU⁺ and HU⁻ strains. Moreover, the published studies leave open the question of whether the complex whose formation is affected by HU incorporates that protein along with the specific binding protein. If it does not, HU would have to work as a "DNA chaperone"[163] perhaps by prebending the DNA to which the specific proteins bind. For such a chaperone activity to significantly affect complex formation, the process by which proteins bind to their targets must be rate-determining. This is not usually the case, at least for the typical kind of in vitro experiment used to implicate HU as a facilitator of protein-DNA interactions. It might be argued that, by changing the rates of competing pathways for association between protein and DNA, a DNA chaperone would let a specific protein form a complex with a given target that is structurally distinct from the one formed in the absence of the chaperone. As far as we know, there is no evidence supporting this possibility in the cases reported above for HU. Perhaps HU is not a DNA chaperone but is incorporated along with the specific protein, presumably altering the stability of the resulting complex (i.e., a thermodynamic effect) and not just its rate of formation (a kinetic effect). One could then invoke the capacity of HU to stabilize loops between multiple copies of the specific protein to account for the effect on complex formation. But, for Lac repressor this possibility has been discounted[161] and it would be unprecedented for CRP and IHF since these proteins have not been reported to form higher multimers. Alternatively, HU could stabilize complexes by making direct contact with the specific protein. While this possibility is plausible for HU modulation of IHF binding[162] (since HU and IHF are homologs and other members of the family tend to make linear array),[164] it is less likely to be a general property that would extend to Lac repressor and CRP. Finally, HU might facilitate binding of the specific protein by preferentially occupying nonspecific DNA that flanks the specific site. In this sense, HU would act as an element of bacterial chromatin (see below). In summary, much work needs to be done before one can understand whether HU assists protein-DNA complexes solely by assisting loop formation or whether there is a more general role for HU involving the modulation of DNA flexibility.

C. Phenotypes of Strains with Multiple Deficiencies

When mutations that eliminate HU or IHF function are combined with each other, cells become much sicker. Most dramatically, cells that are both HU$^-$ and IHF$^-$ cannot be grown at temperatures greater than 40° and, even at the optimal temperature, colonies are very small and doubling time in liquid cultures is slow.[165] These synergistic effects probably are the in vivo counterpart of the partial substitution of IHF and HU for each other that was described above for several in vitro systems. In general, although loss of either HU or IHF produces only a modest phenotype under most laboratory growth conditions, the importance of these proteins often becomes more obvious when other functions are damaged. For example, combination of IHF mutants with alleles of gyrase produces strains with new properties,[166] and mutants that render IHF$^-$ strains inviable have been isolated.[167] Most dramatic is the claim that the triple knockout involving HU, IHF and HNS cannot be constructed, even under conditions where each of the double knockouts is viable.[168] (This is especially surprising because, in HU$^+$ cells, IHF and H-NS seem to counteract each other in the sense that, in two different systems[169,170] the double knockout restores a function that is missing in the IHF single mutant.)

At present we do not know whether one particular cellular function is responsible for the increased defects in multiple mutants or whether these defects represent the incremental accumulation of altered cellular transactions. Two approaches seem to offer good ways to answer this question. First is the construction of conditional mutants of IHF and HU so that acute changes in cell physiology can be monitored following removal of the functional protein. This can be accomplished either by the construction of strains in which expression of IHF and/or HU is under the investigator's control[75,167] or by the construction of mutants with conditional function. The latter method has been pioneered by Sayre and Geiduschek,[171] who made changes in an HU family member (the TF1 protein of bacteriophage SPO1) that were designed to render folding of the protein sensitive to thermal denaturation. That one of their mutants indeed displays a temperature-sensitive phenotype encourages the wider application of this technique. The second approach is to analyze genetic suppressors than can ameliorate the slow growth and/or temperature-restricted phenotype of some multiply mutant strains.

D. Bacterial Chromatin

In eukaryotes, DNA is organized into chromatin, a form that serves to compact the genetic material and render it differentially accessible for transcription and replication.[172] The fundamental building block of eukaryotic chromatin is the nucleosome, a bead-like structure in which DNA is wrapped around the surface of an octamer of histone protein subunits.[173] Eukaryotic cells have a variety of ways to discourage nucleosome occupancy at specific places in the genome, thereby providing the differential accessibility that is essential for developmental processes.[174] Since compaction and accessibility of genetic material are likely to be important for all life forms, it has long been suspected that bacteria have some analogy of eukaryotic chromatin. Bacterial proteins that, like eukaryotic histones, are abundant nonspecific DNA binders have been traditional candidates for the key organizers of such bacterial chromatin.[1] How do HU and IHF fit into this picture?

If a stable regular structure serves as the organizer of bacterial chromatin, it has yet to be discovered. Thus, in the strictest sense, neither HU, IHF nor any other bacterial protein can be histone-like; not only do they fail to share an obvious sequence relationship with eukaryotic histones (for an exception see ref. 175), they cannot be identified as contributing to a isolatable, bead-like structure. Moreover, there is no evidence that the nonspecific binding by HU and IHF would tend to leave important DNA elements preferentially accessible or could be induced to do so by some cellular apparatus. As an extreme alternative to the existence of bacterial histones, one might imagine that the compaction of DNA that is needed to make the *E. coli* chromosome fit into the confines of the bacterial cell is accomplished primarily by small intracellular molecules like magnesium ions, spermidine, etc.[176] According to this view, HU and IHF interactions with DNA merely accomplish a set of specific tasks involving local deformation of elements like promoters, the replication origin, etc. However, since there are tens of thousands of copies of these two proteins,[75] if the majority of them were occupied in this way, the radical view might merge with the more traditional one. Indeed, examination of the state of the nucleoid by fluorescence microscopy indicates that HU is essential for normal compaction.[20,147] However, it is not clear whether the broadening of the nucleoid in cells deficient for HU represents a direct effect of removal of the protein or is an indirect consequence reflecting the dependence of the synthesis of some primary compacting agent on it. To decide whether HU and/or IHF are important contributors to general DNA compaction, it is essential to know the disposition of these proteins within the cell. However, on this point there is much uncertainty. Electron microscopic observations indicate that most of the HU in the cell is at the periphery of the nucleoid, perhaps associated with the translational apparatus.[176] But studies of the distribution of fluorescently tagged HU in permeabilized cells indicate an even distribution throughout the nucleoid.[177] There appear to be no comparable studies for IHF. A guideline for evaluating the importance of HU and IHF for nucleoid structure might be provided by examining the distribution of these proteins in minicells.[178] Minicells are anucleate bodies that contain a normal complement of cytosol, ribosomes, tRNA, etc. but no nucleoid. From the concentration of HU and IHF found in minicells, one should be able to calculate the proportion of these proteins present in the cytosol of whole cells and thus deduce the proportion associated with the nucleoid. This should let one decide whether HU and IHF are present in the nucleoid in sufficient quantities to significantly assist with the process of DNA compaction.

Acknowledgments

I thank G. Chaconas and B. Lavoie for reading an early version of this manuscript and N. Kleckner for her extensive comments on the penultimate version. I am grateful to M. Werner for help in preparing Figure 8.2, and A. Segall and J. Aarons, respectively, for help in the design and execution of Figure 8.4. Finally, I thank Z. Zanata and M. Christensen for the patient and expert preparation of this manuscript.

REFERENCES

1. Drlica K, Rouviére-Yaniv J. Histonelike proteins of bacteria. Microbiol Rev 1987; 51:301-319.

2. Friedman DI. Integration host factor: A protein for all reasons. Cell 1988; 55:545-554.

3. Pettijohn DE. Histone-like proteins and bacterial chromosome structure. J Biol Chem 1988; 263:12793-12796.

4. Schmid MB, Johnson RC. Southern revival—news of bacterial chromatin. The New Biologist 1991; 3:945-950.

5. Freundlich M, Ramani N, Mathew E et al. The role of integration host factor in gene expression in *Escherichia coli*. Mol Microbiol 1992; 6:2557-2563.

6. Oberto J, Drlica K, Rouviére-Yaniv J. Histones, HMG, HU, IHF: Même combat. Biochimie 1994; 76:901-908.

7. Finkel SE, Johnson RC. Mol Microbiol 1992; 6:3257-3265.

8. Crothers DM, Steitz TA. Transcriptional activation by *Escherichia coli* CAP protein. In: McKinght SL, Yamamoto KR, eds. Transcriptional Regulation. Cold Spring Harbor: Cold Spring Harbor Press, 1992:501-534.

9. Ussery DW, Hinton JCD, Jordi BJAM et al. The chromatin-associated protein H-NS. Biochimie 1994; 76:968-980.

10. Calvo JM, Matthews RG. The leucine-responsive regulatory protein, a global regulator of metabolism in *Escherichia coli*.. Microbiol Rev 1994; 58:466-490.

11. Kolb A, Busby S, Buc H et al. Transcriptional regulation by cAMP and its receptor protein. Annu Rev Biochem 1993; 62:749-795.

12. Rouviére-Yaniv J, Gros F. Characterization of a novel, low-molecular-weight DNA-binding protein from *Escherichia coli*.. Proc Natl Acad Sci USA 1975; 72:3428-3432.

13. Nash HA, Mizuuchi K, Weisberg RA et al. Integrative recombination of bacteriophage lambda-the biochemical approach to DNA insertion. In: Bukhari AI, Shapiro JA, Adhya SL, eds. DNA insertion elements, plasmids, and episomes. Cold Spring Harbor: Cold Spring Harbor Press, 1977:363-373.

14. Miller HI, Kikuchi A, Nash HA et al. Site-specific recombination of bacteriophage λ: the role of host gene products. Cold Spring Harb Symp Quant Biol 1979; 43:1121-1126.

15. Craigie R, Arndt-Jovin DJ, Mizuuchi K. A defined system for the DNA strand-transfer reaction at the initiation of bacteriophage Mu transposition: protein and DNA substrate requirements. Proc Natl Acad Sci USA 1985; 82:7570-7574.

16. Geiduschek EP, Schneider GJ, Sayre MH. TF1, a bacteriophage-specific DNA-binding and DNA bending protein. J Struc Biol 1990; 104:84-90.

17. Neilan JG, Lu Z, Kutish GF et al. An African swine fever virus gene with similarity to bacterial DNA binding proteins, bacterial integration host factors, and the *Bacillus* phage SPO1 transcription factor, TF1. Nucleic Acids Res 1993; 21:1496.

18. Rouviére-Yaniv J, Kjeldgaard NO. Native *Escherichia coli* HU protein is a heterotypic dimer. FEBS Letters 1979; 106:297-300.

19. Wada M, Kano Y, Ogawa T et al. Construction and characterization of the deletion mutant of *hupA* and *hupB* genes in *Escherichia coli*. J Mol Biol 1988; 204:581-591.

20. Huisman O, Faelen M, Girard D et al. Multiple defects in *Escherichia coli* mutants lacking HU protein. J Bacteriol 1989; 171:3704-3712.

21. Werner MH, Clore GM, Gronenborn AM et al. Symmetry and asymmetry in the function of *Escherichia coli* integration host factor: Implications for target identification by DNA-binding proteins. Curr Biol 1994; 4: 477-487.

22. Zulianello L, de la Gorgue de Rosny E, van Ulsen P et al. The HimA and HimD subunits of integration host factor can specifically bind to DNA as homodimers. EMBO J 1994; 13:1534-1540.

23. Tanaka I, Appelt K, Dijk J et al. 3-Å resolution structure of a protein with histone-like properties in prokaryotes. Nature 1984; 310:376-381.

24. White SW, Appelt K, Wilson KS et al. A protein structural motif that bends DNA. Proteins: Struct, Funct, and Genet 1989; 5:281-288.

25. Yang CC, Nash HA. The interaction of *E. coli* IHF protein with its specific binding sites. Cell 1989; 57:869-880.

26. Lee EC, Hales LM, Gumport RI et al. The isolation and characterization of mutants of the integration host factor (IHF) of *Escherichia coli* with altered, expanded DNA-binding specificities. EMBO J 1992; 11:305-313.

27. Granston AE, Nash HA Characterization of a set of integration host factor mutants deficient for DNA binding. J Mol Biol 1993; 234:45-59.

28. Yang S, Nash H. Specific photocrosslinking of DNA-protein complexes: identification of contacts between integration host factor and its target DNA. Proc Natl Acad Sci USA 1994; 91;12183-12187.

29. Vis H, Boelens R, Mariani M et al. ^{1}H, ^{13}C, and ^{15}N resonance assignments and secondary structure analysis of the HU protein from *Bacillus stearothermophilus* using two- and three- dimensional double- and triple-resonance heteronuclear magnetic resonance spectroscopy. Biochem 1994; 33:14858-14870.

30. Gualerzi CO, Losso MA, Lammi M et al. Proteins from the prokaryotic nucleoid. Structural and functional characterization of the *Escherichia coli* DNA-binding proteins NS (HU) and H-NS. In: Gualerzi CO, Pon CL, eds. Bacterial Chromatin. Heidelberg: Springer-Verlag, 1986:101-134.

31. Mendelson I, Gottesman M, Oppenheim AB. HU and integration host factor function as auxiliary proteins in cleavage of phage lambda cohesive ends by terminase. J Bacteriol 1991; 173:1670-1676.

32. Skorupski K, Sauer B, Sternberg N. Faithful cleavage of the P1 packaging site (*pac*) requires two phage proteins, PacA and PacB, and two *Escherichi coli* proteins, IHF and HU. J Mol Biol 1994; 243:268-282.

33. Broyles SS, Pettijohn DE. Interaction of the *Escherichia coli* HU protein with DNA. Evidence for formation of nucleosome-like structures with altered DNA helical pitch. J Mol Biol 1986; 187:47-60.

34. Bonnefoy E, Takahashi M, Rouviere-Yaniv J. DNA-binding parameters of the HU protein of *Escherichia coli* to cruciform DNA. J Mol Biol 1994; 242:116-129.

35. Bonnefoy E, Almeida A, Rouviere-Yaniv J. Lon-dependent regulation of the DNA binding protein HU in *Escherichia coli*. Proc Natl Acad Sci USA 1989; 86:7691-7695.

36. Shindo H, Furubayashi A, Shimizu M et al. Preferential binding of *E. coli* histone-like protein HUα to negatively supercoiled DNA. Nucleic Acids Res 1992; 20:1553-1558.

37. Tanaka H, Goshima N, Kohno K et al. Properties of DNA-binding of HU heterotypic and homotypic dimers from *Escherichia coli*. J Biochem 1993; 113:568-572.

38. Winkelman JW, Hatfield GWJ. Characterization of the integration host factor binding site in the *ilv*P$_G$1 promoter region of the *ilvGMEDA* operon of *Escherichia coli*. J Biol Chem 1990; 265:10055-10060.

39. Kur J, Hasan N, Szybalski W. Physical and biological consequences of interactions between integration host factor (IHF) and coliphage lambda late *p'*$_R$ promoter and its mutants. Gene 1989; 81:1-15.

40. Gardner JF, Nash HA. Role of *Escherichia coli* IHF protein in lambda site-specific recombination. A mutational analysis of binding sites. J Mol Biol 1986; 191:181-189.

41. Prentki P, Chandler M, Galas DJ. *Escherichia coli* integration host factor

bends the DNA at the ends of IS*1* and in an insertion hotspot with multiple IHF binding sites. EMBO J 1987; 6:2479-2487.

42. Mengeritsky G, Goldenberg D, Mendelson I et al. Genetic and biochemical analysis of the integration host factor of *Escherichia coli*. J Mol Biol 1993; 231:646-657.

43. Yang S-W, Nash HA. Comparison of protein binding to DNA *in vivo* and *in vitro*: defining an effective intracellular target. EMBO J 1995; (in press)..

44. Wang S, Cosstick R, Gardner JF et al. The specific binding of integration host factor involves both major and minor grooves of DNA. Biochemistry 1995; 34:13082-13090.

45. Morse BK, Michalczyk R, Kosturko LD. Multiple molecules of integration host factor (IHF) at a single DNA binding site, the bacteriophage λ *cos* I1 site. Biochimie 1994; 76:1005-1018.

46. Goodrich JA, Schwartz ML, McClure WR. Searching for and predicting the activity of sites for DNA binding proteins: compilation and analysis of the binding sites for *Escherichia coli* integration host factor (IHF). Nucleic Acids Res 1990; 18: 4993-5000.

47. Boccard F, Prentki P. Specific interaction of IHF with RIBs, a class of bacterial repetitive DNA elements located at the 3' end of transcription units. EMBO 1993; 12:5019-5027.

48. Oppenheim AB, Rudd KE, Mendelson I et al. Integration host factor binds to a unique class of complex repetitive extragenic DNA sequences in *Escherichia coli*. Mol Microbiol 1993; 10:113-122.

49. Bachellier S, Perrin D, Hofnung M et al. Bacterial interspersed mosaic elements (BIMEs) are present in the genome of *Klebsiella*. Mol Microbiol 1993; 7:537-544.

50. Boffini A, Prentki P. Identification of protein binding sites in genomic DNA by two-dimensional gel electrophoresis. Nucleic Acids Res 1991; 19:1369-1374.

51. Rouviére-Yaniv J, Yaniv M. *E. coli* DNA binding protein HU forms nucleosome-like structure with circular double-stranded DNA. Cell 1979; 17:265-274.

52. Wang JC, Giaever GN. Action at a distance along a DNA. Science 1988; 240:300-304.

53. Hodges-Garcia Y, Hagerman PJ, Pettijohn DE. DNA ring closure mediated by protein HU. J Biol Chem 1989; 264:14621-14623.

54. Paull TT, Haykinson MJ, Johnson RC. The nonspecific DNA-binding and -bending proteins HMG1 and HMG2 promote the assembly of complex nucleoprotein structures. Genes Dev 1993; 7:1521-1534.

55. Wu HM, Crothers DM. The locus of sequence-directed and protein-induced DNA bending. Nature 1984; 308:509-513.

56. Zinkel SS, Crothers DM. Comparative gel electrophoresis measurement of the DNA bend angle induced by the catabolite activator protein. Biopolymers 1990; 29:29-38.

57. Thompson JF, Landy A. Empirical estimation of protein-induced DNA bending angles: applications to lambda site-specific recombination complexes. Nucleic Acids Res 1988; 16:9687-9705.

58. Kosturko LD, Daub E, Murialdo H. The interaction of *E. coli* integration host factor and lambda cos DNA: multiple complex formation and protein-induced bending. Nucleic Acids Res 1989; 17:317-334.

59. Yang S-M, Kahn JD, Crothers DM. 1995; (in preparation).

60. Schneider GJ, Sayre MH, Geiduschek EP. DNA-bending properties of TF1. J Mol Biol 1991; 221:777-794.

61. Kim JL, Nikolov DB, Burley SK. Co-crystal structure of TBP recognizing the minor groove of a TATA element. Nature 1993; 365:520-527.

62. Kim Y, Geiger JH, Hahn S et al. Crystal structure of a yeast TBP/TATA-box complex. Nature 1993; 365:512-520.

63. Kahn JD, Crothers DM. Protein-induced bending and DNA cyclization. Proc Natl Acad Sci USA 1992: 89:6343-6347.

64. Teter B, PhD Thesis (University of S. California, 1991).

65. Sun D, Harshey RM, Hurley LH. IHF-induced DNA bending studied by DNA cyclization. 1995; (in preparation).

66. Pontiggia A, Negri A, Beltrame M et al. Protein HU binds specifically to kinked DNA. Mol Microbiol 1993; 7:343-350.

67. Castaing B, Zelwer C, Laval J et al. HU protein of *Escherichia coli* binds specifically to DNA that contains single-strand breaks or gaps. J Biol Chem 1995; 270:10291-10296.

68. Craig NL, Nash HA. *E. coli* integration host factor binds to specific sites in DNA. Cell 1984; 39:707-716.

69. Zulianello L, van Ulsen P, van de Putte P et al. Participation of the flank regions of the IHF protein in the specificity and stability of DNA-binding. J Biol Chem 1995; 270:17902-17907.

70. Lavoie BD, Chaconas G. Site-specific HU binding in the Mu transpososome: conversion of a sequence-independent DNA-binding protein into a chemical nuclease. Genes Dev 1993; 7:2510-2519.

71. Neidhardt FC. Multigene Systems and Regulons. In: Neidhardt FC ed. Escherichia coli and Salmonella typhimurium. Washington, DC: American Society for Microbiology, 1987:1313-1317.

72. Magasanik B, Neidhardt FC. Regulation of carbon and nitrogen utilization. In: Neidhardt FC, ed. Escherichia coli and Salmonella typhimurium. Washington DC: American Society for Microbiology, 1987:1318-1325.

73. Ball CA, Osuna R, Ferguson KC et al. Dramatic changes in Fis levels upon nutrient upshift in *Escherichia coli*. J Bacteriol 1992; 174:8043-8056.

74. Ninnemann O, Koch C, Kahmann R. The *E. coli* fis promoter is subject to stringent control and autoregulation. EMBO 1992; 11:1075-1083.

75. Ditto MD, Roberts D, Weisberg RA. Growth phase variation of integration host factor level in *Escherichia coli*. J Bacteriol 1994; 176:3738-3748.

76. Kohno K, Wada M, Kano Y et al. Promoters and autogenous control of the *Escherichia coli hupA* and *hupB* genes. J Mol Biol 1990; 213:27-36.

77. Bushman W, Thompson JF, Vargas L et al. Control of directionality in lambda site specific recombination. Science 1985; 230, 906-911.

78. Nash HA, Robertson CA, Flamm E et al. Overproduction of *Escherichia coli* integration host factor, a protein with nonidentical subunits. J Bacteriol 1987; 169:4124-4127.

79. Aviv M, Giladi H, Schreiber G et al. Expression of the genes coding for the *Escherichia coli* integration host factor are controlled by growth phase, *rpoS*, ppGpp and by autoregulation. Mol Microbiol 1994; 14:1021-1031.

80. Mechulam Y, Blanquet S, Fayat GJ. Dual level control of the *Escherichia coli pheST-himA* operon expression tRNA^{Phe}-dependent attenuation and transcriptional operator-repressor control by *himA* and the SOS network. J Mol Biol 1987; 197:453-470.

81. Flamm EL, Weisberg RA. Primary structure of the *hip* gene of *Escherichia coli* and of its product, the β subunit of integration host factor. J Mol Biol 1985; 183:117-128.

82. Miller HI, Kirk M, Echols H. SOS induction and autoregulation of the *himA* gene for site-specific recombination in *Escherichia coli*. Proc Natl Acad Sci USA 1981; 78:6754-6758.

83. Peterson KR, Mount DW. Differential repression of SOS genes by unstable LexA41 (Tsl-1) protein causes a "split-phenotype" in *Escherichia coli* K-12. J Mol Biol 1987; 193:27-40.

84. Mahan MJ, Slauch JM, Mekalanos JJ. Selection of bacterial virulence genes that are specifically induced in host tissues. Science 1993; 259:686-688.

85. Neidhardt FC. Chemical composition of *Escherichia coli*. In: Neidhardt FC ed. Escherichia coli and Salmonella typhimurium. Washington DC: American Society for Microbiology, 1987:3-6.

86. von Hippel PH, Revzin A, Gross CA et al. Non-specific DNA binding of genome regulating proteins as a biological control mechanism: 1. The *lac* operon: equilibrium aspects. Proc Nat Acad Sci USA 1974; 71:4808-4812.

87. Lin S-Y, Riggs AD. The general affinity of *lac* repressor for E. coli DNA: implications for gene regulation in procaryotes and eucaryotes. Cell 1975; 4:107-111.

88. Stickle DF, Vossen KM, Riley DA et al. Free DNA concentration in *E. coli* estimated by an analysis of competition for DNA binding proteins. J Theor Biol 1994; 168:1-12.

89. Grosschedl R, Giese K, Pagel J. HMG domain proteins: Architectural elements in the assembly of nucleoprotein structures. Trends Genet 1994; 10:94-100.

90. Richet E, Abcarian P, Nash HA. The interaction of recombination proteins with supercoiled DNA: defining the role of supercoiling in lambda integrative recombination. Cell 1986; 46:1011-1021.

91. Robertson CA., Nash HA. Bending of the bacteriophage λ attachment site by *Escherichia coli* integration host factor. J Biol Chem 1988; 263:3554-3557.

92. Goodman SD, Nash HA. Functional replacement of a protein-induced bend in a DNA recombination site. Nature 1989; 341:251-244.

93. Goodman SD, Nicholson SC, Nash HA. Deformation of DNA during site-specific recombination of bacteriophage λ: replacement of IHF protein by HU protein or sequence-directed bends. Proc Natl Acad Sci USA 1992; 89:11910-11914.

94. Moitoso de Vargas L, Kim S, Landy A. DNA looping generated by DNA bending protein IHF and the two domains of lambda integrase. Science 1989; 244:1457-1461.

95. Kim S, de Vargas LM, Nunes-Düby SE et al. Mapping of a higher order protein-DNA complex: two kinds of long-range interactions in λ attL. Cell 1990; 63:773-781.

96. Nash HA. Bending and supercoiling of DNA at the attachment site of bacteriophage λ. Trends Biochem Sci 1990; 15:222-227.

97. Landy A. Mechanistic and structural complexity in the site-specific recombination pathways of Int and FLP. Curr Opin Genet Devel 1993; 3:699-707.

98. Tang RS, Draper DE. Bulge loops used to measure the helical twist of RNA in solution. Biochemistry 1990; 29:5232-5237.

99. Kustu S, Santero E, Keener J et al. Expression of σ^{54} (*ntrA*)-dependent genes is probably united by a common mechanism. Microbiol Rev 1989; 53:367-376.

100. Hoover TR, Santero E, Porter S et al. The integration host factor stimulates interaction of RNA polymerase with NIFA, the transcriptional activator for nitrogen fixation operons. Cell 1990; 63:11-22.

101. Claverie-Martin F, Magasanik BJ. Positive and negative effects of DNA bending on activation of transcription from a distant site. J Mol Biol 1992; 227:996-1008.

102. Pérez-Martin J, Timmis KN, de Lorenzo VJ. Co-regulation by bent DNA. J Biol Chem 1994; 269:22657-22662.

103. Molina-López J, Govantes F, Santero EJ. Geometry of the process of transcription activation at the σ54-dependent nifH promoter of *Klebsiella pneumoniae*. J Biol Chem 1994; 269:25419-25425.

104. Krause HM, Higgins NP. Positive and negative regulation of the Mu operator by Mu repressor and *Escherichia coli* integration host factor. J Biol Chem 1986; 261:3744-3752.

105. van Rijn PA, van de Putte P, Goosen, N. Analysis of the IHF binding site in the regulatory region of bacteriophage Mu. Nucleic Acids Res 1991; 19:2825-2834.

106. Alazard R, Bétermier M, Chandler M. *Escherichia coli* integration host factor stabilizes bacteriophage Mu repressor interactions with operator DNA in vitro. Mol Microbiol 1992; 6:1707-1714.

107. Betermier M, Rousseau P, Alazard R et al. Mutual stabilisation of bacteriophage Mu repressor and histone like proteins in a nucleoprotein structure. J Mol Biol 1995; 249:332-341.

108. Mizuuchi M, Mizuuchi K. Efficient Mu transposition requires interaction of transposase with a DNA sequence at the Mu operator: implications for regulation. Cell 1989; 58:399-408.

109. Lleung PC, Teplow DB, Harshey. Interaction of distinct domains in Mu transposase with Mu DNA ends and an internal transpositional enhancer. Nature 1989; 338:656-658.

110. Surette MG, Chaconas G. The Mu transpositional enhancer can function in trans: requirement of the enhancer for synapsis but not strand cleavage. Cell 1992; 68:1101-1108.

111. Surette MG, Lavoie BD, Chaconas G. Action at a distance in Mu DNA transposition: an enhancer-like element is the site of action of supercoiling relief activity by integration host factor (IHF). EMBO J 1989; 8:3483-3489.

112. Vologodskii AV, Cozzarelli NR. Conformational and thermodynamic properties of supercoiled DNA. Ann Rev Biophys Biomol Struct 1994; 23:609-643.

113. Kornberg A, Baker TA. DNA replication. Second Edition. New York: WH Freeman & Co, 1992.

114. Polaczek P. Bending of the origin of replication of *E. coli* by binding of IHF at a specific site. The New Biologist 1990; 2:265-271.

115. Filutowicz M, Roll J. The requirement of IHF protein for extrachromosomal replication of the *Escherichia coli oriC* in a mutant deficient in DNA polymerase I activity. The New Biologist 1990; 2:818-827.

116. Roth A, Urmoneit B, Messer W. Functions of histone-like proteins in the initiation of DNA replication at *oriC* of *Escherichia coli*. Biochimie 1994; 6:917-923.

117. Kano Y, Ogawa T, Ogura T et al. Participation of the histone-like protein HU and of IHF in minichromosomal maintenance in *Escherichia coli*. Gene 1991; 103:25-30.

118. Hwang DS, Kornberg AJ. Opening of the replication origin of *Escherichia coli* by DnaA protein with protein HU or IHF. J Biol Chem 1992; 267:23083-23086.

119. Cassler MR, Grimwade JE, Leonard AC. Cell cycle-specific changes in nucleoprotein complexes at a chromosomal replication origin. EMBO J 1995; (in press).

120. Lu M, Campbell JL, Boye E et al. SeqA: a negative modulator of replication initiation in E. coli. Cell 1994; 77:413-426.

121. Boye E, Lyngstadaas A, Løbner-Olesen A et al. Regulation of DNA replication in *Escherichia coli*. In: Fanning E, Knippers R, Winnacher E-L, eds. DNA Replication and the Cell Cycle. Heidelberg: Springer-Verlag, 1992; 15-26.

122. Kleckner N, Chalmers R, Kwon D et al. Tn10 and IS10 transposition and chromosome rearrangements: mechanism and regulation in vivo and in vitro. In: Saedler H, Gierl A, eds. Transposable Elements. Current Topics in Microbiology and Immunology. New York: Springer-Verlag, 1995; (in press).

123. Morisato D, Kleckner N. Tn10 transposition and circle formation in vitro. Cell 1987; 51:101-111.

124. Chalmers RM, Kleckner NJ. Tn*10*/IS*10* Transposase purification, activation, and in vitro reaction. J Biol Chem 1994; 269:8029-8035.

125. Signon L, Kleckner N. Negative and positive regulation of Tn10/IS10-promoted recombination by IHF: two distinguishable processes inhibit transposition off of multicopy plasmid replicons and activate chromosomal events that favor evolution of new transposons. Genes Dev 1995; 9:1123-1136.

126. Goosen N, van de Putte P. The regulation of transcription initiation by integration host factor. Mol Microbiol 1995; 16:1-7.

127. Pereira RF, Ortuno MJ, Lawther RP. Binding of integration host factor (IHF) to the *ilvGp1* promoter of the *ilvGMEDA* operon of *Escherichia coli* K12. Nucleic Acids Res 1988; 16:5973-5989.

128. Pagel JM, Winkelman JW, Adams CW et al. DNA topology-mediated regulation of transcription initiation from the tandem promoters of the *ilvGMEDA* operon of *Escherichia coli.* J Mol Biol 1992; 224:919-935.

129. Huisman O, Errada PR, Signon L et al. Mutational analysis of IS10's outside end. EMBO J 1989; 8:2101-2109.

130. Giladi H, Igarashi K, Ishihama A et al. Stimulation of the phage λ pL promoter by integration host factor requires the carboxy terminus of the α-subunit of RNA polymerase. J Mol Biol 1992; 227:985-990.

131. Giladi H, Gottesman M, Oppenheim AB. Integration host factor stimulates the phage lambda pL promoter. J Mol Biol 1990; 213:109-121.

132. Giladi H, Koby S, Gottesman ME et al. Supercoiling, integration host factor, and a dual promoter system, participate in the control of the bacteriophage λ pL promoter. J Mol Biol 1992; 224:937-948.

133. Higgins NP, Collier DA, Kilpatrick MW et al. Supercoiling and integration host factor change the DNA conformation and alter the flow of convergent transcription in phage mu. J Biol Chem 1989; 264:3035-3042.

134. van Rijn PA, Goosen N, van de Putte P. Integration host factor of *Escherichia coli* regulates early and repressor transcription of bacteriophage Mu by two different mechanisms. Nucleic Acid Res 1988; 16:4595-4605.

135. Mizuuchi K. Transpositional recombination: mechanistic insights from studies of Mu and other elements. Annu Rev Biochem 1992; 61:1011-1051.

136. Mizuuchi M, Baker T, Mizuuchi K. Assembly of the active form of the transposase-Mu DNA complex: a critical control point in Mu transposition. Cell 1992; 70:303-311.

137. Surette MG, Chaconas GJ. A protein factor which reduces the negative supercoiling requirment in the Mu DNA strand transfer reaction is *Escherichia coli* integration host factor. J Biol Chem 1989; 264:3028-3034.

138. Lavoie BD, Chaconas GJ. Immunoelectron microscopic analysis of the A, B, and HU protein content of bacteriophage Mu transpososomes. J Biol Chem 1990; 265:1623-1627.

139. Lavoie BD, Chaconas GJ. A second high affinity HU binding site in the phage Mu transpososome. J Biol Chem 1994; 269:15571-15576.

140. Johnson R. Mechanism of site-specific DNA inversion in bacteria. Curr Opin Genet Devel 1991; 1:412-416.

141. Johnson RC, Bruist MF, Simon MI. Host protein requirements for in vitro site-specific DNA inversion. Cell 1986; 46:531-539.

142. Paull TT, Haykinson MJ, Johnson RC. HU and functional analogs in eukaryotes promote Hin invertasome assembly. Biochimie 1994; 76:992-1004.

143. Haykinson MJ, Johnson RC. DNA looping and the helical repeat in vitro and in vivo: effect of Hu protein and enhancer location on Hin invertasome assembly. EMBO J 1993; 12:2503-2512.

144. Hillyard DR, Edlund M, Hughes KT et al. Subunit-specific phenotypes of *Salmonella typhimurium* HU mutants. J Bacteriol 1990; 172:5402-5407.

145. Wada M, Kutsukake K, Komano T et al. Participation of the *hup* gene product in site-specific DNA inversion in *Escherichia coli*. Gene 1989; 76:345-352.

146. Koch C, Kahmann RJ. Purification and properties of the *Escherichia coli* host factor required for inversion of the G segment in bacteriophage Mu. J Biol Chem 1986; 261:15673-15678.

147. Paull TT, Johnson RC. DNA looping by *Saccharomyces cerevisiae* high mobility group proteins NHP6A/B. J Biol Chem 1995; 270:8744-8754.

148. Bianchi ME. Prokaryotic HU and eukaryotic HMG1: a kinked relationship. Mol Microbiol 1994; 4:1-5.

149. Segall AM, Goodman SD, Nash HA. Architectural elements in nucleoprotein complexes. EMBO J 1994; 13:4536-4548.

150. Nash HA, Robertson CA. Purification and properties of the *Escherichia coli* protein factor required for λ integrative recombination. J Biol Chem 1981; 256:9246-9253.

151. Gamas P, Burger AC, Churchward G et al. Replication of pSC101: effects of mutations in the *E. coli* DNA binding protein IHF. Mol Gen Genet 1986; 204:85-89.

152. Stenzel TT, Patel P, Bastia D. The integration host factor of *Escherichia coli* binds to bent DNA at the origin of replication of the plasmid pSC101. Cell 1987; 49:709-717.

153. Hoyt MA, Knight DM, Das A et al. Control of phage λ development by stability and synthesis of cII protein: role of the viral cIII and host *hflA*, *himA* and *himD* genes. Cell 1982; 31:565-573.

154. Mahajna J, Oppenheim AB, Rattray A et al. Translation initiation of bacteriophage lambda gene cII requires integration host factor. J Bacteriol 1986; 165:167-174.

155. Rabin RS, Collins LA, Stewart V. In vivo requirement of integration host factor for *nar* (nitrate reductase) operon expression in *Escherichia coli* K-12. Proc Natl Acad Sci USA 1992; 89:8701-8705.

156. Schröder I, Darie S, Gunsalus RP. Activation of the *Escherichia coli* nitrate reductase (*narGHJI*) operon by *NarL* and *Fnr* requires integration host factor. J Biol Chem 1993; 268:771-774.

157. Wu Y, Datta PJ. Integration host factor is required for positive regulation of the *tdc* operon of *Escherichia coli*. J Bacteriol 1992; 174:233-240.

158. Weiss DS, Klose KE, Hoover TR et al. Prokaryotic transcriptional enhancers. In: McKinght SL, Yamamoto KR eds. Transcriptional Regulation. Cold Spring Harbor: Cold Spring Harbor Press, 1992:667-694.

159. Pérez-Martín J, de Lorenzo V. Integration host factor (IHF) suppresses promiscuous activation of the σ54-dependent promoter *Pu* of *Pseudomaonas putita*. Proc Natl Acad Sci USA 1995; 92:7277-7281.

160. Ogawa T, Wada M, Kano Y et al. DNA replication in *Escherichia coli* mutants that lack protein HU. J Bacteriol 1989; 171:5672-5679.

161. Flashner Y, Gralla JD. DNA dynamic flexibility and protein recognition: Differential stimulation by bacterial histone-like protein HU. Cell 1988; 54:713-721.

162. Bonnefoy E, Rouviére-Yaniv J. HU, the major histone-like protein of *E. coli*, modulates the binding of IHF to *oriC*. EMBO J 1992; 11:4489-4496.

163. Travers AA, Ner SS, Churchill, MEA. DNA chaperones: a solution to a persistence problem. Cell 1994; 77:167-169.

164. Schneider GJ, Geiduschek EP. Stoichiometry of DNA binding by the bacteriophage SPO1-encoded type II DNA-binding protein TF1. J Biol Chem 1990; 265:10198-10200.

165. Kano Y, Imamoto F. Requirement of integration host factor (IHF) for growth of *Escherichia coli* deficient in HU protein. Gene 1990; 89:133-137.

166. Friedman DI, Olson EJ, Carver D et al. Synergistic effect of *himA* and *gyrB* mutations: evidence that Him functions control expression of *ilv* and *xyl* genes. J Bacteriol 1984; 157:484-489.

167. Roberts DE. Genetic analysis of mutants of *E. coli* affected for Tn10 transposition. (Harvard University, 1986).

168. Yasuzawa K, Hayashi N, Goshima N et al. Histone-like proteins are required for cell growth and constraint of supercoils in DNA. Gene 1992; 122:9-15.

169. Goshima N, Kano Y, Tanaka H et al. IHF supresses the inhibitory effect of H-NS on HU function in the *hin* inversion system. Gene 1994; 141:17-23.

170. Kano Y, Yasuzawa K, Tanaka H et al. Propagation of phage Mu in IHF-deficient *Escherichia coli* in the absence of the H-NS histone-like protein. Gene 1993; 126:93-97.

171. Sayre MH, Geiduschek EP. Construction and properties of a temperature-sensitive mutation in the gene for the bacteriophage SPO1 DNA-binding protein TF1. J Bacteriol 1990; 172:4672-4681.

172. Hayes JJ, Wolffe AP. The interaction of transcription factors with nucleosomal DNA. BioEssays 1992; 14:597-603.

173. Ramakrishnan V. Histone structure. Curr Opin Struct Biol 1994; 4:44-50.

174. Wallrath LL, Lu Q, Granok H et al. Architectural variations of inducible eukaryotic promoters: preset and remodeling chromatin structures. Bioessays 1994; 16:165-170.

175. Sandman K, Perler FB, Reeve JN. Histone-encoding genes from Pyrococcus: evidence for members of the HMF family of archaeal histones in a non-methanogenic Archaeon. Gene 1994; 150:207-208.

176. Dürrenberger M, Bjornsti MA, Uetz T et al. J Bacteriol 1988; 170:4757-4768.

177. Shellman VL, Pettijohn DE. Introduction of proteins into living bacterial cells: distribution of labeled HU protein in *Escherichia coli*. J Bacteriol 1991; 173:3047-3059.

178. Revzin A. Techniques for characterizing nonspecific DNA-protein interactions. In: Reuzin A, ed. The Biology of Nonspecific DNA-Protein Interactions. Ann Arbor: CRC Press, 1990:6-31.

179. Toussaint B, Delic-Attree I, De Sury d'Aspremont R et al. Purification of the integration host factor homolog of *Rhodobacter capsulatus*: cloning and sequencing of the *hip* gene, which encodes the β subunit. J Bacteriol 1993; 175:6499-6504.

180. Kim S, Landy A. Lambda Int protein bridges between higher order complexes at two distant chromosomal loci *att*L and *att*R. Science 1992; 256:198-203.

181. Richet E, Abcarian P, Nash HA. Synapsis of attachment sites during lambda integrative recombination involves capture of a naked DNA by a protein-DNA complex. Cell 1988; 52:9-17.

182. Pérez-Martín J, Rojo F, de Lorezo V. Promoters responsive to DNA bending: A common theme in prokaryotic gene expression. Microbiol Rev 1994; 58:268-290

CHAPTER 9

THE *LAC* AND *GAL* OPERONS TODAY

Sankar Adhya

".....I have done no more than summarize what is considered established ideas in contemporary science." — Jacque Monod

1. INTRODUCTION

In 1961, Jacob and Monod proposed the operon model of gene expression and its negative control primarily from the experimental results obtained by these authors and their colleagues studying the induced synthesis of proteins involved in the utilization of sugar lactose and development of bacteriophage λ from a prophage state in *Escherichia coli*.[1] The experiments and arguments used in formulating the model revolved around genetic analysis of these two systems. To explain the results of the lactose system, they suggested the *théorie de la double négativité*. In the double negative control of the induced synthesis of lactose enzymes, in summary, a repressor protein binds to a DNA element, called an operator, and represses gene expression by *inhibiting* transcription initiation from the promoter which controls a set of contiguous structural genes (operon or transcription unit). When present, an inducer binds to the repressor *inhibiting* the operator-binding activity of the repressor by a mechanism which is called allosteric modification[2] and allows transcription initiation of the operon. By analyzing the regulation of the genes encoding enzymes of arabinose utilization (the *ara* operon) in *E. coli*, Engelsberg and his colleagues, in 1965, proposed that regulation of transcription initiation can also occur through a mechanism of positive control.[3] In this system, the regulatory protein serves the role of an activator, which is required for transcription initiation. In its absence, transcription initiates at a very low rate. As in the case of negative control, the activity of the activator can also be modulated by an allosteric modifier. The activator in the *ara* operon performs positive control only when the sugar arabinose is present. Positive control was later demonstrated to be superimposed also in the *lac* operon of *E. coli*, in which a cAMP receptor protein (CRP, also known as catabolite gene activator protein or CAP), when allosterically changed by binding cAMP, acts as an activator of transcription initiation.[4-7] The positive control of the *lac* operon by the cAMP·CRP complex also explained an earlier observation called the

Regulation of Gene Expression in Escherichia coli, edited by E. C. C. Lin and A. Simon Lynch. © 1996 R.G. Landes Company.

"glucose effect," sometimes referred to as "catabolite repression."[8,9] In the glucose effect, the expression of the *lac* operon cannot be induced by an inducer if glucose is also present in the growth media. It is now known that the intracellular level of cAMP is very low in glucose grown cells, thus not permitting full expression of the *lac* operon even in the absence of a functional repressor.[10] (For the portion of the glucose effect attributable to inducer exclusion see chapter 11). Since the proposal of the two original models of regulation of transcription initiation, negative and positive controls, regulation has also been demonstrated or perceived to occur at almost every step of gene expression. From the study of numerous bacterial and bacteriophage genes and operons, it is clear that operons are structured for such purposes. Regulation has been discovered at the level of transcription elongation and termination, at several levels of translation, as well as at the level of mRNA and protein processing. Genes are also known to be turned on and off by reversibly splitting a functional operon into incomplete parts. Such variation in the mechanisms of gene regulation makes for overall cellular economy, coordination and differentiation and minimizes chaos.

In this chapter, we will describe the regulatory features of the *lac* and galactose (*gal*)[11] operons in *E. coli*, and then provide an overview of the present state of our knowledge of the mechanisms of the superimposed positive and negative controls in the two operons. Several elegant and authoritative articles discussing the critical genetic analyses of the *lac* operon and its regulation have been published previously.[12-15] A summary of the earlier work in the *gal* operon has also been reported.[16]

2. THE *lac* AND *gal* OPERONS ENCODE ENZYMES OF A CONTINUOUS BIOCHEMICAL PATHWAY

The enzymes involved in the uptake and metabolism of lactose and galactose are shown in Figure 9.1. Galactose is one of the metabolic products of lactose. Whereas the role of the proteins for lactose uptake and metabolism—β-galactoside permease and β-galactosidase—encoded in the *lac* operon, is purely catabolic, the enzymes of galactose metabolism—galactose-1-epimerase, galactokinase, galactose-1-phosphateuridyl transferase and uridinodiphosphogalactose epimerase—encoded in the *gal* operon, play an amphibolic role. Some of the intermediates of galactose metabolism are involved in the biosynthesis of complex carbohydrates. The two permeases that import extracellular galactose are encoded by other operons.[17,18] Note that a product of lactose, which is a β-D-galactoside, is β-D-galactose and the substrate for galactokinase is α-D-galactose.[19] Whereas the mutarotation of β-galactose to the α-anomer occurs spontaneously at a slow rate, a newly discovered cistron in the *gal* operon, *galM*, encodes an aldose-1-epimerase (mutarotase) which is largely responsible for the mutarotation in vivo.[20] A mutation in the *galM* gene makes the cell largely deficient in the utilization of the galactose moiety of lactose. The aldose-1-epimerase thus now links the products of the *lac* and *gal* operons in the form of a single metabolic pathway as shown in Figure 9.1. The structural genes of the *lac* and *gal* operons, encoding different enzymes of the pathway, are shown in Figure 9.2.

3. THE REGULATORY CIRCUITS AND THEIR COMPONENTS

The regulatory proteins and their cognate DNA elements, together with the negative and positive controls they bring to the *lac* and *gal* operons (also shown in Fig. 9.2), are briefly described below and

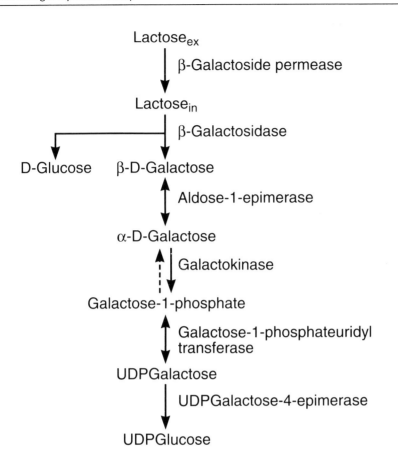

Fig. 9.1 (left). Metabolism of lactose and galactose in Escherichia coli.

Fig. 9.2 (below). The structural and regulatory genes and proteins as well as the DNA control elements of the (A) lac and (B) gal operons. The regulatory circuits are explained in the text. Not drawn to scale.

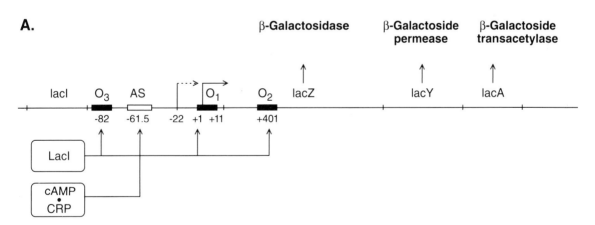

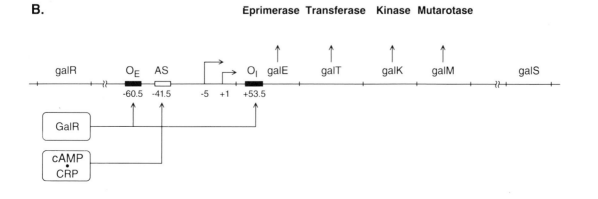

Table 9.1. *Expression of the* lac *operon in different regulatory mutants*

Genotype	Relative amounts of β-galactosidase synthesis	
	-Inducer	+Inducer
Wild Type	< 0.1	100
lacI⁻	140	140
lacIˢ	10	10-20
lacIᵃᵈⁱ	3	100
O_1^C	5-50	100
O_2^C	3.5	100
O_3^C	5.5	100
$O_2^C\ O_3^C$	70	100
$O_1^C\ O_3^C$	100	100
$O_1^C\ O_2^C$	100	100
lacP⁻	< 0.1	6-7
lacP(AS)⁻	< 0.1	2-6
lacP$_{UV5}$	< 0.1	120
lacP$_{UV5}$ cya⁻	< 0.1	100
crp⁻	< 0.1	2
cya⁻	< 0.1	2
crpᵖᶜ	< 0.1	5-10

Data taken from different sources[1,4,5,12,22,111] are normalized arbitrarily and rounded off for this chapter.

Table 9.2. *Expression of the* gal *operon in different regulatory mutants*

Genotype	Relative amounts of galactokinase synthesis	
	- Inducer	+ Inducer
Wild type	5	100
galR⁻	100	300
galS⁻	5	300
galR⁻ galS⁻	300	300
lacIᵃᵈⁱ*	130	300
O_E^C	300	300
O_I^C	70	100
$O_E^C\ O_I^C$	300	300
P1⁻	1-2	10-20
P2⁻	1-2	20-25
cya⁻	2	50
P1⁻ cya⁻	1-2	40-50

Data taken from previous publications[40,54,55,97,101,112,113] and normalized arbitrarily and rounded off for this publication. * For an explanation of the use of lacIᵃᵈⁱ mutation in *gal*, see section 7.

elaborated in later sections. The expression of the *lac* and *gal* operons in the wild-type *E. coli* and various mutants that defined the regulatory components, as discussed below, are shown in Tables 9.1 and 9.2, respectively.

THE *lac* OPERON

The promoter

The *lac* operon is exclusively transcribed in vivo from a single promoter (*P*), mutations which render the operon at least partially inactive.[21,22] The transcription initiation site (+1) from this promoter is subject to both positive control by cAMP·CRP and negative control by Lac repressor (LacI). A second promoter, *P2*, is located 22 bp upstream of *P*.[23] Both closed and open complexes are formed at *P2* efficiently but it is defective in its start site at position -22, which in the RNA is a U.[15] When this U is changed to an A, *P2* becomes active.[24] The native *P2* is not known to help or modulate *lac* transcription in any way.[25,26]

The activation site (AS)

It is the binding site of the cAMP·CRP complex and is located at the -61.5 portion from the start site of transcription (+1).[22,27] The essential part of it contains a dyad symmetry of 16 bp, which is recognized by the cAMP·CRP complex.[28] Mutation or deletion of this site, like a mutation in the promoter, also makes the operon inactive.[22]

cAMP·CRP

CRP, encoded by the gene *crp*, is a dimer with a two-domain structure in each subunit.[6,7,29] The N-domain binds cAMP, a product of adenylate cyclase which is encoded by the *cya* gene, for allosteric changes which make CRP bind to the activation site.[5,30] The C-domain, connected to the N-domain by a hinge, contains a helix-turn-helix DNA binding motif which binds specifically to a half symmetry of the 16 bp activation site. cAMP·CRP bound to the activation site stimulates *lac* transcription (positive control). Mutation of either the *cya* or *crp* genes makes the cell unable to transcribe the *lac* operon.

The operators

The *lac* operon is now known to contain not one, as originally thought, but three operators, O_1, O_2 and O_3 at positions +11, +401 and -82, respectively.[31-33] LacI binds to the three operators and inhibits *lac* transcription (negative control).

Mutation of the different operators results in different degrees of constitutivity. The wild-type O_1 allele contributes most to the repression.[34] The operator sequences contain variously hyphenated 17 bp dyad symmetry.

The Lac repressor

The Lac repressor (LacI) is the product of the *lacI* gene.[1] The purified protein is a tetramer in solution.[35] Each subunit is a two domain protein linked by a hinge.[36] The N-domain contains a helix-turn-helix DNA binding motif specific for a half symmetry in the operator.[37] The tetramer can bind to two operator sites simultaneously.[38] LacI belongs to a family of repressors (GalR-LacI family) with considerable homology to each other.[39] The family includes two repressors of the *gal* operon discussed later (Fig. 9.3). Such homology between repressor molecules permits construction of active hybrid repressors (see below). The C-domain in LacI contains oligomerization as well as inducer binding properties. Allolactose, a metabolic product of lactose, or isopropyl-β-D-thiogalactose, a nonmetabolizable analog of lactose, rather than lactose itself binds to LacI to inhibit its function.[14] Inactivation of the *lacI* gene makes the operon constitutive, while a *lacI* mutant (*lacI^s*) whose product does not bind the inducer is noninducible.[1] LacI is not known to act on any other operon.

THE *gal* OPERON

The two promoters

The *gal* operon is transcribed from two promoters *P*1 and *P*2.[40,41] The two promoters, whose corresponding start sites (+1 and -5, respectively) are separated by 5 bp, are modulated by cAMP·CRP differently. cAMP·CRP is an activator of *P*1 (positive control), but a repressor of *P*2 (negative control). The site (AS) to which cAMP·CRP binds to exert such dual action is located at position -41.5.[42]

The bipartite operators

The two promoters of the *gal* operon are modulated negatively by Gal repressor binding to two operators, O_E and O_I.[43-45] The two operators contain a 16 bp dyad symmetry and are separated by 113 bp. O_E is located upstream to the promoter at position -60.5 from the start site of *P*1 transcription, while O_I is located at position +53.5 (which in fact is within the first structural gene). Both operators participate in repression; mutation of either operator causes at least partial constitutivity.

The Gal repressor

Both *gal* promoters are negatively regulated by Gal repressor (GalR), encoded by the *galR* gene, interacting with O_E and O_I.[43,46,47] Purified GalR is a dimer in solution.[48] Like LacI, each monomer is a two domain protein; the N-domain contains a helix-turn-helix motif which recognizes an operator half-symmetry.[49] The C-domain contains the inducer, D-galactose, or its nonmetabolizable analog, D-fucose, binding site.[50,51] The inducer inhibits the function of GalR as a repressor, providing another example of double negative control. While a null mutation in repressor (*galR$^-$*) makes both promoters constitutive, a mutation (*galR^s*) whose product does not bind to the inducer makes them noninducible.[40,52,53]

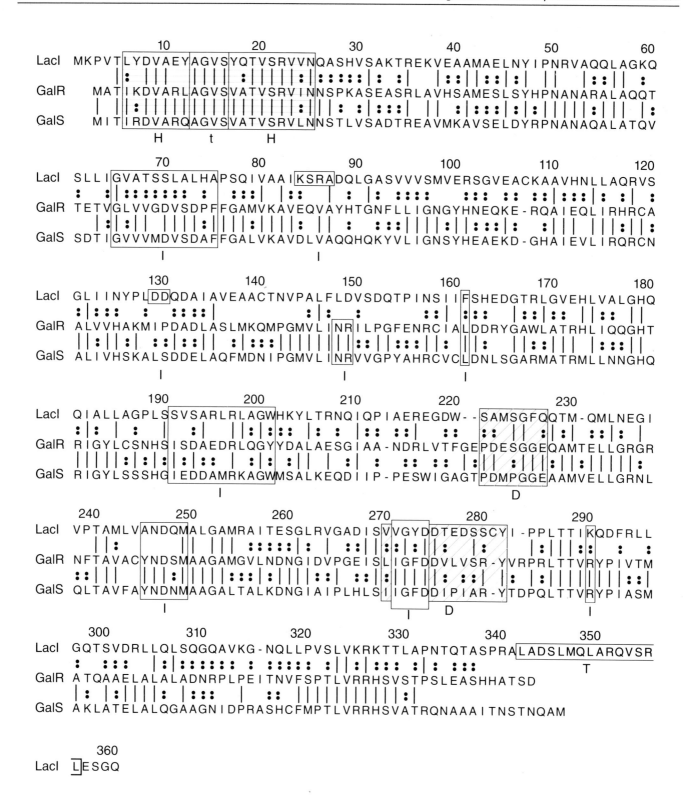

Fig. 9.3. Comparison of the amino acid sequences of the LacI, GalR and GalS proteins.[39] Identity is designated by vertical lines and similarity by colons. Known functional regions are boxed. Helix (H)-turn (t)-helix (H) domain boxes with horizontal lines is the DNA binding motif;[49,60] I (open box), the inducer binding regions;[50,110,114] D (boxes with slanted lines), the dimerization domain;[12,110,115] the region marked T (box shaded) forms an antiparallel four-helix bundle for tetramerization in LacI.[102,103] Such domains do not exist in GalR and GalS.

The isorepressor

The *gal* operon is also negatively regulated by a second repressor, GalS.[54-56] GalS is 85% similar to GalR in amino acid sequence. GalS acts through the same two operators, O_E and O_I (Fig. 9.3). The amount of repression brought out by GalS on the *gal* promoters, however, is very weak. The physiological significance of GalS on the *gal* operon is unknown. The action of GalR is epistatic to that of GalS, i.e., only in the absence of GalR can the cells be further induced for *gal* operon by the addition of inducer. The phenomenon is called ultra-induction. Incidentally, GalR and GalS modulate a few other operons, encoding various galactose transport proteins and genes encoding themselves and thus are part of a regulon.[57] The degree of repression by GalR and GalS on these operons varies. Both GalR and GalS are homologous to LacI, i.e., members of the GalR-LacI family.[39]

Helix swap between LacI and GalR

Because of extensive amino acid homology between the Lac and the Gal repressors (Fig. 9.3), it has been possible to precisely transplant by genetic engineering the helix-turn-helix motif of LacI into GalR and vice versa (Fig. 9.4). Such transplantations, following the principle of similar experiments using Lambdoid phage repressors,[58,59] have resulted in repressors with specificity switched for the operators, as shown both in vitro and in vivo.[60,61] For example, a LacI carrying the helix-turn-helix motif of GalR effectively represses the wild-type *gal* operon and is induced by IPTG.

4. MODULATION OF PROMOTERS BY cAMP·CRP

cAMP and its binding protein, although initially discovered as the critical components of catabolite repression, are now known to modulate a much wider variety of genes (reviewed in ref. 62). The way they modulate the *lac* and *gal* operons is summarized below.

Fig. 9.4. Swap of helix-turn-helix DNA binding motifs between LacI and GalR.[60,61] O^G and O^L, half-operator sites of gal *and* lac *dyad symmetry operators, respectively; GalR (A), LacI (B), GalR-rHL (C) and LacI-rHG (D) represent GalR, LacI, GalR and LacI with the DNA recognition helices of GalR, LacI, LacI and GalR, respectively. Only the amino acids that are boxed were exchanged between GalR and LacI. The dotted lines indicate amino acid base contacts.[60,49]*

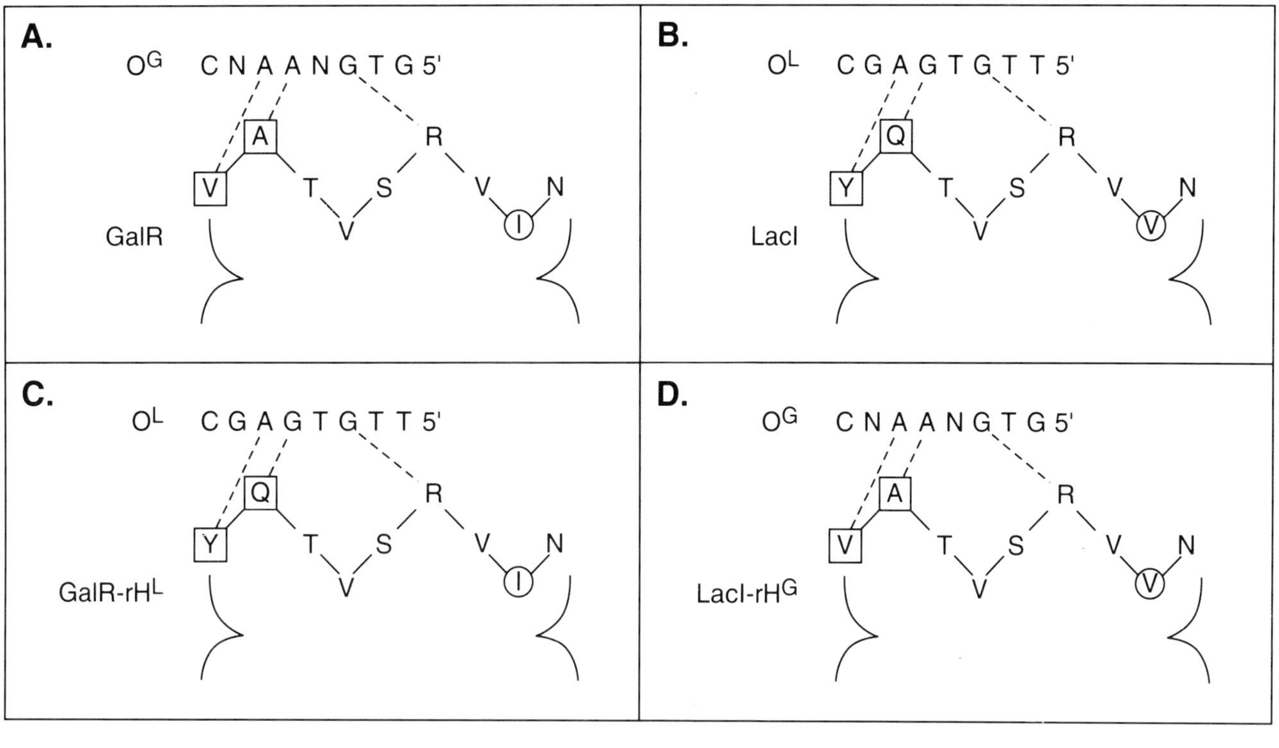

ACTIVATION OF *lac* PROMOTER BY cAMP·CRP

The *lac* promoter belongs to a class of promoters that are intrinsically defective.[63] The -35 and -10 elements of the promoter sequence differ significantly from the consensus.[64] The defect of the promoter is resurrected by cAMP and CRP. As mentioned before, the cAMP·CRP complex binds to the activation site and activates *lac* transcription. Two of the initial steps in the pathway of transcription initiation by RNA polymerase—formation of a weak RNA polymerase promoter complex (closed complex) and isomerization of the closed complex to a more stable form with partial opening of the DNA strands (open complex)—have been measured in the *lac* promoter.[23] The wild-type *lac* promoter forms a closed complex poorly; cAMP·CRP has been shown to stimulate the closed complex formation but not the rate of isomerization, although a role of cAMP·CRP beyond the steps of isomerization has not been tested (see Natural Polarity). The mode of cAMP·CRP action in *lac* promoter activation is briefly discussed below and is elaborated in chapter 12. Although the poor activity of the second *lac* promoter (*P2*) is inhibited by cAMP·CRP in vitro,[23] as mentioned before, there seems to be no biological consequence of the presence of this promoter on *lac* operon expression in vivo.[25]

DUAL CONTROL OF *gal* PROMOTERS BY cAMP·CRP

The two promoters of the *gal* operon (*P1* and *P2*) are separated by 5 bp, which is equivalent to half of a B-DNA helical turn. Although both are active on their own, their regulatory features are somewhat different. First, the intrinsic strength of the *P1* promoter is 2-fold enhanced in vitro by the presence of periodic tracks of 4-6 adenine residues at positions -84.5, -74 and -63 on the DNA.[65] Since the presence of specific A tracks bend DNA toward the face of *P1* to which RNA polymerase binds, the A tracks induced DNA curvature may help some upstream DNA to contact RNA polymerase for stimulating promoter activity.[63,66,67] The *P2* promoter, unlike the *P1* promoter, makes a large amount of aborted as well as pseudotemplated stuttered RNA oligomers (see Control of *P2* by UTP in *gal*) suggesting that RNA polymerase fails to go to the elongating mode efficiently and idles at this promoter.[68,69] As mentioned earlier, cAMP·CRP modulates the two *gal* promoters, which partially overlap and are located on the opposite face of the DNA helix, in opposite directions.[68] The complex stimulates transcription initiation from *P1* about 3-fold and inhibits the synthesis of aborted, stuttered, as well as full length RNA completely (Fig. 9.5). Whereas *P1* stimulation occurs at the level of closed complex formation,[70,71] the level at which cAMP·CRP inhibits *P2* is not clear. The interpretation of the kinetic data for *gal* transcription is problematic due to overlapping of the two promoters.

MECHANICS OF POSITIVE CONTROL BY cAMP·CRP

The activation site of cAMP·CRP complex in the *lac* promoter is located at position -61.5. The DNA-bound activator can affect transcription initiation in two ways: by allosterically altering either (a) RNA polymerase or (b) the *lac* promoter, not necessarily in a mutually exclusive way (see Fig. 9.6).[72,73]

    a. CRP-RNA polymerase contact in the *lac* promoter. Several lines of investigations suggest a direct interaction between cAMP·CRP and RNA polymerase at the *lac* promoter: (i) Free RNA polymerase and cAMP·CRP interact as suggested from fluorescence,[74,75] ultracentrifugation,[76] and immunological experiments.[77] (ii) cAMP·CRP and RNA polymerase

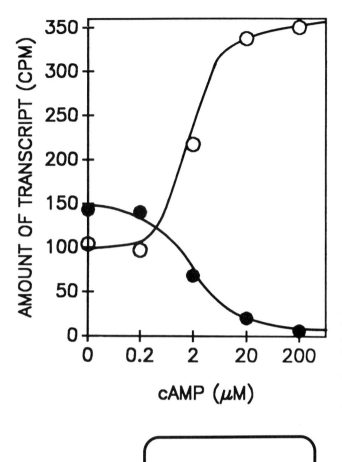

Fig. 9.5. Effect of cAMP concentrations on transcription from the P1 and P2 promoters in gal operon.[68] Open circles, RNA made from P1; closed circles, RNA made from P2.

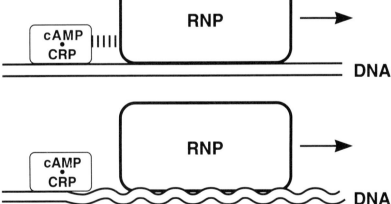

Fig. 9.6. Models of cAMP·CRP action. (Top) The DNA-bound cAMP·CRP makes a direct contact with RNA polymerase; (Bottom) The cAMP·CRP complex induces a structural change in DNA that is transmitted to the promoter.

bind to the *lac* promoter cooperatively.[78-80] Cooperative binding of two proteins to DNA is best explained by a direct protein-protein contact.[81] (iii) A cluster of mutations (positive control or *pc* mutations) located on a surface loop covering amino acids 152 to 166 of CRP affect cooperative binding but not the intrinsic DNA binding.[82-84] These mutations make cAMP·CRP unable to activate transcription at the *lac* promoter. (iv) In experiments where the cAMP·CRP binding site is engineered into different locations upstream of the +1 start site in similar promoters, cAMP·CRP activates transcription only when the binding site is away from the +1 site by an integral number of

B-DNA helical turns (reviewed in ref. 62). cAMP·CRP activates either poorly or not at all in cases in which the cAMP·CRP site is located at intermediate distances. These results argue that cAMP·CRP contacts RNA polymerase for transcription activation since activation occurs only when the two proteins are on the same face of DNA. Additionally, in an experiment using concatenated DNA circles, cAMP·CRP bound on a circle does not activate transcription if the *lac* promoter is located on the second circle.[85] (v) In vitro transcription of the *lac* promoter using reconstituted RNA polymerase holoenzyme has shown that the C-terminus of the α subunit of RNA polymerase is a target of cAMP·CRP action.[86] Reconstituted RNA polymerase carrying deletions of the C-terminus of the α subunit are capable of transcription initiation from several promoters but not transcription activation by cAMP·CRP at the *lac* promoter. Point mutations around amino acid positions 258-265 in the α subunit that prevent CRP activation of the *lac* promoter have also been reported.[87] The C-terminal region of the α subunit has been shown by limited proteolysis to be an independent domain connected to the N-terminal region by a hinge, and has some DNA binding activity.[88] It is likely that these mutations define a region in the α subunit that interacts with the 156-162 loop in CRP. Is the activating patch, called the "158/159 loop," of a specific (the promoter-proximal or the promoter-distal) subunit of DNA-bound cAMP·CRP or that of both subunits responsible for transcription activation? This question has been answered unequivocally as follows:[89] First, heterodimers consisting of a CRP subunit having a mutated activation loop but a wild-type DNA binding motif and a CRP subunit having a wild-type activation loop but DNA binding motif with altered specificity were constructed. Next, the heterodimers were forced to bind to the *lac* promoter in the two possible orientations by the use of hybrid activation sites, consisting of one wild-type half site and one half site that is only compatible with the CRP with altered DNA binding specificity. The results clearly show that transcription activation in *lac* requires the 158/159 loop of only one subunit: the promoter proximal one. For a more in depth description about the interaction between cAMP·CRP and RNA polymerase in several promoters, see Chapter 12.

b. Role of DNA allostery in the *lac* promoter. cAMP·CRP bound to the -61.5 region may activate the *lac* promoter by influencing the conformation of the nearby promoter.[72,73] Creation of single stranded gaps in the space between CRP and RNA polymerase binding sites of the *lac* promoter prevented the action of DNA bound cAMP·CRP.[85] Single stranded gaps at the same region did not affect transcription of a mutant *lac* promoter that is independent of cAMP·CRP (P_{UV5}). Thus, both a protein-protein (cAMP·CRP-RNA polymerase) bridge and a bridge of intact DNA between CRP-DNA and RNA polymerase-promoter complex are involved in transcription activation at *lac*. It has been suggested that CRP induces an allosteric change in the *lac* promoter.[85] The effect of the proposed DNA allostery is very likely to make the promoter better suited for transcription initiation. The purpose of the cAMP·CRP-RNA polymerase contact may well be to anchor RNA polymerase and create (or maintain) torsion so an essential DNA conformational change can be effected (or stabilized). An alternative idea is that the C-terminal domain of α makes contact with both cAMP·CRP and the spacer DNA resulting in a higher level of closed complex formation.[88]

Unlike the *lac* promoter, in which cAMP·CRP binds at position -61.5 to activate transcription, in *gal* it binds to position -41.5 to stimulate transcription from the *P*1 promoter. Although it has been suggested that a second molecule of cAMP·CRP binds to the -61 region of *gal* for transcription activation of P1,[88a] it has been argued that the *lac* promoter and the *gal P*1 promoter are fundamentally different.[62] First, CRP *pc* mutants with defects in the 158/159 loop that fail to act at the *lac* promoter do not show any defect in acting on the *P*1 promoter in *gal*. A second surface exposed loop in CRP spanning the region of amino acids 52 through 56 is required to activate transcription from the position 41.5.[89] Second, the use of "oriented" heterodimers, like the one described above for *lac*, at the -41.5 site in *gal* have revealed that "the 52 loop" of the CRP subunit distal to the +1 position of *P*1 is essential for action from the -41.5 site.[89] Although the 52 loop is on the opposite side from the 158/159 loop on the CRP subunit, the 52 loop of one subunit is adjacent to the 158/159 loop of the other subunit in the dimer which has 2-fold rotational symmetry. The 52 loop may see a region of RNA polymerase different from the one seen by the 158/159 loop.

5. CONTROL OF *P*2 BY UTP IN *gal*

E. coli cells expressing the *gal* operon mostly from the *P*2 promoter grow slowly when using galactose as the sole carbon source when their internal UTP pool is adjusted to high levels.[69] The presence of UTP and UDP sugars in the pathway of galactose metabolism may be the reason for a potential control of the *gal* operon expression by UTP. It has been shown that the step of promoter clearance by RNA polymerase is regulated by UTP in an intriguing way at the *P*2 promoter in *gal*. In vitro, RNA polymerase stays at this promoter and not only makes a large amount of aborted RNA oligomers (Fig. 9.5) but also makes a large amount of pseudotemplated "stuttered" RNA oligomers of the composition $pppAU_n$ (n = 2 to > 20) at high concentrations of UTP, but not at low concentrations of the triphosphate (Fig. 9.7).[69] RNA polymerase makes template encoded RNA at low UTP concentrations. The cAMP·CRP complex which inhibits the synthesis of aborted and full length RNA from *P*2 also represses the "stuttered" RNA synthesis. The apparent correlation between in vitro and in vivo results suggests that the cellular UTP concentration can modulate the expression of the *gal* operon from the *P*2 promoter. Such a control is reminiscent of the regulation of the promoter of *pyrBI* operon, encoding earlier enzymes for UTP biosynthesis.[90] In this system, high UTP concentrations block the promoter clearance step of RNA polymerase and cause the latter to stutter in a similar fashion.

6. NATURAL POLARITY

The molar amount of polypeptides made from a promoter proximal cistron is higher than that from the promoter distal cistron in both *lac* and *gal* operons, a phenomenon termed natural polarity.[91-95]

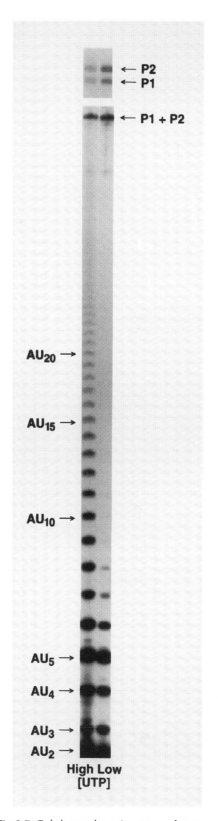

Fig. 9.7. Gel electrophoresis pattern of stuttered RNA synthesis from the P2 promoter in gal.[69] Top (8% gel), full length RNA from P1 and P2 as indicated; bottom (24% gel), RNA oligomers of the composition indicated on the margin from the P2 promoter. UTP concentrations 0.2 mM (left) and 0.02 mM (right).

The molar amount of β-galactosidase polypeptide chains made from the *lac* operon is considerably higher than that of β-galactoside transacetylase. Similarly, the amount of uridinediphosphogalactose-4-epimerase can be about 4-fold higher than that of galactokinase. Several levels have been cited at which natural polarity may originate: transcription termination at intercistronic regions, preferential degradation of promoter distal mRNA, and lower efficiency of translation initiation of the distal cistrons. Whether the cause is one or more of the above mechanisms remains to be determined. Since the natural polarity in the *gal* operon is more severe in the absence of cAMP, cAMP·CRP may play a role in altering the relative expression of the *gal* cistrons.

7. NEGATIVE CONTROL BY REPRESSOR-OPERATOR INTERACTIONS

MECHANISM OF REPRESSION IN *gal*: DNA LOOPING

As mentioned before, each of the two operator elements, O_E and O_I, in *gal* is needed for normal repression of $P1$ and $P2$. Mutation of O_E causes complete derepression of both $P1$ and $P2$, whereas inactivation of O_I results in complete derepression of $P2$ and partial derepression of $P1$.[40,43,96] The mere existence of two spatially separated operators and their need for the action of GalR, suggested that repression requires an interaction between the O_E and O_I bound repressors which will generate a loop of the intervening promoter DNA (Fig. 9.8).[43,45] The requirement of DNA looping for repression of the *gal* promoters has been supported by the following set of experiments:

i. Operator swap: If each *gal* operator, referred to as O_E^+ and O_I^+ in Figure 9.9, is genetically substituted by a *lac* operator (O_E^L- O_I^L), then the two *gal* promoters are completely repressed by LacI and inducible by IPTG (Fig. 9.9).[97] However if O_E^+ or O_I^+ alone is replaced by a *lac* operator (O_E^L - O_I^+ or O_E^+ - O_I^L), normal repression does not follow the occupation of the two operators by respective repressors.

ii. Cooperative binding: Both GalR and LacI bind to the bipartite operators, O_E^+ - O_I^+ and O_E^L - O_I^L, respectively, in a cooperative fashion in vivo.[97] Cooperative binding is best explained by an interaction of DNA bound repressors, i.e., DNA looping.[81] Although GalR does not, LacI shows cooperative binding in vitro to O_E^L- O_I^L DNA.[98-100] The failure to show cooperative binding by GalR in vitro is ascribed to the absence of a missing factor.[96]

iii. Electron microscopy: Electron microscopic studies directly show that a O_E-O_I DNA in the presence of a repressor can generate a DNA loop.[101] When linear DNA fragments with the O_E^L - O_I^L genotype are mixed with LacI, DNA loops of the expected size at the correct location on the DNA fragment are observed frequently. Such studies do not show DNA looping with O_E^G- O_E^G DNA in the presence of GalR (see the section on Modulation of *gal* promoters in the absence of DNA looping: repressor-RNA polymerase contact).

iv. Nonlooping repressor mutant: Tetramer formation in LacI has been shown to depend upon the presence of a helix containing three leucine-heptad repeats at the C-terminal end of the protein.[102,103] The C-terminal helix of each monomer forms an antiparallel four-helix bundle in the tetramer through global hydrophobic packing and not through

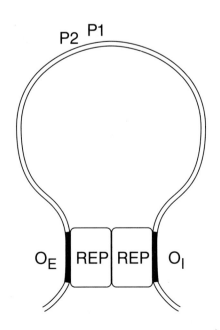

Fig. 9.8. DNA looping of the gal promoters. Repressor dimers bound at O_E and O_I interact to generate a loop of the promoter DNA segment.

direct leucine-leucine contacts.[103] Mutations in the leucines disrupt the global hydrophobic packing. A LacI mutant which does not contain this terminal segment does not tetramerize and fails to repress the *gal* promoters in the O_E^L- O_I^L strain.[101] The mutant Lac repressor also does not form DNA loops as observed under the electron microscope and does not show cooperative binding to the DNA.[99,101]

The requirement of DNA looping for efficient repression of *P1* and *P2* has been confirmed in vitro. Wild-type looping-proficient LacI represses both *P1* and *P2* (both aborted and full length RNA synthesis) in O_E^L- O_I^L DNA, while neither the mutant LacI (in O_E^L- O_I^D) nor wild-type GalR (in wild-type DNA), which are looping deficient, repress the promoters normally.[96] DNA looping prevents transcription of *P1* in both the absence and presence of cAMP·CRP. It has been proposed that a topologically closed DNA loop of 114 bp, which is inflexible to torsional changes, makes the promoters refractory to transcription initiation presumably by resisting DNA unwinding needed for open complex formation.[104]

MODULATION OF *gal* PROMOTERS IN THE ABSENCE OF DNA LOOPING: REPRESSOR-RNA POLYMERASE CONTACT

In the absence of DNA looping, nevertheless, occupation of O_E^+ or O_E^L alone by GalR or LacI, respectively, represses only *P1* (about 4- to 5-fold) and results in an almost two-fold stimulation of *P2*.[96] Occupation of O_I does not affect such regulation. The repression of *P1* by a repressor bound to O_E in the absence of DNA looping is through

Repression

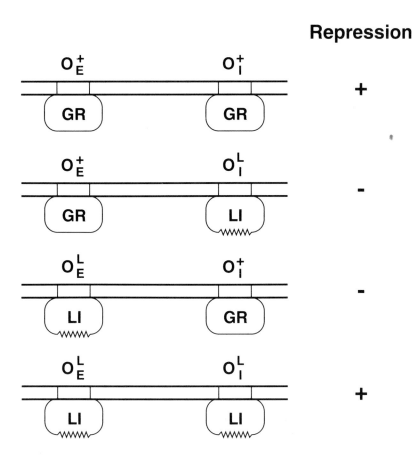

Fig. 9.9. Effect of operator conversion in repression of the gal operon.[97] O_E^+ and O_I^+, wild-type gal operators; O_E^L and OS_I^L, converted to lac operators. GR and LI denote GalR and LacI, respectively.

a direct contact between the repressor with RNA polymerase (Fig. 9.10). First, binding of the repressor to O_E does not prevent RNA polymerase binding to $P1$ or $P2$ showing that the repressor does not sterically block RNA polymerase binding to cause repression.[105] Second, the repressor fails to repress $P1$ and activate $P2$ transcription by a reconstituted RNA polymerase which is missing the C-terminal domain of the α subunit.[105] These results suggest that the repression of $P1$ and the activation of $P2$ require the physical presence of the C-terminal domain of the α subunit although the latter is not required for transcription itself. It has been suggested that the repressor inhibits or stimulates transcription initiation by disabling or stimulating RNA polymerase activity at a post-binding step by modulating the C-terminal α domain.[105] These results provide the first evidence that negative control of transcription initiation occurs by a direct action of repressor on RNA polymerase activity rather than by a passive role of hindering RNA polymerase binding to DNA. These results also oppose a model of RNA polymerase partitioning for the activation of $P2$ by O_E bound repressor in the absence of DNA looping: repressor by inhibiting the binding of RNA polymerase to $P1$ makes more polymerase available for $P2$.[70]

MULTIPARTITE OPERATORS AND MECHANISM OF REPRESSION IN *lac*

Although it was assumed originally that O_1 is the sole operator responsible for LacI mediated repression of the *lac* operon, discovery of the O_2 and O_3 sites required re-investigation of the potential contribution of the latter operators in repression. Systemic genetic analysis has shown that a point mutation of O_1 results in a 5- to 50-fold but not total derepression, and multi-site mutations of O_2 or O_3 cause a 2- to 3-fold loss of repression.[34] Besides, the nonlooping mutant Lac repressor dimer also shows 3-fold loss of repression.[34,101] These results suggest that while O_1 plays a major role, O_2 and O_3 also contribute in bringing about total repression. A cooperative binding or DNA looping between O_1 and O_2 or O_3 increases the stability of repressor binding to O_1 and thus repression.[34] Thus, repressor binding to a point mutation in O_1 can be partially restored by cooperative binding with O_2 or O_3 resulting in only 5- to 50-fold derepression, rather than 500-fold derepression observed with the O_1 mutation without the presence of O_2 and O_3 elements.

The location of the primary operator O_1 (+11) partially overlapping with the promoter originally suggested that LacI inhibits transcription by sterically hindering RNA polymerase binding. This idea is supported by the finding that repressor prevents RNA polymerase-promoter closed complex formation,[106] although other results suggested LacI action at a post RNA polymerase binding step.[107-109]

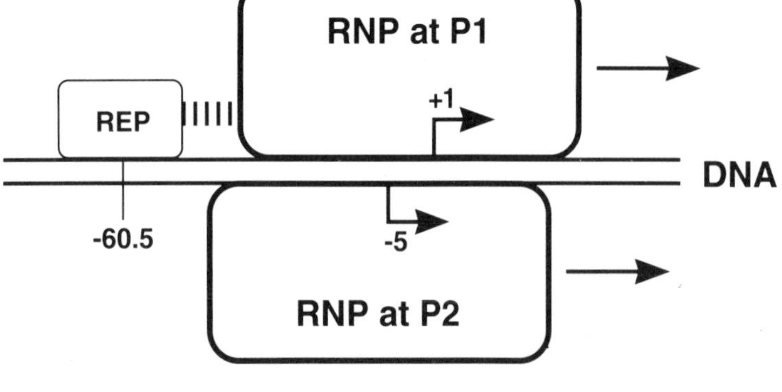

8. EPILOGUE

The original operon model of genetic regulatory mechanisms, although proposed at the genetic and physiological level, had the tempting power of being tested at the level of chemistry. This led and continues to lead scores of new investigators to confirming, modifying, or discovering additional features of gene

Fig. 9.10. Repression of P1 in gal by repressor RNA polymerase contact.[105] Details are given in the text.

regulation, resulting in a wealth of information that has made the subject of gene regulation perhaps the most understood area of molecular biology. Even studying the regulatory mechanisms in the classical operons more than three decades later continues to provide new and fascinating insights about both negative and positive controls. Several noteworthy concepts have emerged during the last decade: communication between multipartite control elements by DNA looping; direct communication between an activator or a repressor and RNA polymerase to directly facilitate or inhibit steps of transcription initiation; and awareness that a regulatory protein has the built-in potential to participate as both a repressor and an activator in such direct fashion. But much remains to be determined at the molecular level: What is the chemical mechanism of transcription initiation? What does a regulatory protein do to RNA polymerase to activate or to repress transcription? How do DNA structural changes regulate transcription? Is there any role of cAMP·CRP beyond the initial steps of transcription? As both *lac* and *gal* operons are units of transcription, RNA synthesis must terminate at their ends. Such signals are yet to be identified. With the current pace of investigations, we will very likely have many of these answers within the next few years.

REFERENCES

1. Jacob F, Monod J. Genetic regulatory mechanisms in the synthesis of proteins. J Mol Biol 1961; 3:318-356.
2. Monod J, Changeux JP, Jacob F. Allosteric proteins and cellular control systems. J Mol Biol 1963; 6:306-329.
3. Engelsberg E, Irr J, Power J et al. Positive control of enzyme synthesis by gene C in the L-arabinose system. J Bacteriol 90; 90:946-957.
4. Schwartz D, Beckwith JR. Mutants missing a factor necessary for the expression of catabolite-sensitive operons in *E. coli*. In: Beckwith JR, Zipser D, eds. The Lac Operon. Cold Spring Harbor, NY: Cold Spring Harbor Laboratory, 1970:417-422.
5. Perlman RL, Pastan I. Pleiotropic deficiency of carbohydrate utilization in an adenyl cyclase deficient mutant of *Escherichia coli*. Biochem Biophys Res Comm 1969; 37:151-157.
6. Emmer M, deCrombrugghe B, Pastan I et al. Cyclic AMP receptor protein of *E. coli*: its role in the synthesis of inducible enzymes. Proc Natl Acad Sci USA 1970; 66:480-487.
7. Zubay G, Schwartz DO, Beckwith JR. The mechanism of activation of catabolite-sensitive genes: a positive control system. Proc Natl Acad Sci USA 1970; 66:104-110.
8. Monod J. The phenomenon of enzymatic adaptation. Growth 1947; 11:223- 289.
9. Magasanik B. Catabolite repression. Cold Spring Harbor Symposium Quant. Biol. 1961; 26:249-256.
10. Makman RS, Sutherland EQ. Adenosine 3',5'-phosphate in *Escherichia coli*. J Biol Chem 1965; 240:1309-1314.
11. Buttin G. Mechanismes regulateurs dans biosynthese des enzymes du metabolism du galactose chex *Escherichia coli* K-12. 1. La biosynthese indreile de la galactokinase el l'induction simultanee de la sequence enzymatique. J Mol Biol 1963; 7:164-182.
12. Miller JH. The *lac*I gene: its role in *lac* operon control and its use as a genetic system. In: Miller JH, Reznikoff WE, esd. The Operon. Cold Spring Harbor, NY: Cold Spring Harbor Laboratory, 1978:31-88.
13. Beckwith J. The operon: an historical account. In: Miller JH, Reznikoff WE, eds. The Operon. Cold Spring Harbor, NY: Cold Spring Harbor Laboratory, 1987:1439-1443.

14. Beckwith J. The Lactose operon. In: Neidhardt FC, Ingraham JL, Low KB, Magasaki, B, Schaecter, M, Umberger, HE, eds. *Escherichia coli* and *Salmonella typhimurium*. ASM, 1987:1444-1452.

15. Beckwith JR. *lac*: The genetic system. In: Miller JH, Reznikoff WE, eds. The Operon. Cold Spring Harbor, NY: Cold Spring Harbor Laboratory, 1970:11-30.

16. Adhya S. The Galactose operon. In: Neidhardt FC, Ingraham FL, Low KB, Magasanik B, Schaecter M, Umberger HE, eds. *Escherichia coli* and *Salmonella typhimurium*. ASM, 1987:1503-1512.

17. Robbins AR, Guzman R, Rotman B. Roles of individual *mgl* gene products in the beta-methylgalactoside transport system of *Escherichia coli* K12. J Biol Chem 1976; 25:3112-3116.

18. Riordan G, Kornberg HL. Proc R Soc Lon (Biol) 1977; 198:401-410.

19. Sherman JR, Adler J. Galactokinase in *Escherichia coli* K-12. J Biol Chem 1963; 238:873-878.

20. Bouffard G, Rudd K, Adhya S. Dependence of lactose metabolism upon mutarotase encoded in the *gal* operon in *Escherichia coli*. J Mol Biol 1994; 244:269-278.

21. Scaife J, Beckwith JR. Mutational alteration of the maximal level of *lac* operon expression. Cold Spring Harbor Symposium Quant. Biol. 1966; 31:403-408.

22. Hopkins JD. A new class of promoter mutations in the lactose operon of *Escherichia coli*. J Mol Biol 1974; 87:715-724.

23. Malan TP, McClure WR. Dual promoter control of the *Escherichia coli* lactose operon. Cell 1984; 39:173-180.

24. Rajendrakumar G, personal communication.

25. Yu X-M, Reznikoff W. Deletion analysis of the *Escherichia coli* lactose promoter *P*2. Nucleic Acids Res 1987; 13:2457-1468.

26. Donnelly CE, Reznikoff WS. Mutations in the *lac P*2 promoter. J Bacteriol 1987; 169:1812-1817.

27. Simpson RB. Interaction of the cAMP receptor protein with the *lac* promoter. Nucleic Acids Res 1980; 8:759-766.

28. Ebright RH, Cossart P, Gicquel-Sanzey B et al. Mutations that alter the DNA sequence specificity of the catabolite gene activator protein of *E. coli*. Nature 1984; 311:232-235.

29. Weber IT, Steitz TA. Structure of a complex of catabolite gene activator protein and cyclic AMP at 2.5Å resolution. J Mol Biol 1987; 198:311-326.

30. Adhya S, Ryu S, Garges S. Role of allosteric changes in cyclic AMP receptor protein function. Subcell Biochem. In: Bigwas BB, Roy S, eds. Proteins: Structure, Function and Engineering. New York: Plenum Press, 1994; 24:303-321.

31. Gilbert W, Gralla J, Majors J et al. Lactose operator sequences. In: Sund H, Blauer G, eds. Protein Ligand Interactions. Perlin: de Gruyter, 1975:193-210.

32. Reznikoff WS, Winter RB, Hurley CK. The location of the repressor binding sites in the *lac* operon. Proc Natl Acad Sci USA 1974; 71:2314-2318.

33. Pfahl M, Gulde V, Bourgeois S. "Second" and "third" operator of *lac* operon: an investigation of their role in the regulatory mechanisms. J Mol Biol 1979; 127:339-344.

34. Oehler S, Eismann ER, Kramer H et al. The operators of the *lac* operon cooperate in repression. EMBO J 1990; 9:973-975.

35. Müller-Hill B, Beyreuter K, Gilbert W. Lac repressor from *Escherichia coli*. Methods Enzymol 1971; 21D:483-487.

36. Geisler N, Weber K. Isolation of amino-terminal fragment of lactose repressor necessary for DNA binding. Biochem 1977; 16:938-943.

37. Ebright RH. Evidence for a contact between glutamine 18 of *lac* repressor and base pair 7 of *lac* operator. Proc Natl Acad Sci USA 1986; 83:303-307.

38. O'Gorman RO, Rosenberg JM, Kallai, OB et al. Equilibrium binding of inducer to *lac* repressor-operator DNA complex. J Biol Chem 1980; 255:10107- 10114.

39. Weickert MJ, Adhya S. A family of bacterial regulators homologous to Gal and Lac repressors. J Biol Chem 1992; 267:15869-15874.

40. Adhya S, Miller W. Modulation of the two promoters of the galactose operon of *Escherichia coli*. Nature 1979; 279:492-494.

41. Musso R, deLauro R, Adhya S et al. Dual control for transcription of the galactose operon by cyclic AMP and its receptor protein at two interspersed promoters. Cell 1977; 12:847-854.

42. Taniguchi T, O'Neill M, deCrombrugghe B. Interaction site of *Escherichia coli* cyclic AMP receptor protein on DNA of galactose operon promoters. Proc Natl Acad Sci USA 1979; 76:5090-5094.

43. Irani M, Orosz L, Adhya S. A control element within a structural gene: the *gal* operon of *Escherichia coli*. Cell 1983; 32:783-788.

44. Fritz H-J, Bicknase H, Gleumes B et al. Characterization of two mutations in the *Escherichia coli gal*E gene inactivating the second galactose operator and comparative studies of repressor binding. EMBO J 1983; 2:2129-2135.

45. Majumdar A, Adhya S. Demonstration of two operator elements in *gal*: in vitro repressor binding studies. Proc Natl Acad Sci USA 1984; 81:6100-6104.

46. Buttin G. Mechanismes regulateurs dans la biosynthese des enzymes du metabolisme du galactose chez *Escherichia coli* K-12. II. Le determinisme de la regulation. J Mol Biol 1963; 7:183-205.

47. Adhya S, Echols H. Glucose effect and the galactose enzymes of *Escherichia coli*: correlation between glucose inhibition of induction and inducer transport. J Bacteriol 1966; 92:601-608.

48. Majumdar A, Rudikoff S, Adhya S. Purification and properties of Gal repressor: *p*L-*gal*R fusion in pKC31 plasmid vector. J Biol Chem 1987; 262:2326-2331.

49. Majumdar A, Adhya S. Probing the structure of Gal operator-repressor complexes. J Biol Chem 1986; 262:13258-13262.

50. Müller-Hill B. Sequence homology between Lac and Gal repressors and three sugar-binding periplasmic proteins. Nature 1983; 302:163-164.

51. Hsieh M, Hensley P, Brenowitz M et al. A molecular model of the inducer-binding domain of the galactose repessor of *Escherichia coli*: J Biol Chem 1994; 269:13825-13835.

52. Saedler H, Gullon A, Fiethen L et al. Negative control of the galactose operon in *E. coli*. Mol Gen Genet 1968; 102:79-88.

53. Zhou Y, Chatterjee S, Roy S et al. The non-inducible nature of superrepressors of the *gal* operon in *Escherichia coli*. J Mol Biol 1995; 253:414-425.

54. Tokeson JPE, Garges S, Adhya S. Further inducibility of a constitutive system: ultrainduction of the *gal* operon. J Bact 1991; 173:2319-2327.

55. Golding A, Weickert MJ, Tokeson JPE et al. A mutation defining ultrainduction of the *Escherichia coli gal* operon. J Bact 1991; 173: 6294-6296.

56. Weickert MJ, Adhya S. Isorepressor of the *gal* regulon in *Escherichia coli*. J Mol Biol 1992; 226:69-83.

57. Weickert MJ, Adhya S. The galactose regulon of *Escherichia coli*. Mol Microbiol 1993; 10:245-251.

58. Wharton RB, Brown EL, Ptashne M. Substituting an α-helix switches the sequence-specific DNA interactions of a repressor. Cell 1984; 38:361-369.

59. Wharton RP, Ptashne M. A new-specificity mutant of 434 repressor that defines an amino acid-based pair contact. Nature 1985; 316:601-605.

60. Lehming N, Savtorius J, Miemöller M et al. The interaction of the recognition helix of *lac* repressor with *lac* operator. EMBO J 1987; 6:3145-3153.

61. Ketter J, personal communication.

62. Kolb A, Busby S, Buc H et al. Transcriptional regulation by cAMP and its receptor protein. Ann Rev Biochem 1993; 62:749-795.

63. Adhya S, Gottesman M, Garges S et al. Promoter resurrection by activators—a minireview. Gene 1993; 132:1-6.

64. Hawley DK, McClure WR. Compilation and analysis of *Escherichia coli* promoter sequences. Nucleic Acids Res 1983; 11:2237-2255.

65. Lavigne M, Herbert H, Kolb A, Buc, H. Upstream curved sequences influence the initiation of transcription at the *Escherichia coli* galactose operon. J Mol Biol 1992; 224:293-306.

66. Plaskon RR, Wartell RM. Sequence distribution associated with DNA curvature are found upstream of strong *E. coli* promoter. Nucleic Acids Res 1987; 15:785-796.

67. Nagaich AK, Bhattacharyya D, Brahmachari SK et al. CA/TG sequence at the 5' end of oligo(A) tracts strongly modulates DNA curvature. J Biol Chem 1994; 269:7824-7833.

68. Choy HE, Adhya S. RNA polymerase idling and clearance in *gal* promoters: use of supercoiled minicircle DNA template made in vivo. Proc Natl Acad Sci USA 1993; 96:472-476.

69. Jin DJ. Slippage synthesis at the *galP2* promoter of *Escherichia coli* and its regulation by UTP concentration and cAMP·cAMP receptor protein. J Biol Chem 1994; 269:17221-17227.

70. Goodrich JA, McClure WR. Regulation of open complex formation at the *Escherichia coli* galactose operon promoters. J Mol Biol 1992; 224:15-29.

71. Herbert M, Kolb B, Buc H. Overlapping promoters and their control in *Escherichia coli*: the *gal* case. Proc Natl Acad Sci USA 1986; 83:2807-2811.

72. Reznikoff WS, Abelson JN. The *lac* promoter. In: Miller JH, Reznikoff WS, eds. The Operon. Cold Spring Harbor, NY: Cold Spring Harbor Laboratory, 1978:221-243.

73. Adhya S, Garges S. Positive control. J Biol Chem 1990; 265:10797-10800.

74. Heyduk T, Lee J, Ebright Y et al. CAP interacts with RNA polymerase in solution in the absence of promoter DNA. Nature 1993; 364:548-549.

75. Pinkey M, Goggelt J. Binding of the cyclic AMP receptor protein of *Escherichia coli* to RNA polymerase. Biochem J 1988; 250:897-902.

76. Blazy B, Takahashi M, Baudras A. Binding of CRP to DNA-dependent RNA polymerase from *E. coli*: modulation by cAMP of the interactions with free and DNA-bound holo. Mol Biol Rep 1980; 6:39-43.

77. Riftina F, DeFalco E, Krakow J. Effects of an anti-alpha monoclonal antibody on interaction of *Escherichia coli* RNA polymerase with *lac* promoters. Biochem 1990; 29:4440-4446.

78. Ren YL, Garges S, Adhya S et al. Cooperative DNA binding of heterologous proteins: evidence for contact between cyclic AMP receptor protein and RNA polymerase. Proc Natl Acad Sci USA 1988; 85:4138-4142.

79. Spassky A, Busby S, Buch H. On the action of the cyclic AMP-cyclic AMP receptor protein complex at the *Escherichia coli* lactose and galactose promoter regions. EMBO J 1984; 3:43-50.

80. Straney D, Straney S, Crothers D. Synergy between *Escherichia coli* CAP protein and RNA polymerase in the *lac* promoter open complex. J Mol Biol 1989; 206:41-57.

81. Hochschild A, Ptashne M. Cooperative binding of λ repressors to sites separated by integral turns of the DNA helix. Cell 1986; 44:681-687.

82. Eschenlauer AC, Reznikoff WS. *Escherichia coli* catabolite activator protein mutants defective in positive control of *lac* operon transcription. J Bacteriol 1991; 173:5024-5029.

83. Bell A, Gaston K, Williams R et al. Mutations that affect the ability of *Escherichia coli* cyclic AMP receptor protein to activation transcription. Nucleic Acids Res 1990; 18:7243-7250.

84. Zhore Y, Zhang X, Ebright RH. Identification of the activating region of CAP: isolation and characterization of mutants of CAP specifically defective in transcription activation. Proc Natl Acad Sci USA 1993; 90:6081-6085.

85. Ryu S, Garges S, Adhya S. An arcane role of DNA in transcription activation. Proc Natl Acad Sci USA 1994; 91:8582-8586.

86. Igarashi K, Ishihama A. Bipartite functional map of the *E. coli* RNA polymerase alpha subunit: involvement of the C-terminal region in transcription activation by cAMP-CRP. Cell 1991; 65:1015-1022.

87. Tang H, Severinov K, Goldfarb A et al. Location, structure and function of the target of a transcription activator protein. Genes Dev 1994; 8:3058-3067.

88. Blatter EE, Tang H, Ross W et al. Domain organization of RNA polymerase α subunit: C-terminal 85 amino acids constitute an independently folded domain capable of dimerization and DNA binding. Cell 1994; 78:889-896.

88a. Shanblatt SH, Revzin A. Interactions of the catabolite activator protein (CAP) at the galactose and lactose promoters of *Escherichia coli* probed by hydroxy radical footprinting. J Biol Chem 1986; 261:10885-10890.

89. Zhou Y, Rendergrast PS, Bell A et al. The functional subunit of a dimeric transcription activator protein depends on promoter architecture. EMBO J 1994; 13:4549-4557.

90. Jin DJ, Turnbough CL Jr. An *Escherichia coli* RNA polymerase defective in transcription due to its overproduction of abortive initiation products. J Mol Biol 1994; 236:72-80.

91. Ullmann A, Joseph E, Danchin A. cyclic AMP as a modulator of polarity in polycistronic units. Proc Natl Acad Sci USA 1979; 76:3194-3197.

92. Wilson DB, Hogness DS. The enzymes of the galactose operon in *Escherichia coli* IV. The frequencies of translation of the terminal cistrons in the operon. J Biol Chem 1969; 244:2143-2148.

93. Adhya S, Garges S. Unpublished results.

94. Zabin I. Cold Spring Harbor Symposia on Quant Biol 1963; 28:431-435.

95. Michels CA, Zipser D. The non-linear relationship between the enzyme activity and structural protein concentration of thiogalactoside transacetylase of *E. coli*. Biochem. Biophys Res. Commun 1969; 34:522-527.

96. Choy HE, Adhya S. Control of *gal* transcription through DNA looping: inhibition of the initial transcribing complex. Proc Natl Acad Sci USA 1992; 90:472-476.

97. Haber R, Adhya S. Interaction of spatially separated protein-DNA complexes for control of gene expression: operator conversions. Proc Natl Acad Sci USA 1988; 86:9683-9687.

98. Brenowitz M, Jamison E, Majumdar A et al. Interaction of the *Escherichia coli* Gal repressor protein with its DNA operators in vitro. Biochem 1990; 29:3374-3383.

99. Brenowitz M, Mandal N, Pickar A et al. DNA-binding properties of a Lac repressor mutant incapable of forming tetramers. J Biol Chem 1991; 266:1281- 1288.

100. Kramer H, Niemoller M, Amouyal M et al. *lac* repressor forms loops with linear DNA carrying two suitable spaced *lac* operators. EMBO J 1987; 6:1481- 1491.

101. Mandal N, Su W, Haber R et al. DNA looping in cellular repression of transcription of the Galactose operon. Genes & Develop 1990; 4:410-418.

102. Alberti S, Oehler S, von Wilcken-Bergmann B et al. Genetic analysis of the leucine heptad repeats of Lac repressor: evidence for a 4-helicas bundle. New Biol 1991; 3:57-62.

103. Friedman AM, Fischmann TO, Steitz TA. Crystal structure of *lac* repressor core tetramer and its implication for DNA looping. Science 1995; 268:1721- 1727.

104. Choy HE, Park SW, Parrack P and Adhya S. Transcription regulation by inflexibility of promoter DNA in a looped complex. Proc Natl Acad Sci USA 1995; 92:7327-7331.

105. Choy H, Park SW, Aki T, Parrack P, Fujita N, Ishihama A, Adhya S. Repression and activation of transcription by Gal and Lac repressors: Involvement of alpha subunit of RNA polymerase. EMBO J 1995; 14:4523-4529.

106. Schlax PJ, Capp MW, Record Jr MT. Inhibition of transcription initiation by Lac repressor. J Mol Biol 1995; 245:331-350.

107. Straney SB, Crothers DM. Lac repressor is a transient gene-activating protein. Cell 1987; 51:699-707.

108. Lee J, Goldfarb A. Lac repressor acts by modifying the initial transcribing complex so that it cannot leave the promoter. Cell 1991; 66:793-798.

109. Brodolin KL, Studitsky VM, Mirzabekov AD. Conformational changes in *E. coli* RNA polymerase during promoter recognition. Nucleic Acids Res 1993; 21:5748-5753.

110. Kleina LG, Miller JH. Genetic studies of the *lac* repressor. XIII. Extensive amino acid replacements generated by the use of natural and synthetic nonsense suppressors. J Mol Biol 1990; 212:295-318.

111. Eschenlauer AC, Reznikoff WS. *Escherichia coli* catabolite gene activator protein mutants defective in positive control of *lac* operon transcription. J Bacteriol 1991; 173:5024-5029.

112. Busby S, Irani M, deCrombrugghe B. Isolation of mutant promoters in the *Escherichia coli* galactose operon using localized mutagenesis on cloned DNA fragments. J Mol Biol 1982; 154:197-209.

113. Busby S, Aiba H, deCrombrugghe B. Mutations in the *Escherichia coli* operon that define two promtoers and the binding site of cyclic AMP receptor protein. J Mol Biol 1982; 154:211-227.

114. Vyas NK, Vyas MN, Quiocho FA. Comparison of the periplasmic receptors for L-arabinose, D-glucose/D-galactose and D-ribose. Structural and functional similarity. J Biol Chem 1991; 266:5226-5237.

115. Chakerian AE, Matthews KS. Characterization of mutations in oligomerization domain of Lac repressor protein. J Biol Chem 1991; 266: 22206-22214.

CHAPTER 10

THE MALTOSE SYSTEM

Winfried Boos, Ralf Peist, Katja Decker and Eva Zdych

INTRODUCTION AND SCOPE

The maltose system consists of a number of genes that are under the control of a single transcriptional activator, the MalT protein. The proteins encoded are localized in different compartments of the cell, the cytoplasm, the periplasm, the cytoplasmic membrane and the outer membrane. This is probably the reason why the maltose system has become so attractive to scientists of different biological interests, and hence has become a model system for the analysis of many important biological functions:

- The genetic approach used to elucidate the machinery of protein secretion employed the maltose system extensively and most *sec* genes have been defined using this system.[1]
- The proteins comprising the complex transport system have become the standard for defining the function of binding protein-dependent ABC transporters in gram-negative bacteria.[2-4]
- The modern methods of *phoA*- and *lacZ*-fusions used to elucidate the two-dimensional topology of membrane proteins were introduced using MalF, one of the intrinsic membrane proteins of the system.[5]
- The λ-receptor, located in the outer membrane, was one of the first proteins recognized as a specific diffusion pore specifically catalyzing the entry of maltodextrins into the periplasm.[6-9]
- The MalT-dependent regulation of the maltose genes was the first system recognized to be controlled by a purely positive acting and inducer-dependent transcriptional activator.[10-12]
- The envelope localized λ-receptor has been used to genetically engineer surface located epitopes for the production of specific antibodies.[13] Similarly, the periplasmic maltose-binding protein (MBP) has become a vehicle for export from *E. coli* of engineered proteins of eukaryotic origin.[14-16]

For the historical perspectives and the biological relevance of the maltose system the reader is referred to the comprehensive review by Schwartz[17] covering the literature up to 1986.

Regulation of Gene Expression in Escherichia coli, edited by E. C. C. Lin and A. Simon Lynch. © 1996 R.G. Landes Company.

The topics in this book deal with the regulatory control systems in *E. coli*. Superficially, the regulation of the maltose system seems to be straightforward. There is MalT, a cytoplasmic transcriptional activator necessary for the transcription of all *mal* genes, which are organized in operons and dispersed over the chromosome. The transcription of *malT* itself, as well as some of the *malT*-dependent *mal* genes, is subject to catabolite repression and is thus dependent on the cAMP·CAP complex. In addition, even though maltodextrins from maltose to maltoheptaose (and probably longer) induce the expression of the *mal* genes when present in the growth medium, implying that they could be recognized by MalT, it is, in fact, only maltotriose which can activate MalT to become an activator inside the cell. Thus, for the geneticist, the *mal* system is controlled by a positive regulator that becomes active by binding the inducer maltotriose.

However, this is only the basic framework of the regulation seen in the maltose system, namely the final step concerning gene expression.

Yet, several other regulatory phenomena have been observed with the maltose system that are not so straightforward and most of which are only understood on a phenomenological level:

- MalK, the ATP-hydrolyzing subunit of the maltose-binding protein-dependent transport system, acts phenotypically as a repressor, reducing the MalT-dependent transcription of the *mal* genes. The mechanism of this repression is unclear.
- The uninduced level of expression, as well as the constitutive expression observed in *malK* mutants, is strongly reduced at high osmolarity, for instance at 250 mM NaCl. It is possible that this phenomenon is related to the observation that in certain *envZ* mutants the over-phosphorylation of OmpR, the response regulator of the two-component regulatory system of osmoregulation, causes repression of *malT*, thus strongly affecting *mal* gene expression.
- *malQ* mutants lacking amylomaltase, one of the key metabolic enzymes of the system, are constitutive for the expression of the maltose system. This constitutivity can be observed only if the strain is not defective in its capability to synthesize glycogen. Thus the synthesis and degradation of glycogen is connected to the expression of the maltose system.
- The uninduced level of *mal* gene expression (for instance in the absence of external maltodextrins) is controlled by the internal synthesis of inducer, presumably maltotriose. The synthesis of inducer is independent of all *mal* genes, but is counteracted by amylomaltase and maltodextrin glucosidase, two maltose metabolizing enzymes. By this *mal* gene-independent pathway, the cytoplasmic degradation of carbohydrates other than maltodextrins can result in the formation of internal inducer, the degradation of trehalose being a prominent example. This indicates the presence of an enzyme responsible for the formation of inducer. The available evidence points to an as yet uncharacterized phosphorylase which will form maltose from glucose and glucose-1-P and subsequently, maltotriose from glucose-1-P and maltose. The gene for this enzyme is as yet unknown. Connected to the production of internal inducer is probably also the derepression of the maltose system, as well as other sugar metabolizing systems, during starvation for glucose.

- At 36 minutes on the *E. coli* chromosome, a gene cluster has been found which compromises *malI*, which encodes a repressor and next to it, *malX* and *malY*, both being controlled (repressed) by the gene product of *malI*. Mutants in *malI*, leading to the constitutive expression of *malY*, result in a strong repression of the *malT*-dependent expression of all *mal* genes. The nature of this repression is not understood, but it is clear that additional components defined by insertion mutants are necessary for MalY-dependent repression to occur. By the same type of analysis, it has been found that MalK and MalY-dependent repression of the *mal* genes occurs by a common mediator.

This review will focus on the *physiological* control circuits affecting *mal* gene expression. The MalT-dependent transcriptional control that has been analyzed on a molecular level in great detail was pioneered by the work of Olivier Raibaud and Evelyn Richet and their co-workers. This topic will only be dealt with on a superficial level necessary for the understanding of the physiology.

Since an important role in the physiological control circuit is performed by a subunit of the transport system, and since the enzymes take part in controlling the level of inducer, we will also present a discussion on the transport system and the metabolic enzymes. In Table 10.1 we summarize the *mal* genes, as well as other genes related to the maltose system, their genetic organization, and the function of their products.

Some of the *mal* genes are organized in clusters. The *malA* region at 75 minutes contains the two divergently oriented operons with *malT* transcribed clockwise and *malP/Q* transcribed counterclockwise.[85-87] Likewise, all maltose transport genes are clustered at 91 minutes in the *malB* region with two divergently organized operons: *malE/F/G* in counterclockwise, and *malK/lamB/malM* in clockwise orientation.[32,88]

THE POSITIVE TRANSCRIPTIONAL ACTIVATOR MalT

THE *malT* GENE, ITS PRODUCT AND ITS EXPRESSION

An early genetic approach identified *malT*, a gene that appears essential for the expression of all maltose-inducible functions.[10,85,89] Since all maltose-inducible operons require *malT* and since *malT* fails to repress the operons in the absence of inducer,[90,91] it seemed clear that the protein encoded by *malT* is a purely positive regulator that is activated by inducer. The *malT* gene was cloned[92] and sequenced,[18] and its product, MalT, purified and characterized.[11] The protein appears monomeric in solution, contains 901 amino acids and exhibits a molecular weight of 103,000. It binds ATP (K_D 0.4 μM)[21] and maltotriose (K_D of about 20 μM)[93], both of which are necessary for transcriptional activation, even though ATP hydrolysis is not required, since the binding of ADP or nonhydrolyzable ATP analogs also stimulates transcription.[21]

Mutations in *malT* (*malT^C*) have been isolated resulting in the constitutive expression of all *mal* genes.[90,93] These mutations center around amino acids 243 and 358 of the polypeptide chain. Whereas the wild-type MalT protein is inactive in the absence of maltotriose, the mutant proteins are active in the absence of maltotriose and exhibit an increased affinity towards this sugar. All mutant proteins can still be stimulated by maltotriose.[93] The last C-terminal 95 amino acids of MalT constitute the DNA binding domain. In this area, MalT exhibits homology to a number of procaryotic transcriptional activators.[22]

Table 10.1. The mal *genes and their products and genes controlling their function*

Gene	min on the chromosome	Gene product	Comment	Reference
malT	75	transcriptional activator	controls transcription of all *mal* genes except the *malI/X/Y* gene cluster	11,12,18-22
malE	91.5	periplasmic maltose-binding protein	binds maltose/maltodextrins with μM affinity	6,23-26
malF	91.5	intrinsic membrane protein of the transport system	forms, together with MalG, the translocator	27-31
malG	91.5	intrinsic membrane protein of the transport system	forms, together with MalF, the translocator	29,32-36
malK	91.5	transport ATPase responsible for energization of transport	peripherally bound to MalF/G on the inside of the plasma membrane; overproduction represses the *mal* genes	37-45
lamB	91.5	receptor for phage λ and specific pore for maltodextrins	efficient pore for other carbohydrates at low concentration	6-9,46-49
malM	91.5	periplasmic protein of unknown function	not essential for transport; contains an Ala-Pro linker region also found in OmpA	50-52
malP	75	maltodextrin phosphorylase	substrates are maltopentaose and larger maltooligosaccharides	53-55
malQ	75	amylomaltase	maltodextrinyl-transferase with maltotriose as smallest substrate	56-60
malS	80	periplasmic α-amylase	cleaves preferentially maltohexaose from the nonreducing end of maltodextrins	7,61,62
malZ	9	maltodextrin glucosidase and γ-cyclodextrinase	cleaves glucose sequentially from the reducing end of maltodextrins; maltotriose is the smallest substrate	63,64
malI	36	repressor for *malX/Y*	not dependent on MalT, inducer not known	65
malX	36	enzyme II of the PTS	transports and phosphorylates glucose; can transport maltose by diffusion	66
malY	36	βC-S lyase (cystathionase)	overproduction represses the *mal* genes	66,67
mac	10	glucose/maltose transacetylase	not MalT-dependent; responsible for exit of maltose as acetyl-maltose, and repression by removing glucose and maltose, the precursors of the inducer	68,69
glgA	75.4	glycogen synthase	ADP-glucose-dependent synthesis of glycogen	70,71
glgC	75.4	ADP-glucose- pyrophosphorylase		70,72
glgP	75.4	glycogen phosphorylase		73
glgB	75.4	branching enzyme		70,74
glgX	75.4	amylase-like enzyme	role in glycogen degradation unclear	75
amyA	42-43	cytoplasmic α-amylase	not MalT-dependent; no apparent role in glycogen degradation	76
galU	26	UDP-glucose pyrophosphorylase	possible origin of cytoplasmic glucose	77
glgS	66	short polypeptide	involved in RpoS-dependent glycogen synthesis	78
glk	52	glucokinase	control of endogenous glucose	79,80; accession number U2249
pgm	15.5	phosphoglucomutase	endogenous induction	81-84

MalT mutant proteins in which this fragment is deleted are negatively dominant over the function of the wild-type protein, indicating that the protein interacts with itself when binding DNA.[18]

The expression of *malT* is not autoregulated but is subject to catabolite repression and, therefore, requires the presence of the cAMP·CAP complex.[86,94,95] Mutations in the control region of the *malT* gene have shown that its expression in the wild type is limited at both the transcriptional and translational levels.[19] The use of deletions introduced upstream of the *malT* promoter revealed that the 120 base pairs upstream of the transcriptional start site, encompassing the binding sites for the polymerase and the cAMP·CAP complex, are sufficient for *malT* expression. However, the deletion of DNA further upstream increased expression of *malT*, indicating the removal of a binding site for a protein reducing expression.[87,96] Whether or not this site is involved in the dominant negative effect of certain mutations in *envZ*[97-99] resulting in an overphosphorylated OmpR protein,[100-102] is still unclear at present. Interestingly, the DNA upstream of the malT promoter whose deletion results in an increase of *malT* expression contains sequences that are highly homologous to sequences identified as binding sites for phosphorylated OmpR.[103] It is clear that mutations in *envZ* affect transcription of *malT*.[104]

THE MalT BOX

The first feature recognized for at least three MalT-dependent promoters was the lack of the usual -35 region, whereas the -10 region corresponds to those of constitutive promoters.[105] Instead, the asymmetric hexanucleotide sequence 5'-GGA(G/T)GA-3', the so called MalT box, was recognized centered at -37.5 or at -38.5 bp upstream of the transcriptional start point.[106-109] In addition, DNA upstream of the -38 region was found to be essential for *mal* gene expression,[108,109] whereas the 30 base pair sequence preceding the transcriptional start point contains only a few positions that are essential for promoter activity.[110]

The second structural feature recognized to be essential for MalT-dependent *mal* gene expression was two additional MalT boxes in direct repeat upstream of the -38 region.[12,111,112] The analysis of these MalT boxes also lead to an extension in the consensus sequence which is now defined as 5'-GGGGA(G/T)GAGG-3'.[111] In contrast to the orientation of the MalT box which is proximal to the transcription initiation site and always oriented 5' to 3' in the direction of transcription, the orientation of the two repeats can be either way. This motif of MalT-dependent promoter structure with three MalT boxes has also been found by sequence analysis upstream of *malS*[62] and *malZ*,[64] two MalT-dependent genes encoding maltodextrin-hydrolyzing enzymes in *E. coli*. By using the decanucleotide 5'-GGGGAGGAGG-3', the MalT-binding consensus box, and nucleotide sequences present in the polylinker of a vector plasmid as connecting sequences, Danot and Raibaud constructed several functional semisynthetic promoters and tested them for MalT-dependent gene expression. These studies confirmed the importance of the structural motif formed by two MalT-binding sites in a direct repeat. This motif is involved in promoter activation either alone or in conjunction with a third MalT-binding site proximal to the transcription start site. In this configuration, the promoters are active irrespective of the orientation of the repeat. Provided that the alignment along the axis of the helix is retained, the distance of the repeat to the proximal MalT-binding site can be varied to some extent. This analysis defines the inducible MalT-dependent promoter that does not require the cAMP·CAP complex.[113] The role of these sites for MalT to contact and activate RNA polymerase has been assessed by mutating each site and measuring the contribution of the remaining sites.[114]

Of the five known MalT-dependent operons in *E. coli*, expression of *malP/Q* and *malZ* is independent of cAMP·CAP, whereas expression of *malK/lamB/malM* and *malE/malF/malG* is dependent on cAMP·CAP. Expression of *malS* is most likely also cAMP·CAP dependent since a putative cAMP·CAP binding site precedes the *malS* gene. The *malK/lamB/malM* and *malE/malF/malG* operons are oriented divergently to each other. Their transcription start sites are 271 base pairs apart. Located between these start sites is a 210 base pair regulatory region which comprises two series of MalT boxes (three MalT boxes, including the repeat, in front of *malK*, and two MalT boxes, the repeat, in front of *malE*) separated by three cAMP·CAP binding sites. Apparently, the entire control region is required for the full expression of both operons. This suggested that multiple copies of MalT and cAMP·CAP form an unique nucleoprotein structure at this regulatory region that is essential for both promoters.[12] The observation that the function of these sites is sensitive to the phase of the DNA helix has lead to a model in which the DNA is wrapped around the complex composed of MalT and cAMP·CAP.[12,115] The state of supercoiling also appears to be important for the effectiveness by which the two operons are transcribed. Not only is relaxed DNA less efficient in transcription initiation, but also the sites occupied by the activator complex are shifted in supercoiled and relaxed DNA.[116]

By a series of elegant experiments, it was found that the effect of cAMP·CAP binding in the intergenic regulatory region between *malE* and *malK* results in a repositioning of MalT binding. In the absence of cAMP·CAP, MalT binds with high affinity to the three MalT boxes upstream of the *malK* transcriptional start site (identified previously). In the presence of cAMP·CAP, MalT-binding is shifted by three nucleotides towards the Pribnow box of the *malK* promoter. The cAMP/CAP effect requires the *malKp*-distal MalT-binding sites.[117] The repositioning is caused by DNA-bending, that can be mimicked when replacing cAMP·CAP by the integration host factor.[118] Recently, it has been observed that Lrp, the leucine responsive protein,[119] also affects the transcription of *malT,* as well as of some *malT*-dependent genes.[120] It is unclear whether this effect of Lrp is the consequence of its direct action on the *mal* gene promoters or an indirect effect, for instance via the level of the cAMP·CAP complex.

THE MALTOSE/MALTODEXTRIN TRANSPORT SYSTEM

The maltose/maltodextrin transport system represents a member of the multicomponent and periplasmic binding protein-dependent ATP-Binding Cassette (ABC) high affinity transport systems of gram-negative enteric bacteria.[2-4,121-123] The substrate recognition site of the system is primarily determined by the soluble binding protein with a high affinity for maltose and maltodextrins (K_D around μM)[6,23] that is located in the periplasm in high concentration (around mM and in 30- to 50-fold molar excess over the intrinsically membrane-bound proteins of the system).[124] This maltose-binding protein (MBP) consists of two nearly symmetrical lobes between which the binding site is formed.[25] Substrate-loaded and substrate-free forms of MBP, as well as other substrate-binding proteins, differ dramatically in their conformation such that substrate bound to the protein no longer has access to bulk solvent; whereas, in the substrate-free form, the lobes are open and the substrate-binding site becomes accessible to the bulk solvent.[125] Both the substrate-loaded and the substrate-free form of MBP have access to the membrane components, MalF and MalG,[126] but only the loaded form is able to initiate substrate translocation through the membrane.[127]

The interaction of MBP with MalF and MalG has been studied by the genetic approach of mutant and suppressor analysis. This study in combination with the knowledge of the crystal structure of MBP has led to the conclusion that one lobe of MBP interacts with MalF, the other with MalG.[26] The application of the *phoA* fusion technology allowed the determination of the two-dimensional topology of both MalF[30,128] and MalG.[34,36] Accordingly, MalF consists of eight membrane spanning α-helical segments (MSS) with one large periplasmic loop between MSS 1 and 2, and both termini of the polypeptide chain protruding into the cytoplasm. MalG, as judged by the same criteria, consists of six MSS, and again, both termini extend into the cytoplasm. Near their C-termini, MalF and MalG carry a sequence conserved in all intrinsic membrane-bound subunits of binding protein-dependent ABC systems.[129] This site might be related to their transport function or may be essential for the binding of the last component of this complex transport machinery, the MalK subunit.

The MalK subunit contains a classical consensus sequence found in all ATP hydrolyzing proteins.[130] It consists of two subsites, the A and the B domain. These two motifs, as well as the sequence around these sites, are conserved in the equivalent subunits of all binding protein-dependent transport systems.[131] In addition, they are found in many proteins of prokaryotic and eukaryotic origin whose function is in the transport of molecules, including polysaccharides, peptides and proteins.[132] MalK and equivalent proteins of other binding protein-dependent transport systems have been characterized not only as ATP-binding proteins, but as ATPases that hydrolyze ATP when embedded in the membrane (or solubilized in detergent) as a complex with MalF and MalG, or the corresponding binding protein triggered by the interaction with substrate-loaded MBP or the corresponding binding protein.[43,121,127,133,134] The active complex consists of two MalK subunits connected by one molecule each of MalF and G.[135] When this complex is reconstituted in liposomes, ATP-dependent active transport of maltose into the liposomes can be measured, demonstrating the function of the MalK dimer as an energy module driving the active transport of maltose. Nothing is known about the mechanism of energy coupling of ATP hydrolysis to the accumulation of maltose. Since neither the substrate nor any of the proteins involved have been seen to become phosphorylated during transport, it has become commonplace to interpret energy coupling as the transfer of the energy gained through ATP hydrolysis to protein-conformational energy in MalF and G, followed by subsequent binding protein-triggered release of the substrate, resulting in the unidirectional translocation of substrate through the membrane. Mutants in *malF* or *malG* have been isolated that no longer require the triggering by substrate-loaded MBP. It is noteworthy that the MalFGK$_2$ complex of these mutants no longer requires the presence of substrate-loaded MBP to stimulate ATPase activity of MalK;[121] ATPase activity has become uncoupled in these mutants, similarly to the uncoupled ATPase activity of the purified wild-type MalK subunit alone.[41] This has led to the notion that the transport system is a signal transduction pathway that begins in the periplasm with the recognition of maltose by the binding protein and ends with the control of MalK ATPase activity in the cytoplasm.[121]

No detailed studies on the synthesis of MalK in comparison to its partners in the membrane, MalF and MalG, are available. Even though the stoichiometric composition of the biochemically active complex has been determined as MalF/MalG/MalK$_2$,[135] it is unclear whether or not MalK is synthesized in excess or, depending on regulatory conditions, in varying amounts. This is not unlikely since *malF* and *malG* are located in a different operon than *malK*, and substantial amounts of MalK have been reported in contrast to its membrane-bound partners, MalF and MalG.[40] As will be discussed later, MalK plays an important role in the MalT-dependent control of *mal* gene expression. It is only natural to see MalK not only as an energy module to drive transport, but also as a control unit of transport activity. Obviously, the availability of ATP will have an influence on transport activity. But since the K$_m$ of MalK for ATP is in the µM range[134] and the physiological concentration of ATP in the mM range, the cells would presumably have to become exhausted of ATP before they would stop transporting maltose.

We postulate an additional level of control for the function of MalK in transport which works by regulating the affinity of MalK to its partners MalF and MalG. We observed that transport of glycerol-3-phosphate (G3P) by the binding protein-dependent Ugp transport system is inhibited by internal P$_i$.[136] The Ugp system exhibits a surprising homology to the maltose system in all subunits in spite of the differences in substrate specificity and regulation.[137] In the attempt to find whether or not UgpC, the MalK analog of the Ugp system, was the target of P$_i$ inhibition, we exchanged MalK by UgpC[45] and tested a possible inhibition of maltose transport by P$_i$. While P$_i$ did not inhibit the UgpC-mediated uptake of maltose, it failed to inhibit uptake of G3P via the Ugp system when UgpC was overproduced. This indicated that P$_i$ inhibition affects the interaction between UgpC and its cognate membrane components. It also suggests the possibility that the degree of association of the ATP hydrolyzing subunit with the membrane components may control transport activity.

The MalK subunit is usually pictured as being peripherally membrane associated through binding to the intrinsic membrane proteins MalF and MalG. This membrane association of MalK may, in fact, be more intimate. Studies with the functionally related binding protein-dependent histidine transport system of *S. typhimurium* have shown that HisP, the MalK analog of the system, is actually accessible from the periplasm,[138] even though the interaction of the cognate periplasmic binding protein, HisJ, takes place with the intrinsic membrane proteins of the system and apparently not with HisP.[139]

Efficient uptake of maltose and particularly longer maltodextrins at low concentrations requires the presence of maltoporin (λ-receptor), the specific diffusion pore for maltodextrins and other carbohydrates, in the outer membrane.[6,7,47,140-142] It is interesting to note that *lamB*, the gene for the λ-receptor, is, together with *malK*, located in a different operon than the remaining genes (*malE malF malG*) of the transport system. Since MalK is involved in the regulation of the system, and since the λ-receptor is also required for the diffusion of carbohydrates other than maltodextrins, a differential regulation of these two operons (in comparison to the other *mal* genes) should be considered. In addition, translational control of *lamB*[143,144] may be evoked when considering the need of carbohydrate uptake under starvation conditions.[142,145]

THE ENZYMES OF THE MALTOSE SYSTEM

AMYLOMALTASE (ENCODED BY *malQ*) IS A DEXTRINYL TRANSFERASE

Amylomaltase[56-58,60] is a dextrinyl transferase that can transfer maltosyl and longer dextrinyl residues onto glucose, maltose and longer maltodextrins.[59] The smallest substrate which amylomaltase recognizes is maltotriose. Acting on maltotriose, it releases glucose from the reducing end, forms a maltosyl-enzyme complex, and transfers the maltosyl residue onto the nonreducing end of an acceptor, be it glucose, maltose or any larger maltodextrin. The paradoxical consequence of this scheme is that maltose itself is not a primary substrate. Pure maltose is not disproportioned into maltodextrins and glucose unless trace amounts of maltodextrins are present acting as a primer in the reaction. When (^{14}C)-labeled glucose is present during this reaction the maltodextrins that are formed from maltose or longer maltodextrins are exclusively labeled in the glucose residue located at the reducing end of the dextrin, thus proving the mechanism of dextrinyl transfer outlined above. It should be stressed that amylomaltase, in contrast to the MalZ enzyme discussed later, is not only able to release glucose but also longer dextrins from the reducing end of its maltodextrin substrates.[59] Even though this will not result in the release of "productive" glucose and only shuffle dextrins of different sizes, the release of dextrins larger than maltotriose is necessary for the formation of more primers when maltose is the only substrate and when only trace amounts of primers are present initially. When maltose is incubated with amylomaltase, the formation of glucose is initially very slow but increases autocatalytically until an equilibrium between glucose and maltodextrins is reached.[59] Since amylomaltase, strictly speaking, is not a hydrolase but a transferase, it follows that the sum of glycosidic linkages of all maltodextrins involved remains constant. Only when glucose is removed from this equilibrium, for instance by glucokinase-mediated phosphorylation to glucose-6-phosphate, does the degradation of maltose continue while forming longer maltodextrins.

It is important to stress that maltose is not a substrate of amylomaltase but only an acceptor in the transfer reaction catalyzed by the enzyme. Thus, it follows that for maltose degradation the cell has to be able to internally produce small amounts of maltodextrins as primers with the minimum size of maltotriose. Yet, amylomaltase is essential for maltose degradation and *malQ* mutants are unable to grow on maltose. *malQ* mutants are not only Mal⁻, their growth is also inhibited by maltose.[85] These mutants accumulate large amounts of free maltose inside the cell.[6] In the presence of an alternative carbon source, mutations arise that are found preferentially in *malT* or in *malK*. Why no mutations in other maltose transport genes arise whose products are essential in transport remains a mystery.[17] As discussed below, the function of MalK as a repressor is possibly related to this phenomenon.

MALTODEXTRIN PHOSPHORYLASE (ENCODED BY *malP*) FORMS GLUCOSE-1-PHOSPHATE BY SEQUENTIAL PHOSPHOROLYSIS OF THE NONREDUCING END GLUCOSE MOIETIES OF LARGER DEXTRINS

As discussed above, amylomaltase disproportionates maltose into glucose and maltodextrins. Since glucose will be removed in vivo by glucokinase to form glucose-6-phosphate, maltodextrins would accumulate. Indeed, *malP malQ*⁺ mutants can grow on maltose. Under these conditions they become very large, are filled with long linear dextrins[146,147] and stain blue with iodine.[81] Surprisingly, they do not transform these dextrins into glycogen, even though the *glgB*-encoded branching enzyme is present. Accumulation of maltodextrins does not occur in the *malP*+ wild-type strain. Maltodextrin phosphorylase[53,54,148] recognizes maltopentaose and longer linear maltodextrins and cleaves glucose-1-phosphate by phosphorolysis from the nonreducing end of the maltodextrin. Obviously, it is important that maltodextrin phosphorylase does not attack maltotetraose and maltotriose, since dextrins of a minimum size are required for full activity of amylomaltase. One is led to wonder why glycogen phosphorylase, the *glgP*-encoded enzyme supposedly involved in the degradation of glycogen and linear dextrins,[73] cannot replace maltodextrin phosphorylase in the utilization of maltose and maltodextrins as a carbon source.

The reaction catalyzed by maltodextrin phosphorylase is, of course, reversible. The rate in the phosphorolysis direction will be stimulated by increasing concentrations of cytoplasmic P_i. In turn, the concentration of internal P_i will vary depending on the availability of external P_i and phosphorus containing organic compounds, as well as the state of induction of the *phoB*-dependent *pho* regulon.[149] There are other connections between the *mal* and the *pho* regulon. Aside from the unusual sequence homology between the proteins of the maltose- and the *phoB*-dependent Ugp transport system for glycerol-3-phosphate[137] and the functional exchangeability between their ATP-binding subunits, MalK and Ugp,[45] there is the common response (repression) of the two systems by dominant mutations in *envZ* resulting in overphosphorylation of OmpR, the response regulator of the two-component regulatory EnvZ/OmpR system involved in osmoregulation.[104]

In addition, the utilization of a fermentable carbon source, such as glucose, trehalose or maltose, under conditions of P_i concentrations below 1 mM in the growth medium requires the derepression of the *pho* regulon. *phoB* mutants do not turn deep red on MacConkey indicator plates (0.3 mM P_i) unless 5mM P_i is added. Also, whereas strains that turn red on these plates derepress the *pho* regulon, nonfermenters that appear pale remain repressed for the *pho* regulon. Apparently, the utilization of a fermentable carbon source requires higher cellular P_i concentrations than that of a noncarbohydrate carbon source present in MacConkey plates.[150]

THE ROLE OF GLUCOKINASE (ENCODED BY *glk*) AND PHOSPHOGLUCOMUTASE (ENCODED BY *pgm*) IN MALTOSE/MALTODEXTRIN METABOLISM

The final products of the combined action of amylomaltase and maltodextrin phosphorylase are glucose and glucose-1-phosphate. Therefore, to funnel these end-products of the specific maltose enzymes into general metabolism, the cells rely on glucokinase for the phosphorylation of glucose to glucose-6-phosphate and on phosphoglucomutase for the transformation of glucose-1-phosphate to glucose-6-phosphate for entry into glycolysis. Mutants unable to phosphorylate glucose due to the lack of glucokinase and enzyme II^Glc (encoded by *ptsG*) of the phosphotransferase (PTS)-mediated phosphorylation, are unable to grow on maltose.[151] Similarly, they are unable to grow on trehalose, which is degraded by an enzyme that produces internal glucose.[152] Therefore, the utilization of internally produced glucose appears to be essential for growth. It is unclear whether the inability to grow on these sugars in the absence of glucose phosphorylation is due to a possible toxic effect of accumulating internal glucose (requiring extrusion) or the insufficient flow of carbon and energy via the phosphoglucomutase pathway alone. The drain of glucose-1-phosphate for the biosynthesis of polysaccharides might be another factor.

pgm mutants lacking phosphoglucomutase activity are able to grow on maltose even though seemingly only one half of the maltose moiety can be used as a carbon and energy source. These strains exhibit a maltose blue phenotype[81,82] caused by the massive production of dextrins due to the accumulation of glucose-1-phosphate and the reversal of the maltodextrin phosphorylase action. The problem created by the accumulation of maltodextrins is eased by the action of maltodextrin glucosidase, the MalZ enzyme (discussed below) which produces glucose from the maltodextrins, followed by glucokinase-dependent phosphorylation to glucose-6-phosphate. Indeed, *pgm malZ* double mutants are unable to grow on maltose. Similarly, *pgm* mutants can grow on galactose,[81] but only when all the enzymes of the maltose system, including MalZ, are present.[84] *pgm* mutants, even null mutations[83] created by insertion elements, still show residual activity when the enzyme activity is assayed by coupling the formation of glucose-6-phosphate from glucose-1-phosphate to the NADP⁺-dependent oxidation by glucose-6-phosphate dehydrogenase. This activity is caused by a sugar-phosphate transferase which cannot only hydrolyze sugar phosphates but transfer the phosphate moiety to another sugar molecule.[153] Thus, in the presence of free glucose, glucose-1-phosphate can be transformed to glucose-6-phosphate, mimicking phosphoglucomutase activity.

MALTODEXTRIN GLUCOSIDASE (ENCODED BY *malZ*), AN ENZYME OF UNCLEAR FUNCTION

The MalZ enzyme was discovered in *malF* or *malG* mutants that transport maltose independently of MBP. In contrast to the wild type, these mutants also transport paranitrophenyl α-maltoside (NPG2) and are able to hydrolyze this compound in vivo.[63] Since amylomaltase is unable to hydrolyze NPG2 and since the observed NPG2 hydrolyzing activity is maltose-inducible and MalT-dependent, it was clear from the start that the novel enzyme must be a member of the maltose regulon. The cloning and sequencing of the *malZ* gene and the isolation and biochemical characterization of the encoded protein revealed an enzyme that hydrolyzed maltoheptaose and smaller maltodextrins to glucose and maltose. The smallest substrate is maltotriose; maltose is not a substrate. In contrast to other glucosidases, the MalZ enzyme removes glucose consecutively from the reducing end of the maltodextrin chain.[64] The deduced amino acid sequence of MalZ reveals homology to cyclomaltodextrinase.[154] In our early experiments we did not observe hydrolysis of α-cyclodextrin. However, after we became aware of the significant homology to cyclodextrinyl transferases, we checked the enzyme for hydrolyzing other cyclodextrins and pullulan. We found that the enzyme does not hydrolyze α-cyclodextrin nor does it hydrolyze pullulan. But γ-cyclodextrin appears to be an excellent substrate forming maltose and glucose. The significance of this finding is unclear. γ-cyclodextrin is not a carbon source for *E. coli*. The compound can neither diffuse through the outer membrane, nor can it be taken up by the maltose transport system. The function of MalZ remains unclear. *malZ* mutants grow normally on maltose and maltodextrins. Also, the induction of the maltose system seems not to be influenced by lack of the MalZ enzyme. *malQ* mutants cannot grow on maltose or maltotriose. This is somewhat surprising since the action of the MalZ enzyme releases glucose (which could be used as a carbon source) from maltotriose. Most likely, the accumulation of maltose resulting from the MalZ-dependent hydrolysis of maltotriose in the *malQ* mutant is toxic. Indeed, the accumulation of maltose from the medium in a *malQ* mutant is growth inhibitory.

In the attempt to isolate mutants in a postulated maltose/maltotriose phosphorylase involved in the endogenous formation of maltotriose, we selected pseudorevertants of *malQ* mutants that were once again able to grow on maltose. The only mutation necessary was found to be in *malZ*, suggesting that the substrate specificity of MalZ had changed to allow the hydrolysis of maltose. Surprisingly, the purified mutant enzyme is not able to hydrolyze purified maltose, and shows the same activity in hydrolyzing maltodextrins as the wild type. However, using (^{14}C) glucose in the presence of unlabeled maltotriose or maltotetraose, the mutant enzyme, unlike the wild-type enzyme, is able to transfer maltodextrins onto the labeled glucose. The mutant MalZ enzyme has acquired a quality that is similar to that of amylomaltase. Yet, the property of amylomaltase to be strictly a transferase is not true for the mutant MalZ enzyme. The latter also remains a hydrolase, allowing the net hydrolysis of maltose to glucose in the presence of maltodextrins.[155] As in the case of amylomaltase, the utilization of maltose by the mutant MalZ enzyme in vivo requires the endogenous formation of maltodextrins that can be used as primers. The MalZ enzyme, mutant or wild type, cleaves the glycosidic linkage of only the last glucose residue at the reducing end, and not, like amylomaltase, also at other positions. The first step in the catalysis of amylomaltase and the MalZ enzyme (release of glucose) is identical. But, whereas the formed maltodextrinyl-enzyme complex (in the case of amylomaltase) can only be transferred onto the hydroxyl group of the C-4 carbon of the nonreducing glucose residue of the acceptor molecule, in the case of MalZ, it can also be transferred to water.

THE PERIPLASMIC α-AMYLASE (ENCODED BY *malS*) PREFERENTIALLY CLEAVES MALTOHEXAOSE FROM LARGER MALTODEXTRINS

The *malS* gene[62] was discovered as a maltose-inducible and MalT-dependent gene by *lacZ* fusion technology.[61] It cleaves maltodextrins but not maltose. Its preferred product released from larger dextrins is maltohexaose. *malS* mutants have no recognizable maltose phenotype. The function of the enzyme is most likely the degradation of longer dextrins that enter the periplasm to shorter dextrins that can be transported by the binding protein-dependent maltose/maltodextrin transport system.[7] Even though the maltose-binding protein, the recognition site of the transport system, binds all maltodextrins from maltose to amylose, only dextrins up to the size of maltohexaose can be transported across the membrane.[156] This is most likely due to the modes of substrate recognition by MBP. Those substrates that can be attached to MBP at the reducing end will be transported, those which are bound within the dextrinyl chain (such as cyclodextrins) are not.[6,23,157-159]

Figure 10.1 shows the pathway by which we picture maltose is degraded into glucose and glucose-1-phosphate by the maltose-specific enzymes.

NONCLASSICAL REGULATORY PHENOMENA

MALK, THE ATP-HYDROLYZING SUBUNIT OF THE TRANSPORT SYSTEM, ACTS PHENOTYPICALLY AS A REPRESSOR

It has been known for a long time that mutations in the transport system, later identified as *malK* mutations, lead to an elevated expression

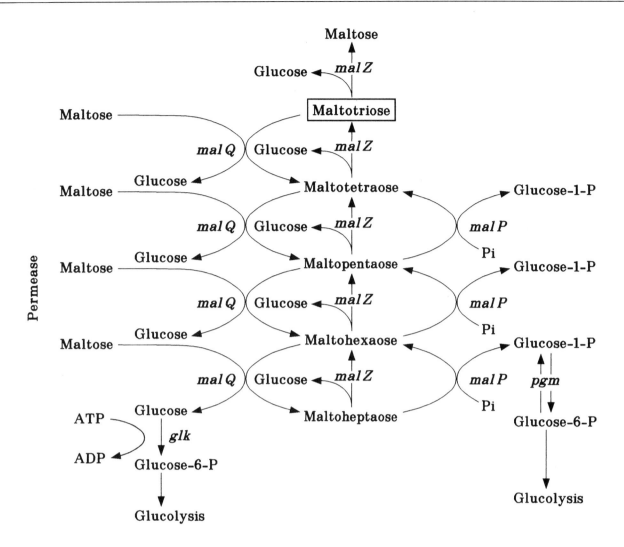

Fig. 10.1. Maltose degradation by the maltose enzymes. The enzymes amylomaltase (malQ), maltodextrin phosphorylase (malP) and maltodextrin glucosidase (malZ) are indicated by their genes. After transport of maltose by the binding protein-dependent ABC transporter a maltosyl, maltotriosyl (and so on) residue is transferred from maltotriose, maltotetraose or maltopentaose (and so on) onto the incoming maltose releasing glucose in the process. Maltopentaose and longer maltodextrins are recognized by the maltodextrin phosphorylase forming glucose-1-phosphate and a maltodextrin that is smaller by one glucosyl residue. Maltodextrin glucosidase recognizes maltotriose and longer maltodextrins (up to maltoheptaose), releasing glucose consecutively from the reducing end of the maltodextrin. This scheme demonstrates that maltose degradation by the maltose enzymes requires the endogenous presence of a maltodextrin primer with the minimal size of maltotriose. The activity of the last maltose enzyme, the periplasmic α-amylase, is not considered in this scheme. Adapted from Decker K et al, J Bacteriol 1993; 175:5655-5665.

of the remaining *mal* genes.[160,161] This phenomenon can be conveniently observed with *malK-lacZ* fusions that no longer exhibit MalK function. These mutants exhibit high and constitutive β-galactosidase activity. An intact MalT activator is required for this constitutivity. The same *malK-lacZ* fusion in a *malK*⁺ merodiploid genetic background, exhibits low β-galactosidase activity that can be induced by maltose.[161] This indicates that MalK acts as a repressor. Also, when plasmid-encoded MalK is overexpressed in the original *malK-lacZ* fusion strain, the β-galactosidase activity is abolished. This effect can also be observed in a wild-type strain where the overexpression of MalK renders the strain unable to grow on maltose even though the strain can grow normally on glycerol.[42] The mutational analysis of MalK has shown that the regulatory function of MalK resides in the C-terminal domain of the protein and that transport and regulatory function of the protein are independent of each other.[44] Surprisingly, a *malT*ᶜ mutation expressing the *mal* genes in the absence of inducer can counteract the repressing effects of overproduced MalK.[42]

At present, it is still unclear how MalK mediates its repressing function. Two possibilities have been excluded. The first was that the C-terminal portion of MalK might act as an enzyme that degrades or inactivates the inducer, maltotriose. This possibility was attractive since *malT*ᶜ mutants, which no longer require inducer or require less inducer,[93] are resistant to the repression mediated by overproduced MalK. However, the ability of bacteria to synthesize maltotriose is not abolished by the overproduction of MalK, nor does the production of maltotriose in the presence of overproduced MalK lead to induction.[84] Thus, MalK does not seem to act at the level of induction. The second possibility was that MalK, as long as it is not engaged in transport, interacts with the MalT protein and inactivates it. The MalT protein has been proposed to exist in an equilibrium of two forms, only one of which would be able to act as a transcriptional activator.[20] Thus, it seemed possible that MalK and inducer would compete for MalT, shifting the equilibrium either with maltotriose to the inducing conformation, or with MalK to the inactive form. However, by using antibodies against MalK or MalT either in the absence or the presence of maltotriose and using strains that overproduced both MalT and MalK, we were unable to biochemically demonstrate complex formation between the two proteins, even though, surprisingly, we found that antibodies against β-galactosidase precipitate a complex of β-galactosidase and MalT, but not MalT alone (Decker and Boos, unpublished observations).

The genetic approach may still be the most promising one in elucidating the mechanism of MalK-mediated repression. By isolating antibiotic resistance insertion mutations on the chromosome, we found insertions in two different genes which abolish the repressing effect of MalK. This clearly indicates that the inactivating effect of MalK on MalT is indirect and may require as many as two additional proteins or their products. The two insertions alone in an otherwise wild-type genetic background result in a level of *mal* gene expression that is 2- to 6-fold over the uninduced wild-type level. The insertions do not map in any of the known *mal* genes. Hopefully, the sequencing of the genes in which the insertions have occurred will reveal the function of the encoded proteins.

The MalK protein is unusually long in comparison to other ATP hydrolyzing subunits of binding protein-dependent ABC transport systems, the difference being a C-terminal extension. On the other hand, UgpC, the corresponding subunit of the Ugp system, is homologous to MalK over its entire length. Indeed, MalK and UgpC can be exchanged in the two systems and complement the transport defect of the heterologous system, but only in the absence of the cognate subunit[45] indicating the lower affinity in forming the heterologous complex. It was of interest whether or not UgpC would also be active as a repressor. While the overproduction of UgpC had no repressing effect on *mal* gene expression, it did reduce the expression of the remaining *ugp* genes (Tommassen, personal communication; Boos, unpublished observation). Thus, the phenomenon of repression exerted by an ATP hydrolyzing subunit of a binding protein-dependent transport system is not a peculiarity of MalK, but may represent a novel form of regulation occurring in binding protein-dependent transport systems.

The MalK subunit is also the target of PTS-mediated inducer exclusion that is brought about by the interaction of nonphosphorylated EIIA^{Gluc} with MalK.[44,162,163] This interaction leads to a reduction in V_{max} without changing the K_m of the transport system.[162] This is reminiscent of the effect of the UgpC-mediated inhibition of the Ugp transport system by internal P_i,[136,149] which we interpret by the dissociation of UgpC from its cognate membrane components. Thus, it appears that the state of MalK (membrane bound or cytosolic) is important for its function in transport and regulation.

THE *malI malX malY* GENE CLUSTER EFFECTS *mal* GENE EXPRESSION

The *malK-lacZ* fusion lacking MalK function and being expressed at high levels has been used to select mutations whose characterization might reveal the cause for the constitutivity or the mechanism of MalK-dependent *mal* gene repression. A chromosomal Tn*10* insertion was isolated that abolished the constitutivity of the *malK-lacZ* fusion. The gene was mapped at 36 minutes on the genetic map, outside any known *mal* operon. It was named *malI* to indicate the expected function of the gene product in the synthesis of the endogenous *mal* gene inducer.[164] The *malI* gene was cloned and sequenced. The deduced amino acid sequence of the *malI* gene product revealed that it is highly homologous to the classical repressor proteins, such as the Lac or Gal repressors,[65] and is likely to encode a repressor itself. An operon containing two genes, *malX* and *malY*, is located next to, and transcribed divergently from, *malI*. This operon, as well as *malI* itself, is repressed by the MalI protein.[66] Therefore, MalI is not an enzyme involved in the synthesis of the endogenous inducer of the *mal* system, but mutations in *malI* lead to the constitutive expression of the *malX malY* operon, which in turn represses the *mal* system. Sequencing revealed that *malX* encodes a protein that shows homology to the enzyme II^{Glc} of the PTS, encoded by *ptsG,* and catalyzing transport of glucose. Indeed, the expression of *malX* could complement *ptsG* mutants for growth on glucose. Yet, the expression of *malX* is not causative in repressing *mal* gene expression. It is *malY*, the distal gene in the operon, that causes repression when overexpressed, either in a *malI* mutant, or when cloned under *tac* promoter control.[66] The sequence of *malY* indicates weak homology to aminotransferases, including the consensus sequence of a pyridoxal phosphate binding site. However, the purified enzyme did not exhibit any aminotransferase activity with glutamate as substrate. Instead, we found βC-S lyase activity (cleavage of a βC-S bond in amino acids) associated with the protein, similar to cystathionase activity.[67] This enzyme, encoded by *metC*, is essential in methionine biosynthesis and cleaves cystathionine to homocysteine, ammonia and pyruvate.[165] Indeed, *metC* mutations can be complemented for growth in the absence of methionine by *malI* mutants or by multicopy plasmids harboring *malY*, even though MalY and MetC do not show significant sequence homology and the quaternary structure of the two proteins is entirely different. MetC is a hexamer and MalY a monomer.

However, the βC-S lyase activity of MalY is not the agent for the repression of the *mal* genes. A mutation in the pyridoxal phosphate binding site of MalY was constructed that resulted in the loss of βC-S lyase activity, even though it was still active in repressing the *mal* genes. In addition, overproduction of the MetC protein had no effect on the regulation of the *mal* genes.[67] Either the MalY-mediated repression is due to protein-protein interaction, thus requiring a certain protein conformation, or the MalY protein exhibits an additional enzymatic activity that is causative for repression.

The mechanism by which MalY interferes with the regulation of the maltose system remains elusive. The final target must be MalT since MalT^c mutants are as insensitive to overproduced MalY as they are to overproduced MalK. As in the case of MalK-dependent repression, the production of endogenous inducer does not seem to be affected by MalY.[84] Also, it was not possible to precipitate a complex of MalY and MalT from cellular extracts with specific antibodies against MalY, even when both proteins were overproduced. As mentioned above, two different antibiotic resistance insertion mutations have been isolated that abolish the effect of overproduced MalK on the expression of the *mal* genes. Surprisingly, one of them also abolished the effect of overproduced MalY. In addition, a third insertion mutation has been isolated that only affects MalY mediated repression. The analysis of this mutation revealed that the action of MalY is competitive in nature. That is, when MalY is synthesized constitutively by the *malI* mutant, or at an uninduced level (no IPTG) from a *tac* promoter-controlled *malY*-carrying plasmid, the insertion mutation allows full expression of a *malK-lacZ* fusion. However, when MalY is overproduced from the plasmid-encoded gene (in the presence of IPTG), the insertion can no longer prevent MalY-mediated repression. Thus, there may be a titration effect. The sequence analysis of the regions adjacent to the insertion proved the insertion to lie in *malQ*. This suggests that the effect of the accumulated endogenous inducer in a *malQ* mutant, i.e. high level of *malK-lacZ* expression is titrable by increasing the amount of *malY*.

THE ENDOGENOUS INDUCER OF THE MALTOSE SYSTEM CAN BE PRODUCED FROM GLYCOGEN OR THE GLYCOGEN SYNTHESIZING ENZYMES

We observed that *malQ* mutants are constitutive for the maltose transport system and other *malT*-dependent enzymes. This constitutivity is dependent on glycogen, or, more correctly, the presence of the glycogen synthesizing enzymes.[84] *malQ* mutants that carry an additional mutation in *glgA* (encoding glycogen synthetase) or *glgC* (encoding ADP-glucose pyrophosphorylase) are no longer constitutive but normally inducible by maltose. Thus, the constitutivity in *malQ* mutants must be caused by the production of endogenous inducer from glycogen, and the function of amylomaltase is to remove the inducer. With the knowledge of the activity of amylomaltase, it is clear that the inducer, most likely maltotriose, is removed and kept at low concentrations by the formation of larger maltodextrins and glucose. Indeed, extracts of *malQ* mutants contain glucose, maltose, maltotriose and larger maltodextrins that are not found, or found in much lower concentrations, in the corresponding *malQ*⁺ strain.[164] Obviously, the turnover of glycogen in the *malQ*⁺ strain, representing a futile cycle, cannot be extensive, since *malQ*⁺ is not selected against when cells are growing on carbon sources other than maltodextrins.

Whereas the removal of endogenous inducer is clearly a function of amylomaltase, it is less clear how inducer is derived from glycogen. *malQ malZ* or *malQ amyA* double mutants are still constitutive in the expression of the *mal* genes, indicating that maltodextrin glucosidase[62,64] or the cytoplasmic amylase[76] are not involved in producing maltotriose from glycogen.[84] A likely candidate is *glgX*, a gene found in the glycogen gene cluster which shows homology to amylase genes.[75]

ENDOGENOUS MALTOTRIOSE CAN BE SYNTHESIZED IN THE ABSENCE OF GLYCOGEN

Even in the absence of glycogen, or, more correctly, in *glgA* and *glgC* mutants defective in the synthesis of glycogen, the *mal* genes can be induced in the absence of external maltodextrins, but under conditions that generate glucose and glucose-1-phosphate (or glucose-6-phosphate) inside the cell. The best example is the metabolism of trehalose, which induces the *mal* genes (in the absence of glycogen) to about 30% of the fully induced or constitutive level of expression.[140] The products of trehalose metabolism are glucose and glucose-6-phosphate.[152] Similarly, when glucose is transported into the cell without phosphorylation (in a *ptsG glk* mutant), the maltose genes become induced. This glucose-dependent induction is, in turn, dependent on glucose-1-phosphate since *pgm* mutants, which lack phosphoglucomutase and, hence, are unable to deliver glucose-1-phosphate by gluconeogenesis, are not induced by glucose. Using exogenous (^{14}C) glucose in such a mutant, the glucose-1-phosphate-dependent formation of internal maltose and maltotriose can be observed. The formation of these maltodextrins is independent of any *mal* gene enzyme since it can be observed in *malT* mutants as well.[84]

It appears that, even in the absence of trehalose metabolism or when no glucose is being transported into the cell, the *mal* genes are endogenously induced by the low level formation of maltotriose. In wild-type strains, this basal expression is low, curbed by the action of the MalK protein, but it becomes significant in *malK-lacZ* fusion strains. Since these fusions are also constitutively expressed in strains lacking glycogen synthesis, it is obvious that endogenous formation of the inducer maltotriose is derived from gluconeogenesis. We have proposed the existence of a maltose/maltotriose phosphorylase that will reversibly synthesize maltose from glucose and glucose-1-phosphate and maltotriose from glucose-1-phosphate and maltose. The gluconeogenic origin of glucose-1-phosphate is clear; less clear is the origin of free internal glucose. There are several possibilities. The first is the hydrolytic function of a phosphoglucotransferase, an enzyme that has been purified some time ago but whose gene has not yet been identified. This enzyme not only transfers the phosphate moiety from glucose-phosphate to another sugar, but also to water, thus forming free glucose.[153] The other possibility is the release of free glucose from UDP-glucose in analogy to the release of galactose from UDP-galactose.[166] Indeed, *galU* mutants lacking UDP-glucose pyrophosphorylase are impaired in the constitutivity of a *malK-lacZ* fusion. Since these mutants are still able to synthesize glycogen, they will not be devoid of endogenous inducer. The test for endogenous induction will have to be repeated in a mutant lacking both pathways of inducer formation.

Figure 10.2 depicts our view of the role and function of maltose/maltotriose phosphorylase postulated to function in the production of endogenous inducer.

malK-lacZ fusions are hardly affected in their constitutive expression in *pgm* mutants lacking phosphoglucomutase. Similarly, the induction of the maltose genes by trehalose takes place to the same extent in *pgm* mutants as in a *pgm*⁺ strain. This is surprising in view of our postulate that inducer formation needs glucose and glucose-1-phosphate. The only reasonable explanation is that small amounts of glucose-1-phosphate will be produced from glucose-6-phosphate byphosphoglucotransferase in the absence of phosphoglucomutase. Since induction only requires micromolar levels of maltotriose[167] but no metabolic flow, the rate of synthesis needed for maintaining this concentration may be very low.

THE ELEVATED EXPRESSION OF THE *mal* GENES DURING GLUCOSE STARVATION IS DUE TO ENDOGENOUS INDUCTION

When *E. coli* is grown in the chemostat with glucose as the limiting carbon source, the maltose system is expressed at elevated levels.[141,142,145] The authors conclude that the observed elevated expression is due to the increased synthesis of endogenous maltotriose as inducer of the system. This is in strong contrast to the situation when the cells are growing logarithmically in batch cultures at high concentrations of glucose. Under these conditions transport of glucose via

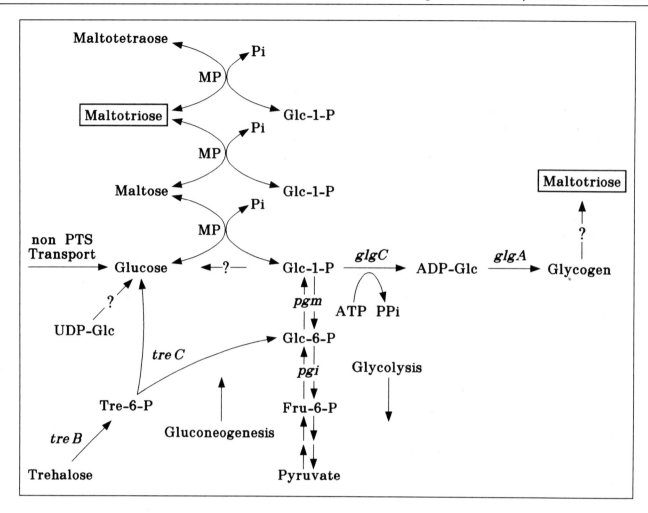

Fig. 10.2. Endogenous formation of maltose and maltotriose and induction of the maltose system by glucose. ADP-glucose pyrophosphorylase (glgC), glycogen synthase (glgA), phosphoglucomutase (pgm), phosphoglucoisomerase (pgi), trehalose-6-phosphate hydrolase (treC) and trehalose transport (treB) are indicated by their genetic designations. MP stands for maltose phosphorylase that is proposed as central enzyme in the formation of endogenous inducer.[84] In contrast to maltodextrin phosphorylase (malP) and glycogen phosphorylase (glgP), this enzyme would recognize free glucose and maltose. It is responsible for the formation of maltose and maltotriose. At present it is unclear how free glucose is generated from glucose-1-phosphate. Phosphoglucotransferase[153] is one possibility, the release from UDP-glucose, in analogy to the release of galactose from UDP-galactose,[166] another one. A second pathway of inducer formation is the degradation of glycogen. A third pathway of inducer formation is the degradation of trehalose.[152] With modification, adapted from Decker K et al, J Bacteriol 1993; 175:5655-5665.

the PTS exerts strong catabolite repression, abolishing the transcription of malT and the malT-dependent genes of the malK lamB malM, as well as the malE malF malG operons. In addition, transport of maltose is reduced by inducer exclusion. This phenomenon is caused by the interaction of unphosphorylated EIIA of the PTS with MalK.[168] Apparently, the induction of high affinity uptake systems under starvation conditions is an emergency measure to ensure scavenging of possible carbon sources.

THE HUNT FOR THE ELUSIVE MALTOSE/MALTODEXTRIN PHOSPHORYLASE

With the intention of proving the existence of the phosphorylase postulated for the endogenous synthesis of maltotriose, we employed several strategies to identify the gene encoding this enzyme:

The first rationale was to count on the reversible nature of the phosphorylase and try to isolate mutants that would overproduce the enzyme allowing the metabolic production of glucose and glucose-1-phosphate from maltose in a malQ mutant lacking amylomaltase, the key enzyme of maltose metabolism. The strain escaped this selection pressure by a mutation in malZ, turning the MalZ protein into an enzyme which exhibited not only hydrolase but also transferase activity, as discussed above.

The second approach used a *malK⁺-lacZ* fusion in which *lacZ* is fused to the last codon of *malK*. In this strain, the *lacZ* fusion is inducible by maltose (the strain is Mal⁺), as well as by trehalose, and forms blue colonies on X-Gal indicator plates with trehalose as the carbon source. According to our scheme, inducer formation requires the function of the proposed phosphorylase. Among mutant white colonies on the indicator plate we expected to find those defective in our phosphorylase. Yet, all mutants isolated in this way carried mutations in *malI*, resulting in *mal* gene repression due to the overproduction of MalY.

The last approach so far was a variation of the second one in which we used *malK-lacZ malK⁺* merodiploid strains in a genetic background that was deleted for the *malI malX malY* gene cluster. Again, we looked for white colonies on X-Gal indicator plates with trehalose as a carbon source after insertion mutagenesis. This screen yielded mostly insertions in *malT*. One mutant could not be complemented with a *malT* harboring plasmid. The insertion was cloned and the adjacent DNA was sequenced. The insertion had occurred in *dnaK*. The same mutation in an otherwise *mal⁺* wild-type strain showed the typical phenotype of *dnaK* mutants.[169] The effect of *dnaK* remains unclear, a role in the assembly of the MalF/G/K₂ complex being a likely possibility.

THE MALTOSE SYSTEM IS UNDER OSMOTIC CONTROL BUT NOT WHEN INDUCED BY MALTOSE

As mentioned above, a *malK-lacZ* fusion that lacks any MalK activity is expressed constitutively due to the presence of endogenous inducer. With increasing osmolarity of the growth medium, the expression of the fusion is abolished. Using a *malK-lacZ* fusion that is merodiploid for *malK* (and can therefore grow on maltose), we found that the fusion is expressed at low levels when grown on glycerol but can be induced by maltose, the induction ratio being similar to the induction ratio of a wild-type maltose system. When the effect of high osmolarity was tested on this fusion strain, we found that the uninduced level was still subjected to repression by high osmolarity, whereas the maltose-induced level was not.[161] The latter is consistent with the observation that high osmolarity does not affect growth on maltose. So far, very little is known about the mechanism by which osmolarity affects the maltose system. The sigma factor RpoS, a central regulator for genes controlled by starvation and osmolarity, is not involved in this osmo-dependence (Kossmann and Boos, unpublished observations). Since only the uninduced level of *mal* gene expression is affected by osmolarity, one may argue that it is the synthesis of endogenous inducer that is inhibited under conditions of high osmolarity. Yet, the *malK-lacZ* fusion is not repressed by high osmolarity in *malQ* mutants, provided that they contain the glycogen-synthesizing enzyme.[161]

PERSPECTIVES

It was the intention of the authors to point out that the understanding of the regulation of a system such as the *E. coli* maltose system is not complete, even when expression of the genes involved is understood at a detailed level and the function of the products can be reproduced in vitro using purified components. Obviously, the physiological state of the cell will play a decisive role in the fine modulation of gene expression. In a "normal situation" such control circuits may go by unnoticed, especially when the cells are growing in a logarithmic fashion. Only the drastic situation mediated by mutations, or the overproduction of certain proteins, or stress (such as high osmolarity or starvation) will unbalance the usually observed control and reveal surprising connections. One of the phenomena observed with the maltose system, the apparent repressor-like activity of the subunit of the cognate transport system, may very well represent a novel regulatory principle, observed also in the Ugp system, for instance. One may argue that some of these nonclassical regulatory circuits are unimportant for the normal growth situation and are only created by a "nonphysiological" mutation. In the case of the MalY effect on *mal* gene expression, this is most likely not the case. The *malI* mutation leading to a derepressed synthesis of MalY and to a strong repression of the *mal* regulon will be equivalent to the induced state of the *malX malY* operon. The problem is that we have not yet identified the inducer of the MalI repressor. The projected functions of the *malX* and *malY* gene products seem to indicate a connection of *mal* gene regulation to the biosynthesis of amino acids. Other cross-regulatory circuits are more obvious. This is certainly the case for the induction of the maltose system by an unrelated sugar, trehalose. There, it is clear that the internal metabolism of trehalose to glucose and glucose-6-phosphate is essential for the production of maltotriose, the inducer of the maltose system. This demonstrates that any situation (for instance starvation) that alters the prevailing concentrations of these products of gluconeogenesis will affect the level of *mal* gene expression.

Hopefully, a closer look at these physiological conditions will give us a better understanding of the complicated regulatory network by which gene expression is controlled.

REFERENCES

1. Schatz PJ, Beckwith J. Genetic analysis of protein export in *Escherichia coli*. Annu Rev Genet 1990; 24:215-248.
2. Shuman HA, Panagiotidis CH. Tinkering with transporters—periplasmic binding protein-dependent maltose transport in *Escherichia coli*. J Bioenerg Biomembrane 1993; 25:613-620.
3. Nikaido H. Maltose transport system of *Escherichia coli*: An ABC-type transporter. FEBS Lett 1994; 346:55-58.
4. Boos W, Lucht JM. Periplasmic binding-protein-dependent ABC transporters In: Neidhardt F, ed. *Escherichia coli* and *Salmonella typhimurium*; cellular and molecular biology. Washington, DC:American Society of Microbiology, 1996; (in press).
5. Manoil C, Beckwith J. A genetic approach to analyzing membrane protein topology. Science 1986; 233:1403-1408.
6. Szmelcman S, Schwartz M, Silhavy TJ et al. Maltose transport in *Escherichia coli* K12. A comparison of transport kinetics in wild-type and λ-resistant mutants with the dissociation constants of the maltose binding protein as measured by fluorescence quenching. Eur J Biochem 1976; 65:13-19.
7. Freundlieb S, Ehmann U, Boos W. Facilitated diffusion of *p*-nitrophenyl-α-D-maltohexaoside through the outer membrane of *Escherichia coli*. J Biol Chem 1988; 263:314-320.
8. Benz R, Francis G, Nakae T et al. Investigation of the selectivity of maltoporin channels using mutant LamB proteins—mutations changing the maltodextrin binding site. Biochim. Biophys. Acta 1992; 1104:299-307.
9. Schirmer T, Keller TA, Wang YF et al. Structural basis for sugar translocation through maltoporin channels at 3.1 ångstrom resolution. Science 1995; 267:512-514.
10. Hatfield D, Hofnung M, Schwartz M. Nonsense mutations in the maltose A region of the genetic map of *Escherichia coli*. J Bacteriol 1969; 100:1311-1315.

11. Richet E, Raibaud O. Purification and properties of the MalT protein, the transcription activator of the *Escherichia coli* maltose regulon. J Biol Chem 1987; 262:12647-53.

12. Raibaud O, Vidal-Ingigliardi D, Richet E. A complex nucleoprotein structure involved in activation of transcription of two divergent *Escherichia coli* promoters. J Mol Biol 1989; 205:471-485.

13. Charbit A, Boulain JC, Ryter A et al. Probing the topology of a bacterial membrane protein by genetic insertion of a foreign epitope; expression at the cell surface. EMBO J 1986; 5:3029-3037.

14. Rodseth LE, Martineau P, Duplay P et al. Crystallization of genetically engineered active maltose-binding proteins, including an immunogenic viral epitope insertion. J Mol Biol 1990; 213:607-611.

15. Martineau P, Guillet JG, Leclerc C et al. Expression of heterologous peptides at two permissive sites of the MalE protein: Antigenicity and immunogenicity of foreign B-cell and T-cell epitopes. Gene 1992; 113:35-46.

16. Szmelcman S, Clement JM, Jehanno M et al. Export and one-step purification from *Escherichia coli* of a MalE-CD4 hybrid protein that neutralizes HIV in vitro. J Acq Immun Defic Syndrome 1990; 3:859-872.

17. Schwartz M. The maltose regulon. In: Neidhardt FC, ed. *Escherichia coli* and *Salmonella typhimurium*: cellular and molecular biology. Washington DC: American Society of Microbiology, 1987; 2:1482-1502.

18. Cole ST, Raibaud O. The nucleotide sequence of the *malT* gene encoding the positive regulator of the *Escherichia coli* maltose regulon. Gene 1986; 42:201-208.

19. Chapon C. Expression of *malT*, the regulator gene of the maltose region in *Escherichia coli*, is limited both at transcription and translation. EMBO J 1982; 1:369-74.

20. Débarbouillé M, Schwartz M. Mutants which make more *malT* product, the activator of the maltose regulon in *Escherichia coli*. Mol Gen Genet 1980; 178:589-95.

21. Richet E, Raibaud O. MalT, the regulatory protein of the *Escherichia coli* maltose system, is an ATP-dependent transcriptional activator. EMBO J 1989; 8:981-987.

22. Vidal-Ingigliardi D, Richet E, Danot O et al. A small C-terminal region of the *Escherichia coli* MalT protein contains the DNA-binding domain. J Biol Chem 1993; 268:24527-24530.

23. Kellerman O, Szmelcman S. Active transport of maltose in *Escherichia coli* K-12. Involvement of a periplasmic maltose-binding protein. Eur J Biochem 1974; 47:139-149.

24. Duplay P, Bedouelle H, Fowler A et al. Sequences of the *malE* gene and of its product, the maltose-binding protein of *Escherichia coli* K12. J Biol Chem 1984; 259:10606-13.

25. Spurlino JC, Lu GY, Quiocho FA. The 2.3-A resolution structure of the maltose- or maltodextrin-binding protein, a primary receptor of bacterial active transport and chemotaxis. J Biol Chem 1991; 266:5202-19.

26. Hor LI, Shuman HA. Genetic analysis of periplasmic binding protein dependent transport in *Escherichia coli*. Each lobe of maltose-binding protein interacts with a different subunit of the MalFGK2 membrane transport complex. J Mol Biol 1993; 233:659-670.

27. Shuman HA, Silhavy TJ, Beckwith JR. Labeling of proteins with β-galactosidase by gene fusion. Identification of a cytoplasmic membrane component of the *Escherichia coli* maltose transport system. J Biol Chem 1980; 255:168-174.

28. Froshauer S, Beckwith J. The nucleotide sequence of the gene for MalF protein, an inner membrane component of the maltose transport system of *Escherichia coli*. Repeated DNA sequences are found in the *malE-malF* intercistronic region. J Biol Chem. 1984; 259:10896-903.

29. Treptow N, Shuman H. Genetic evidence for substrate and periplasmic binding-protein recognition by MalF and MalG proteins, cytoplasmic membrane components of the *Escherichia coli* maltose transport system. J Bacteriol 1985; 163:654-660.

30. Ehrmann M, Boyd D, Beckwith J. Genetic analysis of membrane protein topology by a sandwich gene fusion approach. Proc Natl Acad Sci USA 1990; 87:7574-7578.

31. Covitz KMY, Panagiotidis CH, Hor LI et al. Mutations that alter the transmembrane signaling pathway in an ATP binding cassette (ABC) transporter. EMBO J 1994; 13:1752-1759.

32. Silhavy TJ, Brickman E, Bassford PJ et al. Structure of the *malB* region in *Escherichia coli* K12. II. Genetic map of the *malE,F,G* operon. Mol Gen Genet 1979; 174:249-59.

33. Dassa E, Hofnung M. Sequence of gene *malG* in *Escherichia coli* K12: homologies between integral membrane components from binding protein-dependent transport systems. EMBO J 1985; 4:2287-2293.

34. Dassa E, Muir S. Membrane topology of MalG, an inner membrane protein from the maltose transport system of *Escherichia coli*. Mol Microbiol 1993; 7:29-38.

35. Dassa E. Sequence-function relationships in MalG, an inner membrane protein from the maltose transport system in *Escherichia coli*. Mol Microbiol 1993; 7:39-47.

36. Boyd D, Traxler B, Beckwith J. Analysis of the topology of a membrane protein by using a minimum number of alkaline phosphatase fusions. J Bacteriol 1993; 175:553-556.

37. Bavoil P, Hofnung M, Nikaido H. Identification of a cytoplasmic membrane-associated component of the maltose transport system of *Escherichia coli*. J Biol Chem 1980; 255:8366-8369.

38. Dahl MK, Francoz E, Saurin W et al. Comparison of sequences from the *malB* regions of *Salmonella typhimurium* and *Enterobacter aerogenes* with *Escherichia coli* K12: A potential new regulatory site in the intergenic region. Mol Gen Genet 1989; 218:199-207.

39. Gilson E, Nikaido H, Hofnung M. Sequence of the *malK* gene in *Escherichia coli* K12. Nucl Acids Res 1982; 10:7449-7458.

40. Shuman HA, Silhavy TJ. Identification of the *malK* gene product. A peripheral membrane component of the *Escherichia coli* maltose transport system. J Biol Chem 1981; 256:560-562.

41. Morbach S, Tebbe S, Schneider E. The ATP-Binding cassette (ABC) transporter for maltose/maltodextrins of *Salmonella typhimurium*—characterization of the ATPase activity associated with the purified MalK subunit. J Biol Chem 1993; 268:18617-18621.

42. Reyes M, Shuman HA. Overproduction of MalK protein prevents expression of the *Escherichia coli mal* regulon. J Bacteriol 1988; 170:4598-4602.

43. Panagiotidis CH, Reyes M, Sievertsen A et al. Characterization of the structural requirements for assembly and nucleotide binding of an ATP-binding cassette transporter—The maltose transport system of *Escherichia coli*. J Biol Chem 1993; 268:23685-23696.

44. Kühnau S, Reyes M, Sievertsen A et al. The activities of the *Escherichia coli* MalK protein in maltose transport, regulation and inducer exclusion can be separated by mutations. J Bacteriol 1991; 173:2180-2186.

45. Hekstra D, Tommassen J. Functional exchangeability of the ABC proteins of the periplasmic binding protein-dependent transport systems Ugp and Mal. J Bacteriol 1993; 175:6546-6552.

46. Randall-Hazelbauer L, Schwartz M. Isolation of the bacteriophage lambda receptor from *Escherichia coli*. J Bacteriol 1973; 116:1436-1446.

47. Szmelcman S, Hofnung M. Maltose transport in *Escherichia coli* K12. Involvement of the bacteriophage lambda receptor. J Bacteriol 1975; 124:112-118.

48. Ferenci T, Schwentorat M, Ullrich S et al. Lambda Receptor in the outer membrane of *Escherichia coli* as a binding protein for maltodextrins and starch polysaccacharides. J Bacteriol 1980; 142:521-526.

49. Clement JM, Hofnung M. Gene sequence of the lambda receptor, an outer membrane protein of *Escherichia coli* K-12. Cell 1981; 27:507-514.

50. Gilson E, Rousset J, Charbit A et al. *malM*, a new gene of the maltose regulon in *Escherichia coli* K-12. I. *malM* is the last gene of the *malK-lamB* operon and encodes a periplasmic protein. J Mol Biol 1986; 191:303-311.

51. Rousset JP, Gilson E, Hofnung M. *malM*, a new gene of the maltose regulon in *Escherichia coli* K12. II. Mutations affecting the signal peptide of the MalM protein. J Mol Biol 1986; 191:313-20.

52. Schneider E, Francoz E, Dassa E. Completion of the nucleotide sequence of the 'maltose B' region in *Salmonella typhimurium*: the high conservation of the *malM* gene suggests a selected physiological role for its product. Biochim Biophys Acta 1992; 1129:223-7.

53. Schwartz M, Hofnung M. La maltodextrin phosphorylase d'*Escherichia coli*. Eur J Biochem 1967; 2:132-145.

54. Palm D, Goerl R, Burger KJ. Evolution of catalytic and regulatory sites in phosphorylases. Nature 1985; 313:500-502.

55. Schinzel R, Palm D. *Escherichia coli* maltodextrin phosphorylase: contribution of active site residues glutamate-637 and tyrosine-538 to the phosphorolytic cleavage of alpha-glucans. Biochemistry 1990; 29:9956-9962.

56. Monod J, Torriani AM. De l'amylomaltase d'*Escherichia coli*. Ann Inst Pasteur 1950; 78:65-77.

57. Wiesmeyer H, Cohn M. The characterization of the pathway of maltose utilization by *Escherichia coli*. I. Purification and physical chemical properties of the enzyme amylomaltase. Biochim Biophys Acta 1960; 39:417-426.

58. Wiesmeyer H, Cohn M. The characterization of the pathway of maltose utilization by *Escherichia coli*. II. General properties and mechanism of action of amylomaltase. Biochim Biophys Acta 1960; 39:427-439.

59. Palmer NT, Ryman BE, Whelan WJ. The action pattern of amylomaltase from *Escherichia coli*. Eur J Biochem 1976; 69:105-115.

60. Pugsley AP, Dubreuil C. Molecular characterization of *malQ*, the structural gene for the *Escherichia coli* enzyme amylomaltase. Mol Microbiol 1988; 2:473-479.

61. Freundlieb S, Boos W. α-Amylase of *Escherichia coli*, mapping and cloning of the structural gene, *malS*, and identification of its product as a periplasmic protein. J Biol Chem 1986; 261:2946-2953.

62. Schneider E, Freundlieb S, Tapio S et al. Molecular characterization of the MalT-dependent periplasmic α-amylase of *Escherichia coli* encoded by *malS*. J Biol Chem 1992; 267:5148-5154.

63. Reyes M, Treptow NA, Shuman HA. Transport of *p*-nitrophenyl-a-maltoside by the maltose transport system of *Escherichia coli* and its subsequent hydrolysis by a cytoplasmic maltosidase. J Bacteriol 1986; 165:918-922.

64. Tapio S, Yeh F, Shuman HA et al. The *malZ* gene of *Escherichia coli*, a member of the maltose regulon, encodes a maltodextrin glucosidase. J Biol Chem 1991; 266:19450-19458.

65. Reidl J, Römisch K, Ehrmann M et al. MalI, a novel protein involved in regulation of the maltose system of *Escherichia coli*, is highly homologous to the repressor proteins GalR, CytR, and LacI. J Bacteriol 1989; 171:4888-4899.

66. Reidl J, Boos W. The *malX malY* operon of *Escherichia coli* encodes a novel enzyme II of the phosphotransferase system recognizing glucose and maltose and an enzyme abolishing the endogenous induction of the maltose system. J Bacteriol 1991; 173:4862-76.

67. Zdych E, Peist R, Reidl J et al. MalY of *Escherichia coli* is an enzyme with the activity of a βC-S lyase. J. Bacteriol. 1995; (in press).

68. Boos W, Ferenci T, Shuman HA. Formation and excretion of acetylmaltose after accumulation of maltose in *Escherichia coli*. J Bacteriol 1981; 146:725-732.

69. Brand B, Boos W. Maltose transacetylase of *Escherichia coli*. Mapping and cloning of its structural gene, *mac*, and characterization of the enzyme as a dimer of identical polypeptides with a molecular weight of 20.000. J Biol Chem 1991; 266:14113-14118.

70. Okita TW, Rodriguez RL, Preiss J. Biosynthesis of bacterial glycogen: Cloning of the glycogen enzyme structural genes of *Escherichia coli*. J Biol Chem 1981; 256:6944-6952.

71. Kumar A, Larsen CE, Preiss J. Biosynthesis of bacterial glycogen. Primary structure of *Escherichia coli* ADP-glucose: α-1,4-glucan, 4-glucosyl transferase as deduced from the nucleotide sequence of the *glgA* gene. J Biol Chem 1986; 261:16256-16259.

72. Baecker PA, Furlong CE, Preiss J. Biosynthesis of bacterial glycogen. Primary structure of *Escherichia coli* ADP-glucose synthase as deduced from the nucleotide sequence of the *glgC* gene. J Biol Chem 1983; 258: 5084-5088.

73. Yu F, Jen J, Takeuchi E et al. α-Glucan phosphorylase from *Escherichia coli*. Cloning of the gene and purification and characterization of the protein. J Biol Chem 1988; 263:13706-13711.

74. Baecker PA, Greenberg E, Preiss J. Biosynthesis of bacterial glycogen. Primary structure of *Escherichia coli* 1,4-α-D-glucan 6-α-(1,4-α-D-glucano)-transferase as deduced from the nucleotide sequence of the *glgB* gene. J Biol Chem 1986; 261:8738-8743.

75. Romeo T, Kumar A, Preiss J. Analysis of the *Escherichia coli* glycogen gene cluster suggests that catabolic enzymes are encoded among the biosynthetic genes. Gene 1988; 70:363-376.

76. Raha M, Kawagishi I, Muller V et al. *Escherichia coli* produces a cytoplasmic alpha-amylase, AmyA. J Bacteriol 1992; 174:6644-6652.

77. Weissborn AC, Liu QY, Rumley MK et al. UTP:alpha-D-glucose-1-phosphate uridylyltransferase of *Escherichia coli*; isolation and DNA sequence of the *galU* gene and purification of the enzyme. J Bacteriol 1994; 176:2611-2618.

78. Hengge-Aronis R, Fischer D. Identification and molecular analysis of *glgS*, a novel growth-phase-regulated and *rpoS*-dependent gene involved in glycogen synthesis in *Escherichia coli*. Mol Microbiol 1992; 6:1877-1886.

79. Curtis SJ, Epstein W. Phosphorylation of D-glucose in *Escherichia coli* mutants defective in glucophosphotransferase, mannosephosphotransferase, and glucokinase. J Bacteriol 1975; 122:1189-1199.

80. Fukuda Y, Yamaguchi S, Shimosaka M et al. Cloning of the glucokinase gene in *Escherichia coli* B. J Bacteriol 1983; 156:922-925.

81. Adhya S, Schwartz M. Phosphoglucomutase mutants of *Escherichia coli* K12. J Bacteriol 1971; 108:621-626.

82. Roehl RA, Vinopal RT. New maltose blue mutations in *Escherichia coli* K12. J Bacteriol 1979; 139:683-685.

83. Lu M, Kleckner N. Molecular cloning and characterization of the *pgm* gene encoding phosphoglucomutase of *Escherichia coli*. J Bacteriol 1994; 176:5847-5851.

84. Decker K, Peist R, Reidl J et al. Maltose and maltotriose can be formed endogenously in *Escherichia coli* from glucose and glucose-1-phosphate

independently of enzymes of the maltose system. J Bacteriol 1993; 175:5655-5665.

85. Hofnung M, Schwartz M, Hatfield D. Complementation studies in the maltose-A region of *Escherichia coli* K12 genetic map. J Mol Biol 1971; 61:681-694.

86. Débarbouillé M, Schwartz M. The use of gene fusions to study the expression of *malT* the positive regulator gene of the maltose regulon. J Mol Biol 1979; 132:521-34.

87. Raibaud O, Débarbouillé M, Schwartz M. Use of deletions created in vitro to map transcriptional regulatory signals in the *malA* region of *Escherichia coli*. J Mol Biol 1983; 163:395-408.

88. Raibaud O, Clement JM, Hofnung M. Structure of the *malB* region in *Escherichia coli* K12. III. Correlation of the genetic map with the restriction map. Mol Gen Genet 1979; 174:261-7.

89. Hatfield D, Hofnung M, Schwartz M. Genetic analysis of the maltose A region in *Escherichia coli*. J Bacteriol 1969; 98:559-567.

90. Débarbouillé M, Shuman HA, Silhavy TJ et al. Dominant constitutive mutations in *malT*, the positive regulator of the maltose regulon in *Escherichia coli*. J Mol Biol 1978; 124:359-371.

91. Hofnung M, Schwartz M. Mutations allowing growth on maltose of *Escherichia coli* K 12 strains with a deleted *malT* gene. Mol Gen Genet 1971; 112:117-32.

92. Raibaud O, Schwartz M. Restriction map of the *Escherichia coli malA* region and identification of the *malT* product. J Bacteriol 1980; 143:761-71.

93. Dardonville B, Raibaud O. Characterization of *malT* mutants that constitutively activate the maltose regulon of *Escherichia coli*. J Bacteriol 1990; 172:1846-1852.

94. Chapon C. Role of the catabolite activator protein in the maltose regulon of *Escherichia coli*. J Bacteriol 1982; 150:722-729.

95. Chapon C, Kolb A. Action of CAP on the *malT* promoter in vitro. J Bacteriol 1983; 156:1135-1143.

96. Raibaud O, Vidal-Ingigliardi D, Kolb A. Genetic studies on the promoter of *malT*, the gene that encodes the activator of the *Escherichia coli* maltose regulon. Res Microbiol 1991; 142:937-42.

97. Wanner BL, Sarthy A, Beckwith J. *Escherichia coli* mutant that reduces amounts of several periplasmic and outer membrane proteins. J Bacteriol 1979; 140:229-239.

98. Wandersman C, Moreno F, Schwartz M. Pleiotropic mutations rendering *Escherichia coli* K-12 resistant to bacteriophage TP1. J Bacteriol 1980; 143:1374-83.

99. Granett S, Villarejo M. Regulation of gene expression in *Escherichia coli* by the local anesthetic procaine. J Mol Biol 1982; 160:363-367.

100. Aiba H, Nakasa F, Mizushima S et al. Evidence for the importance of the phosphotransfer between the two regulatory components EnvZ and OmpR in osmoregulation in *Escherichia coli*. J Biol Chem 1989; 264:14090-14094.

101. Waukau J, Forst S. Molecular analysis of the signaling pathway between EnvZ and OmpR in *Escherichia coli*. J Bacteriol 1992; 174:1522-1527.

102. Russo FD, Silhavy TJ. EnvZ controls the concentration of phosphorylated OmpR to mediate osmoregulation of the porin genes. J Mol Biol 1991; 222:567-580.

103. Maeda S, Takayanagi K, Nishimura Y et al. Activation of the osmoregulated *ompC* gene by the OmpR protein in *Escherichia coli*: A study involving synthetic OmpR-binding sequences. J Biochem Tokyo 1991; 110:324-327.

104. Case CC, Bukau B, Granett S et al. Contrasting mechanisms of *envZ* control of *mal* and *pho* regulon genes in *Escherichia coli*. J Bacteriol 1986; 166:706-12.

105. Danot O, Raibaud O. Which nucleotides in the "-10" region are crucial to obtain a fully active MalT-dependent promoter? J Mol Biol 1994; 238:643-648.

106. Bedouelle H, Schmeissner U, Hofnung M et al. Promoters of the *malEFG* and *malK-lam*B operons in *Escherichia coli* K12. J Mol Biol 1982; 161:519-31.

107. Bedouelle H. Mutations in the promoter regions of the *malEFG* and *malK-lamB* operons of *Escherichia coli* K12. J Mol Biol 1983; 170:861-82.

108. Gutierrez C, Raibaud O. Point mutations that reduce the expression of *malPQ*, a positively controlled operon of *Escherichia coli*. J Mol Biol 1984; 177:69-86.

109. Raibaud O, Gutierrez C, Schwartz M. Essential and nonessential sequences in *malPp*, a positively controlled promoter in *Escherichia coli*. J Bacteriol 1985; 161:1201-8.

110. Débarbouillé M, Raibaud O. Expression of the *Escherichia coli malPQ* operon remains unaffected after drastic alteration of its promoter. J Bacteriol 1983; 153:1221-7.

111. Vidal-Ingigliardi D, Richet E, Raibaud O. Two MalT binding sites in direct repeat. A structural motif involved in the activation of all the promoters of the maltose regulons in *Escherichia coli* and *Klebsiella pneumoniae*. J Mol Biol 1991; 218:323-34.

112. Vidal-Ingigliardi D, Raibaud O. 3 Adjacent binding sites for cAMP receptor protein are involved in the activation of the divergent *malEp-malKp* promoters. Proc Natl Acad Sci USA 1991; 88:229-233.

113. Danot O, Raibaud O. On the puzzling arrangement of the asymmetric MalT-Binding sites in the MalT-dependent promoters. Proc Natl Acad Sci USA 1993; 90:10999-11003.

114. Danot O, Raibaud O. Multiple protein-DNA and protein-protein interactions are involved in transcriptional activation by MalT. Mol Microbiol 1994; 14:335-346.

115. Raibaud O. Nucleoprotein structures at positively regulated bacterial promoters: homology with replication origins and some hypotheses on the quaternary structures of the activator proteins in these complexes. Molec Microbiol 1989; 3:455-458.

116. Richet E, Raibaud O. Supercoiling is essential for the formation and stability of the initiation complex at the divergent *malEp* and *malKp* promoters. J Mol Biol 1991; 218:529-42.

117. Richet E, Vidal-Ingigliardi D, Raibaud O. A new mechanism for coactivation of transcription initiation: Repositioning of an activator triggered by the binding of a second activator. Cell 1991; 66:1185-1195.

118. Richet E, Sogaard-Andersen L. CRP induces the repositioning of MalT at the *Escherichia coli malKp* promoter primarily through DNA bending. EMBO J 1994; 13:4558-4567.

119. Calvo JM, Matthews RG. The leucine-responsive regulatory protein, a global regulator of metabolism in *Escherichia coli*. Microbiol Rev 1994; 58:401-425.

120. Tchetina E, Newman EB. Identification of *lrp*-regulated genes by inverse PCR and sequencing: regulation of two *mal* operons of *Escherichia coli* by leucine-responsive regulatory protein. J Bacteriol 1995; 177:2679-2683.

121. Davidson AL, Shuman HA, Nikaido H. Mechanism of maltose transport in *Escherichia coli*. Transmembrane signaling by periplasmic binding proteins. Proc Natl Acad Sci USA 1992; 89:2360-2364.

122. Hengge R, Boos W. Maltose and lactose transport in *Escherichia coli*. Examples of two different types of concentrative transport systems. Biochim Biophys Acta 1983; 737:443-478.

123. Shuman HA. Active transport of maltose in *Escherichia coli* K12. Role of the periplasmic maltose-binding protein and evidence for a substrate recognition site in the cytoplasmic membrane. J Biol Chem 1982; 257:5455-5461.

124. Dietzel I, Kolb V, Boos W. Pole cap formation in *Escherichia coli* following induction of the maltose-binding protein. Arch Microbiol 1978; 118:207-218.

125. Sharff AJ, Rodseth LE, Spurlino JC et al. Crystallographic evidence of a large ligand-induced hinge-twist motion between the two domains of the maltodextrin binding protein involved in active transport and chemotaxis. Biochemistry 1992; 31:10657-10663.

126. Bohl E, Shuman HA, Boos W. Mathematical treatment of the kinetics of binding protein dependent transport systems reveals that both the substrate loaded and unloaded binding proteins interact with the membrane components. J Theor Biol 1995; 172:83-94.

127. Dean DA, Hor LI, Shuman HA et al. Interaction between maltose-binding protein and the membrane-associated maltose transporter complex in *Escherichia coli*. Mol Microbiol 1992; 6:2033-2040.

128. Froshauer S, Green GN, Boyd D et al. Genetic analysis of the membrane insertion and topology of MalF, a cytoplasmic membrane protein of *Escherichia coli*. J Mol Biol 1988; 200:501-11.

129. Saurin W, Köster W, Dassa E. Bacterial binding protein-dependent permeases: Characterization of distinctive signatures for functionally related integral cytoplasmic membrane proteins. Mol Microbiol 1994; 12: 993-1004.

130. Walker JE, Saraste M, Runswik JM et al. Distantly related sequences in the alpha- and beta- subunits of ATP synthase, myosin kinases and other ATP-requiring enzymes and a common nucleotide binding fold. EMBO J 1982; 1:945-951.

131. Higgins CF, Hiles ID, Salmond GPC et al. A family of related ATP-binding subunits coupled to many distinct biological processes in bacteria. Nature 1986; 323:448-450.

132. Higgins CF. ABC transporters—From microorganisms to man. Annu Rev Cell Biol 1992; 8:67-113.

133. Davidson AL, Nikaido H. Overproduction, solubilization, and reconstitution of the maltose transport system from *Escherichia coli*. J Biol Chem 1990; 265:4254-4260.

134. Dean DA, Davidson AL, Nikaido H. The role of ATP as the energy source for maltose transport in *Escherichia coli*. Res Microbiol 1990; 141:348-52.

135. Davidson AL, Nikaido H. Purification and characterization of the membrane-associated components of the maltose transport system from *Escherichia coli*. J Biol Chem 1991; 266:8946-51.

136. Brzoska P, Rimmele M, Brzostek K et al. The *pho* regulon-dependent ugp uptake system for glycerol-3-phosphate in *Escherichia coli* is trans inhibited by Pi. J Bacteriol 1994; 176:15-20.

137. Overduin P, Boos W, Tommassen J. Nucleotide sequence of the ugp genes of *Escherichia coli* K-12: homology to the maltose system. Mol Microbiol 1988; 2:767-75.

138. Baichwal V, Liu DX, Ames GFL. The ATP-binding component of a prokaryotic traffic ATPase is exposed to the periplasmic (external) surface. Proc Natl Acad Sci USA 1993; 90:620-624.

139. Prossnitz E. Determination of a region of the HisJ binding protein involved in the recognition of the membrane complex of the histidine transport system of *Salmonella typhimurium*. J Biol Chem 1991; 266:9673-9677.

140. Klein W, Boos W. Induction of the lambda receptor is essential for the effective uptake of trehalose in *Escherichia coli*. J Bacteriol 1993; 175:1682-1686.

141. Death A, Notley L, Ferenci T. Derepression of LamB protein facilitates outer membrane permeation of carbohydrates into *Escherichia coli* under conditions of nutrient stress. J Bacteriol 1993; 175:1475-1483.

142. Death A, Ferenci T. Between feast and famine: endogenous inducer synthesis in the adaptation of *Escherichia coli* to growth with limiting carbohydrates. J Bacteriol 1994; 176:5101-5107.

143. Hall MN, Gabay J, Débarbouillé M et al. A role for mRNA secondary structure in the control of translation initiation. Nature 1982; 295:616-618.

144. De Smit MH, Van Duin J. Control of translation by mRNA secondary structure in *Escherichia coli*. J Mol Biol 1994; 244:144-150.

145. Notley L, Ferenci T. Differential expression of *mal* genes under cAMP and endogenous inducer control in nutrient-stressed *Escherichia coli*. Mol Microbiol 1995; 16:121-129.

146. Schwartz M. Aspects biochimiques et génétiques du metabolism du maltose chez *Escherichia coli* K12. C R Acad Sci 1965; 260:2613-2616.

147. Schwartz M. Phenotypic expression and genetic localization of mutations affecting maltose metabolism in *Escherichia coli* K 12. Ann Inst Pasteur (Paris) 1967; 112:673-98.

148. Becker S, Palm D, Schinzel R. Dissecting differential binding in the forward and reverse reaction of *Escherichia coli* maltodextrin phosphorylase using 2-deoxyglucosyl substrates. J Biol Chem 1994; 269:2485-2490.

149. Xavier KB, Kossmann M, Santos H et al. Kinetic analysis by in vivo 31P nuclear magnetic resonance of internal Pi during the uptake of *sn*-glycerol-3-phosphate by the *pho*-regulon-dependent Ugp system and the *glp*-regulon-dependent GlpT system. J Bacteriol 1995; 177:699-709.

150. Hartmann A, Boos W. Mutations in *phoB*, the positive gene activator of the *pho* regulon in *Escherichia coli*, affect the carbohydrate phenotype on MacConkey indicator plates. Res Microbiol 1993; 144:285-293.

151. Buhr A, Daniels GA, Erni B. The glucose transporter of *Escherichia coli*. Mutants with impaired translocation activity that retain phosphorylation activity. J Biol Chem 1992; 267:3847-3851.

152. Rimmele M, Boos W. Trehalose-6-phosphate hydrolase of *Escherichia coli*. J Bacteriol 1994; 176:5654-5664.

153. Stevens-Clark JR, Theisen MC, Conklin KA et al. Phosphoramidates. IV. Purification and characterization of a phosphoryl transfer enzyme from *Escherichia coli*. J Biol Chem 1968; 243:4468-4473.

154. Podkovyrov SM, Zeikus JG. Structure of the gene encoding cyclomaltodextrinase from *Clostridium thermohydrosulfuricum* 39E and characterization of the enzyme purified from *Escherichia coli*. J Bacteriol 1992; 174:5400-5405.

155. Peist R, Schneider-Fresenius C, Boos W. A mutation in the *Escherichia coli* MalZ protein changing the enzyme from a hydrolase to a transferase. Unpublished observations, 1995.

156. Ferenci T. The recognition of maltodextrins by *Escherichia coli*. Eur J Biochem 1980; 108:631-636.

157. Ferenci T, Muir M, Lee K-S et al. Substrate specificity of the *Escherichia coli* maltodextrin transport system and its component proteins. Biochim Biophys Acta 1986; 860:44-50.

158. Nikaido H, Pokrovskaya ID, Reyes L et al. Maltose transport system of *Escherichia coli* as a member of ABC transporters. In: Torriani A, Yagil E,

Silver S, eds. Phosphate in Microorganisms; Cellular and molecular biology. Washington, DC: American Society of Microbiology, 1994:91-96.

159. Thieme R, Lay H, Oser A et al. 3-Azi-1-methoxybutyl D-maltooligosaccharides specifically bind to the maltose/maltooligosaccharide-binding protein of *Escherichia coli* and can be used as photoaffinity labels. Eur J Biochem 1986; 160:83-91.

160. Hofnung M, Hatfield D, Schwartz M. *malB* region in *Escherichia coli* K-12: characterization of new mutations. J Bacteriol 1974; 117:40-47.

161. Bukau B, Ehrmann M, Boos W. Osmoregulation of the maltose regulon in *Escherichia coli*. J Bacteriol 1986; 166:884-91.

162. Dean DA, Reizer J, Nikaido H et al. Regulation of the maltose transport system of *Escherichia coli* by the glucose-specific enzyme III of the phosphoenolpyruvate-sugar phosphotransferase system. Characterization of inducer exclusion-resistant mutants and reconstitution of inducer exclusion in proteoliposomes. J Biol Chem 1990; 265:21005-10.

163. van der Vlag J, van Dam K, Postma PW. Quantification of the regulation of glycerol and maltose metabolism by IIA(Glc) of the phosphoenolpyruvate-dependent glucose phosphotransferase system in *Salmonella typhimurium*. J Bacteriol 1994; 176:3518-3526.

164. Ehrmann M, Boos W. Identification of endogenous inducers of the *mal* system in *Escherichia coli*. J Bacteriol 1987; 169:3539-3545.

165. Belfaiza J, Parsot C, Martel A et al. Evolution in biosynthetic pathways: two enzymes catalyzing consecutive steps in methionine biosynthesis originate from a common ancestor and posses a similar regulatory region. Proc Natl Acad Sci USA 1986; 83:867-871.

166. Wu HCP, Kalckar HM. Endogenous induction of the galactose operon in *Escherichia coli* K12. Proc Natl Acad Sci 1966; 55:622-629.

167. Raibaud O, Richet E. Maltotriose is the inducer of the maltose regulon. J Bacteriol 1987; 169:3059-3061.

168. Postma PW, Lengeler JW, Jacobson GR. Phosphoenolpyruvate—carbohydrate phosphotransferase systems of bacteria. Microbiol Rev 1993; 57:543-594.

169. Bukau B, Walker GC. Mutations altering heat shock specific subunit of RNA polymerase suppress major cellular defects of *E. coli* mutants lacking the DnaK chaperone. EMBO J 1990; 9:4027-4036.

THE PHOSPHOENOLPYRUVATE-DEPENDENT CARBOHYDRATE: PHOSPHOTRANSFERASE SYSTEM (PTS) AND CONTROL OF CARBON SOURCE UTILIZATION

Joseph W. Lengeler

INTRODUCTION

Unicellular microorganisms such as *Escherichia coli* must be able to detect changes in their environment and to adapt their metabolism rapidly to external fluctuations. Adaptation of bacterial populations to such changes is in general transient, i.e., the cellular adaptation persists not much longer than the environmental change lasts and will be accommodated during a prolonged change. Prokaryotes monitor their surroundings directly by membrane-bound sensors, and indirectly by intracellular sensors which detect changes in pools of intracellular metabolites that vary as the consequence of extracellular changes. The pools usually correlate with the transport capacity of a cell. Most sensors are linked through complex signal transduction pathways to global regulatory networks. Global control systems, however, regulate metabolic networks, e.g., those involved in carbon source utilization, cellular differentiation processes, and the behavior of bacterial populations. Their activity leads eventually to the adaptation of cells to the change in conditions. Formally speaking, bacterial adaptation and differentiation processes can be viewed in analogy to other transiently acting sensory processes, and the entire bacterial cell may thus be seen as the equivalent of a transiently responding sensory system.

Systematic research on carbon source utilization started in *E. coli* with the pioneering work of J. Monod[1] on diauxic growth and the "glucose effect". During the following years relevant phenomena were described, e.g., inducer exclusion and catabolite repression by B. Magasanik and F. Neidhardt[2,3] and catabolite inhibition by K. Paigen,[4]

Regulation of Gene Expression in Escherichia coli, edited by E. C. C. Lin and A. Simon Lynch. © 1996 R.G. Landes Company.

and essential regulatory systems were discovered, e.g., the cAMP·CAP system by various teams (references in 5), and the phosphoenolpyruvate-dependent carbohydrate: phosphotransferase system (PTS) by E.C.C. Lin[6] and by S. Roseman.[7] The descriptive phase ended with the formulation of a model[8-10] according to which elements of the PTS and of the cAMP·CAP system regulate catabolite repression and inducer exclusion. By now, it is well supported by experimental evidence (for recent reviews see refs. 11, 12), but the model itself has not progressed beyond its original version. However, two concepts have emerged which place carbon source utilization and its control in a new context. According to these, the cAMP·CAP system must be considered as a global regulatory system for all catabolic pathways in *E. coli*,[13] which constitutes a functional unit with the PTS as its signal transduction system.[14,15] The PTS comprises 20 to 30 proteins or functional domains (modules) which communicate among themselves through phosphoryl group transfer with membrane-bound transport systems acting to monitor the environment of the cell. This information is used in single form to control the expression of individual operons and regulons. In an integrated form, the information is used to control the *crp* modulon and through it carbon source utilization. The PTS may be considered a paradigm for the close connection between the transport activity of a cell and cellular control of, for example, carbon source utilization at a global level. Hence, in this chapter we will begin by describing a series of phenomena related to carbon source utilization in *E. coli*, followed by a description of the hierarchical structure of gene regulation and of the role of signal transduction pathways in this control. In the major part of the chapter, we will describe the PTS, its molecular components and its multiple functions related to carbohydrate transport and to the control of carbon source utilization through carbon catabolite repression, inducer exclusion and related phenomena. The major purpose of this review is to summarize relevant data and to discuss them in relation to the new concept. Due to space limitations, older literature[16] will only be included where required to present a coherent picture. The concept of global regulatory and signal transduction systems as well as the PTS will only be presented in summary, and as far as needed to understand the molecular details of carbon source utilization and its control.

REGULATORY PHENOMENA RELATED TO CARBON SOURCE UTILIZATION

Diauxie

When cells of *E. coli* are exposed to pairs of carbon sources at relatively high concentrations (≥ 0.5 mM), usually one is used in preference. Growth occurs in two phases (diauxic growth). The two phases are often separated by a lag period which may last up to one generation time (Fig. 11.1).[17] As suggested originally by Monod,[1] Class A carbon sources, namely glucose, mannitol, N-acetyl-glucosamine, gluconate and glucose-6-phosphate for *E. coli*, exclude Class B substrates (e.g., fructose, β-glucosides, L-sorbose, glucitol, galactose, maltose, lactose, L-arabinose, xylose and glycerol) during the first growth phase from the cells. Subsequent studies[16] showed that Class A substrates indeed prevent the uptake of the first inducing molecules of any Class B substrate and hence induction of the corresponding transport systems and metabolic enzymes. During the growth lag period the cells adapt to the second carbon source.

CARBON CATABOLITE REPRESSION

For enteric bacteria, the best studied Class A substrate is glucose, hence the name "glucose effects" to describe these exclusion phenomena. This name, now considered out of date, was changed into carbon catabolite repression by Neidhardt and Magasanik (references in 2,3). These authors showed that the "glucose effects" were not restricted to glucose and Class A substrates, but could be caused by any carbon source provided its catabolism exceeds the rate of anabolism, e.g., under the conditions of phosphate or nitrogen limitation. In mutants with completely deregulated metabolic enzymes for a Class B substrate (e.g., glycerol), the latter also caused a strong "glucose effect."[18] On the other hand, when the Class A substrates glucose and mannitol were forced to enter cells through the transport systems for the Class B substrates fructose and glucitol, respectively, they behaved like Class B substrates.[17,19] From such data, Magasanik[2] concluded that all growth conditions which cause an excess of catabolism over anabolism produce high levels of catabolic intermediates. One or more of these then inhibit transcription of many (most) inducible operons through carbon catabolite repression.

Carbon catabolite repression, however, is a complex regulatory phenomenon which comprises at least three regulatory mechanisms (references in 3-5, 16). The PTS is central to each of these as will be described in detail below:

i. Permanent catabolite repression

This form of repression, as indicated by the name, decreases the steady state level of enzyme synthesis. It lasts as long as the repressing (Class A) substrate is metabolized rapidly, or as the repressing conditions (excess of catabolism over anabolism) last. Permanent repression affects all inducible operons that code for peripheral catabolic enzymes and for enzymes involved in the quest of food, including those for flagella synthesis, cell motility and chemotaxis. It is usually measured in mutant cells expressing the enzymes under consideration in a constitutive way, e.g., those for the degradation of a Class B substrate. Enzyme levels from cells grown on a repressing carbon source (normally glucose) are compared to the levels from cells grown on a nonrepressing substrate (usually glycerol). This test prevents unwanted effects due to, e.g., inhibition of inducer uptake. Permanent repression rarely decreases enzyme synthesis more than 2- to 3-fold.

ii. Transient catabolite repression

A more severe (up to 95% decrease) type of repression is characteristic for cells undergoing a change in growth rate such as seen during shift-up experiments or during diauxic growth (Fig. 11.2). It may also be caused by nonmetabolizable substrate analogues and is transient (0.1 to 1 doubling).

It is generally believed by now (references in 5,11,20) that both types of repression, though not identical, are closely related to the

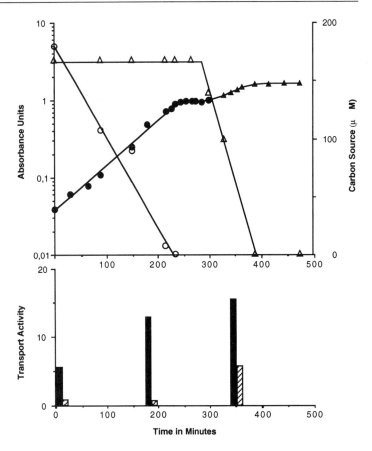

Fig. 11.1. Diauxic growth of E. coli K-12 on D-mannitol and D-glucitol. Cells of strain JWL159 (mtl⁺ gut⁺), pregrown on glucose, were washed and grown on a mixture of 180 μM mannitol and of 170 μM glucitol. Growth on mannitol (●-●) and on glucitol (▲-▲), as well as the disappearance of mannitol (O-O) and of glucitol (Δ-Δ) from the medium was recorded; activities for mannitol (solid bars) and for glucitol (stippled bars) transport (in nmoles per minute per mg of protein) are given for the times indicated. Data redrawn from Fig. 4, Lengeler J et al, J Bacteriol 1972; 112:840-848.

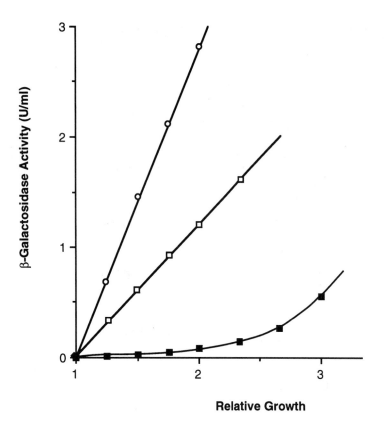

Fig. 11.2. *Permanent and transient catabolite repression in* E. coli *K-12. Cells of strain W3110, inducible for the* lac *operon, were pregrown on glucose or on glycerol. The glucose-grown cells were re-inoculated into glucose + IPTG (□-□; culture 2), the glycerol-grown cells into glycerol medium to which either IPTG (○-○; culture 1), or IPTG + glucose (■-■; culture 3), were added. Beta-galactosidase synthesis in culture 1 compared to culture 2 reflects permanent repression, while the difference between cultures 2 and 3 reflects transient repression and inducer exclusion. Data redrawn from Fig. 4, Paigen K et al, Adv Microb Physiol 1970; 4:251-324.*

bacterial second-messenger (or alarmone) cyclic-3'-5'-adenosine-monophosphate (cAMP), and its intracellular pools: (i) glucose, especially when added in high concentrations to cells growing on poor carbon sources ("shift-up"), causes a rapid drop in intracellular cAMP concentrations;[21] (ii) exogenous cAMP suppresses permanent and transient catabolite repression;[22] (iii) to each carbon source (generation time) corresponds a typical cellular cAMP level which correlates fairly well with the degree of repression.[23] Feast conditions correspond to low cAMP levels and strong catabolite repression; famine conditions, however, to high cAMP levels and weak repression. Earlier studies by various groups (references in 11, 20) led to the discovery of the cAMP-binding protein (CAP) or the cAMP-cAMP receptor protein (CRP) system. It is considered by now as the paradigm of a well-studied global regulatory network.[13] When complexed with cAMP, CAP binds to specific sites located in inducible promoters. If bound to a promoter, CAP activates (or rarely inhibits) transcription by bending the promoter DNA and by modulating transcription initiation (for details see chapter 14). All operons and regulons whose transcription is regulated by this epistatic global control system form the *crp* modulon. For most, catabolite repression thus corresponds to the lack of CAP activation (positive control) and not to a repression in the strict sense (negative control).

iii. Inducer exclusion

If present in high concentrations, many carbon sources or derivatives thereof inhibit the activity of transport systems and of catabolic enzymes for secondary class substrates. Hence, transport, uptake or synthesis of inducing molecules are inhibited ("inducer exclusion")[3] and induction of the corresponding genes is prevented or delayed. The term *catabolite inhibition*[4] has also been proposed to distinguish this indirect "repression" from permanent and transient catabolite repressions which modulate directly enzyme synthesis. Several mechanisms are involved in inducer exclusion, not all involving the PTS. We will discuss these and the two repression mechanisms in more detail after having described the hierarchical structure of bacterial regulatory networks and the PTS in its complexity.

BACTERIAL TRANSPORT SYSTEMS AND GLOBAL REGULATORY NETWORKS FORM A UNIT

In bacteria, genes coding for related functions are often grouped as a transcriptional unit or *operon* which allows the coordinated expression of these genes. The many genes for complex metabolic pathways, perhaps not easily accommodated in a single operon, often map in more than one operon. Provided they are still regulated by a single common regulatory protein (activator or repressor), e.g., MalT for maltose or GlpR for glycerol utilization, such a group of operons is called a *regulon*. This individual regulation of operons and regulons by, e.g.,

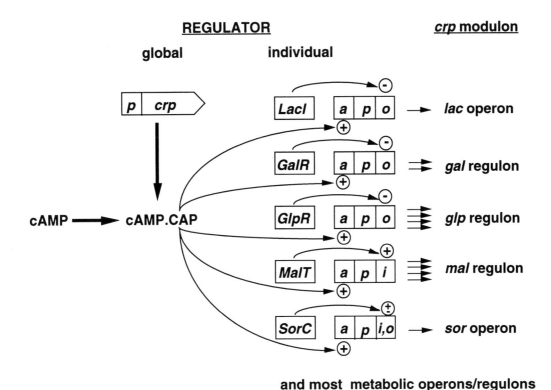

Fig. 11.3. The crp modulon of E. coli K-12. Several operons (single arrow) and regulons (≥2 arrows) represented are controlled by an individual repressor at an operator (o), or by an activator at an initiator (i), or by both (sor operon). Each member of the crp modulon, i.e., most operons involved in the quest of food, is also regulated by the global and dominant regulator cAMP-binding-protein (CAP). CAP (gene crp) forms a complex with the alarmone ("second-messenger") cAMP and binds to a consensus sequence (a) present in all promoters of operons which belong to the crp modulon. The cAMP·CAP complex activates (rarely inhibits) transcription.

the substrates (induction) or by the end products (repression) of the enzymes which they encode, is thus at the same hierarchical level and is in fact identical.

Many, even more complex, bacterial activities oriented towards a common global purpose require coordination of whole blocks or networks of metabolic pathways. The corresponding genes are encoded in dozens of operons and regulons and a higher level of organization is required. The term *modulon*[24] designates such groups of operons and regulons that are not only regulated by individual regulatory proteins, but in addition are under the control of a common global regulator (Fig. 11.3). The global system is epistatic, i.e., superimposed or dominant over the individual regulators. Mutations in the regulator cause pleiotropic phenotypes, that is they affect the expression of all members of a modulon. As an example, the *crp* modulon can be used. Starving conditions are signaled to the cells by increased levels of the alarmone cAMP. During growth in poor media, cAMP·CAP-complexes are abundant and bound to all promoters of the *crp* modulon, catabolite repression is eliminated (not existent). The capacity to synthesize all enzymes of the modulon is thus maximal, although in the absence of an inducer, all systems remain repressed (are not activated) by their individual repressor (activator). The addition of any inducing substrate will cause maximal synthesis, but only of its catabolic enzymes, e.g., the addition of a β-galactoside inducing only the *lac* operon. When grown, in contrast, in rich media, cells have decreased levels of intracellular cAMP. Hence induction remains low even in the presence of an inducer, all members of the modulon being repressed by the dominating CAP-dependent global control system.

A single stimulus, e.g., the addition of glucose to cells of *E. coli* starved for a prolonged time, can trigger a response in more than one modulon and act through different regulatory mechanisms. Under these conditions, members of other global regulatory networks involved in the control of carbon starvation (Cst) and in stringent-relaxed (RelA/SpoT) control are also affected (see chapters 25 and 27). The term *stimulon* has been proposed[25] to designate large groups of operons that respond together to an environmental stimulus, but which are not necessarily controlled through a common regulatory system. The operons rather contribute to the achievement of one common goal, that is selection of the optimal carbon source in the chosen example.

Bacteria monitor their surroundings by an impressive array of extra- and intracellular sensors. Sensors convert a stimulus originating from the environment to a signal. Signals are communicated by transducers to regulators, e.g., those controlling the transcription of a modulon. The response of a culture of *E. coli* to the addition of lactose may be described in terms of a sensory reception system: lactose in the medium (*stimulus*) is sensed by the LacY permease (*sensor*). Lactose is transported through the membrane and converted by β-galactosidase into allolactose (*signal*), the inducer which binds to the *lac* repressor. This *regulator* controls the synthesis of the lactose degrading enzymes which hydrolyze lactose (*response*). If fermented at a rate sufficient to cause carbon catabolite repression, a new equilibrium in enzyme synthesis, enzyme activity and fermentation is reached (*adaptation*).

Many global regulatory systems are connected to specific chemosensors and control their modulon in response to drastic global changes which generate a state of general alarm in a cell. The alarm is often signaled to the cell as changes in specific indicator molecules or *alarmones* such as cAMP and other nucleotide derivatives.[25] Alarmone levels usually reflect the activity of entire metabolic networks, e.g., cAMP pools reflect the activity of all peripheral catabolic pathways. They are detected by intracellular sensors. Classical bacterial sensory systems share a common theme which is best exemplified by the so-called two-component systems comprising a sensor and a response-regulator (Fig. 11.4) (references in 26, 27). Each component contains two functional domains or modules, and the four domains together form the backbone

Fig. 11.4. Two-component signal transduction systems. In its most frequent form, a two-component system consists of a chemosensor and a response-regulator. A stimulus (input) is perceived by a receptor domain which modulates (thin line) the activity of a transmitter kinase. The kinase domain autophosphorylates at a His-residue at the expense of ATP. It phosphorylates the receiver domain of the response-regulator at an Asp-residue, and modulates the fourth or regulator domain. The regulator often controls (output) global regulatory systems involved in metabolism, cellular differentiation and cellular behavior; the system then senses alarmone pools which reflect metabolic network activities.

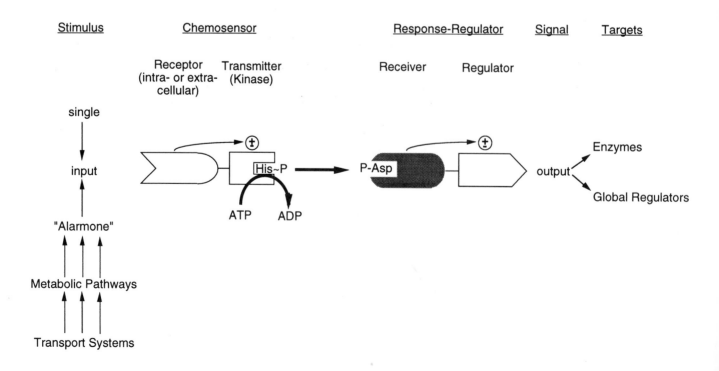

of a signal transduction pathway. Information transfer is by direct protein-protein contact and especially by protein phosphorylation.

Chemosensors, as the name implies, react to chemical stimuli which they recognize through the first or receptor domain. Stimulation normally corresponds to the binding of a chemical, e.g., an inducer or an alarmone, to the receptor ("input module"). To each receptor corresponds a transmitter domain. Both form a complex and often are covalently fused at the peptide level. Together they constitute the first component of a signal transduction pathway. Similar to other protein kinases, the transmitter autophosphorylates in an ATP-dependent reaction at a histidine residue. Depending on the system, stimulation increases or decreases this autophosphorylation. In its phosphorylated and activated form the transmitter communicates with the second component, the response-regulator, of the signal transduction pathway. Communication is through transfer of the phosphoryl group from the histidine residue of the transmitter to an aspartate residue of the third or receiver domain which together with the fourth or regulator domain ("output module") constitutes the response-regulator. Receiver and regulator may again be free or covalently fused, but in each case their activity is modulated through phosphorylation and dephosphorylation.

The regulator controls various cellular functions, e.g., the switch of the flagellar motor and chemotaxis or gene expression. Gene regulation often involves alternative sigma factors or global transcription factors, especially for genes organized in a modulon. Global control of most carbohydrate transport systems and peripheral catabolic pathways in *E. coli* through carbon catabolite repression acts formally, though not in detail, in analogy to two-component systems. In this particular control, the PTS has the role of a very complex sensor and signal transduction system which is linked to the cAMP·CAP transcription regulator through protein phosphorylation.[14]

THE BACTERIAL PTS IS A TRANSPORT AND SIGNAL TRANSDUCTION SYSTEM

The PEP-dependent PTS[12,28,29] is involved in the transport and phosphorylation of a large number of carbohydrates. It must therefore first be viewed as a collection of transport systems present in a single bacterium, e.g., the 16 PTSs of *E. coli*. These systems, first described by Roseman and co-workers as phosphotransferase systems[7] and by Lin and co-workers as carbohydrate transport systems,[30] accumulate their substrates as the corresponding phosphate esters at the expense of PEP. Studies based on amino acid sequences and on the analysis of the structure of PTS proteins revealed that all PTSs contain identical functional domains (or modules) which catalyze similar biochemical reactions (references in 31). The different steps and molecules involved in transport and phosphorylation of a substrate through any PTS are summarized in Figure 11.5. The sequence of reaction begins with the PEP-dependent autophosphorylation of a soluble protein kinase called traditionally Enzyme I (EI). It is equivalent to the catalytic subunit of other targeted protein kinases from pro- and eukaryotic organisms,[14] but it is phosphorylated at a histidine residue thus resembling two-component kinases. EI donates its phosphoryl group to a small Histidine protein (HPr), which is also phosphorylated at a histidine residue (His15). HPr constitutes a universal joint because it donates the phosphoryl group to a large number of targeting subunits, the Enzymes II (EIIs). EI and HPr are encoded in a single *pts* operon (genes *ptsI* and *ptsH*) and constitute the general PTS proteins.

Fig. 11.5. General scheme of a PTS. Each PTS starts with the general protein kinase Enzyme I which autophosphorylates at the expense of PEP. The phosphoryl group is transferred in a fully reversible reaction cascade (His- and Cys-residues), through the soluble ancillary protein (domain)s Histidine protein (HPr), IIA and IIB, to a substrate molecule bound in the membrane-bound and substrate-specific transporter IIC. The process ends with the release of a substrate-phosphate molecule into the cytoplasm and with rephosphorylation of all components. Domains IIA, IIB and IIC together constitute an Enzyme II of a PTS.

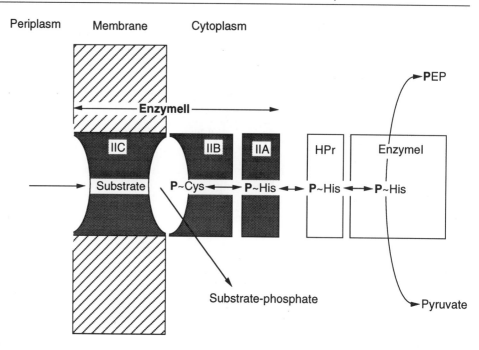

In contrast to EI and HPr which phosphorylate all EIIs of a cell, EIIs are substrate-specific and encoded by individual operons.[12] The different EIIs and the operons which encode them are normally named according to their major substrate, e.g., the mannitol PTS with its EII^{Mtl} is a member of the *mtl* operon. Each EII comprises three functional domains named IIA, IIB and IIC, respectively. Domain IIC corresponds to a membrane-bound transporter. IIC alone folds correctly into the membrane and binds all its substrates with a normal affinity. Normally, however, IIC forms a complex with a cytoplasmic domain IIB, itself complexed with another cytoplasmic domain IIA. Related IIC and IIB domains are normally fused into one protein, while the corresponding IIA domain may be in a free form. As indicated in Figure 11.5, IIAs are phosphorylated by phospho-HPr, again at a histidine residue, and donate the phosphoryl group to their IIB. Most IIBs are phosphorylated at a cysteine residue (few at a histidine, e.g., the mannose and the L-sorbose PTS). All phosphorylation steps from the catalytic subunit EI to the targeting subunits IIBs are fully reversible. The IIC transporter alone is unable to translocate high-affinity substrates with good efficiency. When complexed to a phospho-IIB domain, the transporter catalyzes with a high efficiency the facilitated diffusion of substrates and their subsequent phosphorylation by phospho-IIB. Transport ends with the release of substrate-phosphate molecules into the cytoplasm and the rephosphorylation of all domains at the expense of PEP.

The PTS may, however, also be regarded as a single regulatory unit in which all members communicate through protein-protein interactions and reversible protein phosphorylation.[14] The system is organized in a hierarchical way (Fig. 11.6) with the catalytic unit or kinase EI at the top. Through the universal carrier protein HPr, EI phosphorylates all IIAs of a cell. As expected for a protein interacting with many targeting units (at least 16 different IIAs in *E. coli*), the molecular interactions between HPr and the IIAs are not very specific. According to nuclear magnetic resonance and X-ray crystallography studies,[32] the residues forming the active center of HPr, including His15,

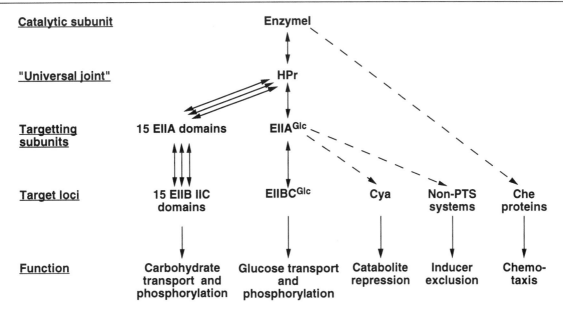

Fig. 11.6. The PTSs of a cell as a hierarchical regulatory unit. The catalytic subunit or kinase Enzyme I is linked through the universal phosphoryl carrier HPr to all IIA domains (targeting subunits) of a cell, and through these to the various target loci and functions indicated. Because all phosphorylation steps are reversible, the various activities of all PTSs are integrated at the level of EI and HPr. This information is used to control carbon catabolite repression indirectly through IIA^{Glc} (formerly called III^{Glc}) and bacterial chemotaxis directly. Modified from Lengeler JW et al, In: Torriani-Gorini AM et al, eds. Phosphate in Microorganisms. Cellular and Molecular Biology. Washington: ASM Press, 1994:192-188.

are clustered at the surface of the molecule. Similar studies revealed for IIA^{Glc} from *E. coli*[33] that two essential His residues are also located in close proximity at the surface of the molecule. This includes His90, the residue phosphorylated by phospho-HPr, and His75. Replacement of His90 with Gln resulted in a IIA^{Glc} that could no longer be phosphorylated by phospho-HPr. The same exchange in His75, however, resulted in a protein that could still be phosphorylated, but could no longer phosphorylate IIB^{Glc}.[34] His75 thus seems to be important in phosphotransfer between IIA^{Glc} and IIB^{Glc}. Both His residues are located within a shallow depression which is formed by hydrophobic residues. This surface structure appears as complementary to the corresponding part of HPr such that phospho-group transfer from His15 of HPr to His90 of IIA^{Glc} may occur within the hydrophobic interface formed by the proteins. The residues involved in the interaction between IIA^{Mtl} and HPr from *E. coli*[35] are also located at the surface of both molecules and surround the catalytic His residues. No consensus with the corresponding residues from IIA^{Glc} is found. This corroborates the hypothesis of a universal, nonspecific protein-protein interaction at this early stage of the hierarchy. The binding structures between IIAs and their IIB domains, in contrast, are highly specific and require individual amino acid to amino acid interactions (references in 31).

IIA^{Glc} OF THE PTS IS CENTRAL TO CARBON CATABOLITE REPRESSION

As expected according to the scheme presented in Figure 11.6, mutants of *E. coli* defective in EI (gene *ptsI*), HPr (gene *ptsH*) or in EI and HPr are unable to grow on PTS carbohydrates. They are, however, also unable to grow on a number of non-PTS carbon sources, e.g., lactose, maltose, melibiose, glycerol, the pentoses, Krebs-cycle intermediates and other Class B components. Their unexpected phenotype resembles the phentotype of Cya$^-$ mutants which are defective in adenylate cyclase (gene *cyaA*) and in the synthesis of cAMP from ATP. This similarity is extended by the observation that growth of Pts$^-$ mutants on non-PTS carbohydrates (for obvious reasons not on PTS carbohydrates!) and growth of Cya⁻ mutants on all carbon sources can be restored by the addition of exogenous cAMP. Later studies showed that adenylate cyclase activity is decreased in Pts⁻ mutants compared to Pts⁺ wild-type cells. Furthermore, synthesis of cAMP in intact cells and adenylate cyclase activity in toluenized cells was inhibited in the presence of a PTS carbohydrate like glucose or its nonmetabolizable analog methyl-α, d-glucopyranoside.[5,21,22]

In a different approach, suppressor mutations were isolated that restored growth of Pts⁻ and Cya⁻ mutants (references in ref. 16). Some of these were found to restore growth only on the carbon source used during selection. These specific mutations affected repressor genes, promoter sequences and other regulatory elements; they caused consistently the constitutive expression of the corresponding genes or restored inducibility under the conditions of carbon catabolite repression. Other mutations restored growth on all carbon sources mentioned above. One class, called *crp**, or correctly *crp* (In), resulted in CAP molecules active in the absence of cAMP. These mutations obviously simulate the addition of exogenous cAMP to Pts⁻ or Cya⁻ mutants. Saier and Roseman[9,10,36] detected first in *Salmonella typhimurium* a new class which they called *crr* (mnemonic for *c*arbohydrate *r*epression *r*esistant) because the mutation restored in leaky *ptsI* mutants growth on several (but not all) non-PTS carbohydrates simultaneously. Similar mutants were detected in *E. coli* and all were shown to affect IIA^{Glc} (formerly called Enzyme III), which functionally also belongs to the glucose PTS and its IICB^{Glc} (gene *ptsG*).[37-39] It could also be shown that the addition of any PTS-carbohydrate, provided its PTS was induced, inhibits the uptake of lactose, maltose, melibiose and glycerol through their specific non-PTS transport system within seconds and minutes. In Crr⁻ mutants this inhibition of transport activity (inducer exclusion) was not seen.

Based primarily on these data, Roseman, Postma, and Saier[8-10] proposed a model which explained the repression (synthesis) as well as the inhibition (activity) observed during carbon catabolite repression, and which postulated IIA^{Glc} of the PTS as the essential regulatory molecule. This model has been confirmed through biochemical and genetic results. It is generally accepted for all enteric bacteria in the form outlined schematically in Figure 11.7: The regulatory molecule IIA^{Glc} can exist in a phosphorylated (P-IIA^{Glc}) and in the unphosphorylated (IIA^{Glc}) form. Its phosphorylation state is determined by the balance between phosphorylation through phospho-HPr (P-HPr) and dephosphorylation through IICB^{Glc}, e.g., during uptake of glucose. Because the phosphorylation steps of all PTS proteins (Fig. 11.5) are fully reversible, uptake of any PTS carbohydrate which dephosphorylates P-HPr will also drain phosphoryl groups from P-IIA^{Glc} and influence the equilibrium. In its phosphorylated form, P-IIA^{Glc} is probably involved in the activation of the enzyme adenylate cyclase and hence, through modulation of intracellular cAMP levels, in the control of the *crp* modulon. Free IIA^{Glc}, however, binds to and inhibits several proteins and enzymes essential in the metabolism of several non-PTS carbohydrates, e.g., the lactose and melibiose permeases, the MalK protein of the maltose transport system, and glycerol kinase in *E. coli*. This rapid inhibition of the uptake of the first inducing molecules (lactose, melibiose, maltose), or of the synthesis of the inducer (sn-glycerol-phosphate) corresponds to the inducer exclusion first proposed by Monod.[1] Enzyme repression and inducer exclusion act in concert during carbon catabolite repression. In cells growing on a non-PTS carbohydrate, all PTS proteins including IIA^{Glc} should be in the phosphorylated form (high cAMP levels, no inducer exclusion). Addition of a PTS carbohydrate, especially of glucose, should lower (transiently) the phosphorylation leading to the deactivation of adenylate cyclase (low cAMP levels, i.e., repression) and to inducer exclusion by IIA^{Glc}. We will discuss the different processes in more detail below.

IIA^{Glc}, THE REGULATION OF ADENYLATE CYCLASE ACTIVITY AND OF INTRACELLULAR cAMP LEVELS

In *E. coli*, cAMP is synthesized from ATP by the enzyme adenylate cyclase (encoded by *cyaA*). It can be hydrolyzed by a phosphodiesterase (encoded by *cpd*), expelled to the medium in an energy-dependent process, and taken up into the cell from the medium.[5] Although adenylate cyclase from *E. coli* has been purified to near homogeneity and the gene *cyaA* has been sequenced, neither its regulation at the level of enzyme synthesis nor of enzyme activity is understood.[40] For cAMP excretion and uptake from the medium, the transport systems involved are not even known (for recent reviews see refs. 11, 20).

As discussed before, the postulate of a direct link between the phosphorylation state of IIA^{Glc} and cAMP synthesis rests basically on indirect results: the similarity in the phenotype of Pts$^-$ and Cya$^-$ mutants which can be suppressed by exogenous cAMP,[22,40] a lowered cAMP

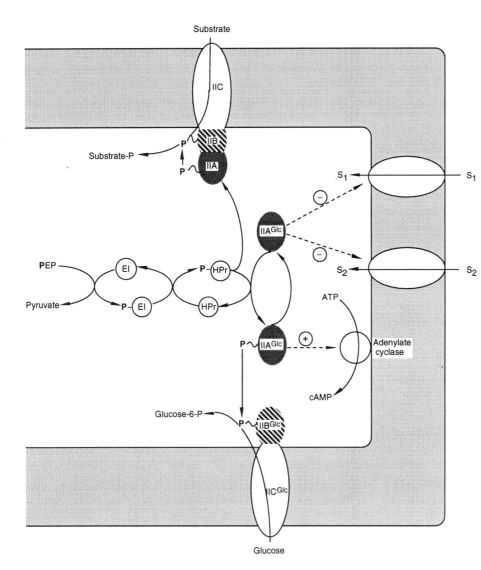

Fig. 11.7. Model of the regulation of carbon catabolite repression through IIA^{Glc} and the PTS. The general PTS proteins Enzyme I (EI) and HPr are shown together with the glucose PTS (domains IIA^{Glc}, IIB^{Glc} and IIC^{Glc}), and an unspecified EII. During uptake of a PTS-carbohydrate, IIA^{Glc} is dephosphorylated directly (during glucose uptake) or indirectly (during uptake through any EII) due to full reversibility of all phosphoryl reactions. Activation (+) of adenylate cyclase by phospho-IIA^{Glc} and inhibition (–) of two different non-PTS uptake systems S1 and S2 (e.g., for lactose, maltose, melibiose, and glycerol), by nonphosphorylated IIA^{Glc} are also indicated. Modified from Postma PW et al, Microbiol Rev 1993 57:543-594.

level in *crr* negative mutants,[41,42] the inhibition of adenylate cyclase activity in toluene treated cells by PTS carbohydrates, provided the corresponding EII is also present,[40,43] and above all the isolation of Crr⁻ mutations as suppressors in Pts⁻ strains.[10,37,39,44] From these results it was concluded that P-IIA^{Glc} might be an activator for adenylate cyclase (alternatively it might inactivate an inhibitor).[41,42] All essays to demonstrate such an activation of adenylate cyclase by adding in vitro P-IIA^{Glc}, or purified EI, HPr, IIA^{Glc} and PEP have failed,[45] or even gave contradictory results.[46] Nor is there any evidence that the cyclase is phosphorylated by P-IIA^{Glc} or the PTS. However, the genetic data strongly indicate IIA^{Glc} to be the (an) essential modulator.

NOT ONLY cAMP LEVELS, BUT ALSO CAP LEVELS ARE ESSENTIAL IN CATABOLITE REPRESSION

The intracellular level of cAMP is essential in permanent and transient repression. Although it is clear that the regulation of adenylate cyclase by IIA^{Glc} plays an essential role, numerous studies suggest that the regulation of cAMP pools is more complex and that cAMP may not be the sole mediator of repression (references in 20,47-49). Thus mutants which lack CAP produce large amounts of cAMP, most of which is excreted into the medium.[50,51] The increased production (~100-fold) appears to require IIA^{Glc} because it is not seen in Crr⁻ mutants.[52,53] With the help of *cya-lacZ* fusion strains, it was shown that the *cya* promoter P2 is negatively controlled by the cAMP·CAP complex.[54] This 4-fold decrease in the transcription rate is not sufficient to explain the large variations in cellular cAMP levels (at least 100-fold) observed in CAP⁺ versus CAP⁻ mutants,[50] during a shift from derepression to repression,[55,56] or the 10- to 20-fold inhibition of adenylate cyclase by PTS-carbohydrates in toluene treated cells.[43] Several mutations in the *cya* gene have been isolated which abolish increased cAMP levels in CAP⁻ mutants.[52,57] Their analysis has not yet revealed the mechanism involved in this activity regulation.

Recent studies by Aiba and co-workers[58,59] resolve part of the discrepancy. These authors confirmed older studies which had indicated that the correlation between, e.g., the expression of β-galactosidase and intracellular cAMP levels was not strict,[23] that catabolite repression can occur in Cya⁻ mutants if they carry *crp*(In) suppressor mutations, and that exogenous cAMP could not suppress permanent catabolite repression completely but could eliminate transient catabolite repression (references in 5, 20). They showed furthermore that the level of CAP does vary appreciably (from 1,400 to 6,600 molecules per cell) and is markedly reduced in the presence of glucose. The reduction is mostly due to a decrease in *crp* mRNA levels (repression) and not to, e.g., an inactivation of the DNA-binding capacity of CAP. Exogenous cAMP does eliminate permanent repression in cells overproducing CAP (5-fold from an F-plasmid encoded and constitutively expressed *crp* gene). Finally, β-galactosidase expression decreases even in the absence of glucose when the expression of CAP is reduced below wild-type levels and then shows a good correlation with the CAP levels. According to these data, transient and permanent catabolite repression can be explained as follows: the addition of glucose to cells growing on a poor carbon source stimulates through an unknown mechanism (modification of CAP?) a rapid excretion of cAMP. This excretion is an energy-dependent process and stimulated by metabolizable carbon sources to various degrees.[56,60] It is (together with inducer exclusion for inducible systems, see below) primarily responsible for transient repression and can be reverted by exogenous cAMP in mM concentrations. The addition of glucose also represses the synthesis of CAP leading (by dilution over several generations) to a lowered (2- to 5-fold) CAP level. Meanwhile, a new cAMP steady-state level (about 2-fold lower) is reached. It is most likely due to the lack of P-IIA^{Glc} and deactivated adenylate cyclase molecules. This reduction in cAMP and in CAP levels is the major cause for permanent catabolite repression. Molecular details of the postulated mechanisms have yet to be elucidated.

IIA^{Glc} AND INDUCER EXCLUSION

As mentioned before, any regulatory process related to carbon catabolite repression which prevents the uptake of inducing molecules or which inhibits inducer synthesis through protein or enzyme inhibition is called inducer exclusion (catabolite inhibition). Several mechanisms may be involved (references in 4,16,61), e.g., competition at the transport level between substrate isomers like glucose/galactose, glucose/fructose, and mannitol/glucitol, and allosteric inhibition of transport systems by accumulated carbohydrate phosphates. Because their effect on gene expression is indirect, only through exclusion of the inducer from the cell, we will restrict their description in this series on gene regulation to mechanisms which involve the PTS proteins HPr and IIA^{Glc} (for a complete review see ref. 12).

The model outlined in Figure 11.7 postulates that IIA^{Glc} in its free form binds to and inhibits several non-PTS transport systems and metabolic enzymes, thus preventing uptake and metabolism of the corresponding carbohydrates. Binding of purified IIA^{Glc} to the lactose permease (LacY),[62-64] the MalK component of the maltose transport system[65,66] and to purified glycerol kinase[66,67] has been demonstrated, as well as its inhibitory effects on the transport systems reconstituted in liposomes (references in 12,61). Only free IIA^{Glc} binds to these proteins and phosphorylation prevents its binding. Mutations have been isolated in crr[39,44,68] and in its target proteins LacY,[69,70] MalK[65,71] and MelB[72,73] that allow growth of ptsI,H mutants on non-PTS carbohydrates. The mutations have been identified by DNA sequencing and alter residues involved in binding, as has been proven through binding studies for LacY[69] and for IIA^{Glc}.[74,75]

The three-dimensional complex between four IIA^{Glc} molecules and the tetrameric glycerol kinase from E. coli has been determined.[76] The contact between both molecules is rather limited and confined to a very hydrophobic structure in which residues 472-481 of the kinase project into the active site of IIA^{Glc} with the phospho-group acceptor His90. It is tempting to speculate that this is another example of an unspecific binding through hydrophobic bonds as found between IIA^{Glc} and HPr. Such a binding may be easily disrupted by introducing a negative charge at His90 of IIA^{Glc} and explain why only IIA^{Glc}, but not P-IIA^{Glc} inhibits glycerol kinase activity.[77] Several mutations in crr which were defective in kinase regulation but which were still active in glucose transport and phosphorylation mapped within the hydrophobic region mentioned thus support such a model.[34,69,74,75]

Three factors, one of them not predicted by the model as outlined in Figure 11.7, are essential in the response of cells to PTS-carbohydrates through carbon catabolite repression: (i) directly the phosphorylation state of IIA^{Glc}, and indirectly (because of the reversibility of all steps) the phosphorylation state of all PTS proteins; (ii) the number of nonphosphorylated IIA^{Glc} molecules relative to the number of target proteins, i.e., their stoichiometry; and (iii) the presence of a substrate for the target protein (e.g., for lactose, maltose or glycerol metabolism) because binding of IIA^{Glc} to the target proteins requires the presence of a substrate.[64,77] In cells growing on poor non-PTS carbohydrates, all PTS proteins will be phosphorylated due to the high free energy of hydrolysis of the phospho group of PEP and to the complete reversibility of all phosphorylation steps. Adenylate cyclase activity will be high and induced systems will be expressed at maximum transcription rates. It had been noted before by Cohn and Horibata[78] that cells of *E. coli,* fully induced or constitutive for the *lac* operon, could not be inhibited by glucose for lactose transport and metabolism whereas they could be inhibited during partial induction. The results were repeated in *S. typhimurium* and extended to the *mal,* the *mel* and *glp* system.[62,66,73] Mutants with lowered levels of IIA^{Glc} escape more easily from inhibition as do cells in which two target proteins, e.g., those for lactose or maltose and for glycerol, compete for IIA^{Glc}.[42,79] Cells of *S. typhimurium* which overexpress IIA^{Glc} from a plasmid-encoded *crr*, in contrast, become hypersensitive to inducer exclusion by glucose.[66] These data together with the original selection procedure for Crr$^-$ mutants as resistant to carbon catabolite repression[36] strongly support the stoichiometric interactions between IIA^{Glc} and its target proteins as postulated in Figure 11.7. Cells (fully induced) of *E. coli* and of *S. typhimurium* contain about 15.000 IIA^{Glc} molecules, 8.000 lactose permease molecules and a similar number of glycerol kinase and MalK proteins, i.e., enough IIA^{Glc} to complex most copies of one target protein. Based on the known K_D-values (5-16 μM for GlpK-IIA^{Glc}) and on recent tests,[80] complete inhibition of glycerol uptake by PTS-carbohydrates was calculated and found in *E. coli* for about four IIA^{Glc} copies per kinase tetramer. This is in excellent agreement with the co-crystal studies mentioned before.[76] The predicted dephosphorylation of IIA^{Glc} (from about 80% to below 10%) during uptake of the nonmetabolizable glucose analog methyl-α, D-glucoside has also been shown for *S. typhimurium.*[77]

CATABOLITE REPRESSION AND INDUCER EXCLUSION ACT IN CONCERT

The regulatory system as outlined (Fig. 11.7) is of a remarkable economy: (i) the complete PTS with its multiple transport systems and proteins, all coupled through reversible phosphorylation steps and organized in strict hierarchy (Fig. 11.6), is used as the sensor and signal transduction system. Its various activities, directly connected to the PEP pool of the cells, accurately reflect the phosphoryl energy pool and the activities of all peripheral catabolic pathways; (ii) among the PTS-proteins, IIA^{Glc} constitutes the direct link between the sensory and the regulatory part of the carbon catabolite control system. IIA^{Glc} is the only IIA of the large glucose-sucrose PTS family, to which it belongs,[12] found in free form and involved in the uptake of several carbohydrates (i.e., sucrose, trehalose, maltose), among them glucose, the preferred carbon source in enteric bacteria. It is also the only IIA encoded together with the general PTS proteins in the *pts* operon.[81] Because it is synthesized in a coordinate way with EI and HPr, it may also be regarded as a general PTS protein.[16,81,82] Large EIIs, like II^{Nag} for N-acetyl-glucosamine and II^{Bgl} for β-glucosides, contain fused IIA^{Glc}-like domains. These domains can replace IIA^{Glc} (in Crr$^-$ mutants) and can be replaced by IIA^{Glc} in glycoside transport and phosphorylation assays.[69,83-85] While in these earlier studies the IIA-like domains failed to restore adenylate cyclase activation and inducer exclusion, a recent study[74] showed that a plasmid encoding II^{Nag} restored inducer exclusion in *crr nagE* mutants from *S. typhimurium.* Similarly, IIA^{Glc} from the II^{Glc} of *Bacillus subtilis* restored inducer exclusion in appropriated

E. coli mutant strains.[86] The discrepancy probably reflects the different expression levels (and hence variations in the stoichiometry) of the various constructs rather than true differences. (iii) IIA^{Glc} binds only in the presence of a substrate to its target proteins. Thus in cells induced to intermediate levels for the *mal* and the *glp* system, the addition of 2-deoxy-glucoside results in a lower inhibition of the maltose transport system when glycerol is also present,[66] i.e., when the glycerol-glycerolkinase complex binds IIA^{Glc} and lowers its concentration in the free form. This arrangement prevents the nonproductive binding of IIA^{Glc} molecules to proteins for which no substrates are available. Lowering of CAP synthesis during repressing conditions[59] similarly may prevent the nonproductive synthesis of this protein; iv) compared to mutants in which the *cya* gene has been deleted, Pts⁻ and Crr⁻ mutants retain a residual level of cAMP synthesis. Different promoters of the *crp* modulon require different concentrations of cAMP (and of cAMP·CAP) for full expression (references in ref. 12). This may explain why Crr⁻ mutants of *E. coli* still grow on some carbon sources (e.g., lactose, maltose, glycerol, melibiose, or xylose), but not on others (e.g., Krebs cycle intermediates and L-rhamnose). Variations in cAMP and CAP levels are also most likely at the base of the different phenotypes observed in mutants carrying various *ptsI, ptsH, crr, cya* and *crp* alleles, and between members of the enteric bacteria, e.g., *E. coli* and *S. typhimurium*.[16,87] A basal level of cAMP synthesis allows growth on some carbon sources with a low requirement for cAMP·CAP among others most PTS carbohydrates. Low cAMP levels are not sufficient for growth on other substrates, often the poor ones, e.g., Krebs cycle intermediates and L-rhamnose; (v) even within the PTS-carbohydrates of *E. coli*, there is a strict hierarchy between dominant (the aldose D-glucose, the aminosugar N-acetyl-glucosamine, and the polyhydric alcohol D-mannitol) and lesser ones (all others).[17,19] The relevant properties are the uninduced basal level of expression of the corresponding genes, the different requirement in cAMP levels for full expression, the sensitivity of the corresponding EIIs towards carbohydrate-phosphate inhibition, and variations in the affinities of the different IIA-domains for HPr (see ref. 16).

CARBON CATABOLITE REPRESSION THROUGH PTS-CONTROL IS PART OF A STIMULON

Regulation of carbon source utilization in *E. coli* is a very complex process. The model, as outlined in Figure 11.7, probably describes the role of the PTS and of IIA^{Glc} in CAP-activation and inducer exclusion in a fairly adequate, but incomplete way (references in 11,48,49). It does not, for example, explain catabolite repression in strains deleted for the *cya* (no cAMP), and especially for the *crp* gene. Neither does it explain the strong repression and inducer expulsion observed in many gram-positive bacteria, (most of) which lack intracellular cAMP (references in ref. 88). Carbon source utilization must be linked to gluconeogenesis, to the carbon starvation response, to ATP and PEP synthesis, to acetyl-phosphate and acetyl-CoA synthesis and at least also to nitrogen metabolism. Under nutrient limitation, cells of *E. coli* respond by multiple alterations commonly known as the starvation/stationary phase response which involves the Cst/Csi (for <u>c</u>arbon <u>s</u>tarvation <u>i</u>nduced) global systems[13,25,89] linked in an ill-defined way with the *crp* modulon. Hence, a stimulon might exist in which a large number of functions reacts in concert to the stimulus "feast/famine," and cooperates in the quest of food and in optimal utilization of the carbon sources.[90] This stimulon would include the PTS as a sensory system, cAMP as a relevant alarmone, and the cAMP·CAP response regulator of the *crp* modulon as one essential global regulatory network among several involved in the response. Other candidate members of the putative stimulon with a link to the PTS are: (i) FruR as a pleiotropic regulator for gluconeogenesis; (ii) the catalytic subunit EI as a regulator for other protein kinases; and (iii) PTS proteins as a link between carbon and nitrogen metabolism.

(i) FruR, product of the gene *fruR*, is a classical repressor of the LacI family,[91,92] i.e., it comprises an amino-terminal helix-turn-helix (HTH) motif, which is connected by a linker to an inducer-binding domain (inducer: fructose 1-phosphate), and a carboxy terminal oligomerization domain.[93] Loss of FruR causes the constitutive expression of the *fruFKA* operon which codes for a protein, called FPr, a fructose-1-phosphate kinase and IIBC^{Fru}.[94-96] FPr contains the IIA^{Fru} domain of the fructose PTS fused to an HPr-like domain.[97] Because this "pseudo-HPr" can functionally replace HPr, provided the *fru* operon is expressed constitutively, *fruR* mutants can be isolated as suppressor mutations in strains lacking HPr. As expected for a repressor, FruR binds to the *fruFo* operator and fructose 1-phosphate releases the complex.[93,98] Tight FruR$^-$ mutants are also unable to express at a high level several enzymes involved in gluconeogenesis, e.g., PEP synthase and PEP carboxykinase, and hence to grow on lactate and pyruvate.[99-101] FruR seems to act like an activator for the expression of the corresponding operons.[101] From DNA binding studies of known FruR controlled operators, a consensus sequence was established. Computer searches revealed the presence of this sequence in numerous functional operons, e.g., those for PEP synthase, PEP carboxykinase and fructose 1,6-bisphosphatase (gluconeogenic enzymes); isocitrate lyase and malate synthase (glyoxalate shunt enzymes), isocitrate dehydrogenase (Krebs cycle); cytochrome d; and even in the *pts* operon itself. FruR could also be shown to bind to these operators.[98,102] Although this seems to imply that FruR is a global regulator in enteric bacteria,[92] the hypothesis is not easily consistent with the major role of FruR as the individual repressor for the *fru* operon modulated by fructose 1-phosphate, and needs further physiological proof.

(ii) Acetate kinase catalyzes the conversion of acetate and ATP into acetyl-phosphate and P$_i$. In vitro, phospho-EI can directly donate its phosphoryl group in a reversible reaction to the kinase, and in the presence of HPr, IIA^{Glc} can thus be phosphorylated at the expense of ATP. Potentially, this reaction could link the PTS to the enzymes of the Krebs cycle, of acetyl-CoA and of acetyl-phosphate metabolism.[103] Acetyl-phosphate has been proposed as the alarmone for this important metabolic bloc and as a potential modulator for many two-component sensory transduction systems (for a review see ref. 104). The physiological relevance has not been shown, however, for any of these reactions.

(iii) Already the first studies on catabolite repression (references in ref. 2, 25) indicated a close connection between the regulation of carbon and nitrogen metabolism. The missing-link has never been found. Recently, it was observed in *Klebsiella pneumoniae* and in *E. coli* that the structural gene *rpoN* for an alternative sigma-54 factor maps in an operon together with genes for a IIA- and an HPr-like protein.[105,106] Sigma-54 is required for a number of genes involved in nitrogen assimilation. Its synthesis is controlled in a negative way by (one of the) gene products from its operon[107] and can be phosphorylated in a PTS-dependent way.[108] Again, as for the two previous systems, the exact physiological role of these gene products in coupling carbon and nitrogen metabolism, is unclear.[109]

Similar to other cells *E. coli* does not correspond to a "bag full of enzymes." It comprises instead a highly organized and sophisticated network of enzyme catalyzed reactions which are controlled in a hierarchical way. Consequently, to understand the cell in its entire complexity we must not only know all its constituents, but also the magnitude of all its metabolic fluxes in a quantitative way. For the PTS and its central role in carbon catabolite control this information is beginning to emerge for the first time.[110] It promises totally new insights into the old phenomena of diauxie, catabolite repression and inducer exclusion, and should pave the way to new flux control theories and a modern metabolic engineering.

CONCLUDING REMARKS

One of the principal regulatory mechanisms for various cellular functions in the eukaryotes is post-translational modification of proteins by protein kinases and phospho-protein phosphatases (for a review see ref. 111). Both enzymes act as catalytic subunits which through targeting-subunits modulate the activity of (usually a large variety of) target proteins. The kinases (or phosphatases) often respond to extracellular stimuli which are sensed by membrane-bound sensors and signaled to the catalytic units in the form of second messengers, e.g., cAMP and cGMP. Only now, more than fifty years after the first thorough analysis of the "glucose effects" in *E. coli* by J. Monod,[1] and thirty years after the first description of the PTS by S. Roseman,[7] and others[30] do we realize how exactly both systems fit into the "eukaryotic" scheme (Fig. 11.8). For a free-living unicellular organism, "outside" normally signifies the physical universe, while for a cell of a multicellular organism, the neighboring cells usually constitute outside. Intercellular communication thus is of vital importance for eukaryotic cells. For *E. coli*, however, the PTS with its various transport

Fig. 11.8. The PTS as a signal transduction pathway in carbon catabolite repression of enteric bacteria. If drawn in analogy to a two-component system, domains IIC,B of a PTS correspond to the receptor, the other domains including the kinase Enzyme I to a complex transmitter. Phospho-IIA^Glc transfers the integrated information from all PTS to the receiver adenylate cyclase (Cya), which synthesizes the alarmone cAMP. This second-messenger modulates the activity of the global regulator CAP for the crp modulon, and through it the many cellular functions involved in the quest of food.

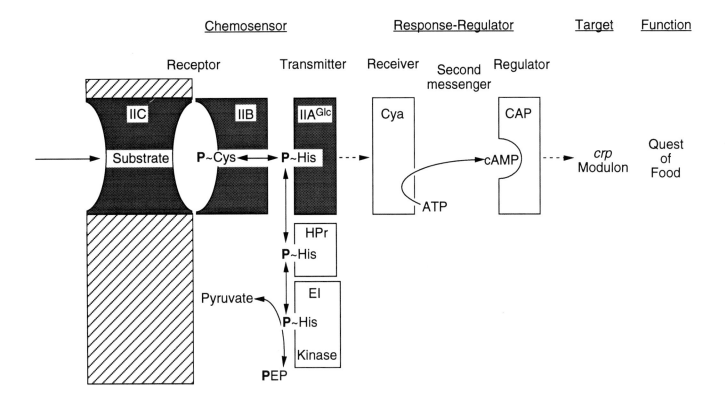

systems constitutes an essential sensory system through which the cell monitors its environment. The information gathered through all PTS proteins is integrated at the level of the catalytic subunit kinase EI and (through HPr) of the signaler IIA^{Glc} (targeting subunits). This information transfer follows a strict hierarchical pattern. IIA^{Glc} controls the activity of a large variety of membrane-bound transport systems (target loci) and synthesis of cAMP. This alarmone ("second messenger") modulates the activity of the global regulator protein CAP and through it the synthesis of all members of the *crp* modulon. Members of this modulon control a plethora of cellular functions (for a complete list see ref. 20). Among them are many which correspond in *E. coli* to cellular differentiation processes, like cell division, cell adhesion and motility, adaptations to long-lasting starvation periods, etc. Bacterial modulons thus correspond to eukaryotic regulatory networks involved at a higher hierarchical level in the global control of cellular differentiation, e.g., systems controlled by the maternal, the gap and the homeo-box genes of *Drosophila*, coupled through protein kinases/phosphatases to signal transduction cascades. This similarity cannot be purely accidental, because seryl-, threonyl- and tyrosyl-specific kinases with a role in bacterial differentiation are found in a rapidly increasing number in the prokaryotic world, and histidyl-kinases, which closely resemble those from bacterial two-component systems, have been detected in the eukaryotes (references in ref. 112). It rather seems to reflect the existence of a universal biochemistry involved in cellular sensory reception, signal transduction, and global control through large regulatory networks, which is found throughout the pro- and the eukaryotic world.

ACKNOWLEDGMENTS

I would like to thank C.-A. Alpert for help with the figures, E. Placke for help in preparing the manuscript and the Deutsche Forschungsgemeinschaft for financial support through SFB171, TP C3/C4.

REFERENCES

1. Monod J. Recherches sur la croissance des cultures bactériennes. Hermann et Cie, Paris. 1942.
2. Magasanik B. Catabolite repression. Cold Spring Harbor Symp Quant Biol 1961; 26:249-256.
3. Magasanik B. Glucose effects: Inducer exclusion and repression. In: Beckwith JR, Zipser D eds. The lactose operon. Cold Spring Harbor, NY: Cold Spring Harbor, 1970; 189-219.
4. Paigen K, Williams B. Catabolite repression and other control mechanisms in carbohydrate utilization. Adv Microb Physiol 1970; 4:251-324.
5. Pastan I, Adhya S. Cyclic adenosine 3'-5'-monophosphate in *Escherichia coli*. Bacteriol Rev 1976; 40:527-551.
6. Lin ECC. The genetics of bacterial transport systems. Annu Rev Genet 1970; 4:225-262.
7. Kundig W, Gosh S, Roseman S. Phosphate bound to histidine in a protein as an intermediate in a novel phospho-transferase system. Proc Natl Acad Sci USA 1964; 52:1067-1074.
8. Postma PW, Roseman S. The bacterial phosphoenolpyruvate:sugar phosphotransferase system. Biochim Biophys Acta 1976; 457:213-257.
9. Saier Jr MH, Roseman S. Sugar transport. Inducer exclusion and regulation of the melibiose, maltose, glycerol, and lactose transport systems by the phosphoenol-pyruvate:sugar phosphotransferase system. J Biol Chem 1976; 251:6606-6615.

10. Saier Jr MH, Roseman S. Sugar transport. The *crr* mutation: its effect on repression of enzyme synthesis. J Biol Chem 1976; 251:6598-6605..

11. Kolb A, Busby S, Buc H et al. Transcriptional regulation by cAMP and its receptor protein. Annu Rev Biochem 1993; 62:749-795.

12. Postma PW, Lengeler JW, Jacobson GR. Phosphoenolpyruvate:carbohydrate phosphotransferase systems of bacteria. Microbiol Rev 1993 57:543-594.

13. Gottesman S. Bacterial regulation: global regulatory networks. Ann Rev Genet 1984; 18:415-441.

14. Lengeler JW, Bettenbrock K, Lux R. Signal transduction through phosphotransferase systems or PTSs. In: Torriani-Gorini AM, Yagil E, Silver S, eds. Phosphate in Microorganisms. Cellular and Molecular Biology. Washington: ASM Press, 1994:192-188.

15. Roseman S, Meadow, ND. Signal transduction by the bacterial phosphotransferase system. J Biol Chem 1990; 265:2993-2996.

16. Postma PW, Lengeler JW. Phosphoenolpyruvate:carbohydrate phosphotransferase system of bacteria. Microbiol Rev 1985; 49:232-269.

17. Kornberg HL. Fine control of sugar uptake by *Escherichia coli*. Symp Soc Exp Biol 1973; 27:175-193.

18. Berman M, Lin ECC. Glycerol-specific revertants of a phosphoenolpyruvate phosphotransferase mutant: suppression by the desensitization of glycerol kinase to feedback inhibition. J Bacteriol 105: 113-120.

19. Lengeler J, Lin ECC. Reversal of the mannitol-sorbitol diauxie in *Escherichia coli*. J Bacteriol 1972; 112:840-848.

20. Botsford JL, Harman JG. Cyclic AMP in prokaryotes. Microbiol Rev 1992; 56:100-122.

21. Pastan I, Perlman RL. Repression of β-galactosidase synthesis by glucose in phosphotransferase mutants of *Escherichia coli*. Repression in the absence of glucose phosphorylation. J Biol Chem 1969; 244:5836-5842.

22. Pastan I, Perlman RL. Cyclic adenosine monophosphate in bacteria. Science 1970; 169:339-344.

23. Epstein W, Rothman-Denes LB, Hesse J. Adenosine 3':5'-cyclic monophosphate as mediator of catabolite repression in *Escherichia coli*. Proc Natl Acad Sci USA 1975; 72:2300-2304.

24. Iuchi S, Lin ECC. *arcA (dye)*, a global regulatory gene in *Escherichia coli*, mediating repression of enzymes in aerobic pathways. Proc Natl Acad Sci USA 1988; 85:1888-1892.

25. Neidhardt FC, Ingraham JL, Schaechter M. Physiology of the bacterial cell. A molecular approach. Sunderland, Massachusetts: Sinauer Associates, Inc. 1990.

26. Parkinson JS. Signal transduction schemes of bacteria. Cell 1993; 73:857-871.

27. Stock J. Phosphoprotein talk. Curr Biol 1993; 3:303-305.

28. Meadow ND, Fox DK, Roseman S. The bacterial phosphoenolpyruvate:glycose phosphotransferase system. Ann Rev Biochem 1990; 59:497-542.

29. Robillard GT, Lolkema JS. Enzymes II of the phosphoenolpyruvate-dependent sugar transport systems: a review of their structure and mechanism of sugar transport. Biochim Biophys Acta 1988; 947:493-519.

30. Tanaka S, Lin ECC. Two classes of pleiotropic mutants of *Aerobacter aerogenes* lacking components of a phosphoenolpyruvate-dependent phosphotransferase system. Proc Natl Acad Sci USA 1967; 57:913-919.

31. Lengeler JW, Jahreis K, Wehmeier UF. Enzymes II of the phosphoenolpyruvate-dependent phosphotransferase systems: their structure and function in carbohydrate transport. Biochim Biophys Acta 1994; 1188:1-28.

32. Sharma S, Georges F, Delbaere LTJ et al. Epitope mapping by mutagenesis distinguishes between the two tertiary structures of the histidine-containing protein HPr. Proc Natl Acad Sci USA 1991; 88:4877-4881.

33. Worthylake D, Meadow ND, Roseman S et al. Three-dimensional structure of the *Escherichia coli* phosphocarrier protein III^{Glc}. Proc Natl Acad Sci USA 1991; 88:10382-10386.

34. Presper KA, Wong CY, Liu L et al. Site-directed mutagenesis of the phosphocarrier protein, III^{Glc}, a major signal-transducing protein in *Escherichia coli*. Proc Natl Acad Sci USA 1989; 86:4052-4055.

35. Van Nuland NAJ, Kroon GJA, Dijkstra K et al. The NMR determination of the IIA^{Mtl} binding site on HPr of the *Escherichia coli* phosphoenolpyruvate-dependent phospho-transferase system. FEBS Lett 1993; 315:11-15.

36. Saier Jr MH, Roseman S. Inducer exclusion and repression of enzyme synthesis in mutants of *Salmonella typhimurium* defective in enzyme I of the phosphoenol-pyruvate:sugar phosphotransferase system. J Biol Chem 1972; 247:972-975.

37. Castro L, Feucht BU, Morse ML et al. Regulation of carbohydrate permeases and adenylate cyclase in *Escherichia coli*. Studies with mutant strains in which Enzyme I of the phosphoenolpyruvate:sugar phosphotransferase system is thermolabile. J Biol Chem 1976; 251:5522-5527.

38. Kornberg HL, Watts PD. Roles of *crr*-gene products in regulating carbohydrate uptake by *Escherichia coli*. FEBS Lett 1978; 89:329-339.

39. Nelson SO, Lengeler J, Postma PW. Role of III^{Glc} of the phosphoenolpyruvate-glucose phosphotransferase system in inducer exclusion in *Escherichia coli*. J Bacteriol 1984; 160:360-364.

40. Peterkofsky A, Svenson I, Amin N. Regulation of *Escherichia coli* adenylate cyclase activity by the phosphoenolpyruvate: sugar phosphotransferase system. FEMS Microbiol Rev 1989; 63:103-108.

41. Feucht BU, Saier MH Jr. Fine control of adenylate cyclase by the phosphoenolpyruvate: sugar phosphotransferase systems in *Escherichia coli* and *Salmonella typhimurium*. J Bacteriol 1980; 141:603-610.

42. Nelson SO, Scholte BJ, Postma PW. Phosphoenolpyruvate:sugar phosphotransferase system-mediated regulation of carbohydrate metabolism in *Salmonella typhimurium*. J Bacteriol 1982; 150:604-615.

43. Harwood JP, Gazdar C, Prasad C et al. Involvement of the glucose enzymes II of the sugar phosphotransferase system in the regulation of adenylate cyclase by glucose in *Escherichia coli*. J Biol Chem 1976; 251:2462-2468.

44. Parra F, Jones-Mortimer MC, Kornberg HL. Phosphotransferase-mediated regulation of carbohydrate utilization in *Escherichia coli* K12: the nature of the *iex (crr)* and *gsr (tgs)* mutations. J Gen Microbiol 1983; 129:337-348.

45. Reddy P, Meadow N, Roseman S et al. Reconstitution of regulatory properties of adenylate cyclase in *Escherichia coli* extracts. Proc Natl Acad Sci USA 1985; 82:8300-8304.

46. Liberman E, Saffen D, Roseman S et al. Inhibition of *E. coli* adenylate cyclase activity by inorganic orthophosphate is dependent on III^{Glc} of the phosphoenolpyruvate:glycose phosphotransferase system. Biochem Biophys Res Commun 1986; 141:1138-1144.

47. Magasanik B, Neidhardt F. Regulation of carbon and nitrogen utilization. In: Neidhardt F, ed. *Escherichia coli* and *Salmonella typhimurium*: Cellular and Molecular Biology. Washington, DC: ASM, 1987:1318-1325.

48. Ullmann A, Danchin A. Role of cyclic AMP in bacteria. Adv Cyclic Nucleotide Res 1983; 15:2-53.

49. Wanner BL, Kodaira R, Neidhardt FC. Regulation of *lac* operon expression: reappraisal of the theory of catabolite repression. J Bacteriol 1978; 136:947-954.

50. Botsford JL, Drexler M. The cyclic 3',5'-adenosine monophosphate receptor protein and regulation of cyclic 3',5'-adenosine monophosphate synthesis in *Escherichia coli*. Mol Gen Genet 1978; 165:47-56.

51. Potter K, Chaloner-Larsson G, Yamazaki H. Abnormally high rate of cyclic AMP excretion from an *Escherichia coli* mutant deficient in cyclic AMP receptor protein. Biochem Biophys Res Commun 1974; 57:379-385.

52. Crasnier M, Danchin A. Characterization of *Escherichia coli* adenylate cyclase mutants with modified regulation. J Gen Microbiol 1990; 136:1825-1831.

53. Den Blaauwen JL, Postma PW. Regulation of cyclic AMP synthesis by enzyme III^{Glc} of the phosphoenolpyruvate:sugar phosphotransferase system in *crp* strains of *Salmonella typhimurium*. J Bacteriol 1985; 164:477-478.

54. Fandl JP, Thorner LK, Artz SW. Mutations that affect transcription and cyclic AMP-CRP regulation of the adenylate cyclase gene (*cya*) of *Salmonella typhimurium*. Genetics 1990; 125:719-727.

55. Joseph E, Bernsley C, Guiso N et al. Multiple regulation of the activity of adenylate cyclase in *Escherichia coli*. Mol Gen Genet 1982; 185:262-268.

56. Saier Jr MH, Feucht BU, McCaman MT. Regulation of intracellular adenosine cyclic 3':5'-monophosphate levels in *Escherichia coli* and *Salmonella typhimurium*. Evidence for energy-dependent excretion of the cyclic nucleotide. J Biol Chem 1975; 250:7593-7601.

57. Crasnier M, Dumay V, Danchin A. The catalytic domain of *Escherichia coli* K-12 adenylate cyclase as revealed by deletion analysis of the *cya* gene. Mol Gen Genet 1994; 243:409-416.

58. Ishizuka H, Hanamura A, Inada T et al. Mechanism of the down-regulation of cAMP receptor protein by glucose in *Escherichia coli*: role of autoregulation of the *crp* gene. EMBO J 1994; 13:3077-3082.

59. Ishizuka H, Hanamura A, Kunimura T et al. A lowered concentration of cAMP receptor protein caused by glucose is an important determinant for catabolite repression in *Escherichia coli*. Molec Microbiol 1993; 10:341-350.

60. Crenon I, Ullmann A. The role of cyclic AMP excretion in the regulation of enzyme synthesis in *Escherichia coli*. FEMS Microbiol Lett 1984; 22:47-51.

61. Saier Jr MH. Protein phosphorylation and allosteric control of inducer exclusion and catabolite repression by the bacterial phosphoenolpyruvate:sugar phosphotransferase system. Microbiol Rev 1989; 53:109-120.

62. Mitchell WJ, Saffen DW, Roseman S. Sugar transport by the bacterial phosphotransferase system. In vivo regulation of lactose transport in *Escherichia coli* by III^{Glc}, a protein of the phosphoenolpyruvate:glycose phosphotransferase system. J Biol Chem 1987; 262:16254-16260.

63. Nelson SO, Wright JK, Postma PW. The mechanism of inducer exclusion. Direct interaction between purified III^{Glc} of the phosphoenolpyruvate:sugar phosphotransferase system and the lactose carrier of *Escherichia coli*. EMBO J 1983; 2:715-720.

64. Osumi T, Saier Jr MH. Regulation of lactose permease activity by the phosphoenol-pyruvate:sugar phosphotransferase system: evidence for direct binding of the glucose-specific enzyme III to the lactose permease. Proc Natl Acad Sci USA 1982; 79:1457-1461.

65. Dean DA, Reizer J, Nikaido H et al. Regulation of the maltose transport system of *Escherichia coli* by the glucose-specific enzyme III of the phosphoenolpyruvate-sugar phosphotransferase system. Characterization of in-

ducer exclusion-resistant mutants and reconstitution of inducer exclusion in proteoliposomes. J Biol Chem 1990; 265:21005-21010.

66. Nelson SO, Postma PW. Interactions in vivo between III^{Glc} of the phosphoenol-pyruvate:sugar phosphotransferase system and the glycerol and maltose uptake systems of *Salmonella typhimurium*. Eur J Biochem 1984; 139:39-34.

67. De Boer M, Broekhuizen CP, Postma PW. Regulation of glycerol kinase by enzyme III^{Glc} of the phosphoenolpyruvate:carabohydrate phosphotransferase system. J Bacteriol 1986; 167:393-395.

68. Scholte BJ, Schuitema ARJ, Postma PW. Characterization of factor III^{Glc} in catabolite repression-resistant (*crr*) mutants of *Salmonella typhimurium*. J Bacteriol 1982; 149:576-586.

69. Postma PW, Broekhuizen CP, Schuitema ARJ et al. Carbohydrate transport and metabolism in *Escherichia coli* and *Salmonella typhimurium*: regulation by the PEP:carbohydrate phosphotransferase system. In: Palmieri F, Quagliariello E, eds. Molecular basis of biomembrane transport. Amsterdam: Elsevier Science Publishers, 1988; 43-52.

70. Wilson TH, Yunker PL, Hansen CL. Lactose transport mutants of *Escherichia coli* resistant to inhibition by the phosphotransferase system. Biochim Biophys Acta 1990; 1029:113-116.

71. Kühnau S, Reyes M, Sievertsen M et al. The activities of the *Escherichia coli* MalK protein in maltose transport, regulation and inducer exclusion can be separated by mutations. J Bacteriol 1991; 174:2180-2186.

72. Kuroda M, De Waard S, Mizushima K et al. Resistance of the melibiose carrier to inhibition by the phosphotransferase system due to substitutions of amino acid residues in the carrier of *Salmonella typhimurium*. J Biol Chem 1992; 267:18336-18341.

73. Okada T, Ueyama K, Niiya S et al. Role of inducer exclusion in preferential utilization of glucose over melibiose in diauxie growth of *Escherichia coli*. J Bacteriol 1981; 146:1030-1037.

74. Postma PW, Van der Vlag J, De Waard JH et al. Enzymes II of the phosphotransferase system: transport and regulation. In: Torriani-Gorini AM, Yagil E, Silver S, eds. Phosphate in Microorganisms. Cellular and Molecular Biology. Washington: ASM Press, 1994; 169-174.

75. Zeng GQ, de Reuse H, Danchin A. Mutational analysis of the enzyme III^{Glc} of the phosphoenolpyruvate phosphotransferase system in *Escherichia coli*. Res Microbiol 1993; 143:251-261.

76. Hurley JH, Worthylake D, Faber HR et al. Structure of the regulatory complex of *Escherichia coli* III^{Glc} with glycerol kinase. Science 1993; 259:673-677.

77. Nelson SO, Schuitema ARJ, Postma PW. The phosphoenolpyruvate:glucose phospho-transferase system of *Salmonella typhimurium*. The phosphorylated form of II^{Glc}. Eur J Biochem 1986; 154:337-341.

78. Cohn M, Horibata K. Physiology of the inhibition by glucose of the induced synthesis of the β-galactosidase-enzyme system of *Escherichia coli*. J Bacteriol 1959; 78:624-635.

79. Saier Jr MH, Novotny MJ, Comeau-Fuhrman D et al. Cooperative binding of the sugar substrates and allosteric regulatory protein (enzyme III^{Glc} of the phosphotransferase system) to the lactose and melibiose permeases in *Escherichia coli* and *Salmonella typhimurium*. J Bacteriol 1983; 155:1351-1357.

80. Van der Vlag J, Van Dam K, Postma PW. Quantification of the regulation of glycerol and maltose metabolism by IIA^{Glc} of the phosphoenolpyruvate-dependent glucose phosphotransferase system in *Salmonella typhimurium*. J Bacteriol 1994; 176:3518-3526.

81. De Reuse H, Danchin A. The *ptsH, ptsI* and *crr* genes of the *Escherichia coli* phosphoenolpyruvate-dependent phosphotransferase system: a complex

operon with several modes of transcription. J Bacteriol 1988; 170: 3827-3837.

82. Fox DK, Presper KA, Adhya S et al. Evidence for two promoters upstream of the *pts* operon: regulation by the cAMP receptor protein regulatory complex. Proc Natl Acad Sci USA 1992; 89:7056-7059.

83. Schnetz K, Rak B. β-glucoside permease represses the *bgl* operon of *Escherichia coli* by phosphorylation of the antiterminator protein and also interacts with glucose-specific enzyme III^{Glc}, the key element in catabolite control. Proc Natl Acad Sci USA 1990; 87:5074-5078.

84. Vogler AP, Broekhuizen CP, Schuitema A et al. Suppression of III^{Glc}-defects by Enzymes II^{Nag} and II^{Bgl} of the PEP: carbohydrate phosphotransferase system. Molec Microbiol 1988; 2:719-726.

85. Vogler AP, Lengeler JW. Complementation of a truncated membrane-bound Enzyme II^{Nag} from *Klebsiella pneumoniae* with a soluble Enzyme III in *Escherichia coli* K12. Mol Gen Genet 1988; 213:175-178.

86. Reizer J, Sutrina SL, Wu L-F et al. Functional interactions between proteins of the phosphoenolpyruvate:sugar phosphotransferase systems of *Bacillus subtilis* and *Escherichia coli*. J Biol Chem 1992; 267:9158-9169.

87. Lévy S, Zeng G-Q, Danchin A. Cyclic AMP synthesis in *Escherichia coli* strains bearing known deletions in the *pts* phosphotransferase operon. Gene 1990; 86:27-33.

88. Steinmetz M. Carbohydrate catabolism: pathways, enzymes, genetic regulation, and evolution. In: Sonenshein AL, Hoch JA, Losick R, eds. *Bacillus subtilis* and other gram-positive bacteria. Washington, DC: ASM 1993; 157-170.

89. Kjellberg S. Starvation in Bacteria. New York & London:Plenum Press, 1993.

90. Chuang, S, Daniels DL, Blattner FR. Global regulation of gene expression in *Escherichia coli*. J Bacteriol 1993;175:2026-2036.

91. Jahreis K, Postma PW, Lengeler JW. Nucleotide sequence of the *ilvH-fruR* gene region of *Escherichia coli* K12 and *Salmonella typhimurium* LT2. Mol Gen Genet 1991; 226:332-336.

92. Vartak NB, Reizer J, Reizer A et al. Sequence and evolution of the FruR protein of *Salmonella typhimurium*: a pleiotropic transcriptional regulatory protein possessing both activator and repressor functions which is homologous to the periplasmic ribose-binding protein. Res Microbiol 1991; 142:951-963.

93. Jahreis K, Lengeler JW. Molecular analysis of two ScrR repressors and of a ScrR-FruR hybrid repressor for sucrose and D-fructose specific regulons from enteric bacteria. Molec Microbiol 1993; 9:195-209.

94. Geerse RH, Ruig, CR, Schuitema ARJ et al. Relationship between pseudo-HPr and the PEP: fructose phosphotransferase system in *Salmonella typhimurium* and *Escherichia coli*. Mol Gen Genet 1986; 203:435-444.

95. Kornberg HL, Elvin CM. Location and function of *fruC*, a gene involved in the regulation of fructose utilization by *Escherichia coli*. J Gen Microbiol 1987; 133:341-346.

96. Reiner AM. Xylitol and D-arabitol toxicities due to derepressed fructose, galactitol and sorbitol phosphotransferases of *Escherichia coli*. J Bacteriol 1977; 132:166-173.

97. Geerse RH, Izzo F, Postma PW. The PEP:fructose phosphotransferase system in *Salmonella typhimurium*: FPr combines Enzyme III^{Fru} and pseudo-HPr activities. Mol Gen Genet 1989; 216:517-525.

98. Ramseier TM, Nègre D, Cortay J-C et al. In vitro binding of the pleiotropic transcriptional regulatory protein, FruR, to the *fru, pps, ace, pts* and *icd* operons of *Escherichia coli* and *Salmonella typhimurium*. J Mol Biol 1993; 234:28-44.

99. Chin AM, Feldheim DA, Saier Jr MH. Altered transcriptional patterns affecting several metabolic pathways in strains of *Salmonella typhimurium* which overexpress the fructose regulon. J Bacteriol 1989; 171:2424-2434.

100. Chin AM, Feucht BU, Saier MH Jr. Evidence for regulation of gluconeogenesis by the fructose phosphotransferase system in *Salmonella typhimurium*. J Bacteriol 1987; 169:897-899.

101. Geerse RH, van der Pluijm J, Postma PW. The repressor of the PEP: fructose phosphotransferase system is required for the transcription of the *pps* gene of *Escherichia coli*. Mol Gen Genet 1989; 218:348-352.

102. Cortay JC, Nègre D, Scarabel M et al. In vitro asymmetric binding of the pleiotropic regulatory protein FruR, to the *ace* operator controlling glyoxylate shunt enzyme synthesis. J Biol Chem 1994; 269: 14885-14891.

103. Fox DK, Meadow ND, Roseman S. Phosphate transfer between acetate kinase and enzyme I of the bacterial phosphotransferase system. J Biol Chem 1986; 261:13498-13503.

104. McCleary WR, Stock JB, Ninfa AJ. Is acetyl phosphate a global signal in *Escherichia coli*? J Bacteriol 1993; 175:2793-2798.

105. Jones DHA, Franklin FCH, Thomas CM. Molecular analysis of the operon which encodes the RNA polymerase sigma factor σ^{54} of *Escherichia coli*. Microbiology 1994; 140:1035-1043.

106. Reizer J, Reizer A, Saier Jr MH et al. A proposed link between nitrogen and carbon metabolism involving protein phosphorylation in bacteria. Protein Sci 1992; 1:722-726.

107. Merrick MJ, Coppard JR. Mutations in genes downstream of the *rpoN* gene (encoding σ^{54}) of *Klebsiella pneumoniae* affect expression from σ^{54}-dependent promoters. Molec Microbiol 1989; 3:1765-1775.

108. Begley GS, Jacobson GR. Overexpression, phosphorylation, and growth effects of ORF162, a *Klebsiella pneumoniae* protein that is encoded by a gene linked to *rpoN*, a gene encoding σ^{54}. FEMS Microbiol Lett 1994; 119: 389-394.

109. Powell BS, Court DL, Inada T et al. Novel proteins of the phosphotransferase system encoded within the *rpoN* operon of *Escherichia coli*. Enzyme IIA^{ntr} affects growth on organic nitrogen and the conditional lethality of an era(Ts) mutant. J Biol Chem 1995; 270:4822-4839.

110. Van der Vlag J, van 't Hof R, Van Dam K et al. Control of glucose metabolism by the enzymes of the glucose phosphotransferase system in *Salmonella typhimurium* Eur J Biochem 1995; 230:170-182.

111. Hubbard MJ, Cohen P. On target with a new mechanism for the regulation of protein phosphorylation. Trends Biochem Sci 1993; 18:172-177.

112 Alex LA, Simon MI. Protein histidine kinases and signal transduction in prokaryotes and eukaryotes. Trends Genet 1994; 10:133-138.

CHAPTER 12

THE CAP MODULON

Stephen Busby and Annie Kolb

THE LONG HISTORY OF CAP

The catabolite gene activator protein (CAP) is a transcription factor found in *Escherichia coli*. It was originally identified as a gene where disruptions suppressed the expression of the lactose (*lac*) operon, but it was rapidly realized that it had a role at a large number of other promoters. The story of the discovery of CAP is fascinating and has been told many times (e.g., by Pastan and Adhya,[1] Ullmann and Danchin[2]). In the mid-sixties it had been shown that glucose repression of *lac* expression could be countered by the inclusion of cyclic AMP (cAMP) in the growth media. In the late sixties, mutants at two unlinked loci that suppressed *lac* expression were identified: the effects of the mutants at just one of the loci (at map position 85 minutes) could be suppressed by added cAMP. The simplest explanation was that the 85 minute mutations identified the gene encoding adenyl cyclase (*cya*), while the other locus, at 73 minutes, mapped a cAMP receptor. This triggered a rush to purify this receptor, and by the early seventies, simple in vitro systems could be used to show cAMP-dependent *lac* transcription. The rush also led to some confusion in nomenclature with the receptor being variously known as CRP (cyclic AMP receptor protein), CAP or CGA (catabolite gene activator protein). Over the past 20 years, the story has unfolded with the advent of recombinant DNA technology leading to the cloning of many CAP-dependent promoters, the sequence of the *crp* gene, the discovery of CAP-induced DNA bending, and finally the structure of CAP protein-cAMP complexes either alone, or bound to a target site (most recently reviewed by Kolb et al[3]). At the start of the story, CAP attracted attention because it was a paradigm for the newly discovered class of gene regulatory DNA-binding proteins. The aim of this review is to outline the global role of CAP in regulating gene expression in *E. coli*, and to sketch out the organization of the different types of targets where CAP intervenes. Additionally, since CAP is one of the best-characterized transcription factors in terms of structure-function relationships, we discuss the way in which the molecular architecture of CAP allows a multiplicity of functions at different targets.

CYCLIC AMP AND GENE EXPRESSION

There is overwhelming evidence that the intracellular level of cAMP controls the activity of CAP.[1-4] First, CAP is a cAMP-binding protein

Regulation of Gene Expression in Escherichia coli, edited by E. C. C. Lin and A. Simon Lynch. © 1996 R.G. Landes Company.

and the specific binding of CAP to target sites in vitro is completely cAMP-dependent. In vivo, CAP activity is dependent on the *cya* locus but added cAMP can compensate for *cya* mutations. Intracellular cAMP levels vary according to the growth conditions: in particular they are depressed by glucose. The observation that the effects of glucose on the synthesis of catabolic enzymes could be countered by the addition of exogenous cAMP led to the supposition that cAMP was the second messenger used to signal the induction of catabolically-repressed enzymes. Thus, higher glucose levels reduce cAMP levels and cause repression (in fact lack of induction) at CAP-dependent promoters. Conversely, glucose starvation leads to increased cAMP levels and hence the possibility of inducing expression at numerous promoters, many of which control enzymes needed for the catabolism of alternative sugars. This simple idea, which gave birth to the name CAP, received crucial support from experiments that showed correlations between the actual intracellular cAMP concentration and β-galactosidase expression during growth in a variety of conditions.[5] However, despite the simplicity of the idea, the precise role of CAP in catabolite repression is rather more complex and cAMP cannot be the only modulator of catabolite repression and the effects of glucose on gene expression. For example, several investigators failed to observe a simple correlation between β-galactosidase expression and cAMP levels,[2] and, in some cases, catabolite repression can occur without cAMP (this occurs in *cya* strains where the *crp* gene carries mutations that confer cAMP-independence on CAP activity).[6] Such observations suggest that other factors must be playing more or less important roles in catabolite repression, and these include the glucose-mediated exclusion of inducers, glucose-induced changes in the structure of catabolite-sensitive promoters (for example via supercoiling) and the existence of other catabolite modulatory factors (reviewed by Botsford and Harman[4]). While it is indisputable that the cAMP·CAP complex plays a role in promoter activation, the precise contribution of cAMP-triggered effects to catabolite repression and glucose effects remains unclear. This problem was reexamined by Aiba and colleagues,[7,8] who made careful measurements of cAMP levels in a variety of conditions and concluded that these fluctuations could not account for the observed changes in β-galactosidase expression. Further, the addition of cAMP failed to completely reverse glucose-induced catabolite suppression of β-galactosidase expression. In contrast, glucose-induced effects were suppressed if CAP was expressed from a constitutive promoter on a multi-copy plasmid, suggesting that fluctuations in CAP levels may be important. Aiba and colleagues showed that CAP levels do indeed vary according to growth conditions and glucose status and concluded that these fluctuations may be as important as changes in cAMP levels in catabolite repression.[7] Figure 12.1 outlines a plausible model sketching the various glucose-induced effects on CAP and cAMP levels.

CAP AS A GLOBAL REGULATOR: THE CAP MODULON

Figure 12.2 gives a simple outline of the different steps involved in gene activation by CAP. First, cAMP binds to CAP causing a conformational change. Second, the cAMP·CAP complex binds to specific sites located at target promoters. Third, bound cAMP·CAP activates transcription.[9]

CAP is involved in the regulation expression of a vast number of *E. coli* genes. Although two-dimensional gel electrophoresis has been used to estimate the extent of this regulation, these studies do not provide quantitative data since they cannot distinguish between direct

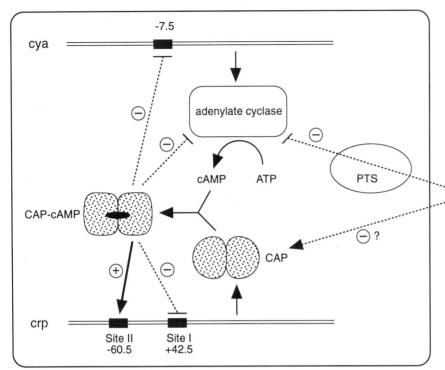

Fig. 12.1. Down regulation of cAMP and CAP levels by glucose. This figure shows that glucose entry into E. coli lowers the intracellular cAMP level by inhibiting adenyl cyclase via the PTS system. Glucose also lowers the CAP level by an unknown mechanism. The reduction in both cAMP and CAP leads to decreased levels of the cAMP·CAP complex. The initial reduction in cAMP·CAP affects the positive autoregulatory circuit at the crp promoter causing a further reduction in crp expression. Adapted from Ishizuka et al, EMBO J 1994; 13: 3077-82.

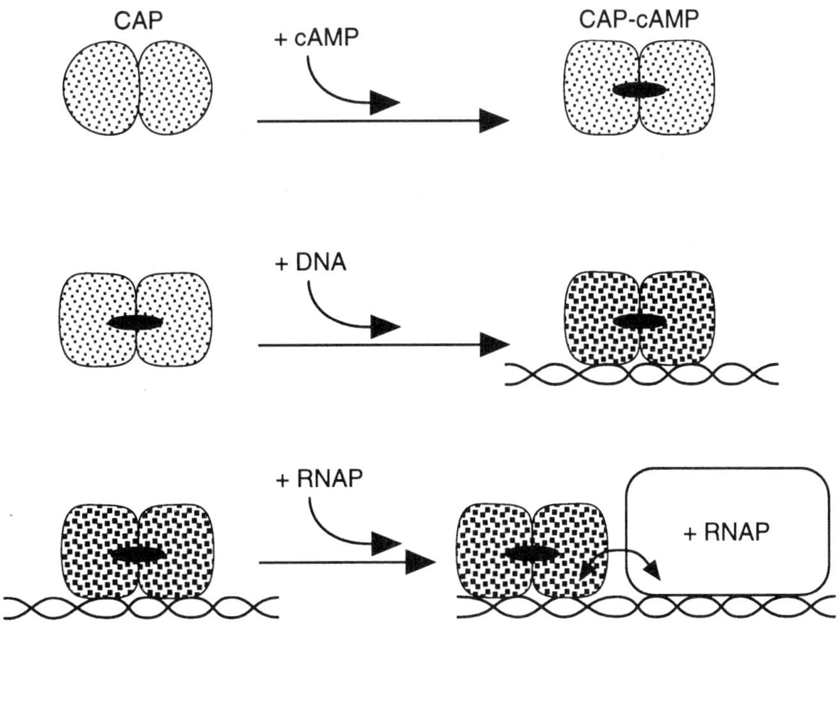

Fig. 12.2. Regulation of transcription by cAMP and CAP. The figure shows that cAMP first binds to CAP causing an allosteric change in the protein. Second, the cAMP·CAP complex recognizes specific targets. Third, the DNA-bound cAMP·CAP can regulate transcription initiation. In the case of the simplest CAP-dependent promoters, the bound cAMP·CAP can recruit RNA polymerase via a direct contact, thereby activating transcription initiation.

and indirect effects. The gels show that CAP is involved in the repression of gene expression as well as in activation.[10,11] Cells lacking CAP or adenyl cyclase grow more slowly than their wild-type counterparts, they show increased resistance to some mutagens and decreased resistance to neutral detergents and other inhibitors.[12] In some conditions crp or cya mutations cause cells to become rounded, although there is little direct evidence for a role of cAMP·CAP in regulating cell division.[13]

The first characterized CAP-dependent promoters were mostly concerned with the control of sugar catabolism and, at one stage, it appeared that the role of CAP was solely to control the use of alternatives to glucose. However it is now clear that CAP controls (or participates in the control of) many noncatabolic functions. Table 12.1 lists some of these functions. It is difficult to see any common strand running through the various "tasks" that CAP performs, although it is possible to rationalize CAP's function as being involved in the expression of proteins that handle adaptation to "particular" (usually poor) growth conditions.[2] A better way to explain the diversity of CAP's functions is to suppose that there are many conditions where intracellular cAMP levels can be sufficient for CAP to be active, and that the cell has evolved to recruit active CAP as a general transcription factor.

Table 12.1 classes CAP-sensitive functions according to the type of gene product controlled. Obviously, the largest group of promoters are those that control catabolic functions, including the catabolism of sugars, amino acids and nucleosides. CAP controls the expression of a substantial number of membrane proteins, some but not all of which are involved in metabolite transport, and of a number of proteins involved in carbon starvation and other stresses. Finally, CAP regulates a substantial group of genes that control operon-specific regulators.

Many promoters where CAP intervenes are controlled by several transcription factors. In most cases CAP acts together with an operon-specific activator or repressor. Thus, CAP activity "interprets" a global signal (glucose levels as reflected by intracellular cAMP), while the operon-specific regulator monitors the level of a specific metabolite which may or may not be present (e.g., lactose and Lac repressor or maltose and MalT protein). In this way, if CAP is active, a target promoter will be "cocked" (because levels of glucose are low) but will not "fire" unless a particular metabolite is present (e.g., lactose for LacI-repressed promoters or maltose for MalT-dependent promoters). In some of these cases, CAP exerts its influence indirectly by regulating the concentration of the regulator without intervening directly at the target. A good example of this is the *melAB* operon, encoding the genes for melibiose catabolism and transport. The *melAB* promoter is totally dependent on MelR, a melibiose-triggered transcription activator. MelR expression, in turn, is totally dependent on CAP and cAMP.[14] Thus MelR is only made during conditions of higher cAMP levels: MelR synthesis then signals readiness to respond to melibiose. In other cases, where a second activator is involved, CAP and the second activator act directly at targets, controlling transcription synergistically.[15]

Figure 12.3 illustrates four of the more common situations where gene expression is controlled by CAP. In the simplest cases, expression of the target gene is directly dependent on CAP because the promoter requires CAP alone for activation. In many of these cases, promoter-specific regulation is assured by repressors. In most examples, these repressors bind independently of CAP (e.g., at the *lac* and *gal* operon promoters), although, at CytR-repressed promoters, CytR binds directly to CAP, which acts as a corepressor (see below). In more complex situations, expression of the target gene is dependent on a gene-specific regulator, and this, in turn, is controlled by CAP (e.g., the *mel* operon). In other cases, expression of the target gene is co-activated by CAP and a second more specific regulator. Finally, in a number of cases CAP acts as a repressor: the simplest case is where CAP binding overlaps the RNA polymerase-binding promoter elements, but there are more complex situations (see below).

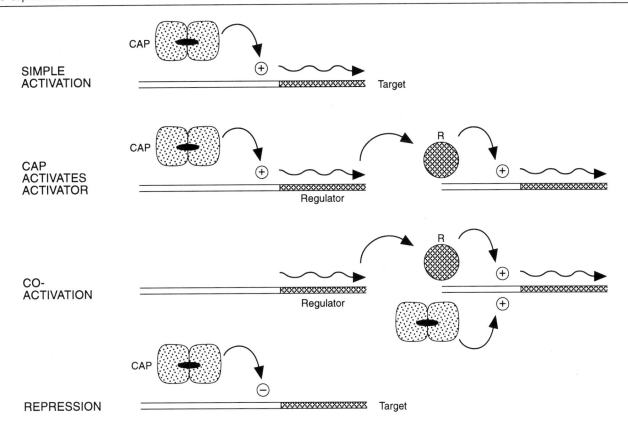

CAP BINDING AT TARGET PROMOTERS AND STRUCTURAL STUDIES

The cAMP·CAP system has proven amenable to genetic, biochemical and biophysical analyses, which have allowed access to the molecular details of CAP interactions at a number of promoters. The target site for cAMP·CAP binding was initially deduced from sequence comparisons of different CAP-dependent promoters,[9,16] and the location of CAP-binding at many promoters has been confirmed by footprinting: the derived consensus is a 22 base pair palindromic sequence, 5' aaaTGTGAtntanaTCACAttt 3', where the two TGTGA motifs are the best-conserved elements at different promoters. However, no known naturally occurring promoter carries the consensus sequence, and CAP-binding sequences always diverge from the consensus. Interestingly, as sites diverge from the consensus, the binding affinity is lowered and higher levels of cAMP are required to trigger CAP binding.[17,18] Thus, in vivo, stronger binding sites will be filled first when the intracellular level of cAMP is raised (say, in response to glucose starvation). The variation in CAP-binding sequence from one promoter to another imposes a hierarchy in expression of different target genes. This was corroborated by the observation that CAP-dependent expression from an artificial promoter containing a consensus CAP-binding sequence was unaffected by fluctuations in cAMP levels in a *cya*⁺ background: presumably the level of cAMP fails to fall below the low level that triggers CAP binding to this site.[19] Thus, although the consensus CAP binding sequence is effective in coupling gene expression to CAP, it is useless as a switch.

Fig. 12.3. Different types of regulation by CAP. Simple activation: cAMP·CAP binds upstream of the target promoter and directly activates transcription initiation (e.g., the lac operon). CAP activates activator: cAMP·CAP activates the expression of a second activator which, in turn, triggers the expression of a set of genes (e.g., the mel operon). Co-activation: both cAMP·CAP and a second activator must bind at the target promoter to trigger expression (e.g., at the malB locus). Repression: in the simplest case of repression cAMP·CAP binds to the promoter and blocks binding of RNA polymerase. Other more complex cases of CAP-dependent repression are discussed in the text.

Table 12.1. Some binding sites for the cAMP·CAP complex

Genes		E	F	Other genes involved in regulation	Distance	Sequences	Ref
ald	aldehyde dehydrogenase	+	C		-59.5	tttTATGAagccctTCACAgaa	89
ansB	asparaginase	+	C	*fnr*	-90.5	tttTGTTAcctgccTCTAActt	57
ansB (*S. typhimurium*) site 1		+	C		-40.5	ttaATCGTatggcgTCACAtta	90
site 2		+			-90.5	tttTGTTAtccatcTCTAAaaa	
araBAD - *araC*	arabinose metabolism	+	C	*araC*	-93.5	aagtGTGAcgccgtGCAAAtaa *	16
		+	R	*araC*	-54.5	ttaTTTGCacggcgTCACActt *	16
araE		+	T	*araC*	-94.5	aatTGGAAtaccaTCACATata	91
araFGH		+	T	*araC*	-41.5	cgaTTGGAtattgcTCTCCtat	56
araJ		+		*araC*	~-89.5	actGGAAAgtacgtTTGCAgtg	56
bglGFB mutant	β-glucoside metabolism	+	R,T,C	*bglG*	-61.5	aacTGTGAgcatggTCATAttt	16
cat	chloramphenicol acetyltransferase	+	M		-44.5	aaaTGAGAcgttgaTCGGCacg	16
cdd site 1	cytidine deaminase	+	C	*cytR*	-41.5	taaTGAGAttcagaTCACAtat	44
site 2		+		*cytR*	-91.5	attTGCGAtgcgtcGCGCAcgc	
site 3		0		*cytR*	-93.5	aaaTTTGCgatgcgTCGCGcat	
cea site 1	colicin E1 production	+	M	*fnr, lexA*	-63.5	tttTTTGAtcgtttTCACAaaa	16
site 2		0		*fnr, lexA*	-250.5	aacTGTGAacgcgaTCTGCctg	92
cir	colicin I receptor, siderophore	+	T		+118.5	agaTGTGAgcgataACCCAttt	93
cpdB	cyclic phosphodiesterase	+	M		-43.5	aacTGTGAtagtgtCATCAttt	94
cpdB (*S. typhimurium*)		+			~-43.5	aacTGTGAtaccgtCAGCGgtt	
crp site 1	CAP	-	R		+42.5	gtaTGCAAaggacgTCACAtta *	16
site 2		+			-60.5	gaaGGCGAcctgggTCATGCgtg	7
crp (*H. influenzae*) site 1		-				aagCGTGAtttacGCGAAgga	95
crp (*S. typhimurium*) site 1		-			~+42.5	gtaTGCAGaggacaTCACAtta	96
cstAB	carbon starvation induced	+	T		-90.5,-98.5,-106.5 –	cggAGTGAtcgagtTAAACattg	97
cya P2	adenylate cyclase	-	R		-7.5	aggTGTTAaattgaTCACGttt *	16
cya (*H. influenzae*)		-				aatTGTGAtttatgTCACAttt	98
cyoABCDE	cytochrome b562-o production	+	M	*fnr, arcA*	-70.5	ataTGTGAcctggcAGCCAaat	99
cytR	nucleoside catabolism repressor	+	R	*cytR*	-64.5	aaaTTCAAtattcaTCACActt	85
dadAX	alanine catabolism	+	C	*dadQ*	-59.5	agaTGTGAgccagcTCACcata	100
deoCABD P2 site 1	deoxyribonucleoside metabolism	+	C	*cytR, deoR*	-40.5	aatTGTGAtgtgtaTCGAAgtg	16
site 2		+		*cytR, deoR*	-93.5	ttaTTTGAaccagaTCGCAtta	16
exuT	galacturonate catabolism	+	C	*exuR, uxuR*		attTGTGAtggctcTCACCttt	101
fadBA	fatty acid catabolism	+	C	*fadR, arcA*	-59.5	gagCGTGAtcagatCGGCAttt	142
flhDC	flagella regulators	+	R		~-72.5	tgcGGTGAaaccgcTAAAAata	102

Gene	Function	+/−	Factor	Regulators	Position	Sequence	Ref
fucPIKR- fucAO site 1	fucose metabolism	+	C,R	fucR		aagTGTGAccgccgtTCATAtta taaTATGAcggcggtTCACActt	103
site 2		−	C	fucR		ttaAAGTGAtggtagtTCACAtaa ttaTGTGActaccaTCACTtta	
site 3						tttTGTGAcctggtTCAACtaa ttaGTTGAaccaggtTCACAaaa	
site 4						taaTGTTGttcctttTCACActa tagTGTGAaaggaacAACAtta	
fur P1	ferric uptake regulation	+	R	fur	−70.5	aaaTGTAAgctgtgCCACGttt	104
fur P2		−			−77.5	−	
galETKM P1	galactose metabolism	+	C,T	galR, galS	−41.5 −36.5	taaTTTATtccatgTCACActt *	16
galETKM P2		−			−37.5	− *	
galE P1 Δ4 mutant		−			−41.5	aaaTGTGAtctagaTCACActg	72
galP	galactose transport	+	R		−41.5	tgaTGTGAtttgctTCACAtct	63
galS	galactose isorepressor	+	R	galS	−42.5	tgcTGTGActcgatTCACGaag	105
gapA P3	glyceraldehyde 3 phosphate dehydrogenase	+	C	rpoH, fis	−41.5	aatCGTGAtgaaaaTCACAttt	106
gapB	inactive gene	+		fruR	−71.5	aagTGTGAtgtgagTCAGAtaa	143
glgS	glycogen synthesis activator	+	R	rpoS	−40.5, −63.5, −72.5	aagTGTGAtcggggACAATata	107
glnA P1 glnAntrBntrC P2	glutamine synthesis	+	M,R	ntrC, ntrB, rpoN ntrC	~−70.5 −185.5	cttTGTGAtcgcttTCACGgag −	108
glpEGR - glpD	glycerol metabolism	+	M,R	arcA, glpR	−63.5 −58.5	tagAGTGAtatgtaTAACAtta taaTGTTAtacataTCACTcta	109
		+	C	arcA, glpR			
glpFK		+	T,M	glpR	−60.5	tttTATGAcgaggcACACAcat	110
glpTQ - glpACB site 1		+	T,M	glpR, fnr	−41.5 −90.5	atgTGTGCggaccaTCACAttt aaaTGTGAtggtccGCACAcat	111
		+	C	glpR, fnr			
site 2		+		glpR, fnr	−91.5 −40.5	aaaCGTGAtttcatGCGTCatt aatGACGCatgaaaTCACGttt	
		+		glpR, fnr			
gutABDMR	glucitol metabolism	+	T,C,R	gutM, gutR	−41.5	tttTGCGAtcaaaaTAACActt	112
hupA	HU-α	+	M	fis	−63.5	aaaCGTGAtttaacGCCTGatt	88, 148
hupB P3	HU-β	+	M	fis, hupA	−83.5	aatAGTGAcctcgcGCAAAatg	88, 148
hutUH (K. aerogenes) site 1	histidine	+	C	nac, hutC	−41.5	ttcCTGTTAaatcctGGCTTgcg *	113
site 2		+			−81.5	aaaCGTGAttgctgACGCAata *	
ilvB	isoleucine-valine metabolism	+	M	relA	~−64.5	aaaCGTGAtcaaccCCTCAatt *	16
lacZYA	lactose metabolism	+	C,T,M	lacR	−61.5	taaTGTGAgttagcTCACTcat *	16
lacO mutant		−			+11.5	aatTGTGAgcggatTCACAttt *	71
malEFG - malK, lamB, malM site 1	maltose metabolism	+	T	malT	−76.5 −195.5	ttcTGTAAcagagaTCACAcaa * ttgTGTGAtctctgTTACAgaa *	51
		+		malT			
site 2		+			−105.5 −166.5	ttaTGTGCgcatctCCACAtta taaTGTGGagatgcGCACAtaa	
site 3		+			−139.5 −132.5	tttTGCAAgcaacaTCACGaaa * tttCGTGAtgttgctTTGCAaaa *	

Cont'd...

Table 12.1. continued

Genes		E	F	Other genes involved in regulation	Distance	Sequences	Ref
malI		+	R		-39.5	tagTGAGGcataaaTCACAtta	114
malPQ		O	C	*malT*	-93.5	tttAAGTGGttgagaTCACATtt	115
malT		+	R		-70.5	aatTGTGAcacagtGCAAAttc *	16
manXYZ	mannose transport	+	T		-40.5	attACGGAtcttcaTCACATaa	116
mccABC	microcin production	+	M	*hns, rpoS*	-59.5	tttTGTGAaataaaTCTATgtt	144
mdh	malate dehydrogenase	+	M		-60.5	taaATTGCgtgactACACATtc	117
melR	melibiose catabolism activator	+	R		-41.5	aacCGTGCtcccacTCGCAgtc	14
mglBAC	methylgalactoside transport	+	T	*galS*	-41.5	atcTGTGAgtgattTCACAgta	118
mtlAD site 1	mannitol metabolism	+	T,C		-58.5	ttaTGTGAttgataTCACAcaa	119-121
site 2					-102.5	tttTGTGAtgaacgTCACGtca	
site 3					-175.5	aaaTGTGAcactacTCACATtt	
site 4					-219.5	tgtTGTGAttcagaTCACAaat	
site 5					-261.5	taaCATGCtgtagaTCACAtca	
nagE -	N-acetylglucosamine metabolism	+	T	*nagC*	-61.5	tttGGTGAcaaaacTCACAAaa *	116
nagBACD		+	C,R	*nagC*	-71.5	tttTGTGAgtttgtGTCACCaaa	
nmpC	porin	+	T	*hns, ompR*	-73.5	aatAGAGAtctactTCACAAaat	122
nupC site 1	transport of nucleosides	+	T	*cytR*	-40.5	tagTGTGTgtcagaTCTCGttt	145
site 2					-89.5	aaaTGTATgacagaTCACTAtt	
nupG site 1	transport of nucleosides	+	T	*cytR, deoR*	-40.5	attTGCCAcaggtaACAAAaaa	123
site 2		+		*cytR, deoR*	-92.5	aaaTGTTAtccacaTCACAAtt	
ompA	outer membrane protein	-	T		-36.5	atgCCTGAcggagtTCACActt *	16
ompB P2	porin regulation	+	R		-71.5	taaCGTGAtcatatCAACAgaa *	124
ompB P1		-			-49.5	—	
osmY (csi-5)	periplasmic protein	-	T	*hip,himA,rpoS,lrp*	-12.5	aatTGTGAtctataTTTAAcaa	125
pac	penicillin acylase	+	M		-68.5	attAGTTAtcgcgcTCACAgtt	126
papBAHCDJKEFG-papI	pili formation	+	R,M	*hns, papB,*	-215.5	ttaTTTGAtgtgtgTCACATtt	60
papI		+	R	*papI, lrp, dam*	-115.5	aaaTGTGAcacacaTCAAATaa	
pBR P4	plasmid copy number regulation	+	R		-42.5	atcTGTGCGgtatttTCACAccg *	16
pdhR aceEF lpd	pyruvate dehydrogenase	+	M	*arcA, fnr*	-82.5	agtTGTTAaaatgtGCACAgtt	127
pep (S. typhimurium)	dipeptidase	+	C		~-60.5	gccTGTGAcacgcgTAACAtca	128
ppiA (rot) P2	peptidyl-prolyl-isomerase	+	M	*cytR*	-42.5	agaGGTGAtttgaTCACGgaa	121,129
P4				*cytR*	-41.5	tttTGTGAtctgttTAAATgtt	

Gene/Promoter	Function	Effect	F	Other genes	Distance	Sequence	Ref.
ptsH1crr P0a P0b P1a P1b	phosphotransferase system	+ - - +	M		-63.5 -60.5 -35.5 -42.5	tttTGTGGcctgctTCAAActt * - tttTATGATttggtTCAATtct -	130, 131
putP P1	proline metabolism	+	T	nac, putA	~-60.5	aaaTGTGAgagagtGCAACctg	132
putA	proline metabolism	+	C	nac, putA	~-50.5	gctTGCTAcgcatgTCACAttt	132
rbsDACBK	ribose metabolism	+	T		-61.5	cgtTTCGAggttgaTCACAttt	133
rhaBAD	rhamnose metabolis	+	C	rhaS	-92.5	aatTGTGAacatcaTCACGttc	134
rhaT	rhamnose transport	+	T			gaaAATGAttatcaATGCCgta	135
rhaT (S. thyphimurium)						aaaCGTGAagttaaTCACTtca	
rec-1 (H. influenzae)	recombination	+	M	himA, hip, lexA	~-178.5 ~-234.5	attTGTGAgccagaTCGTAaat actTCTGAttcatTTACAtca	136
rpoS P2 site 1 site 2	stationary-phase sigma	?	R	rpoS	-62.5 +55.5	agtTGTGAtcaagcCTGCAcaa aacTGCGAccacggTCACAgcg	147
rpoH P5	sigma heat-shock	+	R		~-46.5	actTGTGGataaaaTCACGgtc *	137
sdaCB	serine catabolism	+	T,C	lrp		attTGAGAtcaagaTCACTgat	146
sdhCDAB	succinate dehydrogenase	+	M	arcA		tatCGTGAcctggaTCACTgtt	138
spf	synthesis of spot 42 RNA	-	M		-84.5	tttTGTGAtggctaTTAGAaat	86
tdcABC	threonine metabolism	+	R,C, T	himA, hip, tdcA, tdcR	-40.5	attTGTGAgtggtcGCACAtat	16
tnaAB	tryptophan metabolism	+	C,T		-61.5	gatTGTGAttcgatTCACAttt	16
toxAB	enterotoxin synthesis	+	M		-88.5	aaaCATGAttgacaTCATGttg	139
tsx P2 site 1 site 2	nucleoside uptake	+ +	T	cytR	-40.5 -73.5	aacTGTGAaacgaaACATAttt aaaCGTGAacgcaaTCGATtac	83
udp site 1 site 2	uridine phosphorylase	+ +	M	cytR	-41.5 -93.5	catGGTGAtgagtaTCACGaaa ttaTGTGAtttgcaTCACTttt	140
uhpT	sugar phosphate transport	+	T	uhpA	-103.5	aagCGTGAtacaccTCACCttt	141
uxaCA	galacturonate catabolism	+	C	exuR, uxu R	~-80.5	tttATTGAtctaacTCACGaaa	101
uxuAB	glucuronate metabolism	+	C	exuR, uxuR	~-58.5	tgtTGTGAtgtggtTAACCcaa	16
CONSENSUS						aaaTGTGAtctagaTCACAttt	

CAP binding sites are listed in alphabetical order by the name of the gene(s) they regulate. When the gene does not belong to the *E. coli* genome, the name of the species is given in brackets. For promoters which contain multiple CAP binding sites, the sites are numbered "1", "2"... When a gene is transcribed from multiple promoters, the name of the CAP regulated promoter is given. The effect of CAP is indicated in column E: CAP can activate (+), repress (−), or have no effect (0) on the regulation of the promoter. The genes are classified into four different groups according to their functions (column F): catabolism (C), regulator (R), transporter (T) and miscellaneous (M). Other genes known to be involved in the regulation of the CAP dependent promoters are also listed. The sequences of the 22 bp CAP binding site are aligned according to the 5' TGTGA 3' motif and orientated towards the direction of transcription (i.e. from left to right). In the case where the same CAP binding site regulates two divergent promoters, the two complementary strands of the CAP binding site are bracketed. The distance of the centre of symmetry of the CAP binding site with respect to the transcription start site is also listed. The sites where CAP binding has been shown to regulate transcription in vitro are marked with an asterisk (*).

CAP clearly retards DNA fragments in band shift assays and these have been extensively used to study binding affinities. Anomalies in the mobilities of complexes were interpreted as evidence for CAP-induced DNA distortion and subsequent analyses have suggested that DNA is sharply bent by CAP (for example see ref. 17). Electrostatic calculations suggest that a tract of positive potential running along the protein may attract backbone phosphates flanking CAP-binding sites.[20] This probably explains why CAP binding is stabilized by the presence of bendable sequences flanking CAP sites, and why the sequence information required for optimum binding is greater than that for a protein solely dependent on interactions with a helix-turn-helix motif.[21]

A major contribution to our understanding has come X-ray crystallography studies of cAMP·CAP.[22] The structure of co-crystals of cAMP·CAP, bound to a 30 base pair DNA target, has been determined at 3 angstrom resolution.[23] Each CAP subunit consists of two domains: the larger N-terminal domain containing the cAMP-binding site and the smaller C-terminal domain carrying a helix-turn-helix motif. The structure reveals the CAP dimer binding to DNA with the recognition helix of each subunit penetrating two adjacent major grooves (Fig. 12.4). The structure suggests a model for the molecular basis of binding specificity,[23] and this is largely consistent with previous genetic and biochemical studies (summarized by Gunasekera et al[24]). The principal contacts are due to interactions between the first two residues, R180 and E181, of the recognition helix and the two G:C pairs in the consensus motif TGTGA. The structure shows a sharp bend of over 90° in the complexed DNA, due principally to 45° kinks between the TGT and GA parts of the consensus motif. Thus, the central 10 base pairs are bent towards the protein. Between the kinks, interactions involve eight hydrogen bonds to six DNA phosphates and specific base contacts by E181 and R185. On each side, the 10 base pairs flanking the hinges are engaged in another set of interactions: six backbone phosphates are linked to the protein via hydrogen bonds and ionic interactions while R180 is hydrogen bonded to the first G of the TGTGA motif.

Fig. 12.4. Skeleton Structure of the CAP-DNA complex. The figure shows the backbone structure of two CAP monomers with the DNA-binding domains penetrating into two adjacent major grooves. The DNA helix axis is shown as a black line running down the middle of the DNA helix and illustrates the CAP-induced bend. Adapted with permission from Schultz SC et al, Science 1991; 253: 1001-7.

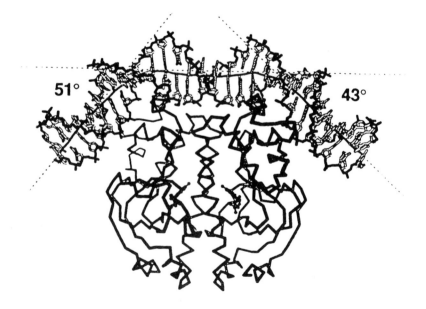

The X-ray structural analysis has provided a wealth of information on the organization of CAP and specific binding to DNA. However the mechanism by which cAMP triggers binding to specific DNA sites remains obscure, since useful crystals are obtained only in the presence of cAMP. Some clues concerning the action of cAMP have come from the study of substitutions which render CAP active in the absence of cAMP (reviewed by Kolb et al[3]). A cluster of these substitutions falls around the "hinge" between the C and D helices. From the location of these substitutions it is possible to deduce that cAMP, upon binding to the large domain of each subunit, triggers a conformational change via this hinge, which results in a reorientation of the DNA-binding domains such that the dimer has increased affinity for specific binding sites.

ACTIVATION BY CAP AT "SIMPLE" PROMOTERS

The simplest function of CAP is positive regulation of transcription and there are a number of well-characterized promoters where CAP alone is sufficient to activate transcription initiation by RNA polymerase. The paradigm for this is the *lac* promoter, and detailed in vitro studies have shown that the cAMP-induced binding of one CAP dimer centered between base pairs -61 and -62 (-61.5) with respect to the transcript start is sufficient to activate the promoter.[25] Kinetic and footprint analyses show that CAP promotes the initial binding of RNA polymerase to the promoter and that RNA polymerase binds adjacent to the bound CAP.[26,27] CAP contacts RNA polymerase directly via a 7 amino-acid surface-exposed β-turn (residues 156-162: activating region 1).[28] Genetic and biochemical analysis shows that activating region 1 in the downstream subunit of the CAP dimer makes contact with a site located in the C-terminal part of the α subunit of RNA polymerase (contact site 1).[28-30] The simplest model suggests that, at the *lac* promoter, CAP recruits the α subunit of RNA polymerase to bind just downstream, and that this contact guides the RNA polymerase into place such that correct contacts with both the -10 and -35 regions of the promoter can then be made.[31]

A long-standing puzzle concerns the diversity in the location of the CAP-binding sites at "simple" promoters where CAP alone is sufficient for activation. This is most likely explained by the observation that the C-terminal domain of the α subunit of RNA polymerase, which contains contact site 1, can fold as an autonomous domain, and is tethered to the N-terminal part of the α subunit by a flexible linker.[32] Results with an artificial promoter system, in which a consensus CAP-binding sequence was cloned at different distances upstream from the *melR* -10 region, showed that a single bound CAP dimer could activate transcription when bound around -41, -61 (as at the *lac* promoter) and -71 but not at intermediate positions.[33] Additionally, in some cases, activation from distances further upstream (e.g., around -81 and -91) may be possible, but these results are less conclusive.[34] The activation of promoters dependent on a single bound CAP dimer upstream of -61 requires the same contact between activating region 1 (in CAP) and contact site 1 (in the C-terminal part of α) that is needed at the *lac* promoter: most likely, the flexibility in α allows the C-terminal domain contact site to reach CAP bound at a number of different positions.[35]

At a number of promoters, including the prototype *galP1*, the CAP-binding site is centered near -41 and, thus, CAP overlaps with promoter elements that bind RNA polymerase (these are known as Class II promoters in contrast to Class I promoters such as the *lac* promoter where CAP binds further upstream).[33,36] Footprint analysis shows that RNA polymerase binds both upstream and downstream of CAP at a number of Class II promoters, the upstream protection being due to α, which is displaced from its normal binding site by CAP.[37] In this situation, activating region 1 in the upstream subunit of the CAP dimer interacts with α, which is located just upstream of the CAP dimer.[38] However, paradoxically, the C-terminal region of α is not essential for CAP-dependent activation at promoters such as *galP1* (in contrast to the situation at the *lac* promoter).[39] A simple explanation for this is that the displaced α subunit at promoters such as *galP1* is inhibitory, but that this is overcome by α binding to activating region 1 in the upstream subunit of the CAP dimer.[38] Thus, at Class II promoters, although α makes contact with CAP, there must be other essential CAP-RNA polymerase contacts needed for transcription activation.

A clue to the nature of these alternative CAP-polymerase contacts has come from the study of substitutions at E96 and K52 in CAP, that appear to improve regions that can make contact with RNA polymerase only when CAP is bound around -40, and are active only in the downstream subunit of the CAP dimer.[40,41] Thus, at promoters where CAP binds around -41, transcription activation by CAP may involve three activating regions, with region 1 functional just in the upstream subunit and the two other functional regions in the downstream subunit (reviewed in Busby et al[42]). Examination of the position of these three activating regions in the CAP structure shows that, although they are far apart in the monomer, they are displayed on adjacent faces in the CAP dimer (Fig. 12.5). This suggests a simple model in which, as polymerase makes contact upstream of CAP, it "stretches" along one side of the CAP dimer and can make contact with the three activating regions that are displayed on the two different adjacent faces. It is likely that the CAP-induced DNA bending facilitates these interactions and the subsequent upstream contact of RNA polymerase with upstream sequences. The downstream subunit of the CAP dimer is likely to make at least one direct contact with the C-terminal part of sigma, which is known to bind directly to the -35 hexamer at promoters.[36,43]

Although most promoters that are simply dependent on CAP for activation carry just one CAP-binding site, there are a number of cases where such promoters carry multiple CAP sites.[3] For example, at the *cdd* promoter, optimal activation requires bound CAP at two sites centered at -91 and -41: deletion of the upstream CAP-binding site reduces CAP-dependent expression 8-fold.[44] The mechanism of synergy between tandemly bound CAP dimers has been investigated at two sets of semisynthetic promoters. In one case CAP dimers were centered near -61 and -93[45] while in the other case they were centered around -41 and -90.[46] In both instances the evidence suggests that both bound CAP dimers make contact with RNA polymerase. This argues that RNA polymerase contains multiple contact sites for CAP and that these sites can accommodate contacts with CAP dimers bound at different positions.

ACTIVATION BY CAP AT COMPLEX PROMOTERS

Many CAP-dependent promoters are also regulated by a second transcription activator. Usually the operon-specific activator binds close to the RNA polymerase binding site with CAP binding further upstream. There is great variety in the organization of such promoters and diversity in the position of the CAP-binding site.[3,15,47]

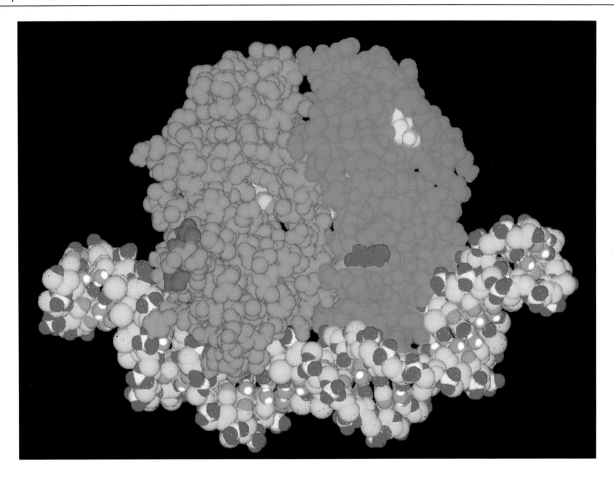

The best understood case is the *malK* promoter which requires maltose-induced MalT binding to both proximal and distal sites and CAP binding to a number of sites in between (Fig. 12.6).[48] In the absence of CAP, maltotriose-MalT binds to three sites upstream of *pmalK* but the position of the sites is such that transcription activation cannot take place. CAP binding triggers a re-positioning of MalT, which, in turn, triggers transcription initiation and ensures that *malK* expression is co-regulated by CAP and MalT.[49] In this situation, CAP makes no direct contact with RNA polymerase and activating region 1 is not involved. The CAP sites can be replaced by IHF binding or by DNA with a suitably positioned static bend, suggesting that DNA bending is the prerequisite for repositioning MalT.[50] Interestingly, the same array of bound MalT and CAP also activates initiation at the divergent *malE* promoter, but in this case, the bound CAP is closer to the *malE* transcription start and contacts with bound polymerase may be possible.[51]

The promoters of the arabinose regulon provide a second illustration of cooperation between CAP and a second activator, in this case, AraC. Despite similarities with the *mal* promoters, the situation is more complex as AraC can behave both as a repressor and an activator of transcription, in the absence and presence of arabinose, respectively.[52] In the best-studied case, the *araBAD* promoter, activation is dependent on arabinose triggering AraC binding to the *I1* and *I2* sites: the *I2* site overlaps the -35 region of the promoter and it is the occupation of this site that is responsible for transcription activation.[52,53] In the absence of arabinose, AraC occupies *I1* and the upstream *O2* site, forming a repression loop. cAMP·CAP binds just upstream of *I1* and

Fig. 12.5. Adjacent faces of the CAP dimer display different residues. The crystallographic coordinates of the CAP-DNA complex were obtained from the Brookhaven Protein data bank (accession code 1CGP). The image was generated and recorded by Virgil Rhodius on a Silicon Graphics 4D120 using the program Quanta (MSI). The figure shows a side-on view of the space-filling model of the CAP dimer bound to a consensus DNA site, with T158 and H159 highlighted in purple and K52 and E96 highlighted in red and yellow, respectively, on the adjacent face of the neighboring subunit. At Class II promoters, RNA polymerase can make contacts with the region around H159 in the upstream subunit, and the two regions around K52 and E96 in the downstream subunit.

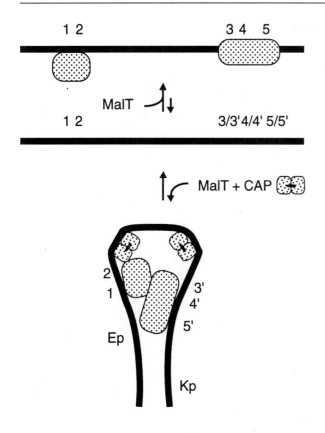

Fig. 12.6. Coactivation at the malE-malK promoters: CAP repositions MalT. MalT binds to sites 1-5 in the absence of CAP, but occupation of sites 3-5 prevents transcription activation. cAMP·CAP binding forces MalT to bind to sites 3',4' and 5', which overlap sites 3, 4 and 5. This results in increased occupancy of sites 1 and 2 and the resulting nucleoprotein interacts productively with RNA polymerase and expression from both the malE and malK promoters is induced. Adapted from Richet E et al, EMBO J 1994; 13: 4558-67.

its role appears 2-fold.[54,55] First, it helps break the *I1-O2* repression loop, and, second, it appears to help AraC occupy *I2*. Thus, in this case, it primarily plays the role of repositioning the operon-specific activator, albeit in a more complex fashion than at p*malK*. However, at the *araFG* promoter, which is also dependent on both CAP and AraC, the action of CAP is somewhat different.[56] In this case CAP binds around -41, as at Class II promoters, and the activity of the promoter is totally dependent on CAP (in contrast to the *araBAD* promoter). However expression is also dependent on AraC which binds upstream. It is most likely that AraC provides an extra contact for RNA polymerase, similar to the situation found at promoters with tandem CAP-binding sites. An interesting parallel is found at the *E. coli ans* promoter that is coregulated by FNR and CAP.[57] In this case FNR binds around -41 and CAP binds around -91. In this situation FNR behaves as a Class II-type activator and CAP presumably makes contact with upstream elements of RNA polymerase.

Two simple principles to explain the phenomenon of coregulation have emerged from study of the promoters of the maltose and arabinose operons. In some cases activator 1 binds in a nonproductive mode and requires activator 2 to reposition it in a productive mode (e.g., the *malK* and *araBAD* promoters). In other cases, there is no apparent repositioning of (or cooperative binding between) activators, but RNA polymerase needs to make at least two contacts with activators (e.g., the *araFG* and possibly the *malE* promoter). In these latter cases, it is interesting to consider why one single activator is not sufficient as at the "simple" promoters detailed above. In most cases, one of the activators binds around -40 and must lead to displacement of the RNA polymerase α subunits. We suppose that, at simple promoters, these displaced and, thus, potentially inhibitory α subunits are bound either to the single activator or to α-binding tracts of DNA sequence located upstream. However, in cases where neither of these options are possible, a second activator may be required to make a contact to relieve inhibition.

Studies of the *pap* regulon, encoding pyelonephritis-associated pili in a number of pathogenic *E. coli* strains, suggest a third simple mechanism for co-regulation. In this case, the expression of the genes encoding pilus functions is dependent on two activators, PapB and PapI, as well as CAP, which, surprisingly, binds around 215 base pairs upstream from the *papB* transcription start.[58,59] However, the requirement for cAMP·CAP is lost in host backgrounds carrying disruptions in the gene encoding H-NS,[60] an abundant low-specificity DNA-binding protein that plays a role in the structure of the bacterial chromosome.[61] These observations suggest that H-NS normally represses *pap* operon expression, but that bound H-NS (or the structure created by bound H-NS) is disrupted by CAP. Thus, in this case, CAP is acting to relieve repression rather than as a true transcription activator. Interestingly, a further control in the *pap* system is exerted via DAM/adenine methylation of two GATC sites in the *papB-papI* intergenic region: this regulation depends on Lrp but is independent of CAP.[62]

It is important to stress that most naturally occurring promoters are subject to regulation by multiple factors, and that CAP is most commonly either involved in co-activating with a second activator or acting in tandem with a repressor (for example, at the *lac*[25] or *gal*[63] promoters). However, there are also a number of cases where CAP-dependent promoters are subject to dual activation and expression can be triggered either by CAP or by a second activator (in contrast to co-activation where both activators are needed for optimal expression). The simplest examples of this are promoters dependent either on CAP or FNR. This is possible because the CAP binding consensus sequence, 5'aaaTGTGAtntanaTCACAttt 3', resembles the consensus for FNR binding, 5'aaaT*TT*GAtntanaTCA*A*Attt 3'.[64] Promoters carrying hybrid sequences such as 5'aaaT*G*TGAtntanaTCA*A*Attt 3' can be activated by either CAP or FNR, and are coupled into both modulons.[65,66]

CAP AS A REPRESSOR AND A CO-REPRESSOR

Remarkably, CAP is responsible for turning off as many genes as it activates. In the simplest scenario, which is found very rarely, CAP-binding at target promoters blocks the access of RNA polymerase to promoter elements. The best example of this is the *cya* P2 promoter, where CAP overlaps the -10 region.[67,68] Clearly however, CAP binding to any promoter is likely to repress, if it is not located at a position where it can activate. Examples are the *galP2* and *ompA* promoters.[69,70] Misplaced CAP sites have been engineered into both the *lacUV5* and *galP1* promoters to create derivatives that were strongly repressed by CAP.[71,72] Using a synthetic CAP site, Aiba and colleagues[73] showed that a CAP site cloned downstream of a promoter led to CAP-dependent repression, but that the repression became less as the CAP site was moved farther from the transcript start.

One of the most puzzling promoters is the *crp* gene promoter itself which is autoregulated by CAP binding to a site centered around 42 base pairs downstream of the transcript start point.[74-76] In this case, repression is not simple, as the downstream-bound CAP promotes RNA polymerase binding to a second divergent promoter that runs counter to the *crp* promoter. It is clear that repression is due to the presence of RNA polymerase at this second promoter rather than just CAP binding. Thus, in this case, CAP-dependent repression is a consequence of CAP-dependent activation at an alternative promoter which then occludes the target promoter. It is likely that there are many other cases of overlapping promoters being essential in maintaining repression: this is discussed at length by Kolb et al.[3]

A further level of complexity in the mechanism of repression is found in the family of promoters that is coordinately co-repressed by CAP and CytR. The CytR repressor controls the expression of at least ten operons involved in nucleoside and deoxynucleoside uptake and degradation.[77,78] All *CytR* promoters are positively controlled by CAP which binds around -41 (as at Class II promoters). In most cases, a second CAP-binding site is located upstream around -93, but CAP-binding to the upstream site is not essential for activation. Transcription initiation is blocked by the simultaneous binding of CAP and CytR to the promoter.[78,79] In vitro studies show that CytR has weak affinity for its operator, but in the presence of tandemly bound CAP dimers, located around -41 and -93, the CytR dimer binds tightly (Fig. 12.7).[80] Thus CytR recognizes the "array" of tandemly bound CAP dimers separated by 52-53 base pairs, and CAP is a co-repressor along with CytR. This explains the puzzle of the apparently redundant tandem CAP sites at CytR-dependent promoters and the mystery of why CytR binds so poorly to target promoters. The stability of the nucleoprotein complex is thus mediated by direct protein-protein interactions between liganded CAP subunits and the CytR repressor. The

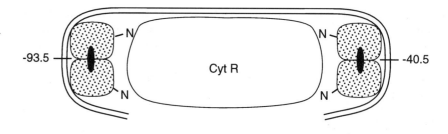

Fig. 12.7. CytR and cAMP·CAP bound at the deoP2 promoter. The CytR dimer is sandwiched between two CAP dimers. The sharp bend of the DNA around CAP allows contact of CytR with segments in the N-terminal domain: these segments have been pinpointed by substitutions in CAP that suppress repression by CytR. Note that CytR contacts DNA in the region that would be contacted by the α subunits of RNA polymerase (in the absence of CytR). Adapted from Søgaard-Andersen L et al, Mol Microbiol 1991; 5: 969-75.

CytR inducer, cytosine, does not affect the intrinsic affinity of CytR for operator DNA but abolishes the cooperativity with CAP. Moreover some substitutions on the face of CAP opposite to the DNA binding region (H17R, C18R, V108A, P110S) severely decrease CytR repression without affecting CAP-dependent activation.[81] It is likely that these substitutions identify the surface of CAP that makes specific contact with CytR.

The paradigm for a CytR-repressed promoter is *deoP2*; in this case, the distance between the center of the two CAP-binding sites is 53 base pairs.[79,82] CytR regulation depends crucially on the distance between the two CAP sites, as CytR is sandwiched between the two bound CAP dimers. At the *cdd* promoter, the two CAP-binding sites are separated by only 51 base pairs, but CytR provokes a repositioning of the distal-bound CAP to a weaker binding site located 2 bp further upstream.[44] At the *tsxP2* promoter, the two CAP-binding sites are separated by only 33 bp, but CytR provokes a relocation of the downstream CAP dimer by 20 bp.[83] In this case, the downstream CAP dimer is shifted from a position around -40, where it can activate transcription initiation, to a weaker site around -20, where it represses *tsxP2*. Finally, at the *cytR* promoter, formation of the autoregulatory repression complex involves only one bound CAP dimer (around -64) and CytR bound immediately downstream.[84,85] Interestingly, in this case, the repression is relatively weak, presumably because interactions between CytR and a single bound CAP are commensurately feebler. In this case CytR binding antagonizes the CAP-induced bend in the promoter DNA.

In summary, we now know about several situations where CAP participates directly in repression of promoters. In the most simple cases bound CAP directly blocks the access of RNA polymerase to promoter elements. In other cases bound CAP promotes occupation of a secondary promoter and it is this occupation that represses the target promoter. In the most complicated cases, CAP acts as a corepressor. A further scenario is found at the *spf* promoter: in this case the CAP-binding site overlaps the site for a gene-specific activator, and, thus, CAP binding prevents the activation process by interfering with activator binding.[86]

CAP: PARADIGM OR ARTIFACT?

The history of CAP is a story of increasing complexity. Although, CAP started as a "simple" transcription factor with a "simple" role at the *lac* promoter, we now see it as a truly "global" factor with diverse functions at a wide range of promoters. Perhaps the most fascinating aspect for molecular biologists has been in trying to fit the different functions to the molecular structures present in CAP. Looking into the future, we can expect to gain some insights from the study of CAP homologs in other organisms. Presumably these homologs will retain some functions found in *E coli* CAP but will lack others. The multiplicity of functions of CAP is very reminiscent of some eucaryotic transcription factors.[87] Although procaryotic nucleoprotein structures are clearly simpler than their eucaryotic counterparts, it seems likely that the principles governing their assembly and regulation are going to be similar.

We finish by highlighting some of the many remaining areas of ignorance, which range from problems of molecular biology to physiology. The crucial molecular biology problem remains: How does CAP exactly works. Great progress has been made in studies in vitro at model promoters and we now have a credible idea of how the different RNA polymerase subunits are arranged, and how they contact CAP in initiation complexes. Over the next years we can expect structural information for several RNA polymerase subunits. The crucial question is whether this information can be put together to give a mechanism, and whether this mechanism can then be applied to less well-characterized promoters. At the physiology end of the subject, we also face tough questions: What is the exact role of CAP, how are cAMP levels fixed, what is their significance and is there a role for CAP in cell division? Finally, the question of the *E coli* folded chromosome[88] remains a puzzle: Does CAP fit in somewhere alongside HU, H-NS, Lrp, Fis and IHF? Is the interior of *E. coli* like a Swiss watch or like a busy city center? Our hope is that this chapter will convince readers that CAP is far from being just a curious artifact and will provoke fresh new insights which will lead to new paradigms.

ACKNOWLEDGMENTS

We are grateful to many friends and colleagues for communicating their results prior to publication. S.B. is an E.P.A. Cephalosporin Fund Research Fellow.

REFERENCES

1. Pastan I, Adhya S. Cyclic adenosine 5'-monophosphate in *Escherichia coli*. Bacteriol Reviews 1976; 40:527-51.
2. Ullmann A, Danchin A. Role of cyclic AMP in bacteria. Adv Cyclic Nucleotide Res 1983; 15:1-54.
3. Kolb A, Busby S, Buc H et al. Transcriptional regulation by cAMP and its receptor protein. Annu Rev Biochem 1993; 62:749-95.
4. Botsford JL, Harman JG. Cyclic AMP in prokaryotes. Microbiol Rev 1992; 56:100-22.
5. Epstein W, Rothman-Denes LB, Hesse J. Adenosine 3' 5'-cyclic monophosphate as mediator of catabolite repression of *Escherichia coli*. Proc Natl Acad Sci USA 1975; 72:2300-4.
6. Harman JG, McKenney K, Peterkofsky A. Structure-function analysis of three cAMP-independent forms of the cAMP receptor protein. J Biol Chem 1986; 261:16332-9.
7. Ishizuka H, Hanamura A, Kunimura T et al. A lowered concentration of cAMP receptor protein caused by glucose is an important determinant for catabolite repression in *E. coli*. Mol Microbiol 1993; 10:341-350.
8. Ishizuka H, Hanamura A, Inada T et al. Mechanism of the down-regulation of cAMP receptor protein by glucose in *Escherichia coli*: role of autoregulation of the *crp* gene. EMBO J 1994; 13:3077-82.
9. de Crombrugghe B, Busby S, Buc H. cAMP receptor protein: role in transcription activation. Science 1984; 224:831-38.
10. Mallick U, Herrlich P. Regulation of synthesis of a major outer membrane protein: cyclic AMP represses Escherichia coli protein III synthesis. Proc Natl Acad Sci USA 1979; 76:5520-23.

11. Botsford JL. Analysis of protein expression in response to osmotic stress in *Escherichia coli*. FEMS Microbiol Lett 1990; 72:335-360.

12. Kumar S. Properties of adenyl-cyclase and cyclic adenosine monophosphate receptor protein-deficient mutants of *Escherichia coli*. J Bacteriol 1976; 125:545-555.

13. D'Ari R, Jaffe A, Bouloc P et al. Cyclic AMP and cell division in *Escherichia coli*. J Bacteriol 1988; 170:65-70.

14. Webster C, Gaston K, Busby S. Transcription from the *Escherichia coli* *mel*R promoter is dependent on the cyclic AMP receptor protein. Gene 1988; 68:297-305.

15. Raibaud O, Schwartz M. Positive control of transcription initiation in bacteria. Annu Rev Genet 1984; 18:173-206.

16. Berg OG, von Hippel PH. Selection of DNA binding sites by regulatory proteins. II. The binding of specificity of cyclic AMP receptor protein to recognition sites. J Mol Biol 1988; 200:709-23.

17. Kolb A, Spassky A, Chapon C et al. On the different affinities of CRP at the *lac*, *gal* and *mal*T promoter regions. Nucleic Acids Res 1983; 11:7833-52.

18. Kolb A, Busby S, Herbert M et al. Comparison of the binding sites for the *Escherichia coli* cAMP receptor protein at the lactose and galactose promoters. EMBO J 1983; 2:217-22.

19. Gaston K, Kolb A, Busby S. Binding of the *Escherichia coli* cyclic AMP receptor protein to DNA fragments containing consensus nucleotide sequences. Biochem J 1989; 261:649-53.

20. Warwicker J, Engelman BP, Steitz TA. Electrostatic calculations and model building suggest that DNA bound to CAP is sharply bent. Proteins 1987; 2:283-9.

21. Gartenberg M, Crothers D. DNA sequence determinants of CAP-induced bending and protein binding affinity. Nature 1988; 333:824-29.

22. Weber IT, Steitz TA. Structure of a complex of catabolite gene activator protein and cyclic AMP refined at 2.5 Å resolution. J Mol Biol 1987; 198:311-26.

23. Schultz SC, Shields GC, Steitz TA. Crystal structure of a CAP-DNA complex: the DNA is bent by 90°. Science 1991; 253:1001-7.

24. Gunasekera A, Ebright YW, Ebright RH. DNA sequence determinants for binding of the *Escherichia coli* catabolite gene activator protein. J Biol Chem 1992; 267:14713-20.

25. Reznikoff WS. The lactose operon-controlling elements: a complex paradigm. Mol Microbiol 1992; 6:2419-22.

26. Spassky A, Busby S, Buc H. On the Action of the cyclic AMP-cyclic AMP receptor protein complex at the *Escherichia coli* lactose and galactose promoter regions. EMBO J 1984; 3:43-50.

27. Malan TP, Kolb A, Buc H et al. Mechanism of CRP-cAMP activation of *lac* operon transcription initiation: activation of the P1 promoter. J Mol Biol 1984; 180:881-909.

28. Ebright RH. Transcription activation at Class I CAP-dependent promoters. Mol Microbiol 1993; 8:797-802.

29. Zou C, Fujita N, Igarashi K et al. Mapping the cAMP receptor protein contact site on the alpha subunit of *Escherichia coli* RNA polymerase. Mol Microbiol 1992; 6:2599-605.

30. Chen Y, Ebright YW, Ebright RH. Identification of the target of a transcription activator protein by protein-protein photocrosslinking. Science 1994; 265:90-92.

31. Kolb A, Igarashi K, Ishihama A et al. *E. coli* RNA polymerase, deleted in the C-terminal part of its alpha-subunit, interacts differently with the cAMP-CRP complex at the *lac*P1 and at the *gal*P1 promoter. Nucleic Acids Res 1993; 21:319-26.

32. Blatter EE, Tang H, Ross W et al. Domain organization of RNA polymerase alpha subunit: the C-terminal 85 amino acids constitute an independently folded domain capable of dimerization and DNA binding. Cell 1994; 78:889-96.

33. Gaston K, Bell A, Kolb A et al. Stringent spacing requirements for transcription activation by CRP. Cell 1990; 62:733-43.

34. Ushida C, Aiba H. Helical phase dependent action of CRP: effect of the distance between the CRP site and the -35 region on promoter activity. Nucleic Acids Res 1990; 18:6325-30.

35. Zhou Y, Merkel TJ, Ebright RH. Characterization of the activating region of *Escherichia coli* catabolite gene activator protein (CAP). II. Role at Class I and Class II CAP-dependent promoters. J Mol Biol 1994; 243:603-610.

36. Ishihama A. Protein-protein communication within the transcription apparatus. J Bacteriol 1993; 175:2483-9.

37. Attey A, Belyaeva T, Savery N et al. Interactions between the cyclic AMP receptor protein and the alpha subunit of RNA polymerase at the *Escherichia coli* galactose operon P1 promoter. Nucleic Acids Res 1994; 22:4375-4380.

38. Zhou Y, Pendergrast S, Bell A et al. The functional subunit of a dimeric transcription activator protein depends on promoter architecture. EMBO J 1994; 13:4549-57.

39. Igarashi K, Hanamura A, Makino K et al. Functional map of the alpha subunit of *Escherichia coli* RNA polymerase: two modes of transcription activation by positive factors. Proc Natl Acad Sci USA 1991; 88:8958-62.

40. Williams R, Bell A, Sims G et al. The role of two surface exposed loops in transcription activation by the *Escherichia coli* CRP and FNR proteins. Nucl Acids Res 1991; 19:6705-12.

41. West D, Williams R, Rhodius V et al. Interactions between the *Escherichia coli* cyclic AMP receptor protein and RNA polymerase at Class II promoters. Mol Microbiol 1993; 10:789-797.

42. Busby S, Kolb A, Buc H. The *E. coli* cyclic AMP receptor protein. Nucleic Acids and Mol Biol (Lilley D, Eckstein F, eds) 1995; 9:177-191.

43. Kumar A, Grimes B, Fujita N et al. Role of the sigma 70 subunit of *Escherichia coli* RNA polymerase in transcription activation. J Mol Biol 1994; 235:405-413.

44. Holst B, Søgaard-Andersen L, Pedersen H et al. The cAMP-CRP/CytR nucleoprotein complex in *Escherichia coli*: two pairs of closely linked binding sites for the cAMP-CRP activator complex are involved in combinatorial regulation of the *cdd* promoter. EMBO J 1992; 11:3635-43.

45. Joung JK, Le L, Hochschild A. Synergistic activation of transcription by *Escherichia coli* cAMP receptor protein. Proc Natl Acad Sci USA 1993; 90:3083-87.

46. Busby S, West D, Lawes M et al. Transcription activation of the *Escherichia coli* cAMP receptor protein. J Mol Biol 1994; 241:341-352.

47. Gottesman S. Bacterial regulation: global regulatory networks. Annu Rev Genet 1984; 18:415-41.

48. Raibaud O, Vidal-Ingigliardi D, Richet E. A complex nucleoprotein structure involved in activation of transcription of two divergent *Escherichia coli* promoters. J Mol Biol 1989; 205:471-85.

49. Richet E, Vidal-Ingigliardi D, Raibaud O. A new mechanism for coactivation of transcription: repositioning of an activator triggered by the binding of a second activator. Cell 1991; 66:1185-95.

50. Richet E, Søgaard-Andersen L. CRP induces the repositioning of MalT at the *E. coli malKp* promoter primarily through DNA bending. EMBO J 1994; 13:4558-67.

51. Vidal-Ingigliardi D, Raibaud O. Three adjacent binding sites for cAMP receptor protein are involved in the activation of the divergent *malEp-malKp* promoters. Proc Natl Acad Sci USA 1991; 88:229-233.

52. Lobell RB, Schleif RF. DNA looping and unlooping by AraC protein. Science 1990; 250:528-32.

53. Carra JH, Schleif RF. Variation of half-site organization and DNA looping by AraC protein. EMBO J 1993; 12:35-44.

54. Lobell RB, Schleif RF. AraC-DNA looping: orientation and distance-dependent loop breaking by the cyclic AMP receptor protein. J Mol Biol 1991; 218:45-54.

55. Lee N, Francklyn C, Hamilton EP. Arabinose-induced binding of AraC protein to *araI₂* activates the *araBAD* operon promoter. Proc Natl Acad Sci USA 1987; 84:8814-18.

56. Hendrickson W, Stoner C, Schleif R. Characterization of the *Escherichia coli araFGH* and *araJ* promoters. J Mol Biol 1990; 215:497-510.

57. Jennings MP, Beacham IR. Co-dependent positive regulation of the *ansB* promoter of *Escherichia coli* by CRP and the FNR protein: a molecular analysis. Mol Microbiol 1993; 9:155-64.

58. Baga M, Göransson M, Norman S et al. Transcriptional activation of a Pap pilus virulence operon from uropathogenic *Escherichia coli*. EMBO J 1985; 4:3887-93.

59. Göransson M, Forsman K, Nilsson P et al. Upstream activating sequences that are shared by two divergently transcribed operons mediate cAMP-CRP regulation of pilus-adhesin in *Escherichia coli*. Mol Microbiol 1989; 3:1557-65.

60. Forsman K, Sonden B, Göransson M et al. Antirepression function in *Escherichia coli* for the cAMP-cAMP receptor protein transcriptional activator. Proc Natl Acad Sci USA 1992; 89:9880-4.

61. Higgins CF, Hinton JCD, Hulton CSJ et al. Protein H1: a role for chromatin structure in the regulation of bacterial gene expression and virulence? Mol Microbiol 1990; 4:2007-12.

62. Braaten BA, Nou X, Kaltenbach LS et al. Methylation patterns in pap regulatory DNA control pyelonephritis-associated pili phase variation in *E. coli*. Cell 1994; 76:577-88.

63. Weickert MJ, Adhya S. The galactose regulon of *Escherichia coli*. Mol Microbiol 1993; 10:245-51.

64. Spiro S, Guest JR. FNR and its role in oxygen-regulated gene expression in *Escherichia coli*. FEMS Microbiol Reviews 1990; 75:399-428.

65. Bell AI, Gaston KL, Cole JA et al. Cloning of binding sequences for the *Escherichia coli* transcription activators, FNR and CRP: location of bases involved in discrimination between FNR and CRP. Nucleic Acids Res 1989; 17:3865-74.

66. Eiglmeier K, Honoré N, Iuchi S et al. Molecular genetic analysis of FNR-dependent promoters. Mol Microbiol 1989; 3:869-78.

67. Aiba H. Transcription of the *Escherichia coli* adenylate cyclase gene is negatively regulated by cAMP-cAMP receptor protein. J Biol Chem 1985; 260:3063-70.

68. Mori K, Aiba H. Evidence for negative control of *cya* transcription by cAMP and cAMP receptor protein in intact *Escherichia coli* cells. J Biol Chem 1985; 260:14838-42.

69. Musso RE, Di Lauro R, Adhya S et al. Dual control for transcription of the galactose operon by cyclic AMP and its receptor protein at two interspersed promoters. Cell 1977; 12:847-54.

70. Movva RN, Green P, Nakamura K et al. Interaction of cAMP receptor protein with *ompA*, a gene for a major outer membrane protein of *Escherichia coli*. FEBS Lett 1981; 128:186-90.

71. Irwin N, Ptashne M. Mutants of the catabolite activator protein of *Escherichia coli* that are specifically deficient in the gene-activating function. Proc Natl Acad Sci USA 1987; 84:8315-9.

72. Bell A, Gaston K, Williams R et al. Mutations that alter the ability of the *Escherichia coli* cyclic AMP receptor protein to activate transcription. Nucleic Acids Res 1990; 18:7243-50.

73. Morita T, Shigesada K, Kimizuka F et al. Regulatory effect of a synthetic CRP recognition sequence placed downstream of a promoter. Nucleic Acids Res 1988; 16:7315-32.

74. Aiba H. Autoregulation of the *Escherichia coli crp* gene: Crp is a transcriptional repressor of its own gene. Cell 1983; 32:141-9.

75. Hanamura A, Aiba H. Molecular mechanism of negative autoregulation of *Escherichia coli crp* gene. Nucleic Acids Res 1991; 19:4413-19.

76. Hanamura A, Aiba H. A new aspect of transcriptional control of the *Escherichia coli crp* gene: positive autoregulation. Mol Microbiol 1992; 6:2489-97.

77. Valentin-Hansen P. DNA sequences involved in expression and regulation of *deo*R and *cyt*R and cAMP-CRP-controlled genes in *E. coli*. In: Glass R, Spicek J, eds. Gene manipulation and Expression. London: Croom Helm-G.B., 1985:273-88.

78. Søgaard-Andersen L, Valentin-Hansen P. Protein-protein interactions in gene regulation: the cAMP-CRP complex sets the specificity of a second DNA-binding protein, the CytR repressor. Cell 1993; 75:557-66.

79. Søgaard-Andersen L, Møllegaard NE, Douthwaite SR et al. Tandem DNA bound cAMP-CRP complexes are required for transcriptional repression of the *deo*P2 promoter by the CytR repressor in *Escherichia coli*. Mol Microbiol 1990; 4:1595-1601.

80. Pedersen H, Søgaard-Andersen L, Holst B et al. Heterologous cooperativity in *Escherichia coli*. J Biol Chem 1991; 266:17804-8.

81. Søgaard-Andersen L, Mironov AS, Pedersen H et al. Single amino acid substitutions in the cAMP receptor protein specifically abolish regulation by the CytR repressor in *Escherichia coli*. Proc Natl Acad Sci USA 1991; 88:4921-25.

82. Søgaard-Andersen L, Pedersen H, Holst B et al. A novel function of the cAMP-CRP complex in *Escherichia coli*: cAMP-CRP functions as an adaptor for the CytR repressor in the *deo* operon. Mol Microbiol 1991; 5:969-75.

83. Gerlach P, Søgaard-Andersen L, Pedersen H et al. The cyclic AMP (cAMP)-cAMP receptor protein complex functions both as an activator and as a corepressor at the *tsx-p*$_2$ promoter of *Escherichia coli* K-12. J Bacteriol 1991; 173:5419-30.

84. Gerlach P, Valentin-Hansen P, Bremer E. Transcriptional regulation of the *cyt*R repressor gene of *Escherichia coli*: autoregulation and positive control by the cAMP/CAP complex. Mol Microbiol 1990; 4:479-488.

85. Pedersen H, Søgaard-Andersen L, Holst B et al. cAMP-CRP activator complex and the CytR repressor protein bind co-operatively to the cytRP promoter in *Escherichia coli* and CytR antagonizes the cAMP-CRP-induced DNA bend. J Mol Biol 1992; 227:396-406.

86. Polayes DA, Rice PW, Garner MM et al. Cyclic AMP-cyclic AMP receptor protein as a repressor of transcription of the *spf* gene of *Escherichia coli*. J Bacteriol 1988; 170:3110-4.

87. Ptashne M. A genetic switch. Palo-Alto, Calif., Oxford-G.B.: Cell Press & Blackwell Scientific Publications 2nd ed. 1986.

88. Drlica K, Rouvière-Yaniv J. Histonelike proteins of bacteria. Microbiol Reviews 1987; 51:301-19.

89. Hidalgo E, Chen Y-M, Lin ECC et al. Molecular cloning and DNA sequencing of the *Escherichia coli* K-12 *ald* gene encoding aldehyde dehydrogenase. J Bacteriol 1991; 173:6118-23.

90. Jennings MP, Scott SP, Beacham IR. Regulation of the *ansB* gene of *Salmonella enterica*. Mol Microbiol 1993; 9:165-72.

91. Stoner C, Schleif R. The *araE* low affinity L-arabinose transport promoter: cloning, sequence, transcription start site and DNA binding sites of regulatory proteins. J Mol Biol 1983; 171:369-81.

92. Eraso JM, Weinstock GM. Anaerobic control of colicin E1 production. J Bacteriol 1992; 174:5101-9.

93. Griggs DW, Kafka K, Nau CD et al. Activation of expression of the *Escherichia coli cir* gene by an iron-independent regulatory mechanism involving cyclic AMP-cyclic AMP receptor protein complex. J Bacteriol 1990; 172:3529-33.

94. Liu J, Beacham IR. Transcription and regulation of the *cpdB* gene in *Escherichia coli* K12 and *Salmonella typhimurium* LT2: evidence for modulation of constitutive promoters by cyclic AMP-CRP complex. Mol Gen Genet 1990; 222:161-5.

95. Chandler MS. The gene encoding cAMP receptor protein is required for competence development in *Haemophilus influenzae* Rd. Proc Natl Acad Sci USA 1992; 89:1626-30.

96. Cossart P, Groisman EA, Serre M-C et al. *crp* genes of *Shigella flexneri*, *Salmonella typhimurium*, and *Escherichia coli*. J Bacteriol 1986; 167:639-46.

97. Schultz JE, Matin A. Molecular and functional characterization of a carbon starvation gene of *Escherichia coli*. J Mol Biol 1991; 218:129-40.

98. Dorocicz IR, Williams PM, Redfield RJ. The *Haemophilus influenzae* adenylate cyclase gene: cloning, sequence, and essential role in competence. J Bacteriol 1993; 175:7142-9.

99. Minagawa J, Nakamura H, Yamato I et al. Transcriptional regulation of the cytochrome b_{562}-o complex in *Escherichia coli*: gene expression and molecular characterization of the promoter. J Biol Chem 1990; 265:11198-203.

100. Lobocka M, Hennig J, Wild J et al. Organization and expression of the *Escherichia coli* K-12 *dad* operon encoding the smaller subunit of D-amino acid dehydrogenase and the catabolic alanine racemase. J Bacteriol 1994; 176:1500-10.

101. Blanco C, Mata-Gilsinger M. Identification of cyclic AMP-CRP binding sites in the intercistronic regulatory *uxaCA-exuT* region of *Escherichia coli*. FEMS Microbiol Lett 1986; 33:205-9.

102. Bartlett DH, Frantz BB, Matsumura P. Flagellar transcriptional activators FlbB and FlaI: gene sequences and 5' consensus sequences of operons under FlbB and FlaI control. J Bacteriol 1988; 170:1575-81.

103. Lu Z, Lin ECC. The nucleotide sequence of *Escherichia coli* genes for L-fucose dissimilation. Nucleic Acids Res 1989; 17:4883-4.

104. de Lorenzo V, Herrero M, Giovannini F et al. Fur (ferric uptake regulation) protein and CAP (catabolite-activator protein) modulate transcription of *fur* gene in *Escherichia coli*. Eur J Biochem 1988; 173:537-46.

105. Weickert MJ, Adhya S. Control of transcription of *gal* repressor and isorepressor genes in *Escherichia coli*. J Bacteriol 1993; 175:251-8.

106. Charpentier B, Branlant C. The *Escherichia coli gapA* gene is transcribed by the vegetative RNA polymerase holoenzyme $E\sigma^{70}$ and by the heat shock RNA polymerase $E\sigma^{32}$. J Bacteriol 1994; 176:830-9.

107. Hengge-Aronis R, Fischer D. Identification and molecular analysis of *glgS*, a novel growth-phase-regulated and *rpoS*-dependent gene involved in glycogen synthesis in *Escherichia coli*. Mol Microbiol 1992; 6:1877-86.

108. Reitzer LJ, Magasanik B. Expression of *glnA* in *Escherichia coli* is regulated at tandem promoters. Proc Natl Acad Sci USA 1985; 82:1979-83.

109. Choi Y-L, Kawase S, Kawamukai M et al. Regulation of *glpD* and *glpE* gene expression by a cyclic AMP-cAMP receptor protein (cAMP-CRP) complex in *Escherichia coli*. Biochem Biophys Acta 1991; 1088:31-35.

110. Weissenborn DL, Wittekindt N, Larson TJ. Structure and regulation of the *glpFK* operon encoding glycerol diffusion facilitator and glycerol kinase of *Escherichia coli* K-12. J Biol Chem 1992; 267:6122-31.

111. Larson TJ, Cantwell JS, van Loo-Bhattacharya AT. Interaction at a distance between multiple operators controls the adjacent, divergently transcribed *glpTQ-glpACB* operons of *Escherichia coli* K-12. J Biol Chem 1992; 267:6114-21.

112. Yamada M, Saier MH Jr. Glucitol-specific enzymes of the phosphotransferase system in *Escherichia coli*: nucleotide sequence of the *gut* operon. J Biol Chem 1987; 262:5455-63.

113. Osuna R, Janes KJ, Bender RA. Roles of catabolite activator protein sites centered at -81.5 and -41.5 in the activation of the *Klebsiella aerogenes* histidine utilization operon *hutUH*. J Bacteriol 1994; 176:5513-24.

114. Reidl J, Römisch K, Ehrmann M et al. MalI, a novel protein involved in regulation of the maltose system of *Escherichia coli*, is highly homologous to the repressor proteins GalR, CytR, and LacI. J Bacteriol 1989; 171:4888-99.

115. Danot O, Raibaud O. Multiple protein-DNA and protein-protein interactions are involved in the transcriptional activation by MalT. Mol Microbiol 1994; 14:335-46.

116. Plumbridge J, Kolb A. CAP and Nag repressor binding to the regulatory regions of the *nagE-B* and *manX* genes of *Escherichia coli*. J Mol Biol 1991; 217:661-79.

117. Vogel RF, Entian K-D, Mecke D. Cloning and sequence of the *mdh* structural gene of *Escherichia coli* coding for malate dehydrogenase. Arch Microbiol 1987; 149:36-42.

118. Hogg RW, Voelker C, von Carlowitz I. Nucleotide sequence and analysis of the *mgl* operon of *Escherichia coli* K12. Mol Gen Genet 1991; 229:453-9.

119. Davis T, Yamada M, Elgort M et al. Nucleotide sequence of the mannitol (*mtl*) operon in *Escherichia coli*. Mol Microbiol 1988; 2:405-12.

120. Wang MX, Church GM. A whole genome approach to in vivo DNA-protein interactions in *E. coli*. Nature 1992; 360:606-9.

121. Hale BW, van der Woude MW, Low DA. Analysis of nonmethylated GATC sites in the *Escherichia coli* chromosome and identification of sites that are differentially methylated in response to environmental stimuli. J Bacteriol 1994; 176:3438-41.

122. Coll JL, Heyde M, Portalier R. Expression of the *nmpC* gene of *Escherichia coli* K-12 is modulated by external pH. Identification of *cis*-acting regulatory sequences involved in this regulation. Mol Microbiol 1994; 12:83-93.

123. Munch-Petersen A, Jensen N. Analysis of the regulatory region of the *Escherichia coli nupG* gene, encoding a nucleoside-transport protein. Eur J Biochem 1990; 190:547-51.

124. Huang L, Tsui P, Freundlich M. Positive and negative control of *ompB* transcription in *Escherichia coli* by cyclic AMP and the cyclic AMP receptor protein. J Bacteriol 1992; 174:664-70.

125. Lange R, Barth M, Hengge-Aronis R. Complex transcriptional control of the σ^s-dependent stationary-phase-induced and osmotically regulated *osmY* (*csi-5*) gene suggests novel roles for Lrp, cyclic AMP (cAMP) receptor protein-cAMP complex, and integration host factor in the stationary-phase response of *Escherichia coli*. J Bacteriol 1993; 175:7910-7.

126. Valle F, Gosset G, Tenorio B et al. Characterization of the regulatory region of the *Escherichia coli* penicillin acylase structural gene. Gene 1986; 50:119-22.

127. Quail MA, Haydon DJ, Guest JR. The *pdhR-aceEF-lpd* operon of *Escherichia coli* expresses the pyruvate dehydrogenase complex. Mol Microbiol 1994; 12:95-104.

128. Conlin CA, Hakensson K, Liljas A et al. Cloning and nucleotide sequence of the cyclic AMP receptor protein-regulated *Salmonella typhimurium pepE* gene and crystallization of its product, an α-aspartyldipeptidase. J Bacteriol 1994; 176:166-72.

129. Nørregaard-Madsen M, Mygind B, Pedersen R et al. The gene encoding the periplasmic cyclophilin homolog, PPIase A, in *E. coli*, is expressed from four promoters, three of which are activated by the cAMP-CRP complex and negatively regulated by the CytR repressor protein. Mol Microbiol 1994; 14:989-997.

130. de Reuse H, Kolb A, Danchin A. Positive regulation of the expression of the *Escherichia coli pts* operon: identification of the regulatory regions. J Mol Biol 1992; 226:623-35.

131. Ryu S, Garges S. Promoter switch in the *Escherichia coli pts* operon. J Biol Chem 1994; 269:4767-72.

132. Nakao T, Yamato I, Anraku Y. Mapping of the multiple regulatory sites for *putP* and *putA* expression in the *putC* region of *Escherichia coli*. Mol Gen Genet 1988; 214:379-88.

133. Bell AW, Buckel SD, Groarke JM et al. The nucleotide sequences of the *rbsD*, *rbsA*, and *rbsC* genes of *Escherichia coli* K12. J Biol Chem 1986; 261:7652-8.

134. Egan SM, Schleif RF. A regulatory cascade in the induction of *rhaBAD*. J Mol Biol 1993; 234:87-98.

135. Tate CG, Muiry JAR, Henderson PJF. Mapping, cloning, expression, and sequencing of the *rhaT* gene, which encodes a novel-L-rhamnose-H$^+$ transport protein in *Salmonella typhimurium* and *Escherichia coli*. J Biol Chem 1992; 267:6923-32.

136. Zulty JJ, Barcak GJ. Structural organization, nucleotide sequence, and regulation of the *Haemophilus influenzae rec-1$^+$* gene. J Bacteriol 1993; 175:7269-81.

137. Nagai H, Yano R, Erickson JW et al. Transcriptional regulation of the heat shock regulatory gene *rpoH* in *Escherichia coli*: involvement of a novel catabolite-sensitive promoter. J Bacteriol 1990; 172:2710-5.

138. Wood D, Darlison MG, Wilde RJ et al. Nucleotide sequence encoding the flavoprotein and hydrophobic subunits of the succinate dehydrogenase of *Escherichia coli*. Biochem J 1984; 222:519-34.

139. Gibert I, Villegas V, Barbé J. Expression of heat-labile enterotoxin genes is under cyclic AMP control in *Escherichia coli*. Current Microbiol 1990; 20:83-90.

140. Walton L, Richards CA, Elwell LP. Nucleotide sequence of the *Escherichia coli* uridine phosphorylase (*udp*) gene. Nucleic Acids Res 1989; 17:6741.

141. Merkel TJ, Nelson DM, Brauer CL et al. Promoter elements required for positive control of transcription of the *Escherichia coli uhpT* gene. J Bacteriol 1992; 174:2763-70.

142. Black PN, DiRusso CC. Molecular and biochemical analyses of fatty acid transport, metabolism, and gene regulation in *Escherichia coli*. Biochim Biophys Acta 1994; 1210:123-45.

143a. Alefounder PR, Perham RN. Identification, molecular cloning and sequence analysis of a gene cluster encoding the Class II fructose 1,6-bisphosphate aldolase, 3-phosphoglycerate kinase and a putative second glyceraldehyde 3-phosphate dehydrogenase of *Escherichia coli*. Mol Microbiol 1989; 3:723-32.

143b. Charpentier B., Branlant C. Manuscript in preparation.

144a. González-Pastor JE, San Millán JL, Moreno F. The smallest known gene. Nature 1994; 369:281.

144b. González-Pastor JE, Kolb A, Moreno F. Manuscript in preparation.

145. Craig JE, Zhang Y, Gallagher MP. Cloning of the *nupC* gene of *Escherichia coli* encoding a nucleoside transport system, and identification of an adjacent insertion element, IS*186*. Mol Microbiol 1994; 11:1159-68.

146. Shao ZQ, Lin RT, Newman EB. Sequencing and characterization of the *sdaC* gene and identification of the *sdaCB* operon in *Escherichia coli* K12. Eur J Biochem 1994; 222:901-7.

147. Takayanagi Y, Tanaka K, Takahashi H. Structure of the 5' upstream region and the regulation of the *rpoS* gene of *Escherichia coli*. Mol Gen Genet 1994; 243:525-31.

148. Claret L, Rouvière-Yaniv J. Manuscript in preparation.

CHAPTER 13

REGULATION OF NITROGEN ASSIMILATION

Boris Magasanik

The preferred single nitrogen source for *Escherichia coli* and other enteric bacteria is ammonia, in at least 1 mM concentration. Although a number of other nitrogen containing compounds for example arginine, proline, glutamine, glutamate and aspartate in the case of *E. coli*, and in addition dinitrogen, nitrate, histidine and urea in the case of *Klebsiella pneumoniae* can be used, none support a growth rate as fast as that attained when ammonia serves as the source of nitrogen. The expression of genes subject to nitrogen regulation (Ntr) can be activated during growth on any one of these growth rate limiting sources of nitrogen. In the case of some of these Ntr genes or operons, nitrogen limitation is sufficient to elicit the response. Other operons, such as those comprising the genes for the degradation of histidine and proline are in addition subject to repression which is overcome by the addition of the particular nitrogen source to the medium. Thus in these cases, both the presence of the inducer and nitrogen limitation are required for the response (reviewed in Magasanik).[1]

The extracellular stimulus for the response is a drop in the ammonia concentration below the level that allows the NADP-linked glutamate dehydrogenase to catalyze the synthesis of glutamate from α-ketoglutarate, ammonia and NADPH. Because of its unfavorable equilibrium, this enzyme fails to provide glutamate when the ammonia concentration of the medium is reduced to below 1 mM. The only other important reaction in which ammonia participates is the synthesis of glutamine from glutamate with the concomitant conversion of ATP to ADP and P_i, a reaction catalyzed by glutamine synthetase. It is this enzyme which becomes the sole entryway for ammonia in nitrogen-depleted cells. It can fulfill this role because of the presence of another enzyme, glutamate synthase, which catalyzes the transfer of the amide group of glutamine to the keto group of α-ketoglutarate using NADPH as reducing agent to produce two molecules of glutamate, one from glutamine and the other from α-ketoglutarate. Together glutamine synthetase and glutamate synthase catalyze the net synthesis of glutamate from α-ketoglutarate, ammonia and NADPH and the concomitant hydrolysis of ATP which ensures a favorable equilibrium for this reaction sequence.[1]

The amino nitrogen of glutamate is the source of 85% of all cellular nitrogen and the amide nitrogen of glutamine provides the remaining

Regulation of Gene Expression in Escherichia coli, edited by E. C. C. Lin and A. Simon Lynch. © 1996 R.G. Landes Company.

15%. It is therefore apparent that in cells grown with an excess of ammonia, glutamine synthetase is responsible for only 15% of the ammonia taken up, but in cells grown with ammonia limitation, glutamine synthetase becomes the sole agent of ammonia assimilation. It is therefore not surprising that the immediate response of the cell to ammonia deprivation is a sharp increase in the rate of transcription of *glnA*, the structural gene for glutamine synthetase. Subsequently, depending on the composition of the nitrogen deficient medium, the transcription of other nitrogen regulated genes is initiated (reviewed in Reitzer and Magasanik, and Magasanik.)[2,3]

THE *GLNALG(GLNA NTRBC)* OPERON

The structural gene for glutamine synthetase *glnA* is a member of the complex *glnALG* operon. The products of *glnG(ntrC)* and of *glnL(ntrB)* are, respectively, nitrogen regulators I and II (NR$_I$ and NR$_{II}$) which have the primary responsibility for the regulation of transcription of the *glnALG* operon as well as of other nitrogen regulated genes.

The *glnALG* operon is equipped with three promoters (Fig. 13.1). The *glnAp1* and *glnAp2* promoters are located upstream of *glnA* and *glnLp* is located between *glnA* and *glnL* just downstream from a terminator which arrests approximately 80% of the transcripts that have passed through *glnA*. In cells growing with an excess of ammonia, transcription is initiated by σ^{70}-RNA polymerase at *glnAp1* and at *glnLp*. The product of *glnG*, NR$_I$, is a protein capable of binding to specific sites on the DNA which overlap the *glnAp1* and *glnLp* promoters. In this manner NR$_I$ negatively regulates its own synthesis together with the synthesis of NR$_{II}$ as well as the synthesis of glutamine synthetase.[3]

A drop in the level of ammonia causes NR$_{II}$ to phosphorylate NR$_I$. NR$_I$-phosphate in turn activates the initiation of transcription at *glnAp2*, resulting in greatly increased synthesis of glutamine synthetase, and of NR$_I$.[4] Apparently, the binding of NR$_I$ to the site overlapping *glnLp* does not block the elongation of the transcript initiated at *glnAp2*. The increased intracellular concentration of NR$_I$ results in the complete arrest of transcription initiation at *glnAp1* and *glnLp*, so that *glnAp2* alone serves as promoter of the *glnALG* operon.[5]

THE σ^{54}-DEPENDENT PROMOTER

Transcription at *glnAp2* is initiated by σ^{54}-RNA polymerase.[6,7] In addition a number of other genes and operons subject to Ntr, for example the *nif* (nitrogen fixation) genes of *K. pneumoniae*, have σ^{54}-dependent promoters. It was the fact that the mutations in a gene (called *glnF*) not linked to *glnA* could prevent the increase in *glnA* expression and that of genes subject to Ntr during nitrogen starvation, as well as the discovery of a nucleotide sequence consensus in the promoter region of nitrogen-regulated genes differing from that of other promoters, that led to the recognition of the product of this gene, now renamed *rpoN*, as the previously unknown sigma factor σ^{54}.

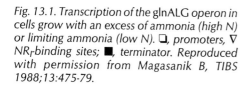

Fig. 13.1. Transcription of the glnALG operon in cells grow with an excess of ammonia (high N) or limiting ammonia (low N). ☐, promoters, ▽ NR$_I$-binding sites; ■, terminator. Reproduced with permission from Magasanik B, TIBS 1988;13:475-79.

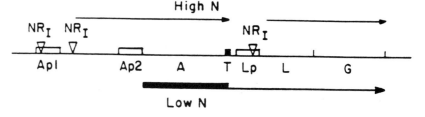

Nevertheless, σ^{54} is not a specific component of the nitrogen regulation system since the transcription of genes unrelated to nitrogen metabolism, like those for formic hydrogen lyase, is also initiated by σ^{54}-RNA polymerase (reviewed in Kustu et al).[8]

The consensus sequence for σ^{54}-dependent promoters located 10 bp from the transcriptional start site is TGGPyPuPyPu——PyPyGCA/T. The critical elements are GG and GC separated by exactly 10 base pairs.[9] The affinity of the promoter for σ^{54}-RNA polymerase is increased by the presence of T in the four positions preceding the GC pair.[10]

In cells grown with an excess of nitrogen, σ^{54}-RNA polymerase is bound to the *glnAp2* promoter as part of a closed complex.[11] The transformation of the closed to the open complex requires NR_I-phosphate which exerts its effect on the promoter from the two strong binding sites overlapping the *glnAp1* promoter, which are located 100 bp upstream from the transcriptional start site. The binding sites can be moved to positions as far as 1000 bp upstream or downstream from the promoter without significant impairment of the ability of the activator to catalyze the transition of the closed σ^{54}-RNA polymerase promoter complex to the open complex. They are therefore the prokaryotic equivalent of eukaryotic enhancers.[12,13]

It has since been found that all σ^{54}-dependent promoters require activators bound to enhancers for the transition of the closed to the open complex. This is in contrast to σ^{70}-dependent promoters where this transcription can be accomplished without the help of an activator.

TRANSCRIPTIONAL ENHANCERS

The NR_I-specific transcriptional enhancers are always found in pairs and share a characteristic sequence of nucleotides. The consensus sequence reveals an inverted repeat consisting of 17 bp: TGCACCA---TGGTGCA.[14] The strong binding sites upstream from *glnAp2* differ in a single base pair from the consensus. The NR_I-binding sites at other Ntr promoters have less homology to the consensus and lower affinity. As a result the NR_I present in low concentration in cells grown with an excess of nitrogen (approximately five molecules per cell) is adequate for full activation of transcription at *glnAp2*, but not at other NR_I-activated promoters. Activation at these promoters must await the increase in the intracellular concentration of NR_I to approximately 70 molecules per cell which results from the increase in the rate of transcription of the *glnALG* operon when the *glnAp2* promoter is activated. Consequently, the first response of the cell to ammonia deficiency is increased synthesis of glutamine synthetase which allows the cell to utilize ammonia present in low concentration in the medium followed by increased synthesis of enzymes and permeases that allow the cell to use alternative sources of nitrogen.[3]

PHOSPHORYLATION OF NR_I

Although unphosphorylated NR_I binds to its enhancers, only phosphorylated NR_I is able to activate transcription. The agent of phosphorylation is NR_{II}, the product of *glnL*. NR_{II} is also responsible for the dephosphorylation of NR_I-phosphate when combined with the small protein P_{II}, the product of *glnB*, a gene unlinked to *glnALG*.[4] NR_{II} is enabled to phosphorylate NR_I when nitrogen deprivation causes the enzyme uridylyltransferase (UTase), the product of still another unlinked gene, *glnD*, to uridylylate P_{II}.[3] This control of P_{II} will be described in a later section.

Evidence for the role of NR_{II} in the regulation of the initiation of transcription came from experiments demonstrating that in a transcription system using highly purified components, σ^{54} RNA polymerase, NR_I and NR_{II} were essential for the initiation of transcription at *glnAp2*,[6] that incubation of NR_I with NR_{II} and ATP resulted in the phosphorylation of NR_I and that the addition of P_{II} to the transcription system resulted in the dephosphorylation of NR_I-phosphate and the arrest of transcription initiation at *glnAp2*.[4] The involvement of NR_{II} in the dephosphorylation of NR_I-phosphate was demonstrated by the existence of a mutant form of NR_{II}, capable of phosphorylating NR_I but in whose presence P_{II} failed to bring about the dephosphorylation of NR_I-phosphate.[4]

However, in apparent contradiction to the postulated role of NR_{II} was the observation that deletion of *glnL*, the structural gene for NR_{II}, did not prevent initiation of transcription at *glnAp2* in intact cells, although transitions from lack of transcription initiation to rapid initiation and vice versa in cells with functional NR_{II} were very fast and in cells lacking NR_{II}, very slow.[5] Furthermore, NR_{II} was not required for the activation of transcription in a cell free transcription-translation system.[7] An explanation for these apparent contradictions was the discovery that acetylphosphate, a normal metabolite whose intracellular concentration increases when the rate of glucose catabolism exceeds the rate of biosynthesis, can bring about the phosphorylation of NR_I in the absence of NR_{II}. In cells with functional NR_{II}, acetylphosphate appears to be without effect since NR_{II} is responsible for the rapid phosphorylation of NR_I as well as for the rapid dephosphorylation of NR_I-phosphate in response to the nitrogen state of the cell.[15]

NR_I/NR_{II} AS TWO-COMPONENT PARADIGM

In the course of determining the deduced amino acid sequence of NR_I and NR_{II} derived from a species of *Bradyrhizobium*, Ausubel and his co-workers discovered in 1986 that amino-terminal domains of a number of regulatory proteins were homologous to that of NR_I and that each of these proteins had a partner whose carboxy-terminal domain was homologous to that of NR_{II}.[16] The simultaneous discovery, mentioned earlier, that NR_{II} phosphorylates NR_I, suggested that this type of regulation in which a modulator or sensor, such as NR_{II}, phosphorylates a response regulator, such as NR_I, is the general property of these two-component systems.[4] In the succeeding years many of these systems have been identified.

Most of the modulators are membrane proteins, which appear to phosphorylate their partners in response to extracellular signals.[17] NR_{II}, however, is a cytoplasmic protein and as mentioned before, its ability to phosphorylate NR_I or to dephosphorylate NR_I-phosphate is determined by the absence or presence of the unmodified P_{II} protein.

The reaction of NR_{II} with ATP results in the transfer of the γ-phosphate of ATP to a histidine residue of NR_{II},[18] subsequently identified as the histidine in position 139 of NR_{II}, located in the homologous domain of all modulators in a fully conserved position.[19] The phosphate is then transferred to an aspartate residue of NR_I,[18] located in position 54, a fully conserved residue in the high homology amino-terminal domain of all response regulators.[20] The ability to phosphorylate this aspartate does not reside in the modulator but in the response regulator, as shown by the fact that treatment of NR_I (or of other response regulators) with acylphosphates such as acetylphosphate, carbamyl phosphate or phosphoramidate, but not with ATP, results in the phosphorylation of this aspartate residue.[21] As mentioned earlier, phosphorylation of NR_I by acetylphosphate in NR_{II}-deficient cells accounts for the activation of transcription at *glnAp2*.[15] The role of NR_{II} is to catalyze its own phosphorylation by ATP in order to present NR_I with an exceptionally good phosphate donor.

NR_I-phosphate, like some other response regulator phosphate,s has an autophosphatase activity.[18,22] However, the rate of its dephosphorylation is greatly increased by NR_{II}, P_{II} and ATP. Initially, P_{II} in considerable excess over NR_{II} was found to be required, but it has now been shown that the presence of glutamate reduces the necessary concentration of P_{II} to one equimolar with that of NR_{II}.[23]

The dimeric NR_I is composed of three domains which are homologous to the domains of other activators of transcription at σ^{54}-dependent promoters. The regulatory N-terminal domain (residues 1-120), contains the site of phosphorylation, the aspartate residue at position 54. The central domain, which appears to be responsible for the interaction of NR_I-phosphate with σ^{54}-RNA polymerase contains an ATP-binding motif (residues 168-175) and the C-terminal domain (residues 380-469) contains a helix-turn-helix motif characteristic for proteins capable of binding to specific DNA sites.[17,24]

Mutant proteins lacking the C-terminal domain are monomeric, fail to bind to the enhancer sites, but can be phosphorylated; mutant proteins lacking the N-terminal domain are dimeric, can bind to DNA, but cannot be phosphorylated. All three domains of NR_I are required for the activation of transcription.[25]

Not all activators for σ^{54}-dependent promoters have homologous N-terminal domains. For example, the NifA protein responsible for the activation of transcription at promoters for *nif* genes and operons, is highly homologous to NR_I in its central domain, contains a C-terminal domain with a helix-turn-helix motif responsible for its ability to bind to DNA, but has a totally different N-terminal domain.[26] This protein is able to activate transcription in its unphosphorylated form and in fact, there is no reason to believe that it is ever phosphorylated.[27] Its ability to activate transcription is negatively regulated by the NifL protein in response to nitrogen levels and oxygen tension.[28]

ACTIVATION OF TRANSCRIPTION

The fact that the response regulator bound to enhancer sites far from the binding site for the RNA polymerase is highly effective in the activation of transcription initiation suggested that the flexibility of the intervening DNA enabled the response regulator to contact the σ^{54}-RNA polymerase bound to the promoter.[12] The existence of the resulting DNA loops has been visualized in electron micrographs in the case of *glnAp2*.[29] However, this flexibility of the DNA is not in all cases adequate for the contact between the response regulator and the σ^{54}-RNA polymerase: as first discovered in the study of transcription initiation at the *nifH* promoter by the product of the *nifA* gene bound to its enhancer, another protein, integration host factor (IHF), known for its ability to bend DNA, must bind to a site located between this enhancer and the promoter to allow transcription to be initiated; a binding site for IHF in this location is found in many other σ^{54}-dependent systems.[30] In the case of *glnHp2*, a promoter for an operon whose products play a role in the transport of glutamine, activation in supercoiled DNA is stimulated by, but not dependent on, IHF.[31] In this case it could be shown that IHF bound to its site was only effective when all three proteins, NR_I-phosphate, IHF and σ^{54}-RNA polymerase, were bound to the same face of the DNA helix even though the distances separating them could be altered; however, moving anyone of the three binding sites to the opposite face of the DNA helix did not affect transcription in the absence of IHF, but converted IHF from a potentiator into a powerful inhibitor of open complex formation. Apparently, this placement caused IHF to bend the DNA so as to prevent contact between response regulator and polymerases (Fig. 13.2).[32] This example demonstrates the ability of a DNA binding protein, such as IHF, to play a positive or negative role in transcription initiation without contact with either the activator or the RNA polymerase.

The exact role of the enhancers was not immediately apparent, since NR$_I$-phosphate could activate transcription at *glnAp2* in intact cells or in a cell-free system on DNA templates lacking the enhancers when present in high concentration.[12,13] In addition, although both NR$_I$ and NR$_I$-phosphate could bind to enhancers, only NR$_I$-phosphate was able to activate the initiation of transcription at *glnAp2*.[4,13] Furthermore, NR$_I$ was shown to acquire upon phosphorylation the ability to hydrolyze ATP to ADP and P$_i$, a reaction required for the activation of transcription.[33,34]

It has now become clear that the enhancers play a double role. Their first role is to tether the NR$_I$-phosphate in the vicinity of the promoter, thus increasing its local concentration. This was demonstrated by the ability of NR$_I$-phosphate in low concentration to activate open complex formation at *glnAp2* when the enhancer and the promoter were located on separate DNA rings of a singly linked catenane, but not when the two rings were unlinked.[35] Their second role which requires pairing of the enhancers, is to increase the probability of a cooperative interaction of two NR$_I$-phosphate dimers to form a tetramer. This conformational change results in the acquisition of ATPase activity and of the ability to activate the isomerization of the closed σ^{54}-RNA polymerase-*glnAp2* complex to the open complex. The experimental

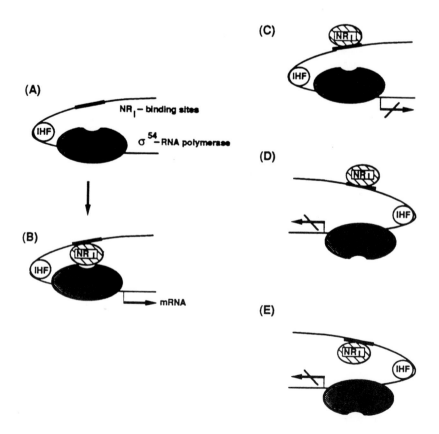

Fig. 13.2. Schematic model for the stimulation or inhibition of glnHp2 transcription activation by IHF. The indentation in the oval that designates σ^{54}-RNA polymerase indicates a receptor site for NR$_I$. By bending the DNA, IHF stimulates transcription activation when the binding sites for NR$_I$, IHF and σ^{54}-RNA polymerase (shaded) are placed correctly on the DNA helix (A) and (B). When the binding sites for NR$_I$ (C) or for IHF (D), or for both (E) are placed on the other side of the DNA helix with respect to the RNA polymerase, IHF becomes an inhibitor of transcription. Reproduced with permission from Claverie-Martin F et al, J Mol Biol 1992; 227:996-1008.

evidence for this role of paired enhancers was provided by the observation that though the phosphorylation of NR_I did not improve its affinity for a single enhancer, it greatly increased its affinity for the paired enhancers. Furthermore, on a template containing *glnAp2* and paired enhancers, the ability of NR_I-phosphate to activate the initiation of transcription increased with the occupation of the enhancers, while on a similar template with a single enhancer even its full occupation by NR_I-phosphate was not effective: rather, the concentration of NR_I-phosphate had to be increased beyond this level to allow transcription to be initiated.[14] Apparently, the increased concentration allowed a second molecule of NR_I-phosphate to interact with the one bound to the single enhancer to produce the tetramer able to hydrolyze ATP and to catalyze the formation of the open complex. Additional strong evidence in support of this hypothesis is the observation that a mutant form of NR_I-phosphate lacking the ability to bind to DNA can be used to endow NR_I-phosphate bound to a single enhancer with the ability to activate the initiation of transcription at *glnAp2*.[36]

A recent study led to the surprising result that it is possible to replace the enhancers by a segment of DNA with an entirely different sequence. This segment consists of many repeats of 5A residues separated by five other nucleotides and introduces locally a superhelical structure. This sequence-induced superhelical DNA-segment is as effective as the paired enhancers in binding NR_I-phosphate and in stimulating the oligomerization which endows it with ATPase activity and the ability to activate the initiation of transcription of *glnAp2*. It is possible that the repeated bent DNA tracts in the superhelical structure enhance its affinity for NR_I-phosphate, a protein able to bend DNA;[29] and the tight packing of the DNA in this structure may then facilitate the oligomerization of NR_I-phosphate.[37]

RESPONSE TO NITROGEN AVAILABILITY

The heart of the Ntr system is the *glnALG* operon. The three products of these genes, respectively glutamine synthetase, NR_{II} and NR_I, together with the small protein P_{II} (the product of *glnB*) and the enzyme UTase/UR (the product of *glnD*) catalyzing the uridylylation and deuridylylation of P_{II}, are generally responsible for the changes in the expression of genes in response to the availability and quality of the nitrogen source.[3] In addition, the products of the *nac* and *nifA* genes link the expression of genes whose products can supply the cell with ammonia by the degradation of certain nitrogen containing compounds or by the fixation of atmospheric nitrogen to the central components of the Ntr system.[38,39]

In cells growing with an excess of ammonia, the enzyme glutamine synthetase is present in relatively low concentration and, in part, in an inactive form. The inactivation results from the adenylylation of approximately one half of the 12 subunits comprising the enzyme. The enzyme adenylyltransferase (ATase), the product of the *glnE* gene, is stimulated by P_{II} to adenylylate glutamine synthetase. At the same time, P_{II} also interacts with NR_{II} to stimulate the dephosphorylation of NR_I-phosphate, greatly reducing any initiation of transcription of the *glnALG* operon at the *glnAp2* promoter. A reduction of the ammonia concentration of the growth medium to below 1 mM, results in diminished glutamate synthesis by glutamate dehydrogenase, diminished glutamine synthesis by glutamine synthetase and consequently in a low intracellular concentration of glutamine and a high intracellular concentration of α-ketoglutarate. The change in the ratio of α-ketoglutarate to glutamine activates the UTase; as a result, P_{II} is converted to P_{II}-UMP, which activates the ATase to deadenylylate glutamine synthetase, and which therefore fails to interfere with the phosphorylation of NR_I by NR_{II}. Consequently, transcription is now initiated at the σ^{54}-dependent *glnAp2* promoter, resulting in an increase in the intracellular level of active glutamine synthetase and of NR_I-phosphate.[2]

In summary, the extracellular stimulus, a drop in the ammonia concentration, has caused the sensor, glutamine synthetase, to generate the intracellular signal, reduction in the concentration of glutamine; this signal is transmitted by the transducers UTase/UR and P_{II} to the modulator NR_{II}, which in turn activates the response regulator NR_I.[3]

As a result of the increase in the expression of the *glnALG* operon the level of NR_I rises sufficiently to activate the transcription of σ^{54}-dependent promoters associated with enhancer pairs of lower affinity for NR_I than that associated with *glnAp2*. Among these promoters are those for the *glnHPQ* and the *hisYQMP* operons whose products are, respectively, components of the uptake systems for glutamine and for histidine.[40,41] In addition, the increase in the intracellular level of NR_I activates the σ^{54}-dependent promoters of the *nac* gene and of the *nifLA* operon. The expression of these genes has been studied in *Klebsiella*. *E. coli* has a functional *nac* gene but lacks the *nif* genes; *S. typhimurium* lacks both *nac* and *nif*. The *nifA* product is the activator of transcription at the σ^{54}-dependent promoters of the *nif* genes encoding the components of the nitrogen fixation system and the *nifL* product reduces its activity in response to the presence of oxygen and a high level of ammonia.[39] The *nac* product is the activator of transcription of the *hut* and *put* operons, whose products are responsible for the degradation of histidine and proline, respectively, and of the *ure* genes encoding urease.[38] The expression of the *hut* and *put* genes is also negatively regulated by repressors whose effects are overcome specifically by the presence of histidine or proline, respectively, in the growth medium. In addition, the expression of these operons can also be activated in the absence of the product of the *nac* gene by the catabolite activator protein (CAP) charged with cyclic AMP. Consequently, the expression of these genes requires the presence of the inducer, and either a deficiency of nitrogen or a deficiency of carbon and energy.[40]

Addition of ammonia to cells growing in a nitrogen-deficient medium, reverses the process of activation. The rise in the level of glutamine causes UTase/UR to remove the uridylyl groups from P_{II}-UMP. Consequently, P_{II} stimulates ATase to adenylylate glutamine synthetase; this prevents the loss of glutamate that would result from the rapid conversion of glutamate to glutamine by the high level of glutamine synthetase in the nitrogen-starved cells.[42] P_{II} reacts with NR_{II} to bring about the dephosphorylation of NR_I-phosphate and consequently blocks the initiation of transcription at σ^{54}-dependent promoters.[2] Eventually, the growth of these cells in the ammonia-containing medium lowers the levels of the regulated enzymes and permeases and adjusts the expression of the corresponding genes to that appropriate for growth in the new environment.

REFERENCES

1. Magasanik, B. 1982. Genetic control of nitrogen assimilation in bacteria. Annu Rev Genet 1982; 16:135-68.
2. Reitzer L, Magasanik B. Ammonia assimilation and the biosynthesis of glutamine, glutamate, aspartate, asparagine, L-alanine, and D-alanine. In: Neidhardt FC, ed. *Escherichia coli* and *Salmonella typhimurium*: Cellular and Molecular Biology. Washington: American Society for Microbiology, 1987:302-407.

3. Magasanik B. Reversible phosphorylation of an enhancer binding protein regulates the transcription of bacterial nitrogen utilization genes. TIBS 1988;13:475-79.

4. Ninfa AJ, Magasanik B. Covalent modification of the *glnG* product, NR$_I$, by the *glnL* product, NR$_{II}$, regulates the transcription of the *glnALG* operon in *Escherichia coli.*. Proc Natl Acad Sci USA 1986; 83:5909-13.

5. Reitzer LJ, Magasanik B. Expression of *glnA* in *Escherichia coli* is regulated at tandem promoters. Proc Natl Acad Sci USA 1985; 82:1979-83.

6. Hunt TP, Magasanik B. Transcription of *glnA* by purified *Escherichia coli* components: core RNA polymerase and the products of *glnF, glnG,* and *glnL*. Proc Natl Acad Sci USA 1985; 82:8453-57.

7. Hirschman J, Wong PK, Sei K, Keener J, Kustu S. Products of nitrogen regulatory genes *ntrA* and *ntrC* of enteric bacteria activate *glnA* transcription *in vitro*: evidence that the *ntrA* product is a σ factor. Proc Natl Acad Sci USA 1985; 82:7525-29.

8. Kustu S, Santero E, Keener J et al. Expression of σ^{54}*(ntrA)*-dependent genes is probably united by a common mechanism. Microbiol. Rev. 1989; 53:367-76.

9. Ausubel FM. Regulation of nitrogen fixation genes. Cell 1954; 37:5-6.

10. Cannon W, Buck M. Central domain of the positive control protein NifA and its role in transcriptional activation. J Mol Biol 1992; 225:271-86.

11. Sasse-Dwight S, Gralla JD. Probing the *Escherichia coli glnALG* upstream activation mechanism *in vivo*. Proc Natl Acad Sci USA 1988; 85:8934-38.

12. Reitzer LJ, Magasanik B. Transcription of *glnA* in *Escherichia coli* is stimulated by activator bound to sites far from the promoter. Cell 1985; 45:785-92.

13. Ninfa AJ, Reitzer LJ, Magasanik B. Initiation of transcription at the bacterial *glnAp2* promoter by purified *E. coli* components is facilitated by enhancers. Cell 1987; 50:1039-46.

14. Weiss V, Claverie-Martin F, Magasanik B. Phosphorylation of nitrogen regulator I of *Escherichia coli* induces strong cooperative binding to DNA essential for activation of transcription. Proc Natl Acad Sci USA 1992; 89:5088-92

15. Feng J, Atkinson MR, McCleary W et al. Role of phosphorylated metabolic intermediates in the regulation of glutamine synthetase synthesis in *Escherichia coli*. J Bacteriol 1992; 174:6061-70.

16. Nixon BT, Ronson CW, Ausubel FM. Two-component regulatory systems responsive to environmental stimuli share strongly conserved domains with the nitrogen assimilation regulatory genes *ntrB* and *ntrC*. Proc Natl Acad Sci USA 1986; 83:7850-54.

17. Stock JB, Ninfa AJ, Stock AM. Protein phosphorylation and regulation of adaptive responses in bacteria. Microbiol Rev 1989; 53:450-90.

18. Weiss V, Magasanik B. Phosphorylation of nitrogen regulator I (NR$_I$) of *Escherichia coli*. Proc Natl Acad Sci USA 1988; 85:8919-23.

19. Ninfa AJ, Bennett RL. Identification of the site of autophosphorylation of the bacterial kinase/phosphatase NR$_{II}$. J Biol Chem 1991; 266:6888-93.

20. Sanders DA, Gillece-Castro BL, Burlingame AL et al. Phosphorylation site of NtrC, a protein phosphatase whose covalent intermediate activates transcription. J Bacteriol 1992; 174:5117-22.

21. Lukat GS, McCleary WR, Stock AM et al. Phosphorylation of bacterial response regulator proteins by low molecular weight phosphate donors. Proc Natl Acad Sci USA 1992; 89:718-22.

22. Keener J, Kustu S. Protein kinase and phosphoprotein phosphatase activities of nitrogen regulatory proteins NTRB and NTRC of enteric bacteria: roles of the conserved amino-terminal domain of NTRC. Proc Natl Acad Sci USA 1988; 88:4976-80.

23. Liu J, Magasanik B. Activation of the dephosphorylation of nitrogen activator NR_I-phosphate of *Escherichia coli*. J Bacteriol 1995, in press.

24. Marett E, Segovia L. The σ^{54} bacterial enhancer-binding protein family: mechanism of action and phylo-genetic relationship of their functional domain. J Bacteriol 1993; 175:6067-74.

25. North AK, Klose KE, Stedman KM, Kustu S. Prokaryotic enhancer-binding proteins reflect eukaryotic-like modularity: the puzzle of nitrogen regulatory protein C. J Bacteriol 1993; 175:4267-73.

26. Drummond M, Whitty P, Wootton J. Sequence and domain relationships of ntrC and *nifA* from *Klebsiella pneumoniae*: homologies to other nitrogen regulatory proteins. EMBO J 1986; 5:441-47.

27. Lee HS, Berger DK, Kustu S. Activity of purified NIFA, a transcriptional activator of nitrogen fixation genes. Proc Natl Acad Sci USA 1993; 90:2266-70.

28. Berger DK, Narberhaus F, Kustu S. The isolated catalytic domain of NIFA, a bacterial enhancer-binding protein, activates transcription *in vitro*: activation is inhibited by NIFL. Proc Natl Acad Sci USA 1994; 91:103-07.

29. Su W, Porter S, Kustu S, Echols H. DNA looping and enhancer activity: association between DNA-bound NtrC activator and RNA polymerase at the bacterial *glnA* promoter. Proc Natl Acad Sci USA 1990; 87:5504-08.

30. Hoover TR, Santero E, Porter S, Kustu S. The integration host factor stimulates the interaction of RNA polymerases with NIFA, the transcriptional activator of nitrogen fixation operons. Cell 1990; 63:11-22.

31. Claverie-Martin F, Magasanik B. Role of integration host factor in the regulation of the *glnHp2* promoter of *Escherichia coli*. Proc Natl Acad Sci USA 1991; 88:1631-35.

32. Claverie-Martin F, Magasanik B. Positive and negative effects of DNA bending on activation of transcription from a distant site. J Mol Biol 1992; 227:996-1008.

33. Popham DL, Szeto D, Keener J et al. Function of a bacterial activator protein that binds to transcriptional enhancers. Science 1989; 243:629-35.

34. Weiss DS, Batut J, Klose KE et al. The phosphorylated form of the enhancer-binding protein NTRC has an ATPase activity that is essential for the activation of transcription. Cell 1991; 67:155-67.

35. Wedel A, Weiss D, Popham D et al. A bacterial enhancer functions to tether a transcriptional activator near a promoter. Science 1990; 248:486-90.

36. Porter SC, North AK, Wedel AB et al. Oligomerizatoon of NTRC at the *glnA* enhancer is required for transcriptional activation. Genes & Dev 1993; 7:2258-73.

37. Brahms G, Brahms S, Magasanik B. A sequence-induced superhelical DNA-segment serves as transcriptional enhancer. J Mol Biol 1995, in press.

38. Bender RA. The role of the NAC protein in the nitrogen regulation of *Klebsiella aerogenes*. Mol Microbiol 1991;5:2575-80.

39. Merrick MJ. Nitrogen control of the *nif* regulon in *Klebsiella pneumoniae*: involvement of the *ntrA* gene and analogies between *ntrC* and *nifA*. EMBO J 1983; 2:39-44.

40. Magasanik B. Neidhardt FC. Regulation of carbon and nitrogen utilization. In: Neidhardt FC, ed. *Escherichia coli* and *Salmonella typhimurium*: Cellular and Molecular Biology, vol 2. Washington: American Society for Microbiology, 1987:1318-25.

41. Nohno T, Saito T, Hong JS. Cloning and complete nucleotide sequence of the *Escherichia coli* glutamine permease operon (*glnHPQ*). Mol Gen Genet 1986; 205:260-67.

42. Kustu S, Hirschman J, Burton D et al. Covalent modification of bacterial glutamine synthetase: physiological significance. Mol Gen Genet 1984; 197:309-17.

CHAPTER 14

HISTORY OF THE PHO SYSTEM

Annamaria Torriani-Gorini

In Milan, I collaborated with the biochemist Luigi Gorini, with whom I subsequently shared my life. We were studying the resistance of *E. coli* to a new antibiotic "penicillin"[1] and its effect on the metabolism of acidoproteolytic bacteria.[2] In 1948 we went to Paris. I joined Jacques Monod at the Institut Pasteur on studies of regulation of carbon metabolism. Thus began my studies in the field of cellular regulatory processes, whose rules were being explored and characterized by André Lwoff, Max Delbrück, Herman Kalckar, Jacques Monod, Mike Doudoroff, Luigi Gorini, and Sol Spiegelman.

I started by studying maltose metabolism. After the discoveries of the Cori ester (glucose-1-phosphate), Doudoroff's sugar phosphoryl cycle, and the phosphorylases, the prevailing opinion was that the other polysaccharides, like starch and glycogen, were produced in a similar way. How surprising it was for us to observe the synthesis from maltose of a starch-like polysaccharide without phosphate intervention! This led to the discovery of the enzyme amylomaltase, now known to be the product of the *malQ* gene.[3,4] The enzyme was induced by maltose and repressed by glucose. One of the key questions was whether the enzyme, required to metabolize the carbon source, was destroyed by proteolysis as soon as the inducer (maltose in this case) was eliminated from the medium? Such a hypothesis was advanced by the Finnish Nobelist, Atturi I. Virtanen, and his collaborator, U. Winkler, in 1949.[5] On the basis of their observation that in contrast to constitutive proteolytic enzymes, which retain their activities during nitrogen starvation, "adaptive enzymes (e.g., lactase), which are necessary only in definite nutritional conditions seem, as a rule, to decrease powerfully or to disappear entirely with the lowering of the N-content of the cells." Monod found this statement worth checking experimentally since it related directly to the mechanism of metabolic regulation.

In order to test the hypothesis, we compared the level of an inducible and a constitutive enzyme in cells growing in conditions of excess and starvation levels of the substrate/inducer. The enzymes we chose were the inducible amylomaltase and a constitutive phosphatase.[6] I realized that the hypothesis could not be properly tested using phosphate starvation. During P_i starvation, the constitutive phosphatase (with

Regulation of Gene Expression in Escherichia coli, edited by E. C. C. Lin and A. Simon Lynch. © 1996 R.G. Landes Company.

a pH optimum of 4.0) was superseded by a new enzyme with an alkaline pH optimum of 8.2.)[7,8] It seemed possible that the alkaline phosphatase (AP) was repressed by phosphate. Monod was not convinced by my results, as they contradicted the prevailing notion of regulation, namely, as induction by substrate.

Phosphate metabolism, however, was coming to the forefront: another independent group provided definitive evidence that the synthesis of alkaline phosphatase was not in fact induced by P_i, but repressed by it.[9] Soon thereafter this enzyme also became the vehicle for studying nonsense suppression and therein the nature of the genetic code by C. Levinthal, A. Garen, S. Garen, F. Rothman, H. Echols and me at MIT.

Although phenotypic suppression prevented Levinthal from accomplishing his project on colinearity, the isolation and mapping of many other alkaline phosphate-negative mutants identified *phoA* as the structural gene.[10] Peptide analysis showed that *phoA* specified a single chain and, further, that the active enzyme was a dimer.[11] This nonspecific phosphomonoesterase turned out to be located in the periplasm.[12] In recent years, the school of J. Beckwith has exploited alkaline phosphatase gene fusions as a method of probing the topology of transmembrane proteins.[13]

Among the alkaline phosphatase-negative mutants isolated, some were found to be defective in *phoB* (or R_{1a}) which is closely linked to *phoA* at minute 9 of the *E. coli* map. In vitro experiments indicated the need for PhoB as a positive regulatory element,[14] which was later confirmed.[15] Mutants in which alkaline phosphatase expression was insensitive to P_i repression at first suggested that a repressor is also involved in wild-type cells growing at concentrations of P_i sufficient to prevent the synthesis and function of the *phoB* product. Some of these constitutive mutations mapped in a genetic site closely linked to *phoB* and were referred to as *phoR*, or R_1b (R for repressibility).[16] We now know that PhoR is the P_i-sensing histidine kinase and that PhoB is the cognate regulator of a two-component signal transduction system.[17-19] The "constitutivity" observed in *phoR* mutants was later shown to be due to the *phoM* (later renamed *CreC*) gene product, another sensor kinase for PhoB, because a *phoR phoM* (*creC*) double mutant was also alkaline phosphatase negative.[19]

Since mutants with *phoA* insensitive to P_i repression could be selected on solid medium containing glycerol-3-phosphate as the sole carbon and energy source in the presence of P_i, a large group was readily collected. Genetic mapping revealed yet another apparent regulatory locus, R_2 (now *pst* for phosphate specific transport) at minute 83. This locus was subsequently resolved into *phoS* (now *pstS*), *pstC*, *pst2* (now *pstA*) *pstB*, and *phoT*$_{35}$ (now *phoU*), arranged in counterclockwise order as an operon.[20-23] Although mutations in these genes allow *phoA* to escape P_i repression by attenuating inflow of the signal molecule P_i, none of them are unconditionally defective in P_i transport. This is because *E. coli* has another transport system for P_i encoded by *pit* (phosphate inorganic transport).[29] Expression of this gene is apparently insensitive to P_i and is therefore always transcriptionally active. The K_t of the transport protein, however, is high (9-25 μM), and under conditions of P_i starvation (10^{-7}M) this transport is ineffectual. The work of Willsky and Malamy clarified most of these intricacies.[25]

It is worth noting that *E. coli* has two systems for the transport of glycerol phosphate: GlpT for providing carbon and energy, and Ugp for providing phosphate.[26,27] Only *ugp* is under P_i control like *pst*.

How the starvation signal is transduced is particularly intriguing in the Pho regulon. The signal of P_i starvation requires the functioning of Pst. This exceptional double function was demonstrated by Cox et al, with mutants of *pstA*, *pstB* or *pstC* defective in P_i transport but not for alkaline phosphatase repression.[28,29] A peculiarity of the Pst operon is the presence of five genes[30] instead of the four normally present in other bacterial transport systems. PstS is an abundant periplasmic protein with a very high affinity ($Km = 10^{-7}$ M) for exogenous P_i or the P_i liberated by the secreted phosphatases and nucleotidases. PstS delivers its sequestered P_i to the cytoplasm through a membrane channel (PstC-PstA) possibly with the help of the ATP-driven PstB. We recently showed by immunoblot of a crude extract that PstB binds 8-azido-^{32}P-ATP (Torriani and Chan, unpublished).

The function of the fifth protein, PhoU, remains unsolved. Mutants deleted in *phoU* are constitutive for the synthesis of alkaline phosphatase. It is possible that during P_i transport by the Pst system in a phosphate-poor medium, PhoU might be signaling to modulate the activity of the sensor kinase PhoR. However, the PhoU protein may have other functions since a missense mutant has a pleiotropic phenotype.[31,32,34] Does P_i act as an external effector via *phoR* alone? Rosenberg and collaborators arrived at an affirmative answer from studies of the phosphate transport system.[20] By analyzing the expression of a *phoA-lacZ* fusion we found that P_i, when generated in the cytoplasm from glycerol-3-phosphate (transported by the GlpT system), had no effect on the expression of the *phoA* promoter. But this cytoplasmic P_i, if excreted to the periplasm by exchange with glycerol-3-phosphate (*glpD* mutant) did repress *phoA*.[33] The following chapter will consider recent further developments of our knowledge of the Pho modulon.

REFERENCES

1. Gorini L, Torriani A. Biochemistry of *Escherichia coli* and the production of penicillinase. Nature 1947; 16:332-333.

2. Gorini L, Torriani A. Action de la penicillinase sur l'activité proteolytique des bacteries acido-proteolytiques. Biochim Biophys Acta 1948; 2:226-238.

3. Monod J, Torriani A. De l'amylomaltase d'*Escherichia coli*. Ann Inst Pasteur (Paris) 1950; 78:65-78.

4. Schwartz M. Expression phénotypique et génétique de mutations affectant le metabolisme du maltose chez *Escherichia coli* K12. Ann Inst Pasteur (Paris) 1967; 112:673-702.

5. Virtanen AI, Winkler U. Effect of decrease in protein content of cells on the proteolytic enzyme system. Acta Chem Scand 1949; 3:272-278.

6. Roche J, van Thoai N. Phosphatase alkaline. Adv Enzymol 1950; 10:83-122.

7. Torriani A. Effect of inorganic phosphate in the formation of phosphatases by *E. coli*. Fed Proc 1959; 18:339.

8. Torriani A. Effect of inorganic phosphate in the formation of phosphatases by *E. coli*. Biochim Biophys Acta 1960; 38:460-469.

9. Horiuchi T, Horiuchi S, Mizuno D. A possible negative feedback phenomenon controlling formation of alkaline phosphorous esterase in *Escherichia coli*. Nature 1959; 183:1529-1530.

10. Rothman F, Byrne R. Fingerprint analysis of alkaline phosphate of *E. coli K12*. J Mol Biol 1963; 6:330-340.

11. Schlesinger MJ, Barrett K. The reversible dissociation of alkaline phosphatase of *E. coli*. J Biol Chem 1965; 240:4248-4292.

12. Schlesinger MJ, Reynolds JA, Schlesinger S. Formation and localization of the alkaline phosphatase of *Escherichia coli*. Ann N Y Acad Sci 1969; 166:368-379.

13. Derman AL, Beckwith J. *Escherichia coli* alkaline phosphatase fails to acquire disulfide bonds when retained in the cytoplasm. J Bacteriol 1991; 173:7719-7722.

14. Dohan FCJr, Rubman RH, Torriani A. In vitro synthesis of *E. coli* alkaline phosphatase monomers. J Mol Biol 1971; 58:469-471.

15. Inouye H, Pratt C, Beckwith J et al. Alkaline phosphatase synthesis in a cell-free system using DNA and RNA templates. J Mol Biol 1977; 110:75-87.

16. Garen A, Otsuji N. Isolation of a protein specified by a regulator gene. J Mol Biol 1964; 8:841-852.

17. Makino D, Shinagawa H, Nakata A. Regulation of the phosphate regulon in *Escherichia coli* K12: regulation and role of the regulatory gene *phoR*. J Mol Biol 1985; 8:231-240.

18. Makino K, Shinagawa H, Amemura M et al. Signal transduction in the phosphate regulon of *Escherichia coli* involves phosphotransfer between PhoR and PhoB proteins. J Mol Biol 1989; 210:551-559.

19. Wanner BL, Latterell P. Mutants affected in alkaline phosphatase expression: evidence for multiple positive regulators for the phosphate regulon in *Escherichia coli*. Genetics 1980; 96:242-266.

20. Surin BP, Rosenberg H, Cox GB. Phosphate specific transport system of *Escherichia coli*: nucleotide sequence and gene-polypeptide relationship. J Bacteriol 1985; 161:189-198.

21. Echols H, Garen A, Garen S et al. Genetic control of repression of alkaline phosphatase in *E. coli*. J Mol Biol 1961; 3:425.

22. Zuckier G, Torriani A. Genetic and physiological test of three phosphate-specific transport mutants of *E. Coli*. J Bacteriol 1981; 145:1249-1256.

23. Nakata A, Amemura M, Makino K et al. Genetic and biochemical analysis of the phosphate-specific transport system in *E. coli*. In: Torriani-Gorini A, Rothman F, Silver S et al, eds. Phosphate metabolism and cellular regulation in microorganisms. Washington: American Society for Microbiology, 1987:150-155.

24. Rosenberg H, Gerdes RG, Chegwidden K. Two systems for the uptake of phosphate in *Escherichia coli*. J Bacteriol 1977; 131:505-511.

25. Willsky GR, Malamy MH. Characterization of two genetically separable inorganic phosphate transport systems in *Escherichia coli*. J Bacteriol 1980; 144:356-365.

26. Lin ECC. Glycerol dissimilation and its regulation in bacteria. Ann Rev Microbiol 1976; 30:535-578.

27. Argast M, Ludke D, Silhavy TU et al. A second transport system for *sn*-glycerol-3-phosphate in *Escherichia coli*. J Bacteriol 1978; 136:1070-1083.

28. Cox GB, Webb D, Godovan-Zimmerman J et al. Arg220 of the PstA protein is required for phosphate transport through the phosphate-specific transport system in *E. coli* but not for alkaline phosphatase repression. J Bacteriol 1988; 170:2283-2286.

29. Cox GB, Webb D, Rosenberg H. Specific amino acid residues in both the PstB and PstC proteins are required for phosphate transport by the *E. coli* Pst system. J Bacteriol 1989; 171:1531-1534.

30. Surin BP, Cox GB, Rosenberg H. Molecular studies on the phosphate-specific transport system of *Escherichia coli*. In: Torriani-Gorini A, Rothman FG, Silver S et al, eds. Phosphate metabolism and cellular regulation in microorganisms. Washington: American Society for Microbiology, 1987:145-149.

31. Muda M, Rao NN, Torriani A. The role of PhoU in phosphate transport and alkaline phosphatase regulation. J Bacteriol 1992; 174:8057-8064.

32. Steed PM, Wanner BL. Use of the *rep* technique for allele replacement to construct mutants with deletions of the *pstSCAB-phoU* operon: evidence

of a new role for the PhoU protein in the phosphate regulon. J Bacteriol 1993; 175:6797-6809.

33. Rao NN, Roberts MF, Torriani A et al. Effect of *glpT* and *glpD* mutations on expression of the *phoA* gene in *Escherichia coli*. J Bacteriol 1993; 175:74-79.

34. Zuckier G, Ingenito E, Torriani A. Pleiotropic effects of alkaline phosphatase regulatory mutations *phoB* and *phoT* on anaerobic growth of and polyphosphate synthesis in *Escherichia coli*. J Bacteriol 1980; 143:934-941.

ARE THE MULTIPLE SIGNAL TRANSDUCTION PATHWAYS OF THE PHO REGULON DUE TO CROSS TALK OR CROSS REGULATION?

Barry L. Wanner, Weihong Jiang, Soo-Ki Kim, Sayaka Yamagata, Andreas Haldimann and Larry L. Daniels

INTRODUCTION

The phosphate (Pho) regulon includes a large number of co-regulated genes whose expression is controlled by the environmental (extracellular) inorganic phosphate (P_i) level. Altogether 11 Pho regulon promoters for a total of 38 different genes have now been characterized in *E. coli* or closely related bacteria (Fig. 15.1). The expression of most of them has been shown to require the transcription factor PhoB that upon phosphorylation activates transcription by binding to Pho Box sequences within the respective promoter region of each of these genes or operons. The *E. coli* Pho regulon currently comprises 31 genes that are arranged in eight unlinked transcriptional units. At least 13 of these genes are also believed to exist in *Salmonella typhimurium*, although 16 others are known to be absent. In addition, *S. typhimurium* has seven Pho regulon genes that are absent in *E. coli*. Most of the corresponding gene products have roles in the use of various phosphorus (P) compounds as sole P sources for growth, while the roles of others are unknown.[1]

Studies on the *E. coli* Pho regulon emanated from early observations[3,4] that the synthesis of bacterial alkaline phosphatase (the *phoA* gene product) increased many hundred-fold under conditions of P_i limitation. Shortly thereafter a large number of regulatory mutants were isolated. A personal account of these early developments is given in chapter 14. We now know that P_i control of the Pho regulon is a paradigm of a signal transduction pathway in which occupancy of a cell surface receptor(s) regulates gene expression in the cytoplasm.[5] This

Fig. 15.1. Sequenced genes and operons belonging to the Pho regulon. Only those genes known to be activated by PhoB are shown. All are present in E. coli except phnR to phnX, which are present only in S. typhimurium. The phoBR, phoE, pstSCAB-phoU and ugpBAECQ loci exist in both E. coli and S. typhimurium. The phoA-psiF and phnC to phnP operons are absent in S. typhimurium. Whether phoH and psiE exist in S. typhimurium has not been determined. Arrows denote mRNA transcripts. Squares indicate locations of Pho Box sequences. Whether phospho-PhoB activates only phnW or both phnW and phnR by binding to the single Pho Box within the phnW to phnR intergenic region has not been established. C-P, carbon-phosphorus; Pn, phosphonate; psi, phosphate-starvation-inducible gene. Additional information and literature citations have been given elsewhere.[1,2]

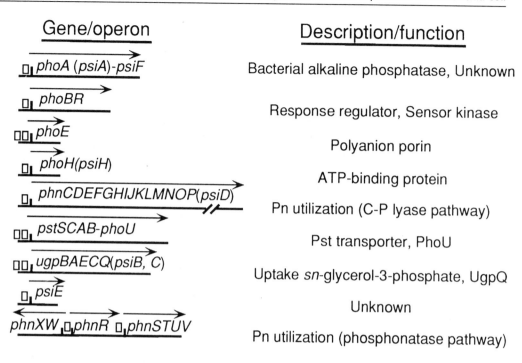

Gene/operon

phoA (psiA)-psiF

phoBR

phoE

phoH(psiH)

phnCDEFGHIJKLMNOP(psiD)

pstSCAB-phoU

ugpBAECQ(psiB, C)

psiE

phnXW   phnR   phnSTUV

Description/function

Bacterial alkaline phosphatase, Unknown

Response regulator, Sensor kinase

Polyanion porin

ATP-binding protein

Pn utilization (C-P lyase pathway)

Pst transporter, PhoU

Uptake *sn*-glycerol-3-phosphate, UgpQ

Unknown

Pn utilization (phosphonatase pathway)

regulation is mediated by the activities of a membrane localized sensor kinase (PhoR) and its partner response regulator (PhoB), a transcriptional activator whose activity is enhanced by phosphorylation. These proteins are members of the large superfamily of two-component regulatory systems that are prevalent in bacteria,[6,7] and that may be involved in signal transduction in eucaryotic cells as well.[8,9]

P_i control of the Pho regulon involves a process of transmembrane signaling in which cells detect environmental P_i and regulate gene expression in response to the extracellular P_i level. This signaling pathway requires in addition to PhoR an intact phosphate-specific transporter (the ABC family PstSCAB system) and a protein called PhoU. It is believed that regulation is brought about via phosphorylation of PhoB by PhoR under conditions of P_i limitation and (presumably) via dephosphorylation of phospho-PhoB by PhoR (perhaps together with a Pst component or PhoU) when P_i is in excess. Further, it is now clear that two additional controls (one requiring the sensor kinase CreC [originally called PhoM] and the other requiring acetyl phosphate) may also lead to phosphorylation of PhoB in vivo, at least in some mutants.[10] Whether these signaling pathways under certain conditions have a role in the regulation of the Pho regulon in wild-type cells has not been established. On the one hand, the activation of the Pho regulon by CreC and acetyl phosphate may be an example of highly efficient in vivo cross talk.[6] On the other hand, activation by these signaling pathways may be indicative of cross regulation that may be important for the overall global control of cell growth and metabolism and for regulatory connections between different signaling pathways.[11]

In this chapter, we will briefly describe how we arrived at our current understanding of each of these Pho regulon controls. We would like to point out at the onset that the regulatory components involved in these controls were identified in genetic studies long before protein phosphorylation was recognized as a gene regulatory mechanism in bacteria. Those genetic studies showed long ago that Pho regulon control was complex because it involved multiple levels of both negative and positive controls. The results of those studies fit extraordinarily well with present concepts about the involvement of signal transduction pathways and protein phosphorylation in Pho regulon control. It is perhaps even more surprising to see in retrospect how well many of the same results also matched with earlier hypotheses later proven to be wrong, at least in regard to the Pho regulon. It will be interesting to see how well our current ideas hold up as we gain further information about this regulatory system in the years to come. Various aspects in this chapter concerning P_i control, transmembrane signal transduction and these multiple controls of the Pho regulon have recently been reviewed elsewhere.[1,5,12,13]

GENES FOR P_i CONTROL OF THE PHO REGULON

Early genetic studies on the regulation of the Pho regulon led to the identification of two loci (originally named R1 and R2), in which mutations resulted in constitutive or uninducible expression of *phoA*. It was especially interesting that mutations of the R1 locus blocked repression, induction or both; mutations of the R2 locus blocked only repression.[14,15] These studies were carried out at a time when the Jacob-Monod operon model of negative control was favored as a means to describe all genetic regulatory mechanisms. Therefore, in order to explain the apparent dual (negative and positive) role of the R1 locus, it was proposed that R1 specified the formation of an endogenous inducer and that this inducer was required as a precursor for synthesis of a cytoplasmic repressor encoded by the R2 locus.[14] Even though this mechanism was later shown to be incorrect, some important interpretations of these early studies remain consistent with current models.

Subsequent studies have revealed that the R1 locus encodes the *phoBR* operon near 9 minutes on the *E. coli* chromosome, that the R2 locus encodes the *pstSCAB-phoU* operon near 83 minutes and that PhoR has a dual regulatory role.[16] As it has been known for some time, the Pho regulon is subject to positive control by the transcription factor PhoB. We now use the terms inhibition and activation to describe Pho regulon control in order to avoid confusion with classic mechanisms of repression and induction that are due to repressor-DNA interactions.[1] Additional early studies of the Pho regulon also established that both inhibition and activation are active processes.[16] Those R1 mutations that blocked both inhibition (repression) and activation (induction) were called R1a mutations; these correspond to null *phoR* mutations. In agreement, PhoR is now thought to activate PhoB by phosphorylation in response to P_i limitation and to inhibit PhoB (presumably by dephosphorylation of phospho-PhoB) when P_i is in excess. Those R1 mutations that blocked only inhibition (repression) were called R1b mutations; these correspond to missense *phoR* mutations that are now called constitutively active *phoR* alleles. A constitutively active PhoR is believed to activate PhoB by phosphorylation even in the absence of a signal for P_i limitation. Those R1 mutations that blocked only activation (induction) were called R1c mutations; these correspond to null *phoB* mutations. A null *phoB* mutant is unable to activate gene expression due to the role of PhoB as the transcriptional activator of the Pho regulon. Mutations of the R2 locus abolished only inhibition (repression); null mutations of any gene of the *pstSCAB-phoU* operon prevent signaling by extracellular P_i.[17]

Evidence for control of the Pho regulon by extracellular P_i was provided by the discovery that mutations of the high affinity PstSCAB transporter mapped to the R2 locus and, like other R2 mutations, resulted in high level activation of the Pho regulon.[18] Further, intracellular P_i levels (ca. 10 mM) are unchanged under conditions of P_i

limitation.[18,19] More importantly, the processes of P_i transport and inhibition can be uncoupled; a site-directed mutation of the channel protein PstA abolishes P_i transport without affecting inhibition of Pho regulon gene expression by extracellular P_i.[20] As expected, this mutation is without effect on activation under conditions of P_i limitation. Mutants with a defect in the PstSCAB system are able to grow using P_i as a P source due to the presence of the low affinity-high velocity P_i transporter, Pit. Mutations blocking only the Pit transporter are without effect on Pho regulon control.[21]

R1a mutants correspond to null *phoR* alleles and abolish both inhibition and activation by PhoR. Yet, these *phoR* mutations result in low level activation of the Pho regulon. Typically, an R1a mutant synthesizes about 30% as much PhoA when P_i is in excess as a wild-type strain makes under conditions of P_i limitation (Table 15.1). Also, the amount synthesized by an R1a mutant is unaffected by P_i.[15] It was later shown that the low level activation observed in an R1a mutant is due to CreC,[22] since this synthesis is abolished by a null mutation in *creC*, which is now known to lie in the *creABCD* operon. CreC acts as an alternative sensor kinase that is capable of efficient phosphorylation of PhoB,[23] at least in the absence of PhoR. Although activation of the Pho regulon due to CreC is unaffected by P_i levels, it is highly regulated by carbon and energy sources.[1] In the past, activation by CreC was thought to result in constitutive synthesis. The effects of carbon and energy sources on activation by CreC were apparently overlooked in earlier studies because they were frequently carried out with strains carrying the *creC510* mutation,[24] which results in constitutive signaling.[1]

TRANSMEMBRANE SIGNALING BY ENVIRONMENTAL P_i

P_i control of the Pho regulon involves two aspects: inhibition and activation, both of which are active processes. The expression of Pho regulon genes is inhibited when P_i, the preferred P source, is in excess; the expression of these genes is activated many hundred-folds under conditions of P_i limitation. Inhibition requires PhoR, all four components of the Pst system (PstS, PstC, PstA and PstB) and a protein called PhoU; activation requires only PhoR and PhoB. Because multiple components are required for P_i inhibition and many of these (in particular those of the Pst system) are known to interact, it is reasonable to suggest that the process of P_i inhibition involves an inhibition complex and that P_i control is therefore dependent on multiple protein-protein interactions within such a complex. Accordingly, activation results from a subtle change in this complex leading to the formation of an activation complex. These complexes are expected to be located in the cytoplasmic membrane, as illustrated in Figure 15.2. The existence of membrane-associated inhibition and activation complexes containing the Pst components, PhoU and PhoR is consistent with features of these components. Both the Pst transporter and PhoR are localized in the cytoplasmic membrane. Furthermore, the membrane-association of PhoR is required

Table 15.1. Effects of **phoB,** **phoR,** **creC** *and* **phoU** *mutations on expression of the Pho regulon*

Genotype[a]	PhoA synthesis[b]		Phenotype
	Excess P_i	Limited P_i	
Wild-type	0.4	172	Inhibited/Activated
phoR (R1a allele)	42.1	47.0	Low level activated
phoR69 (R1b allele)	116	176	High level activated
phoB (R1c allele)	0.1	0.1	Negative
phoU35 (R2 allele)	185	168	High level activated
phoR creC	1.6	1.7	Negative
creC	0.3	185	Inhibited/Activated
phoR69 creC	142	162	High level activated

[a] All mutations except *phoR69* and *phoU35* correspond to null alleles. The *phoR69* and *phoU35* alleles result in the T220N and A147E missense changes,[1] respectively.
[b] Cells were grown in glucose MOPS medium with 2.0 mM (excess) or 0.1 mM (limited) P_i for at least 10 generations before sampling.[16] Units are nmoles of *o*-nitrophenol made per minutes per cell O. D.$_{420}$.

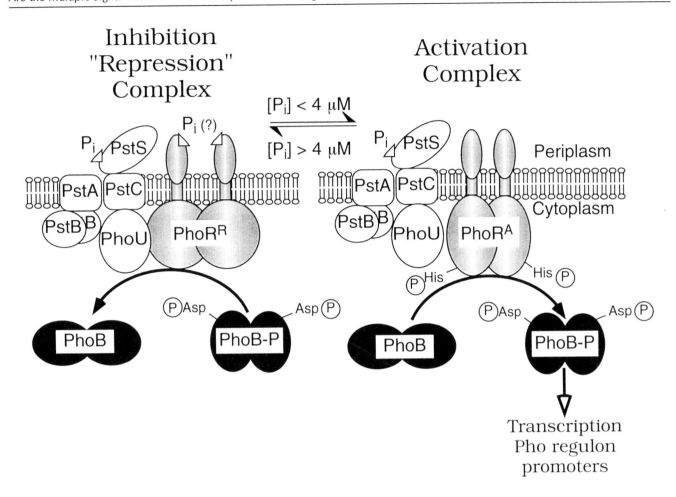

Fig. 15.2. Transmembrane signal transduction by environmental P_i. The P_i binding site on PstS and the hypothetical P_i regulatory site on PhoR are shown as small triangles. To account for the dual role of PhoR in Pho regulon control,[15,22] it was proposed that PhoR exists in two forms. These were named PhoR^A and PhoR^R long before it was understood how they may act.[22] Other versions of the same basic model have been described in detail elsewhere.[1,5,12,13,50] The inhibition complex was originally called the repression complex while no activation complex was indicated.[50] Symbols: PhoB, response regulator; PhoB-P, phospho-PhoB; PhoR^A, PhoR activation (autophosphorylated) form; PhoR^R, PhoR repression (inhibition) form; PhoU, phoU gene product; PstA and PstC, integral membrane channel proteins of Pst transporter; PstB, traffic ATPase of Pst system; PstS, periplasmic P_i binding protein of Pst transporter.

for inhibition. In addition, PhoU appears to be a peripheral cytoplasmic membrane protein, even though it lacks features of a membrane protein. PhoU can be associated with the membrane via an interaction with the Pst system, PhoR or both. Protein-protein interactions among these components can involve interactions between respective cytoplasmic domains, membrane domains (such as membrane-spanning helices) or both.[5]

The inhibition (repression) complex in Figure 15.2 contains all components of the Pst system, PhoU and PhoR, because all of these are required for inhibition. By analogy to other members of the membrane sensor kinase protein family, PhoR is thought to act as the P_i sensor in the process of inhibition. Whether it detects P_i directly via a regulatory site (as shown) or indirectly solely via an interaction with the Pst system or PhoU is unknown. It was previously proposed that a P_i regulatory site may exist on PhoR.[5] The presence of such a site would allow the Pst system to be fully saturated with P_i when in excess and yet to be able to bind P_i for the purpose of P_i uptake without causing inhibition under conditions of P_i limitation. If there is a regulatory site on PhoR, then a Pst mutation can abolish inhibition because P_i saturation of the Pst system can be necessary to facilitate P_i binding to the regulatory site. Such a regulatory site can include residues of the membrane spanning region near the periplasmic face of the membrane (the interfacial membrane layer[25]) because PhoR can have only a few amino acids exposed to the periplasm. Other mechanisms for P_i signaling that do not involve a P_i regulatory site on PhoR have been described elsewhere.[5] Regardless of which component(s) senses P_i, the process of inhibition probably involves the dephosphorylation of phospho-PhoB when P_i is in excess. PhoR^R can act alone or together with PhoU or a Pst component to facilitate the dephosphorylation of phospho-PhoB. Alternatively, PhoU or a Pst component may interact with phospho-PhoB, and PhoR^R can somehow stimulate one of them to cause dephosphorylation. Either way PhoR is absolutely required for the process of P_i inhibition. In the absence of PhoR, two other signaling pathways—involving CreC or acetyl phosphate—lead to high level activation of PhoB.[10] Further, constitutively active *phoR* missense alleles are specifically blocked in the process of inhibition.[1]

The activation complex can consist of PhoR^A alone, or it can contain all components of the inhibition complex in an altered conformation in which PhoR^A is formed. In the former case, P_i limitation can lead to formation of PhoR^A upon its release from the inhibition complex. In the latter case, a subtle change within the complex can lead to formation of PhoR^A. In regard to the overall process of Pho regulon control by P_i, it should be mentioned that P_i limitation also leads to increased amounts of the Pst components, PhoU and PhoR, as well as of PhoB. Therefore, a stoichiometric mechanism can be partly responsible for the interconversion of PhoR^R and PhoR^A. For example, under conditions of P_i limitation the number of PhoR molecules can exceed the number of complexes containing P_i and excess PhoR may be released as PhoR^A.[5] Regardless, none of the Pst components or PhoU is required for activation. Deletions of these genes (like the original R2 mutations) abolish inhibition, resulting in activation by PhoR^A.[17] In any case, PhoR^R is expected to predominate when the extracellular P_i is in excess (greater than about 4 μM) and PhoR^A is expected to predominate under conditions of P_i limitation (less than about 4 μM). Accordingly, PhoR^R can act by facilitating the dephosphorylation of phospho-PhoB while PhoR^A can correspond to autophosphorylated PhoR, a phosphoryl transferase and PhoB kinase. Phospho-PhoB in turn leads to transcriptional activation of each of the Pho regulon promoters.[1]

We purposely described the mechanisms of P_i sensing and of signal transduction across the cytoplasmic membrane in rather general terms above, because the mechanisms involved are still poorly understood. In contrast, we know much more about the biochemistry of transcriptional activation. This understanding has resulted from the discovery that PhoR and PhoB share sequence similarities at the protein level to the respective family members of the superfamily of signaling kinases and partner response regulators in bacteria.[6,7] Like other signaling kinases, a truncated form of PhoR (lacking its N-terminal membrane spanning segments) has been shown to be autophosphorylated by ATP in vitro[26] on His-213[27] and phospho-PhoR has been shown to rapidly transfer the phosphoryl group to Asp-53 of PhoB.[28] Also, even though both PhoB and phospho-PhoB bind DNA in a site-specific manner,[26,29] DNA binding is greatly enhanced by phosphorylation. In this regard, it should be mentioned that PhoB had been shown to act as a transcriptional activator in vitro long before phosphorylation was thought to be necessary for transcriptional activation by PhoB.[30,31] However, in retrospect those extracts activated transcription poorly. Further, as indicated above, it is reasonable to suppose that PhoR (perhaps together with PhoU or a Pst component) facilitates the dephosphorylation of phospho-PhoB, although no such activity has yet been demonstrated.[1]

GENES FOR P_i INDEPENDENT CONTROLS
OF THE Pho REGULON

Three signaling pathways lead to activation of the Pho regulon, all of which lead to activation via phosphorylation of PhoB. One pathway is regulated by P_i and leads to activation by PhoR. The other two are P_i independent. One of these leads to activation by CreC; the other leads to activation by acetyl phosphate.[12] Evidence for activation of the Pho regulon by these P_i independent controls resulted from the isolation of *creC* mutants, the finding of which was entirely accidental. They were uncovered in a search aimed at finding mutants that interfered with localization of PhoA to the periplasm.[32] A null *phoR* mutant was (fortuitously) used in that study in order to facilitate screening. The strain was also *phoA*. Mutants were sought that remained PhoA negative following introduction of a *phoA⁺* gene on a transducing phage. In that study we anticipated finding both regulatory and nonregulatory mutants. We expected regulatory mutants to be mutated in *phoB* and to be identifiable as ones with lesions that mapped in *phoB*. We anticipated that mutants altered in localization may reveal themselves as nonregulatory mutants with lesions that mapped elsewhere. We unexpectedly found a new class of regulatory mutations with lesions in a new locus called *creC*. Our finding of *creC* mutations was especially surprising because no such regulatory mutants had been previously isolated, in spite of the large number of mutants that had been previously characterized. Yet, nearly one-half of our regulatory mutants had lesions in *creC*; the others had lesions in *phoB*. In addition, the genetic characterization of these *creC* mutations gave unexpected results.

CreC was shown to be responsible for activation of PhoB in the absence of PhoR; and activation by CreC was shown to be inhibited by PhoR when P_i is in excess[22] (Fig. 15.3). No *creC* mutation had been previously uncovered because CreC is without an apparent effect on Pho regulon control in the presence of PhoR. This dominant effect of PhoR was responsible for our unexpected results in the genetic characterization of *creC* mutants. We had tested our regulatory mutants for lesions linked to the *phoBR* region in two ways; we used them as donors and we used them as recipients, in order to map the mutations by P1 transduction. Because we obtained no mutant transductants when using the *creC* mutants as donors of the *phoBR* region, the mutants appeared to have mutations unlinked to *phoB*. Yet, we obtained wild-type transductants when using the mutants as recipients of the *phoBR* operon. Because the latter crosses were done using a *phoR⁺* donor, we suspected that PhoR and CreC were able to activate PhoB alone and that they acted in an analogous manner. This hypothesis

Fig. 15.3. An early model for activation of the Pho regulon by PhoR and CreC. Regulation of the Pho regulon involves inhibition by PhoRᴿ when P_i is in excess and activation by PhoRᴬ under conditions of P_i limitation. In the absence of PhoR, expression of the Pho regulon results from activation by CreC. This model was originally proposed in order to explain the dual role of PhoR and how CreC is able to substitute for only one of these roles.[22] PhoRᴿ was shown to inhibit phoA expression because at that time it was believed that PhoR may directly cause inhibition, via an association with PhoB.[38]

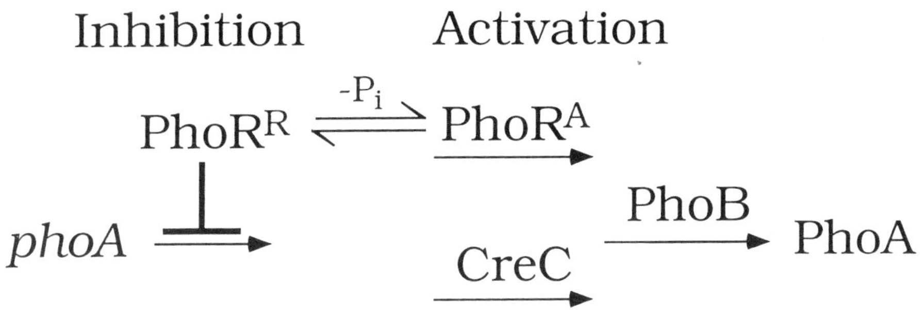

turned out to be correct. It is important to realize that the isolation of *creC* mutations or the similar roles of PhoR and CreC may have gone unrecognized, if we had not carried out both sets of crosses.

Accordingly, one pathway for activation of the Pho regulon requires PhoR and PhoB and another requires CreC and PhoB (Fig. 15.3). The expression of the Pho regulon is inhibited by PhoR when P_i is in excess and activated by PhoR under conditions of P_i limitation. Both of these processes are abolished by a null *phoR* allele (an R1a mutation) that leads to low level expression (Table 15.1, lines 1 and 2). In contrast, the *phoR69* allele (the original R1b mutation[15]) essentially abolishes P_i inhibition and leads to high level activation (line 3). Activation due to this mutation is increased somewhat upon P_i limitation. Under all conditions, activation requires PhoB because expression is completely abolished by a null *phoB* allele (an R1c mutation;[14] line 4). Also, the *phoU35* allele (an R2 mutation) results in high level activation (line 5). The low level expression of a null *phoR* mutant results from activation by CreC because it is abolished in a *phoR creC* mutant (lines 2 and 6). Activation by CreC is unaffected by P_i. Further, there is no apparent effect due to CreC in the presence of PhoR (line 7). Apparently, the *phoR69* allele produces a constitutively active gene product because it leads to high level activation even in the absence of CreC (line 8).

The finding of R1a and R1b mutations provided the first evidence of gene regulation by a positive control mechanism,[15,33] although it was many years later before the first convincing evidence of positive control was obtained (in the arabinose system,[34,35] as reviewed elsewhere[36]). Ironically, the original finding of *phoB* (R1c) mutations was initially interpreted in terms of a model for negative control, in which the R1 locus was proposed to specify the synthesis of an endogenous inducer.[14] It was more than ten years later when new *phoB* mutations were isolated that the R1 locus was shown to encode two gene products (PhoB and PhoR) which were proposed to be involved in positive control of the Pho regulon.[37,38] One criteria for determining whether a regulatory system is subject to negative or positive control involves the characterization of deletions of the regulatory gene. Evidence of positive control in the Pho regulon resulted from the isolation of Δ*phoB* mutants that blocked activation of *phoA*.[39] At this time, the concept of positive control had also been widely accepted. Studies carried out using various *phoB*, *phoR* and *pstSCAB-phoU* mutations showed that *phoA* is part of a regulon in which the synthesis of multiple gene products is co-regulated by these loci.[38,40]

Acetyl phosphate was found to activate the Pho regulon in the absence of PhoR or CreC.[10] This finding resulted from genetic studies aimed at demonstrating that PhoR and CreC are involved in positive control. Under most growth conditions, acetyl phosphate is synthesized from acetyl CoA and P_i by phosphotransacetylase (Pta) and acetyl phosphate and ADP are converted to acetate and ATP by acetate kinase (AckA; Fig. 15.4). Second-site revertants of *phoR creC* mutants were isolated in order to determine how PhoR and CreC activate PhoB. It was surprising that *phoR creC* mutants carrying missense, nonsense or deletion mutations gave rise to frequent pseudorevertants.[41] Many of these grew extremely poorly and were difficult to study; a few auxotrophs were also isolated.[5,13] Many years later (when the *ackA* sequence was deposited in GenBank in 1989), we discovered that some of these second-site revertants were mutated in the structural gene for acetate kinase (*ackA*).[10] The *ackA* mutants grew extremely poorly and gave

A) The phosphotransacetylase (Pta)-acetate kinase (AckA) pathway

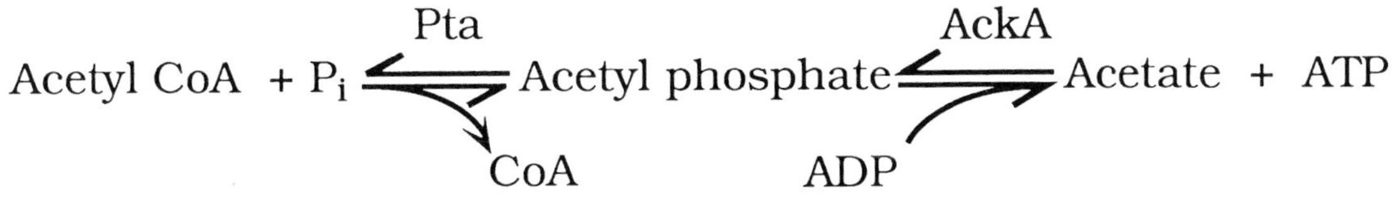

$$\text{Acetyl CoA} + P_i \underset{\text{Pta}}{\overset{}{\rightleftharpoons}} \text{Acetyl phosphate} \underset{\text{AckA}}{\overset{}{\rightleftharpoons}} \text{Acetate} + \text{ATP}$$

CoA   ADP

B) Mutational effects on activation of the Pho regulon in *phoR creC* mutants

Genotype	Carbon source (PhoA synthesis, sp act)		
	Glucose	Pyruvate	Acetate
Wild-type	0.2	104	0.6
AckA⁻	422	409	1.4
Pta⁻	0.2	0.2	43.7
Δ(*ackA pta*)	0.2	0.2	0.3

rise to compensatory mutants, many of which had lesions in the adjacent gene (*pta*) encoding phosphotransacetylase. These results pointed to an involvement of acetyl phosphate in activation of the Pho regulon.[10]

We were able to show that the regulatory consequences of *ackA* and *pta* mutations resulted from effects on acetyl phosphate synthesis because the Pta-AckA pathway is freely reversible in vivo (Fig. 15.4A). It operates in the direction of ATP synthesis during growth on glucose or pyruvate and in the direction of acetyl CoA synthesis during growth on acetate. The Pta-AckA pathway is nonessential during growth on glucose, pyruvate or acetate. An alternative pathway for acetate metabolism leads to the formation of acetyl CoA via acetyl CoA synthetase, whose synthesis is inducible by acetate. Accordingly, we tested for effects of *ackA* and *pta* mutations on activation of the Pho regulon in the absence of PhoR and CreC (Fig. 15.4B).[10] In an otherwise wild-type *phoR creC* mutant, genes of the Pho regulon are expressed at a low basal level during growth on glucose or acetate as sole carbon sources (Fig. 15.4B). In contrast, the expression of the Pho regulon is activated about 500-fold during growth on pyruvate. These results are consistent with activation due to acetyl phosphate. During growth on glucose or acetate, acetyl phosphate levels are expected to be low. Acetyl phosphate levels are expected to be low on glucose because during growth on glucose, acetyl CoA primarily enters biosynthesis and only a small amount enters the Pta-AckA pathway. Acetyl phosphate levels are expected to be low on acetate because acetate is an energy poor carbon source. In contrast, acetyl phosphate levels are expected to be high on pyruvate because pyruvate is an energy rich carbon source. Both acetyl phosphate and acetyl CoA levels had been previously shown to be greatly elevated during growth on pyruvate.[42] Further, an *ackA* mutation results in activation of the Pho regulon during growth on glucose; it also leads to an enhanced activation during growth on pyruvate. These effects are attributable to accumulation of acetyl phosphate because under these conditions an *ackA* mutation prevents its breakdown. Accordingly, no effect of an *ackA* mutation is expected during growth on acetate. In agreement, only a small effect is observed on acetate. A

Fig. 15.4. Evidence for activation of the Pho regulon by acetyl phosphate. (A) The Pta-AckA pathway. This pathway operates in the direction of acetate and ATP synthesis during growth on glucose or pyruvate and in the reverse direction during growth on acetate. (B) Mutational effects on activation of the Pho regulon in phoR creC mutants. The amounts of PhoA specific activity are given for phoR creC mutants that are otherwise wild-type, ackA, pta or Δ(ackA pta). Strains were assayed following growth on glucose, pyruvate or acetate MOPS medium containing excess P_i.[10] All effects are due to activation of PhoB because synthesis is abolished by a null phoB mutation. All effects are also inhibited by PhoR in the presence of excess P_i (data not shown).

small effect can be due to low level accumulation of acetyl phosphate being synthesized from acetyl CoA. In addition, a *pta* mutation has no effect during growth on glucose, whereas it abolishes activation during growth on pyruvate. This is expected because a *pta* mutation blocks the synthesis of acetyl phosphate under these conditions. Compelling evidence that activation is due to acetyl phosphate comes from the effect of a *pta* mutation during growth on acetate. Under these conditions, a *pta* mutation is expected to result in accumulation of acetyl phosphate by preventing its breakdown. As predicted, a *pta* mutation results in activation of the Pho regulon during growth on acetate, whereas a Δ(*ackA pta*) mutation abolishes activation under all conditions.

ACTIVATION BY CReC AND ACETYL PHOSPHATE

The finding that PhoR leads to activation by phosphorylation of PhoB originated from the observation that PhoB and PhoR belong to families of partner proteins (called two-component regulatory systems) that share sequence similarities at the protein level to other members of the same family.[6,43] One family includes PhoB and the other includes PhoR. The PhoB family has highly conserved response regulatory or receiver domains; the PhoR family has highly conserved sensory or transmitter domains.[7] The finding of these sequence similarities was especially important because in nitrogen control the activation of the PhoB homolog NtrC (NR$_I$) by the PhoR homolog NtrB (NR$_{II}$) was shown to result from phosphorylation of NtrC by NtrB, as reviewed in chapter 13. It was later shown that PhoB is phosphorylated by PhoR and that phospho-PhoB is an even better transcriptional activator than PhoB.[26]

CreC was also shown to belong to the sensor kinase protein family. This was especially significant because it suggested that PhoR and CreC were likely to activate PhoB via a common biochemical mechanism. Hence, activation of the Pho regulon by CreC may result from cross talk to PhoB.[6] In agreement, CreC was shown to phosphorylate PhoB as well as its partner protein CreB (formerly called PhoM-Orf2).[23] Accordingly, CreC and CreB comprise a two-component regulatory system that probably controls the expression of an unknown set of target genes.

CreC and CreB are encoded together with CreA and CreD in the *creABCD* operon (Fig. 15.5).[1] CreA is a periplasmic protein and CreD is an inner membrane protein, of unknown functions. The gene organization near the *creABCD* operon is interesting. In particular, the location of *robA* is suggestive of regulatory interactions between the *creABCD* operon and *robA*, or their gene products. RobA was isolated as an *oriC* DNA-binding protein.[44] In spite of this, its in vivo role is uncertain. Clearly, RobA is nonessential since extensive deletions of the *robA-creABCD* region exist.[41] Also, a *robA* insertion has no striking phenotype.[44] RobA is probably a regulatory protein. It has a highly conserved helix-turn-helix motif as well as other strong sequence similarities in common with a number of regulatory proteins, including: AdaA, MarA, MelR, RhaS, SoxS and TetD[44] (data not shown). As the synthesis of many of these is regulated by a gene product(s) encoded by an adjacent, divergently transcribed gene or operon (as described for SoxR control of SoxS in chapter 21). CreC and CreB may regulate *robA* expression. Curiously, *arcA* coding for the response regulator of the ArcB-ArcA modulon (chapter 18) lies immediately downstream of the *creABCD* operon, although no regulatory connection is inferred by this gene arrangement.

It was originally thought that CreC-mediated activation led to constitutive, i.e., unregulated, expression of the Pho regulon. This turned out to be incorrect. Instead, activation resulting from CreC is highly regulated, especially by the carbon source.[1,45] The constitutivity observed in earlier studies was due to the presence of the *creC510* (formerly called *pho-510*) mutation in strains used in those studies.[24,46] The *creC510* allele results in a R77P change within a CreC domain expected to lie within the periplasm (Fig. 15.5B) and, on the basis of its phenotype, probably results in constitutive signaling. Unlike PhoR, CreC has a large periplasmic domain and may therefore (like many other sensor proteins containing large periplasmic domains) respond to an extracellular signal by binding a ligand(s)

A. Gene organization near *creABCD* operon

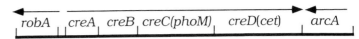

B. Proposed topology of sensor kinase CreC

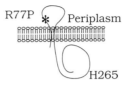

C. The *ackA-pta* operon

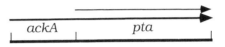

in the periplasm. Accordingly, activation of the Pho regulon by CreC is expected to occur in response to this signal. Yet, in spite of its topology, CreC appears to respond to a cellular metabolite. This is because activation by CreC is highly regulated by the carbon and energy source, even though CreC does not appear to respond to a particular carbohydrate or growth condition per se. Instead, many different carbohydrates (especially glucose) lead to strong activation of the Pho regulon via CreC. These and other results[10,24,45] imply that CreC responds to a catabolite that is a normal metabolic intermediate whose level changes under various growth conditions. Since CreC may respond to an extracellular ligand, such a metabolite may be secreted when made in large amounts. However, there is no direct evidence in support of this notion.

The finding that acetyl phosphate leads to activation of the Pho regulon indicated that acetyl phosphate leads to the phosphorylation of PhoB.[10,11] This is because phosphorylation is the only mechanism known for activation of PhoB. Two ways of how acetyl phosphate can activate PhoB were considered.[10] It was suggested that acetyl phosphate leads to activation of an unknown sensor (presumably a PhoR or CreC homolog) that in turn activates PhoB by phosphorylation. It was also suggested that acetyl phosphate can directly activate PhoB by acting as a chemical phosphorylating agent (which is equivalent to acetyl phosphate acting as a substrate if PhoB is an autokinase). No sensor for acetyl phosphate is yet known. Also, our attempts to find mutants lacking a putative sensor were unsuccessful. Furthermore, the subsequent findings that acetyl phosphate acts as a chemical phosphorylating agent in the in vitro autophosphorylation of other response regulators as well as of PhoB provide evidence in favor of the latter possibility.[1] Yet, it should be pointed out that in vitro studies alone do not prove how acetyl phosphate acts in vivo. In this regard, the concentrations of acetyl phosphate used in those in vitro studies do not appear to be physiological ones. Also, our inability to find mutants lacking an acetyl phosphate sensor does not show that none exist. Our failure to identify mutations of a sensor kinase that block activation of PhoB by acetyl phosphate may instead indicate the presence of multiple signaling kinases capable of activating PhoB under

Fig. 15.5. Genes involved in P_i independent controls of Pho regulon. (A) Gene organization near creABCD operon. Acronyms: robA, right oriC binding protein;[44] creABCD, catabolite regulatory operon;[47] arcA, aerobic respiration control.[48] (B) Proposed topology of the sensor kinase CreC. The site of the creC510 (R77P) mutation and probable site of autophosphorylation (H265) are shown. Strains carrying the creC510 mutation probably also carry a mutant form of robA, the robA1 (E15K) allele.[1] (C) The ackA-pta operon. Two promoters are indicated to account for the lack of polarity due to ackA insertions on pta expression.[1,10] Measurements of mRNA synthesis indicate that ackA and pta are co-transcribed from a strong promoter preceding ackA and that pta is also transcribed from a weaker promoter preceding pta.[49]

those conditions, or that mutations of the hypothetical acetyl phosphate sensor may be deleterious. Therefore, whether acetyl phosphate acts via an unknown sensor kinase or as a direct phosphorylating agent in vivo remains an open question. Importantly, acetyl phosphate levels are subject to considerable variation in vivo, making acetyl phosphate an attractive candidate as an effector molecule,[10] regardless of how it acts.

CROSS TALK, CROSS REGULATION AND A HYPOTHESIS

The terms cross talk and cross regulation have been used in order to describe two different kinds of interactions between sensor kinases and response regulators.[11] Because these proteins share sequence similarities at the protein level with other members of the same family, sensors and regulators are probably structurally and functionally similar to other sensors and regulators, respectively. Accordingly, cross reactivities can result in biochemical reactions in which sensors of similar sequence are able to phosphorylate noncognate regulators.

Cross talk refers to those interactions that are believed to be due to nonspecific interactions or noise and that are unlikely to be of physiological importance.[11] Interactions indicative of cross talk are especially likely to occur among noncognate proteins in those cases where one of the interacting proteins is present in abnormal abundance. A few examples of interactions likely to be due to cross talk were cited previously. Numerous additional examples have now been found.

The term cross regulation was adopted in order to distinguish those interactions that are likely to be of biological significance from those interactions due to cross talk.[11] Cross regulation refers only to those instances in which significant regulatory effects occur when the interacting proteins are present in normal amounts. The term cross regulation was also intended as a more general term; it refers to any control of a response regulator of one two-component regulatory system by a different regulatory system. Yet, cross regulation may always involve the phosphorylation (or dephosphorylation) of a response regulator, because these are the only mechanisms known to affect the activity of a response regulator. Cross regulation can involve the phosphorylation of a response regulator by a noncognate sensor kinase (or a chemical phosphorylating agent such as acetyl phosphate); cross regulation can involve a different covalent modification (such as an adenylylation or acetylation); or cross regulation can involve binding of an effector molecule.

A hypothesis was presented that the Pho regulon is subject to cross regulation by CreC and acetyl phosphate, in addition to its normal P_i control by PhoR.[11] Cross regulation by these systems is seen under normal growth conditions in response to signals thought to lead to activation of CreC or increased levels of acetyl phosphate, respectively. Also, regulatory effects due to CreC or acetyl phosphate are observed when each of the respective genes is present in single copy and expressed from its normal promoter. Activations by CreC and acetyl phosphate are also independent of each other, although additive effects are apparent.[10] In spite of this, cross regulation by these controls is observed only in null *phoR* mutants. Activation due to all signaling pathways is inhibited by PhoR when P_i is in excess (Fig. 15.6). There is also no apparent requirement for CreC or acetyl phosphate during P_i limitation, because under these conditions PhoR alone is capable of high level activation of the Pho regulon. Therefore, it is uncertain what role, if any, CreC or acetyl phosphate has in the presence of PhoR. In order for them to have a role in Pho regulon control in wild-type cells, a mechanism must exist for preventing inhibition by PhoR in

A. Inhibition in presence of excess P_i

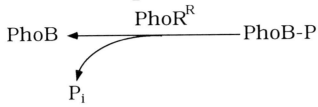

B. Inhibition of activation due to CreC

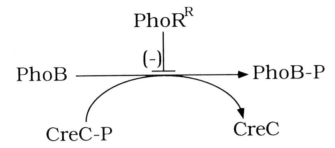

C. Inhibition of activation due to acetyl phosphate

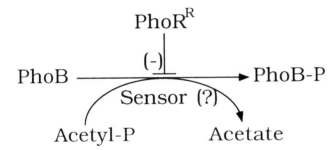

the presence of P_i. Some ways in which CreC or acetyl phosphate can lead to activation of the Pho regulon in the presence of PhoR and excess P_i are considered in the next section.

IS THERE EVIDENCE FOR CROSS REGULATION?

The Pho regulon provides the most compelling evidence of cross regulation in the control of a two-component regulatory system, where cross regulation has been proposed to be an important mechanism for global control of PhoB.[11] It is likely that to maximize growth yield or rate a cell be able to coordinate many diverse branches of metabolism. This coordination is likely to involve many transcriptional controls, as well as other controls. It is also likely to involve many input signals. Different genes, operons or regulons are subject to specific (individual) controls. They can also share controls to create an overlapping network in order to coordinate particular metabolic processes and eventually for global regulation. In the case of the Pho regulon, cross regulation can provide a regulatory coupling(s) between Pho regulon gene expression and central pathways of carbon and energy metabolism, many of which are also formally pathways of P_i metabolism. Accordingly, activation of PhoB by CreC and acetyl phosphate can be a means for cross regulation of Pho regulon promoters by these central pathways.

Fig. 15.6. Inhibition of PhoB activation by PhoR. (A) Inhibition in presence of excess P_i. Inhibition by PhoR requires excess P_i and in addition a functional Pst transporter and PhoU (not shown). Under these conditions PhoR^R can cause inhibition by dephosphorylation of phospho-PhoB, as depicted. Alternatively, PhoR^R can (perhaps via a tight association with PhoB) cause inhibition by preventing phosphotransfer to PhoB, similar to how inhibition is depicted in B and C. (B) Inhibition of activation due to CreC. Activation due to CreC is inhibited by PhoR^R in the presence of excess P_i. (C) Inhibition of activation due to acetyl phosphate. Likewise, activation due to acetyl phosphate is inhibited by PhoR^R. Whether acetyl phosphate leads directly to the phosphorylation of PhoB or acts via a sensor kinase is unknown. Results leading to these interpretations have been summarized elsewhere.[1,5,10,11,13]

Further, cross regulation can be important only under certain growth conditions.

The primary reason for proposing that cross regulation by protein phosphorylation of response regulators can be a global regulatory mechanism is a teleological one,[11] in accordance with the connotation for teleonomy. It was based on the discovery that conditions resulting in increased levels of acetyl phosphate lead to activation of PhoB.[10] Acetyl phosphate is an intermediate of the Pta-AckA pathway, a pathway of carbon, energy and P_i metabolism. Importantly, the overall process of P_i assimilation involves two steps. The first step involves uptake of environmental P_i into the cell; the second step involves the incorporation of intracellular P_i into ATP, the primary phosphoryl donor in metabolism (Fig. 15.7A). The finding that increased levels of acetyl phosphate lead to activation of the Pho regulon indicated that a regulatory link exists between pathways of substrate-level phosphorylation and activation of PhoB. Hence, cross regulation involving acetyl phosphate can result in phosphorylation of PhoB due to an increased level of acetyl phosphate. Since the control of a pathway is likely to respond to its end product, cross regulation involving acetyl phosphate can actually detect the ATP-to-acetyl phosphate ratio, with a lowered ratio leading to activation of PhoB. Similarly, cross regulation involving CreC can result in activation of PhoB in response to a signal from a different central pathway. While it is unknown what signal leads to activation by CreC, it is reasonable to suppose that this signal is coupled to a central pathway, on the basis of the effects various carbon and energy sources have on activation by CreC (unpublished data).[10,45]

Accordingly, the Pho regulon is subject to three controls,[10,11] each of which leads to its activation by phosphorylation of PhoB (Fig. 15.7C). One of these is its normal control by P_i; both of the others involve cross regulation. Its control by P_i responds to extracellular P_i in a manner that is regulated by the PstSCAB system and PhoU, leading to phosphorylation of PhoB by PhoR. One system for cross regulation responds to an unknown catabolite in a manner that is coupled to a central pathway, leading to phosphorylation of PhoB by CreC. The other system for cross regulation responds to ATP synthesis in a manner that is regulated by the Pta-AckA pathway, leading to activation of PhoB by acetyl phosphate. In this context, it is unimportant whether the in vivo activation of PhoB by acetyl phosphate is due to a direct phosphorylation reaction or an unknown sensor kinase. Accordingly, the normal regulation of the Pho regulon is coupled to the first step of P_i metabolism (P_i uptake) while both forms of cross regulation of the Pho regulon are coupled to subsequent steps in P_i metabolism (incorporation of P_i into ATP).

The expression of the Pho regulon is subject to inhibition by PhoR when P_i is in excess. How then does cross regulation by CreC or acetyl phosphate occur in the presence of PhoR? Two simple ways for how this can come about in wild-type cells have been previously mentioned.[11] There are probably more. One way involves PhoU. Under normal conditions in the presence of excess P_i, PhoU may interfere with phosphorylation of PhoB by PhoR but not with the dephosphorylation of phospho-PhoB by PhoR. If under certain conditions PhoU (perhaps via an association with PhoR) were to interfere with both PhoR reactions, phosphorylation of PhoB can result from cross regulation by CreC or acetyl phosphate. Another way for how cross regulation can act in wild-type cells involves the destruction of PhoR by proteolysis. Under conditions of P_i inhibition, PhoR is expected to be made in

A. Pathways of P assimilation from environmental P_i

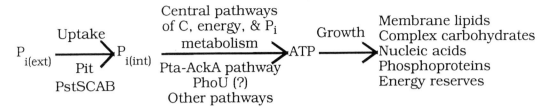

B. Pta-AckA pathway for incorporation of P_i into ATP

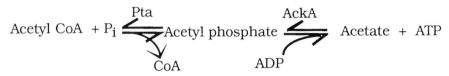

C. Multiple controls of Pho regulon and cross regulation

Signal	Control	Sensor or effector molecule	Regulator
$P_{i(ext)}$	PstSCAB, PhoU	PhoR	
Unknown catabolite	Central pathway	CreC (PhoM)	PhoB
ATP synthesis	Pta-AckA pathway	Acetyl phosphate — Unknown sensor(s) ?	

very low amounts. These small amounts may be susceptible to proteolysis under certain conditions. Under these conditions, cross regulation resulting in phosphorylation of PhoB by CreC or acetyl phosphate is expected to restore the levels of PhoR by increasing PhoR synthesis. Accordingly, cross regulation can be especially important only under certain growth conditions, or during shifts from one condition to another.

Further, it should also be mentioned that most studies on Pho regulon control by CreC and acetyl phosphate have concerned only the expression of *phoA*. Cross regulation of a different promoter can occur in the presence of PhoR with excess P_i. If different amounts of phospho-PhoB are required for transcriptional activation of individual promoters, then preferential effects on a particular promoter(s) can result. It has also not been tested whether cross regulation affects basal level expression of Pho regulon promoters. It may be important to regulate the basal level of Pho regulon promoters and cross regulation may control basal level gene expression.

OVERVIEW AND PROSPECTS FOR FUTURE STUDIES

In summary, we have discussed how genetic studies showed the Pho regulon to be subject to both multiple positive controls (by PhoB and PhoR) and negative controls (by PhoR, the Pst system and PhoU) and how these controls fit into current ideas about the role of signaling pathways and protein phosphorylation in transcriptional activation of the Pho regulon. It is now generally accepted that activation under conditions of P_i limitation is due to phosphorylation of PhoB by PhoR, resulting in transcriptional activation by phospho-PhoB. Although it

Fig. 15.7. P assimilation and control of the Pho regulon. (A) Pathways of P assimilation from environmental P_i.[1,47] The primary systems for P_i uptake are the low affinity-high velocity P_i transporter Pit and the high affinity-low velocity, binding protein-dependent P_i specific transporter PstSCAB. Principal routes for incorporation of P_i into ATP involve glyceraldehyde-3-phosphate dehydrogenase and phosphoglycerate kinase in glycolysis, succinyl-coenzyme A synthetase in the TCA cycle, the F_1F_0 ATP synthase, the Pta-AckA pathway and others. In addition, PhoU may have a role in intracellular P_i metabolism.[17] Symbols: PhoU, phoU gene product; $P_{i(ext)}$, extracellular P_i; $P_i^{(int)}$, intracellular P_i. (B) Pta-AckA pathway for incorporation of P_i into ATP.[10] Symbols: AckA, acetate kinase; Pta, phosphotransacetylase. (C) Multiple controls of Pho regulon and cross regulation.[11] The solid arrow indicates normal control by P_i, the PstSCAB system, PhoU and the sensor kinase PhoR. Dashed arrows indicate cross regulation by an unknown catabolite (presumably) from a central pathway and the sensor kinase CreC or by a signal for ATP synthesis via the Pta-AckA pathway and acetyl phosphate, which may involve an unknown sensor kinase.

is also accepted that inhibition under conditions of excess P_i involves PhoR, the Pst system and PhoU, very little is known about how inhibition is brought about (probably resulting in dephosphorylation of phospho-PhoB). How is P_i detected? Is P_i detected solely by the Pst system? Does a P_i regulatory site exist on PhoR? How is the signal transmitted across the membrane? Do PhoR, the Pst system and PhoU interact? If so, how do they interact? What is the role of PhoU? How is phospho-PhoB dephosphorylated?

In addition, we have discussed how genetic studies showed the Pho regulon to be subject to two additional controls (by CreC and acetyl phosphate) that lead to its activation in the absence of PhoR. While the discovery of these signaling pathways provided insights into the mechanism of PhoB activation by phosphorylation, it remains uncertain whether CreC or acetyl phosphate have a role in Pho regulon control under certain conditions in the presence of PhoR. The normal role of CreC and acetyl phosphate in gene regulation are also unknown. CreC is the sensor kinase of the CreC-CreB two-component regulatory system. What does CreC sense and what are the target genes of this system? Does acetyl phosphate act as an effector molecule? Does a sensor kinase detect acetyl phosphate or does sensor (or PhoB) use acetyl phosphate as a substrate? If a sensor is involved, what is the role of that sensor(s)?

Searching for answers to these and other questions is likely to continue to fascinate investigators in the future. It is hoped that we will then gain a much better understanding of Pho regulon control as well as the role, if any, of its apparent cross regulation by CreC and acetyl phosphate. Of course, it may turn out that activation by CreC and acetyl phosphate are shown to be solely due to cross talk resulting from nonspecific interactions among noncognate proteins. Nevertheless, studying them may reveal important basic information on how a sensor kinase of one two-component regulatory system is able to recognize and interact specifically with a regulator of that system, in order to avoid interactions between noncognate sensor and regulator proteins.

ACKNOWLEDGMENTS

This laboratory is supported by NIH grant GM35392 and NSF grant MCB 9405929.

REFERENCES

1. Wanner BL. Phosphorus assimilation and control of the phosphate regulon. In: Neidhardt FC, Curtiss RI, Gross CA et al, eds. *Escherichia coli* and *Salmonella typhimurium* cellular and molecular biology. 2nd ed, chapter 84. Washington, DC: Am Soc Microbiol,1996; (in press).

2. Jiang W, Metcalf WW, Leeks KS et al. Molecular cloning, mapping and regulation of Pho regulon genes for phosphonate breakdown by the phosphonatase pathway of *Salmonella typhimurium* LT2. J Bacteriol 1995; 177:6411-26.

3. Horiuchi T, Horiuchi S, Mizuno D. A possible negative feedback phenomenon controlling formation of alkaline phosphomonoesterase in *Escherichia coli*. Nature 1959; 183:1529-30.

4. Torriani A. Influence of inorganic phosphate in the formation of phosphatases *by Escherichia coli*. Biochim Biophys Acta 1960; 38:460-9.

5. Wanner BL. Signal transduction and cross regulation in the *Escherichia coli* phosphate regulon involving the phosphate sensor PhoR, the catabolite regulatory sensor CreC, and acetyl phosphate. In: Hoch JA, Silhavy

TJ, eds. Two-component signal tranduction. Washington, DC: Am Soc Microbiol, 1995:203-21.

6. Ronson CW, Nixon BT, Ausubel FM. Conserved domains in bacterial regulatory proteins that respond to environmental stimuli. Cell 1987; 49:579-81.

7. Parkinson JS, Kofoid EC. Communication modules in bacterial signaling proteins. Annu Rev Genet 1992; 26:71-112.

8. Swanson RV, Alex LA, Simon MI. Histidine and aspartate phosphorylation: Two-component systems and the limits of homology. Trends Biochem Sci 1994; 19:485-90.

9. Herskowitz I. MAP kinase pathways in yeast: for mating and more. Cell 1995; 80:187-97.

10. Wanner BL, Wilmes-Riesenberg MR. Involvement of phosphotransacetylase, acetate kinase, and acetyl phosphate synthesis in the control of the phosphate regulon in *Escherichia coli*. J Bacteriol 1992; 174:2124-30.

11. Wanner BL. Minireview. Is cross regulation by phosphorylation of two-component response regulator proteins important in bacteria? J Bacteriol 1992; 174:2053-8.

12. Wanner BL. Gene regulation by phosphate in enteric bacteria. J Cell Biochem 1993; 51:47-54.

13. Wanner BL. Multiple controls of the *Escherichia coli* Pho regulon by the P_i sensor PhoR, the catabolite regulatory sensor CreC, and acetyl phosphate. In: Torriani-Gorini A, Yagil E, Silver S, eds. Phosphate in Microorganisms, Cellular and Molecular Biology. Washington, DC: Am Soc Microbiol, 1994:13-21.

14. Garen A, Echols H. Genetic control of induction of alkaline phosphatase synthesis in *E. coli*. Proc Natl Acad Sci USA 1962; 48:1398-402.

15. Garen A, Echols H. Properties of two regulating genes for alkaline phosphatase. J Bacteriol 1962; 83:297-300.

16. Wanner BL. Phosphate regulation of gene expression in *Escherichia coli*. In: Neidhardt FC, Ingraham J, Low KB et al. eds. *Escherichia coli* and *Salmonella typhimurium* cellular and molecular biology, Volume 2. Washington, DC: Am Soc Microbiol, 1987:1326-33.

17. Steed PM, Wanner BL. Use of the *rep* technique for allele replacement to construct mutants with deletions of the *pstSCAB-phoU* operon: evidence of a new role for the PhoU protein in the phosphate regulon. J Bacteriol 1993; 175:6797-809.

18. Willsky GR, Bennett RL, Malamy MH. Inorganic phosphate transport in *Escherichia coli*: involvement of two genes which play a role in alkaline phosphatase regulation. J Bacteriol 1973; 113:529-39.

19. Shulman RG, Brown TR, Ugurbil K et al. Cellular applications of ^{31}P and ^{13}C nuclear magnetic resonance. Science 1979; 205:160-6.

20. Cox GB, Webb D, Godovac-Zimmermann J et al. Arg-220 of the PstA protein is required for phosphate transport through the phosphate-specific transport system in *Escherichia coli* but not for alkaline phosphatase repression. J Bacteriol 1988; 170:2283-6.

21. Willsky GR, Malamy MH. Characterization of two genetically separable inorganic phosphate transport system in *Escherichia coli*. J Bacteriol 1980; 144:356-65.

22. Wanner BL, Latterell P. Mutants affected in alkaline phosphatase expression: evidence for multiple positive regulators of the phosphate regulon in *Escherichia coli*. Genetics 1980; 96:242-66.

23. Amemura M, Makino K, Shinagawa H et al. Cross talk to the phosphate regulon of *Escherichia coli* by PhoM protein: PhoM is a histidine protein kinase and catalyzes phosphorylation of PhoB and PhoM-open reading frame 2. J Bacteriol 1990; 172:6300-7.

24. Wanner BL. Control of *phoR*-dependent bacterial alkaline phosphatase clonal variation by the *phoM* region. J Bacteriol 1987; 169:900-3.

25. White SH, Wimley WC. Peptides in bilayers: structural and thermodynamic basis for partitioning and folding. Curr Opin Struct Biol 1994; 4:79-86.

26. Makino K, Shinagawa H, Amemura M et al. Signal transduction in the phosphate regulon of *Escherichia coli* involves phosphotransfer between PhoR and PhoB proteins. J Mol Biol 1989; 210:551-9.

27. Shinagawa H, Makino K, Yamada M et al. Signal transduction in the phosphate regulon of *Escherichia coli*: dual functions of PhoR as a protein kinase and a protein phosphatase. In: Torriani-Gorini A, Yagil E, Silver S, eds. Phosphate in Microorganisms. Washington, DC: Am. Soc Microbiol, 1994:285-9.

28. Makino K, Amemura M, Kim S-K et al. Mechanism of transcriptional activation of the phosphate regulon in *Escherichia coli*. In: Torriani-Gorini A, Yagil E, Silver S, eds. Phosphate in Microorganisms, Cellular and Molecular Biology. Washington, DC: Am Soc Microbiol, 1994:5-12.

29. Makino K, Shinagawa H, Amemura M et al. Regulation of the phosphate regulon of *Escherichia coli*: activation of *pstS* transcription by PhoB protein in vitro. J Mol Biol 1988; 203:85-95.

30. Inouye H, Pratt C, Beckwith J et al. Alkaline phosphatase synthesis in a cell-free system using DNA and RNA templates. J Mol Biol 1977; 110:75-87.

31. Pratt C. Kinetics and regulation of cell-free alkaline phosphatase synthesis. J Bacteriol 1980; 143:1265-74.

32. Wanner BL, Sarthy A, Beckwith JR. *Escherichia coli* pleiotropic mutant that reduces amounts of several periplasmic and outer membrane proteins. J Bacteriol 1979; 140:229-39.

33. Adhya S. Obituary. Harrison Echols (1933-1993). Cell 1993; 73:833-4.

34. Englesberg E, Irr J, Power J et al. Positive control of enzyme synthesis by gene C in the L-arabinose system. J Bacteriol 1965; 90:946-57.

35. Englesberg E, Sheppard D, Squires C et al. An analysis of "revertants" of a deletion mutant in the *C* gene of the L-arabinose gene complex in *Escherichia coli* B/r: isolation of initiator constitutive mutants (I^c). J Mol Biol 1969; 43:281-98.

36. Schleif R. 91. The L-Arabinose operon. In: Neidhardt FC, Ingraham JL, Low KB et al. eds. *Escherichia coli* and *Salmonella typhimurium*. Cellular and Molecular Biology, Volume 2. Washington, D.C.: Am Soc Microbiol, 1987:1473-81.

37. Bracha M, Yagil E. A new type of alkaline phosphatase-negative mutants in Escherichia coli K12. Mol Gen Genet 1973; 122:53-60.

38. Morris H, Schlesinger MJ, Bracha M et al. Pleiotropic effects of mutations involved in the regulation of *Escherichia coli* K-12 alkaline phosphatase. J Bacteriol 1974; 119:583-92.

39. Brickman E, Beckwith J. Analysis of the regulation of *Escherichia coli* alkaline phosphatase synthesis using deletions and φ80 transducing phages. J Mol Biol 1975; 96:307-16.

40. Willsky GR, Malamy MH. Control of the synthesis of alkaline phosphatase and the phosphate-binding protein in *Escherichia coli*. J Bacteriol 1976; 127:595-609.

41. Wanner BL, Bernstein J. Determining the *phoM* map location in *Escherichia coli* K-12 by using a nearby transposon Tn*10* Insertion. J Bacteriol 1982; 150:429-32.

42. Hunt AG, Hong J-S. A micromethod for the measurement of acetyl phosphate and acetyl coenzyme A. Anal Biochem 1980; 108:290-4.

43. Nixon BT, Ronson CW, Ausubel FM. Two-component regulatory systems responsive to environmental stimuli share strongly conserved domains

with the nitrogen assimilation regulatory genes *ntrB* and *ntrC*. Proc Natl Acad Sci USA 1986; 83:7850-4.

44. Skarstad K, Thny B, Hwang DS et al. A novel binding protein of the origin of the *Escherichia coli* chromosome. J Biol Chem 1993; 268:5365-70.

45. Wanner BL, Wilmes MR, Young DC. Control of bacterial alkaline phosphatase synthesis and variation in an *Escherichia coli* K-12 *phoR* mutant by adenyl cyclase, the cyclic AMP receptor protein, and the *phoM* operon. J Bacteriol 1988; 170:1092-102.

46. Wanner BL, Wilmes MR, Hunter E. Molecular cloning of the wild-type *phoM* operon in *Escherichia coli* K-12. J Bacteriol 1988; 170:279-88.

47. Wanner BL. Phosphorus assimilation and its control of gene expression in *Escherichia coli*. In: Hauska G, Thauer R, eds. The molecular basis of bacterial metabolism. Heidelberg: Springer-Verlag, 1990:152-63.

48. Iuchi S, Lin ECC. *arcA* (*dye*), a global regulatory gene in *Escherichia coli* mediating repression of enzymes in aerobic pathways. Proc Natl Acad Sci USA 1988; 85:1888-92.

49. Kakuda H, Hosono K, Shiroishi K et al. Identification and characterization of the *ackA* (acetate kinase A)-*pta* (phosphotransacetylase) operon and complementation analysis of acetate utilization by an *ackA-pta* deletion mutant of *Escherichia coli*. J Biochem (Tokyo) 1994; 116:916-22.

CHAPTER 16

THE FNR MODULON AND FNR-REGULATED GENE EXPRESSION

John R. Guest, Jeffrey Green, Alistair S. Irvine and Stephen Spiro

INTRODUCTION

In recent years it has been realized that the elaborate network of interacting metabolic processes operating in living bacteria is not maintained simply by controlling enzyme activities in response to specific metabolites (substrates, end-products and allosteric effectors) or by controlling enzyme synthesis via specific regulatory proteins that activate or repress relevant transcriptional units (genes, operons or regulons) in response to the corresponding metabolites (coeffectors). There is yet another tier of complexity imposed by global regulators which control families of transcriptional units in response to general metabolic or environmental factors. These families have been called regulatory networks or modulons. By belonging to one or more such modulons, the pattern of gene expression can be adapted to that required for a specific metabolic mode or physiological state. A major current challenge is to understand how multiple regulatory factors exert their various effects on a single transcriptional unit, and whether such interactions are sufficient to establish and maintain the complex coordinated metabolic networks operating under diverse physiological conditions, or whether other factors such as the regulation of regulatory gene expression, make a significant contribution to the overall process.

The FNR modulon of *Escherichia coli* represents a family of genes and operons whose expression is regulated in response to anaerobiosis by FNR, the anaerobic transcriptional regulator. Evidence for the existence of this global regulatory system derives from the isolation in several laboratories of pleiotropic mutants lacking the ability to use fumarate, nitrate or nitrite for anaerobic respiratory growth.[1] Indeed, the FNR designation stems from the combined defects in anaerobic fumarate and nitrate reduction exhibited by these mutants. The nature of the defect was much debated until the *fnr* gene was cloned and sequenced, whereupon it became clear that its product is a transcriptional regulator resembling CAP (the catabolite gene activator protein,

Regulation of Gene Expression in Escherichia coli, edited by E. C. C. Lin and A. Simon Lynch. © 1996 R.G. Landes Company.

also known as CRP, the cyclic-AMP receptor protein).[1] Thereafter, the CAP system has served as a very useful model for much of the work on FNR. Progress has been relatively slow, especially at the in vitro level, due to the insolubility of the FNR protein and the poor yields from early expression systems. However, it is now abundantly clear that assumptions concerning the close structure-function relationships between FNR and CAP were justified. Thus FNR operates the master switch between aerobic and anaerobic metabolism by ensuring that oxygen is used in preference to alternative electron acceptors, in much the same way that CAP ensures that glucose is metabolized in preference to other carbohydrates. This review will cover aspects of the role and mode of action of FNR that are of special interest to the authors. Comprehensive reviews of anaerobic gene expression and details of individual members of the FNR modulon, can be found elsewhere.[1-7]

THE METABOLIC ARENA

E. coli is a metabolically versatile chemoheterotroph which can use a variety of growth substrates under aerobic and anaerobic conditions, deriving energy from either respiration or fermentation. Respiration involves membrane-bound proton-translocating electron transport chains in which metabolic energy is conserved by coupling substrate oxidation to the reduction of either oxygen (aerobic respiration) or some other exogenous electron acceptor such as, nitrate, nitrite, fumarate, trimethylamine-N-oxide, dimethylsulfoxide or tetrahydrothiophene 1-oxide (anaerobic respiration). In contrast, fermentation involves the formation of endogenous electron acceptors during a redox-balanced dismutation of the substrate, and energy is conserved primarily by substrate-level phosphorylation. The metabolic mode adopted depends on the nature of the growth substrate and the availability of oxygen and alternative electron acceptors. The growth substrate is typically a respirable and fermentable carbohydrate (preferably glucose), or a nonfermentable substrate which has to be respired (e.g., glycerol or lactate). In either case the substrate serves as the major carbon and energy source during growth in minimal medium. However, *E. coli* can use H_2 as the oxidizable substrate, in which case the electron acceptor, fumarate or malate, serves as the carbon source.

Examples of the basic metabolic modes are shown for glucose catabolism in Figure 16.1. The glycolytic conversion of glucose to pyruvate is common to all modes, whereafter the routes diverge. During aerobic respiration with nonrepressing glucose concentrations, pyruvate is oxidatively decarboxlyated by the pyruvate-inducible pyruvate dehydrogenase (PDH) complex, the resulting acetyl units are totally oxidized in the citric acid cycle, and the reducing equivalents are transferred to oxygen via the aerobic electron transport chain with cytochrome *o* and cytochrome *d* as the terminal oxidases. This generates the highest yield of utilizable energy (Fig. 16.1). Under anaerobic conditions, the PDH complex is partially repressed and inhibited, and the conversion of pyruvate to acetyl-CoA increasingly depends on pyruvate formate-lyase (PFL), which is induced and activated. Most of the citric acid cycle enzymes are repressed and the cycle becomes noncyclic, primarily due to the severe anaerobic repression of the 2-oxoglutarate dehydrogenase complex. The anaerobic repression of succinate dehydrogenase (SDH) and fumarase A (FumA) is paralleled by the induction of their anaerobic counterparts, fumarate reductase (FRD) and fumarase B (FumB). The metabolic flow in the C_4-dicarboxylate sector of the cycle is reversed and this affords some energy conservation via the fumarate reductase system under certain circumstances. When the citric acid cycle is repressed, the residual activities perform essentially anabolic functions and most of the substrate carbon flows elsewhere to produce less oxidized products. During anaerobic respiration with nitrate the major product is acetate (Fig. 16.1). Pyruvate is metabolized by the PDH complex and PFL, and energy conservation involves the nitrate reductase system which accepts reducing equivalents via ubiquinone and menaquinone.[8,9] A similar situation exists with fumarate as electron acceptor except that pyruvate metabolism is probably more dependent on PFL, fumarate reductase only accepts reducing equivalents via menaquinone, and the reduction of formate may rely more on the constitutive FDH_O and basal FDH_N activities (Fig. 16.1). The yield of utilizable energy (4ATP and 8H per mol of glucose) reflects the incomplete oxidation of the substrate. During fermentation, acetyl-CoA is generated nonoxidatively by PFL, and in the absence of an exogenous electron acceptor, redox-balance is accomplished by reducing approximately half of the acetyl-CoA to ethanol (Fig. 16.1). The products of this simplified mixed acid fermentation therefore include acetate, ethanol and formate, which is interconvertible with H_2 and CO_2 by the formate hydrogenlyase system.[10] The formate hydrogenlyase system is induced by the accumulation of formate,[11,12] so it is not expressed during aerobic respiration, because there is no PFL, or during anaerobic respiration, where formate is readily oxidized by

Aerobic Respiration

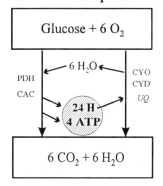

$\Delta G_o' = -2830$ kJ/mol ; $E_0' = +820$ mV

Nitrate Respiration

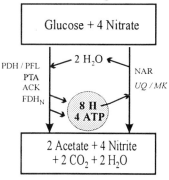

$\Delta G_o' = -858$ kJ/mol ; $E_0' = +420$ mV

Fumarate Respiration

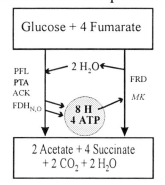

$\Delta G_o' = -550$ kJ/mol ; $E_0' = +31$ mV

Fermentation

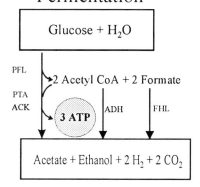

$\Delta G_o' = -218$ kJ/mol ; $E_0' = -412$ mV

Fig. 16.1. The metabolic modes of Escherichia coli. Schematic and simplified representations of the pathways of glucose catabolism during aerobic respiration, anaerobic respiration with nitrate and fumarate as electron acceptors, and fermentation. Each route starts with the glycolytic conversion of glucose to pyruvate. The potential yields of utilizable energy are indicated by the relative amounts of ATP and reducing equivalents [H] produced. The free energies ($\Delta G_o'$, kJ/mol glucose) for each overall reaction and the redox potentials of the electron acceptors, are indicated. Abbreviations: ACK, acetate kinase; ADH, alcohol dehydrogenase; CAC, citric acid cycle; CYD, cytochrome d; CYO, cytochrome o; FDH, formate dehydrogenase; FHL, formate hydrogenlyase; FRD, fumarate reductase; MK, menaquinone; NAR, nitrate reductase; PDH, pyruvate dehydrogenase complex; PFL, pyruvate formate-lyase; PTA, phosphotransacetylase; UQ, ubiqinone.

exogenous electron acceptors (nitrate and presumably fumarate). Other minor fermentation products include succinate and D-lactate which vary in amount depending on the growth conditions. The utilizable energy (3ATP per mol of glucose) comes solely from substrate-level phosphorylation in glycolysis, and from the conversion of one mol of acetyl-CoA to acetate.

The yields of utilizable energy (ATP + H) for the different metabolic modes are consistent with the free energy changes calculated for the overall reactions (Fig. 16.1) and with the redox-potentials of the electron acceptors (E_o', mV): oxygen (+820); nitrate (+420); nitrite (+374); DMSO (+160); TMAO (+130); fumarate (+31); acetyl-CoA (-412). When given the choice, *E. coli* clearly uses the most energetically favorable process. Thus aerobic respiration is preferred to nitrate, nitrite and fumarate respiration, nitrate respiration is preferred to fumarate and DMSO respiration, and it seems likely that when present, fumarate is used in preference to endogenously generated electron acceptors (acetyl-CoA and pyruvate) in the hydrogen-evolving fermentation. However, this hierarchy is not imposed by a single regulatory system which senses substrate potential energies or the redox potentials of available electron acceptors, nor is the same hierarchy adopted by all organisms.[2] On the contrary, the switch between different metabolic modes is imposed by four major global regulators responding to different environmental stimuli. They operate at the transcriptional level although there is some fine control at the post-translational level. For example, PFL activity is regulated by a reversible activation-deactivation system,[13] and what appears to be a superfluous anaerobic synthesis of the PDH complex may be modulated by the inhibitory effects of NADH on lipoamide dehydrogenase.[14] Likewise, the anaerobic nitrate and fumarate transport systems are reversibly inhibited by oxygen as well as being repressed at the transcriptional level.[15]

The most significant global regulators in this metabolic arena (and their primary spheres of influence) are: ArcA (aerobic respiration); FNR (anaerobic respiration); NarL and NarP (nitrate and nitrite respiration); and FhlA (fermentative formate metabolism and hydrogen evolution). However, there is considerable overlap between these spheres of influence (modulons) such that some operons belong to more than one modulon. This complexity is further compounded by the fact that regulators often function as activators and repressors with different target operons. There are also very significant overlaps with the modulons controlled by: CAP and CreC-PhoB (catabolite repression); SoxRS and OxyR (oxidative stress); Fur (ferric uptake); and RpoN (NtrA or σ^{54}; nitrogen source). Gene expression is also effected by the underlying effects of anaerobicity and osmolarity on DNA supercoiling, which will not be considered here.

FNR activates and represses target genes in response to anaerobiosis, and the FNR modulon contains a variety of genes concerned with anaerobic metabolism (see below). The ArcA protein is the response regulator of a two-component signal transduction system in which anaerobic stress in the aerobic respiratory chain, or the accumulation of fermentation intermediates and products (e.g., pyruvate, acetate, D-lactate and NADH) caused by anaerobiosis, is apparently detected and signaled by the membrane-bound sensor, ArcB.[5,16] The regulator is converted to its phosphorylated form which functions as an anaerobic repressor of citric acid cycle, glyoxylate cycle and other aerobic enzymes, e.g., L-lactate dehydrogenase,[4] cytochrome *o*,[17] Mn-superoxide dismutase[18] and possibly the PDH complex.[4,14] ArcA also functions as an anaerobic activator for cytochrome *d*[19] and PFL.[20] Since ArcA like some other response regulators, may be phosphorylated by acetyl-phosphate, it is possible that this compound provides a direct route for signaling redox stress.[21,22] The NarXL and NarQP systems are analogous sensor-regulator pairs which respond to exogenous nitrate and nitrite with different effector and target specificities.[7,23] The NarL regulator activates the respiratory nitrate reductase, the NADH-dependent nitrite reductase, formate dehydrogenase$_N$, and the nitrite transporter, but represses the formate-linked nitrite reductase, the respiratory fumarate and DMSO reductases (but not TMAO reductase), the anaerobic dicarboxylate transport systems and alcohol dehydrogenase, in response to nitrate. Other regulatory events are initiated by NarL in response to nitrite (e.g., repression of the *nrf*-encoded nitrite reductase$_F$) and by NarP in response to nitrate and nitrite.[7] The FhlA protein induces the synthesis of formate hydrogenlyase in response to formate accumulation during fermentation.[11,12,24] The corresponding modulon encodes the formate dehydrogenase$_H$ (*fdhF*) and hydrogenase-3 (*hyc*) components of formate hydrogenlyase, and other hydrogenase activities (*hypA-E*). The *hyp-fhlA* and *hyc* genes comprise the divergent *fhl* operon which is positively autoregulated by FhlA and the FhlA-dependent promoters are unusual in using σ^{54}-RNA polymerase having the *rpoN*-encoded sigma factor, rather than σ^{70}-RNA polymerase.

THE FNR MODULON

The FNR modulon comprises a family of transcriptional units whose expression is regulated in response to anaerobiosis by FNR. It contains 29 known transcriptional units (70 genes) that are mainly concerned with anaerobic metabolic processes. Current members are listed in Table 16.1 according to whether they are activated or repressed by FNR. The supporting evidence for FNR-dependent regulation is indicated together with the locations of the FNR-sites (where known) and other regulatory factors that affect their expression. They are distributed among four of the six recently-defined functional categories[25] of *E. coli* gene products: 53 (I, intermediary metabolism); 3 (II, biosynthesis of small molecules); 13 (V, cellular processes); and 1 (VI, other functions). Clearly most of the FNR modulon is concerned with maximizing the capacity for anaerobic energy generation. In fact, all but one member (*cea*) are directly involved in anaerobic metabolism or in the transport or synthesis of substrates and cofactors used by other members of the modulon. Interestingly, the genes encoding anaerobic transport systems for formate, dicarboxylic acids and nitrite are located within or close to related transcriptional units (*focA-pfl*, *aspA-dcuA*, *dcuB-fumB*, *narXL-narK-narGHJI*). It is also interesting that *arcA* belongs to the FNR modulon, not simply because anaerobic *arcA* expression is amplified 4-fold by FNR, but this subservience elevates FNR to the highest ranking regulator of anaerobic gene expression.

The ultimate criterion for membership requires that FNR binds site-specifically to a target site(s) *and* activates or represses transcription in vitro. Unfortunately, this has only been achieved with a few promoters (*ndh, fnr, narX*) and with the semisynthetic FNR-regulated *FFpmelR* promoter.[26-29] Such in vitro studies have confirmed the general importance of predicted FNR-sites centered at about -41.5 in positively regulated promoters and elsewhere in negatively regulated promoters, but also cast doubt on the functional significance of other potential sites, e.g., *nirBDC* -89.5 (see below). Evidence based on enzyme and *lacZ* fusion activities in *fnr* mutants and the effects of multicopy *fnr* plasmids is less rigorous, because the observed changes may be of secondary rather than direct origin, or else due to invasion of a physiologically

Table 16.1. The FNR modulon

Genes and functions		FNR-sites	Evidence	Other global regulators
A. FNR-activated				
aeg-46.5	Putative periplasmic nitrate reductase	-64.5	D E F	NarL(-) NarP(+)
ansB	L-Asparaginase II	-41.5, -74.5?	C D E F	CAP(+)
arcA	ArcA	-82.5	D E F	ArcA(+)
aspA-dcuA	L-Aspartase and dicarboxylate transport	ND	D	NarL(-)CAP(+)
cea	Colicin E1	-64.5 to -31.5 (5 potential sites)	E F	LexA(-) CAP(+)
[*cydAB*]	Cytochrome *d*	-54.5	D E F	ArcA(+)
dcuB-fumB	Dicarboxylate transport and fumarase B	ND	D E F	NarL(-)
dmsABC	DMSO reductase	-49.5	D E, F	NarL(-)
fdnGHI	Formate dehydrogenase-N	-42.5, -97.5	B C D E F	NarL,P(+) ArcA(-)
feoAB	Iron(II) transport	ND	D E F	Fur(-)
FFpmelR	Semi-synthetic fusion	-41.5	A B C D E F	
focA-pfl	Formate transport and pyruvate formate-lyase	P6 -41.5, (-380.5) P7 -59.5	B C D E F	ArcA(+) NarL(-) IHF
frdABCD	Fumarate reductase	-45.5 or -46.5	D E, F	NarL(-)
glpABC	Anaerobic glycerol-3-P dehydrogenase	-40.5	D	NarL?
glpTQ	Glycerol-3-P transport	-91.5	D	NarL? CAP(+)
hypBCDE-fhlA	Hydrogenase activities and formate regulation	-42.5	E F	RpoN
narGHJI	Nitrate reductase	-41.5	B C D E F	NarL(+) IHF
narK	Nitrite extrusion protein	-41.5, -79.5	C D E F	NarL(+) IHF Fis
nikA-E (hydC)	Nickel transport	ND	D E F	
nirBDC	NADH-dependent nitrite reductase	-41.5, -89.5?	B C E	NarL(+)
nrfABCDEFG	Formate-linked nitrite reductase (cyt c552)	-42.5	D E F	NarL(-) NarP(+)
nrdD	Anaerobic ribonucleotide reductase	ND	E F	
	Molybdate reductase	ND	E	
B. FNR-repressed				
[*cyoABCDE*]	Cytochrome *o*	ND	D E	ArcA(-) CAP(-)
fnr	FNR	-0.5, -103.5	A B C D E	
hemA	Glutamyl-tRNA dehydrogenase	-23.5	D	ArcA(+) IHF
narX	NarX	-106.5, -75.5, +107.5	A D F	NarL(+)
ndh	NADH dehydrogenase II	-50.5, -94.5	A B D E F	RpoN
[*pdhR-aceEF- lpd*]	Pyruvate dehydrogenase complex and regulator	-49.5	F	ArcA?
[*sodA*]	Mn-Superoxide dismutase	-35.5 or -32.5	D E F	SoxRS,Q(+) Fur(-), ArcA(-) IHF

Members of the FNR modulon are listed alphabetically according to whether they are activated or repressed by FNR. The positions of FNR-site centers are defined relative to the transcriptional start sites, where known. The evidence for a functional FNR-site(s) in each promoter is categorized as follows: A, in vitro transcription; B, in vitro footprinting; C, mutational analysis; D, *lacZ* fusion analysis; E, altered expression in *fnr* mutant; F, prediction from nucleotide sequence. ND, not determined. Uncertain members are enclosed by square brackets ([ ]). Roles for other global regulators are indicated, those for NarL and NarP refer only to the effects of nitrate. (A complete citation list for the modulon compiled in September 1994 can be obtained from the authors).

insignificant site by the amplified regulator. This type of evidence should thus be treated with caution.

The PDH complex represents an interesting case where the *pdhR-aceEF-lpd* operon is controlled by the pyruvate-responsive repressor (PdhR) and there is a putative FNR-site centered at -49.5.[14] The PDH complex is utilized aerobically and during nitrate respiration but is not essential under all anaerobic conditions, where its metabolic role is increasingly adopted by PFL (Fig. 16.1). *A priori*, the *pdhR-aceEF-lpd* operon might have been expected to exhibit reciprocal regulatory responses to those of the *focA-pfl* operon[30] and be repressed anaerobically by FNR. However, *lacZ*-fusion studies indicated that this is not the case although a multicopy *fnr⁺* plasmid caused severe repression under aerobic and anaerobic conditions. So, it is not certain whether this operon is a member of the FNR modulon. The alcohol dehydrogenase gene (*adhE*) appears not to be regulated by FNR despite the presence of a putative FNR-site.[31]

Most members of the FNR modulon are members of other modulons (Table 16.1). Especially interesting are those such as *cyoA-E*, *cydAB*, *sodA* and *focA-pfl*, which were thought to be regulated by FNR and ArcA, sometimes with conflicting results.[3,17,19] However, it is now realized that *arcA* belongs to the FNR modulon,[16,32] so the effects of FNR on the ArcA-activated *cyd*, and the ArcA-repressed *cyo* and *sodA* promoters are likely to be indirect. Dual regulation by FNR and ArcA may be used to fine-tune gene expression in response to anoxia. Hence, with the initial onset of anaerobiosis ArcA is activated, and if these conditions persist or become more anaerobic, FNR is activated leading in turn to the up-regulation of ArcA and intensification of its effects.[32]

THE FNR PROTEIN AND RELATIONSHIPS WITH CAP

By comparing the amino acid sequences of FNR and CAP it was predicted that all of the secondary structural elements in CAP are retained in FNR, including the helix-turn-helix motif in the DNA-binding domain, the series of β-strands in the nucleotide-binding domain and the major helix at the dimer interface (Fig. 16.2). However, there are some important differences in FNR. First, isolated FNR is monomeric (M_r 28,000 by gel filtration, 30,000 by SDS-PAGE) rather than dimeric, although it appears to be dimeric when bound to DNA (see below). Nine N-terminal residues are removed when the protein is exposed to periplasmic or membrane-bound proteases yielding products of M_r 26,823 ± 6 and 26,901 ± 1 (by electrospray mass spectrometry).[35] Secondly, the residues that interact with cyclic AMP in CAP are not conserved and there is no evidence for an interaction between cyclic AMP and FNR.[36] Thirdly, a 'patch' of residues in a surface-exposed β_9-β_{10} loop of CAP that makes an activating contact with the α-subunit of RNA polymerase (RNAP) is not conserved in FNR, although its role may be adopted by residues in the β_3-β_4 loop in FNR.[37] Another potential activation patch in the β_4-β_5 loop is conserved in both proteins.[38] These discrepancies may be related to the different locations of regulator-binding sites in different promoters. Fourth, FNR has a cysteine-rich N-terminal extension, which contains three of the four essential cysteine residues (C20, C23, C29 and C122).[39,40] Finally, isolated FNR is associated with a variable amount of iron (0.02-1.10 atoms per monomer).[26] It has a weak absorption maximum at 420 nm but this appeared not to be due to an iron-sulfur center because the absorbance is not correlated with the iron content and no acid-labile sulfur was detected.[26] A high iron content can be maintained by purification in the presence of 0.1 mM ferrous ammonium sulfate, or restored by incubating apo-FNR with ferrous iron and β-mercaptoethanol.[a] The iron content is inversely related to the reactive sulfhydryl content, indicating that the cysteine residues serve as iron ligands. However, only one of the five proteins containing single cysteine substitutions (C122A) is substantially deficient in iron.[35] The N-terminal region is thus envisaged as contributing to an iron-binding sensory domain, which initiates a redox-mediated conformational change in FNR resulting in the activation or repression of transcription. Despite the differences between FNR and CAP, particularly with respect to its built-in sensor, the mode of FNR action is likely to be similar to that of CAP.

The finding that iron-containing FNR (holo-FNR) activates and represses transcription in vitro represented a very significant advance, because it showed that the FNR protein could be isolated in a functional form.[27] The conversion of inactive apo-FNR to active holo-FNR by treatment with ferrous ions and β-mercaptoethanol was later demonstrated[41] and this has posed important questions concerning the nature of the bound iron.

When analyzed by nonreducing SDS-PAGE, purified FNR contains significant amounts of two species, M_r 27,000 and 30,000, the former (FNR$_{27}$) being converted to the latter (FNR$_{30}$) upon treatment with sulfhydryl compounds.[35] The inactive cysteine mutant proteins were fixed in one form (FNR$_{27}$) except for C122A which had the FNR$_{30}$ mobility. It was therefore suggested that FNR$_{27}$ could represent a modified form possibly containing an intramolecular disulfide bond linking C122 to one of the other cysteine residues. The relative abundance of FNR$_{27}$ in whole cells increases from 14% in anaerobic bacteria to 40% in aerobic bacteria but the physiological significance of this is unclear.[35] There is no evidence for significant amounts of a stable or covalently linked dimeric form.

The FNR protein can now be purified in a single chromatographic step from genetically-amplified sources in suitable amounts for biochemical studies (10 mg per liter of culture).[a] Even in the presence of detergents and salt, it comes out of solution when concentrated. The amplified protein behaves as an unstable dimer (average M_r 40,000) during gel filtration in the presence of ferrous iron, which suggests that the incorporation of iron promotes dimerization. The iron contents of isolated or reconstituted FNR can be as high as 2.7 Fe atoms per monomer, and acid-labile sulfur contents of 0.07 or 0.25 S atoms per monomer are observed after rapid aerobic or anaerobic purification.[a] Furthermore, iron reconstitution is associated with a broad absorbance maximum at 380 to 440 nm (ε_{380} 2 400 M^{-1} cm^{-1}) under aerobic conditions, and maxima at 315 nm (ε_{315} 3 700 M^{-1} cm^{-1}) and 420 nm (ε_{420} 300 M^{-1} cm^{-1}) under anaerobic conditions, as if the protein contains an unstable iron-sulfur center. It would therefore appear that FNR contains its own redox-sensor, probably in the form of a redox-sensitive iron-sulfur center in the N-terminal region.

THE DNA-BINDING SPECIFICITY OF FNR

The close relationship between FNR and CAP extends to their DNA binding sites. The 22 bp FNR-site consensus deduced from sequence comparisons, is a partial palindrome (--**TTGAT**--**ATCAA**--) containing a **TTGA** half-site motif instead of the **GTGA** motif in the analogous CAP-site consensus (Fig. 16.2). This relationship has been confirmed in vivo by showing that the regulator specificities of natural and synthetic promoters can be interconverted by replacing **T**(**A**) with **G**(**C**) and vice versa, at the critical symmetrical positions in the

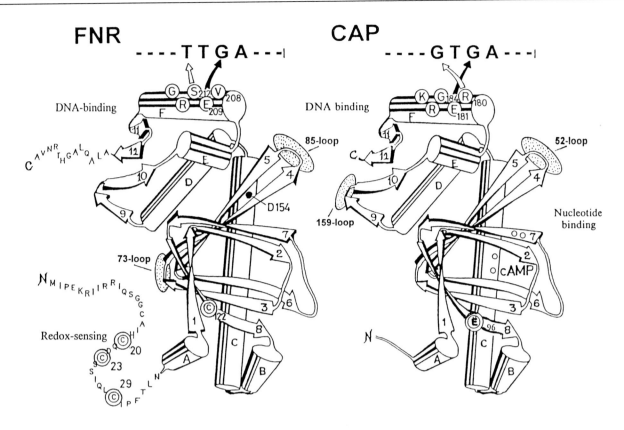

Fig. 16.2. Predicted structure of the FNR monomer based on the structure of CAP. The DNA-binding domain containing the helix-turn-helix motif (α_E-α_F), the nucleotide-binding or allosteric domain containing a series of anti-parallel β-strands (β_1-β_8), and the dimer interface (α_C), and the putative redox-sensing domain of FNR containing four essential cysteine residues, are indicated. The FNR and CAP half-sites with the consensus core motifs are shown and the conserved (filled arrow) and discriminatory (open arrow) interactions are indicated.[33,34] The site of the D154A substitution in an FNR* protein and the loops making contact with RNAP during transcription activation are also indicated.

half-site motifs.[1] The DNA-recognition specificities of the two regulators have also been interconverted by appropriate amino acid substitutions in the DNA-binding face of the recognition helices (α_F).[34] The proposed specificity-conferring interactions are the conserved interaction between E209$_{FNR}$ (E181$_{CAP}$) and the **G-C** base pair common to both core motifs and a discriminatory interaction between S212$_{FNR}$ and the unique **T-A** base pair in the FNR-site, which replaces that between R180$_{CAP}$ and the corresponding **G-C** base pair in the CAP-site (Fig. 16.2). The role of another conserved interaction involving R213$_{FNR}$ (R185$_{CAP}$) and the common **G-C** base pair in sequence discrimination is still uncertain.[33] The corresponding binding-site motifs can therefore be denoted -E--SR (FNR) and RE---R (CAP). The successful construction of FNR hybrids that activate CAP-dependent promoters in response to anoxia and CAP hybrids that activate FNR-dependent promoters in response to glucose starvation, amply confirms the early predictions concerning the close structural and functional relationships between the two regulators.

A powerful in vitro tool for studying protein:DNA interactions is the electrophoretic-mobility-shift or gel-retardation assay. Unfortunately, it has proved very difficult to demonstrate specific binding between isolated FNR and target DNA using this approach. Non-specific binding was observed between the semisynthetic *FFpmelR* promoter and wild-type FNR (K_d 10^{-7} M)[26] but enhanced binding has recently been obtained using a synthetic 48 bp DNA target with FNR protein (D154A) from an *fnr** mutant.[42] In contrast, DNase I footprinting has proved far more successful for studying site-specific FNR-binding at several positively-regulated (*FFpmelR*, *pflP6*, *fdn*, *narGHJI*, *nirB*) and negatively-regulated (*ndh* and *fnr*) promoters.[26,27,a] Representative DNase I footprints are shown in Fig. 16.3. The protected regions (24-33 bp)

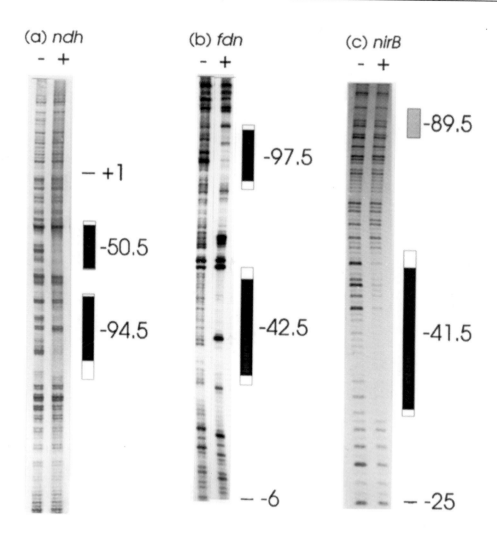

Fig. 16.3. DNase I footprints of representative FNR-dependent promoters. The regions protected by FNR are shown for: (a) the negatively regulated ndh promoter (-50.5 and -94.5); (b) the positively regulated fdn promoter (-42.5 and -97.5); and (c) the positively regulated nirB promoter (-41.5) where a potential site centered at -89.5 is not protected. The positions of the predicted 22 bp FNR-sites (black bars) are indicated within the protected regions (open bars).

overlap the predicted sites and show that two FNR monomers are bound. The *ndh* and *fdn* promoters have two protected regions centered at -50.5 and -94.5 (*ndh*) or -42.5 and -97.5 (*fdn*), indicating that multiple FNR-sites may be important in FNR-dependent gene expression (Fig. 16.4). This contrasts with the *nirB* promoter, where two FNR-sites centered at -41.5 and -89.5 are predicted, but protection occurs only at the -41.5 site (Fig. 16.3) which is known to be essential for FNR-dependent expression.[43] The *ansB* promoter likewise has two predicted FNR-sites (-41.5 and -74.5; Fig. 16.2) and only the physiologically significant downstream site is protected by FNR.[a]

Other important conclusions arising from the footprinting studies are that anaerobiosis (or the presence of reducing agents) and holo-FNR are not essential for specific binding to target DNA. However, although the protection patterns are unaffected by iron content, the DNA-binding affinities (K_d 10^{-7} to 10^{-8} M) are reduced about 2-fold by iron depletion, and DNA binding is very significantly impaired or abolished with mutant proteins lacking the essential cysteine residues, particularly the iron-deficient C122A protein.[26,27,35]

IN VITRO TRANSCRIPTION ACTIVATION AND REPRESSION

FNR-dependent transcription regulation was first demonstrated in vitro using holo-FNR to activate or repress the respective *FFpmelR* or

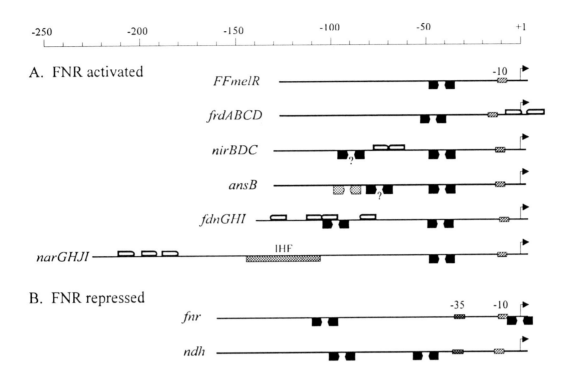

Fig. 16.4. Promoter regions of representative FNR-regulated genes or operons. The positions of specific regulatory sites and consensus -35 and -10 sequences (▨) are located relative to the transcriptional start sites (arrows) in linear forms with a scale in base-pairs. Regulator binding sites are denoted by the following motifs: FNR, ◼; CAP/FNR, ▨; NarL, ▷.

ndh promoters.[27] It was later shown that a high iron content is absolutely essential for both in vitro regulatory activities.[41,44,a] The C122A mutant protein, which lacks iron, is also incapable of mediating FNR-regulated transcription in vitro. A requirement for iron has likewise been inferred from the inhibitory effects of chelating agents in in vitro studies on the autorepression of the fnr gene.[29] In the same context, an in vitro switch has been devised in which inactive apo-FNR could be reactivated by preincubation with ferrous iron and β-mercaptoethanol.[41] The reactivation was accompanied by the incorporation of iron (and it could be prevented and reversed by adding chelating agents).[a] These observations show that FNR-dependent FFpmelR transcription can be switched on and off in vitro but this should not necessarily be regarded as mimicking the natural FNR-mediated switch. Permanganate footprinting has also been used both in vivo and in vitro, to show that holo-FNR is essential for efficient open-complex-formation at the FFpmelR promoter.[44]

TRANSCRIPTIONAL ORGANIZATION OF REPRESENTATIVE PROMOTERS

Successful transcription activation generally requires contact between a DNA-bound activator and RNAP, in order to generate an effective ternary complex at an otherwise inadequate promoter site. There are many ways in which a successful complex (or caged structure[45]) can be assembled. For example, CAP binds at a variety of sites in different promoters (-41.5, galP1 and melR; -61.5, lacP1; -70.5, malT; -103, uhpT) where it makes different contacts with RNAP and elicits quite distinct modes of transcription activation.[46,47] Furthermore, relatively minor changes in the position of a regulator binding site can affect the contact with RNAP such that an activator becomes a repressor. The positions of FNR-site in FNR-regulated genes are listed in Table 16.1, and Fig. 16.4 shows the regulatory organization of representative promoters.

(A) TRANSCRIPTIONAL ACTIVATION BY FNR

In common with most positively regulated promoters, FNR-activated promoters generally have a good -10 sequence but a poor -35 sequence. However, there is usually a 'compensatory' FNR-site centered at about -41.5 which allows FNR to activate transcription initiation via RNAP (Table 16.1). This means that most of the FNR-activated promoters resemble Class II CAP-dependent promoters, such as *galP1* and *melR*. This type of organization may thus be regarded as the basic transcriptional unit for FNR-mediated activation. The situation is complicated by the presence of sites for other regulatory factors upstream of the basic unit (Fig. 16.4), or by the presence of additional upstream FNR-sites which seem to be associated with FNR-mediated repression (see below).

(B) TRANSCRIPTIONAL REPRESSION BY FNR

Repression can occur when a regulatory protein binds directly to the promoter region in such a way as to interfere with the binding or activation of RNAP. It can also occur by cooperative interaction between repressors bound at sites upstream and downstream of the promoter, or in the case of CytR by a cooperative interaction with CAP molecules bound at upstream sites. However, a different mechanism seems to operate in FNR-mediated anaerobic repression. The promoters which have been most studied, *ndh*, *fnr* and *narX*, have good -35 and -10 sequences but apart from containing more than one FNR-site, there isn't a common pattern (Table 16.1, Fig. 16.4). The *ndh* promoter has FNR-sites centered at -50.5 and -94.5, the *fnr* promoter has a site that overlaps the transcription start (-0.5) and an upstream site (-103.5), whereas the divergent *narX/narK* promoter has two upstream sites and a downstream site (-106.5, -75.5 and +107.5) relative to *narX* transcription. In each case the site that is furthest upstream is particularly important because its deletion severely impairs FNR-mediated repression.[28,29] Repression of the *ndh* promoter is not simply a matter of promoter occlusion. On the contrary, it would seem that an FNR dimer binds at each of the two sites in order to inhibit open-complex formation, presumably by a mechanism that involves direct interaction with RNAP and the abolition of essential upstream DNA:RNAP contacts.[28] It has likewise been concluded for the *fnr* and *narX* promoters that there are essential factor-independent activation signals (possibly RNAP-DNA contacts) which are blocked when active FNR is bound at the most upstream site.[29] Also relevant is the fact that of the two protected FNR-sites in the *fdn* promoter (Figs. 16.3 and 16.4), only the downstream site centered at -42.5 is essential for FNR-dependent activation, whereas the upstream site at -97.5 is thought to be involved in down-regulating *fdn* expression in the absence of nitrate.[48] Thus the presence of multiple FNR-sites, including a far upstream site where important DNA:RNAP contacts can be disrupted, may be an essential feature of FNR-mediated repression.

(C) MULTIFACTORIAL REGULATION

For many genes, the basic pattern of FNR-mediated activation and repression is modified because they are members of other modulons (Table 16.1). The convergence of regulators responding to different stimuli at a single promoter poses challenging questions concerning their interaction with RNAP and with each other. Such interactions presumably provide the mechanisms for establishing the priorities for utilizing different electron acceptors, ensuring that the most energetically efficient metabolic process is operative in a given environment.

From the arrangement of different regulatory sites in FNR-dependent promoters several mechanisms can be envisaged, including: competition for overlapping binding sites; cooperative DNA-binding; independent interactions between each regulator and RNAP; and modulation of the activity of one regulator that contacts RNAP by another that makes no direct contact with RNAP. Some specific examples will now be considered.

Dual regulation by FNR and NarL

Most anaerobic respiratory genes are regulated by FNR and the NarXL two-component sensor-regulator system (or its counterpart NarQP) in response to anaerobiosis and nitrate, respectively.[7,23] NarL acts as a classical repressor when its binding site is close to the transcription start, e.g., in the *frdA* (Fig. 16.4) and *dmsA* promoters. More interesting is the way in which FNR and NarL control transcription activation in response to anaerobiosis and nitrate. Here there are conflicting reports on the need for a direct interaction between FNR and NarL in different systems. Such an interaction provides a plausible explanation for the lack of nitrate-inducible *narG* expression in an *fnr* mutant,[49] and for later observations that anaerobic *nirB* expression is lowered in the absence of nitrate in a *narL* mutant and when the NarL-site is deleted or altered.[50] Studies with a *narG-lacZ* reporter in some *fnr** mutants further indicated that interaction between FNR and NarL is needed for maximal aerobic expression.[51] On the other hand, FNR-independent expression of *narG* is observed when the -35 region is replaced by the consensus sequence, and the resulting activity is enhanced by NarL, indicating that a FNR:NarL interaction is not essential.[52] Other evidence shows that IHF is required for FNR- and NarL-dependent *narG* expression, offering the possibility that DNA-bending by IHF facilitates direct contact between FNR, NarL and RNAP (Fig. 16.4).[53] This leaves a confusing picture for dual control by FNR and NarL, and highlights the need for further direct experimentation as well as the problem of interpreting observations made with restructured promoters. Some mechanistic diversity would of course be expected in view of the different binding-site relationships (Fig. 16.4) and the impact of other accessory proteins such as IHF.

FNR and ArcA interactions

The prime regulators of aerobic and anaerobic gene expression, FNR and ArcA, coordinate the response to anoxia by what are essentially reciprocal anaerobic activation and repression mechanisms. They are intimately linked through the genes controlled by both regulators (Table 16.1) and by the recent demonstration that *arcA* expression is anaerobically-activated by FNR.[32] The FNR dependence of *arcA* presumably serves to fine-tune the coordinated regulation of aerobic and anaerobic metabolism. Both respond to anaerobiosis, but whereas FNR seems to respond at a redox-potential ($E_o' = > +0.4$ V) poised between oxygen and nitrate respiration,[54] ArcA may sense the concentrations of relevant metabolites[16] and a decrease in $\Delta\mu_{H}+$.[55] A DNA-binding site has not been identified for ArcA, but the existence of such sites is by no means excluded. The dual action of FNR and ArcA on the expression of some genes (*sodA* in the absence of Fur, *cyo* and *cyd*) may be explained by FNR regulation of *arcA* but in other cases (*focA-pfl* and *arcA*) direct FNR:ArcA interaction is implied.[32] The nature of the interaction has not been defined but a model involving IHF-induced DNA-bending and a transcription complex containing two FNR dimers, ArcA and RNAP, has been proposed for *focA-pflP6* and *P7*.[20]

FNR and CAP

The FNR and CAP binding sites are sufficiently similar to envisage joint occupancy as a real possibility. Indeed, overlapping sites seem to be important in the regulation of genes that are members of both the FNR and CAP modulons. The colicin E1 gene (*cea*) is subject to FNR-dependent activation and catabolite repression, and the *cea* promoter contains several potential FNR-sites the best of which is a potential CAP-site centered at -64.5.[56] The anaerobic induction of *ansB* (asparaginase II) is mediated by an FNR-site at -41.5 (Fig. 16.4) and catabolite repression is mediated through a CAP/FNR hybrid site at -91.5.[57] There is a second potential FNR-site at -74.5 which seems to be of little physiological significance and is not protected by FNR in DNase I footprinting reactions.[a] Both FNR and CAP are required for maximum expression and the way in which they combine to achieve this is unknown. One possibility, which parallels that for FNR and NarL at the *fdn* promoter,[48] is that FNR down-regulates anaerobic *ansB* expression by occupying both sites until glucose is exhausted, whereafter cAMP-CAP displaces FNR from the hybrid site to allow maximum expression.

POTENTIAL FNR CONTACTS WITH RNA POLYMERASE AND DNA-BENDING

An interesting difference between CAP and FNR is that CAP-sites are located at a variety of positions relative to the transcription start site, e.g., -41.5 (Class II) or -61.5, -70.5, and -81.5 (Class I), whereas FNR-sites are generally located at about -41.5, in positively controlled promoters. However, FNR can activate promoters, albeit weakly, when the FNR-site is relocated at -61.5 in the semisynthetic *FFpmelR* promoter or when the CAP-site of *lacP1* is mutated to an FNR-site.[43,58] Transcriptional activation involves protein-protein interactions between the bound activator and RNAP, and mutational analyses have recently shown that there are three activation patches in CAP that are essential for contacting RNAP in the positive control process.[46] One of these, the 159-loop in the β_9-β_{10} turn of the *downstream* subunit of the CAP dimer (Fig. 16.2) is thought to contact the C-terminal region of the α-subunit of RNAP when CAP is bound at a Class I promoter. This contact is also important at Class II promoters where CAP is bound closer to the RNAP binding site, but here it is the 159-loop of the *upstream* subunit of the CAP dimer that contacts the α-subunit of RNAP. At Class II promoters, other contacts involving the E96 region in β_8[59] and the 52-loop situated in the β_4-β_5 turn of CAP (Fig. 16.2) and RNAP are necessary. These are only active in the *downstream* subunit of CAP and may thus contact the C-terminal part of the σ-subunit of RNAP.[60] The amino acid sequence of the 159-loop of CAP is not conserved in FNR and there is evidence to suggest that this region is not involved in transcription activation by FNR.[37] However, comparable mutational studies have indicated that the 85-loop of FNR provides an important activation contact (Fig. 16.2).[37,38] This loop is analogous to the 52-loop of CAP involved in Class II promoter activation, and its importance is consistent with the apparent similarity between FNR-dependent promoters and Class II CRP-dependent promoters. No evidence for an activation patch in FNR equivalent to the E96 region of CAP has been obtained, but it may be significant that there is a glutamate residue (E123) at the corresponding position in FNR and that it is adjacent to an essential cysteine, C122 (Fig. 16.2). The integrity of the 85-loop (like the 52-loop) is not essential for positive control when the regulator binding site is located at -61.5. Nevertheless, recent studies with the semisynthetic FNR-dependent *FF+20pmelR* promoter have now identified a site in FNR (S73) which is essential for activation from the upstream position.[37] This site in the putative β3-β4 turn of FNR (Fig. 16.2) is needed with both classes of FNR-dependent promoters so it is conceivable that it performs an analogous function to that of the 159-loop of CAP. That FNR interacts with the C-terminal region of the α-subunit of RNAP has been inferred from studies with *rpoA* mutants and the FNR-dependent Class II *pepT* promoter of *S. typhimurium*.[61] Thus it can be concluded that members of the CAP-FNR family share at least two potential activation patches, one represented by the 52-loop in CAP and the 85-loop in FNR, important solely with Class II promoters, and the other represented by the 159-loop in CAP (possibly replaced by the 73-loop in FNR) which is important with promoters of Class I and Class II. Further mutational studies should, in the absence of direct structural information, allow the respective contact sites in RNAP to be defined for each type of CAP- and FNR- regulated promoter.

CAP-induced DNA bending is thought to play a significant role in transcription activation at CAP-dependent promoters and, in view of the predicted similarity between CAP and FNR, it seemed likely that this property would apply to FNR.[33] Using a gel shift assay developed with FNR* protein (D154A), it has now been shown that target DNA is bent up to 92° by both FNR* and wild-type FNR, despite the 10-fold lower affinity of the latter.[79] The presence of hypersensitive sites within the FNR-protected regions of DNase I footprints (Fig. 16.2) further shows that FNR induces DNA bending.[a] These hypersensitive sites correspond to those observed in some CAP footprints and may thus define sites of FNR-induced kink formation in DNA. Since DNA bending can itself activate transcription, it seems likely that FNR induced bending and distortion participate in transcription regulation, either by facilitating direct contact between RNAP and FNR, or by allowing RNAP to contact -10 and -35 regions as well as important upstream regions of promoter DNA. Deletion of upstream sequences from the FNR repressible promoters *ndh, narX* and *fnr* resulted in inhibition of expression both in vivo and in vitro, suggesting that RNAP makes important contacts in the upstream regions (around -100) of these promoters and that FNR may regulate transcription at least in part by preventing or altering these contacts.[28,29]

THE MYSTERY OF REDOX-SENSING

The process of redox-sensing and signal transduction is a mystery because holo-FNR, purified from aerobically grown bacteria under aerobic conditions, can activate and repress transcription in vitro without needing an anaerobic stimulus.[27] This suggests that at least a fraction of the purified protein is in the active conformation and that oxygen per se is not directly responsible for FNR inactivation (unless the added ferrous ions afford protection during purification). It is also consistent with the observation that the regulatory switch can be mediated by hexacyanoferrate III (E_o' +520 mV) in the complete absence of oxygen.[54] Based on current evidence, several plausible mechanisms can be proposed for the sensing and response to anaerobiosis in FNR-mediated transcriptional regulation.

(A) FERRIC-FERROUS REDOX CYCLING

One working hypothesis is that FNR contains an iron cofactor which cycles between ferric (inactive) and ferrous (active) states under aerobic and anaerobic conditions, respectively (Fig. 16.5). A role for

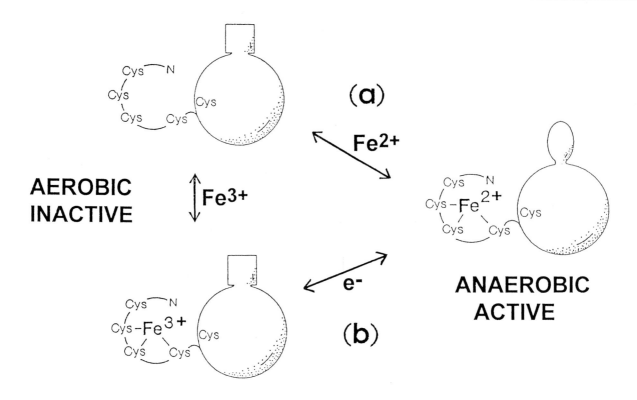

Fig. 16.5. Two models for the mode of action of FNR. Anaerobic activation could involve the conversion of FNR to the transcriptionally active form by (a) assembly of a ferrous iron containing cofactor, or (b) reduction of a cysteine-bound ferric iron cofactor. Only one subunit of the active DNA-bound dimer is shown. Recent studies suggest that the functional form of FNR may contain up to two reduced [4Fe 4S] clusters per dimer and that the anaerobic regulatory switch might be mediated by a very significant increase in DNA-binding affinity that accompanies assembly or reduction of the cluster.[80,a] Oxygen and chelating agents produce iron-depleted apo-FNR.

iron in regulating FNR activity was originally inferred from in vivo studies with chelating agents prompted by analogous studies with the NifA system of *B. japonicum*.[62] These studies showed that iron-chelators mimic oxygen in their effects on both the expression of FNR-regulated *lacZ* fusions[63] and the reactivities of FNR sulfhydryl groups in intact bacteria.[64] Moreover, these effects were best prevented by added ferrous iron. It was also shown that depending on the availability of oxygen and iron, the active and inactive forms of FNR are readily interconverted in the absence of de novo protein synthesis.[65] The iron content of *E. coli* is not significantly different during aerobic or anaerobic growth[66] but there could be differences in the ratios of ferric to ferrous iron which might be sensed and adopted by FNR. The switch could thus be initiated by reduction of a ferric iron cofactor (or iron-sulfur center) in the sensory domain (Fig. 16.5a). Although this model lacks direct experimental confirmation, it is envisaged that cofactor reduction could involve FNR dimers that are already bound to target DNA (as preinduction complexes) or the reduction could initiate simultaneous DNA-binding and activation. In either case, consequent conformational changes in FNR might promote effective contacts with RNAP leading to open-complex formation and transcription activation, or to repression.

Further support for this model comes from the isolation of *fnr** mutants in which FNR is fixed in a partially active state under aerobic conditions.[40,51] One class having single alterations in the putative sensory domain could be explained if the changes (D22G, D22S, L28H, or an insertion of S between residues 17 and 18) either mimic the conformational changes associated with reduction, or affect the tuning or stability of the iron center such that the allosteric switch is initiated at a higher redox potential. Another class has single alterations in the putative dimer interface (e.g., D154A), which may enhance dimer formation or stability. More recently, physicochemical studies with an FNR* protein (L28H D154A) combining both types of alteration, have significantly advanced our understanding of the iron component of FNR.[80] Preparations of this protein were shown to contain an oxygen- and redox-sensitive polynuclear iron-sulfur cluster, with a stoichiometry of about one [3Fe 4S] or [4Fe 4S] center per 10 monomers. Moreover, the affinity for target DNA was up to 40-fold greater than for wild-type FNR. Anaerobic reduction of FNR* with dithionite increased the DNA binding affinity 10-fold relative to protein that had been exposed to oxygen,[80] and this factor might be increased to 50- or 100-fold, were the protein to contain a full complement of one iron-sulfur center per dimer or monomer (respectively). The increased DNA binding affinity associated with the reduction of FNR* suggests that the redox state of the iron-sulfur center governs DNA binding, and thus provides a plausible regulatory mechanism which should likewise apply to the wild-type protein. Indeed, the wild-type FNR protein has been shown to contain iron and acid-labile sulfur, and to exhibit a uv-visible spectrum consistent with the presence of an iron-sulfur center (see above).[a] It will now be important to define the number of iron-sulfur centers per functional dimeric protein and to determine whether each center is liganded to one or both subunits.

(B) Uptake and Release of Iron

In a related mechanism, the switch could be activated by the reversible uptake of iron, so that the iron cofactor (assembled iron-sulfur center) is associated solely with the anaerobic (active) form of FNR (Fig. 16.5). This is supported by the variable iron content of isolated FNR[26,80] and by the increased reactivity of FNR sulfhydryl in aerobic bacteria.[64] Since iron is not essential for site-specific DNA-binding, it is possible that pre-induction complexes containing apo-FNR and RNAP are assembled at FNR-regulated promoters, waiting to be activated by the incorporation of ferrous iron (Fig. 16.5b). The recent activation of apo-FNR by preincubation with Fe^{2+} and β-mercaptoethanol could be regarded as providing in vitro support for this mechanism of transcriptional activation, without excluding the redox-cycling mechanism.[41]

(C) Monomer-Dimer Transition

Although isolated FNR is predominantly monomeric, it appears to be dimeric when bound to target DNA.[26] It is not known whether a preformed dimer is bound or whether the dimer is formed cooperatively on the DNA. In either case, dimerization could be an important step in the anaerobic activation of FNR. Recent support for activation by dimerization has come from studies with a class of *fnr** mutants altered in the putative dimer-interface domain.[42] One such FNR* protein (D154A) has a higher molecular mass (M_r 45,000) and a higher affinity for target DNA than the wild-type protein. This type of FNR* protein could bypass the normal sensory pathway by being fixed in the active (DNA-binding) dimeric form, rather than being converted to the inactive monomer under aerobic conditions. Studies with dominant negative mutants (where the mutant subunits inhibit wild-type activity in mixed dimers) provide further support for this view.[42] The D154A class of FNR* protein may adopt the same active conformation as that which is ultimately adopted by FNR* proteins with substitutions in the N-terminal sensory domain. The combination of N-terminal and dimer interface substitutions produces an FNR* (L28H D154A) which is substantially dimeric (M_r 66,000) and has an increased DNA binding affinity.[80] Thus the D154A type of FNR* protein might have the conformation that is normally induced via the sensory domain in response to the incorporation of iron, or the reduction of the iron cofactor.

(D) PROTEIN MODIFICATION AND A REDOX SIGNALING EFFECTOR

Covalent modification is commonly used to regulate the function or activity of proteins. Clues for such a mechanism have been sought in studies with wild-type and cysteine-substituted FNR proteins. For example, the reversible formation of an intramolecular disulfide bond would provide a plausible redox-sensing mechanism.[1] This mechanism is consistent with the coexistence of two forms of FNR in whole cells and in purified preparations.[35] However, it would follow that replacing the relevant cysteine residues should fix FNR in an active (dithiol) conformation rather than cause the observed inactivation.[39,40] Also, since it is generally accepted that the cytoplasmic proteins of *E. coli* do not contain disulfide bonds, reversible disulfide-bond formation seems an unlikely mechanism, unless there is a specific enzyme catalyzing the reversible formation of a disulfide in FNR. No phosphorylated derivative of FNR has been detected, but electrospray mass spectrometry has indicated that purified FNR contains a covalently-modified component bearing an unidentified substituent of mass 78 ± 7 Da.[35] It was speculated that this might reflect the reductive acylation of an intramolecular disulfide bond. However, in view of the recent evidence that active FNR contains an iron-sulfur center,[80,a] it is possible that the additional mass is due to the retention of S (as persulfides) or an Fe-S adduct, by a fraction of the isolated protein. It is also significant that the iron-deficient C122A and FNR-574 (Δ2-30) proteins lack the modifying substituent(s), whereas the inactive proteins that retain iron (C20S and C23G) are modified.[35,a]

(E) AEROBIC ACTIVITY OF FNR

The possibility that FNR participates in aerobic transcription regulation was raised by the observation that cellular levels of eight polypeptides are reduced in an *fnr* mutant but only under aerobic conditions.[67] In addition, the activation of the secondary *hyp* promoter located in the *hypA* coding region is FNR-dependent under both aerobic and anaerobic conditions,[24] and the autorepression of the *fnr* and *hlyX* genes is not dependent on anaerobiosis, suggesting both regulators can function aerobically.[68,69] Since apo-FNR is known to bind target DNA aerobically and may thus form RNAP-FNR-DNA pre-induction complexes awaiting activation by anaerobiosis,[41] it is quite conceivable that the special geometry or some other feature of specific promoters might allow their activation by the aerobic form of FNR. It has been long known that a high glucose concentration permits some aerobic derepression of fumarate reductase synthesis, just as it prevents the aerobic derepression of the 2-oxoglutarate dehydrogenase complex.[70,71] However, it is not known whether FNR is required for activating the *frd* operon under these conditions. Further study of the aerobic activities of FNR could provide useful clues about its mode of action.

(F) CONCLUSIONS

Clearly there is still a lot to be learned about redox-sensing and signal transduction in FNR. Two factors seem to be particularly important in FNR-mediated transcription regulation: the activation of FNR (by the reduction or assembly of an iron-sulfur center); and the relationship between dimer formation and DNA-binding affinity. It is envisaged that such events are communicated allosterically to RNAP in the transcriptional complex and this in turn leads to the activation or repression of target genes.

Assuming that FNR has its own redox-sensor, the immediate source of the electrons responsible for its reduction is unknown. FNR might interact with a component of the aerobic electron transport chain and thus respond directly to alterations in electron flow during anaerobic stress. There is evidence for such an interaction in *Paracoccus denitrificans*[72] and possibly also in *E. coli* in view of the (weak but significant) aerobic derepression of nitrate reductase in ubiquinone-deficient mutants.[73] Alternatively, FNR might respond to changes in the intracellular ratio of ferric to ferrous ions that may accompany an aerobic-anaerobic shift. Any mechanism of FNR action must account for the ready interconversion of active and inactive forms. The transfer of electrons to oxygen via a component of the electron transport chain (e.g., ubiquinone) or the release and oxidation of the ferrous iron cofactor under aerobic conditions could account for the reversible inactivation of FNR. All of these uncertainties highlight the very pressing need to reproduce in vitro, the aerobic-anaerobic switch and thereby seek out the routes of FNR reduction, oxidation and signal transduction.

STRUCTURAL AND FUNCTIONAL HOMOLOGS OF FNR

FNR is a member of a growing family of regulatory proteins having structures that are predicted to resemble CAP.[74,75] Excluding the CAP proteins, some 19 FNR homologs have so far been identified in a variety of gram-negative and gram-positive species (Table 16.2). They perform diverse regulatory functions and comparisons could shed light on unresolved questions concerning the *E. coli* protein. The FNR homologs can be classified according to their domain organization, cysteine-residue conservation, and type of DNA-recognition motif, as summarized in Figure 16.6. The proteins included in each class are as follows: **FNR** (FNR$_{Ec}$, FNR$_{St}$, FNR$_{Vf}$, EtrA, HlyX, FnrA, ANR, BTR, FixK$_{Bj}$, FnrN and AadR); **FLP** (FLP$_{Ll}$ and FLP$_{Lc}$); **FixK** (FixK$_{Rm}$, FixK$_{Ac}$, Orf4, and NNR); **PrfA** (PrfA$_{Li}$ and PrfA$_{Lm}$); and **FNR$_{Bs}$** (FNR$_{Bs}$). This structural classification is not mirrored by any obvious functional grouping, indeed regulators of nitrogen fixation are found in different classes.

The DNA binding domain is the most highly conserved region, and where studied, the homologs appear to recognize DNA sequences that are very similar to the *E. coli* FNR-site consensus.[75] The residues that make specific contacts with the DNA target sites are known only for FNR$_{Ec}$ and CAP.[34] Nevertheless, in many cases the site-specific DNA binding motif of FNR$_{Ec}$ (-E--SR) is conserved in the putative DNA-recognition helices of most of the relatives (Fig. 16.6). The exceptions are FNR$_{Bs}$, where the corresponding motif more closely resembles that of CRP (**RE---R**), and the two PfrA proteins, where the motif (----SR) retains only one of the specificity-conferring residues of the FNR motif and the palindromic PrfA-sites do not align precisely with either FNR- or CRP-sites.[72]

Thirteen of the homologs (including FNR$_{Ec}$) exhibit partial or complete conservation of the four essential cysteine residues that are thought to be involved in signal recognition in FNR$_{Ec}$ (Fig. 16.6). There is some degree of variation in both the number and spacing of the cysteine residues in the N-terminal clusters, despite the fact that at least one insertion between C20 and C23 severely affects FNR activity.[40] Interestingly, the FNR-like proteins of the gram-positive species are organized differently. FLP$_{Ll}$ and FLP$_{Lc}$ have only one cysteine

Table 16.2. FNR homologs

Protein	Organism	Function regulated	Distance from FNR
FNR$_{St}$ (OxrA)	*Salmonella typhimurium*	Anaerobic respiration	0.01
FNR$_{Vf}$	*Vibrio fischeri*	Luminescence	0.16
EtrA	*Shewanella putrefaciens*	Anaerobic respiration	0.28
HlyX	*Actinobacillus pleuropneumoniae*	Hemolysin biosynthesis	0.36
FnrA	*Pseudomonas stutzeri*	Arginine fermentation	0.64
ANR	*Pseudomonas aeruginosa*	Denitrification and arginine fermentation	0.67
BTR	*Bordetella pertussis*	Hemolysin biosynthesis?	1.36
FixK$_{Bj}$	*Bradyrhizobium japonicum*	N$_2$ fixation and denitrification	1.55
FnrN	*Rhizobium leguminosarum*	Nitrogen fixation	1.72
FixK$_{Rm}$	*Rhizobium meliloti*	Nitrogen fixation	1.82
AadR	*Rhodopseudomonas palustris*	Aromatic acid degradation	1.83
Orf4	*Rhizobium* IC3342	Not known	1.84
FixK$_{Ac}$	*Azorhizobium caulinodans*	Nitrogen fixation	1.95
FLP$_{Ll}$	*Lactococcus lactis*	Not known	2.09
FLP$_{Lc}$	*Lactobacillus casei*	Not known	2.12
CAP	*Escherichia coli*	Catabolite repression	2.23
NNR	*Paracoccus denitrificans*	Nitrite and NO respiration	2.62
FNR$_{Bs}$	*Bacillus subtilis*	Anaerobic respiration	2.66
PrfA$_{Lm}$	*Listeria monocytogenes*	Virulence	2.86
PrfA$_{Li}$	*Listeria ivanovii*	Virulence	2.88

The evolutionary distances from FNR were estimated by PROTDIST in the PHYLIP package[76] using an alignment for FNR and 19 homologs. The *S. typhimurium* sequence is from Genbank (entry U05668). Other sequences can be found elsewhere,[75,77,78] except for two that were kindly provided prior to publication, FNR$_{Bs}$ (P Glaser) and FLP$_{Ll}$ (MJ Gasson, DO Gostick, HJ Griffin and JR Guest).

residue near the N-terminus but retain the central cysteine residue (C122), and in FNR$_{Bs}$ the sensory domain is split such that four cysteine residues are clustered near the C-terminus and neither of the two remaining cysteines align with the central conservation site (Fig. 16.6). Conservation of a cysteine cluster and a central cysteine residue in most of the proteins suggests that they contain polynuclear iron-sulfur centers and have a common mechanism of signal recognition. The structural variations might point to differences in the redox-potentials of the redox-sensitive iron centers, or in some cases, to the presence of different metal cofactors. In the naturally occurring homologs which lack the critical cysteine residues (FixK of *R. meliloti*, FixK of *A. caulinodans* and Orf4 of *Rhizobium* IC3342) it appears that expression of the protein rather than its activity is oxygen-regulated. In this context it is interesting that six homologs, including three that lack the essential cysteine residues, have alanine residues at the site of the FNR* substitution (D154A). However, three others (FixK of *B. japonicum*, FnrN, and AadR) do not, indicating that the D154A substitution is not necessarily associated with redox-independent activity, but it could signify a dimeric quaternary structure.

There is a degree of sequence conservation in the region of FNR that has been implicated in transcriptional activation but there are also some notable differences. For example, substituting D86 severely affects positive control[37] but this residue is replaced by a positively charged residue (arginine) in the rhizobial proteins, some of which retain the ability to activate transcription in *E. coli*. Members of the CRP-FNR family clearly exhibit different combinations of signal and target recognition specificity. Hopefully further structure-function studies will reveal the underlying molecular bases of these variations.

Several functionally related regulatory proteins have been described. These include NifA, a redox- and iron-dependent regulator of nitrogen fixation in *B. japonicum.*[62] Like FNR it has four essential cysteine residues and could well contain a similar redox-sensitive iron-sulfur center. Indeed, the FNRs and NifA would seem to be the regulatory protein counterparts of the oxygen-labile dehydratase enzymes (e.g., aconitase A and fumarase A) which contain unstable [4Fe 4S] centers that are likewise reactivated by Fe^{2+} and a sulfhydryl compound. In *E. coli,* the superoxide stress response is initiated by the SoxR regulator, which contains a stable iron-sulfur center needed for transcription activation but not for DNA-binding.[81] In contrast, OxyR, the peroxide stress regulator of *E. coli* is activated via a cysteine sulfhydryl,[82] and the oxygen-responsive sensor (FixL) of the FixLJ two-component regulatory system for nitrogen fixation in *R. meliloti* contains a hemoglobin-like sensory domain.[83] Clearly, iron and iron-containing prosthetic groups have been recruited to perform essential roles in transcription regulation as well as in enzymes and electron or oxygen carriers.

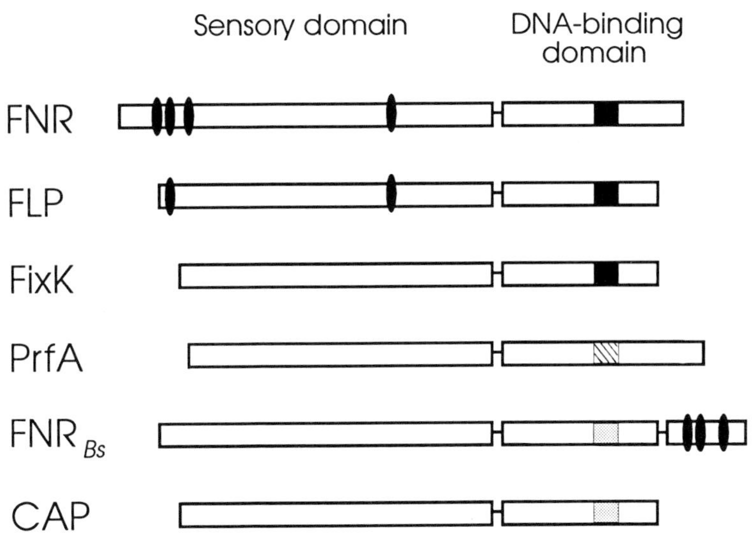

Fig. 16.6. Structural organization in the CRP-FNR family of regulatory proteins. Domain organization and other features of CRP, FNR and FNR-related regulators. The proteins are grouped according to the conservation of cysteine residues and binding-site motifs in putative DNA recognition helices, relative to FNR$_{Ec}$ (see text). Cysteine residues that are essential in FNR$_{Ec}$ and appear to be conserved in other proteins are marked (I). The three types of DNA-recognition motif are denoted thus: FNR-like, -E--SR (filled box); CRP-like, RE--R (stippled box): and PfrA---SR (hatched box). Unaligned cysteine residues of unknown importance, including two in CAP and three in FNR$_{Bs}$, one of which is in the C-terminal cysteine cluster, are not shown.

CONCLUDING REMARKS

The existence of X-ray structures of CAP complexed with DNA and with its coeffector, as well as the extensive literature on CAP function, have provided the basis for exploring the structure-function relationships of FNR and its homologs. Clear parallels have been established with FNR but there remain significant differences with respect to regulator:RNAP interactions, protein:DNA contacts, and the overall process of signal sensing and response. FNR is not a good candidate for addressing general problems of transcription activation and repression, even though some of its disadvantages could be overcome by the effective use of mutant proteins (e.g., various classes of FNR* protein). However, FNR does pose specific problems that deserve attention.

An FNR structure would provide invaluable information but numerous crystallization trials with the wild-type protein have so far been unsuccessful, as have nmr-spectroscopic studies with the cysteine-containing segment.[a] In this respect, the dimeric FNR* protein or the seemingly homogeneous cysteine mutants might be more suitable candidates for structural studies.

The interaction of FNR with RNAP presents specific problems in transcriptional activation and repression. FNR operates at Class II-like promoters and appears less able to use FNR-sites at other locations. There are few natural Class I equivalents (Table 16.1) and compared with CAP, FNR seems to use different positive control contacts. FNR-mediated repression likewise differs from the classical model in which the binding or progression of RNAP is blocked by the regulator. In contrast it appears to be an active process requiring multiple upstream FNR-sites without involving other proteins. Here, studies with 'negative control' mutants could be as enlightening as those with their 'positive-control' counterparts. The role of DNA bending in all of these processes also deserves further attention.

The ability to sense and respond to specific signals is a fundamental property of all transcription regulators. The structure of the putative redox-sensing domain with and without iron, and its relation to the allosteric and DNA-binding domains may prove the key to understanding redox (oxygen) sensing and response. Further studies with the more amenable FNR* proteins that are less sensitive to oxidative inactivation, should point the way to elucidating the fundamental property of this class of regulators. Contributions from epr and Mössbauer spectroscopy and studies with model peptides, should be illuminating but a complete three-dimensional structure and a functional in vitro switch are the 'holy grails' of FNR research, even though mutational and in vitro studies on the interactions between FNR, DNA, RNAP and other factors at a variety of different promoters, offer productive diversions in the immediate future.

ACKNOWLEDGMENTS

Support from The Wellcome Trust and the Biochemical and Biological Sciences Research Council is gratefully acknowledged. The Krebs Institute is a BBSRC-designated Center for Molecular Recognition Studies.

REFERENCES

1. Spiro S, Guest JR. FNR and its role in oxygen-regulated gene expression in *Escherichia coli*. FEMS Microbiol Rev 1990; 75:399-428.

2. Unden G, Becker S, Bongaerts J et al. Oxygen regulated gene expression in facultatively anaerobic bacteria. Antonie van Leeuwenhoek 1994; 66, 3-23.

3. Gunsalus RP. Control of electron flow in *Escherichia coli*: coordinated transcription of respiratory pathway genes. J Bacteriol 1992; 174: 7069-7074.

4. Lin ECC, Iuchi S. Regulation of gene expression in fermentative and respiratory systems in *Escherichia coli* and related bacteria. Annu Rev Genet 1991; 25:361-387.

5. Iuchi S, Lin ECC. Adaptation of *Escherichia coli* to redox environments by gene expression. Mol Microbiol 1993; 9:9-15.

6. Stewart V. Nitrate respiration in relation to facultative metabolism in enterobacteria. Microbiol Rev 1988; 52:190-232.

7. Stewart V. Nitrate regulation of anaerobic respiratory gene expression in *Escherichia coli*. Mol Microbiol 1993; 9:425-434.

8. Lin ECC, Kuritzkes DR. Pathways for anaerobic electron transport. In: Neidhardt FC, ed. *Escherichia coli* and *Salmonella typhimurium* Cellular and Molecular Biology. Washington: Amer Soc Microbiol, 1987:201-221.

9. Kaiser M, Sawers G. Pyruvate formate-lyase is not essential for nitrate respiration by *Escherichia coli*. FEMS Microbiol Lett 1994; 117:163-168.

10. Clark DP. The fermentation pathways of *Escherichia coli*. FEMS Microbiol Rev 1989;63:223-234.

11. Rossmann R, Sawers G, Bock A. Mechanism of regulation of the formate-hydrogen-lyase pathway by oxygen, nitrate and pH: definition of the formate regulon. Mol Microbiol 1991; 5:2807-2814.

12. Sauter M, Bohm R, Bock A. Mutational analysis of the operon (*hyc*) determining hydrogenase 3 formation in *Escherichia coli*. Mol Microbiol 1992; 6:1523-1532.

13. Knappe J, Sawers G. A radical-chemical route to acetyl-CoA: the anaerobically induced pyruvate-formate-lyase system of *Escherichia coli*. FEMS Microbiol Rev 1990; 75:383-398.

14. Quail MA, Haydon DJ, Guest JR. The *pdhR-aceEF-lpd* operon of *Escherichia coli* expresses the pyruvate dehydrogenase complex. Mol Microbiol 1994; 12:95-104.

15. Engel P, Krämer R, Unden G. Transport of C$_4$-dicarboxylates by anaerobically-grown *Escherichia coli*: energetics and mechanism of exchange, uptake and efflux. Eur J Biochem 1994; 174:605-614.

16. Iuchi S, Aristarkhov A, Dong JM et al. Effects of nitrate respiration on expression of the arc-controlled operons encoding succinate dehydrogenase and flavin-linked L-lactate dehydrogenase. J Bacteriol 1994; 176:1695-1701.

17. Cotter PA, Gunsalus RP. Contribution of the *fnr* and *arcA* gene products in coordinate regulation of cytochrome *o* and *d* oxidase (*cyoABCDE* and *cydAB*) genes in *Escherichia coli*. FEMS Microbiol Lett 1992; 91:31-36.

18. Hassan HM, Sun HCH. Regulatory roles of Fnr, Fur and Arc in expression of manganese-containing superoxide dismutase in *Escherichia coli*. Proc Natl Acad Sci USA 1992; 89:3217-3221.

19. Fu HA, Iuchi S, Lin ECC. The requirement of *arcA* and *fnr* for peak expression of the *cyd* operon in *Escherichia coli* under microaerobic conditions. Mol Gen Genet 1991; 226:209-213.

20. Sawers G. Specific transcriptional requirements for positive regulation of the anaerobically inducible *pfl* operon by ArcA and FNR. Mol Microbiol 1993; 10:737-747.

21. McCleary R, Stock JB, Ninfa AJ. Is acetyl phosphate a global signal in *Escherichia coli*? J Bacteriol 1993; 175: 2793-2798.

22. Wanner BL, Wilmes-Riesenberg MR. Involvement of phosphotransacetylase, acetate kinase and acetyl phosphate synthesis in control of the phosphate regulon in *Escherichia coli*. J Bacteriol 1993; 174: 2124-2130.

23. Rabin RS, Stewart V. Dual response regulators (NarL and NarP) interact with dual sensors (NarX and NarQ) to control nitrate-and nitrite-regulated gene expression in *Escherichia coli*-K12. J Bacteriol 1993; 175:3259-3268.

24. Lutz S, Jacobi A, Schlensog V et al. Molecular characterization of an operon (*hyp*) necessary for the activity of the three hydrogenase isoenzymes in *Escherichia coli*. Mol Microbiol 1991; 5:123-135.

25. Riley M. Functions of the gene products of *Escherichia coli*. Microbiol Rev 1993; 57:862-952.

26. Green J, Trageser JM, Six S et al. Characterization of the FNR protein of *Escherichia coli*, an iron-binding transcriptional regulator. Proc Roy Soc Lond B 1991; 244:137-144.

27. Sharrocks AD, Green J, Guest JR. FNR activates and represses transcription in vitro. Proc Roy Soc Lond B 1991; 245:219-226.

28. Green J, Guest JR. Regulation of transcription at the *ndh* promoter of *Escherichia coli* by FNR and novel factors. Mol Microbiol 1994; 12:433-444.

29. Takahashi K, Hattori T, Nakanishi T et al. Repression of in vitro transcription of the *Escherichia coli fnr* and *narX* genes by FNR protein. FEBS Lett 1994; 340: 59-64.

30. Suppmann B, Sawers G. Isolation and characterization of hypophosphite-resistant mutants of *Escherichia coli*: identification of the FocA protein, encoded by the *pfl* operon, as a putative formate transporter. Mol Microbiol 1994; 11:965-982.

31. Chen Y-M, Lin ECC. Regulation of the *adhE* gene, which encodes ethanol dehydrogenase in *Escherichia coli*. J Bacteriol 1991; 173:8009-8013.

32. Compan I, Touati D. Anaerobic activation of *arcA* transcription in *Escherichia coli*: roles of Fnr and ArcA. Mol Microbiol 1994; 11:955-964.

33. Schultz SC, Shields GC, Steitz TA. Crystal structure of a CAP-DNA complex: the DNA is bent by 90. Science 1991; 253:1001-1007.

34. Spiro S, Gaston KL, Bell AI et al. Interconversion of the DNA-binding specificities of two related transcription regulators, CRP and FNR. Mol Microbiol 1990; 4:1831-1838.

35. Green J, Sharrocks AD, Green B et al. Properties of FNR proteins substituted at each of the five cysteine residues. Mol Microbiol. 1993; 8:61-68.

36. Unden G, Duchene A. On the role of cyclic AMP and the FNR protein in *E. coli* growing anaerobically. Arch Microbiol 1987; 147:195-200.

37. Bell A, Busby S. Location and orientation of an activating region in the *Escherichia coli* transcription factor, FNR. Mol Microbiol 1994; 11:383-390.

38. Williams R, Bell A, Sims G et al. The role of two surface exposed loops in transcription activation by the *Escherichia coli* CRP and FNR proteins. Nucl Acids Res 1991; 19:6705-6712.

39. Sharrocks AD, Green J, Guest JR. In vivo and in vitro mutants of FNR, the anaerobic transcriptional regulator of *E. coli*. FEBS Lett 1990; 270:119-132.

40. Melville SB, Gunsalus RP. Mutations in *fnr* that alter anaerobic regulation of electron transport-associated genes in *Escherichia coli*. J Biol Chem 1990; 265:18733-18736.

41. Green J, Guest JR. Activation of FNR-dependent transcription by iron: an in vitro switch for FNR. FEMS Microbiol Lett 1993; 113:219-222.

42. Lazazzera BA, Bates DM, Kiley PJ. The activity of the *Escherichia coli* transcription factor FNR is regulated by a change in oligomeric state. Genes Dev 1993; 7:1993-2005.

43. Bell A, Cole JA, Busby SJW. Molecular genetic analysis of an FNR-dependent anaerobically inducible *Escherichia coli* promoter. Mol Microbiol 1990; 4:1753-1763.

44. Green J, Guest JR. A role for iron in transcriptional activation by FNR. FEBS Lett 1993; 329:55-58.

45. Adhya S, Gottesman M, Garges S. Promoter resurrection by activators - a minireview. Gene 1993; 132:1-6.

46. Kolb A, Busby S, Buc H et al. Transcriptional regulation by cAMP and its receptor protein. Ann Rev Biochem 1993; 62:747-795.

47. Collado-Vides J, Magasanik B, Gralla JD. Control site location and transcriptional regulation in *Escherichia coli*. Microbiol Rev 1991; 55:371-394.

48. Li J, Stewart V. Localization of upstream sequence elements required for nitrate and anaerobic induction of *fdn* (formate dehydrogenase-N) operon expression in *Escherichia coli* K-12. J Bacteriol 1992; 174:4935-4942.

49. Stewart V. Requirement of Fnr and NarL functions for nitrate reductase expression in *Escherichia coli* K-12. J Bacteriol 1982; 151:1320-1325.

50. Tyson KL, Bell AI, Cole JA et al. Definitioin of nitrite and nitrate response elements at the anaerobically inducible *Escherichia coli* nirB promoter—interactions between FNR and NarL. Mol Microbiol 1993; 7:151-157.

51. Kiley PJ, Reznikoff W. Fnr mutants that activate gene expression in the presence of oxygen. J Bacteriol 1991; 173:16-22.

52. Walker MS, Demoss JA. Role of alternative promoter elements in transcription from the *nar* promoter of *Escherichia coli*. J Bacteriol 1992; 174:1119-1123.

53. Schröder I, Darie S, Gunsalus RP. Activation of the *Escherichia coli* nitrate reductase (*narGHJI*) operon by NarL and Fnr requires integration host factor. J Biol Chem 1992; 268:771-774.

54. Unden G, Trageser M, Duchene A. Effect of positive redox potentials (>+400mV) on the expression of anaerobic respiratory enzymes of *Escherichia coli*. Mol Microbiol 1990; 4:315-319.

55. Bogachev AV, Murtazina RA, Skulacher VP. Cytochrome *d* induction in *Escherichia coli* growing under unfavourable conditions. FEBS Lett 1993; 336:75-78.

56. Eraso JM, Weinstock GM. Anaerobic control of colicin E1 production. J Bacteriol 1992; 174:5101-5109.

57. Jennings MP, Beacham IR. Co-dependent positive regulation of the *ansB* promoter of *Escherichia coli* by CRP and the FNR protein: a molecular analysis. Mol Microbiol 1993; 9:155-164.

58. Zhang X, Ebright RH. Substitution of 2 base pairs (1 base pair per DNA half-site) within the *Escherichia coli lac* promoter DNA site for catabolic gene activator protein places the lac promoter in the FNR regulon. J Biol Chem 1990; 265:12400-12403.

59. West D, Williams R, Rhodius V et al. Interactions between the *Escherichia coli* cyclic AMP receptor protein and RNA polymerase at Class II promoters. Mol Microbiol 1993; 10: 789-797.

60. Kumar A, Grimes B, Fujita N et al. Role of sigma [70] subunit of *Escherichia coli* RNA polymerase in transcription activation. J Mol Biol 1994; 235:405-413.

61. Lombardo MJ, Bagga D, Miller CG. Mutations in *rpoA* affect expression of anaerobically regulated genes in *Salmonella typhimurium*. J Bacteriol 1991; 173:7511-7518.

62. Fischer HM, Bruderer T, Hennecke H. Essential and non-essential domains in the *Bradyrhizobium japonicum* NifA protein: identification of indispensible cysteine residues potentially involved in redox activity and/or metal binding. Nucl Acids Res 1988; 16:2207-2224.

63. Spiro S, Roberts R, Guest JR. FNR-dependent repression of the *ndh* gene of *Escherichia coli* and metal ion requirement for FNR regulated gene expression. Mol Microbiol 1989; 3:601-608.

64. Trageser M, Unden G. Role of cysteine residues and of metal ions in the regulatory functioning of FNR, the transcriptional regulator of anaerobic respiration in *Escherichia coli*. Mol Microbiol 1989; 3:593-599.

65. Engel P, Trageser M, Unden G. Reversible interconversion of the functional state of the gene regulator FNR from *Escherichia coli* in vivo by O_2 and iron availability. Arch Microbiol 1991; 156:463-470.

66. Niehaus F, Hantke K, Unden G. Iron content and FNR-dependent gene regulation in *Escherichia coli*. FEMS Microbiol Lett 1991; 84:319-324.

67. Sawers RG, Zehelein E, Bock A. Two-dimensional gel electrophoretic analysis of *Escherichia coli* proteins: influence of various anaerobic growth conditions and the *fnr* gene product on cellular protein composition. Arch Microbiol 1988; 149: 240-244.

68. Pascal M-C, Bonnefoy V, Fons M et al. Use of gene fusions to study the expression of *fnr*, the regulatory gene of anaerobic electron transfer in *Escherichia coli*. FEMS Microbiol Lett 1986; 36:35-40.

69. Soltes GA, MacInnes JI. Regulation of gene expression by the HlyX protein of *Actinobacillus pleuropneumoniae*. Microbiol 1994; 140: 839-845.

70. Spencer M, Guest JR. Isolation and properties of fumarate reductase mutants of *Escherichia coli*. J Bacteriol 1973; 114: 563-570.

71. Amarasingham CR, Davis BD. Regulation of -ketoglutarate dehydrogenase formation in *Escherichia coli*. J Biol Chem 1965; 240: 3664-3668.

72. Kucera I, Matchova I, Spiro S. Respiratory inhibitors activate an FNR-like regulatory protein in *Parracoccus denitrificans*. Biochem Mol Biol Int 1994; 32:245-250.

73. Giordano G, Grillet L, Rosset R et al. Characterization of an *Escherichia coli* mutant that is sensitive to chlorate when grown aerobically. Biochem J 1978; 176: 553-561.

74. Irvine AS, Guest JR. *Lactobacillus casei* contains a member of the CRP-FNR family. Nucl Acids Res 1993; 21:753.

75. Spiro S. The FNR family of transcriptional regulators. Antonie van Leeuwenhoek 1994; 66, 23-36.

76. Felsenstein J. PHYLIP (Phylogeny Inference Package) Version 3.5c, 1993. University of Washington, Seattle.

77. Lampdis R, Gross R, Sokolovic Z et al. The virulence regulator of *Listeria ivanovii* is highly homologous to PfrA from *Listeria monocytogenes* and both belong to the Crp-Fnr family of transcription regulators. Mol Microbiol 1994; 13:141-151.

78. Van Spanning RJ, Anthonius M, De Boer PN et al. Nitrite and nitric oxide reduction in *Paracoccus denitrificans* is under the control of NNR, a regulatory protein that belongs to the FNR family of transcriptional activators. FEBS Lett 1995; in press.

79. Ziegelhoffer EC, Kiley PJ. In vitro analysis of a constitutively active mutant form of the *Escherichia coli* global transcription factor FNR. J Mol Biol 1995; 245: 351-361.

80. Khoroshilova N, Beinert H, Kiley PJ. Association of a polynuclear iron-sulfur center with a mutant FNR protein enhances DNA-binding. Proc Natl Acad Sci USA 1995; 92: 2499-2505.

81. Hidalgo E, Demple B. An iron-sulphur center essential for transcriptional activation by the redox-sensing SoxR protein. EMBO J 1994; 13: 138-146.

82. Kullick I, Toledano MB, Tartaglia LA et al. Mutational analysis of the redox-sensitive transcriptional regulator OxyR: regions important for oxidation and transcriptional activation. J Bacteriol 1995; 177: 1275-1284.

83. Gilles-Gonzalez MA, Ditta GS, Helinski DR. A haemoprotein with kinase activity encoded by the oxygen sensor of *Rhizobium meliloti*. Nature 1991; 350: 170-172.

NOTE

[a] Unpublished work by the authors and colleagues.

CHAPTER 17

THE NAR MODULON SYSTEMS: NITRATE AND NITRITE REGULATION OF ANAEROBIC GENE EXPRESSION

Andrew J. Darwin and Valley Stewart

INTRODUCTION

B acterial survival often depends on the most efficient exploitation of the available resources. This is particularly true for the vital process of energy generation and the enteric bacterium *Escherichia coli*, with the ability to grow with or without oxygen, has evolved mechanisms to ensure that it takes the most favorable energy-generating option available. This extensively studied organism synthesizes two cytochrome oxidases for respiration at high or low oxygen concentrations. However, when cultivated without aeration the organization of the central metabolic pathways actually typifies that of anaerobes. This is no surprise since the environment of the mammalian intestine is anaerobic. Under these anaerobic conditions the organism generates energy by mixed-acid fermentation and/or by respiration with several terminal electron acceptors. The yield of energy from anaerobic respiration is related to the midpoint potential of the electron acceptor couple.[1] The relatively high midpoint potentials of the nitrate/nitrite and nitrite/ammonium couples result in the preference of nitrate and nitrite over the other anaerobic electron acceptors, dimethylsulfoxide (DMSO), trimethylamine *N* oxide (TMAO) and fumarate. A global regulatory system has evolved to regulate anaerobic gene expression in response to nitrate and nitrite availability. Early observations indicated that this Nar (Nitrate reductase) regulatory system fitted into a two-component regulator class. However, further studies have revealed an unexpected degree of complexity.

This complex regulatory system consists of dual homologous membrane-bound sensor proteins (NarX and NarQ) and dual homologous DNA-binding response regulators (NarL and NarP). Recent observations have revealed that the system regulates anaerobic gene expression in response to two electron acceptors: nitrate and its reduction product,

Regulation of Gene Expression in Escherichia coli, edited by E. C. C. Lin and A. Simon Lynch. © 1996 R.G. Landes Company.

nitrite.[2,3] Operons that are known to be members of the Nar modulon are listed in Table 17.1. This review will describe how the Nar system has been characterized to date. We will discuss current ideas regarding sensor protein function, interactions between the sensors and response regulators and how two transcriptional regulatory proteins control the expression of common target operons. The Nar system is not involved in the regulation of genes required for the utilization of nitrate as a nitrogen source for biosynthesis. In enteric bacteria this process is mediated by a different set of enzymes that are under separate genetic control from the enzymes involved in anaerobic nitrate respiration.[4,5]

ANAEROBIC RESPIRATION

The pathways of anaerobic respiration are composed of distinct dehydrogenases, a common quinone pool and distinct reductases. Electron donors include formate, hydrogen, NADH, glycerol-3-phosphate, succinate and lactate. The respiratory reductases include nitrate reductase (encoded by the *narGHJI* operon), formate-dependent nitrite reductase (*nrfABCDEFG*), DMSO reductase (*dmsABC*), TMAO reductase (*torCAD*), and fumarate reductase (*frdABCD*). Another nitrate reductase encoded by the *narZYWV* operon is constitutively expressed at relatively low levels.[6] In addition, sequence analysis suggests that the *nap-ccm* (*aeg-46.5*) operon[7] encodes a periplasmic nitrate reductase homologous to the NapAB enzyme of *Alcaligenes eutrophus*.[8,9] Nitrite is also reduced by an NADH-dependent nitrite reductase encoded by the *nirBDC* operon,[10] which apparently functions to detoxify nitrite and to regenerate NAD+.

The synthesis of most anaerobic respiratory enzymes is subject to two levels of global control. First, the enzymes are synthesized only during anaerobic growth, and this control is mediated by the transcriptional regulatory protein Fnr. Regulation of gene expression by Fnr is described in detail in the chapter by Guest et al in this book. The ArcA/ArcB two-component system is also involved in the anaerobic regulation of several genes in *E. coli*,[11] (see also the chapter by Lynch and Lin) although most of the operons described in our review are under Fnr and not ArcA/ArcB control. The second level of global control for anaerobic respiratory genes responds to nitrate. Nitrate induces the synthesis of the components of the formate-nitrate respiratory chain, formate dehydrogenase-N (encoded by the *fdnGHI* operon)[12] and nitrate reductase,[13] and represses the synthesis of other respiratory chain components such as fumarate reductase[14] and formate-dependent nitrite reductase.[15] Nitrite also plays a role in the regulation of anaerobic respiratory gene expression, as discussed below.

Our current model to explain the nitrate and nitrite regulation of some target operons is presented in Figure 17.1. In response to the presence of nitrate, two homologous, membrane-bound sensor proteins (NarX and NarQ) are able to activate (phosphorylate) either of two homologous DNA-binding response regulators (NarL and NarP). In the presence

Table 17.1. Known NarL- and NarP-regulated operons

NarL	NarP	Operon	Function	References
+[a]	Ø	*narGHJI*	Nitrate reductase	13, 27
+	Ø	*narK*	Nitrite extrusion	19, 58
+	+	*fdnGHI*	Formate dehydrogenase-N	12, 27
+	+	*nirBDC*	NADH-nitrite reductase	31, 32
+	+	*nuoA-N*	NADH dehydrogenase I	69
+	+	*narXL*	Nitrate/nitrite regulation	19, 23, 58
+	?	*modABCD*	Molybdate uptake	70
±[b]	+	*nrfABCDEFG*	Formate-nitrite reductase	27, 32
–	Ø	*frdABCD*	Fumarate reductase	14, 27
–	+	*nap-ccm*	Periplasmic nitrate reductase	7, 27
–	–	*pfl*	Pyruvate-formate lyase	71, 72
–	?	*adhE*	Alcohol dehydrogenase	73, 74
–	?	*dmsABC*	DMSO reductase	75

[a] +, positive regulation (activation); –, negative regulation (repression); Ø, no significant effect; ?, not yet examined
[b] NarL protein represses *nrfA* operon expression in response to nitrate, and activates *nrfA* operon expression in response to nitrite. Modified from ref. 3.

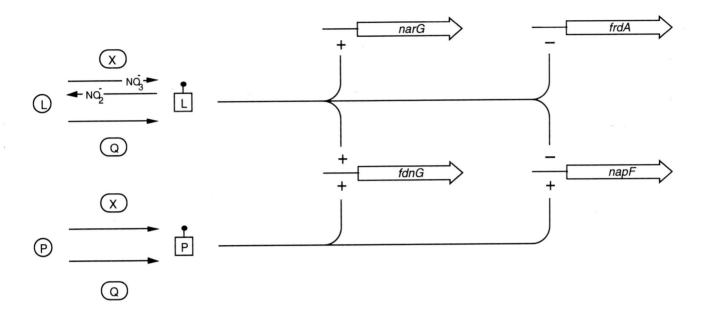

Fig. 17.1. Simplified model for nitrate and nitrite regulation. Open arrows indicate protein-coding regions and their direction of transcription. +, Positive regulation (activation); –, negative regulation (repression). The NarL and NarP proteins are phosphorylated by the NarX and NarQ sensor proteins and bind to target operon control regions to activate or repress transcription. Note that the NarL protein antagonizes NarP-dependent activation of napF operon expression rather than acting as a classical repressor, as is the case at frdA. In the presence of nitrate, the NarX and NarQ proteins phosphorylate both NarL and NarP. In the presence of nitrite, the NarQ protein phosphorylates NarL and NarP, whereas the NarX protein phosphorylates NarP but primarily dephosphorylates NarL.

of nitrite the NarQ protein also phosphorylates NarL and NarP, whereas the NarX protein apparently phosphorylates NarP but primarily dephosphorylates NarL. The phosphorylated response regulators interact with target operon control regions to activate or repress transcription. In some cases transcription is regulated by NarL alone (e.g., *narG* operon expression) whereas the expression of other operons is controlled by the concerted action of both the NarL and NarP proteins (e.g., *nap-ccm* operon expression). The experimental observations which have led us to propose this model will be discussed below.

THE CHARACTERIZATION OF THE NAR REGULATORY SYSTEM

The reduction of chlorate (ClO_3^-) to the toxic product chlorite (ClO_2^-) is catalyzed by nitrate reductase. Consequently early genetic studies of nitrate reductase focused on the isolation of chlorate resistant mutants.[16] A few such mutants appeared to be specifically defective in nitrate reductase synthesis. Others exhibited pleiotropic defects due to failure to synthesize the molybdenum cofactor (which is an essential cofactor for nitrate reductase and several other enzymes).[17,18] One mutant retained basal nitrate reductase activity which was no longer inducible by nitrate. This mutation was termed *narL* and further studies demonstrated that the *narL* gene encodes a specific nitrate-responsive positive regulatory element.[13] The *narL* gene is adjacent to the *narGHJI* operon at 27 minutes on the genetic map. The *narL* gene product is also required for the nitrate induction of *fdnGHI* operon expression and for nitrate repression of *frdABCD* operon expression.[12,14] Therefore, the NarL protein functions both as a transcriptional activator and repressor.

Genetic studies of the *narL* region of the chromosome led to the identification of a closely linked gene, *narX*.[19] Deduced amino acid sequences predicted that the NarL protein was a response regulator of a two-component system and that the NarX protein was its cognate membrane-bound sensor.[20,21] However, this latter prediction was confounded by the observation that insertions in the *narX* gene failed to abolish NarL-dependent nitrate regulation but instead had only a subtle regulatory phenotype.[19,22] As a result of these observations Egan and Stewart investigated the effects of in-frame (nonpolar) *narX* deletions on nitrate regulation.[23] These deletions had no effect on the nitrate induction of a Φ(*fdnG-lacZ*) fusion or the nitrate repression of a Φ(*frdA-lacZ*) fusion. Other experiments demonstrated that the *narX* and *narL* genes form the complex *narXL* operon. Polar insertions in *narX* reduce, but do not abolish, *narL* expression.[23]

Egan and Stewart went on to postulate that, since nonpolar *narX* deletions have no effect on nitrate regulation, there must be another nitrate sensor that can substitute for NarX protein function.[23] Subsequently, genetic screens for insertions that abolish nitrate regulation in a *narX* null strain led to the identification of the second sensor protein, encoded by the *narQ* gene at 53 minutes on the genetic map.[24,25] Either *narX*[+] or *narQ*[+] is sufficient for essentially normal nitrate-dependent activation or repression of target operons by the NarL protein, whereas *narX narQ* double null strains phenotypically resemble *narL* null strains.[24]

After the discovery of the NarL, NarX and NarQ proteins further observations suggested that another protein was involved in the nitrate regulation of anaerobic respiratory gene expression. Formate-dependent nitrite reductase (Nrf) activity is repressed by nitrate and this repression is relieved in a *narL* null strain.[26] However, in a *narL* null strain, Nrf activity is not constitutive but rather nitrate-inducible. Similarly, expression of the *nap-ccm* locus is induced efficiently by nitrate in a *narL* null strain.[7] Furthermore, the NarL-independent nitrate induction of *nap-ccm* operon expression is abolished in a *narX narQ* double null strain.[27] All of these observations suggested the existence of a second response regulator protein which is activated by the NarX or NarQ sensor proteins. A genetic search led to the identification of the *narP* gene which is located immediately adjacent to the *nap-ccm* locus.[8,27] Insertions in *narP* abolish the nitrate-induction of *nap-ccm* operon expression in a *narL* null strain.[27]

DUAL TWO-COMPONENT REGULATORY SYSTEMS

The *narL* and *narP* genes are predicted to encode proteins which share homology with the response regulator family of proteins. Their amino-terminal segments contain conserved Asp and Lys residues which are predicted to constitute the phosphoryl group accepting site. In addition, conserved features of the MalT-LuxR-FixJ-RcsA family of DNA-binding proteins[28,29] are present in the carboxyl-termini of both the NarL and NarP proteins. The most striking observation is that the NarL and NarP proteins are more similar to each other than either is to any other protein.

DNA sequence analysis of the *narX* and *narQ* genes indicates that their protein products are structurally similar to the methyl-accepting chemotaxis proteins. Their amino-terminal regions are predicted to have two membrane spanning regions separated by a periplasmic loop. The periplasmic domains of the NarX and NarQ proteins are largely dissimilar with the exception of a short stretch in which 15 of 17 consecutive amino acid residues are identical between the two proteins.[25] The cytoplasmic carboxyl-terminal regions of the two proteins share similarity with the histidine protein "kinase" sensors.[30] Most notably a conserved His residue, which is the presumed site of autophosphorylation, is present in the NarX and NarQ proteins. As with the NarL and NarP proteins, the NarX and NarQ proteins are more similar to each other than either is to any other protein.

The homology of the NarL/NarP and NarX/NarQ proteins is a strong indication that each pair of proteins arose from a common ancestor by gene duplication. The question is, why are two sensors and two response regulators required for nitrate regulation of anaerobic respiratory gene expression? The recently discovered significance of nitrite in the Nar regulatory system offers a clue to the answer.

THE ROLE OF NITRITE IN THE NAR REGULATORY SYSTEM

It had been known for some time that several of the operons in the Nar modulon are regulated by both nitrite and nitrate. The *fdnG* and *narG* operons are weakly induced by nitrite.[12] Also, the *nrfA* operon is efficiently induced by nitrite.[15,26] Furthermore, the *nirB* operon is induced by nitrite and this induction is abolished in a *narL* null strain.[31] These observations suggested that both nitrite and nitrate regulation of gene expression is mediated by the Nar regulatory system.

Recently the effects of nitrite on the expression of several members of the Nar modulon were investigated.[27] Expression of most of the operons is induced to some extent by nitrite, and for the *nrfA* and *nap-ccm* operons nitrite is the most efficient inducer. Induction of gene expression by nitrite can be mediated by both the NarL and NarP proteins. The weak nitrite induction of *narG* operon expression is completely NarL-dependent.[27] Normal nitrite induction of *nrfA* operon expression occurs in both *narL⁺ narP* and *narL narP⁺* strains, but is abolished in a *narL narP* double null strain.[27,32]

A striking observation was that in a *narX* null strain the weak nitrite induction of *narG* and *fdnG* operon expression is increased such that nitrite is as efficient an inducer as is nitrate.[27] In addition nitrate and nitrite are equally efficient repressors of *frdA* operon expression in a *narX* null strain. However, the nitrite regulation of *narG*, *fdnG* and *frdA* operon expression is not significantly affected by a *narQ* null allele.[27] These observations provided the first indication of a clear difference in the functions of the NarX and NarQ sensor proteins. In the presence of nitrite the NarX protein (but not the NarQ protein) negatively regulates NarL protein activity (Fig. 17.1). In a *narX* null strain this negative regulation is lost but the NarL protein is still efficiently activated (phosphorylated) by the NarQ protein. This effect results in efficient NarL-dependent nitrite regulation of the *narG*, *fdnG* and *frdA* operons. Thus, the negative influence of the NarX protein is essential for normal nitrite regulation of these operons.

THE SENSOR PROTEINS

The relatively recent discovery of the *narQ* gene has meant that most of the genetic studies to date have focused on *narX* gene function. Therefore, much of the following section will deal with studies of NarX protein function and on interactions between the NarX and NarL proteins.

(A) IN VITRO STUDIES

The phosphoryl group transfer from the NarX and NarQ proteins to the NarL protein has been investigated in vitro.[33,34] Full length and amino-terminal truncated versions of the NarX protein and an amino-truncated version of the NarQ protein autophosphorylate when incubated with ATP. Furthermore, the phosphorylated NarX and NarQ proteins can efficiently transfer the phosphoryl group to the NarL protein. Phospho-NarL alone is relatively long-lived but the addition of the NarX protein significantly accelerates the rate of phospho-NarL dephosphorylation.[33,34] This observation supports the interpretation of genetic studies which had indicated that the NarX protein is able to negatively regulate (dephosphorylate) the NarL protein under some conditions.[2,27] An amino-truncated derivative of the NarQ protein has a much less significant effect on the rate of phospho-NarL dephosphorylation than the amino-truncated version of NarX;[34] again this is consistent with genetic studies.[27] The effects of alkaline and acidic conditions on the stability of phospho-NarL and phospho-NarX are consistent with the formation of phosphoaspartyl and phosphohistidyl residues, respectively.[33] Autophosphorylation of the full-length NarX protein, phosphoryl group transfer to the NarL protein or dephosphorylation of phospho-NarL in the presence of the NarX protein are unaffected by the presence of nitrate.[33] These observations indicate that the solubilized NarX protein has lost the ability for ligand-dependent signaling.

(B) The *narX* (H399Q) Mutation

In all sensor proteins studied (with the exception of CheA)[30,35] a conserved histidine residue is the site of autophosphorylation. A site-specific change of His to Gln (H399Q) was constructed to investigate the importance of this conserved residue in NarX protein function.[24]

NarL-dependent nitrate regulation of the *narG*, *fdnG* and *frdA* operons is dependent on the NarL-kinase activity of either the NarX or NarQ protein.[24] This nitrate regulation is abolished in *narX narQ* double null strains and in *narX*(H399Q) *narQ* null strains.[3,24] Therefore, the NarX(H399Q) protein has lost its ability to phosphorylate NarL and resembles the *narX* null allele in this respect.

The phospho-NarL phosphatase activity of the NarX protein in response to nitrite is lost in *narX* null *narQ*⁺ strains and consequently nitrite becomes an efficient effector of *narG*, *fdnG* and *frdA* operon expression.[27] However, the nitrite regulation of these operons is little affected by the *narX*(H399Q) allele,[3] indicating that the NarX(H399Q) protein retains its ability to dephosphorylate phospho-NarL in response to nitrite.

In a separate study, the His-399 residue of NarX was substituted with Gln (H399Q), Glu (H399E) or Lys (H399K).[36] In this study, the *narL* and *narX* genes were present on multicopy plasmids, which resulted in severely aberrant regulation: elevated basal expression of a Φ(*narG-lacZ*) fusion, and reduced basal expression of a Φ(*frdA-lacZ*) fusion. The mutant NarX proteins were unable to activate the NarL protein in response to nitrate in vivo, and also unable to autophosphorylate in vitro.[36] In addition, it was concluded that the NarX(H399Q) and NarX(H399K) proteins are also deficient in phospho-NarL phosphatase activity, both in vivo and in vitro. In contrast, the NarX(H399E) protein clearly retained phospho-NarL phosphatase activity in vivo, although it was severely deficient in this activity in vitro. However, we believe that the in vivo results reveal that both the NarX(H399K) and NarX(H399Q) proteins retain the ability to dephosphorylate NarL in vivo, although not as efficiently as the wild-type NarX protein. For example, plasmids expressing the *narX* (H399K) and *narX* (H399Q) alleles result in a significant increase in basal expression of a Φ(*frdA-lacZ*) fusion due to their negative effect (phosphatase activity) on the NarL protein, which represses Φ(*frdA-lacZ*) expression.[36] Thus, both studies indicate that the NarX(H399Q) protein retains significant phospho-NarL phosphatase activity in vivo.[3,36]

Taken together, these results indicate that the conserved His residue is required for the NarL-kinase but not the phospho-NarL phosphatase function of the NarX protein.[3,36] The inability of the NarX(H399Q) protein to autophosphorylate presumably explains the loss of NarL-kinase activity. Apparently, the conserved His residue is not required for phospho-NarL phosphatase activity. These observations are congruent with similar studies on the role of the conserved His residue in NtrB sensor protein function.[37]

(C) *NARX** MUTATIONS

Altered function alleles of the *narX* gene (*narX** mutations) have been isolated independently by screening for constitutive repression of *frdA* operon expression and constitutive *fdnG* operon expression in the absence of nitrate.[38,39] These constitutive *narX* alleles apparently result in the synthesis of NarX proteins which activate (phosphorylate) NarL in the presence or absence of the inducing signal, nitrate. Several alleles were isolated with different single amino acid substitutions in a region of shared similarity with the methyl accepting chemotaxis proteins. Amino acid substitutions in this region of the Tsr protein (the serine chemoreceptor) affect the signal transduction process.[40] This central linker region immediately follows the second transmembrane region of the NarX protein. Complementation analysis of these *narX** alleles in partial diploid strains revealed that they are recessive to *narX*+, indicating the loss of a function.[39] Initial interpretations were that these NarX* proteins had lost their phospho-NarL phosphatase activity.[24,39] This interpretation is supported by the observation that in both *narQ*+ and *narQ* null strains the *narX**(A224V) and *narX**(E208K) alleles cause increased nitrite induction of *narG* and *fdnG* operon expression in comparison to *narX*+ strains.[3] This aberrant nitrite regulation resembles that of *narX* null *narQ*+ strains in which the NarL-kinase activity of the NarQ protein results in strong nitrite induction of the *narG* and *fdnG* operons in the absence of the antagonistic phospho-NarL phosphatase activity of the NarX protein. However, efficient nitrite induction of *narG* and *fdnG* operon expression occurs in *narX** *narQ* null strains which suggests that the NarX* proteins retain their NarL-kinase activity. Interestingly, the nitrite induction in *narX**(A224V) strains is independent of NarQ protein function whereas nitrite induction in *narX**(E208K) strains was augmented by the NarL-kinase activity of NarQ.[3] Thus, it appears that the NarL-kinase function of NarX*(E208K) is somewhat deficient in comparison to that of NarX*(A224V), consistent with previous interpretations.[24,38]

THE RESPONSE REGULATORS

The Nar modulon comprises a set of operons that can be divided into two general classes: those directly regulated by the NarL protein alone, and those regulated by both the NarL and NarP proteins (Table 17.1, Fig. 17.1).[27] To date no operon has been identified which is regulated by the NarP protein but not the NarL protein. The expression of virtually all operons in the Nar modulon is induced by the Fnr protein in response to anaerobiosis. Most of the operons have apparent Fnr binding sites (consensus, TTGAT-N$_4$-ATCAA) in the -40 to -50 region with respect to their transcription initiation sites (Fig. 17.2).

Sequence comparisons and mutational analysis have identified a consensus sequence for a NarL-binding site, TACYNMT (where Y = C or T and M = A or C).[31,41-43] More recent analysis of interactions between the NarL protein and target operon control regions in vitro has confirmed the importance of these NarL heptamers and has identified additional NarL-binding sites which were not apparent from sequence inspection alone.[43,44] The NarL binding sites of different control

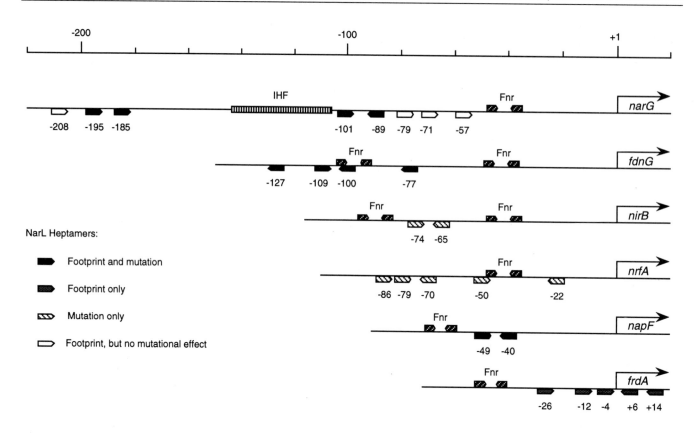

Fig. 17.2. Promoter-regulatory regions for Nar-regulated operons. Modified from other sources.[31,34,42] Scale is in base pairs. Arrows denote transcription initiation sites. Fnr protein binding sites are shown as dark hatched inverted arrows. NarL heptamer sequences are indicated by their position with respect to the transcription initiation site. Heptamers denoted by black arrows have been identified by both mutational analysis and by DNase I footprinting. Heptamers denoted by gray arrows have been identified by DNase I footprinting only. Heptamers denoted by light hatched arrows have been identified by mutational analysis only. Note that mutations in the -86 heptamer of the nrfA operon control region had no effect on nitrate or nitrite regulation in a narL+ narP+ strain.[29] The IHF binding site in the narG control region is denoted by a striped box. Modified from ref. 3.

regions have a wide diversity with respect to their number, location, orientation and spacing (Fig. 17.2).[2] This section will summarize current information on the architecture of some NarL- and NarP- sensitive control regions. NarL and NarP heptamers are designated by the position of the central nucleotide of the heptamer with respect to the transcription initiation site.

THE *narG* OPERON CONTROL REGION

An Fnr binding site centered at position -41.5 with respect to the transcription initiation site is essential for anaerobic induction of *narG* operon expression.[45-47] Deletion and mutational analysis identified the -195 heptamer which is essential for NarL-dependent nitrate induction.[41,47] Subsequent footprinting experiments confirmed that the NarL protein specifically interacts with the -195 heptamer and also identified two other heptamers in the -200 region and five additional heptamers in the -104 to -54 region (Fig. 17.2).[43,44] Mutational analysis has confirmed that at least the -89 heptamer is required for full nitrate induction of *narG* operon expression.[43,44] However, deletion of the -208 heptamer has no effect on nitrate regulation.[47]

The spacing between the -200 region and the transcription initiation site, together with the geometry of this region, are critical for nitrate induction.[41,47] The identification of the NarL heptamers in the -104 to -54 region offers part of the explanation for these observations. However, it was also demonstrated that the sequence-specific DNA bending protein, integration host factor (IHF), is required for full nitrate induction of *narG* operon expression.[48,49] The IHF protein specifically interacts with the -106 to -144 region of the *narG* operon control region in vitro (Fig. 17.2).[44,48,49] One idea is that the IHF protein bends the control region resulting in a looped DNA structure which would bring the -200 region into closer proximity with the transcription initiation site.[44] This may allow NarL molecules bound at the -200 region to interact with those at the -104 to -54 region or with RNA polymerase. However, the precise roles of the DNA-protein and protein-protein interactions in the *narG* operon transcription initiation complex remain to be determined. The architecture of the *narK* control region resembles that of the *narG* operon.[43,50]

THE *FDNG* OPERON CONTROL REGION

Although the patterns of *narG* and *fdnG* operon expression are indistinguishable in wild-type strains[12] their control regions have very different architectures (Fig. 17.2). The *fdnG* operon control region has two Fnr protein binding sites; one centered at -42.5 which is essential for anaerobic induction and another at -97.5 which is apparently involved in repressing *fdnG* operon expression in the absence of nitrate. The mechanism of this Fnr-mediated repression is currently unknown. Deletion, mutational and footprint analysis has identified four NarL-heptamers in the -130 to -74 region (Fig. 17.2).[42,43] Each of these heptamers is required for full nitrate induction of *fdnG* operon expression. The IHF protein does not play a role in *fdnG* operon expression and does not bind to the control region in vitro.[48] Finally, the NarP protein mediates weak nitrate and nitrite induction of *fdnG* operon expression,[27] although NarP binding sites have not yet been identified.

THE *FRDA* OPERON CONTROL REGION

The *frdA* operon control region has an Fnr-binding site centered at -46 (Fig. 17.2).[51] Footprint analysis has identified at least five NarL heptamers in the -29 to +17 region.[43] The location of these NarL binding sites is consistent with the NarL-dependent nitrate repression of *frdA* operon expression.[14]

THE *NIRB* OPERON CONTROL REGION

nirB operon expression is activated by the NarL protein in response to nitrate or nitrite and by the NarP protein in response to nitrate but not nitrite.[26,31,32] Extensive mutational analysis has identified an Fnr binding site centered at -41.5 which is essential for anaerobic induction of *nirB* operon expression.[52-54] Mutational analysis has also identified NarL heptamers -65 and -74 which are involved in nitrate and nitrite induction (Fig. 17.2).[31] Single nucleotide substitutions in either of these heptamers abolish NarL-dependent nitrite induction and NarP-dependent nitrate induction, whereas NarL-dependent nitrate induction is abolished only in constructs with mutations in both heptamers.[31,32] Clearly these results indicate that the NarL and NarP proteins both bind to the -65 and -74 heptamers, although this conclusion awaits support from footprint analysis.

THE *NRFA* OPERON CONTROL REGION

In wild-type cells the *nrfA* operon is activated by the NarL and/or NarP proteins in response to nitrite and repressed by the NarL protein in response to nitrate.[15,26,32] In a *narL* null strain *nrfA* operon expression is induced in response to both nitrate and nitrite by the NarP protein.[27,32] Mutational analysis has identified heptamers -79 and -70 which are required for NarP-dependent nitrate and nitrite induction and for NarL-dependent nitrite induction (Fig. 17.2).[32] Two additional heptamers centered at positions -50 and -22 are required for

NarL-dependent nitrate repression;[32] these heptamers flank the Fnr binding site which is centered at position -41.5 (Fig. 17.2).

The *nrfA* operon is the only member of the Nar modulon that is induced by one effector (nitrite) and repressed by the other (nitrate). The regulation of the *nrfA* operon is thought to be dependent on the relative affinities of phospho-NarL and phospho-NarP for the activation (-79 and -70) and repression (-50 and -22) heptamers and on the level of phospho-NarL in the cell.[32] Nitrite is postulated to result in relatively low levels of phospho-NarL[27] which binds to the activation heptamers with a high affinity. The increased level of phospho-NarL in the presence of nitrate would also result in occupancy of the repression heptamers, resulting in repression of *nrfA* operon expression. Presumably, phospho-NarP binds only to the activation heptamers in the presence of nitrite or nitrate, leading to NarP-dependent nitrite induction (and nitrate induction in a *narL* null strain). In vitro analysis of the interactions of the NarL and NarP proteins with the *nrfA* operon control region is clearly required to test these ideas.

THE *NAP-CCM* OPERON CONTROL REGION

Induction of *nap-ccm* operon expression in response to nitrate or nitrite is solely dependent on the NarP protein, whereas the NarL protein antagonizes this activation.[7,27] DNA sequence analysis identified heptamers -49 and -40 (Fig. 17.2) [55] and mutational analysis has demonstrated that these heptamers are essential for NarP-dependent activation.[56] Mutational analysis has also demonstrated that an apparent Fnr binding site centered at -64.5 is involved in anaerobic induction of *aeg-46.5* operon expression.[56] Footprinting experiments have indicated that both phospho-NarP and phospho-NarL specifically interact with the -49 and -40 heptamers.[56] The NarL protein is unable to activate *aeg-46.5* operon expression and is thought to antagonize NarP-dependent activation by competition for the common DNA-binding site.

HIERARCHY OF NarL-MEDIATED REGULATION

A number of observations have led to the hypothesis that the *narG*, *fdnG* and *frdA* operons are differentially sensitive to NarL-dependent regulation. First, insertions in the *narX* gene, which reduce *narL* expression due to polarity, cause a slight decrease in nitrate induction of *narG* operon expression but almost eliminate nitrate repression of *frdA* operon expression.[19,23] Also, insertions in the *narX* gene reduce the nitrate induction of *fdnG* operon expression more than that of *narG* expression.[12] Secondly, constitutive alleles of *narX* and *narL* cause relatively high-level Φ(*narG-lacZ*) expression, intermediate Φ(*fdnG-lacZ*) expression and little or no repression of Φ(*frdA-lacZ*) expression.[39,57] Finally, in the presence of nitrite, when the level of phospho-NarL is relatively low, NarL-dependent induction of Φ(*narG-lacZ*) expression is higher than that of Φ(*fdnG-lacZ*) expression and there is no repression of Φ(*frdA-lacZ*) expression.[27] All of these observations can be explained by postulating that the various control regions have different affinities for phospho-NarL and are, therefore, differentially sensitive to its concentration. Relatively low levels of phospho-NarL cause significant induction of *narG* operon expression whereas higher levels of phospho-NarL are required for *frdA* operon repression. The explanation for this differential sensitivity to phospho-NarL is currently unclear but it could reflect different NarL-binding site affinities or, given the different architectures of the *narG* and *fdnG* control regions, different modes of transcription activation.

The nitrate induction of *narXL* operon expression,[19,23,58] coupled with the varying sensitivities of the *narG*, *fdnG* and *frdA* control regions to the availability of phospho-NarL, might aid in the fine tuning of target operon expression. When nitrate becomes available, nitrate reductase and formate dehydrogenase-N are synthesized and assembled into an active respiratory chain. As this is accomplished, and with continued nitrate availability, the next priority would be to cease synthesis of alternative anaerobic respiratory enzymes such as fumarate reductase. The basal levels of NarL and NarX might be sufficient for the first phase (formate-nitrate oxidoreductase synthesis), whereas elevated levels of NarL and NarX, resulting from nitrate induction of *narXL* operon expression, might be required for full repression of *frdA* operon expression. Thus, these priorities would be established in the correct order.

INDIRECT NITRATE REGULATION OF GENE EXPRESSION

The nitrate regulation of some genes is not directly mediated by the Nar regulatory system. This section will address two separate examples of indirect nitrate regulation of gene expression. In both cases it has been shown that the presence of nitrate per se is not sufficient for the observed regulation, but rather it is nitrate metabolism which causes the effects on gene expression.

(A) NITRATE REPRESSION OF FORMATE-HYDROGEN LYASE SYNTHESIS

During anaerobic growth in the absence of external electron acceptors, formate is oxidized by the formate-hydrogen lyase (FHL) complex, which consists of formate dehydrogenase-H, several redox carriers and hydrogenase-3.[59] Three different operons are required for the formation of FHL: the *fdhF* gene which encodes formate dehydrogenase-H,[60] the *hyc* operon which encodes structural components of hydrogenase-3,[61] and the *hyp* operon which encodes proteins involved in the post-translational formation of all three hydrogenase isoenzymes.[62] Expression of these genes requires the *rpoN* gene product (the sigma factor, σ^{54})[63] and is induced by anaerobiosis and formate but inhibited by nitrate.[64] However, the nitrate inhibition can be partially overcome by adding formate to the growth medium.[64]

It was demonstrated that the nitrate inhibition of $\Phi(fdhF\text{-}lacZ)$ and $\Phi(hycB\text{-}lacZ)$ expression is partially relieved in a *narG* null mutant which lacks the major nitrate reductase.[22] However, nitrate inhibition of $\Phi(frdA\text{-}lacZ)$ expression, which is directly mediated by the NarL protein, is not relieved in a *narG* null strain.[22] Thus, nitrate respiration is required for the nitrate inhibition of genes encoding the components of FHL. Subsequently, it was demonstrated that the formate induction of *fdhF*, *hyc* and *hyp* gene expression is mediated by the FhlA protein.[65] It was proposed that the oxygen, nitrate, and pH effects on the expression of the *fdhF*, *hyc* and *hyp* genes are all mediated through the intracellular formate concentration which is sensed by the FhlA protein.[66] For example, nitrate induces synthesis of the formate-nitrate oxidoreductase pathway, encoded by the *fdnG* and *narG* operons (see above). The increase in formate-dependent nitrate respiration leads to a decrease in the intracellular formate concentration, reducing the activity of FhlA and, therefore, FHL synthesis. Thus, the nitrate inhibition can be relieved by the addition of formate or by a *narG* null allele which inactivates the formate-nitrate oxidoreductase pathway and prevents the drain on the intracellular formate pool.

(B) ANAEROBIC NITRATE INDUCTION OF *SDH* AND *LLD* (*LCT*) OPERON EXPRESSION

Succinate dehydrogenase (encoded by the *sdhCDAB* operon) and L-lactate dehydrogenase (encoded by *lldD*) are involved in aerobic respiration. (The symbol *lld* [L-lactate dehydrogenase] was recently proposed by E. C. C. Lin in conformity with *dld* which encodes D-lactate dehydrogenase). Indeed, *sdh* and *lld* operon expression is strongly repressed during anaerobic growth. However, the anaerobic growth rate of cells on either succinate or L-lactate as the sole carbon and energy source is increased by the addition of nitrate (but not other electron acceptors such as fumarate and TMAO) to the growth medium.[67] This suggests that succinate dehydrogenase and L-lactate dehydrogenase activities can be recruited for anaerobic nitrate respiration. Consistent with this conclusion was the observation that the anaerobic repression of Φ(*sdh-lac*) and Φ(*lld-lac*) expression is partially relieved by nitrate.[67]

The 14- to 17-fold nitrate induction of Φ(*sdh-lac*) and Φ(*lld-lac*) expression during anaerobic growth is diminished, but not abolished, in a *narL* null mutant. Conversely, the nitrate induction is completely abolished in a *moe* mutant which is deficient in molybdenum cofactor synthesis and, therefore, unable to produce a functional nitrate reductase enzyme.[67] Thus, it appears that the nitrate respiration process is required for the induction of Φ(*sdh-lac*) and Φ(*lld-lac*) expression.

The anaerobic repression of *sdh* and *lld* expression is mediated by the ArcA/ArcB two-component system.[11] In vitro experiments suggest that the ArcB protein senses (among other unknown cellular changes) the abundance of metabolites such as D-lactate, pyruvate, acetate and NADH,[68] the concentrations of which are likely to increase with the lack of an effective exogenous electron sink. It is postulated that nitrate respiration leads to a decrease in the concentration of these metabolites which in turn decreases the ArcB/ArcA mediated repression of *sdh* and *lld* operon expression. This situation is clearly analogous to the proposed mechanism for nitrate regulation of FHL synthesis described above.

CONCLUDING REMARKS

Why have such complex regulatory mechanisms evolved for nitrate and nitrite regulation of anaerobic respiratory gene expression?

The reduction of the preferred anaerobic electron acceptor, nitrate, leads to the production of another electron acceptor, nitrite. Nitrate respiring cells should, therefore, experience a constant shift of nitrate and nitrite concentrations. As these concentrations change the organism must synthesize appropriate levels of the corresponding enzymes. Some enzymes are used primarily when nitrate concentrations are high, such as nitrate reductase (*narG* operon product) and formate dehydrogenase-N (*fdnG* operon product), whereas others, such as DMSO reductase (*dmsA* operon product) and fumarate reductase (*frdA* operon product) are used in the absence of nitrate (or when the nitrate concentration is relatively low). This positive and negative control of gene expression in response to nitrate could be mediated by a "simple" two-component regulatory system. Indeed for the *narG*, *fdnG* and *frdA* operons, the NarX sensor protein and the NarL response regulator are sufficient for normal nitrate regulation.[27] We think that it is the requirement for interactive nitrate and nitrite regulation of other operons that has led to the evolution of a more complex regulatory system. Respiratory nitrite reductase (*nrfA* operon product) is synthesized in the absence of nitrate but only if nitrite is present. Conversely, NADH-nitrite reductase (*nirB* operon product) is synthesized when nitrate and/or nitrite concentrations are high. A better appreciation of the complexity of Nar-mediated regulation is limited because most studies have used cultures grown with excess nitrate or nitrite. More subtle experimentation may be required to reveal the significance of the complexity of Nar-mediated regulation in the context of selective advantage.

The Nar regulatory system provides many challenges for future investigations. We are still a long way from a full understanding of the mechanisms of transmembraneous signal transduction, interaction between the sensors and response regulators, and the activation of transcription by the NarL and NarP proteins.

ACKNOWLEDGMENTS

We appreciate the open exchange of ideas and information with Steve Busby and Jack DeMoss. We thank E. C. C. Lin for his editorial suggestions. We are grateful to our colleagues in the laboratory for their continued interest and support.

REFERENCES

1. Gennis RB, Stewart V. Respiration. In: Neidhardt FC, Ed. *Escherichia coli* and *Salmonella typhimurium*: cellular and molecular biology. Second edition. Washington, DC: ASM Press, 1995: (in press).

2. Stewart V. Nitrate regulation of anaerobic respiratory gene expression in *Escherichia coli*. Mol Microbiol 1993; 9:425-434.

3. Stewart V, Rabin RS. Dual sensors and dual response regulators interact to control nitrate- and nitrite- responsive gene expression in *Escherichia coli*. In: Hoch JA, Silhavy TJ, eds. Two-component Signal Transduction. Washington, DC: ASM Press, 1995:233-252.

4. Lin JT, Goldman BS, Stewart V. The *nasFEDCBA* operon for nitrate and nitrite assimilation in *Klebsiella pneumoniae* M5al. J Bacteriol 1994; 176:2551-2559.

5. Goldman BS, Lin JT, Stewart V. Identification and structure of the *nasR* gene encoding a nitrate- and nitrite-responsive positive regulator of *nasFEDCBA* (Nitrate assimilation) operon expression in *Klebsiella pneumoniae* M5al. J Bacteriol 1994; 176:5077-5085.

6. Blasco F, Iobbi C, Ratouchniak J et al. Nitrate reductases of *Escherichia coli* sequence of the second nitrate reductase and comparison with that encoded by the *narGHJI* operon. Mol Gen Genet 1990; 222:104-111.

7. Choe M, Reznikoff WS. Anaerobically expressed *Escherichia coli* genes identified by operon fusion techniques. J Bacteriol 1991; 173:6139-6146.

8. Richterich P, Lakey N, Gryan G et al. Unpublished DNA sequence. Genbank accession number U00008, 1993.

9. Siddiqui RA, Warnecke Eberz U, Hengsberger A et al. Structure and function of a periplasmic nitrate reductase in *Alcaligenes eutrophus* H16. J Bacteriol 1993; 175:5867-5876.

10. Harborne NR, Griffiths L, Busby SJW et al. Transcriptional control translation and function of the products of the five open reading frames of the *Escherichia coli nir* operon. Mol Microbiol 1992; 6:2805-2813.

11. Iuchi S, Lin ECC. Adaptation of *Escherichia coli* to redox environments by gene expression. Mol Microbiol 1993; 9:9-15.

12. Berg BL, Stewart V. Structural genes for nitrate-inducible formate dehydrogenase in *Escherichia coli* K-12. Genetics 1990; 125:691-702.

13. Stewart V. Requirement of Fnr and NarL functions for nitrate reductase expression in *Escherichia coli* K-12. J Bacteriol 1982; 151:1320-1325.

14. Iuchi S, Lin ECC. The *narL* gene product activates the nitrate reductase operon and represses the fumarate reductase and trimethylamine *N*-oxide reductase operons in *Escherichia coli*. Proc Natl Acad Sci USA 1987; 84:3901-3905.

15. Darwin A, Hussain H, Griffiths L et al. Regulation and sequence of the structural gene for cytochrome c_{552} from *Escherichia coli*: not a hexahaem but a 50 kDa tetrahaem nitrite reductase. Mol Microbiol 1993; 9:1255-1265.

16. Stewart V. Nitrate respiration in relation to facultative metabolism in enterobacteria. Microbiol Rev 1988; 52:190-232.

17. Stewart V, MacGregor C. Nitrate reductase in *Escherichia coli* K-12: involvement of *chlC*, *chlE*, and *chlG* loci. J Bacteriol 1982; 151:788-799.

18. Rajagopalan KV, Johnson JL. The pterin molybdenum cofactors. J Biol Chem 1992; 267:10199-10202.

19. Stewart V, Parales J. Identification and expression of genes *narL* and *narX* of the *nar* (nitrate reductase) locus in *Escherichia coli* K-12. J Bacteriol 1988; 170:1589-1597.

20. Nohno T, Noji S, Taniguchi S et al. The *narX* and *narL* genes encoding the nitrate-sensing regulators of *Escherichia coli* are homologous to a family of prokaryotic two-component regulatory genes. Nucl Acids Res 1989; 17:2947-2957.

21. Stewart V, Parales J, Merkel S. Structure of genes *narL* and *narX* of the *nar* (nitrate reductase) locus in *Escherichia coli* K-12. J Bacteriol 1989; 171:2229-2234.

22. Stewart V, Berg BL. Influence of *nar* (nitrate reductase) genes on nitrate inhibition of formate-hydrogen lyase and fumarate reductase synthesis in *Escherichia coli* K-12. J Bacteriol 1988; 170:4437-4444.

23. Egan SM, Stewart V. Nitrate regulation of anaerobic respiratory gene expression in *narX* deletion mutants of *Escherichia coli* K-12. J Bacteriol 1990; 172:5020-5029.

24. Rabin RS, Stewart V. Either of two functionally redundant sensor proteins, NarX and NarQ, is sufficient for nitrate regulation in *Escherichia coli* K-12. Proc Natl Acad Sci USA 1992; 89:8419-8423.

25. Chiang RC, Cavicchioli R, Gunsalus RP. Identification and characterization of *narQ*, a second nitrate sensor for nitrate-dependent gene regulation in *Escherichia coli*. Mol Microbiol 1992; 6:1913-1923.

26. Page L, Griffiths L, Cole JA. Different physiological roles of two independent pathways for nitrite reduction to ammonia in enteric bacteria. Arch Microbiol 1990; 154:349-354.

27. Rabin RS, Stewart V. Dual response regulators (NarL and NarP) interact with dual sensors (NarX and NarQ) to control nitrate- and nitrite-regulated gene expression in *Escherichia coli* K-12. J Bacteriol 1993; 175:3259-3268.

28. Kahn D, Ditta G. Modular structure of FixJ: homology of the transcriptional activator domain with the -35 binding domain of sigma factors. Mol Microbiol 1991; 5:987-997.

29. Stout V, Torres-Cabassa A, Maurizi MR et al. RcsA, an unstable positive regulator of capsular polysaccharide synthesis. J Bacteriol 1991; 173:1738-1747.

30. Parkinson JS, Kofoid EC. Communication modules in bacterial signaling proteins. Annu Rev Genet 1992; 26:71-112.

31. Tyson KL, Bell AI, Cole JA et al. Definition of nitrite and nitrate response elements at the anaerobically inducible *Escherichia coli* nirB promoter: interactions between FNR and NarL. Mol Microbiol 1993; 7:151-157.

32. Tyson KL, Cole JA, Busby SJW. Nitrite and nitrate regulation at the promoters of two *Escherichia coli* operons encoding nitrite reductase: identification of common target heptamers for both NarP- and NarL-dependent regulation. Mol Microbiol 1994; 13:1045-1055.

33. Walker MS, DeMoss JA. Phosphorylation and dephosphorylation catalyzed in vitro by purified components of the nitrate sensing system, NarX and NarL. J Biol Chem 1993; 268:8391-8393.

34. Schröder I, Wolin CD, Cavicchioli R et al. Phosphorylation and dephosphorylation of the NarQ, NarX, and NarL proteins of the nitrate-dependent two-component regulatory system of *Escherichia coli*. J Bacteriol 1994; 176:4985-4992.

35. Stock JB, Ninfa AJ, Stock AM. Protein phosphorylation and regulation of adaptive responses in bacteria. Microbiol Rev 1989; 53:450-490.

36. Cavicchioli R, Schröder I, Constanti M et al. The NarX and NarQ sensor-transmitter proteins of *Escherichia coli* each require two conserved histidines for nitrate-dependent signal transduction to NarL. J Bacteriol 1995; 177:2416-2424.

37. Atkinson MR, Ninfa AJ. Mutational analysis of the bacterial signal-transducing protein kinase/phosphatase nitrogen regulator II (NR$_{II}$ or NtrB). J Bacteriol 1993; 175:7016-7023.

38. Kalman LV, Gunsalus RP. Nitrate-independent and molybdenum-independent signal transduction mutations in *narX* that alter regulation of anaerobic respiratory genes in *Escherichia coli*. J Bacteriol 1990; 172:7049-7056.

39. Collins LA, Egan SM, Stewart V. Mutational analysis reveals functional similarity between NARX, a nitrate sensor in *Escherichia coli* K-12 and the methyl-accepting chemotaxis proteins. J Bacteriol 1992; 174: 3667-3675.

40. Ames P, Parkinson JS. Transmembrane signaling by bacterial chemoreceptors: *E. coli* transducers with locked signal output. Cell 1988; 55:817-826.

41. Dong XR, Li SF, DeMoss JA. Upstream sequence elements required for NarL-mediated activation of transcription from the *narGHJI* promoter of *Escherichia coli*. J Biol Chem 1992; 267:14122-14128.

42. Li J, Stewart V. Localization of upstream sequences required for nitrate and anaerobic induction of formate dehydrogenase-N operon expression in *Escherichia coli* K-12. J Bacteriol 1992; 174:4935-4942.

43. Li J, Kustu S, Stewart V. In vitro interaction of nitrate-responsive regulatory protein NarL with DNA target sequences in the *fdnG*, *narG*, *narK* and *frdA* operon control regions of *Escherichia coli* K-12. J Mol Biol 1994; 241:150-165.

44. Walker MS, DeMoss JA. NarL-phosphate must bind to multiple upstream sites to activate transcription from the *narG* promoter of *Escherichia coli*. Mol Microbiol 1994; 14:633-641.

45. Walker MS, DeMoss JA. Role of alternative promoter elements in transcription from the *nar* promoter of *Escherichia coli*. J Bacteriol 1992; 174:1119-1123.

46. Walker MS, DeMoss JA. Promoter sequence requirements for Fnr-dependent activation of transcription of the *narGHJI* operon. Mol Microbiol 1991; 5:353-360.

47. Li S-F, DeMoss JA. Location of sequences in the *nar* promoter of *Escherichia coli* required for regulation by Fnr and NarL. J Biol Chem 1988; 263:13700-13705.

48. Rabin RS, Collins LA, Stewart V. In vivo requirement of integration host factor for nitrate reductase (*nar*) operon expression in *Escherichia coli* K-12. Proc Natl Acad Sci USA 1992; 89:8701-8705.

49. Schröder I, Darie S, Gunsalus RP. Activation of the *Escherichia coli* nitrate reductase (*narGHJI*) operon by NarL and Fnr requires integration host factor. J Biol Chem 1993; 268:771-774.

50. Bonnefoy V, DeMoss JA. Identification of functional *cis*-acting sequences involved in regulation of *narK* gene expression in *Escherichia coli*. Mol Microbiol 1992; 6:3595-3602.

51. Eiglmeier K, Honore N, Iuchi S et al. Molecular genetic analysis of FNR-dependent promoters. Mol Microbiol 1989; 3:869-878.

52. Jayaraman PS, Gaston KL, Cole JA et al. The *nirB* promoter of *Escherichia coli*: location of nucleotide sequences essential for regulation by oxygen, the FNR protein, and nitrite. Mol Microbiol 1988; 2:527-530.

53. Jayaraman PS, Cole JA, Busby SJW. Mutational analysis of the nucleotide sequence at the FNR-dependent *nirB* promoter in *Escherichia coli*. Nucl Acids Res 1989; 17:135-146.

54. Bell AI, Cole JA, Busby SJW. Molecular genetic analysis of an Fnr-dependent anaerobically inducible *Escherichia coli* promoter. Mol Microbiol 1990; 4:1753-1764.

55. Choe M, Reznikoff WS. Identification of the regulatory sequence of anaerobically expressed locus *aeg-46.5*. J Bacteriol 1993; 175:1165-1172.

56. Darwin AJ, Stewart V. Nitrate and nitrite regulation of the Fnr-dependent *aeg-46.5* promoter of *Escherichia coli* K-12 is mediated by competition between homologous response regulators (NarL and NarP) for a common DNA-binding site. J Mol Biol 1995; 251:15-29.

57. Egan SM, Stewart V. Mutational analysis of nitrate regulatory gene *narL* in *Escherichia coli* K-12. J Bacteriol 1991; 173:4424-4432.

58. Darwin AJ, Stewart V. Expression of the *narX, narL, narP* and *narQ* genes of *Escherichia coli* K-12: Regulation of the regulators. J Bacteriol 1995; 177:3865-3869.

59. Sauter M, Böhm R, Böck A. Mutational analysis of the operon (*hyc*) determining hydrogenase 3 formation In *Escherichia coli*. Mol Microbiol 1992; 6:1523-1532.

60. Zinoni F, Birkmann A, Stadtman TC et al. Nucleotide sequence and expression of the selenocysteine-containing polypeptide of formate dehydrogenase (formate-hydrogen-lyase-linked) from *Escherichia coli*. Proc Natl Acad Sci USA 1986; 1986:4650-4654.

61. Böhm R, Sauter M, Böck A. Nucleotide sequence and expression of an operon in *Escherichia coli* coding for formate hydrogenylase components. Mol Microbiol 1990; 4:231-244.

62. Lutz S, Jacobi A, Schlensog V et al. Molecular characterization of an operon (*hyp*) necessary for the activity of the three hydrogenase isoenzymes In *Escherichia coli*. Mol Microbiol 1991; 5:123-136.

63. Birkmann A, Sawers G, Böck A. Involvement of the *ntrA* gene product in the anaerobic metabolism of *Escherichia coli*. Mol Gen Genet 1987; 210:535-542.

64. Birkmann A, Zinoni F, Sawers G et al. Factors affecting transcriptional regulation of the formate- hydrogen-lyase pathway of *Escherichia coli*. Arch Microbiol 1987; 148:44-51.

65. Schlensog V, Böck A. Identification and sequence analysis of the gene encoding the transcriptional activator of the formate hydrogenlyase system of *Escherichia coli*. Mol Microbiol 1990; 4:1319-1328.

66. Rossmann R, Sawers G, Böck A. Mechanism of regulation of the formate-hydrogenlyase pathway by oxygen, nitrate, and pH: definition of the formate regulon. Mol Microbiol 1991; 5:2807-2814.

67. Iuchi S, Aristarkhov A, Dong JM et al. Effects of nitrate respiration on expression of the Arc-controlled operons encoding succinate dehydrogenase and flavin-linked L-lactate dehydrogenase. J Bacteriol 1994; 176:1695-1701.

68. Iuchi S. Phosphorylation/dephosphorylation of the receiver module at the conserved aspartate residue controls transphosphorylation activity of histi-

dine kinase in sensor protein ArcB of *Escherichia coli*. J Biol Chem 1993; 268:23972-23980.

69. Bongaerts J, Zoske S, Weidner U et al. Transcriptional regulation of the proton translocating NADH dehydrogenase genes (*nuoA-N*) of *Escherichia coli* by electron acceptors, electron donors and gene regulators. Mol Microbiol 1995; 16:521-534.

70. Miller JB, Scott DJ, Amy NK. Molybdenum-sensitive transcriptional regulation of the *chlD* locus of *Escherichia coli*. J Bacteriol 1987; 169: 1853-1860.

71. Sawers G, Böck A. Anaerobic regulation of pyruvate-formate lyase from *Escherichia coli* K-12. J Bacteriol 1988; 170:5330-5336.

72. Kaiser M, Sawers G. Nitrate repression of the *Escherichia coli pfl* operon is mediated by the dual sensors NarQ and NarX and the dual response regulators NarL and NarP. J Bacteriol 1995; 177:3647-3655.

73. Kalman LV, Gunsalus RP. The *frdR* gene of *Escherichia coli* globally regulates several operons involved in anaerobic growth in response to nitrate. J Bacteriol 1988; 170:623-629.

74. Chen Y-M, Lin ECC. Regulation of the *adhE* gene, which encodes ethanol dehydrogenase in *Escherichia coli*. J Bacteriol 1991; 173:8009-8013.

75. Cotter PA, Gunsalus RP. Oxygen, nitrate, and molybdenum regulation of *dmsABC* gene expression in *Escherichia coli*. J Bacteriol 1989; 171:3817-3823.

REGULATION OF AEROBIC AND ANAEROBIC METABOLISM BY THE ARC SYSTEM

A. Simon Lynch and Edmund C. C. Lin

INTRODUCTION

During the course of the evolution of microbial bioenergetics, progenitors of *E. coli* probably acquired in a stepwise temporal fashion metabolic pathways for fermentation, anaerobic respiration, and aerobic respiration for the process of energy transduction.[1,2] The acquisition of these different pathways, combined with the apparent development of successive layers of regulatory mechanisms to control them, enables *E. coli* to exploit adroitly environmental energy sources to their greatest possible advantage. To this end, a key strategy is to channel electrons from donors to terminal acceptors such that the overall potential difference is maximized for any given growth condition. Conspicuous consequences are improved growth rate and/or yield for any given carbon and energy source.

At the level of transcription, the adaptive responses are coordinated by a group of global regulators: the Fnr protein and the proteins of the two-component signal transduction systems, ArcA/ArcB, NarL/NarX, and NarP/NarQ. This chapter focuses on the role of the Arc system, and its apparent critical role in ensuring the preferential utilization of molecular oxygen as the terminal electron acceptor with the highest midpoint redox potential (E'°=+820 mV).

IDENTIFICATION OF THE *arc* GENES

Some thirty years ago it was established that the tricarboxylic acid (TCA) cycle of *E. coli* is likely to be highly operative only in aerobically grown cells, since the activity levels of a number of TCA enzymes were found to be 2- to 20-fold higher in extracts prepared from aerobically grown cells in comparison to those prepared from cells grown anaerobically.[3-8] In contrast, the activity levels of a number of proteins that function in anaerobic electron transport were observed to be higher in anaerobically grown cells.[5,6] Subsequently, the list of proteins whose cellular levels are influenced by molecular oxygen concentration in the environment was greatly expanded by comparisons of cell extracts

Regulation of Gene Expression in Escherichia coli, edited by E. C. C. Lin and A. Simon Lynch. © 1996 R.G. Landes Company.

using two-dimensional electrophoresis.[9] While these combined studies did not reveal whether differences in levels (or activities) in cells were attributable to induction or repression of synthesis, they did lay a biochemical foundation for future studies of transcriptional regulation. In an attempt to identify a global system involved in the metabolic adaptation to anaerobiosis, Iuchi and Lin[10] exploited the knowledge that the activity level of succinate dehydrogenase (now known to be encoded by the *sdhCDAB* operon) was lowered by anaerobic growth[3,5] and employed a merodiploid Φ(*sdh-lac*), *sdh*⁺ strain to identify *trans*-acting mutants that expressed *sdh* at high levels anaerobically. Numerous mutants that anaerobically expressed up to 20-fold elevated levels of both β-galactosidase and succinate dehydrogenase were isolated. They also expressed elevated anaerobic activity levels of many additional enzymes involved in aerobic metabolism including numerous flavodehydrogenases, enzymes of the TCA cycle and the glyoxylate shunt, a member of the fatty acid degradation pathway, and cytochrome *o*.[10,11] Surprisingly, the anaerobic levels of some of the enzymes studied exceeded those determined from aerobically grown wild-type cells.

Genetic mapping of the mutations identified two putative regulatory loci: *arcA* at the minute 0 region of the genome[10] and *arcB* at minute 69.5.[11] The gene symbol, *arc*, an acronym for <u>a</u>erobic <u>r</u>espiration <u>c</u>ontrol, was chosen in light of the multiple effects that these genes exerted on aerobic metabolism.

THE *arcA* GENE

The *arcA* mutations mapped immediately counterclockwise to the *thr* locus. Since the previously characterized *dye* gene also mapped at minute 0, it became important to determine whether or not *arcA* is allelic with *dye*. The latter was so named because insertion and deletion mutations at the locus conferred hypersensitivity to the phenothiazine dyes methylene blue and toluidine blue.[12-14] While such *dye* mutations do not appear to have any effects on the lipopolysaccharide structure of the cell wall, they do appear to result in significant changes in the protein compositions of both the inner and outer membranes, a property which was proposed to account for their altered sensitivity to redox dyes and other agents.[14]

The *dye* gene, it should be noted, was originally called *msp* (for <u>m</u>ale-<u>s</u>pecific <u>p</u>hage) because mutations in it were observed to confer resistance of Hfr strains to RNA bacteriophage, presumably a consequence of the observed lack of formation of F pili.[15,16] The same gene was also discovered in three independent studies of deficiencies in DNA donor activity of F⁺ cells, resulting in the designations *fexA* for <u>F</u> expression,[17] *sfrA* for <u>s</u>ex <u>f</u>actor <u>r</u>egulation,[18,19] and *cpxC* for <u>c</u>onjugation <u>p</u>lasmid e<u>x</u>pression.[20,21] However, it is now known that *seg* mutations, which affect F plasmid replication (and thereby daughter plasmid <u>seg</u>regation) map to the nearby *dnaK* gene and not the *dye* locus as previously thought.[22]

When *arcA* mutants were tested for two *dye* mutant phenotypes, toluidine blue sensitivity and resistance to infection by the male-specific M13 phage, the results suggested that the same gene had once again been rediscovered. Plasmid complementation analysis definitively showed that *arcA* is allelic with *dye*.[10]

The 238 amino acid open reading frame encoding ArcA was originally sequenced as the *dye* gene by Drury and Buxton,[23] and the predicted molecular weight of 27.3 kDa is close to the size of the cytoplasmic protein estimated by SDS/PAGE analysis.[14] It was also noted that the predicted sequence of the protein showed an overall level of

28% identity with that of the *ompR* gene product, largely due to a stretch of 19 centrally located amino acids which are identical with the exception of a single residue.[23] On the basis of this sequence homology, and data pertaining to the mechanism of OmpR/EnvZ mediated regulation of the *ompF* and *ompC* genes, Iuchi and Lin[10] proposed that the ArcA protein corresponded to a pleiotropic regulator involved in mediating cellular adaptations to changes in environmental aerobiosis, the activity of which was dependent on a signal transmitted by a sensor protein localized in the cytoplasmic membrane. The possibility that the latter was encoded by the newly discovered *arcB* gene was suggested, since *arcA* was known to be the sole member of its operon.[23]

Further comparisons of the ArcA sequence with other bacterial proteins have revealed a high level of homology with a class of proteins referred to as the 'response regulator' units of a large family of prokaryotic two-component signal transduction systems. Within this large family of proteins, ArcA belongs to the OmpR-like sub-family which characteristically has an amino-terminal response regulator domain and a carboxy-terminal domain with a helix-turn-helix DNA binding motif.[24-31]

In the amino-terminal domain of ArcA, three charged residues (Glu,[10] Asp[11] and Asp[54]) correspond to residues which form an 'acidic pocket' in the crystal structure of the homologous CheY protein;[30-37] Lys[103] of ArcA corresponds to an invariant lysine residue found in the carboxy-terminus of the regulator domain of all response regulator proteins (shown diagrammatically in Fig. 18.1). By analogy with studies of other two-component response regulators, Asp[54] of ArcA is predicted to serve as the phospho-acceptor residue.

Transcriptional regulation of the *arcA* gene has been investigated using strains containing Φ(*arcA-lacZ*) fusions, the expression of which showed 4-fold anaerobic elevation.[38] This anaerobic induction is absolutely dependent on Fnr, maximal in *arcA*+ strains, and due (in part) to the Fnr-dependent activation of a second *arcA* promoter which is inactive in aerobically grown cells. These results led to the suggestion that ArcA enhances transcriptional activation mediated by the Fnr protein. While this notion apparently contradicts the results of an earlier study which suggested simple autoregulation by ArcA,[39] the discrepancy may be accounted for by differences in the penetrance of the *arcA* and *fnr* alleles used and by differences in the structures of the gene fusions employed.[38] If *arcA* is indeed a member of the Fnr modulon then, in addition to direct effects of Fnr on anaerobic gene expression, two distinct mechanisms are apparent wherein Fnr may indirectly affect anaerobic gene expression via effects on the Arc system: by elevation of ArcA levels in anaerobic cells, and through FNR-mediated metabolic responses that change the concentrations of certain anaerobic metabolites which affect ArcB function (see later sections).

THE *arcB* GENE

Of seventy independent mutants isolated in the original screen of Iuchi and Lin[10] that showed elevated anaerobic levels of β-galactosidase specified by a Φ(*sdh-lac*) fusion element and increased activity levels of succinate dehydrogenase, five defined a second complementation group mapping to minute 69.5 on the chromosome.[11] Mutations at this newly discovered locus not only elevated anaerobic activity levels of enzymes involved in aerobic metabolism, but also significantly increased sensitivity to toluidine blue and methylene blue, as had previously been found with the *arcA* mutants. The gene was therefore designated as

ArcA protein

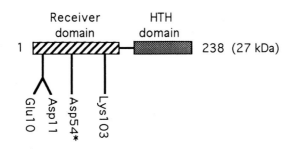

ArcB protein

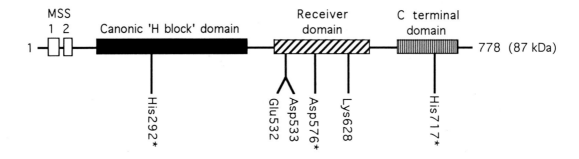

Fig. 18.1. Diagrammatic representations of the proposed domain structures of ArcA and ArcB. MSS: Membrane spanning segments; HTH: Helix-turn-helix DNA binding domain. Asterisks indicate known sites of phosphorylation. Refer to relevant sections of the text for additional details.

arcB.[11] The notion that ArcB plays a role in negative control (i.e., anaerobic repression) of target genes involved in aerobic metabolism arose from studies of a strain containing an *arcB* deletion, which had highly elevated anaerobic levels of aconitase and L-lactate dehydogenase.[11] In accordance with the proposal that ArcB communicates with ArcA in a manner analogous to that of the OmpR/EnvZ system,[40] over-expression of the *arcA* gene was observed to be epistatic over null *arcB* mutations.[11]

Complementation of the dye-sensitivity phenotype of *arcB* mutants was used as a selection in cloning of the gene, the nucleotide sequence of which was subsequently determined and found to contain an open-reading frame encoding a protein of 778 amino acids with a predicted molecular weight of 87.9 kDa.[41] Analysis of the deduced sequence of ArcB revealed significant homology between a central region of about 200 amino acid residues to the conserved 'H-box' domain of a family of proteins that serve as the sensor components of two-component signal transduction systems.[25,27,29,30] Within this family ArcB belongs to a sub-group referred to as 'hybrid kinases', possessing a receiver domain in addition to a histidine kinase transmitter domain (located on the amino-terminal side of the receiver domain). The sub-group currently has seven members, including the RscC, BarA and EvgS proteins of *E. coli*.[25,27,42-44] As with other members of this sub-group, no helix-turn-helix motif is apparent in the receiver domain of ArcB. In the transmitter domain of ArcB, His[292] is expected to be the autophosphorylation site. In the receiver domain, Glu[532], Asp[533], and Asp[576] should constitute the canonical 'acidic pocket', with Lys[628] as the invariant downstream residue (see Fig. 18.1).

Further sequence comparisons have recently revealed a limited but significant level of homology between the carboxy-terminal domain of ArcB (and other hybrid kinases) and an amino-terminal domain of two unorthodox transmitter proteins (CheA and FrzE), which all include a single invariant histidine residue corresponding to His[717] in ArcB.[44-46]

The size of the ArcB protein inferred from the *arcB* gene sequence is compatible with estimates of the size of the native protein determined by SDS/PAGE analysis of ^{35}S-labeled protein extracts prepared from UV-treated maxicells harboring multicopy *arcB*[+] plasmids.[41] These studies additionally indicated that ArcB is localized to the cytoplasmic membrane. As only two potential membrane-spanning segments (Phe[23]-Val[50] and Ser[58]-Val[77]) are apparent in the sequence of the protein[41] (see Fig. 18.1), it seems likely that the bulk of the protein is cytoplasmically located. The periplasmically exposed region of the protein may be as short as only seven residues, in contrast to the rather extensive periplasmic regions of many other membrane-localized sensor proteins. Hence the transmembrane segments of ArcB, rather than its short periplasmic connecting region, probably serve to receive the stimulus. Interactions of the periplasmic segment with a ligand of low molecular weight, however, cannot be readily dismissed. Alternatively, the transmembrane segments may serve a merely structural role in anchoring the protein to the cytoplasmic membrane in which case the stimulus receptor region(s) must lie within the cytoplasmic portion of the protein (see later sections for additional discussion).

In contrast to the case of *arcA*, studies utilizing a Φ(*arcB-lac*) fusion indicated no respiratory regulation of *arcB* at the transcriptional level.[47]

IN VIVO STUDIES OF *arc* MUTANTS

The functions of the ArcB transmitter and receiver domains were explored in experiments using strains with a null *arcB* chromosomal mutation and harboring a plasmid-borne *arcB* allele encoding a truncated ArcB protein with either Phe[516] or Gly[517] at the carboxy terminus. In these strains the aerobic/anaerobic ratio of Φ(*sdh-lac*) expression was reduced by a factor of up to 13-fold in comparison to an isogenic strain expressing an *arcB*[+] plasmid-borne allele.[47] Hence the receiver domain appears to have a critical modulatory effect on the rate of intermolecular transfer of the phosphoryl group from His[292] of the transmitter domain of ArcB to ArcA. Effective anaerobic repression of an Φ(*lctD-lac*) fusion similarly depends on the presence of an intact receiver domain.[41]

Genetic truncation of ArcB from the amino-terminus by deletion of amino acids 1-128 resulted in a cytosolic localization of the mutant protein[48] in support of the notion that the amino-terminus is responsible for anchoring the wild-type protein in the cytoplasmic membrane.

Site-directed mutagenesis studies of ArcB have also been used to test aspects of the signaling pathway, with the design of the ArcB mutants based on biochemical and genetic studies of other two-component signal transduction systems. Substitution of the canonic phosphoryl-donor residue (His[292]) in the transmitter domain of ArcB with a glutamine residue, yielded an allele that failed to complement null *arcB* mutations: specifically, a plasmid-borne *arcB*[His292Gln] allele was unable to repress anaerobically a Φ(*sdh-lac*) fusion in an *arcB* null, *arcA*[+] host. A similar defect was observed with plasmids expressing *arcB*[Asp533Ala] and *arcB*[Asp576Ala] alleles.[47] The possibility that either of the two cysteine residues in the ArcB protein (Cys[180] or Cys[241]) play a role in stimulus (redox) sensing has been discounted since either residue can be substituted by glycine without significant effects on the anaerobic repression of target operons.[47]

The *arcB* gene was recently identified as a multicopy suppressor of a defect in the activation of a Φ(*ompF-lacZ*) fusion, or a Φ(*ompC-lacZ*) fusion, in an *ompR*[+] strain bearing a deletion in the *envZ* gene (encoding the cognate sensor kinase for OmpR).[43] The gene encoding the BarA hybrid kinase was similarly identified as a multicopy suppressor. A subsequent study revealed that the multicopy suppression phenomenon was only dependent on expression of the carboxy-terminal 139 residues of the ArcB protein, or the carboxy-terminal 118 residues of the BarA protein,[44] thus excluding roles for the canonic transmitter and receiver domains. The importance of the conserved histidine residue in the carboxy terminal domains of ArcB and BarA to the phenomenon was demonstrated in experiments which showed that no suppression was observed when His[717] of full-length ArcB was substituted with a leucine residue, or when His[861] of full-length BarA was substituted with an arginine residue. However, *arcB* null strains harboring a single-copy plasmid containing the *arcB*[His717Leu] allele were found to have no detectable phenotypic alterations associated with changes in patterns of *ompC* and *ompF* expression, and to be wild-type with regard to anaerobic repression of a Φ(*lctD-lac*) fusion and toluidine blue resistance.[44] Possible physiological function(s) of the conserved carboxy terminal domain of ArcB is discussed in later sections.

IN VITRO PHOSPHORYLATION STUDIES

The conclusion from in vivo studies that His[292] of ArcB is the sole residue capable of undergoing autophosphorylation is supported by in vitro experiments. Substitution of His[292] by glutamine abolishes autophosphorylation of the protein in everted membrane vesicle preparations in the presence of γ-^{32}P-ATP.[47,49] Moreover, whereas rapid phosphorylation of a purified soluble form of ArcB protein (genetically truncated by 128 amino-terminal amino acids) occurs in the presence of γ-^{32}P-ATP, no phosphorylation of ArcB[129-778] bearing a His[292]Gln substitution occurs.[47,49]

Several observations support the notion that the γ-phosphoryl group of ATP is transferred first to His[292] and is then intermolecularly transferred to Asp[576], allowing for subsequent re-autophosphorylation of His[292]. Firstly, purified ArcB[129-778] labeled by incubation with γ-^{32}P-ATP shows a biphasic loss of the ^{32}P label at 43°C in a pH 12.5 buffer.[48] This is indicative of the presence of both histidyl phosphoryl and aspartyl phosphoryl groups, as it is known that aspartyl phosphate links are significantly more labile under these conditions than histidyl phosphate links. Secondly, overall phosphorylation of ArcB[129-778] is greatly reduced when an Asp[576]Ala substitution is introduced.[47,49] Finally, although ArcB[129-778] catalyzes phosphorylation of ArcB[His292Gln] in everted membrane vesicles, it is unable to phosphorylate ArcB[His292Gln/Asp576Ala] protein under a similar condition.[49] Intermolecular phosphoryl transfer between protomers of oligomeric forms of other two-component sensor proteins has been reported.[29,50]

In the presence of limiting concentrations of purified ArcB[129-778] a linear initial rate of ArcA phosphorylation occurs in the presence of ATP, with the rate of phosphoryl-ArcA (ArcA-P) formation proportional to the concentration of input ArcB[129-778].[48] Intact ArcB in everted membrane preparations rapidly phosphorylates ArcA, in contrast to ArcB variants lacking the receiver domain which retain only weak ArcA kinase activity.[49] On the other hand, ArcB^{Asp576Ala} mutant protein (possessing the receiver domain but lacking its canonic Asp^{576} phospho-acceptor residue) in everted membrane preparations is virtually unable to transphosphorylate ArcA; an ArcB^{Asp533Ala} mutant is similarly defective in this respect.[49] The receiver domain of ArcB therefore plays a critical role in the Arc signaling pathway by controlling the rate of signal transmission.[47-49,51]

Studies of purified ArcB$^{129-778}$ and intact ArcB in everted membrane vesicle preparations both indicated that certain compounds produced in fermentation reactions (D-lactate, acetate, pyruvate, or NADH) increase the level of ArcB phosphorylation by inhibiting an autophosphatase activity.[49] In the framework of the Arc signaling model of Iuchi and Lin[47] this effect should enhance net transphosphorylation of Asp^{576} in the receiver domain, facilitating re-phosphorylation of His^{292}, and consequently phosphoryl transfer from His^{292} to Asp^{54} of ArcA. On the basis of these results, Iuchi[49] proposed that ArcB responds in vivo to the intracellular concentrations of these metabolites.

Although purified ArcB$^{639-778}$ does not undergo autophosphorylation in the presence of γ-^{32}P-ATP, efficient phosphorylation of the protein is catalyzed by cytoplasmic membrane preparations containing intact ArcB.[44] Furthermore, this labeling is dependent on the presence of His^{717} in the ArcB$^{639-778}$ component. Interestingly, phospho-ArcB$^{639-778}$ acts as a kinase for purified OmpR protein, or intact ArcB in cytoplasmic membrane preparations.[44] Hence, it seems that His^{717} can both act as a phospho-acceptor and phospho-donor residue. The physiological relevance of these observations, however, remains to be tested.

Available in vitro and in vivo data on the overall signaling pathway of the Arc system can be accommodated by the model depicted in Figure 18.2. According to this model, the receiver domain of ArcB acts like a competitive inhibitor of the receiver domain of ArcA and thereby functions in 'gating' the activity of a dimeric form of ArcB protein, such that it is only active as an ArcA kinase under the appropriate conditions. Additional support for the model is provided by recent studies which indicate that purified ArcB$^{78-778}$ protein is dimeric in solution, and that ArcA dimerization is strongly promoted by phosphorylation in vitro (J. M. Berger, A. S. Lynch, and E. C. C. Lin, unpublished observations).

The possibility that the receiver domains of hybrid kinases,[29] or the conserved C-terminal domains of sensors like ArcB, BarA and BvgS,[44] may serve a role as targets for physiological 'cross-talk' between, or 'cross-regulation' of, other two-component systems have recently been suggested. For the Arc system, possible routes of phosphoryl transfer are indicated in Figure 18.3. Such cross-communication could diminish the flow of phosphoryl groups from ArcB to ArcA by competing for the His^{292} phosphoryl group. Furthermore, by changing the phosphorylation state of the regulatory receiver domain, or the C-terminal transmitter domain, the ArcA kinase activity may be modulated. Parenthetically, it should be noted that the CpxA sensor protein was originally thought to be the cognate kinase for the ArcA (SfrA) protein,[24] based on the similar effects that mutations in these genes have on conjugative properties of F$^+$ cells.[18,20,21,52-55] However, with the subsequent discovery of the *arcB* sensor gene, combined with characterization of *arcB* and *cpxA* mutants in terms of aerobic respiratory phenotypes and conjugative properties,[56,57] it became apparent that the CpxA/ArcA pairing was erroneous. It is now evident that the true cognate partner of CpxA is encoded by a gene (designated *cpxR*) located upstream

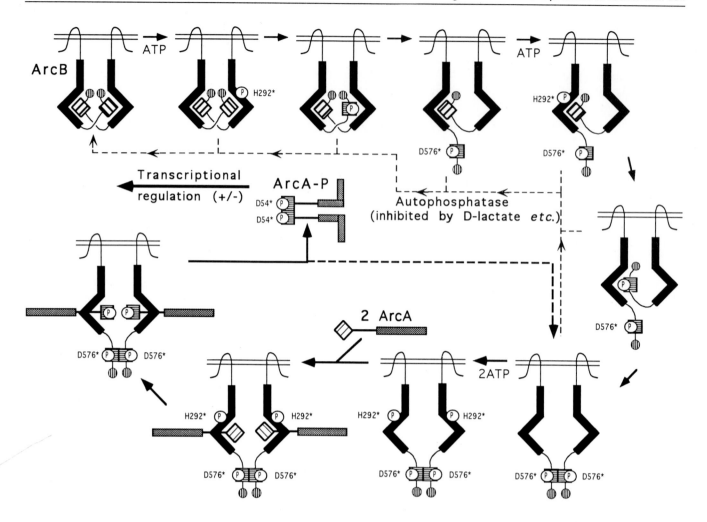

Fig. 18.2. Model for ArcB signaling. Refer to relevant sections of the text for details of data in support of the model.

of, and overlapping with, the *cpxA* gene.[58,59] Genetic studies clearly indicate that CpxR and CpxA do indeed form a cognate pair[60,61] although a clear function for the proteins has yet to be established.

THE ARC MODULON

The term modulon was proposed to designate a family of operons that do not share a specific transcriptional regulator but yet are under the control of a global regulator.[10,62] Genetic and biochemical studies so far have identified thirty operons as members of the Arc modulon (see Table 18.1). Several recent reviews of overall regulations of electron and carbon flow have recently been published.[63-67]

In general the *arc* system mediates anaerobic repression. However, two cases of anaerobic activation are known: *cydAB* (encoding cytochrome *d* oxidase) and *pfl* (encoding pyruvate-formate lyase). Positive control of these operons is readily comprehensible, since their gene products play vital cellular roles when cells are deprived of oxygen.[73-75,85,86,111] Cytochrome *d* oxidase levels are maximal under microaerobic conditions when the function of the enzyme becomes most useful to the cell, and there is evidence indicating that ArcA and Fnr are co-regulators of the operon. The notion that positive control of *cydAB* is mediated by the Arc system is supported by the observation that complete blockage of induction of the heat shock response occurs in strains bearing null *arcA* or *arcB* mutations.[71] The inability of the *arcA1* mutation to abolish expression of a Φ(*cyd-lac*) fusion under

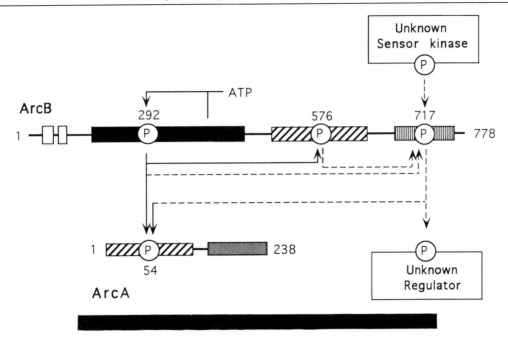

Fig. 18.3. Possible routes of phosphoryl group transfer involving the Arc components. Refer to relevant sections of the text for further details. Adapted from Ishige et al, EMBO J 1994; 13:5195-5202.

microaerobic and anaerobic conditions in a previous study[75] is probably due to slight leakiness of this allele with regard to this phenotype (S. Iuchi and E. C. C. Lin, unpublished data). Although an 'Fnr box' is present in the promoter region of cydAB, no definitive data are available to determine (i) whether Fnr control is positive or negative, and (ii) to what extent the lack of Fnr in fnr mutants indirectly affects the kinase activity of ArcB by altering anaerobic concentrations of compounds such as D-lactate etc.[49,73-75,111]

As described in a preceding section, early mutant studies implicated ArcA function in expression of F plasmid phenotypes, with one study leading to the gene designation, sfrA.[18] It is now known that ArcA/SfrA is required for maximal activation of the promoter of the F plasmid traY operon, encoding proteins required for F pili biogenesis.[54,57] However, the status of traY as a member of the Arc modulon requires clarification for a number of reasons; for instance, there are no significant differences between aerobically and anaerobically grown cells in their abilities to form F pili, and anoxia has only a slight effect on the activity of the traY promoter.[54] Furthermore, using a Φ(sdh-lac) fusion as a reporter of Arc related activity and a Φ(traY-lacZ) fusion as a reporter of Sfr function, Silverman and co-workers have found discordant effects of arc and sfr mutant alleles on the expression of the two reporters.[109] The arcA1 mutation, previously shown to be as effective as a deletion in enabling anaerobic expression of sdh[10]and which was characterized as an eight amino acid insertion mutation between Ala[33] and Thr[34] [109] reduces expression of the Φ(traY-lacZ) fusion by only 40%. Similarly a null arcB mutation (arcB1), resulting from creation of a translational stop signal at codon 240,[47] significantly derepresses anaerobic Φ(sdh-lac) expression, but has little effect on the Sfr phenotype. In contrast, a Val[203]Met substitution in ArcA (previously designated as the sfrA5 mutation)[18] essentially abolishes Sfr function (as monitored by expression of the Φ[traY-lacZ] fusion) without detectably altering Arc function (as monitored by expression of the Φ[sdh-lac] fusion). Whether these data are best interpreted as indicating a separate role for the arcA gene product (i.e., outside of the Arc modulon) awaits further experimental clarification.

Table 18.1. The Arc modulon

Targets		Effects	References
Gene(s)	**Enzyme/Function**		
aceB	Isocitrate lyase	(–)	10
acn	Aconitase	(–)	10
arcA	ArcA	+	38,39
cob	Cobalamin biosynthesis	+	68-70
cydAB	Cytochrome d oxidase	+	71-79
cyoABCDE	Cytochrome o oxidase	–	10,73,74,77
fadB	3-Hydroxyacyl CoA dehydrogenase	(–)	10
fdnGHI	Formate dehyrogenase N	(–)	10,80-82
focA- pfl	Formate transport and Pyruvate formate lyase	+	83-89
fumA	Fumarase A (aerobic)	(–)	90
glpD	Glycerol-3-phosphate dehydrogenase (aerobic)	–	91
gltA	Citrate synthase	–	10,92
hemA	Glutamyl-tRNA dehydrogenase	+	93
hyaA-F	Hydrogenase 1	+	94-97
icd	Isocitrate dehydrogenase	(–)	10
lctPRD	L-Lactate permease, regulator & L-lactate dehydrogenase	–	10,98,99
mdh	Malate dehydrogenase	(–)	10
nuoA-N	NADH:quinone oxidoreductase	–	100
pdhR-aceEF -lpd	Pyruvate dehydrogenase complex and regulator	–	10,101
pdu	Propanediol degradation	+	70
pocR	Positive regulator of cob & pdu	+	70
sdhCDAB	Succinate dehydrogenase	–	10,98,102
sodA	Mn-Superoxide dismutase	–	103-108
sucAB	α-ketoglutarate dehydrogenase	(–)	10
sucCD	Succinate thiolkinase	(–)	10
traY	F plasmid DNA transfer functions	+	18,19,21,52,109,110
ND	D-Amino acid dehydrogenase	(–)	10

All data refer to studies of *E.coli* or *S.typhimurium*.
+   indicates positive control (transcriptional activation of gene/locus).
–   indicates negative control (transcriptional repression of gene/locus).
(+) indicates provisional positive control based on assay of relevant enzyme activity in extracts of wild-
    type and mutant cells.
(–) indicates provisional negative control based on assay of relevant enzyme activity in extracts of
    wild-type and mutant cells.

ArcA DNA BINDING

Despite the availability of the nucleotide sequences corresponding to many of the promoters of operons that comprise the Arc modulon, no conserved *cis* element (i.e., an ArcA, or ArcA-P DNA binding site consensus) has been identified by comparisons. However, should such an element resemble in form the imperfectly conserved heptameric repeat characterized for the homologous NarL response regulator protein,[112-116] (see also chapter 17 by Darwin and Stewart) mere sequence inspection may not be an adequate method to identify and define such a motif.

A more effective way of defining an 'ArcA box' should be through in vitro DNA binding studies, and such studies (of the *sodA* promoter) were first reported by Tardat and Touati.[104] Transcription of *sodA*, encoding the manganese-containing superoxide dismutase (MnSOD), is repressed by the *arcA* gene product during anaerobic growth.[103,105]

Using soluble cell extracts prepared from cells over-expressing the ArcA protein, an ArcA-dependent DNAse I footprint of approximately 65 base-pairs encompassing the -35 and -10 elements of the *sodA* promoter was observed.[104] Attempts to purify the over-expressed ArcA protein by several combinations of gel filtration, ion exchange, and affinity chromatography resulted in the loss of site-specific DNA binding activity as the purification proceeded. Chromatographic separation of the extracts solely by gel filtration resulted in a fraction containing a monomeric form of the ArcA protein which failed to bind DNA, and fractions corresponding to the void volume of the column (containing material with an estimated molecular weight of > 300 kDa) which retained DNA binding activity. Since the DNAse I footprint observed is significantly larger than would be expected for a monomeric (or even dimeric) 27 kDa DNA binding protein, it was concluded that the ArcA protein present in the extracts used was binding the *sodA* promoter in a high-molecular weight complex.[104]

As aerobically grown cells were used for preparation of the extracts, and as similar results were obtained with extracts prepared from an *arcB* null mutant strain, it was further concluded that the predominant form of the ArcA protein present in the extracts was unphosphorylated. However, the possibility remains that the active *sodA* DNA binding species present in the extracts is a minor component corresponding to ArcA-P despite aerobic growth of the cultures used in preparation of the extracts. Moreover, even in extracts of *arcB* mutants, some ArcA-P may well be present as a result of autophosphorylation by compounds such as acetyl phosphate (and see below).[117-119] The possibility also remains that binding of ArcA (or possibly ArcA-P) to the *sodA* promoter is dependent on the presence of another protein(s).

Site-specific DNA binding activity of purified ArcA protein at the *pfl* promoter, as determined by both electrophoretic mobility shift assays (EMSA) and DNAse I footprinting assays, has recently been reported.[88] In a 531-bp DNA fragment encompassing the P_6 and P_7 promoters of the *pfl* operon, four ArcA protected sites were identified. Three of these, ranging in size from 25-66 bp, were observed only at very high ArcA concentrations (0.1 mM) and correspond to (i) two sites that span the transcription start points and -10 regions of P_6 and P_7 and which partially overlap putative Fnr binding sites,[85] and (ii) a site which overlaps a previously characterized integration host factor (IHF) binding site located between the two promoters.[87] The fourth site, with highest affinity for ArcA, extends over 94 bp and is also located in the interpromoter region. The DNA binding activity of the purified ArcA protein was found to be significantly stimulated by prior incubation with either carbamoyl phosphate or acetyl phosphate. The data were interpreted as being consistent with a hypothesis that a higher-order nucleoprotein complex, comprising several proteins (including ArcA, Fnr and IHF), is required to activate transcription from the multiple promoters of the *pfl* operon.

Studies of the binding of ArcA and ArcA-P to the transcriptional regulatory regions of the *lctPRD* and *cydAB* operons, and to the *gltA-sdhCDAB* intergenic region have also been performed. In all cases, phosphorylation of purified ArcA by prior incubation with carbamoyl phosphate or acetyl phosphate led to a significant enhancement of site-specific DNA binding. From results of chemical cross-linking experiments, it was apparent that phosphorylation of the ArcA protein stimulates its dimerization. Furthermore, analysis of nucleoprotein complexes formed between DNA and the phosphorylated and unphosphorylated forms of the protein by EMSA assays indicated that (phosphorylation-promoted) dimerization significantly alters the DNA binding properties of the protein (A. S. Lynch and E. C. C. Lin, unpublished data).

Where ArcA-P is expected to act as a repressor (i.e., *gltA-sdhCDAB*, and *lctPRD*), the phosphorylated protein protected extensive regions of DNA from DNAse I digestion. These protected regions encompass known promoter elements and sites to which transcriptional activators probably bind. Hence, repression would appear to be mediated by effective sequestration of *cis*-controlling elements required for basal (or activated) transcription into an ArcA-P-containing nucleoprotein complex. In a study of *cydAB*, for which ArcA-P is expected to function as a transcriptional activator, discrete ArcA-P binding sites are apparent in the promoter region (A. S. Lynch and E. C. C. Lin, unpublished data); further studies are aimed at elucidating how ArcA-P binding at these sites may stimulate transcription from the *cydAB* promoter(s).

As a whole, preliminary results from studies of ArcA are consistent with those obtained with other response regulator proteins. For instance, the OmpR response regulator appears to bind DNA with a 10-fold lower efficiency in the unphosphorylated form,[120] and oligomerization is the apparent mechanism underlying the phosphorylation-dependent enhancement of DNA binding.[121] Phosphorylation of VanR, a response regulator protein (belonging to the OmpR sub-family) of *Enterococcus faecium* BM4147, apparently increases its affinity for an element in the *vanH* promoter by a factor of five hundred.[122] Lastly, in the case of the NarL response regulator, DNAse I footprints of the control region of the *fdnG* operon (and a number of other regulatory sites) were only observed in one study with the phosphorylated form of a 'constitutive' mutant protein with a Val[88]Ala substitution. One possibility is that this amino acid substitution specifically enhances phosphorylation of the protein by acetyl phosphate.[114] Specific binding of wild-type NarL-P to the transcriptional regulatory region of the *narG* operon has also been reported.[115]

THE Arc STIMULUS

It is postulated that ArcB becomes progressively activated as a kinase during transition from aerobic to microanaerobic growth and remains in the activated state during anaerobic growth. As a consequence, the concentration of ArcA-P becomes elevated leading to repression of some loci (e.g., the *sdh* operon) and activation of others (e.g., the *pfl* operon). While it is apparent that the levels of enzymes regulated by the Arc system vary, depending on the degree of O_2 availability to the cell, there is evidence indicating that this compound is not the direct signal or stimulus for the ArcB kinase. For instance, although aerobic growth results in the highest levels of $\Phi(sdh\text{-}lac)$ expression, anaerobic growth with NO_3^- or fumarate serving as the terminal electron acceptor also results in elevated expression. Thus, the degree of derepression seems to vary with the oxidizing power (E'°) of the terminal electron acceptor.[10] Also, the expression of a $\Phi(cyo\text{-}lac)$ fusion in a strain deleted for both the *cyo* (encoding the terminal oxidase with low O_2 affinity) and *cyd* (encoding the terminal oxidase with high O_2 affinity) operons becomes practically insensitive to the presence of O_2. Even single deletion of either the *cyo* or the *cyd* operon greatly impairs aerobic expression of the $\Phi(cyo\text{-}lac)$ fusion. It was therefore suggested that ArcB indirectly senses the O_2 tension by monitoring the level of an electron transport component in reduced form or that of a nonautoxidizable compound.[74] In this regard, Nystrom[123] recently noted that during growth under aeration, the menaquinone (serving as the redox adapter in anaerobic respiration) level in the cells increases 4.4-fold during transition from logarithmic growth to stationary phase, while ubiquinone (predominately used during aerobic respiration) decreases 3.2-fold.[124-126] However, studies of a *ubiD* mutant, blocked in the biosynthesis of ubiquinone, suggest that this electron carrier is not involved directly in ArcB signaling.[127]

As described in a preceding section, Iuchi[49] has proposed that ArcB activity is influenced by the intracellular levels of certain metabolites that accumulate during anoxia. However, the validity of this hypothesis requires corroboration from in vivo experiments. Nonetheless, it is tempting to speculate that the metabolites serve (in an effector capacity) as cytosolic signals of the metabolic status of the cell, with their accumulation lowering the threshold sensitivity of the ArcB kinase for the primary stimulus (apparently by inhibiting the intrinsic autophosphatase activity of ArcB). On the other hand, if these reduced compounds truly do represent the only and direct chemical stimuli for the ArcB kinase, it is not clear why the protein is anchored in the cytoplasmic membrane.

Recent studies of *cyd* expression in wild-type and *arc* mutant strains in response to treatments of whole cells with the protonophorous uncoupler pentachlorophenol, or agents that oxidize the respiratory chain, have led to the suggestion that a decrease in electrochemical proton potential, rather than reduction of a respiratory chain component, is responsible for the observed induction of *cyd* expression.[76] Hence, it is proposed that the ArcB sensor acts as a 'protometer' for the cell.[76]

Whatever the primary signal for ArcB proves to be, a number of genetic studies indicate that *arcA* has distinct physiological roles in the regulation also of aerobic gene expression. Firstly, the expression pattern of the *hemA* gene (encoding glutamyl-tRNA dehydrogenase) indicates that whereas Fnr functions as a repressor of transcription during anaerobic growth, *arcA* is required for normal activation of *hemA* expression during both anaerobic and aerobic growth.[93] Secondly, a functional *arcA* gene is required for activation of expression of the *cyd* operon aerobically and anaerobically in *fnr* mutant cell, whereas Fnr-mediated repression in anaerobic wild-type cells is dependent on *arcA*.[73,75,111] Thirdly, a recent investigation of the regulation of expression of the *gltA* gene, encoding citrate synthase, led to the suggestion that the *arcA* gene product serves as a repressor in both aerobic and anaerobic cells.[92] Finally, as described in previous sections, a potential aerobic role for *arcA* in the regulation of expression of the F plasmid *traY* promoter has been proposed.[109]

Full aeration of cultures is difficult to achieve in the laboratory and indeed was not attempted in many studies of 'aerobic' roles of the ArcA protein. Even when maximal oxygenation is achieved, we propose that the ArcB protein is always partially active as a kinase for ArcA, and that the activity increases as the O_2 concentration decreases. Therefore, for any given nutrient condition, the ArcA-P/ArcA ratio in the cell is expected to rise as O_2 concentrations decrease. Differential patterns of expression of the members of the Arc modulon would arise as a consequence of differences in the intrinsic affinity of ArcA-P for its DNA binding sites.

FUTURE STUDIES

Although studies of the Arc system have been underway for almost two decades, much remains to be elucidated. The nature of physical and/or chemical factors that modulate ArcB kinase and autophosphatase activities in vivo need to be clarified. In addition, the physiological importance of ArcB-independent routes of ArcA phosphorylation in vivo need to be identified and assessed. In particular, the possibility of aerobic induction of certain members of the Arc modulon as a consequence of autophosphorylation of ArcA (at the expense of low molecular weight phosphoryl donor compounds) should be addressed. Such alternative routes have recently been proposed as a way to modulate cellular metabolism during enteric residence,[118] and during the glucose starvation response.[123] In vitro, differences in the transcriptional regulatory activities of the phosphorylated and unphosphorylated forms of the ArcA protein, indicated by results of in vivo studies, require further explorations at a mechanistic level. These studies should include characterization of differences in the DNA binding properties of the two forms of the protein and investigation of potential interactions with other transcriptional regulatory proteins. Finally, possible physiological interactions of the carboxy-terminal domain of ArcB with other signal transduction components warrant investigation. The existence of such 'cross-talk' or 'cross-regulation' may significantly augment our understanding of integrative metabolic controls that enable the cell to coordinate its responses to an ever-changing environment.

References

1. Gest H. The evolution of biological energy-transducing systems. FEMS Microbiol Lett 1980; 7:73-77.
2. Wilson TH, Lin ECC. Evolution of membrane bioenergetics. J Supramol Structure 1980; 13:421-446.
3. Hirsch CA, Rasminsky M, Davis BD et al. A fumarate reductase in *Escherichia coli* distinct from succinate dehydrogenase. J Biol Chem 1963; 238:3770-3774.
4. Amarasingham CR, Davis BJ. Regulation of α-ketoglutarate dehydrogenase formation in *Escherichia coli*. J Biol Chem 1965; 240:3664-3668.
5. Gray CT, Wimpenny JWT, Hughes DE et al. Regulation of metabolism in facultative bacteria. 1. Structural and functional changes in *Escherichia coli* associated with shifts between the aerobic and anaerobic states. Biochim Biophys Acta 1966; 117:22-32.
6. Gray CT, Wimpenny JWT, Mossman MR. Regulation of metabolism in facultative bacteria. II. Effects of aerobiosis, anaerobiosis and nutrition on the formation of Krebs cycle enzymes in *Escherichia coli*. Biochim Biophys Acta 1966; 117:33-41.
7. Hino S, Maeda M. Effect of oxygen on the development of respiratory activity in *Escherichia coli*. J Gen Appl Microbiol 1966; 12:247-265.
8. Cavari BZ, Avi-Dor Y, Grossowicz N. Induction by oxygen of respiration and phosphorylation of anaerobically grown *Escherichia coli*. J Bacteriol 1968; 96:751-759.
9. Smith MW, Neidhardt FC. Proteins induced by aerobiosis in *Escherichia coli*. J Bacteriol 1983; 154:344-350.
10. Iuchi S, Lin ECC. arcA (dye), a global regulatory gene in *Escherichia coli* mediating repression of enzymes in aerobic pathways. Proc Natl Acad Sci USA 1988; 85:1888-1892.
11. Iuchi S, Cameron DC, Lin ECC. A second global regulator gene (arcB) mediating repression of enzymes in aerobic pathways of *Escherichia coli*. J Bacteriol 1989; 171:868-873.
12. Buxton RS, Drury LS. Cloning and insertional inactivation of the *dye* (sfraA) gene, mutation of which affects sex factor F expression and dye sensitivity of *Escherichia coli* K12. J Bacteriol 1983; 154:1309-1314.
13. Buxton RS, Drury LS, Curtis CAM. Dye sensitivity correlated with envelope protein changes in *dye* (sfraA) mutants of *Escherichia coli* K12 defective in the expression of the sex factor F. J Gen Microbiol 1983; 129:3363-3370.

14. Buxton RS, Drury LS. Identification of the *dye* gene product, mutational loss of which alters envelope protein composition and also affects sex factor F expression in *Escherichia coli* K-12. Mol Gen Genet 1984; 194:241-247.

15. Buxton RS, Hammer-Jespersen K, Hansen TD. Insertion of bacteriophage lambda into the *deo* operon of *Escherichia coli* K-12 and isolation of plaque-forming *lambda deo⁺* transducing bacteriophages. J Bacteriol 1978; 136:668-681.

16. Roeder W, Somerville RL. Cloning the *trpR* gene. Mol Gen Genet 1978; 176:361-368.

17. Lerner T, Zinder N. Chromosomal regulation of sexual expression in *Escherichia coli*. J Bacteriol 1979; 137:1063-1065.

18. Beutin L, Achtman M. Two *Escherichia coli* chromosomal cistrons, *sfrA* and *sfrB*, which are needed for expression of F factor *tra* functions. J Bacteriol 1979; 139:730-737.

19. Beutin L, Manning P, Achtman M et al. *sfrA* and *sfrB* products of *Escherichia coli* K-12 are transcription control factors. J Bacteriol 1981; 145:840-844.

20. McEwen J, Silverman P. Chromosomal mutations of *Escherichia coli* that alter expression of conjugative plasmid functions. Proc Natl Acad Sci USA 1980; 77:513-517.

21. Silverman P, Nat K, McEwen J et al. Selection of *Escherichia coli* K-12 chromosomal mutants that prevent expression of F-plasmid functions. J Bacteriol 1980; 143:1519-1523.

22. Ezaki B, Ogura T, Mori H et al. Involvement of DnaK protein in mini-F plasmid replication: temperature-sensitive *seg* mutations are located in the *dnaK* gene. Mol Gen Genet 1989; 218:183-189.

23. Drury LS, Buxton RS. DNA sequence analysis of the *dye* gene of *Escherichia coli* reveals amino acid homology between the Dye and OmpR proteins. J Biol Chem 1985; 260:4236-4242.

24. Ronson CW, Nixon BT, Ausubel FM. Conserved domains in bacterial regulatory proteins that respond to environmental stimuli. Cell 1987; 49:579-581.

25. Parkinson JS, Kofoid EC. Communication modules in bacterial signaling proteins. Ann Rev Genet 1992; 26:71-112.

26. Volz K. Structural conservation in the CheY superfamily. Biochemistry 1993; 32:11741-11753.

27. Alex LA, Simon MI. Protein histidine kinases and signal transduction in prokaryotes and eukaryotes. Trends Biochem Sci 1994; 10:133-139.

28. Pao GM, Tam R, Lipschitz LS et al. Response regulators: structure, function and evolution. Res Microbiol 1994; 145:356-362.

29. Swanson RV, Alex LA, Simon MI. Histidine and aspartate phosphorylation: two-component systems and the limits of homology. Trends Biochem Sci 1994; 19:485-490.

30. Stock JB, Surette MG, Levit M et al. Two-component signal transduction systems: structure-function relationships and mechanisms of catalysis. In: Hoch JA, Silhavy TJ, eds. Two-component signal transduction. Washington: ASM Press, 1995:25-52.

31. Volz K. Structural and functional conservation in response regulators. In: Hoch JA, Silhavy TJ, eds. Two-component signal transduction. Washington: ASM Press, 1995:53-64.

32. Stock AM, Mottonen JM, Stock JB et al. Three-dimensional structure of CheY, the response regulator of bacterial chemotaxis. Nature 1989; 337:745-749.

33. Volz K, Matsumura P. Crystal structure of *Escherichia coli* Che Y refined at 1.7 Å resolution. J Biol Chem 1991; 266:15511-15519.

34. Bourret RB, Drake SK, Chervitz SA et al. Activation of the phospho-signaling protein Che Y. II. Analysis of the activated mutants by ^{19}F NMR and protein engineering. J Biol Chem 1993; 268:13089-13096.

35. Bruix M, Pascual J, Santoro J et al. ^{1}H- and ^{15}-N-NMR assignment and solution structure of the chemotactic *Escherichia coli* Che Y protein. Eur J Biochem 1993; 215:573-585.

36. Drake SK, Bourret RB, Luck LA et al. Activation of the phosphosignaling protein Che Y. I. Analysis of the phosphorylated conformation by ^{19}F NMR and protein engineering. J Biol Chem 1993; 268:13081-13088.

37. Stock AM, Martinez-Hackert E, Rasmussen BF et al. Structure of the Mg^{2-}-bound form of CheY and mechanism of phosphoryl transfer in bacterial chemotaxis. Biochemistry 1993; 32:13375-13380.

38. Compan I, Touati D. Anaerobic activation of *arcA* transcription in *Escherichia coli*: roles of Fnr and ArcA. Mol Microbiol 1994; 11:955-964.

39. Park SJ, Cotter PA, Gunsalus RP. Autoregulation of the *arcA* gene of *Escherichia coli*. Am Soc Microbiol Abstr 1992; 92:207(Abstract)

40. Slauch JM, Garrett S, Jackson DE et al. EnvZ functions through OmpR to control porin gene expression in *Escherichia coli* K-12. J Bacteriol 1988; 170:439-441.

41. Iuchi S, Matsuda Z, Fujiwara T et al. The *arcB* gene of *Escherichia coli* encodes a sensor-regulator protein for anaerobic repression of the *arc* modulon. Mol Microbiol 1990; 4:715-727.

42. Stout V, Gottesman S. RcsB and RcsC: a two-component regulator of capsule synthesis in *Escherichia coli*. J Bacteriol 1990; 172:659-669.

43. Nagasawa S, Ishige K, Mizuno T. Novel members of the two-component signal transduction genes in *Escherichia coli*. J Biochem 1993; 114:350-357.

44. Ishige K, Nagasawa S, Tokishita S-I et al. A novel device of bacterial signal transducers. EMBO J 1994; 13:5195-5202.

45. Uhl MA, Miller JF. Autophosphorylation and phosphotransfer in the *Bordetella pertussis* BvgAS signal transduction cascade. Proc Natl Acad Sci USA 1994; 91:1163-1167.

46. Uhl MA, Miller JF. *Bordetella pertussis* BvgAS virulence control system. In: Hoch JA, Silhavy TJ, eds. Two-component signal transduction. Washington: ASM Press, 1995:333-350.

47. Iuchi S, Lin ECC. Mutational analysis of signal transduction by ArcB: a membrane sensor protein for anaerobic expression of operons involved in the central aerobic pathways in *Escherichia coli*. J Bacteriol 1992; 174:3972-3980.

48. Iuchi S, Lin ECC. Purification and phosphorylation of the Arc regulatory components of *Escherichia coli*. J Bacteriol 1992; 174:5617-5623.

49. Iuchi S. Phosphorylation/dephosphorylation of the receiver module at the conserved aspartate residue controls transphosphorylation activity of histidine kinase in sensor protein ArcB of *Escherichia coli*. J Biol Chem 1993; 263:23972-23980.

50. Yang Y, Inouye M. Intermolecular complementation between two defective mutant signal-transducing receptors of *Escherichia coli*. Proc Natl Acad Sci USA 1991; 88:11057-11061.

51. Lin ECC, Iuchi S. Role of protein phosphorylation in the regulation of aerobic metabolism by the Arc system in *Escherichia coli*. In: Torriani-Gorini A, Yagil E, Silver S, eds. Phosphate in microorganisms: cellular and molecular biology. Washington, D.C.: American Society for Microbiology, 1994:290-295.

52. Silverman P. Host cell-plasmid interactions in the expression of DNA donor activity by F^{+} strains of *Escherichia coli* K-12. Bioassays 1985; 2:254-259.

53. Albin R, Weber R, Silverman PM. The Cpx proteins of *Escherichia coli* K12: immunologic detection of the chromosomal *cpxA* gene product. J Biol Chem 1986; 261:4698-4705.

54. Silverman P, Wichersham E, Harris R. Regulation of the F plasmid *traY* promoter by host and plasmid factors. J Mol Biol 1991; 218:119-128.

55. Gaudin H, Silverman P. Contributions of promoter context and structure to regulated expression of the F plasmid *traY* promoter in *Escherichia coli* K12. Mol Microbiol 1993; 8:335-342.

56. Iuchi S, Furlong D, Lin ECC. Differentiation of *arcA, arcB,* and *cpxA* mutant phenotypes of *Escherichia coli* by sex pilus formation and enzyme regulation. J Bacteriol 1989; 171:2889-2893.

57. Silverman PM, Tran L, Harris R et al. Accumulation of the F plasmid TraJ protein in *cpx* mutants of *Escherichia coli.* J Bacteriol 1993; 175:921-925.

58. Dong J-M, Iuchi S, Kwan H-S et al. The deduced amino acid sequence of the cloned *cpxR* gene suggests the protein is the cognate regulator for the membrane sensor, CpxA, in a two-component signal transduction system of *Escherichia coli.* Gene 1993; 136:227-230.

59. Plunkett G, Burland V, Daniels DL et al. Analysis of the *Escherichia coli* genome. III. DNA sequence of the region from 87.2 to 89.2 minutes. Nuc Acids Res 1993; 21:3391-3398.

60. Danese PN, Snyder WB, Cosma CL et al. The Cpx two-component signal transduction pathway of *Escherichia coli* regulates transcription of the gene specifying the stress-inducible periplasmic protease, DegP. Genes & Development 1995;9:387-398.

61. Snyder WB, Davis LJB, Danese PN et al. Overproduction of NlpE, a new outer membrane lipoprotein, suppresses the toxicity of periplasmic LacZ by activation of the Cpx signal transduction pathway. J Bacteriol 1995; 177:4216-4223.

62. Neidhardt FC, Ingraham JL, Schaechter M. Physiology of the bacterial cell. A molecular approach. Sinauer Associates, Inc. Sunderland, MA:1990. pp. 382-383.

63. Gunsalus RP. Control of electron flow in *Escherichia coli*: coordinated transcription of respiratory pathway genes. J Bacteriol 1992; 174: 7069-7074.

64. Guest JR, Russell GC. Complexes and complexities of the citric acid cycle in *Escherichia coli.* Curr Topics Cell Reg 1992; 33:231-247.

65. Iuchi S, Lin ECC. Adaptation of *Escherichia coli* to respiratory conditions: regulation of gene expression. Cell 1991; 66:5-7.

66. Iuchi S, Lin ECC. Adaptation of *Escherichia coli* to redox environments by gene expression. Mol Microbiol 1993; 9:9-15.

67. Gunsalus RP, Park S-J. Aerobic-anaerobic gene regulation in *Escherichia coli* control by the ArcAB and Fnr regulons. Res Microbiol 1994; 145:437-450.

68. Andersson DI, Roth JR. Redox regulation of the genes for cobinamide biosynthesis in *Salmonella typhimurium.* J Bacteriol 1989; 171:6734-6739.

69. Andersson DI. Involvement of the Arc system in redox regulation of the Cob operon in *Salmonella typhimurium.* Mol Microbiol 1992; 6: 1491-1494.

70. Ailion M, Bobik TA, Roth JR. Two global regulatory systems (Crp and Arc) control the cobalamin/propanediol regulon of *Salmonella typhimurium.* J Bacteriol 1993; 175:7200-7208.

71. Wall D, Delaney JM, Fayet O et al. *arc*-dependent thermal regulation and extragenic suppression of the *Escherichia coli* cytochrome *d* operon. J Bacteriol 1992; 174:6554-6562.

72. Frey B, Janel G, Michelson U et al. Mutations in the *Escherichia coli fnr* and *tgt* genes: control of molybdate reductase activity and the cytochrome *d* complex by *fnr*. J Bacteriol 1989; 171:1524-1530.

73. Cotter PA, Chepuri V, Gennis RB et al. Cytochrome *o (cyoABCDE)* and *d (cydAB)* oxidase gene expression in *Escherichia coli* is regulated by oxygen, pH, and the *fnr* gene product. J Bacteriol 1990; 172:6333-6338.

74. Iuchi S, Chepuri V, Fu H-A et al. Requirement for terminal cytochromes in generation of the aerobic signal for the *arc* regulatory system in *Escherichia coli*: study utilizing deletions and *lac* fusions of *cyo* and *cyd*. J Bacteriol 1990; 172:6020-6025.

75. Fu H-A, Iuchi S, Lin ECC. The requirement of ArcA and Fnr for peak expression of the *cyd* operon in *Escherichia coli* under microaerobic conditions. Mol Gen Genet 1991; 226:209-213.

76. Bogachev AV, Murtazina RA, Skulachev VP. Cytochrome *d* induction in *Escherichia coli* growing under unfavorable conditions. FEBS Lett 1993; 336:75-78.

77. Cotter PA, Gunsalus RP. Contribution of the *fnr* and *arcA* gene products in coordinate regulation of cytochrome *o* and *d* oxidase (*cyoABCDE* and *cydAB*) genes in *Escherichia coli*. FEMS Microbiol Lett 1992; 91:31-36.

78. Georgiou CD, Dueweke TJ, Gennis RB. Regulation of expression of the cytochrome *d* terminal oxidase in *Escherichia coli*. J Bacteriol 1988; 170:961-966.

79. Fang H, Gennis RB. Identification of the transcriptional start site of the *cyd* operon from *Escherichia coli*. FEMS Microbiol Lett 1993; 108:237-242.

80. Berg BL, Stewart V. Structural genes for nitrate-inducible formate dehydrogenase in *Escherichia coli* K-12. Genetics 1990; 125:691-702.

81. Birkmann A, Zinoni F, Sawers G et al. Factors affecting transcriptional regulation of the formate-hydrogen-lyase pathways of *Escherichia coli*. Arch Microbiol 1987; 148:44-51.

82. Li J, Stewart V. Localization of upstream sequence elements required for nitrate and anaerobic induction of *fdn* (formate dehydrogenase-N) operon expression in *Escherichia coli* K-12. J Bacteriol 1992; 174:4935-4942.

83. Sawers G, Böck A. Anaerobic regulation of pyruvate formate-lyase from *Escherichia coli* K-12. J Bacteriol 1988; 170:5330-5336.

84. Wong KK, Suen KL, Kwan HS. Transcription of *pfl* is regulated by anaerobiosis, catabolite repression, pyruvate, and *oxrA*: *pfl*::Mu dA operon fusions of *Salmonella typhimurium*. J Bacteriol 1989; 171:4900-4905.

85. Sawers G. Specific transcriptional requirements for positive regulation of the anaerobically inducible *pfl* operon by ArcA and FNR. Mol Microbiol 1993; 10:737-747.

86. Sawers G, Suppmann B. Anaerobic induction of pyruvate formate-lyase gene expression is mediated by the ArcA and FNR proteins. J Bacteriol 1992; 174:3474-3478.

87. Sirko A, Zehelein E, Freundlich M et al. Integration host factor is required for anaerobic pyruvate induction of *pfl* operon expression in *Escherichia coli*. J Bacteriol 1993; 175:5769-5777.

88. Drapal N, Sawers G. Purification of ArcA and analysis of its specific interaction with the *pfl* promoter-regulatory region. Mol Microbiol 1995; 16:597-607.

89. Kaiser M, Sawers G. Nitrate repression of the *Escherichia coli pfl* operon is mediated by the dual sensors NarQ and NarX and the dual regulators NarL and NarP. J Bacteriol 1995; 177:3647-3655.

90. Woods SA, Guest JR. Differential roles of the *Escherichia coli* fumarases and *fnr*-dependent expression of fumarase B and aspartase. FEMS Microbiol Lett 1987; 48:219-224.

91. Iuchi S, Cole ST, Lin ECC. Multiple regulatory elements for the *glpA* operon encoding anaerobic glycerol-3-phosphate dehydrogenase and the

glpD operon encoding aerobic glycerol-3-phosphate dehydrogenase in *Escherichia coli*: further characterization of respiratory control. J Bacteriol 1990; 172:179-184.

92. Park S-J, McCabe J, Turna J et al. Regulation of the citrate synthase (*gltA*) gene of *Escherichia coli* in response to anaerobiosis and carbon supply: role of the *arcA* gene product. J Bacteriol 1994; 176:5086-5092.

93. Darie LS, Gunsalus RP. Effect of heme and oxygen availability on *hemA* gene expression in *Escherichia coli*: role of the *fnr*, *arcA*, and *himA* gene products. J Bacteriol 1994; 176:5270-5276.

94. Jamieson DJ, Sawers RG, Rugman PA et al. Effects of anaerobic regulatory mutations and catabolite repression on regulation of hydrogen metabolism and hydrogenase isoenzyme composition in *Salmonella typhimurium*. J Bacteriol 1986; 168:405-411.

95. Sawers RG, Ballantine SP, Boxer DH. Differential expression of hydrogenase isoenzymes in *Escherichia coli* K-12: evidence for a third isoenzyme. J Bacteriol 1985; 164:1324-1331.

96. Jamieson DJ, Higgins CF. Two genetically distinct pathways for transcriptional regulation of anaerobic gene expression in *Salmonella typhimurium*. J Bacteriol 1986; 168:389-397.

97. Brondsted L, Atlung T. Anaerobic regulation of the hydrogenase 1 (*hya*) operon of *Escherichia coli*. J Bacteriol 1994; 176:5423-5428.

98. Iuchi S, Aristarkhov A, Dong J-M et al. Effects of nitrate respiration on expression of the Arc-controlled operons encoding succinate dehydrogenase and flavin-linked L-lactate dehydrogenase. J Bacteriol 1994; 176:1695-1701.

99. Dong J-M, Taylor JS, Latour DJ et al. Three overlapping *lct* genes involved in L-lactate utilization by *Escherichia coli*. J Bacteriol 1993; 175:6671-6678.

100. Bongaerts J, Zoske S, Weidner U et al. Transcriptional regulation of the proton translocating NADH dehydrogenase genes (*nuoA-N*) of *Escherichia coli* by electron acceptors, electron donors and gene regulators. Mol Microbiol 1995; 16:521-534.

101. Quail MA, Haydon DJ, Guest JR. The *pdhR-aceEF-lpd* operon of *Escherichia coli* expresses the pyruvate dehydrogenase complex. Mol Microbiol 1994; 12:95-104.

102. Park S-J, Tseng C-P, Gunsalus RP. Regulation of succinate dehydrogenase (*sdhCDAB*) operon expression in *Escherichia coli* in response to carbon supply and anaerobiosis: role of ArcA and Fnr. Mol Microbiol 1995; 15:473-482.

103. Tardat B, Touati D. Two global regulators repress the anaerobic expression of MnSOD in *Escherichia coli*::Fur (ferric uptake regulation) and Arc (aerobic respiration control). Mol Microbiol 1991; 5:455-465.

104. Tardat B, Touati D. Iron and oxygen regulation of *Escherichia coli* MnSOD expression: competition between the global regulators Fur and ArcA for binding to DNA. Mol Microbiol 1993; 9:53-63.

105. Hassan HM, Sun HC. Regulatory roles of Fnr, Fur, and Arc in expression of manganese-containing superoxide dismutase in *Escherichia coli*. Proc Natl Acad Sci USA 1992; 89:3217-3221.

106. Compan I, Touati D. Interaction of six global transcription regulators in expression of manganese superoxide dismutase in *Escherichia coli* K-12. J Bacteriol 1993; 175:1687-1696.

107. Privalle CT, Kong SE, Fridovich I. Induction of manganese-containing superoxide dismutase in anaerobic *Escherichia coli* by diamide and 1,10-phenanthroline: sites of transcriptional regulation. Proc Natl Acad Sci USA 1993; 90:2310-2314.

108. Beaumont MD, Hassan HM. Characterization of regulatory mutations causing anaerobic derepression of the *sodA* gene in *Escherichia coli* K12:

cooperation between *cis*- and *trans*-acting regulatory loci. J Gen Microbiol 1993; 139:2677-2684.

109. Silverman PM, Rother S, Gaudin H. Arc and Sfr functions of the *Escherichia coli* K-12 *arcA* gene product are genetically and physiologically separable. J Bacteriol 1991; 173:5648-5652.

110. Silverman PM, Wickersham E, Rainwater S et al. Regulation of the F-plasmid *traY* promoter in *Escherichia coli* K12 as a function of sequence context. J Mol Biol 1991; 220:271-279.

111. Cotter PA, Darie S, Gunsalus RP. The effect of iron limitation on expression of the aerobic and anaerobic electron transport pathways genes in *Escherichia coli*. FEMS Microbiol Lett 1992; 79:227-232.

112. Stewart V. Nitrate regulation of anaerobic respiratory gene expression in *Escherichia coli*. Mol Microbiol 1993; 9:425-434.

113. Tyson KL, Bell AI, Cole JA et al. Definition of nitrite and nitrate response elements at the anaerobically inducible *Escherichia coli* *nirB* promoter: interactions between FNR and NarL. Mol Microbiol 1993; 7:151-157.

114. Li J, Kustu S, Stewart V. In vitro interaction of nitrate-responsive regulatory protein NarL with DNA target sequences in the *fdnG*, *narG*, *narK* and *frdA* operon control regions of *Escherichia coli* K-12. J Mol Biol 1994; 241:150-165.

115. Walker MS, DeMoss JA. NarL-phosphate must bind to multiple upstream sites to activate transcription from the *narG* promoter of *Escherichia coli*. Mol Microbiol 1994; 633:641

116. Tyson KL, Cole JA, Busby SJW. Nitrite and nitrate regulation at the promoters of two *Escherichia coli* operons encoding nitrite reductase: identification of common target heptamers for both NarP- and NarL- dependent regulation. Mol Microbiol 1994; 13:1045-1055.

117. Wanner BL. Is cross regulation by phosphorylation of two-component response regulator proteins important in bacteria? J Bacteriol 1992; 174:2053-2058.

118. McCleary WR, Stock JB, Ninfa AJ. Is acetyl phosphate a global signal in *Escherichia coli*? J Bacteriol 1993; 175:2793-2798.

119. McCleary WR, Stock JB. Acetyl phosphate and the activation of two-component response regulators. J Biol Chem 1994; 269:31567-31572.

120. Aiba H, Mizuno T, Mizushima S. Transfer of phosphoryl group between two regulatory proteins involved in osmoregulatory expression of the *ompF* and *ompC* genes in *Escherichia coli*. J Biol Chem 1989; 264:8563- 8567.

121. Nakashima K, Kanamuru K, Aiba H et al. Signal transduction and osmoregulation in *Escherichia coli*. J Biol Chem 1991; 266:10775-10780.

122. Holman TR, Wu Z, Wanner BL et al. Identification of the DNA-binding site for the phospshorylated VanR protein required for vancomycin resistance in *Enterococcus faecium*. Biochemistry 1994; 33:4625-4631.

123. Nystrom T. The glucose-starvation stimulon of *Escherichia coli* induced and repressed synthesis of enzymes of central metabolic pathways and role for acetyl phosphate in gene expression and starvation survival. Mol Microbiol 1994; 12:833-843.

124. Poole RK, Ingledew WJ. The pathways of electrons to oxygen. In: Neidhardt FC, Ingraham JL, Low KB et al. *Escherichia coli* and *Salmonella typhimurium*: cellular and molecular biology. Washington, D.C. American Society for Microbiology, 1987:170-200.

125. Wissenbach U, Ternes D, Unden G. An *Escherichia coli* mutant containing only demethylmenaquinone, but no menaquinone: effects on fumarate, dimethylsulfoxide, trimethylamine N-oxide and nitrate respiration. Arch Microbiol 1992; 158:68-73.

126. Wallace BJ, Young IG. Role of quinones in electron transport to oxygen and nitrate in *Escherichia coli*. Studies with a *ubiA menA* double quinone mutant. Biochim Biophys Acta 1977; 461:84-100.

127. Iuchi S, Lin ECC. Signal transduction in the Arc system for control of operons encoding aerobic respiratory enzymes. In: Hoch JA, Silhavy TJ, eds. Two-component signal transduction. Washington: ASM Press, 1995:223-232.

CHAPTER 19

THE PORIN REGULON: A PARADIGM FOR THE TWO-COMPONENT REGULATORY SYSTEMS

James M. Slauch and Thomas J. Silhavy

INTRODUCTION

The major outer membrane proteins, OmpF and OmpC, serve as passive diffusion pores that allow small hydrophilic molecules to cross the outer membrane of *Escherichia coli* K-12. The total amount of OmpF plus OmpC remains essentially constant. However, the relative amount of each protein is differentially regulated at the transcriptional level by the two-component regulatory system composed of the inner membrane sensory component, EnvZ, and the transcriptional regulator, OmpR. This regulation is in response to the osmolarity of the growth medium. OmpF is predominantly produced in media of low osmolarity, whereas OmpC is preferentially synthesized in media of high osmolarity.

Several reviews on the regulation of OmpF and OmpC or the regulation of outer membrane proteins have been previously published.[1-5] Here we have concentrated on the transcriptional regulation of *ompF* and *ompC* by OmpR and EnvZ in response to media osmolarity. We have covered the material from a historical point of view, focusing on how our current understanding of the porin regulon has evolved. Although osmoregulation has been studied extensively, the porin genes are also regulated in response to a number of other parameters.[6-25] The mechanisms behind these various regulatory networks are not completely understood, but it is clear that some of these factors are not mediated through OmpR and EnvZ and that many act at the post-transcriptional level. These include, but are not limited to, those factors that function through the anti-sense RNA *micF*.[6,21,26-33]

Several aspects of the porin regulon make it especially interesting. First, regulation occurs not in response to a particular molecule, but rather in response to a colligative property of the environment. This

Regulation of Gene Expression in Escherichia coli, edited by E. C. C. Lin and A. Simon Lynch. © 1996 R.G. Landes Company.

is in striking contrast to more familiar systems such as the lactose or arabinose operons. *E. coli* regulates the porin genes in response to the osmolarity of the growth medium while carefully and precisely balancing the osmotic strength of the cytoplasm with that of the external milieu. Second, the regulation of the porin genes is unusual in that it does not work by an on/off switch, as in the regulation of gene expression by λ or Lac repressor, but rather involves the gradual and differential regulation of two promoters. Although this type of differential regulation has not been widely studied, it is almost certainly not unique and, therefore, understanding the nature of this differential regulation may be important for understanding other differentially regulated systems.

OmpR and EnvZ belong to a large family of two-component systems involved in the adaptation and response to environmental parameters. These systems are found throughout the prokaryotes and are involved in processes ranging from chemotaxis to the control of virulence factors of both plant and animal pathogens (for review see refs. 34,35,36,37). The regulation of the porin genes in response to osmolarity serves as a paradigm for these other systems, and the lessons learned will shed light on a variety of biologically and medically relevant control mechanisms.

BACKGROUND

Gram-negative bacteria are surrounded by an outer membrane that serves as a barrier between the cell and its environment. Accordingly, the organisms have evolved sophisticated mechanisms to adjust the properties of the outer membrane in response to changes in environmental conditions. Although protection from harsh environments is crucial, the cell must be able to communicate with its surroundings to take up nutrients and excrete waste products. In addition to specific transport proteins, such as those involved in the transport of iron compounds, maltodextrins, or nucleosides,[38] the cell also produces nonspecific pores in the outer membrane that allow the passive diffusion of small hydrophilic molecules. These pores are formed by trimers of the porin proteins.

Two porin proteins, OmpF and OmpC, are expressed under most conditions and a third protein, PhoE, is induced under conditions of phosphate starvation.[38] The three porin proteins show considerable homology at both the protein and the DNA level.[39] Although the pores formed by OmpF or OmpC are generally nonspecific, the diffusion limit of the two pores for sugars and peptides is about 600 daltons. However, electrical conductance studies show that the size of the pores formed by homotrimers of OmpF versus homotrimers of OmpC differ slightly with the OmpF pore having a diameter of 1.16 nm and the OmpC pore a diameter of 1.08 nm. Thus, although the exclusion limit of the two pores is approximately the same, the rate of diffusion can differ drastically between the two, with OmpF being more efficient.[38]

The differences in diffusion rate through the two porin homotrimers has led Nikaido and Vaara[38] to make the teleonomic argument that the cell produces the type of pore that is best adapted to a given situation. *E. coli* is normally found in two very different environments. In the mammalian intestinal tract, where nutrients and detergents are at a relatively high concentration, the cell produces the smaller, more protective OmpC pore. Whereas, when the cell finds itself in sewers or fresh water (where nutrients are limiting), it preferentially produces the larger, more efficient OmpF pore. Thus, the pores in the outer membrane act as molecular sieves that allow the cell to selectively communicate with its environment.[38]

THE HISTORY OF PORIN REGULATION

EARLY BIOCHEMICAL ANALYSIS OF MAJOR OUTER MEMBRANE PROTEINS

Biochemical analysis of the outer membrane of *E. coli* reveals a limited number of major protein species. Schnaitman[40] originally observed what he thought was one protein with an apparent molecular weight of approximately 44,000 daltons on sodium dodecyl sulfate (SDS)-polyacrylamide gels that accounted for at least 70% of the total protein in the outer membrane. Subsequently, with improved techniques, it was shown by several laboratories that the "major outer membrane protein" of *E. coli* K-12 actually consisted of approximately four species with apparent molecular weights of 36,000 to 44,000 daltons.[41-44] Three of these correspond to OmpF, OmpC and OmpA, a major structural protein. It was subsequently shown that the outer membrane proteins were quite insensitive to denaturation and, therefore, ran anomalously on polyacrylamide gels. After boiling in SDS, it could be shown that OmpF and OmpC lost their normal β-sheet structure and became random coils.[45] This complete denaturation, along with adding urea to the SDS-polyacrylamide gels, allowed complete separation. However, consistent results between various labs were not always obtained and therefore confusion as to the actual number of proteins persisted for several years.

During the course of trying to characterize the outer membrane proteins, it became apparent that the relative amount of the various proteins was affected by the strain background, and media composition.[46] Lugtenberg et al[47] noted that the relative amount of OmpF and OmpC changed depending on the media in which the cells were grown, whereas production of OmpA was relatively constant. Van Alphen and Lugtenberg[48] subsequently showed that it was the osmolarity of the media which affected the relative ratio of the porin proteins. Upon addition of NaCl, KCl, or sucrose to the growth media, there was a drastic increase in OmpC with a corresponding decrease in OmpF production. These authors performed a detailed kinetic study of this phenomenon; the best study done to date. Upon a shift from high to low osmolarity, OmpC production ceases completely, while there is a dramatic increase in OmpF synthesis. This switch in synthesis is rapid; the experiments show no lag time. After approximately 1.5 generations in the new medium, synthesis of both porins resumes to a level which will maintain the proper ratio in steady state conditions. The analogous experiment where the cells were shifted from low to high osmolarity gave the comparable result; production of OmpF ceased while production of OmpC increased, with a subsequent shift in production after 1.5 generations to new steady state levels.

Kawaji et al[49] investigated the influence of carbohydrates of different molecular size on the fluctuation of porin proteins. They found that it required a 3-fold higher concentration of substances smaller than approximately 600 daltons to cause the same degree of change in OmpF and OmpC production as that caused by osmolytes with a molecular weight greater than 600 daltons. Nakae[50] had found that 600 daltons were the limit for diffusion through reconstituted pores formed by OmpF and therefore Kawaji et al concluded that the difference in effect of the various solutes was due to whether the solute could or could not enter the periplasm.

Early Genetic Analysis of the Porin Regulon

The history of the genetics of porin regulation in *E. coli* is long and sometimes circuitous. In the end, it is realized that three genetic loci are involved in the synthesis of the major outer membrane proteins OmpF and OmpC: *ompF* at 21', *ompC* at 47' and *ompB* at 74' on the *E. coli* chromosome. The *ompB* locus encodes two regulatory proteins, OmpR and EnvZ, required for the production of OmpF and OmpC.

Mutations which affected the synthesis of the major outer membrane porin proteins were first isolated by selecting tolerance to colicins, which are small bacteriocidal agents. A tolerant cell is defined as being resistant to the killing action of the colicin but still capable of binding the agent. The first known mutants in the porin regulon were obtained by Nomura and Witten,[51] who isolated several classes of mutants tolerant to colicin K or E2. Mutants in their class TolI, tolerant to colicin K, have a phenotype consistent with mutations in *ompF*.[5] Colicin E2 can use either OmpF or OmpC and, therefore, selecting mutants with this agent often yielded *ompB* mutants, affected in the synthesis of both OmpF and OmpC. Davies and Reeves[52] performed a more exhaustive study, isolating approximately 300 mutants which were tolerant, or resistant (i.e., not able to bind) to various colicins. Many of these mutants were similar to the classes isolated by Nomura and Witten.[51] Likewise, Foulds and Barrett[53] isolated mutants tolerant to bacteriocin JF246, and then categorized the mutants with respect to their sensitivity to eight different colicins. The mutations which mapped to 21' (*ompF*) were termed *tolF*. Mutations in *ompC* apparently have no known effect on resistance or tolerance to known colicins in the absence of secondary mutation in *ompF* or *ompB*.

Because of the role of porins in the general diffusion of substances across the outer membrane, mutants in the porin genes were often isolated as being defective in the uptake of various compounds. For example, Reeve and Doherty[54] isolated mutants which were partially resistant to chloramphenicol. One of these mutants, termed *cmlB*, mapped to 21', and was subsequently shown by Foulds[55] to be in the same gene as *tolF* mutations. Mutants that affect porin synthesis have also been isolated as increasing resistance to Cu^+ and Ag^+ ions.[56-58] Von Meyenburg[58] isolated mutants of *E. coli* B/r which had a decreased affinity for glucose. These mutants were shown to have an apparent 20- to 500-fold increase in the K_m for many carbohydrates as well as the uptake of phosphate and several amino acids. These mutations were termed *kmt* and were shown to map to *ompB*. These results demonstrated the overall role of porins in the interaction of the cell with its environment.

The general defect in outer membrane permeability caused by the lack of porins was also observed by Beachem et al.[59,60] These authors isolated mutants termed *cry* which were "cryptic" for the activity of various periplasmic enzymes. For example, whole cells exhibited dramatically decreased alkaline phosphatase activity. However, if the cells were first treated with EDTA, activity was restored. These observations were explained by proposing that in the absence of porin proteins, external substrate is not available to the periplasmic enzyme. If outer membrane structure is disrupted slightly by the removal of divalent cations, then substrate can enter the periplasm and be acted upon by the enzyme. One of the *cry* mutations was ultimately shown to be in *ompB*.[60,61]

OmpF and OmpC are exposed to the surface of the cell and constitute a major proportion of total exposed protein. Not surprisingly, these proteins serve as receptors for the binding of a myriad of bacteriophages. Studies with these various phages provided a large number of mutants in the porin regulon. Hancock and Reeves[62] isolated mutants resistant to any one of 42 virulent bacteriophages isolated from a sewage treatment farm near Adelaide, Australia. Mutants resistant to phage K20 were termed *ktw*, and subsequently shown to map to *ompF*.[63] Sensitivity to phage K20 has been used extensively as an assay for OmpF production.

Mutants specifically affected in OmpC production were first isolated by Schmitges and Henning as being resistant to bacteriophage TuIb.[64] Although these mutations were not mapped, the authors did note that the mutants were missing OmpC in the outer membrane. The authors called OmpF and OmpC protein Ia and Ib, analogous to the designation 1a and 1b used by Schnaitman,[43] representing the two recently resolved species of what was thought to be a single species, protein I. Schmitges and Henning[64] isolated the proteins Ia and Ib and compared their properties on isoelectric focusing gels and their CnBr cleavage pattern. Because of the high level of similarity between the two proteins, these authors concluded that the OmpF and OmpC represent "essentially the same polypeptide" and they proposed that the two proteins might arise from modification of a single protein or from two almost identical genes. Due to the complicated nature of porin genetics and the biochemical similarity of OmpF and OmpC, distinguishing between these two models required three years and caused much confusion.

Chai and Foulds[65] discovered that previously isolated *ompF* mutants, termed *tolF*, were missing protein Ia in the outer membrane. Starting with this mutant, these authors isolated strains resistant to phage TuIb or PA-2. The mutations conferring phage resistance mapped to *ompB* and caused loss of protein Ib in the outer membrane. The *ompB* mutations in an *ompF*+ background conferred an OmpF+ OmpC− phenotype. These authors concluded that the mutations were working independently and that protein Ia (OmpF) and protein Ib (OmpC) were encoded by two independent genes. However, the simplest interpretation of their data required that the structural gene for OmpC mapped to *ompB*.

The first mutations specifically mapped to *ompC* were isolated by Verhoef et al[66] using a new phage, called Me1, which was isolated as growing specifically on strains which contain OmpC in the outer membrane. Starting with an OmpF+, OmpC+ strain, they isolated Me1 resistant mutants and mapped several, but not all, of their mutations to the *ompC* gene which they termed *meo*.

Mutations in *ompC* were also isolated by Bassford et al.[67] These authors isolated mutants resistant to PA-2 host range phage and mapped the mutations, termed *par* to 21'. These mutants were missing protein 1b, but produced normal amounts of protein 1a which was induced in media of low osmolarity. These authors also noted that known *ompB* mutations conferred PA-2 resistance. Following up on the observation of Chai and Foulds,[65] Bassford et al noted that *tolF* mutants lacked protein 1a but produced protein 1b. When they isolated *par* mutants in a *tolF* background, they found that the strain now produced normal amounts of what they thought was protein 1a. These results, along with biochemical characterization of 1a and 1b lead the authors to conclude that neither *tolF* or *par* represented the structural gene for protein 1 but rather *ompB* was the structural gene for both proteins 1a and 1b and the *tolF* and *par* loci are responsible for modification of this protein to give the two electrophoretically separable species.

This was consistent with the model originally proposed by Schmitges and Henning.[64]

OmpF⁻, OmpC⁻ mutants are pleiotropically decreased in the uptake of nutrients[58] and are therefore at a considerable disadvantage. Accordingly, revertants which produce a porin arise rapidly. This is the most likely explanation for the fact that the *tolF, par* double mutant constructed by Bassford et al[67] produced what they thought was OmpF. Indeed, Pugsley and Schnaitman[68] confirmed that *tolF, par* double mutants lack porin, and characterized second site suppressor mutations which result in the production of "new membrane proteins". These suppressor mutations map to three genetic loci, *nmpA, nmpB* and *nmpC*. It is now known that mutations in *nmpA* and *nmpB* cause the production of PhoE,[69] whereas *nmpC* is the structural gene for a porin produced from a cryptic phage.[70] Either protein provides porin function for the cell.

Henning et al[71] also reported that a particular mutant which was defective in the production of protein Ia and Ib produced a new, but biochemically similar protein, designated Ic (PhoE). The mutation giving rise to this protein did not map to *ompF, ompC* or *ompB*. Amino-terminal sequence analysis showed that the first eight amino acids of proteins Ia and Ic were identical. The authors concluded that this result supports the hypothesis that *ompB* is the one and only structural gene for protein I, which is subsequently modified to give the various species. The authors did not comment on the fact that the third amino acid of species Ib, the sequence of which is also presented, was clearly different from both Ia and Ic.

Ichihara and Mizushima[72] also performed amino-terminal sequence analysis on OmpF and OmpC and realized that there were several differences. They concluded that the proteins were actually different and therefore were the product of different but similar genes.

Genetic analysis by Sarma and Reeves[73] revealed that various classes of tolerant mutations isolated by Davies and Reeves[52] cluster at 74'. In fact, they named this locus *ompB*. Because the *tolF, par* double mutant reported by Bassford et al,[67] produced an OmpF-like species, Sarma and Reeves continued to believe that *ompB* was the structural gene for protein I (OmpF and OmpC).

The model of porin regulation, which ultimately proved correct, was first presented by Verhoef et al[61] with additional data in an accompanying paper by van Alphen et al.[74] These authors noted that mutations in *tolF* confer an OmpF⁻, OmpC⁺ phenotype and mutations at *meoA* confer an OmpF⁺, OmpC⁻ phenotype whereas a mutation at *ompB* can confer all variations of phenotypes (OmpF⁺, OmpC⁻; OmpF⁻, OmpC⁺; OmpF⁻, OmpC⁻; OmpF⁺⁺, OmpC⁻). After confirming the biochemical results of Ichihara and Mizushima[72] these authors concluded that *tolF* and *meoA* are the structural genes for OmpF and OmpC, respectively, and that *ompB* is a regulatory locus. The authors noted that this also explains the regulation of OmpF and OmpC with respect to media composition.

This model was ultimately confirmed by Hall and Silhavy,[75] by constructing operon fusions of the *ompC* promoter to the *lacZ* gene, (encoding the cytoplasmic protein β-galactosidase). Because of the ease with which β-galactosidase can be assayed, this technique provided a way to quantitate transcriptional activity. Hall and Silhavy showed that β-galactosidase activity produced from the *ompC'-lacZ⁺* fusion was controlled by the *ompB* locus and by changes in media osmolarity in the same fashion as the OmpC protein. This left little doubt that the *ompC* locus was a structural gene under the control of the regulatory locus, *ompB*. The authors also showed that *ompC* was not autoregulatory by demonstrating that the activity of the *ompC'-lacZ⁺* fusion was not influenced by the presence of a wild-type *ompC* gene.

Hall and Silhavy[76] also isolated fusions of the lactose operon to *ompF*. As in the case of *ompC*, quantitative analysis of the β-galactosidase activity produced from *ompF'-lacZ⁺* operon fusions showed that *ompF* is transcriptionally controlled in response to changes in media osmolarity. The authors also constructed protein fusions of *ompF* and *lacZ*. Although the absolute level of β-galactosidase activity produced from the *ompF* operon and protein fusions differed, the relative response to changes in media osmolarity was the same. These results suggest that translational control, at least at the level of translation initiation, does not play a large role in the osmoregulation of OmpF.

Hall and Silhavy[76] further showed that the introduction of *ompB101*, a mutation which confers an OmpF⁻, OmpC⁻ phenotype, into the *ompF'-lacZ⁺* fusion strains caused a dramatic decrease in β-galactosidase activity and, significantly, the residual activity is no longer regulated in response to osmolarity. These results are consistent with the hypothesis that the porin genes are transcriptionally controlled in a positive fashion by the *ompB* locus and it is this locus that is also responsible for regulation of the porin genes in response to osmolarity.

Using the *ompF'-lacZ⁺* and *ompC'-lacZ⁺* operon fusions isolated in the previous studies, Hall and Silhavy[77] performed a detailed genetic analysis of the *ompB* locus using six previously isolated *ompB* alleles which confer a variety of porin phenotypes. Fine structure mapping and exhaustive complementation analysis led the authors to conclude that the *ompB* locus consisted of two genes which they named *ompR* and *envZ*. The mutations which existed in *ompR* confer either an OmpF⁻, OmpC⁻ phenotype or an OmpF constitutive, OmpC⁻ phenotype. Mutations in *envZ* confer an OmpF⁻, OmpC constitutive phenotype. In addition, these *envZ* mutations are pleiotropic; they decrease the expression of a number of gene encoding envelope proteins.

Based on these results, Hall and Silhavy[77] proposed a model for the regulation of the porin genes. They proposed that EnvZ was an envelope protein which senses changes in osmolarity and controls the equilibrium between two forms of OmpR; one form activates *ompF* and the other form activates *ompC*. Although this model is simplistic, it is essentially correct and consistent with our current understanding of porin regulation.

THE STRUCTURE OF THE *OmpB* LOCUS

DNA Sequence Analysis

A number of studies reported the cloning of the *ompB* locus. First, Taylor et al[78] cloned *ompR* by complementation of the *ompR101* mutation (OmpF⁻, OmpC⁻) using an *ompF'-lacZ⁺* operon fusion and showed the gene specified a 29 kilodalton (kDa) protein. Similar results were obtained by Mizuno et al.[79] However, because these authors chose different restriction enzymes, they also succeeded in cloning *envZ*, as shown by complementation. They could show that the *ompB* clone produced a 29 kDa protein corresponding to OmpR, but could not detect a product corresponding to EnvZ.

The nucleotide sequence of *ompB*[80,81] showed that *ompR* and *envZ* are transcribed as an operon with *ompR* promoter proximal. Amino-terminal protein sequencing of the 29 kDa OmpR protein isolated from polyacrylamide gels confirmed the start of the *ompR* open reading frame. However, because of a missing base in the sequence, the open reading frame predicted a protein of 32.5 kDa and the stop codon for *ompR* was assigned to a region that is now known to be in the *envZ* open reading frame. This error led to the misidentification of the initiation codon for *envZ*.

Comeau et al[82] corrected the sequence; the *ompR* open reading frame did indeed predict a protein of 28.5 kDa. In addition, linker insertion analysis suggested that the termination codon of *ompR* and the initiation codon of *envZ* actually overlap. Immunoprecipitation of the wild-type EnvZ revealed a protein of 50 kDa.

ompR AND *envZ* ARE TRANSLATIONALLY COUPLED

As mentioned above, even in clones which greatly overexpress OmpR, wild-type EnvZ is difficult to detect. The overlapping translation termination, initiation sequence, AUGA, at the junction between *ompR* and *envZ*, suggests that the two are translationally coupled. Translation initiation of *envZ* is the result of reinitiation by ribosomes that have terminated after translation of *ompR*.

Liljestrom[83] made protein fusions of *ompR* and *envZ* to *lacZ* to monitor translation of the two genes. The results indicate that translation of *ompR* is approximately 8-fold more efficient than translation of *envZ*. It was also shown that nonsense mutations in the very 3' end of *ompR* are absolutely polar, demonstrating that the translation of *envZ* requires complete translation of *ompR*.

THE STRUCTURE OF THE *ompF* AND *ompC* GENES

DNA SEQUENCE ANALYSIS

The *ompF* gene was cloned independently by Mutoh et al[84] and Tommassen et al.[85] The sequence of *ompF*[86] predicts a protein of 340 amino acids synthesized as a precursor with a 22 amino acid signal sequence. A sequence resembling an *rho* independent terminator is located just 3' to the putative translation termination codon, consistent with the transcription of a monocistronic message. The authors noted that the putative promoter region is strikingly AT rich, although the start site of transcription was not clearly defined in this study.

The start site of transcription of the *ompF* gene was determined by Inokuchi et al[87] and by Taylor et al[88] by performing S1 nuclease protection experiments. The latter authors also isolated mutations in the -10 and -35 regions of the promoter which caused decreased expression of an *ompF'-'lacZ* protein fusion. Mutations in the -10 of the *ompF* promoter were also isolated by Dairi et al[89] by selecting OmpR independent expression from the *ompF* promoter. These results are all consistent and pinpoint the start site of transcription with respect to the initiation codon, indicating that the *ompF* mRNA contains a long untranslated leader sequence of 111 bases. The *ompF* message has an extremely long half life of approximately 15 minutes[90] compared to 3 to 5 minutes for most *E. coli* messages.[91] The relatively long untranslated leader sequence might be involved in stabilizing the message.[92]

The *ompC* gene was cloned and sequenced by Mizuno et al.[93] As expected from previous biochemical studies, the predicted amino acid sequence of OmpC was highly homologous to both OmpF and PhoE. The presence of an apparent *rho* independent terminator downstream of the coding sequence suggests that the *ompC* message, like that of *ompF*, is monocistronic. The authors noted no apparent homology between the promoter regions of *ompF* and *ompC*, i.e., there was no obvious OmpR binding site. The start site of transcription of *ompC* was mapped by Ikenaka et al.[94] The *ompC* mRNA, like that of *ompF*, contains a long untranslated leader sequence, in this case 82 bases.

THE *ompF* PROMOTER

Inokuchi et al[87] constructed deletions of the *ompF* promoter region in vitro, starting with transcriptional fusions of the *ompF* promoter to the gene conferring tetracycline resistance. Promoter activity in vivo was measured by the degree to which the construct could confer resistance to tetracycline. Although this study is complicated because of the lack of quantitation, it is clear that the DNA from approximately -112 to +18 from the start site of transcription is critical for OmpR-dependent transcription of *ompF*.

Direct binding of OmpR to the *ompF* promoter region was first demonstrated by Jo et al[95] by gel retardation assay. These authors performed the assay using purified OmpR and DNA fragments containing various regions of the *ompF* promoter. The results are consistent with the deletion analysis performed previously[87] and demonstrate that the region of the promoter required for OmpR-dependent activation of the promoter coincides with the region required for specific binding of the OmpR protein.

DNAse I footprinting studies were carried out by Norioka et al[96] and Mizuno et al.[97] Norioka et al[96] showed that OmpR bound specifically to the region from -60 to -100, consistent with deletion analysis.[87] Mizuno et al[97] showed that as the concentration of OmpR was increased, the protein bound to the sequences from -41 to -59 in addition to the region between -63 to -93. The authors noted that this is a rather large region of DNAse I protection compared to other sequence specific DNA binding proteins. For example, the cyclic-AMP binding protein dimer protects only 25 base pairs,[98] whereas OmpR seems to protect up to 60 base pairs. This suggested that multiple OmpR molecules are bound. Indeed, there is genetic evidence for multimerization.[77] Mizuno et al note in this study, however, that OmpR behaves as a monomer in solution (see below).

Binding of OmpR to the *ompF* promoter in vivo was shown by Tsung et al[99] using dimethyl sulfate. The authors examined the *ompF* promoter region cloned onto a high copy number plasmid and showed that various G residues from -96 to -46 were protected in vivo, consistent with the previous in vitro footprint experiments.[96,97]

Mizuno[100] showed that there is an inherent static bend in the helix of the *ompF* promoter. This is apparently due to four short stretches of dA-dT base pairs separated by 21 base pairs, or two turns of the DNA helix. This type of sequence specific bending has been observed in other systems, such as the λ origin of DNA replication.[101] Mizuno showed that disruption of the spacing between these sequences altered the physical properties of the DNA. The sequence is located between -70 and -102, overlapping the region required for activation of the promoter by OmpR and EnvZ.

Kato et al[102] attempted to directly address the importance of the bent DNA by making mutations in the dA-dT sequences and monitoring their effect on both in vitro binding of OmpR and in vivo activity of a *ompF'-'lacZ* fusion. Although the results are complicated, there is a correlation between the activity of the fusion and the ability of OmpR to bind in vitro. In addition, similar to the deletion studies,[87] mutations at -99, for example, have a great effect on the activity of the fusion but no detectable effect on OmpR binding, consistent with the hypothesis that sequences in this region enhance the binding of OmpR but are not critical for recognition.

The studies outlined above show that the region between approximately -100 to -40 is important for OmpR dependent activation of *ompF*. The fact that OmpR specifically binds this region of DNA also implied that OmpR binding to the promoter region is necessary for activation. However, this region is not sufficient for proper osmoregulation of *ompF*. Ostrow et al[103] monitored promoter activity from *ompF'-lacZ+* operon fusions that were carried on λ specialized transducing phage integrated in single copy in the chromosome. By comparing constructs containing various regions of the *ompF* promoter, and measuring the β-galactosidase activity produced from cells grown in media of high and low osmolarity, Ostrow et al concluded that a region upstream of -240 was required. Huang et al[104] recently reported the identification of an upstream site at -351 to -384 from the start site of transcription. OmpR clearly footprints the region and an insertion mutation in the site results in OmpR-dependent constitutive expression of the gene. These studies were complicated by the fact that this site is in the upstream *asnS* gene, which is essential for viability.[105]

THE *ompC* PROMOTER

The regions of the *ompC* promoter that are required for OmpR and EnvZ dependent transcriptional regulation were delineated by Mizuno and Mizushima.[106] These authors started with a fusion of the entire promoter region, mRNA leader sequence and the sequences for the first four amino acids of the OmpC signal sequence fused in frame to *lacZ*. Deletion analysis of this construct revealed that the region between -94 to +1 from the start site of transcription was critical for OmpR-dependent promoter activity. These authors then mutagenized this region, screening for mutations which decreased OmpR-dependent transcription. Although every mutant they isolated contained more than one mutation, there were two interesting classes of mutants which suggested that three ten base pair repeats (tGaAaCATcT), located on the same face of the helix, each separated by two helical turns, might be important for transcriptional activation by OmpR and EnvZ.

In a series of reports, Maeda, Mizuno and colleagues have performed a detailed analysis of the *ompC* promoter region, concentrating on the three ten base pair repeats described above. First, Maeda et al[107] showed that deletion of either the most promoter distal or the most promoter proximal of the repeats greatly decreased OmpR-dependent transcription. The authors then increased the distance between the three repeats and the -35 and -10 regions of the promoter. They found that OmpR-dependent activation was maintained only when the repeats were moved a near integral number of turns of the DNA helix, suggesting that OmpR must be bound on the correct face of the helix in order to activate the promoter.

In the second report, Maeda and Mizuno[108] show that the upstream region can activate the promoter in an OmpR-dependent fashion in either orientation, again provided that the binding site is on the correct face of the helix. Finally, Maeda and Mizuno[109] show that the most distal ten base pair repeat can independently activate the promoter provided it is positioned correctly with respect to the -35 and -10 regions. In this report, they also provide evidence that OmpR can independently bind each of the three sequences containing the repeats. However, the data suggest that binding to the most distal repeat influences binding to the other two repeats, suggesting cooperative binding.

The OmpR Binding Sites

Both the *ompF* and *ompC* promoters are dependent on OmpR and EnvZ for activation, yet early analysis of these two promoters suggested that they have little in common. The *ompC* promoter seems relatively straight forward. Evidence suggests that three ten base pair repeats are important for OmpR-dependent transcriptional regulation and these sequences seem to simply position OmpR correctly with respect to the -10 and -35 sequences and, therefore, the polymerase. The *ompF* promoter on the other hand, is more complicated. OmpR binds differentially to a large region of the promoter including an area at -350. In addition, sequence induced bends play a role in osmoregulation of the promoter.

As outlined above, Mizuno and Mizushima[106] pointed out three 10 bp repeats in the *ompC* promoter and characterized mutations that affect these sequences. Analogous sequences are also found in the *ompF* promoter,[104,110] including two repeats in the -350 region. Indeed, Maeda et al[111] constructed synthetic oligonucleotides (20-mers) that contain these apparent motifs and showed that one such sequence can function in OmpR-dependent transcriptional activation in an orientation-independent fashion. They also showed, similar to their results in the *ompC* promoter,[108,109] that the sequence must be properly positioned with respect to the DNA helix.[111] Alignment of the eight repeats from *ompF* and *ompC* gives the "consensus" sequence shown in Figure 19.1. Note that the length of this motif is not clear. For example, although several more upstream bases were required to get OmpR-dependent transcription in the experiments of Maeda et al,[111] there is clear evidence that OmpR does not interact with these base pairs in all of the repeats.[104] For the repeat at - 90 in the *ompF* promoter, the region immediately upstream was suggested by Kato et al[102] to be important for OmpR binding to the DNA but not directly involved in recognition. The A/T richness of the sequence makes the consensus somewhat less convincing. However, this sequence is repeated in all of the OmpR binding sites and the spacing between the invariant Gs in the repeats is 20 or 21 bp in the primary binding sites in *ompF* and *ompC* and 18 bp in the *ompF* -350 site. Interestingly, Aoyama and Oka[112] point out similar sequences in promoters regulated by PhoB and VirG, proteins that are homologous to OmpR (see below). Although these data are suggestive, alternative binding motifs have been proposed.[99]

Many of the early OmpR binding experiments outlined above were performed with purified OmpR that was nonphosphorylated (see below) and therefore had low affinity for the DNA. Hence, most of these footprinting experiments required large molar concentrations of OmpR relative to the DNA on the order of 20-30:1. In addition, few of the experiments included the binding site at -350. Clearly a more systematic analysis is required.

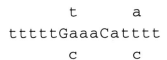

Fig. 19.1. Consensus alignment of potential OmpR binding sites. The lower case bases in the main sequence are present in at least five out of eight repeats. The upper case G and C are invariant. In those cases where a single base is not present in at least five cases, the bases are listed in order of relative representation from top to bottom.

THE ROLES OF OmpR AND EnvZ

EnvZ is an Inner Membrane Sensor Protein

To be consistent with the proposed role of EnvZ as an environmental sensor of media osmolarity, Hall and Silhavy[77] postulated that EnvZ is an envelope protein. This indeed proved to be the case. The nucleotide sequence of *envZ* revealed that there are two long hydrophobic stretches of amino acids in the protein predicting a structure similar to the chemoreceptors with an amino-terminal periplasmic domain bordered by the two putative membrane spanning segments. The carboxy-terminal half of EnvZ is in the cytoplasm where it is free to perform its role in the transcriptional regulation of the porin genes.

The fact that EnvZ is an inner membrane protein was first demonstrated by Liljestrom[113] who showed that EnvZ was associated with the inner membrane after fractionation of the cell on sucrose gradients and immunoblotting with EnvZ antisera. Forst et al[114] performed a similar experiment. After overproducing EnvZ from a plasmid, the protein was shown to be in the inner membrane by immunoprecipitating various membrane fractions. Forst et al[114] also created protein fusions of *envZ* to β-lactamase, an antibiotic resistance gene which is only functional when localized to extra-cytoplasmic locations. Fusions which positioned β-lactamase to the putative periplasmic portions of EnvZ were found to be highly active, whereas fusions in which β-lactamase was joined to the carboxy-terminal portion of EnvZ had decreased activity. These results are consistent with EnvZ being an inner membrane protein with the topology described above.

The original mutations which mapped in the *ompB* locus conferred a variety of porin phenotypes. Hall and Silhavy[77] demonstrated that the mutations in *ompR* conferred either an OmpF⁻, OmpC⁻, or an OmpF constitutive, OmpC⁻ phenotype, whereas the mutations in *envZ* conferred an OmpF⁻, OmpC constitutive phenotype. In addition, the *envZ* mutants (e.g., *envZ473*) were pleiotropic, causing a decrease in production of a number of envelope proteins that are not normal members of the porin regulon. These include the proteins of the maltose regulon, and PhoA and PhoE, the last two being members of the phosphate regulon, normally induced upon phosphate starvation.[61,115-117] This is similar to the phenotype observed in wild-type cells upon addition of low concentrations of the anesthetic procaine.[118-121]

These pleiotropic *envZ* alleles are codominant with wild type, suggesting that none of the alleles represent the *envZ* null phenotype.[77,122] Taken together, these results suggested that *envZ* coded for an essential protein involved in the production of a number of envelope proteins.[77] To address this question directly, Garrett et al[123] isolated *envZ* amber mutations in an *ompB* merodiploid background so that mutants would survive even if *envZ* was essential. The authors subsequently showed that a haploid *envZ* amber strain is viable, and therefore that EnvZ is not essential.

Once it was realized that *envZ* was not an essential gene, Garrett et al[124] isolated chromosomal deletions that removed the entire *ompB* operon and surrounding chromosome. These deletion mutations confer an OmpF⁻, OmpC⁻ phenotype identical to the phenotype conferred by *ompR* null mutations. In addition, when these deletion strains are lysogenized with a specialized λ transducing phage carrying the wild-type *ompR* but not *envZ*, the phenotype is identical to the phenotype conferred by the *envZ* amber mutations. These results show that the amber mutants truly reflect the phenotype of an *envZ* null. Mizuno and Mizushima[125] subsequently isolated *envZ* deletion mutations confirming the results of Garrett et al.[124]

The *envZ* amber mutant synthesizes a small amount of OmpF and virtually no OmpC (OmpF⁻/⁺, OmpC⁻), indicating that EnvZ is required for the proper expression of both porin genes. However, expression of the maltose operon and *phoA* is unaffected in the *envZ* amber strain, indicating that EnvZ is not normally involved in the expression of these genes. Garrett et al[123] also showed that the *envZ* amber strains are completely resistant to low concentrations of procaine, consistent with the hypothesis that the effect of procaine is mediated through EnvZ. These results suggest that the pleiotropic phenotypes are the result of an alteration of EnvZ, either by mutation or the addition of procaine, such that it interferes with the expression of genes that it does not normally regulate (see below).

The model proposed by Hall and Silhavy[77] suggested that EnvZ functions only as a sensor, and that it functions in the transcriptional regulation of the porin genes by signaling to OmpR. We tested this aspect of the model by constructing an *ompR101, envZ473* double mutant.[126] The *ompR101* mutation is an in-frame deletion[127] and thus allows normal expression of the downstream *envZ* gene. We could show that all of the pleiotropic phenotypes conferred by *envZ473* were alleviated in the *ompR* null background. In addition, we could show that under certain conditions OmpR could function independently of EnvZ to activate porin gene expression. These results were consistent with the role of EnvZ as a sensor of osmolarity that communicates information to OmpR.

OmpR Works Both Positively and Negatively to Control Transcription of *ompF*

Although work to this point had helped to clarify the roles of OmpR and EnvZ and the *ompF* and *ompC* promoters in the regulation of the porin regulon, the question of how OmpR and EnvZ differentially regulate the porin genes remained. Because EnvZ is required for activation of both porin genes,[123-125] the switch that controls the differential expression of *ompF* and *ompC* could not be simply the presence or absence of a signal from EnvZ. To address this issue, we isolated and characterized a series of *ompR* missense mutations that decreased the expression of either *ompF* or *ompC*, but not both.[128] Diploid analysis of the various *ompR* mutations along with tests of epistasis using known *envZ* alleles was performed to probe the nature of the genetic switch that controls porin gene expression. Two themes emerged. First, decreased expression of *ompC* was accompanied by a loss of osmoregulation of *ompF*. Second, a dominant OmpC-constitutive phenotype was always accompanied by a dominant OmpF- phenotype. This was observed whether the phenotype was conferred by a mutant *ompR*, a mutant *envZ* functioning through OmpR, or wild-type proteins in high osmolarity. Because of the generality of this phenotype, we suggested that it reflects the normal regulation of the porin genes and that OmpR functions in both a positive and a negative fashion to regulate the expression of *ompF*.[128] This implied that there must be both positive and negative sites in the *ompF* promoter region at which OmpR works to activate or inhibit transcription. The genetic arguments that led to these conclusions were similar in concept to those used by Englesberg et al[129] in their classic studies of the *ara* operon.

Using the above data, along with previous results from a variety of labs, we formulated a working model for porin regulation.[128] The model proposed that OmpR normally exists in at least three functionally distinct states. Conversion among these states is mediated by the inner membrane signal transducer, EnvZ.

The first state is the null state. The existence of such a null form in our model is mandated by the phenotype of envZ nonsense and deletion mutations. In the absence of EnvZ, transcription at both porin genes is greatly reduced.[123-125] It had been shown that OmpR is stable in the absence of EnvZ.[130] Therefore, there must be a form of OmpR that is largely nonfunctional.

The second state is the low osmolarity state. We proposed that this form of OmpR is responsible for activating expression of *ompF* and does so by binding to a site(s) (positive site) in the promoter region. This form does not bind to the *ompC* promoter and consequently has no effect on *ompC* expression.

The third state is the high osmolarity state. We proposed that this state activates *ompC* expression by binding to a site(s) in the *ompC* promoter. In addition, this form of OmpR negatively regulates *ompF* expression by binding to a site(s) (negative site) in the *ompF* promoter region.

According to this model, the key to osmoregulation is the dominance of the high osmolarity form of OmpR to the low osmolarity form. To regulate porin expression, the cell need only regulate the amount of the high osmolarity form because this form of OmpR both activates *ompC* and inhibits *ompF* expression. Although this model provided no explanation for the molecular difference between the functionally distinct states of OmpR, it was consistent with all of the known genetic and biochemical data.

The model proposed above suggests that there are both positive and negative acting sites in the *ompF* promoter that account for the differential expression of the gene. As outlined above, studies of the *ompF* promoter region reveal a complex pattern of DNA-protein interaction. DNAse I protection studies revealed that OmpR binds to the region between -40 and -100 from the start site of transcription.[97] More recent studies reveal that OmpR also binds to the region from approximately -350 to -380 from the start site of transcription,[104] and this region is required for proper osmoregulation.[103,104] DNA footprinting studies with mutant OmpR proteins suggest that OmpR2 proteins, which we have shown can no longer negatively regulate *ompF*,[128] can only bind to a subset of the sites, namely the region from -60 to -100,[97] whereas OmpR3 proteins, which always repress *ompF* expression,[97] show the same footprint pattern as wild-type OmpR.[97] Taking all of these results into consideration, we proposed that the negative site(s) required for osmoregulation of *ompF* included the site from -40 to -60. The site from -350 to -380, not included in the footprint studies of Mizuno et al,[97] is also involved in the negative regulation of *ompF*; mutations in this region result in loss of negative regulation.[104] Indeed, Rampersaud et al[131] have recently shown in vitro that OmpR preferentially binds to the region from approximately -60 to -100 in the *ompF* promoter. As the concentration of OmpR was increased, the protein bound to the -40 to -60 site and subsequently to the site at -350. Binding to these secondary sites was dependent on binding to the -60 to -100 site. Moreover, these authors showed that OmpR472, a mutant in the OmpR2 class, binds poorly to both the -40 to -60 site and the -350 site.

Further analysis of the *ompF* promoter suggested that it has a complicated three dimensional structure. Mizuno[100] has shown that the promoter region is naturally bent and has localized this effect to the two sets of T residues at -80 and -100, separated by two turns of the DNA helix. Furthermore, Tsui et al[132] have shown that integration host factor (IHF) is required for proper regulation of *ompF*. It has subsequently been shown that IHF binds both at -170 and -60 from the start site of *ompF* transcription.[133] (IHF is a small histone-like protein that performs a role in recombination, DNA replication and gene regulation (for review see ref. 134). It appears that IHF bends DNA into the correct conformation to allow various protein-protein interactions to occur. Interestingly, IHF⁻ strains are constitutive for *ompF* expression.[132] Based on the fact that the *ompF* promoter is naturally bent[100] and that there is an OmpR binding site at position -350 that is involved in repression of *ompF*, we speculated[3] that the negative regulation of the *ompF* promoter involves a DNA loop structure similar to that proposed for other systems.

In order to substantiate various aspects of our model for differential regulation of the *ompF* promoter, we isolated and characterized *cis*-acting mutations that affected negative regulation by OmpR and, accordingly, conferred OmpR-dependent constitutive expression on *ompF*.[110] The two mutations deleted one of the dA-dT base pairs in the -80 region and directly affected the DNA bend,[100] suggesting that the conformation of the promoter is required for proper osmoregulation.[110] All of the results with the *ompF* promoter are consistent with the hypothesis that in high osmolarity, OmpR interacts with all of the sites in the *ompF* promoter resulting in a looped structure that represses transcription.[104,110]

PHOSPHORYLATION AND SIGNAL TRANSDUCTION

OmpR and EnvZ belong to a large family of two-component regulatory systems[135] involved in environmental sensing and adaptive responses ranging from chemotaxis to the control of gene expression (for review see refs. 34, 35, 36, 37). These systems are found throughout the prokaryotes. The EnvZ-like proteins function as sensory components and OmpR-like proteins are effector components.

The homology in the sensor class of molecules in usually localized to the carboxy-terminal portion including an invariant histidine (amino acid 243 in EnvZ). Many of these proteins, like EnvZ, appear to be membrane receptors and the homologous segments correspond to the cytoplasmic domain which interacts with the cognate effector protein.[34-36,136] In the case of the effectors, the homology, first pointed out by Stock et al,[136] is found in the amino-terminus of the protein. In OmpR, this homology extends from the amino-terminus through approximately amino acid 120. The amino acids corresponding to Asp-12, Asp-55, and Lys-105 in OmpR are always conserved.[34-37]

The best characterized of the two-component systems, both genetically and biochemically, are the ones involved in nitrogen regulation, chemotaxis, phosphate regulation, and porin gene expression in *E. coli*. Hence, these systems are paradigms for environmental sensing and signal transduction systems throughout the prokaryotes, and results from these systems have provided insights into the mechanism of signal transduction. Ninfa and Magasanik,[137] studying nitrogen regulation in *E. coli*, showed that NRII (EnvZ homolog) phosphorylates NRI (OmpR homolog) which then activates expression of the *glnALG* operon. This provided the first clue to the mechanism by which the sensor proteins communicate environmental information to the effector proteins. We now know that phosphorylation mediates signal transduction in all two-component systems.[34-37]

Because of the membrane location of EnvZ, and the fact that it is present at only ten copies per cell, a biochemical analysis of this protein has been difficult. These problems were circumvented with the use of *envZ* alleles constructed in vitro, that lack the amino-terminal portion. These truncated proteins are often called EnvZ*. One allele in this class, *envZ115*, results in a truncated form of EnvZ in which the first 38 amino acids of the protein are replaced with the first eight amino acids of LacZ.[138] As a result of this construction, EnvZ lacks one of its two putative membrane spanning segments and is no longer properly localized to the membrane. A plasmid carrying the *envZ115* allele complements an *envZ* null mutation and allows expression of both *ompF* and *ompC*, indicating that EnvZ115 must retain the ability to communicate with OmpR.[139]

Biochemical analysis using EnvZ115 and similarly truncated proteins has provided evidence that protein phosphorylation is involved in signal transduction in the Omp system.[139-141] When incubated in the presence of ATP, EnvZ phosphorylates itself (autophosphorylation). Addition of OmpR results in the rapid dephosphorylation of phosphorylated EnvZ (EnvZ-P) and the appearance of phosphorylated OmpR (OmpR-P). Moreover, it was shown that EnvZ stimulates the ability of OmpR to activate transcription from the *ompF* promoter in vitro, thus supporting the physiological relevance of the phosphotransfer reaction.[142] EnvZ is also involved in the dephosphorylation of OmpR-P.[140,143,144] Addition of EnvZ115 and ATP (or nonhydrolyzable derivatives of ATP) to OmpR-P results in the dephosphorylation of OmpR and the appearance of inorganic phosphate.[144] Wild-type EnvZ, either in an immune complex[142] or in membrane vesicles[145] is also capable of carrying out these various reactions. Thus, EnvZ catalyzes both the phosphorylation and dephosphorylation of OmpR.

Based on the homologies to the Che (chemotaxis) and Ntr (nitrogen) systems,[146-148] it was predicted that the phosphorylated amino acid in EnvZ is the conserved His-243 residue and the phosphorylated amino acid in OmpR is Asp-55. Biochemical and mutational analysis appears to confirm these predictions. Roberts et al[149] isolated a 25 kDa proteolytic fragment of EnvZ* that corresponds to the carboxy-terminal domain from amino acids 215-240. This fragment behaves as a dimer and is efficiently autophosphorylated. Proteolytic digestion of the phosphorylated protein yielded one phosphorylated peptide that included His-243 and this residue was apparently modified. In addition, site-directed mutagenesis changing His-243 to Arg resulted in a completely nonfunctional protein.[150] As described above, there are two conserved Asp residues in OmpR at positions 12 and 55. By analogy to the crystal structure of CheY,[151] these Asp residues are together in an "acidic pocket." Both of these residues have been altered by site-directed mutagenesis.[150] Mutations at position 12 result in a significant decrease in function, but the protein can still be phosphorylated in vitro. In contrast, mutations at position 55 result in a complete loss of function and the inability of the protein to be phosphorylated by EnvZ in vitro. These results indicate that Asp-55 is the site of phosphorylation,[150] and this is consistent with the fact that CheY is known to be phosphorylated at the analogous position, Asp-57.[152,153]

A common mechanism of phosphotransfer among the two-component regulators is supported by various examples of "cross-talk." For example, EnvZ can phosphorylate the regulator protein of the nitrogen system, NRII, and can thereby stimulate the ability of NRII to activate transcription from the *glnAp2* promoter. Similarly, CheA, the kinase in the chemotaxis system, can phosphorylate OmpR and stimulate OmpR-dependent transcription from the *ompF* promoter in vitro.[144] These results and earlier studies[154] argue for a common phosphotransfer mechanism in the Ntr, Che, and Omp systems and are consistent with the idea that EnvZ differentially regulates the expression of the porin genes simply by phosphorylating OmpR.

Phosphorylation of an effector component by a kinase from a different two-component system (cross-talk) may allow global regulation in response to the environment.[37] This is an intriguing possibility. However, the only in vivo examples of cross-talk are seen in cases where the cognate kinase/phosphatase is mutant.[155] Results obtained under these conditions must be judged carefully. Because EnvZ is both an OmpR kinase and an OmpR-P phosphatase, it seems likely that EnvZ can tightly control the level of OmpR phosphorylation and, therefore, cross-talk in the presence of EnvZ is probably not significant to porin regulation under most conditions (see below).

THE CONVERGENCE OF GENETIC AND BIOCHEMICAL ANALYSIS

The model described in a previous section proposes three states of OmpR: a null state that is largely nonfunctional; a low osmolarity state that activates *ompF* expression; and a high osmolarity state that activates *ompC* expression and represses *ompF* expression.[128] These states represent different functional forms of OmpR that were defined genetically. EnvZ mediates the conversion of OmpR into the various states by phosphorylation, but it was unclear whether the molecular difference between the high and low osmolarity states was quantitative or qualitative. Work over the last several years has provided the resolution to this problem and all the evidence suggests that the difference is simply quantitative.

A number of experiments are consistent with the hypothesis that both the low and high osmolarity states of OmpR are phosphorylated. First, phosphorylation of OmpR increases its ability to bind both the *ompF* and *ompC* promoter regions.[156] Second, mutants of *ompR* that confer an OmpF constitutive, OmpC⁻ phenotype (the low osmolarity state) are strictly dependent on EnvZ, and thus presumably on phosphorylation for function.[128] Third, mutants of *envZ* or *ompR* that confer a OmpC constitutive, OmpF⁻ phenotype (the high osmolarity state) are dominant.[128] The analysis of these mutant proteins in vitro indicates that they result in the accumulation of phosphorylated OmpR.[143] Taken together, these results support the view that both the low and the high osmolarity states of OmpR are phosphorylated.

Forst et al[157] examined the level of OmpR-P in vivo under various conditions. They detected an increase in OmpR-P when cells were grown in high osmolarity versus low osmolarity. Although this result could potentially be complicated by general osmolarity effects on protein stability,[158] they also detected an increase in OmpR-P in *envZ11* cells grown in low osmolarity. This pleiotropic *envZ* allele confers a phenotype that resembles high osmolarity and these results are consistent with the high osmolarity state representing an increased level of OmpR-P.

EnvZ both phosphorylates and dephosphorylates OmpR. However, most of the *envZ* mutants isolated up to this time had been of the pleiotropic variety, conferring an OmpC-constitutive OmpF⁻ phenotype, i.e., high OmpR-P. Russo and Silhavy[159] sought *envZ* mutations that conferred a variety of phenotypes. One class of these mutations behave constitutively as they make EnvZ phosphatase. These alleles are co-dominant to both wild-type *envZ* and the pleiotropic *envZ473*, and the phenotype conferred by these mutations is identical to the phenotype of an *ompR* null mutation. This demonstrates that the nonphoshorylated form of OmpR is completely nonfunctional and that the residual activity seen in an *envZ* null strain is due to crosstalk. We suspect, but have not proven, that the residual activity is the result of OmpR phosphorylation by acetyl phosphate.[160-163] Thus, there are *envZ* missense mutations that can result in either hyperphosphorylation of OmpR or complete dephosphorylation of OmpR-P, and these mutations represent the extremes of *envZ* phenotypes. This supports the hypothesis that EnvZ simply controls the level of OmpR-P.

Russo and Silhavy[159] present an argument based on Michaelis-Menton enzyme kinetics that a change in the concentration of OmpR-P is sufficient to explain the regulation of porin gene expression in response to media osmolarity. Their model assumes that EnvZ possesses both kinase and phosphatase activities and that it is the ratio of these activities that changes in response to osmolarity; as the osmolarity increases, the concentration of OmpR-P increases. Differential regulation of *ompF* and *ompC* is achieved by differences in affinity of the various binding sites in the porin gene promoters. The activating sites for *ompF* are high affinity sites, whereas the activating sites for *ompC* and the repressing sites for *ompF* are low affinity sites. Russo and Silhavy show that a difference in affinity for these sites of as little as 20-fold is sufficient to account for the differential regulation observed.

In a second paper, Russo and Silhavy[164] present further arguments based on enzyme kinetics that the level of OmpR-P is surprisingly independent of the absolute concentration of either OmpR or EnvZ. Their arguments are based on the following assumptions: (1) the kinase and phosphatase reactions are carried out by the same protein, namely EnvZ; (2) the kinase reaction is saturated with respect to substrate, while the phosphatase is not. This analysis explains results that were seemingly paradoxical. Overproduction of OmpR in the absence or presence of single copy *envZ* results in a phenotype that resembles high osmolarity, i.e., high OmpR-P,[126,165] whereas overproduction of both proteins results in normal osmoregulation.[165,166] In addition, overproduction of EnvZ in the presence of single copy *ompR* also results in normal osmoregulation.[145] The high osmolarity phenotype conferred by overproduction of OmpR in the absence of EnvZ is apparently the result of nonspecific phosphorylation, probably by acetyl phosphate. Thus the first assumption of Russo and Silhavy[164] breaks down in that the kinase and phosphatase reactions are not carried out by a single protein. This aberrant phenotype of OmpR overproduction has led to the misinterpretation of data on several occasions.[141]

We do not yet know which of the activities of EnvZ is regulated by osmolarity; autophosphorylation, OmpR kinase, or OmpR-P phosphatase. The analysis of Russo and Silhavy[164] suggests that the kinase activity is regulated. This could be accomplished by regulating autophosphorylation, as in the case of the chemotaxis kinase CheA.[167] Once phosphorylated, EnvZ is a kinase. Yang and Inouye[168] have shown that EnvZ autophosphorylation occurs in a dimer by transphosphorylation of the individual subunits. Perhaps high osmolarity stimulates this transphosphorylation. Inouye and colleagues[169-171] have argued that it is the phosphatase activity which is regulated. However, their conclusions are based on studies with chimeric receptors (see below). Resolution of this question will likely require identification of the signal EnvZ senses.

THE SENSING OF OSMOLARITY

Evidence suggests that EnvZ is the sensor of media osmolarity, but the actual signal to which EnvZ responds remains elusive. However, we can use our limited knowledge to rule out certain possibilities. For example, we know that the signal is not osmolarity per se, because glycerol, which changes the osmolarity yet freely penetrates the cell, has no effect on porin regulation. We use the term osmoregulation for lack of a better description. Indeed, any compound that does not freely penetrate the cell will work to alter porin gene expression with increasing concentration.[49]

An important aspect of the signal is that it must be maintained in steady state in low or high osmolarity. Therefore, EnvZ is not responding to changes in turgor pressure, for example, because the cell has a mechanism(s) to readjust intracellular osmolarity such that turgor remains constant independent of the media osmolarity.[172] This is in contrast to several other so-called osmoregulated genes, such as the *kdp* operon, whose gene products are involved in the uptake of K⁺ ions in response to increasing osmolarity. Expression of the operon is transitory and it has been proposed that regulation is in response to decreasing turgor pressure.[173] Once the internal K⁺ concentration is high enough to re-establish turgor, *kdp* is no longer expressed.

EnvZ is an integral inner membrane protein with a periplasmic domain.[113,114] Several lines of evidence indicate that osmolarity is sensed by the periplasmic domain. First, the truncated *envZ** alleles described above produce a protein that remains in the cytoplasm. A plasmid carrying such a mutation (*envZ115*) complements an *envZ* null mutation allowing expression of both *ompF* and *ompC*. However, the strain no longer responds to media osmolarity. Indeed, point mutations that alter the periplasmic domain alter osmoregulation as well.[159,174] Second, because EnvZ so closely resembles the chemoreceptors, chimeric genes that specify functional receptors can be constructed. Utsumi et al[175] and Baumgartner et al[176] replaced the periplasmic domain of EnvZ with the corresponding domain of Tar and Trg, respectively. Ligand induced activation by signals recognized by the chemoreceptors now allow stimulation of *ompC* expression. These results demonstrate that a periplasmic domain (Tar or Trg) can activate the cytoplasmic domain of EnvZ.

MDO (membrane derived oligosaccharide) was an attractive candidate for the signal sensed by EnvZ.[11] MDO, a large polyanionic compound, is made in dilute media to maintain the osmolarity of the periplasm by establishing a Donnan potential across the outer membrane.[177] However, it has been shown that mutations that completely prevent synthesis of MDO have no effect on the osmoregulation of the porin genes.[178] This rules out MDO as the signal, but perhaps the cell also uses other compounds to maintain periplasmic osmolarity and turgor pressure and this aspect of cellular osmoregulation could still be involved in modulating EnvZ.

Kawaji et al[49] observed that it required a 3-fold higher concentration of a compound that can penetrate the outer membrane to give the same response in porin regulation as a compound that is excluded from the periplasm. This could indicate that EnvZ is actually sensing alteration of the cell wall in response to changes in osmolarity. Perhaps EnvZ monitors tension exerted on the peptidoglycan.

In order to elicit a response, the signal must be transduced to the cytoplasmic domain of EnvZ, and this process must involve the two transmembrane segments. Tokishita et al[179] performed localized mutagenesis on the two domains. They isolated mutations in either transmembrane domain that resulted in both an OmpC constitutive, OmpF- phenotype as well as an OmpF constitutive, OmpC- phenotype. In vitro, these phenotypic classes of mutant EnvZ proteins behaved as expected, having decreased phosphatase activity or decreased kinase activity, respectively. Although these mutations are potentially useful, they do not provide clear insights regarding mechanism.

As mentioned above, hybrid proteins containing the carboxy-terminus of EnvZ and the transmembrane and periplasmic domain of the Tar[175] or Trg chemoreceptor[176] are functional. Sequence comparison reveals significant similarity between Tar, Trg, and EnvZ only in transmembrane domain 1 and the cytoplasmic stretch located immediately after the second transmembrane domain. This sequence similarity along with the functional similarity could suggest that these regions may be involved in signal transduction.[175,176]

FUNCTIONAL DOMAINS OF OMPR

Simplistically, the OmpR protein has four activities: it is phosphorylated and dephosphorylated by EnvZ; it binds DNA; and it interacts with RNA polymerase (see below). As described above, the phosphorylated residue in OmpR is most likely Asp-55. The phosphorylation apparently changes the conformation of the protein and this allows OmpR to bind DNA with higher affinity,[106,110,156,180,181] perhaps by causing an oligomerization.[182]

Mutations that apparently affect the conformational change elicited by phosphorylation were obtained as suppressors of the Asp-55 to Gln mutation.[183,184] These suppresser mutations, Gly-94 to Ser, Tyr-83 to Ala, and Tyr-102 to Cys, allow OmpR to function in the absence of phosphorylation. Nakashima et al[182] have also isolated mutations in this region (Glu-96 to Ala and Arg-115 to Ser) that prevent oligomerization and DNA binding in response to phosphorylation. Mutations changing Gly-94 to Asp or Glu-111 to Lys behave in a similar fashion.[185,186] Perhaps this region of OmpR is involved in the conformational change that leads to the ability to oligomerize and bind DNA upon phosphorylation.

The carboxy-terminal half of OmpR is capable of binding to both the *ompF* and *ompC* promoters.[187,188] Although this localizes the DNA binding domain of OmpR, the molecular aspects of this binding remain a mystery. We have argued that *ompR2* mutations, which confer an OmpF-constitutive, OmpC⁻ phenotype, are defective in DNA binding.[128] These mutations are all recessive, consistent with loss of function mutations. Russo et al[189] examined a series of these mutations to identify those that were specifically defective in DNA binding. Their analysis was based on the levels of *ompF* and *ompC* expression in various *envZ* backgrounds (various levels of OmpR-P) predicted from previous kinetic arguments.[159] Two mutations, Val-203 to Met and Arg-220 to Met, seem to specifically affect DNA binding. The Val-203 to Met mutation (*ompR472*) was isolated by Verhoef et al[61] and was used by Hall and Silhavy[76] to originally define the *ompR2* class. Changing Val-203 to Gln also results in a DNA binding phenotype. However, this mutation is apparently less dramatic than Val-203 to Met. The mutation to Gln is suppressible by *envZ* mutations that confer an OmpC-constitutive, OmpF⁻ phenotype (hyper-OmpR-P[190]), whereas the mutation to Met is not.[128] Interestingly, Val-203 lies in a potential helix proposed to be involved in DNA binding by analogy and similarity to a number of transcription factors, mostly from eukaryotes.[191] Thus, this region is the best candidate for being specifically involved in DNA binding.

Obviously more studies are required to determine the molecular mechanism of specific DNA binding by OmpR. Given the homology of the DNA binding, carboxy-terminal portion of OmpR with a wide variety of important regulatory proteins,[34-37] and the lack of well established DNA binding motifs, the elucidation of the mechanism of DNA binding will no doubt be interesting and important. The carboxy terminus of OmpR has recently been crystallized,[192] and the solution of the structure should help to clarify these issues.

Pratt and Silhavy[193] have identified several mutations in OmpR that specifically affect activation of transcription. These mutations involve amino acids Arg-42, Pro-179, Glu-193, Ala-196, and Glu-198. The Glu-193 mutation was originally isolated by Slauch and Silhavy[128] as having reduced *ompC* expression and classified by Russo et al[189] as an activation mutant. Mutations in these amino acids result in proteins that are defective in activation of both *ompF* and *ompC*. Both of these defects are codominant. The mutant proteins are, however, capable of repressing *ompF* expression and retain the ability to bind DNA, as measured in an in vivo assay of an artificial promoter. These mutant proteins also require EnvZ (phosphorylation) for function.[193] Thus, these amino acids may define regions of OmpR that are required for interaction with RNA polymerase to activate transcription (see below).

INTERACTIONS BETWEEN OMPR-P AND THE TRANSCRIPTION MACHINERY

Activation of a given promoter simply means enhancing the rate at which RNA polymerase initiates transcription. Although the holoenzyme (core and σ) is necessary for transcription initiation, it is not always sufficient. Many promoters, such as *ompF* and *ompC*, require an auxiliary transcriptional activator to permit a sufficient rate of initiation. It is commonly believed that most positive transcriptional activators must interact directly with the RNA polymerase to accomplish this activation, although an effect of simply changing the conformation of the DNA in the promoter region such that it is recognizable by holoenzyme cannot always be ruled out. In the case of OmpR and EnvZ, the interaction with the polymerase seems to be specific and involve the α subunit. The evidence for this hypothesis is outlined below and reviewed in Russo and Silhavy.[194]

As described previously, certain alleles of *envZ*, such as *envZ473* and *envZ11*, confer an OmpF⁻, OmpC-constitutive phenotype. In addition, these mutations are pleiotropic in that they affect the expression of a number of genes which are not normally regulated by OmpR and EnvZ, such as *malT*, *phoA*, and *phoE*. In order to better understand the nature of the pleiotropic phenotype conferred by this class of *envZ* mutations, Garrett and Silhavy[195] isolated extragenic suppressors of *envZ473* that simultaneously alleviated all of the pleiotropic phenotypes. These mutations, termed *sez* (suppressor of *envZ*) all mapped between *rpsL* and *aroE*, at approximately 72.5' on the *E. coli* chromosome. Complementation analysis showed that the *sez* mutations were recessive and were located in the α operon.[195] This operon is in the large ribosomal gene cluster and encodes five genes. Four of these genes encode ribosomal proteins and one gene, *rpoA*, encodes the α subunit of RNA polymerase.[196] Subsequent DNA sequence analysis has localized the *sez* mutations to *rpoA*.[197]

The effect of *sez* on porin expression is class specific. These mutations suppressed both pleiotropic alleles, *envZ473* and *envZ11*, but have no apparent effect on the porin phenotype of *ompR* or *envZ* null mutations. Garrett and Silhavy[195] suggested that these mutations affect an interaction, either direct or indirect, between the mutant EnvZ and the α subunit of RNA polymerase and that this interaction, although altered, reflects the normal interaction between EnvZ and the transcriptional machinery involved in the regulation of the porin genes.

Further support for the role of the α subunit of RNA polymerase in the regulation of porin gene expression comes from the work of Matsuyama and colleagues. A suppressor of the pleiotropic allele, *envZ11*, was isolated and shown to map in *ompR*.[198] This suppressor was given the allele number *ompR77* and was shown to be allele specific. It suppressed the porin and pleiotropic phenotypes of *envZ11*, but behaved like wild-type *ompR* in the presence of any other *envZ* allele, including *envZ160*, which confers an OmpF⁻, OmpC constitutive phenotype similar to *envZ11*, but is not pleiotropic. The authors did not test if *ompR77* could suppress *envZ473*. Matsuyama et al concluded that OmpR and EnvZ interact with each other in the transcriptional control of the porin genes.

Upon transducing the double mutant, *ompR77, envZ11* into various genetic backgrounds, a mutation, termed *szr*, was fortuitously discovered that no longer allowed the suppression of *envZ11* by *ompR77*, i.e., in the triple mutant, *envZ11, ompR77, szr*, the phenotype was that of *envZ11* alone.[199] Genetic mapping and complementation analysis showed conclusively that *szr* mapped in *rpoA*, the gene encoding the α subunit of RNA polymerase. The *szr* mutation, renamed *rpoA77*, apparently had no effect on the porin phenotype conferred by wild-type *ompR* and *envZ* or by *envZ11* alone.

These results suggested that there is a specific interaction between the mutant EnvZ and OmpR and the α subunit of RNA polymerase. However, these *rpoA* mutations had an apparent phenotype only in the presence of a pleiotropic *envZ* allele. Therefore, it was questionable whether this putative interaction between OmpR and the polymerase reflected the normal situation or was of significance only in strains containing the pleiotropic *envZ* alleles that result in high levels of OmpR-P. In order to distinguish between these possibilities, we isolated *rpoA* mutations that specifically affected porin gene regulation in an *ompR⁺ envZ⁺* background.[197] Many of these mutations mapped to the C-terminus of α. Characterization of these *rpoA* mutations in various *ompR* and *envZ* backgrounds showed that they behaved in an allele specific fashion. Sharif and Igo[200] have subsequently isolated more *rpoA* mutations that behave in an allele-specific fashion with respect to OmpR. All of these results suggest that OmpR interacts in a very specific fashion with the carboxy terminus of α to control transcriptional regulation of the porin genes.

Biochemical support for the interaction between OmpR and the carboxy terminus of α comes from the work of Igarashi et al.[201] The carboxy-terminal domain of α can be deleted without affecting the ability of the protein to assemble into a functional holoenzyme. The mutant polymerase still initiates properly at promoters such as *lacUV5*, but can no longer be stimulated by certain transcription activators (for review, see ref. 194). Igarashi et al[201] demonstrated that OmpR cannot activate the RNA polymerase containing truncated subunits.

Although we can conclude that OmpR interacts with the carboxy terminus of α, we do not understand the mechanistic significance of this contact. Does it simply serve to facilitate binding of RNA polymerase to the promoter, or does it alter an activity of this complex enzyme? Evidence to date suggests the latter.[202] We note that the *rpoA* mutations described affect both positive and negative regulation by OmpR,[195,197,200] and we have argued that both OmpR and RNA polymerase are present in the looped complex that represses *ompF* transcription.[110] Accordingly, further analysis may shed light on the mechanism of both activation and repression.

SUMMARY AND CONCLUSIONS

The work summarized in this chapter has provided a reasonably clear picture of the mechanism by which EnvZ and OmpR regulate transcription of the porin structural genes in response to changes in media osmolarity (Fig. 19.2). EnvZ is a receptor kinase/phosphatase. The periplasmic domain of this membrane protein senses changes in media osmolarity, and this induces a conformational change that is passed through the transmembrane segments to the cytoplasmic domain resulting in changes in enzymatic activity. In media of low osmolarity, the ratio of kinase/phosphatase activity is low; in media of high osmolarity this ratio increases.

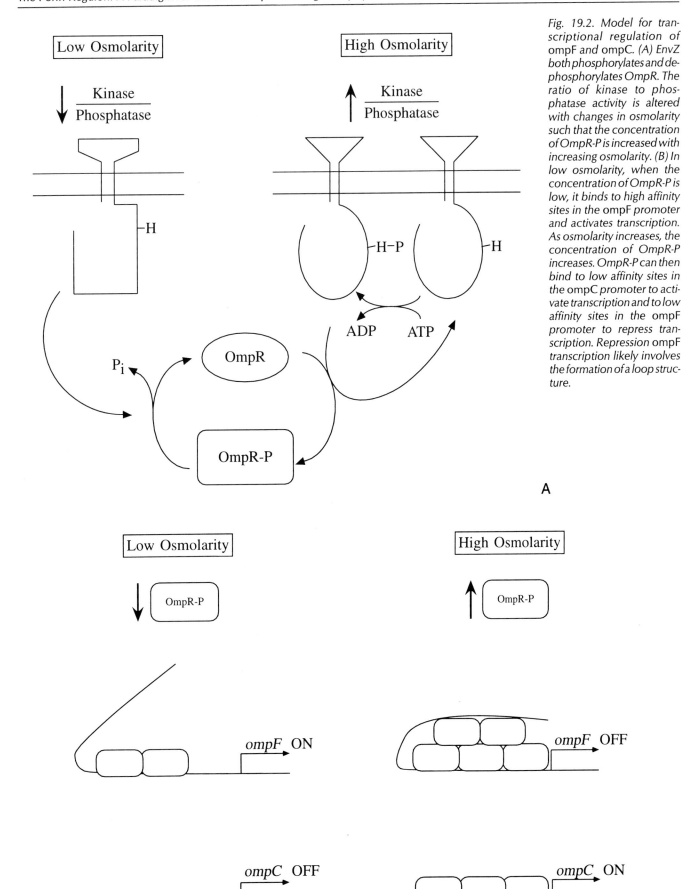

Low Osmolarity

$\dfrac{\text{Kinase}}{\text{Phosphatase}}$

—H

High Osmolarity

$\dfrac{\text{Kinase}}{\text{Phosphatase}}$

H–P   —H

ADP   ATP

P_i

OmpR

OmpR-P

Fig. 19.2. Model for transcriptional regulation of ompF and ompC. (A) EnvZ both phosphorylates and dephosphorylates OmpR. The ratio of kinase to phosphatase activity is altered with changes in osmolarity such that the concentration of OmpR-P is increased with increasing osmolarity. (B) In low osmolarity, when the concentration of OmpR-P is low, it binds to high affinity sites in the ompF promoter and activates transcription. As osmolarity increases, the concentration of OmpR-P increases. OmpR-P can then bind to low affinity sites in the ompC promoter to activate transcription and to low affinity sites in the ompF promoter to repress transcription. Repression ompF transcription likely involves the formation of a loop structure.

A

Low Osmolarity

OmpR-P

High Osmolarity

OmpR-P

ompF ON

ompF OFF

ompC OFF

ompC ON

B

Unphosphorylated OmpR has little if any activity. Phosphorylation stimulates the ability of this protein to bind at specific sites in the porin gene promoters. When OmpR-P concentrations within the cell are low (low osmolarity), the phosphorylated protein binds only to the high affinity sites at *ompF* activating expression of this gene. When OmpR-P concentrations increase (high osmolarity), the phosphorylated protein binds to low affinity sites at *ompC* activating expression of this gene. Under these conditions OmpR-P also binds to low affinity sites at *ompF*. Binding at these widely spaced sites induces a DNA looping that represses expression of *ompF*.

OmpR-P controls the activity of the cellular transcription machinery by direct interaction with the carboxy-terminal domain of the α subunit. In this regard, OmpR resembles other transcriptional activators such as CAP (CRP). However, OmpR can also repress transcription, and the evidence suggests that direct contact with α is required for this function as well.

Our current view of the porin regulon required more than twenty years of work from a variety of different labs. As we have described, the evolution of this model was not straightforward. There were many pitfalls along the way, and at times the problem seemed to defy explanation. In the end though we gained far more than originally anticipated. Studies with EnvZ and OmpR provided some of the first evidence that membrane receptor kinases control gene expression in prokaryotes, and they helped establish the basic properties of two-component regulatory systems. It is now widely appreciated that two-component systems control gene expression in response to a variety of different environmental parameters in all bacteria, and the porin regulon has become a paradigm for this regulatory mechanism. Studies with OmpR also helped establish that a transcriptional regulatory protein can control the activity of RNA polymerase by direct interaction with the subunit. This represents a major advance in our understanding of gene regulation. In recent years it has become apparent that many transcriptional activators share this property as well.

Despite this progress, much remains to be done. We do not understand how EnvZ senses osmolarity or how it transduces this information across the membrane. In addition, we do not understand the mechanistic significance of the interaction between OmpR and the subunit. We think it likely that further study of this model system will again provide insights of broad general interest.

REFERENCES

1. Forst SA, Delgado J, Inouye M. DNA-binding properties of the transcription activator (OmpR) for the upstream sequences of *ompF* in *Escherichia coli* are altered by *envZ* mutations and medium osmolarity. J Bacteriol 1989; 171:2949-2955.

2. Hall MN, Silhavy TJ. The *ompB* locus and the regulation of the major outer membrane porin proteins of *Escherichia coli* K12. J Mol Biol 1981; 146:23-43.

3. Igo MM, Slauch JM, Silhavy TJ. Signal transduction in bacteria: kinases that control gene expression. New Biologist 1990; 2:5-9.

4. Mizuno T, Mizushima S. Signal transduction and gene regulation through the phosphorylation of two regulatory components: the molecular basis for the osmotic regulation of the porin genes. Mol Microbiol 1990; 4:1077-1082.

5. Reeves P. The genetics of outer membrane proteins. In: Inouye M, ed. Bacterial outer membranes: biogenesis and function. New York, NY: John Wiley & Sons, 1979:255-291.

6. Andersen J, Delihas N, Ikenaka K et al. The isolation and characterization of RNA coded by the *micF* gene in *Escherichia coli*. Nucleic Acids Res 1987; 15:2089-2101.

7. Andersen J, Forst SA, Zhao K et al. The function of *micF* RNA: *micF* RNA Is a major factor in the thermal regulation of OmpF protein in *Escherichia coli*. J Biol Chem 1989; 264:17961-17970.

8. Catron KM, Schnaitman CA. Export of protein in *Escherichia coli*: a novel mutation in *ompC* affects expression of other major outer membrane proteins. J Bacteriol 1989; 169:4327-4334.

9. Click EM, Schnaitman CA. Export-defective *lamB* protein is a target for translational control caused by *ompC* porin overexpression. J Bacteriol 1989; 171:616-619.

10. Diedrich DL, Fralick JA. Relationship between the OmpC and LamB proteins of *Escherichia coli* and its influence on the protein mass of the outer membrane. J Bacteriol 1982; 149:156-160.

11. Fiedler W, Rotering H. Properties of *Escherichia coli* mutants lacking membrane-derived oligosaccharides. J Biol Chem 1988; 263:14684-14689.

12. George AM, Levy SB. Amplifiable resistance to tetracycline, chloramphenicol, and other antibiotics in *Escherichia coli*: involvement of a non-plasmid-determined efflux of tetracycline. J Bacteriol 1983; 155:531-540.

13. George AM, Levy SB. Gene in the major cotransduction gap of the *Escherichia coli* K-12 linkage map required for the expression of chromosomal resistance to tetracycline and other antibiotics. J Bacteriol 1983; 155:541-548.

14. Graeme-Cook KA, May G, Bremer E et al. Osmotic regulation of porin expression: a role for DNA supercoiling. Mol Microbiol 1989; 3:1287-1294.

15. Heyde M, Lazzaroni JC, Magnouloux-Blanc B et al. Regulation of porin gene expression over a wide range of extracellular pH in *Escherichia coli* K-12: influence of a *tolA* mutation. FEMS Microbiol Lett 1988; 52:59-66.

16. Higgins CF, Dorman CJ, Stirling DA et al. A physiological role for DNA supercoiling in the osmotic regulation of gene expression in *S. typhimurium* and *E. coli*. Cell 1988; 52:569-584.

17. Huang L, Tsui P, Freundlich M. Positive and negative control of *ompB* transcription in *Escherichia coli* by cyclic AMP and the cyclic AMP receptor protein. J Bacteriol 1992; 174:664-670.

18. Lazzaroni JC, Fognini-Lefebvre N, Portalier RC. Effects of *lkyB* mutations on the expression of *ompF*, *ompC* and *lamB* porin structural genes in *Escherichia coli* K-12. FEMS Microbiol Lett 1986; 33:235-239.

19. Lundrigan MD, Earhart CF. Gene *envY* of *Escherichia coli* K-12 affects thermoregulation of major porin expression. J Bacteriol 1984; 157:262-268.

20. McEwen J, Sambucetti L, Silverman PM. Synthesis of outer membrane proteins in *cpxA cpxB* mutants of *Escherichia coli* K-12. J Bacteriol 1983; 154:375-382.

21. Misra R, Reeves PR. Role of *micF* in the *tolC*-mediated regulation of OmpF, a major outer membrane protein of *Escherichia coli* K-12. J Bacteriol 1987; 169:4722-4730.

22. Murgier M, Pages C, Lazdunski C et al. Translational control of *ompF*, *ompC*, and *lamB* genetic expression during lipid biosynthesis inhibition of *Escherichia coli*. FEMS Microbiol Lett 1982; 13:307-311.

23. NiBhriain N, Dorman CJ, Higgins CF. An overlap between osmotic and anaerobic stress responses: a potential role for DNA supercoiling in the coordinate regulation of gene expression. Mol Microbiol 1989; 3:933-942.

24. Pugsley AP. A mutation in the *dsbA* gene coding for periplasmic disulfide oxidoreductase reduces transcription of the *Escherichia coli ompF* gene. Mol Gen Genet 1993; 237:407-411.

25. Turnowsky F, Fuchs K, Jeschek C et al. *envM* genes of *Salmonella typhimurium* and *Escherichia coli*. J Bacteriol 1989; 171:6555-6565.

26. Aiba H, Matsuyama S, Mizuno T et al. Function of *micF* as an antisense RNA In osmoregulatory expression of the *ompF* gene in *Escherichia coli*. J Bacteriol 1987; 169:3007-3012.

27. Andersen J, Delihas N. *micF* RNA binds to the 5' end of *ompF* mRNA and to a protein from *Escherichia coli*. Biochemistry 1990; 29:9249-9256.

28. Coyer J, Andersen J, Forst SA et al. *micF* RNA In *ompB* mutants of *Escherichia coli*: different pathways regulate *micF* RNA levels in response to osmolarity and temperature change. J Bacteriol 1990; 172:4143-4150.

29. Matsuyama S, Mizushima S. Construction and characterization of a deletion mutant lacking *micF*, a proposed regulatory gene for OmpF synthesis in *Escherichia coli*. J Bacteriol 1985; 162:1196-1202.

30. Mizuno T, Chou MY, Inouye M. A comparative study on the genes for three porins of the *Escherichia coli* outer membrane. J Biol Chem 1983; 258:6932-6940.

31. Mizuno T, Chou MY, Inouye M. A unique mechanism regulating gene expression: translational inhibition by a complementary RNA transcript (micRNA). Proc Natl Acad Sci USA 1984; 81:1966-1970.

32. Ramani N, Hedeshian M, Freundlich M. *micF* antisense RNA has a major role in osmoregulation of OmpF in *Escherichia coli*. J Bacteriol 1994; 176:5005-5010.

33. Schnaitman CA, McDonald GA. Regulation of outer membrane protein synthesis in *Escherichia coli* K-12: deletion of *ompC* affects expression of the OmpF protein. J Bacteriol 1984; 159:555-563.

34. Albright LM, Huala E, Ausubel FM. Prokaryotic signal transduction mediated by sensor and regulatory pairs. Annu Rev Genet 1989; 23:311-336.

35. Bourret RB, Borkovich KA, Simon MI. Signal transduction pathways involving protein phosphorylation in prokaryotes. Annu Rev Biochem 1991; 60:401-441.

36. Hoch JA, Silhavy TJ. Two-component signal transduction. Washington, DC: American Society for Microbiology, 1995.

37. Stock JB, Ninfa AJ, Stock AM. Protein phosphorylation and the regulation of adaptive response in bacteria. Microbiol Rev 1989; 53:450-490.

38. Nikaido H, Vaara M. Outer Membrane. In: Neidhardt FC, ed. *Escherichia coli* and *Salmonella typhimurium*: cellular and molecular biology. Washington, DC: American Society for Microbiology, 1987:7-22.

39. Mizuno T, Chou MY, Inouye M. DNA sequence of the promoter region of the *ompC* gene and amino acid sequence of the signal peptide of pro-OmpC protein of *Escherichia coli*. FEBS Lett 1983; 151:159-164.

40. Schnaitman C. Protein composition of the cell wall and cytoplasmic membrane of *Escherichia coli*. J Bacteriol 1970; 104:890-901.

41. Hindennach I, Henning U. The major proteins of the *E. coli* outer cell envelope membrane: preparative isolation of all major membrane proteins. Eur J Biochem 1975; 59:207-213.

42. Lugtenberg B, Meijers J, Peters R et al. Electrophoretic resolution of the major outer membrane protein of *Escherichia coli* K-12 into four bands. FEBS Lett 1975; 58:254-258.

43. Schnaitman CA. Outer membrane proteins of *Escherichia coli*. IV. Differences in outer membrane proteins due to strain and cultural differences. J Bacteriol 1974; 118:454-464.

44. Uemura J, Mizushima S. Isolation of outer membrane proteins of *Escherichia coli* and their characterization on polyacrylamide gel. Biochim Biophys Acta 1975; 413:163-176.

45. Nakamura K, Mizushima S. Effects of heating in dodecyl sulfate solution on the conformation and electrophoretic mobility of isolated major outer

membrane proteins from *Escherichia coli* K-12. J Biochem 1976; 80:1411-1422.

46. Schnaitman CA. Outer membrane proteins of *Escherichia coli*. III. Evidence that the major protein of *Escherichia coli* O111 outer membrane consists of four distinct polypeptide species. J Bacteriol 1974; 118:442-453.

47. Lugtenburg B, Peters R, Bernheimer H et al. Influence of cultural conditions and mutations on the composition of the outer membrane proteins of *Escherichia coli*. Mol Gen Genet 1976; 147:251-262.

48. van Alphen W, Lugtenberg B. Influence of osmolarity of the growth medium on the outer membrane protein pattern of *Escherichia coli*. J Bacteriol 1977; 131:623-630.

49. Kawaji H, Mizuno T, Mizushima S. Influence of molecular size and osmolarity of sugars and dextrans on the synthesis of outer membrane proteins O-8 and O-9 of *Escherichia coli* K-12. J Bacteriol 1979; 140:843-847.

50. Nakae T. Identification of the outer membrane protein of *E. coli* that produces transmembrane channels in reconstituted vesicle membranes. Biochem Biophys Res Commun 1976; 71:877-884.

51. Nomura M, Witten C. Interaction of colicins with bacterial cells. III. Colicin-tolerant mutants of *Escherichia coli*. J Bacteriol 1967; 94: 1093-1111.

52. Davies JK, Reeves P. Genetics of resistance to colicins in *Escherichia coli* K-12: cross-resistance among colicins of group A. J Bacteriol 1975; 123:102-117.

53. Foulds J, Barrett C. Characterization of *Escherichia coli* mutants tolerant to bacteriocin JF246: two new classes of tolerant mutants. J Bacteriol 1973; 116:885-892.

54. Reeve ECR, Doherty P. Linkage relationships of two genes causing partial resistance to chloramphenicol in *Escherichia coli*. J Bacteriol 1968; 96:1450-1451.

55. Foulds J. *tolF* locus in *Escherichia coli*: chromosomal location and the relationship to locus *cmlB* and *tolD*. J Bacteriol 1976; 128:604-608.

56. Lutkenhaus JF. Role of a major outer membrane protein in *Escherichia coli*. J Bacteriol 1977; 131:631-637.

57. Pugsley AP, Schnaitman CA. Identification of three genes controlling production of new membrane pore proteins in *Escherichia coli* K-12. J Bacteriol 1978; 135:1118-1129.

58. von Meyenburg K. Transport-limited growth rates in a mutant of *Escherichia coli*. J Bacteriol 1971; 107:878-888.

59. Beacham IR, Haas D, Yagil E. Mutants of *Escherichia coli* "cryptic" for certain periplasmic enzymes: evidence for an alteration of the outer membrane. J Bacteriol 1977; 129:1034-1044.

60. Beacham IR, Kahana R, Levy L et al. Mutants of *Escherichia coli* K-12 "cryptic," or dificient in 5'-nucleotidase (uridine diphosphate-sugar hydrolase) and 3'-nucleotidase (cyclic phosphodiesterase) activity. J Bacteriol 1973; 116:957-964.

61. Verhoef C, Lugtenberg B, van Boxtel R et al. Genetics and biochemistry of the peptidoglycan-associated proteins b and c of *Escherichia coli* K-12. Mol Gen Genet 1979; 169:137-146.

62. Hancock REW, Reeves P. Bacteriophage resistance in *Escherichia coli* K-12: general pattern of resistance. J Bacteriol 1975; 121:983-779.

63. Hancock REW, Davies JK, Reeves P. Cross-resistance between bacteriophages and colicins in *Escherichia coli* K-12. J Bacteriol 1976; 126:1347-1350.

64. Schmitges CJ, Henning U. The major proteins of *Escherichia coli* outer cell-envelope membrane: heterogeneity of protein I. Eur J Biochem 1976; 63:47-52.

65. Chai TJ, Foulds J. *Escherichia coli* K-12 *tolF* mutants: alteration in protein composition of the outer membrane. J Bacteriol 1977; 130:781-786.

66. Verhoef C, de Graaff PJ, Lugtenberg EJJ. Mapping of a gene for a major outer membrane protein of *Escherichia coli* K-12 with the aid of a newly isolated bacteriophage. Mol Gen Genet 1977; 150:103-105.

67. Bassford PJ, Diedrich DL, Schnaitman CL et al. Outer membrane proteins of *Escherichia coli* VI. protein alteration in bacteriophage-resistant mutants. J Bacteriol 1977; 131:608-622.

68. Pugsley AP, Schnaitman CA. Outer membrane proteins of *Escherichia coli* VII. Evidence that bacteriophage-directed protein 2 functions as a pore. J Bacteriol 1978; 133:1181-1189.

69. Tommassen J, Lugtenberg B. Outer membrane protein e of *Escherichia coli* K-12 is co-regulated with alkaline phosphatase. J Bacteriol 1980; 143:151-157.

70. Highton PJ, Chang Y, Marcotte WRJ et al. Evidence that the outer membrane protein gene *nmpC* of *Escherichia coli* K-12 lies within the defective qsr' prophage. J Bacteriol 1985; 162:256-252.

71. Henning U, Schmidmayr W, Hindennach I. Major proteins of the outer cell envelope membrane of *Escherichia coli* K-12: multiple species of protein I. Mol Gen Genet 1977; 154:293-298.

72. Ichihara S, Mizushima S. Characterization of major outer membrane proteins O-8 amd O-9 of *Escherichia coli* K-12. J Biochem 1978; 83:1095-1100.

73. Sarma V, Reeves P. Genetic locus (*ompB*) affecting a major outer-membrane protein in *Escherichia coli* K-12. J Bacteriol 1977; 132:23-27.

74. van Alphen L, Lugtenberg B, van Boxtel R et al. *meoA* is the structural gene for outer membrane protein c of *Escherichia coli* K12. Mol Gen Genet 1979; 169:147-155.

75. Hall MN, Silhavy TJ. Transcriptional regulation of *Escherichia coli* K-12 major outer membrane protein 1b. J Bacteriol 1979; 140:342-350.

76. Hall MN, Silhavy TJ. Genetic analysis of the major outer membrane proteins of *Escherichia coli*. Annu Rev Genet 1981; 15:91-142.

77. Hall MN, Silhavy TJ. Genetic analysis of the *ompB* locus in *Escherichia coli* K-12. J Mol Biol 1981; 151:1-15.

78. Taylor RK, Hall MN, Enquist L et al. Identification of OmpR: a positive regulatory protein controlling expression of the major outer membrane matrix porin proteins of *Escherichia coli* K-12. J Bacteriol 1981; 147:255-258.

79. Mizuno T, Wurtzel ET, Inouye M. Cloning of the regulatory genes (*ompR* and *envZ*) for the matrix proteins of the *Escherichia coli* outer membrane. J Bacteriol 1982; 150:1462-1466.

80. Mizuno T, Wurtzel ET, Inouye M. Osmoregulation of gene expression. II. DNA sequence of the *envZ* gene of the *ompB* operon of *Escherichia coli* and characterization of its gene product. J Biol Chem 1982; 257:13692-13698.

81. Wurtzel ET, Chou MY, Inouye M. Osmoregulation of gene expression. I. DNA sequence of the *ompR* gene of the *ompB* operon of *Escherichia coli* and characterization of its gene product. J Biol Chem 1982; 257:13685-13691.

82. Comeau DE, Ikenaka K, Tsung KL et al. Primary characterization of the protein products of the *Escherichia coli ompB* locus: structure and regulation of synthesis of the OmpR and EnvZ proteins. J Bacteriol 1985; 164:578-584.

83. Liljestrom P. Structure and expression of the *ompB* operon of *Salmonella typhimurium* and *Escherichia coli*, Thesis Dissertation. University of Helsinki,Helsinki, Finland 1986;

84. Mutoh N, Nagasawa T, Mizushima S. Specialized transducing bacteriophage lambda carrying the structural gene for a major outer membrane matrix protein of *Escherichia coli* K-12. J Bacteriol 1981; 145:1085-1090.

85. Tommassen J, van der Lay P, van der Ende A et al. Cloning of *ompF*, the structural gene for an outer membrane pore protein of *E. coli* K12: physical localization and homology with the *phoE* gene. Mol Gen Genet 1982; 185:105-110.

86. Inokuchi K, Mutoh N, Matsuyama S et al. Primary structure of the *ompF* gene that codes for a major outer membrane protein of *Escherichia coli* K-12. Nucleic Acids Res 1982; 10:6957-6968.

87. Inokuchi K, Furukawa H, Nakamura K et al. Characterization by deletion mutagenesis in vitro of the promoter region of *ompF*, a positively regulated gene of *Escherichia coli*. J Mol Biol 1984; 178:653-668.

88. Taylor RK, Garrett S, Sodergren E et al. Mutations that define the promoter of *ompF*, a gene specifying a major outer membrane porin protein. J Bacteriol 1985; 162:1054-1060.

89. Dairi T, Inokuchi K, Mizuno T et al. Positive control of transcription initiation in *Escherichia coli*: a base substitution at the pribnow box renders *ompF* expression independent of a positive regulator. J Mol Biol 1985; 184:1-6.

90. Cohen SP, McMurry LM, Levy SB. *marA* locus causes decreased expression of OmpF porin in multiple-antibiotic-resistant (Mar) mutants of *Escherichia coli*. J Bacteriol 1988; 170:5416-5422.

91. Pederson S, Reeh S, Friesen JD. Functional mRNA half lives in *E. coli*. Mol Gen Genet 1978; 166:329-336.

92. Emory SA, Belasco JG. The *ompA* 5' untranslated RNA segment functions in *Escherichia coli* as a growth-rate-regulated mRNA stabilizer whose activity is unrelated to translational efficiency. J Bacteriol 1990; 172:4472-4481.

93. Mizuno T, Chou MY, Inouye M. Regulation of gene expression by a small RNA transcript (micRNA) in *Escherichia coli* K-12. Proc Jpn Acad Sci 1983; 59:794-797.

94. Ikenaka K, Ramakrishnan G, Inouye M et al. Regulation of the *ompC* gene of *Escherichia coli*. Involvement of three tandem promoters. J Biol Chem 1986; 261:9316-9320.

95. Jo YL, Nara F, Ichihara S et al. Purification and characterization of the OmpR protein, a positive regulator involved in osmoregulatory expression of the *ompF* and *ompC* genes in *Escherichia coli*. J Biol Chem 1986; 261:15253-15256.

96. Norioka S, Ramakrishnan G, Ikenaka K et al. Interaction of a transcriptional activator, OmpR, with reciprocally osmoregulated genes, *ompF* and *ompC*, of *Escherichia coli*. J Biol Chem 1986; 261:17113-17119.

97. Mizuno T, Kato M, Jo YL et al. Interaction of OmpR, a positive regulator, with the osmoregulated *ompC* and *ompF* genes of *Escherichia coli*. Studies with wild-type and mutant OmpR proteins. J Biol Chem 1988; 263:1008-1012.

98. de Crombrugghe B, Busby S, Buc H. Cyclic AMP receptor protein: role in transcription activation. Science 1984; 224:831-838.

99. Tsung K, Brissette RE, Inouye M. Identification of the DNA-binding domain of the OmpR protein required for transcriptional activation of the *ompF* and *ompC* genes of *Escherichia coli* by in vivo DNA footprinting. J Biol Chem 1989; 264:10104-10109.

100. Mizuno T. Static bend of DNA helix at the activator recognition site of the *ompF* promoter in *Escherichia coli*. Gene 1987; 54:57-64.

101. Zahn K, Blattner FR. Sequence-induced DNA curvature at the bacteriophage lambda origin of replication. Nature 1985; 317:451-453.

102. Kato M, Aiba H, Mizuno T. Molecular analysis by deletion and site-directed mutagenesis of the *cis*- acting upstream sequence involved in activation of the *ompF* promoter in *Escherichia coli*. J Biochem (Tokyo) 1989; 105:341-347.

103. Ostrow KS, Silhavy TJ, Garrett S. *cis*-acting sites required for osmoregulation of *ompF* expression in *Escherichia coli* K-12. J Bacteriol 1986; 168:1165-1171.

104. Huang KJ, Schieberl JL, Igo MM. A distant upstream site involved in the negative regulation of the *Escherichia coli ompF* gene. J Bacteriol 1994; 176:1309-1315.

105. Yamamoto M, Nomura M, Ohsawa H et al. Identification of a temperature-sensitive asparaginyl-transfer ribonucleic acid synthetase mutant of *Escherichia coli*. J Bacteriol 1977; 132:127-131.

106. Mizuno T, Mizushima S. Characterization by deletion and localized mutagenesis in vitro of the promoter region of the *Escherichia coli ompC* gene and importance of the upstream DNA domain in positive regulation by the OmpR protein. J Bacteriol 1986; 168:86-95.

107. Maeda S, Ozawa Y, Mizuno T et al. Stereospecific positioning of the *cis*-acting sequence with respect to the canonical promoter is required for activation of the *ompC* gene by a positive regulator, OmpR, in *Escherichia coli*. J Mol Biol 1988; 202:433-441.

108. Maeda S, Mizuno T. Activation of the *ompC* gene by the OmpR protein in *Escherichia coli* The *cis*-acting upstream sequence can function in both orientations with respect to the canonical promoter. J Biol Chem 1988; 263:14629-14633.

109. Maeda S, Mizuno T. Evidence for multiple OmpR-binding sites in the upstream activation sequence of the *ompC* promoter in *Escherichia coli*: a single OmpR- binding site is capable of activating the promoter. J Bacteriol 1990; 172:501-503.

110. Slauch JM, Silhavy TJ. *cis*-acting *ompF* mutations that result in OmpR-dependent constitutive expression. J Bacteriol 1991; 173:4039-4048.

111. Maeda S, Takayanagi K, Nishimura Y et al. Activation of the osmoregulated *ompC* gene by the OmpR protein in *Escherichia coli*: a study involving synthetic OmpR-binding sequences. J Biochem (Tokyo) 1991; 110:324-327.

112. Aoyama T, Oka A. A common mechanism of transcriptional activation by the three positive regulators, VirG, PhoB, and OmpR. FEBS Lett 1990; 263:1-4.

113. Liljestrom P. The EnvZ protein of *Salmonella typhimurium* LT-2 and *Escherichia coli* K-12 is located in the cytoplasmic membrane. FEMS Microbiol Lett 1986; 36:145-150.

114. Forst S, Comeau D, Norioka S et al. Localization and membrane topology of EnvZ, a protein involved in osmoregulation of OmpF and OmpC in *Escherichia coli*. J Biol Chem 1987; 262:16433-16438.

115. Case CC, Bukau B, Granett S et al. Contrasting mechanisms of *envZ* control of *mal* and *pho* regulon genes in *Escherichia coli*. J Bacteriol 1986; 166:706-712.

116. Wandersman C, Moreno F, Schwartz M. Pleiotropic mutations rendering *Escherichia coli* K-12 resistant to bacteriophage TP1. J Bacteriol 1980; 143:1374-1383.

117. Wanner BL, Sarthy A, Beckwith J. *Escherichia coli* pleiotropic mutant that reduces amounts of several periplasmic and outer membrane proteins. J Bacteriol 1979; 140:229-239.

118. Granett S, Villarejo M. Selective inhibition of carbohydrate transport by the local anesthetic procaine in *Escherichia coli*. J Bacteriol 1981; 147:289-296.

119. Granett S, Villarejo M. Regulation of gene expression in *Escherichia coli* by the local anesthetic procaine. J Mol Biol 1982; 160:363-367.

120. Pages JM, Lazdunski C. Transcriptional regulation of *ompF* and *lamB* genetic expression by local anesthetics. FEMS Microbiol Lett 1982; 15:153-157.

121. Pugsley AP, Conrard DJ, Schnaitman CA et al. In vivo effects of local anesthetics on the production of major outer membrane proteins by *Escherichia coli*. Biochim Biophys Acta 1980; 599:1-12.

122. Taylor RK, Hall MN, Silhavy TJ. Isolation and characterization of mutations altering expression of the major outer membrane porin proteins using the local anaesthetic. J Mol Biol 1983; 166:273-282.

123. Garrett S, Taylor RK, Silhavy TJ. Isolation and characterization of chain-terminating nonsense mutations in a porin regulator gene, *envZ*. J Bacteriol 1983; 156:62-69.

124. Garrett S, Taylor RK, Silhavy TJ et al. Isolation and characterization of delta *ompB* strains of *Escherichia coli* by a general method based on gene fusions. J Bacteriol 1985; 162:840-844.

125. Mizuno T, Mizushima S. Isolation and characterization of deletion mutants of *ompR* and *envZ*, regulatory genes for expression of the outer membrane proteins OmpC and OmpF in *Escherichia coli*. J Biochem 1987; 101:387-396.

126. Slauch JM, Garrett S, Jackson DE et al. EnvZ functions through OmpR to control porin gene expression in *Escherichia coli* K-12. J Bacteriol 1988; 170:439-441.

127. Nara F, Matsuyama S, Mizuno T et al. Molecular analysis of mutant *ompR* genes exhibiting different phenotypes as to osmoregulation of the *ompF* and *ompC* genes of *Escherichia coli*. Mol Gen Genet 1986; 202:194-199.

128. Slauch JM, Silhavy TJ. Genetic analysis of the switch that controls porin gene expression in *Escherichia coli* K-12. J Mol Biol 1989; 210:281-292.

129. Englesberg E, Squires C, Meronk F, Jr. The L-arabinose operon in *Escherichia coli* B/r: a genetic demonstration of two functional states of the product of a regulatory gene. Proc Natl Acad Sci USA 1969; 62:1100-1107.

130. Forst S, Delgado J, Rampersaud A et al. In vivo phosphorylation of OmpR, the transcription activator of the *ompF* and *ompC* genes in *Escherichia coli*. J Bacteriol 1990; 172:3473-3477.

131. Rampersaud A, Harlocker SL, Inouye M. The OmpR protein of *Escherichia coli* binds to sites in the *ompF* promoter region in a hierarchical manner determined by its degree of phosphorylation. J Biol Chem 1994; 269:12559-12566.

132. Tsui P, Helu V, Freundlich M. Altered osmoregulation of *ompF* in integration host factor mutants of *Escherichia coli*. J Bacteriol 1988; 170:4950-4953.

133. Ramani N, Huang L, Freundlich M. In vitro interactions of integration host factor with the *ompF* promoter- regulatory region of *Escherichia coli*. Mol Gen Genet 1992; 231:248-255.

134. Friedman DI. Integration host factor: a protein for all reasons. Cell 1988; 55:545-554.

135. Nixon BT, Ronson CW, Ausubel FM. Two-component regulatory systems responsive to environmental stimuli share strongly conserved domains with the nitrogen assimilation regulatory genes *ntrB* and *ntrC*. Proc Natl Acad Sci USA 1986; 83:7850-7854.

136. Stock A, Koshland DE, Jr., Stock J. Homologies between the *Salmonella typhimurium* CheY protein and proteins involved in the regulation of chemotaxis, membrane protein synthesis, and sporulation. Proc Natl Acad Sci USA 1985; 82:7989-7993.

137. Ninfa AJ, Magasanik B. Covalent modification of the *glnG* product, NRI, by the *glnL* product, NRII, regulates the transcription of the *glnALG* operon in *Escherichia coli*. Proc Natl Acad Sci USA 1986; 83:5909-5913.

138. Garrett S. Ph.D. Thesis. The Johns Hopkins University, Baltimore, MD 1986.

139. Igo MM, Silhavy TJ. EnvZ, a transmembrane environmental sensor of *Escherichia coli* K-12, is phosphorylated in vitro. J Bacteriol 1988; 170:5971-5973.

140. Aiba H, Mizuno T, Mizushima S. Transfer of phosphoryl group between two regulatory proteins involved in osmoregulatory expression of the *ompF* and *ompC* genes in *Escherichia coli*. J Biol Chem 1989; 264:8563-8567.

141. Forst S, Delgado J, Ramakrishnan G et al. Regulation of *ompC* and *ompF* expression in *Escherichia coli* in the absence of *envZ*. J Bacteriol 1988; 170:5080-5085.

142. Igo MM, Ninfa AJ, Silhavy TJ. A bacterial environmental sensor that functions as a protein kinase and stimulates transcriptioanl activation. Genes and Development 1989; 3:598-605.

143. Aiba H, Nakasai F, Mizushima S et al. Evidence for the physiological importance of the phosphotransfer between the two regulatory components, EnvZ and OmpR, in osmoregulation in *Escherichia coli*. J Biol Chem 1989; 264:14090-14094.

144. Igo MM, Ninfa AJ, Stock JB et al. Phosphorylation and dephosphorylation of a bacterial transcriptional activator by a transmembrane receptor. Genes Dev 1989; 3:1725-1734.

145. Tokishita S, Yamada H, Aiba H et al. Transmembrane signal transduction and osmoregulation in *Escherichia coli*: II. The osmotic sensor, EnvZ, located in the isolated cytoplasmic membrane displays its phosphorylation and dephosphorylation abilities as to the activator protein, OmpR. J Biochem (Tokyo) 1990; 108:488-493.

146. Hess JF, Bourret RB, Simon MI. Histidine phosphorylation and phosphoryl group transfer in bacterial chemotaxis. Nature 1988; 336:139-143.

147. Stock AM, Wylie DC, Mottonen JM et al. Phospho-proteins involved in bacterial signal transduction. Cold Spring Harbor Symp Quant Biol 1988; 53:49-57.

148. Weiss V, Magasanik B. Phosphorylation of nitrogen regulator I (NRI) of *Escherichia coli*. Proc Natl Acad Sci USA 1988; 85:8919-8923.

149. Roberts DL, Bennett DW, Forst SA. Identification of the site of phosphorylation on the osmosensor, EnvZ, of *Escherichia coli*. J Biol Chem 1994; 269:8728-8733.

150. Kanamaru K, Aiba H, Mizuno T. Transmembrane signal transduction and osmoregulation in *Escherichia coli*: I. Analysis by site-directed mutagenesis of the amino acid residues involved in phosphotransfer between the two regulatory components, EnvZ and OmpR. J Biochem (Tokyo) 1990; 108:483-487.

151. Stock AM, Mottonen JM, Stock JB et al. Three-dimensional structure of CheY, the response regulator of bacterial chemotaxis. Nature 1989; 337:745-749.

152. Bourret RB, Hess JF, Simon MI. Conserved aspartate residues and phosphorylation in signal transduction by the chemotaxis protein CheY. Proc Natl Acad Sci USA 1990; 87:41-45.

153. Sanders DA, Gellece-Castro BL, Stock AM et al. Identification of the site of phosphorylation of the chemotaxis response regulator protein, CheY. J Biol Chem 1989; 264:21770-21778.

154. Ninfa AJ, Ninfa EG, Lupas AN et al. Crosstalk between bacterial chemotaxis signal transduction proteins and regulators of transcription of the Ntr regulon: evidence that nitrogen assimilation and chemotaxis are controlled by a common phosphotransfer mechanism. Proc Natl Acad Sci USA 1988; 85:5492-5496.

155. Wanner BL, Wilmes MR, Young DC. Control of bacterial alkaline phosphatase synthesis and variation in an *Escherichia coli* K-12 *phoR* mutant by adenyl cyclase, the cyclic AMP receptor protein, and the *phoM* operon. J Bacteriol 1988; 170:1092-1102.

156. Aiba H, Nakasai F, Mizushima S et al. Phosphorylation of a bacterial activator protein, OmpR, by a protein kinase, EnvZ, results in stimulation of its DNA-binding ability. J Biochem (Tokyo) 1989; 106:5-7.

157. Forst S, Inouye M. Environmentally regulated gene expression for membrane proteins in *Escherichia coli*. Ann Rev Cell Biol 1988; 4:21-42.

158. Kohno T, Roth J. Electrolyte effects on the activity of mutant enzymes in vivo and in vitro. Biochemistry 1979; 18:1386-1392.

159. Russo FD, Silhavy TJ. EnvZ controls the concentration of phosphorylated OmpR to mediate osmoregulation of the porin genes. J Mol Biol 1991; 222:567-580.

160. Feng J, Atkinson MR, McCleary W et al. Role of phosphorylated metabolic intermediates in the regulation of glutamine synthetase synthesis in *Escherichia coli*. J Bacteriol 1992; 174:6061-6070.

161. Lukat GS, McCleary WR, Stock AM et al. Phosphorylation of bacterial response regulator proteins by low molecular weight phospho-donors. Proc Natl Acad Sci U S A 1992; 89:718-722.

162. McCleary WR, Stock JB. Acetyl phosphate and the activation of two-component response regulators. J Biol Chem 1994; 269:31567-31572.

163. Wanner BL, Wilmes-Riesenberg MR. Involvement of phosphotransacetylase, acetate kinase, and acetyl phosphate synthesis in control of the phosphate regulon in *Escherichia coli*. J Bacteriol 1992; 174:2124-2130.

164. Russo FD, Silhavy TJ. The essential tension: opposed reactions in bacterial two-component regulatory systems. Trends Microbiol 1993; 1:306-310.

165. Liljestrom P, Maattanen PL, Palva ET. Cloning of the regulatory locus *ompB* of *Salmonella typhimurium* LT-2 II. Identification of the *envZ* gene product, a protein involved in the expression of the porin proteins. Mol Gen Genet 1982; 188:190-194.

166. Brissette RE, Tsung KL, Inouye M. Intramolecular second-site revertants to the phosphorylation site mutation in OmpR, a kinase-dependent transcriptional activator in *Escherichia coli*. J Bacteriol 1991; 173:3749-3755.

167. Ninfa EG, Stock A, Mowbray S et al. Reconstitution of the bacterial chemotaxis signal transduction system from purified components. J Biol Chem 1991; 266:9764-9770.

168. Yang Y, Inouye M. Intermolecular complementation between two defective mutant signal- transducing receptors of *Escherichia coli*. Proc Natl Acad Sci U S A 1991; 88:11057-11061.

169. Jin T, Inouye M. Ligand binding to the receptor domain regulates the ratio of kinase to phosphatase activities of the signaling domain of the hybrid *Escherichia coli* transmembrane receptor, Taz1. J Mol Biol 1993; 232:484-492.

170. Yang Y, Inouye M. Requirement of both kinase and phosphatase activities of an *Escherichia coli* receptor (Taz1) for ligand-dependent signal transduction. J Mol Biol 1993; 231:335-342.

171. Yang Y, Park H, Inouye M. Ligand binding induces an asymmetrical transmembrane signal through a receptor dimer. J Mol Biol 1993; 232:493-498.

172. Ingraham J. Effect of temperature, pH, water activity, and pressure on growth. In: Neidhardt FC, ed. *Escherichia coli* and *Salmonella typhimurium*: cellular and molecular biology. Washington, DC: American Society for Microbiology, 1987:1543-1554.

173. Laimins LA, Rhoads DB, Epstein W. Osmotic control of *kdp* operon expression in *Escherichia coli*. Proc Natl Acad Sci USA 1981; 78:464-468.

174. Tokishita S, Kojima A, Aiba H et al. Transmembrane signal transduction and osmoregulation in *Escherichia coli*. Functional importance of the periplasmic domain of the membrane-located protein kinase, EnvZ. J Biol Chem 1991; 266:6780-6785.

175. Utsumi R, Brissette RE, Rampersaud A et al. Activation of bacterial porin gene expression by a chimeric signal transducer in response to aspartate. Science 1989; 245:1246-1249.

176. Baumgartner JW, Kim C, Brissette RE et al. Transmembrane signaling by a hybrid protein: communication from the domain of chemoreceptor Trg that recognizes sugar-binding proteins to the kinase/phosphatase domain of osmosensor EnvZ. J Bacteriol 1994; 176:1157-1163.

177. Kennedy EP. Membrane-derived oligosaccharides. In: Neidhardt FC, ed. *Escherichia coli* and *Salmonella typhimurium*: cellular and molecular biology. Washington, DC: American Society for Microbiology, 1987:672-679.

178. Geiger O, Russo FD, Silhavy TJ et al. Membrane-derived oligosaccharides affect porin osmoregulation only in media of low ionic strength. J Bacteriol 1992; 174:1410-1413.

179. Tokishita S, Kojima A, Mizuno T. Transmembrane signal transduction and osmoregulation in *Escherichia coli*: functional importance of the transmembrane regions of membrane-located protein kinase, EnvZ. J Biochem (Tokyo) 1992; 111:707-713.

180. Forst S, Delgado J, Inouye M. Phosphorylation of OmpR by the osmosensor EnvZ modulates expression of the *ompF* and *ompC* genes in *Escherichia coli*. Proc Natl Acad Sci USA 1989; 86:6052-6056.

181. Waukau J, Forst S. Molecular analysis of the signaling pathway between EnvZ and OmpR in *Escherichia coli*. J Bacteriol 1992; 174:1522-1527.

182. Nakashima K, Kanamaru K, Aiba H et al. Signal transduction and osmoregulation in *Escherichia coli*. A novel type of mutation in the phosphorylation domain of the activator protein, OmpR, results in a defect in its phosphorylation-dependent DNA binding. J Biol Chem 1991; 266: 10775-10780.

183. Brissette RE, Tsung KL, Inouye M. Suppression of a mutation in OmpR at the putative phosphorylation center by a mutant EnvZ protein in *Escherichia coli*. J Bacteriol 1991; 173:601-608.

184. Kanamaru K, Mizuno T. Signal transduction and osmoregulation in *Escherichia coli*: a novel mutant of the positive regulator, OmpR, that functions in a phosphorylation-independent manner. J Biochem (Tokyo) 1992; 111:425-430.

185. Bowrin V, Brissette R, Inouye M. Two transcriptionally active OmpR mutants that do not require phosphorylation by EnvZ in an *Escherichia coli* cell-free system. J Bacteriol 1992; 174:6685-6687.

186. Brissette RE, Tsung K, Inouye M. Mutations in a central highly conserved non-DNA-binding region of OmpR, an *Escherichia coli* transcriptional activator, influence its DNA-binding ability. J Bacteriol 1992; 174:4907-4912.

187. Kato M, Aiba H, Tate S et al. Location of phosphorylation site and DNA-binding site of a positive regulator, OmpR, involved in activation of the osmoregulatory genes of *Escherichia coli*. FEBS Lett 1989; 249:168-172.

188. Tate S, Kato M, Nishimura Y et al. Location of DNA-binding segment of a positive regulator, OmpR, involved in activation of the *ompF* and *ompC* genes of *Escherichia coli*. FEBS Lett 1988; 242:27-30.

189. Russo FD, Slauch JM, Silhavy TJ. Mutations that affect separate functions of OmpR the phosphorylated regulator of porin transcription in *Escherichia coli*. J Mol Biol 1993; 231:261-273.

190. Harlocker SL, Rampersaud A, Yang WP et al. Phenotypic revertant mutations of a new OmpR2 mutant (V203Q) of *Escherichia coli* lie in the *envZ* gene, which encodes the OmpR kinase. J Bacteriol 1993; 175:1956-1960.

191. Suzuki M. Common features in DNA recognition helices of eukaryotic transcription factors [published erratum appears in EMBO J 1993 Oct;12(10):4042]. EMBO J 1993; 12:3221-3226.

192. Kondo H, Miyaji T, Suzuki M et al. Crystallization and X-ray studies of the DNA-binding domain of OmpR protein, a positive regulator involved in activation of osmoregulatory genes in *Escherichia coli*. J Mol Biol 1994; 235:780-782.

193. Pratt LA, Silhavy TJ. OmpR mutants specifically defective for transcriptional activation. J Mol Biol 1994; 243:579-594.

194. Russo FD, Silhavy TJ. Alpha: the Cinderella subunit of RNA polymerase. J Biol Chem 1992; 267:14515-14518.

195. Garrett S, Silhavy TJ. Isolation of mutations in the alpha operon of *Escherichia coli* that suppress the transcriptional defect conferred by a mutation in the porin regulatory gene *envZ*. J Bacteriol 1987; 169: 1379-1385.

196. Jaskunas SR, Burgess RR, Nomura M. Identification of a gene for the alpha subunit of RNA polymerase at the *str-spc* region of the *Escherichia coli* chromosome. Proc Natl Acad Sci USA 1975; 72:5036-5040.

197. Slauch JM, Russo FD, Silhavy TJ. Suppressor mutations in *rpoA* suggest that OmpR controls transcription by direct interaction with the alpha subunit of RNA polymerase. J Bacteriol 1991; 173:7501-7510.

198. Matsuyama SI, Mizuno T, Mizushima S. Interaction between two regulatory proteins in osmoregulatory expression of *ompF* and *ompC* genes in *Escherichia coli*: a novel *ompR* mutation suppresses pleiotropic defects caused by an *envZ* mutation. J Bacteriol 1986; 168:1309-1314.

199. Matsuyama S, Mizushima S. Novel *rpoA* mutation that interferes with the function of OmpR and EnvZ, positive regulators of the *ompF* and *ompC* genes that code for outer-membrane proteins in *Escherichia coli* K-12. J Mol Biol 1987; 195:847-853.

200. Sharif TR, Igo MM. Mutations in the alpha subunit of RNA polymerase that affect the regulation of porin gene transcription in *Escherichia coli* K-12. J Bacteriol 1993; 175:5460-5468.

201. Igarashi K, Hanamura A, Makino K et al. Functional map of the alpha subunit of *Escherichia coli* RNA polymerase: two modes of transcription activation by positive factors. Proc Natl Acad Sci U S A 1991; 88:8958-8962.

202. Tsung K, Brissette RE, Inouye M. Enhancement of RNA polymerase binding to promoters by a transcriptional activator, OmpR, in *Escherichia coli*: its positive and negative effects on transcription. Proc Natl Acad Sci U S A 1990; 87:5940-5944.

CHAPTER 20

THE LEUCINE\LRP REGULON

Elaine B. Newman and Rongtuan Lin

I. INTRODUCTION

The leucine/Lrp regulon is a recently described global response governed by a transcriptional regulator called the leucine-responsive regulatory protein (Lrp), a small basic protein (pI 9.2) composed of two identical subunits of molecular weight 18 800 daltons.[1] Its sequence resembles only one other *E. coli* protein, AsnC, a positive regulator of *asnA* (the structural gene of asparagine synthetase A), with which it has 25% amino acid similarity. Cells deficient in Lrp show altered transcription of between 35 and 75 genes,[2,3] increasing the expression of some and decreasing that of others.[4-7] At some target promoters the presence of L-leucine in the growth medium modifies Lrp action, as observed with classic regulators, whereas at other LRP-regulated promoters leucine has little or no effect. Several reviews of the regulon have recently appeared.[4,6,8,9]

Knowledge of the Lrp regulon arose from studies of the regulation of several individual phenomena—*ilvIH* binding protein,[10] "leucine resistance,"[11] the regulation of branched-chain amino acid transport operons,[12] the regulation of oligopeptide permease,[13] the regulation of the *pap* operon,[14] as well as more general studies in the Newman laboratory on the role of leucine as an inducer of *E. coli* enzymes unrelated to leucine metabolism. When R. T. Lin et al demonstrated that a single factor, then called Rbl, regulated these phenomena,[5] it was clear that leucine (via Lrp) regulates a great deal of *E. coli* metabolism. The common physiological thread linking Lrp-regulated genes remains unknown, as does the reason it is governed by leucine. Indeed as we continue to accumulate information and understand more about Lrp regulation, the real function of Lrp (if any) eludes us. The fact that proteins homologous to *E. coli* Lrp have been described in a variety of other bacteria indicates that it is likely to have an important function nonetheless.[15-21]

Whereas the physiological reason that leucine regulates transcription of so many genes is unclear, it is yet more difficult to understand why a large group of genes is affected by Lrp, but not by leucine. Some genes are not transcribed at physiologically significant levels if Lrp is not present.[5] However, there is little evidence that the rate of *lrp* transcription itself varies greatly, except between rich and poor media.[5] In these cases it is clear that the presence of the protein is required to

Regulation of Gene Expression in Escherichia coli, edited by E. C. C. Lin and A. Simon Lynch. © 1996 R.G. Landes Company.

establish transcription, but it is not clear that Lrp is involved in regulating the actual rate of *lrp* transcription. In these cases, by its influence on DNA structure, Lrp may set the stage for other factors.[4]

In fact any of the DNA-binding proteins that are involved in establishing chromosome structure will appear to be global regulators if transcription is compared in wild-type cells and in mutants deficient in the particular factor. In *hns* mutants, transcription of many genes is altered.[22] However there is no evidence of variation in Hns level during metabolism.

Lrp seems to act both as a specific regulator in response to leucine, and as a structural element in establishing DNA conformation. Perhaps it will turn out that many other proteins also play a role in establishing the environment of the cell, in addition to their currently understood roles as chemical catalysts, membrane and ribosome components, and DNA-binding proteins.

II. THE LEUCINE-RESPONSIVE REGULATORY PROTEIN

Lrp is composed of 163 residues, including all amino acids except tryptophan. A stretch of residues centered at position 40 may represent a helix-turn-helix motif.[1] Wild-type Lrp and Lrp-1, a slightly altered form of the protein, were first purified by a series of steps including the identification of the Lrp-containing fractions by their ability to retard a DNA fragment corresponding to sequences located immediately upstream of the *ilvIH* operon, which is positively regulated by Lrp.[1] Subsequently, it was established that the DNA sequence of the *lrp* gene agrees with the first 38 N-terminal amino acids of the purified protein.[1,23] The Lrp molecule is thought to contain, proceeding from the N- to the C-terminus, three domains: those for DNA-binding (40%), for leucine-binding, and for transcription activation.[23]

Lrp is a moderately abundant protein.[1] Immunoassays using anti-Lrp antibodies suggest that Lrp makes up 0.1% of cellular protein in cells grown in glucose minimal medium. However, this estimate may be low since it does not take into account molecules bound tightly to DNA. Comparisons with known proteins on 2D gels suggested a similar figure (0.1 to 0.2%). This corresponds to about 3,000 molecules per cell, an estimate close to that of the number of CAP molecules per cell. Such a number is consistent with the notion that Lrp influences transcription of multiple genes.

Purified Lrp binds well to double-stranded DNA containing an operator site, even in the presence of a 1,000-fold excess of calf thymus DNA, but it does not bind to single-stranded operator DNA.[24] As might be anticipated from the notion that Lrp is an important element in establishing DNA structure, it has been demonstrated by a circular permutation assay that a single Lrp dimer bends its target DNA by an estimated 52° and that two dimeric molecules bound to adjacent sites result in a 135° bond.[24] Wang and Calvo[24] suggest that Lrp plays an architectural role in facilitating the assembly of a nucleoprotein complex regulating transcription.

III. REGULATION OF LRP SYNTHESIS

Lrp regulates its own transcription. In minimal glucose medium, expression of an φ(*lrp-lacZ*) fusion is lowered 2- to 3-fold by the presence of a single chromosomal *lrp*⁺ gene and 10-fold by the presence of a functional *lrp*⁺ gene on a multicopy plasmid.[5,25] The presence of leucine in the medium has no effect on expression. The effect of Lrp on transcription involves the binding of Lrp to a site between 32 and 80 base pairs upstream of the transcription start site, as indicated by gel mobility shifts, DNA footprinting, and deletion studies.[25] In addition, mutations in that region have been shown to alter the response to Lrp, strongly suggesting that autogenous regulation is a direct effect of Lrp binding to this site.[25]

The autogenous regulation, however, represents only a minor effect compared to the effect of growth in rich medium, which reduces *lrp* transcription over 10-fold.[5] This Lrp-independent decrease in lrp transcription[5] can be mainly ascribed to amino acids effects, as it is also observed in medium containing 1% casamino-acids, with or without glucose (C. Sears, R.T. Lin, and E.B. Newman, unpublished results). Furthermore, a large part of this effect can be produced by addition of either the α-ketoglutarate family or the oxalacetate family of amino acids. L-threonine alone reduced transcription 50% (C. Sears and E. B. Newman, unpublished results).

The expression of *lrp* also depends significantly on the carbon source. In cells grown with more oxidized substrates (acetate, succinate or pyruvate), transcription was diminished to 50%. Cells grown on glycerol or a variety of sugars all showed the same level of *lrp* transcription, namely, about 80% of that on glucose. The physiological significance of these variations of Lrp levels seems to be very limited, since no corresponding pattern could be seen in the transcription of 5 *lrp*-regulated target operons in glucose, succinate, or acetate grown cells (C. Sears and E. B. Newman, unpublished results).

However, even if the amount of Lrp made in enriched and minimal media is different, Lrp may still be an important metabolic regulator in all media, especially if the binding affinity of Lrp varies at different promoters. The different levels of *lrp* expression may indicate different settings of the cell's metabolic network, whereby systems with low affinity for Lrp are significantly regulated by it only during growth in minimal medium[26] with the level of exogenous leucine superimposing an additional level of complexity to the picture.

Studies with a plasmid with the Lrp gene under the control of the *araBAD* promoter have shown that the affinity of promoters for Lrp in vivo varies a great deal. In particular, promoters like *gcv* and *gltD* have such high affinity for Lrp that they would likely be saturated even in LB (Liang Tao and E. B. Newman, unpublished results).

IV. TARGET OPERONS OF LRP AND MUTANT PHENOTYPES

All genes that to date have been shown to be regulated by Lrp are listed in Table 20.1. These have been revealed by several experimental protocols: assaying leucine-induced genes (those previously known and those identified by making random insertions of λp*lac*mu); identifying proteins whose levels are altered in an *lrp* mutant, as determined on 2D gels; and identifying Lrp-dependent genes by determination of the DNA sequences flanking λp*lac*mu insertions. Among 50 transformants bearing such random inserts, levels of β-galactosidase activity in six clones were regulated by Lrp, again indicating that Lrp probably influences transcription of a large number of genes (J. Zhang and E. B. Newman, unpublished results).

1. EFFECTS OF A TOTAL LOSS OF LRP: METABOLISM OF *LRP* MUTANTS

Despite the many changes in expression of its genes, the *lrp* mutant can grow in glucose-minimal medium, albeit slowly, indicating that the biochemical pathways are still sufficiently well organized to allow for reasonably efficient growth. However, the reorganization of

Table 20.1. E. coli operons regulated by Lrp

Operon	Map location Min[a]	[e]lrp:lrp+	[f]leu:-Leu	References
Operons at which Lrp activates transcription:				
ilvIH	1.8	0.03	0.05	24,74
serA	63.0	0.16	0.5	5
sdaC	60.2			5,75
leuABCD	1.8	0.09	[b]	5,65
gltBDF	69.8	0.03	0.45	2,35
gcvTHP	62.6	0.05	1.0	5
pntAB	35.4	0.2	0.25	26
ompF	20.7			2
malT	75.3	0.5	1.0	27
malEFG	91.5	0.5	0.67	27
malK[c]	91.5	0.16	0.67	27
lacZYA	8.0	1:1.5	0.91	27
papBA	n.d.	0.03 or 0.004	0.91	14,32,73
fanABC	plasmid	0.01	0.1	14
sfaA	n.d.	0.12	3.0	76
daaABCDE	n.d.	0.12 or 0.02	0.5	68,76
fimB	n.d.	n.d.	n.d.	57
Operons at which Lrp represses transcription:				
sdaA	41.0	8	5	5
glyA	54.8	4	n.d.	[d]
kbl-tdh	81.2	20	8	5,51
oppABCDF	27.7	2.5	n.d.	13,15,33
lysU	93.8	22	4	50,77
livJ	76.3	85	0.01	5,78,79
livKHMG	76.3	9	n.d.	71,78
lrp	19.9	2	1	5,25
ompC	47.7		n.d.	2
fae	plasmid	3	1	80
osmY	n.d.	5	n.d.	18

[a] According to Bachmann.[81]

[b] The strains were leucine auxotrophs; induction in limiting leucine.

[c] Includes malKlamBmalM

[d] M. SanMartano and E.B. Newman, unpublished results

[e] lrp:lrp+ denotes the ratio of expression of the operon in mutant and wild-type cells.

[f] Leu:-Leu denotes the ratio of expression with/without leucine in the medium in the case of an Lrp+ strain?

the cell's metabolism makes it highly sensitive to further perturbations—either environmental, or by addition of further mutations.

Limited availability of L-serine and L-leucine

Growth of the *lrp* mutant in glucose minimal medium is measurably slower than that of wild-type strains (84 minutes doubling time compared to 58 minutes). This defect is alleviated by adding L-serine and L-leucine to the medium.[5,27] These phenomena might be expected from the fact that Lrp is an activator of the first gene in the biosynthetic pathways of both amino acids;[3,5,28] seemingly, reduced Lrp-independent expression of both genes assures some synthesis in the *lrp* mutant. It should be noted that the *lrp* mutant, for reasons not clear, has a strong tendency to accumulate secondary mutations. The mutant allele therefore must be retransduced frequently to maintain the original genetic background (E.B. Newman, unpublished observations).

Predicting the consequences of decreased serine biosynthesis in the *lrp* mutant is complicated by the fact that changes in other areas of metabolism may either exacerbate or compensate for the decreased expression from *serA*. On the one hand, in the absence of Lrp, the L-serine-degrading enzyme, L-serine deaminase 1, is increased 6-fold, enough to allow it to use exogenous L-serine as sole carbon and energy source.[5] However even that might not reduce the L-serine pool, since L-serine deaminase has a remarkably high K_m for L-serine.[21,29] On the other hand, L-serine may be produced by an alternative route, or at least spared, because synthesis of glycine from threonine via L-threonine dehydrogenase is increased.

The *lrp* mutant is probably starved for L-serine and/or leucine in other environmental conditions. It cannot grow in glucose-minimal medium at 42°C and it is also unable to grow anaerobically on glucose at 37°C.[27] In both cases, growth is restored by the addition of L-serine to the medium or by increasing the number of copies of the *serA* gene.[27]

Synthesis of one-carbon units

One-carbon metabolism is greatly altered in the *lrp* mutant. There are two biosynthetic routes to methylene-tetrahydrofolic acid (mTHF): the first is via serine hydroxymethyl transferase (encoded by *glyA*), which catalyzes the formation of glycine and mTHF from L-serine and THF, and which can also be used as a pathway of glycine synthesis during growth on glucose; the second is via the glycine cleavage enzyme complex (Gcv), forming mTHF from the α carbon of glycine, and simultaneously releasing CO_2 and NH_3. Since the second pathway is absent in the *lrp* mutant, it apparently relies on the first pathway for mTHF synthesis.

In wild-type cells growing in glucose minimal medium, the glycolytic intermediate 3-phosphoglycerate is the direct precursor of L-serine, glycine, and one-carbon units. The cell's requirement for one-carbon units is greater than its need for glycine, and the glycine cleavage enzyme complex permits it to equilibrate these syntheses by overproducing glycine and cleaving the excess.[30,31] In the *lrp* mutant, this scheme is radically altered because the *gcv* operon requires Lrp activation for physiologically significant expression.[5] Neither a *lrp* mutant nor a *glyA* mutant[30] can derive one-carbon units or nitrogen from glycine, and all mTHF must be formed by serine hydroxymethyl transferase. Indeed an *lrp*, *glyA* double mutant cannot grow in minimal medium, presumably because it cannot make C1 units for biosynthesis.[5]

Nitrogen metabolism

The *lrp* mutant grows well with ammonium as nitrogen source, at least when supplied at high levels. However, it is deficient in glutamate synthase[32] and therefore cannot use compounds such as arginine or ornithine as organic nitrogen sources. This deficiency in *gltBDF* expression is probably also responsible for the aspartate/glutamate requirement of the *lrp*, *pnt* double mutant, although, why a deficiency in pyridine nucleotide transhydrogenase should cause this, is unclear.

Cellular appendages and solute transport

The relation of the *lrp* mutant with its environment is greatly changed with respect to both its appendages and its solute transport capabilities. Almost all operons coding for pili or fimbriae seem to be affected by Lrp, which intervenes in the phase variation mechanisms. In most cases studied, the *lrp* mutant is unable to make the appendage, suggesting that adhesion to animal tissues would be severely handicapped in the absence of Lrp.[14,33]

Lrp lowers the levels of two high-affinity leucine uptake systems (one also active on isoleucine, valine, threonine, and alanine) and an oligopeptide uptake system, but elevates those of L-serine, maltose, and lactose permeases. Because the cell has so many alternative routes of substrate uptake, the deficiencies have no great effect on growth. For example, the *lrp* mutant uses L-serine well as a carbon source. An additional transport defect is indicated by the resistance of the *lrp* mutant to toxic tripeptides.[15,34]

Timing of Initiation of DNA synthesis unaffected

Among important cellular processes that are not affected by Lrp, it is interesting to note that the timing of DNA replication initiation is not affected in the *lrp* mutant.[35]

2. GROWTH OF *E. COLI* WITH REDUCED LRP LEVELS

Growth in a rich medium, such as LB broth, greatly lowers the Lrp level. In addition, the cell is well provided with leucine, which decreases the efficacy of Lrp. In such an environment, activation or repression of many of the Lrp target operons is expected to be weakened, unless the corresponding promoters have relatively high affinity for the regulator. The binding affinity of Lrp for only a single promoter has been carefully determined.[26] The in vivo assessment of affinities of other promoters now being carried out may help to clarify this issue (L. Tao and E. B. Newman, unpublished results).

Lrp is a particularly strong activator of certain promoters (e.g., *leuABCD*, *serA*, *gcv*, *papBA*, and *gltBDF*). Unless another factor compensates for its absence, these would not be significantly expressed in LB-grown cells. These observations led to the suggestion that Lrp activates biosynthesis, and represses degradation of amino acids.[2,5]

3. REDUCED TOLERANCE OF *LRP* MUTANTS FOR OTHER MUTATIONS

Wild-type cells are remarkably tolerant to changes in their physical environment and to the accumulation of deleterious mutations in genes that are apparently not essential for growth under laboratory conditions. In glucose minimal medium supplemented with L-serine and the three branched chain amino acids at 37°C, the *lrp* mutant grows almost as well as its parent.[5,27] However, the mutant appears to have lost the metabolic flexibility of its parent, as indicated by its reduced tolerance of secondary mutations. For example, an *lrp relA* double mutant has an absolute requirement for leucine and serine in order to grow at a normal rate, suggesting that the reduced expression of *leuA* and *serA* no longer suffices. Similarly, an *lrp pnt* double mutant is hindered in nitrogen metabolism and requires one of glutamate, glutamine, aspartate or asparagine to grow, although mutants defective in only one of the two loci are prototrophic.[27] Similarly, while the cell can tolerate either an *lrp* or a *glyA* mutation, the *lrp glyA* double mutant cannot grow in minimal medium, presumably because it cannot make one-carbon units.

4. AN *LRP* MUTATION FACILITATES GROWTH OF A *METK* MUTANT

The *metK* gene product, S-adenosylmethionine synthetase, catalyzes the formation of S-adenosylmethionine, the direct methyl donor in most methylation reactions. Though this is the only methyl donor known, it was thought to be a dispensable reaction, because *metK* mutants were believed to grow in glucose minimal medium. In fact, *metK* mutants do not grow in minimal medium unless a mutation in the *lrp* gene is also present or the deficiency is suppressed in some other way.[5] This would suggest that Lrp represses the synthesis of some MetK substitutes.

V. LRP AS A CHROMOSOME ORGANIZER

Lrp is known as a global regulator; however, we have suggested that it may be more suitable to think of it as a determinant of higher order DNA structure.[4,6] From this it follows that changes in DNA structure resulting from altered Lrp levels may be the actual mechanism wherein at least some of the altered gene expression is effected.

During evolution the cell had to develop a system for efficiently packaging its DNA. In eukaryotes such packaging involves histones, and in bacteria packaging probably also involves basic DNA-binding proteins. Until recently, however, some of these have been thought of exclusively as specific regulators rather than as chromosome organizers, e.g., CAP, IHF, H-NS, and Lrp. A structural role for some basic proteins such as HU and Fis, which bind to DNA with no apparent site specificity but which significantly affect DNA structure,[36,37] seems likely. However, even those proteins that bind specifically to certain sites may have effects on DNA structure when they bind, and indeed many have been shown to bend DNA.[24,38,39]

The conformation of DNA in vivo must be dependent on the effects of large numbers of different protein molecules binding to it. Indeed, most of the surface of DNA may be coated by proteins. Hence Lrp, as a small basic DNA-bending protein, at a relatively high copy number per cell (approx. 3000), is probably one of the determinants of chromosome structure. This would also be true for other binding proteins which are present in large numbers, e.g., CAP, which is usually thought of only in its role as a regulatory protein. Its regulatory role is clearly understood—in part because the known actions of CAP depend on an effector cAMP. However no search has been made for CAP-regulated genes which are not affected by cAMP.

It is clear that Lrp is also a regulator, especially for leucine-controlled operons. Lrp probably binds DNA at a large number of sites with varying affinity, and its effect may vary according to the affinity of binding and the position of the binding site relative to the start site for transcription. Thus Lrp may play a dual role in cell physiology: as a specific transcriptional regulator of certain operons and as a less specific DNA wrapping and organizing protein at other locations.

As the cell has evolved, chromosome structure has likely evolved too. The establishment of a new Lrp-like DNA-binding protein during evolution might therefore have changed the characteristics of the cell greatly, certainly more than would be expected by appearance of a new sequence specific regulator. It has been suggested that Lrp in effect converts the cells' metabolism from that needed inside a host to that needed for independent growth in impoverished media.[2,3,5] In evolution, such a transition could have happened quite abruptly with the advent of an Lrp-like protein.

Since their roles in chromosome organization depend to a large extent on their ability to bind DNA in a nonsequence-specific manner, the various DNA-organizing proteins may be to some extent interchangeable. If Lrp, IHF, and H-NS fulfill similar DNA organizational functions, overproduction of Lrp in a mutant lacking one of the other proteins might improve the growth properties of the cell. Conversely, a double mutant lacking two of the proteins might be more deleterious to the cell than one would predict from the characteristics of each single mutant. The synergistic impairment observed when *lrp* is coupled with various other mutations may reflect overlapping organizational functions of this sort.

VI. MOLECULAR ASPECTS OF LRP INTERACTIONS AT INDIVIDUAL PROMOTERS

1. THE *ILVIH* PROMOTER

Normal regulation of *ilvIH* requires several hundred base-pairs upstream of the promoter: 331 bp suffice to give normal regulation of transcription initiation and activation by leucine, whereas 200 bp do not.[24,40] Two in vitro transcription start sites have been located, as directed by two promoters: P1, located 31 bp upstream of the ATG initiation codon, and P2, located some 60 bp further upstream. In vitro, transcription from P2 is repressed and P1 is activated 2.7-fold by Lrp (half-maximal activation at 15nM Lrp). No in vivo transcription from P2 has been reported. Adding leucine during growth decreases transcription (from P1) 4- to 7-fold. In the *lrp* mutant, transcription from P1 is decreased even further (7- to 14-fold).[41]

Lrp binding to the *ilvIH* upstream region has been studied both by gel retardation and by footprinting,[10,24,42,43] and Lrp is thought to bind cooperatively to several sites in order to activate transcription.[42] The Lrp-binding sites comprise a region (-255 to -215) which contains two high affinity Lrp binding sites, and a lower affinity downstream region (-101 to -56) which contains four Lrp binding sites.[10,42] Both promoters are located in the downstream region.[10,42] In gel retardation studies, the downstream region (-141 to -60) is 50% saturated at 45 nM Lrp. Although the upstream region is not used for in vivo transcription, it is 50% saturated at a much lower concentration, 8 nM. It is thought that transcriptional activation requires binding to the downstream low-affinity sites, but that this is effected by Lrp binding at the upstream sites.[42] Detailed studies showed cooperative binding of Lrp to the two upstream sites (1 and 2), and to three of the four downstream sites (designated 3,4 and 5), but not to the 6th. Palindromic motifs were found in sites 2 and 6, and a consensus derived from these six sites corresponds to the sequence 5'-AGAATtttATTCT-3'(see below). This view of the overall structure of the *ilvIH* promoter is supported by preliminary mutational studies in which each site was mutated in turn, and the effect on in vivo expression from an ϕ(*ilvIH-lac*) fusion determined.[42]

Lrp has been shown by circular permutation studies to bend DNA carrying a single binding site, *ilvIH* (site 2) by 52°, with the bending centered at 5'-TTTT-3' flanked by a four-base palindromic motif. It bends a DNA fragment carrying two sites (1 and 2) by 135°.[24] Combined with the results of footprinting studies, these results suggest that DNA might be wrapped around multiple Lrp proteins, or that Lrp binding to multiple sites might loop the DNA. The requirement for a stereospecific alignment between the promoter and various *cis*-acting sequences supports the idea of a multi-protein complex at this promoter.[44] This also supports the idea that Lrp may be a general determinant of chromosome structure.[4,6]

2. THE *papBA* PROMOTER: GATC METHYLATION

The *pap* operon includes some 11 genes involved in producing P-pili, including two regulatory genes, *papI* and *papB*. Fimbrial synthesis by the *pap* operon is subject to phase variation, controlled by Dam-directed adenine methylation of two GATC sequences, GATC1 and GATC2 located 102 base pairs apart upstream of the operon.[32] In the ON phase, GATC1 is unmethylated and in the OFF phase, GATC2 is unmethylated.[32] Expression of the *pap* operon requires two positive regulators, PapI and Lrp.[14] Both GATC sites lie within Lrp binding sites. Lrp can bind at the GATC2 site irrespective of its methylation state and, when bound, prevent its methylation. PapI alone does not seem to bind but, in association with Lrp, it promotes binding to the GATC1 site, provided it is unmethylated or hemimethylated. Expression of *papA* requires PapI-Lrp binding here, and this binding keeps the GATC1 site unmethylated, presumably by blocking access to Dam methylase. The switch from OFF to ON requires DNA synthesis, making both sites hemimethylated, with competition between Dam and PapI-Lrp for GATC1.

The ability of Lrp to bend DNA underlies an interesting model which suggests that in phase-on cells, a multiprotein complex bends the DNA and positions RNA polymerase on looped DNA, such that both *papBA* and *papI* can be transcribed.[45] In phase-off cells, methylation prevents binding of Lrp and such a transcription complex is unable to form.

It seems likely that the structure and regulation of the *pap* promoter is very different from that of *ilvIH*. However, this elegant methylation mechanism is not even typical of the fimbrial genes. Phase variation of the *fim* genes is regulated not by methylation but by DNA inversion, and Lrp is involved in regulating that switch too. At least one other gene, orf489, appears to be regulated both by Lrp and methylation.[27,46]

3. THE *serA* PROMOTER

The serA promoter differs from *ilvIH* in that its two promoters are both used in vivo. P1, 45 bp from the translation start site, is activated by Lrp and used in wild-type cells grown in glucose-minimal medium.[47] P2, 93 bp further upstream, is repressed by Lrp, and therefore only used in *lrp* mutants or in growth medium in which Lrp is made at a low level, e.g., in LB broth.

The Lrp-binding sites at *serA* are currently less defined than those at characterized *ilvIH* but they do not appear to correspond exactly. An upstream high affinity site in *serA* (-155 to -81) is in much the same position as the downstream low-affinity *ilvIH* site. The *serA* low affinity site is less well localized, but downstream of -82. Binding of Lrp at the upstream site would be expected to block P2, leaving P1 free to be transcribed. This is similar to *ilvIH*, where binding at the low affinity site would block P2.

4. PRELIMINARY STUDIES AT OTHER PROMOTERS: *GLTBDF, LYSU, GCV* AND *TDH*

Lrp binding to a fragment of the *gltBDF* gene (-322 to +344) has been demonstrated,[31] with leucine acting by increasing the apparent dissociation constant. Therefore at high concentrations of Lrp, leucine has no effect. Deletions upstream of the *tdh* promoter define a segment from -69 to -44 bp as the target for leucine regulation, since deletion of this region makes the operon unresponsive to Lrp.[48] Similarly a sequence between -313 and -169 is involved in regulation of *gcv* by Lrp.[49]

At *lysU*, Lrp binds to a fragment (-51 to -158),[50] with leucine reducing this binding at concentrations similar to the concentrations effective at the *ilvIH* upstream (high affinity) site. The *lysV* footprint indicates probable multiple binding site(s) covering the -60 to -158 region, so that this promoter might be of the *ilvIH* type.

5. LRP INTERACTIONS WITH OTHER REGULATORY FACTORS

In vitro experiments for defining promoter function generally use one or two regulatory proteins and a fragment of DNA. From such studies various models for Lrp action have been proposed, including the ones cited at the *ilvIH* and *pap* promoters.[24,44,45] Clearly, this approach is likely to be too simplistic in that it underestimates the complexity of in vivo regulatory interactions. A suggestion that Hns modulates Lrp activity[22] has been confirmed by studies in our laboratory, showing slightly reduced expression of a φ(*lrp-lacZ*) fusion in an *hns* mutant. It seems likely that we will see complexes of Lrp with other regulatory proteins, perhaps forming different transcription complexes depending on the regulatory responses involved.

6. LRP CONSENSUS DNA BINDING SEQUENCE

Three groups have proposed closely related consensus sequences for Lrp binding. Rex et al[51] proposed an asymmetric 12 bp sequence, 5'-TTTATTCtNaAT-3', upstream of the transcription start of the *kbl-tdh* gene as an Lrp consensus binding site. They also found this consensus in both orientations upstream of other Lrp-regulated genes. Deletion of this sequence located upstream of the *kbl-tdh* operon resulted in high level constitutive expression, independent of leucine and Lrp. This strongly suggests that it is indeed part of the operator site recognized by Lrp, at least for some genes.[51]

Wang and Calvo,[42] in a detailed analysis of the six Lrp-binding sites upstream of the *ilvIH* operon, found that those with strongest affinity for Lrp have a symmetric sequence with a consensus 5'-AGAATTTTATTCT-3'.

Using an efficient method for detection of common motifs in DNA sequences, H. Margalit and co-workers analyzed a set of 23 gene sequences whose transcription is under Lrp control and identified the putative consensus 5'-(g/a)(g/c)nnnTTTATtCTgG-3'. The core of this consensus, TTTATtCT, is compatible with the two consensus sequences described above; but there are significant differences in the flanking regions.

The consensus sequences suggested in these studies are so AT-rich as to suggest that Lrp binds to AT-rich, bent or curved sequences and thereby recognizes the tertiary structure of DNA rather than simply a specific DNA sequence. This would explain why the AT-rich sequence is not always found at an appropriate position relative to the transcription start site,[51] since the DNA might be bent in other ways.

APPENDIX

KNOWN OPERONS THAT ARE REGULATED BY LRP

A. Biosynthetic operons
1. *ilvIH* (isoleucine/valine biosynthesis). The *ilvIH* gene product codes for one of three isoenzymes catalyzing the first step of branched-chain amino acid synthesis.[1,21,23]
2. *serA* (L-serine biosynthesis). The *serA* gene product, phosphoglycerate dehydrogenase, is the first enzyme specific to L-serine biosynthesis.
3. *leuABCD* (leucine biosynthesis).
4. *glyA* (glycine biosynthesis). Gene product, serine hydroxymethyl transferase, is required for glycine biosynthesis from L-serine.[52]
5. *gltBDF* (glutamate synthase). Gene product required for assimilation of organic nitrogen.

B. Degradative Operons
1. *sdaA* (L-serine deaminase 1). Gene product allows degradation of L-serine to pyruvate in cells growing in minimal medium.[53,54]
2. *kbl-tdh* (threonine degradation). Gene products, threonine dehydrogenase and AKB-CoA lyase, allow conversion of threonine to glycine, and are not usually expressed in glucose-minimal medium, threonine dehydrogenase and AKB-CoA lyase.[5,8,48,55,56]
3. *lacZ* (β-galactosidase). Gene product degrades lactose to glucose and galactose.[11]

C. Other Metabolic Operons
1. *gcvTHP* (glycine cleavage). Gene product converts glycine to N5,10-methylene-tetrahydrofolate, NH_3 and one carbon units.[3]

2. *pntAB* (pyridine nucleotide transhydrogenase). Gene product nucleotide transhydrogenase interconverts NADP and NADPH, permitting the cell to produce NADPH independently of the pentose phosphate shunt, or to use NADH to extrude protons.[26,57]

3. *lysU* (lysyl-tRNA synthetase).Gene product is one of the two lysyl-tRNA synthetases, LysU and LysS.[31,50,58-60]

D. Transport Operons

1. *livJ, livKHMGF* (leucine transport). Gene products constitute the two transport systems with high leucine affinity, *livJ* and *livKHMGF*.[3,11,36,46,61-63]

2. *sdaC* (L-serine transport). Gene product may be one of the multiple L-serine transport systems.[55,64]

3. *oppABCD* (oligopeptide transport).Gene products constitute the oligopeptide transport system.[12,13,65]

4. *malEFG, malK-lamB-malM, malT* (maltose transport). Gene products constitute the maltose uptake system.[11,66]

5. *ompF* and *ompC* (outer membrane porins). Gene products, the OmpC and OmpF, permit passage of small molecules into the periplasm.[36,67]

E. Operons Coding for Pili and Fimbriae

All the fimbrial genes which have been tested show regulation by Lrp, including *pap*,[9,14] *sfa*,[25] *daa*,[25,68] and *fim*[38,69] operons, and the *E. coli* plasmid *fan*[70] and *fae*[40] operons.

F. Other Operons

1. *osmY*. Encodes a periplasmic protein of 18 kDa of unclear function induced on entry into stationary phase.[34,71]

2. *orf489*. An unidentified open reading frame.[3,11,72,73]

REFERENCES

1. Willins DA, Ryan CW, Platko JV et al. Characterization of Lrp, an *Escherichia coli* regulatory protein that mediates a global response to leucine. J Biol Chem 1991; 266:10768-74.

2. Ernsting BR, Atkinson MR, Ninfa AJ et al. Characterization of the regulon controlled by the leucine responsive regulatory protein in *Escherichia coli*. J Bacteriol 1992; 174:1109-18.

3. Lin RT, D'Ari R, Newman EB. λplacMu insertions in genes of the leucine regulon: extension of the regulon to genes not regulated by leucine. J Bacteriol 1992; 174:1948-55.

4. D'Ari R, Lin RT, Newman EB. The leucine-responsive regulatory protein: more than a regulator? Trends Biochem Sci 1993; 18:260-63.

5. Lin R, D'Ari R, Newman EB. The leucine regulon of*Escherichia coli*: a mutation in *rblA* alters expression of leucine-dependent metabolic operons. J Bacteriol 1990; 172:4529-35.

6. Newman EB, D'Ari R, Lin RT. The leucine-Lrp regulon in E. coli: a global response in search of a raison d'être. 1992 Cell 68:617-19.

7. Platko JV, Willins DA, Calvo JM. The *ilvIH* operon of *Escherichia coli* is positively regulated. J Bacteriol 1990; 172:4563-70.

8. Calvo JM, Matthews RG. Leucine-responsive regulatory protein- a global regulator of metabolism in *Escherichia coli*. Microbiol Rev 1994; 58:466-98.

9. Newman EB, Lin RT. The leucine-responsive reulatory protein, a global regulator of gene expression in *E. coli*. Annu Rev Microbiol 1995; 49:in press.

10. Ricca E, Aker DA, Calvo JM. A protein that binds to the regulatory region of the *Escherichia coli ilvIH* operon. J Bacteriol 1989; 171:1658-64.

11. Templeton BA, Savageau MA. Transport of biosynthetic intermediates: regulation of homoserine and threonine uptake in *Escherichia coli*. J Bacteriol 1974; 120:114-20.

12. Anderson JJ, Quay SC, and Oxender DL. Mapping of two loci affecting the regulation of branched-chain amino acid transport in *Escherichia coli* K-12. J. Bacteriol 1976; 126:80-90.

13. Austin EA, Andrews JC, Short SA. Selection, characterization and cloning of *oppI*, a regulator of the *E. coli* oligopeptide permease operon. Abstr Mol Genet Bacteria Phages, abstr. Cold Spring Harbor, NY: Cold Spring Harbor Laboratory, 1989:153.

14. Braaten BA, Platko JV, van der Woude MW et al. Leucine-responsive regulatory protein controls the expression of both the *pap* and *fan* pili operons in *Escherichia coli*. Proc Natl Acad Sci USA 1992; 89:4250-4.

15. Andrews JC, Blevins TC, Short SA. Regulation of peptide transport in *Escherichia coli*: induction of the *trp*-linked operon encoding the oligopeptide permease. J Bacteriol 1986; 165:428-33.

16. Arst Jr HN, Integrator gene in *Aspergillus nidulans*. Nature 1994; 262:231-4.

17. Ishizuka H, Hanamura A, Kunimura T et al. A lowered concentration of cAMP receptor protein caused by glucose is an important determinant for catabolite repression in *Escherichia coli*. Mol Microbiol 1994; 6:2489-95.

18. Lange R, Barth M, Hengge-Aronis R. Complex transcriptional control of the sS-dependent stationary-phase-induced and osmotically regulated *osmY* (cis-5) gene suggests novel roles for Lrp, cyclic AMP (cAMP) receptor protein-cAMP complex, and integration host factor in the stationary-phase response of *Escherichia coli*. J Bacteriol 1993; 175:7910-17.

19. Madhusudhan KT, Lorenz D, Sokatch JR. The *bkdR* gene of *Pseudomonas putida* is required for expression of the *bkd* operon and encodes a protein related to Lrp of *Escherichia coli*. J Bacteriol 1993; 175:3934-40.

20. Pizer LI. Glycine synthesis and metabolism in *Escherichia coli*. J Bacteriol. 1967; 89:1145-50.

21. Su H, Lang BF, Newman EB. L-Serine degradation in *Escherichia coli* K-12. Cloning and sequencing of the *sdaA* gene. J Bacteriol 1989; 171:5095-102.

22. Levinthal M, Lejeune P, Danchin A. The H-NS protein modulates the activation of the *ilvIH* operon of *Escherichia coli* K12 by Lrp, the leucine regulatory protein. Mol Gen Genet 1994; 242:736-43.

23. Platko JV, Calvo JM. Mutations affecting the ability of *Escherichia coli* Lrp to bind DNA, activate transcription, or respond to leucine. J Bacteriol 1993; 175:1110-17.

24. Wang Q, Calvo JM. Lrp, a major regulatory protein in *Escherichia coli*, bends DNA and can organize the assembly of a higher-order nucleoprotein structure. EMBO J 1993; 12:2495-501.

25. Wang Q, Wu J, Friedberg D et al. Regulation of the *Escherichia coli lrp* gene. J Bacteriol 1994; 176:1831-9.

26. Ernsting BR, Denninger JW, Blumenthal RM et al. Regulation of the *gltBDF* operon of *Escherichia coli*: how is a leucine-insensitive operon regulated by the leucine-responsive regulatory protein? J Bacteriol 1993; 175:7160-9.

27. Ambartsoumian G, D'Ari R, Lin RT et al. Altered amino acid metabolism in *lrp* mutants of *Escherichia coli* and their derivatives. Microbiol 1994; 140:1737-44.

28. Tchetina E, Newman EB. Identification of Lrp-regulated genes by inverse PCR and sequencing: an insert in *E. coli malF* is regulated by leucine-responsive regulatory protein. J Bacteriol 1995; (in press)

29. Su H, Moniakis J, Newman EB. Use of gene fusions of the structural gene *sdaA* to purify L-serine deaminase 1 from *Escherichia coli* K-12. Eur J biochem 1993; 211:521-7.

30. Newman EB, Batist G, Fraser J et al. The use of glycine as nitrogen source by *Escherichia coli* K-12. Biochim Biophys Acta 1976; 421:97-105.

31. Newman EB, Miller B, Kapoor V. Biosynthesis of single-carbon units in *Escherichia coli* K-12. Biochim Biophys Acta 1974; 338:529-39.

32. Friedberg D., Platko JV, Tyler B, Calvo JM. The amino acid sequence of Lrp is highly conserved in four enteric microorganisms. J Bacteriol 1995; 177:(in press).

33. Braaten BA, Blyn LB, Skinner BS et al. 1991. Evidence for a methylation-blocking factor (*mbf*) locus involved in *pap* pilus expression and phase variation in *Escherichia coli*. J Bacteriol 1991; 173:1789-1800.

34. Andrews JC, Short SA. *opp-lac* operon fusions and transcriptional regulation of the *Escherichia coli trp*-linked oligopeptide permease. J Bacteriol 1986; 165:434-42.

35. Smith DW, Stine WB, Svitil AV et al. *Escherichia coli* cells lacking methylation-blocking factor (leucine-responsive regulatory protein) have precise timing of initiation of DNA replication in the cell cycle. J Bacteriol 1992; 174:3078-82.

36. Drlica K, Rouviere-Yaniv J. Histonelike proteins of bacteria. Microbiol Rev 1987; 51:301-19.

37. Rouviere-Yaniv J, Yaniv M, Germond JE. *E. coli* DNA binding protein HU forms nucleosome-like structure with circular double-stranded DNA. Cell 1979; 17:265-74.

38. Freundlich M, Ramani N, Mathew E et al. The role of integration host factor in gene expression in *Escherichia coli*. Mol Microbiol 1992; 6:2557-63.

39. Schultz SC, Shields GC, Steitz TA. Crystal structure of a CAP-DNA complex: the DNA is bent by 90=B0. Science 1991; 253:1001-7.

40. Haughn GW, Squires CH, De Felice M et al. Unusual organization of the *ilvIH* promoter in *Escherichia coli*. J Bacteriol 1985; 163:186-98.

41. Willins DA, Calvo JM. In vitro transcription from the *Escherichia coli ilvIH* promoter. J Bacteriol 1992; 174:7648-55.

42. Wang Q, Calvo JM. Lrp, a global regulatory protein of *Escherichia coli*, binds co-operatively to multiple sites and activates transcription of *ilvIH*. J Mol Biol 1993; 229:306-18.

43. Wang Q, Sacco M, Ricca E et al. Organization of Lrp-binding sites upstream of *ilvIH* in *Salmonella typhimurium*. Mol Microbiol 1993; 7:883-91.

44. Sacco M, Ricca E, Marasco R et al. A stereospecific alignment between the promoter and the *cis*-acting sequence is required for Lrp-dependent activation of *ilvIH* transcription in *Escherichia coli*. FEMS Microbiol Lett 1993; 107:331-6.

45. Van der Woude MW, Braaten BA, Low DA. Evidence for global regulatory control of pilus expression in *Escherichia coli* by Lrp and DNA methylation: model building based on analysis of pap. Mol Microbiol 1992; 6:2429-35.

46. Hale WB, van der Woude MW, Low DA. Analysis of nonmethylated GATC sites in the *Escherichia coli* chromosome and identification of sites that are differentially methylated in response to environmental stimuli. J Bacteriol 1994; 176:3438-41.

47. Lin R, Characterization of the leucine/Lrp regulon in *Escherichia coli* K-12. Thesis, Concordia University, Montreal, Quebec, Canada, 1992.

48. Ravnikar PD, Somerville RL. Genetic characterization of a highly efficient alternative pathway of serine biosynthesis in *Escherichia coli*. J Bacteriol 1987; 169:2611-17.

49. Wilson RL, Stauffer GV. DNA sequence and characterization of *gcv*A, a *lys*R family regulatory protein for the *Escherichia coli* glycine cleavage enzyme system. J Bacteriol. 1994; 176:2862-8.

50. Lin R, Ernsting B, Hirshfield IN et al. The *lrp* gene product regulates expression of *lysU* in *Escherichia coli* K-12. J Bacteriol 1992; 174:2779-84.

51. Rex JH, Aronson BD, Somerville RL. The *tdh* and *serA* operons of *Escherichia coli*: mutational analysis of the regulatory elements of leucine-responsive genes. J Bacteriol 1991; 173:5944-53.

52. Perez-Martin J, Rojo F, de Lorenzo V. Promoters responsive to DNA bending: a common theme in procaryotic gene expression. Microbiol Rev 1994; 58:268-90.

53. Shao ZQ, Newman EB. Sequencing and characterization of the *sdaB* gene from *Escherichia coli* K-12. Eur J Biochem 1993; 212:777-84.

54. Siegele DA, Kolter R. Life after log. J Bacteriol 1992; 174:345-8.

55. Ferrario M, Ernsting BR, Borst DW et al. The leucine-responsive regulatory protein of *Escherichia coli* negatively regulates transcription of *ompC* and *micF* and positively regulates translation of *ompF*. J Bacteriol 1995; 177:103-13.

56. Raina S, Missiakas D, Baird L et al. Identification and transcriptional analysis of the *Escherichia coli htrE* operon which is homologous to pap and related pilin operons. J Bacteriol 1993; 175:5009-21.

57. Gally DL, Bogan JA, Eisenstein BI et al. Environmental regulation of the fim switch controlling type 1 fimbrial phase variation in *Escherichia coli* K-12: effects of temperature and media. J Bacteriol 1993; 175:6186-93.

58. Isenberg S, Newman EB. Studies on L-serine deaminase in *Escherichia coli* K-12. J Bacteriol 1974; 118:53-8.

59. Leveque F, Gazeau M, Fromant M et al. Control of *Escherichia coli* lysyl-tRNA synthetase expression by anaerobiosis. J Bacteriol 1991; 173:7903-10.

60. Nakamura Y, Ito R. Control and function of lysyl-tRNA synthetases: diversity and co-ordination. Mol Microbiol 1993; 10:225-31.

61. Jamieson DJ, Higgins CF. Anaerobic and leucine-dependent expression of a peptide transport gene in *Salmonella typhimurium*. J Bacteriol 1984; 160:131-6.

62. Kawaji H, Mizuno T, Mizushima S. Influence of molecular size and osmolarity of sugars and dextrans on the synthesis of outer membrane proteins O-8 and O-9 of *Escherichia coli* K-12. J Bacteriol 1979; 140:843-7.

63. Nystrom T, Neidhardt FC. Cloning, mapping and nucleotide sequencing of a gene encoding a universal stress protein in *Escherichia coli*. Mol Microbiol 1992; 6:3187-98.

64. Guardaiola J, De Felice M, Klopotowski T et al. Multiplicity of isoleucine, leucine, and valine transport systems in *Escherichia coli* K-12. J Bacteriol 1974; 117:382-92.

65. Anderson JJ, Oxender DL. *Escherichia coli* transport mutants lacking binding protein and other components of the branched-chain amino acid transport systems. J Bacteriol 1977; 130:384-92.

66. Bedouelle H, Schmeissner E, Hofnung M et al. Promoters of the *malEFG* and *malK-lamB* operons in *Escherichia coli* K12. J Mol Biol 1982; 161:519-31.

67. Ito K, Kawakami K, Nakamura Y. Multiple control of *Escherichia coli* lysyl-tRNA synthetase expression involves a transcriptional repressor and a translational enhancer element. Proc Natl Acad Sci USA 1993 90:302-6.

68. Bilge SS, Apostol Jr JM, Fullner KJ et al. Transcriptional organization of the F1845 fimbrial adhesin determinant of *Escherichia coli*. Mol Microbiol 1993; 7:993-1006.

69. Blomfield IC, Calie PJ, Eberhardt KJ et al. Lrp stimulates phase variation of type 1 fimbriation in *Escherichia coli* K-12. J Bacteriol 1993; 175:27-36.

70. Lawther RP, Calhoun DH, Adams CW et al. Molecular basis of valine resistance in *Escherichia coli* K-12. Proc Natl Acad Sci USA 1981; 78:922-5.

71. Landick R, Anderson JJ, Mayo MM et al. Regulation of high-affinity leucine transport in *Escherichia coli*. J Supramol Struct 1980; 14:527-37.

72. Gerolimatos B, Hanson RL. Repression of *Escherichia coli* pyridine nucleotide transhydrogenase by leucine. J Bacteriol 1978; 134:394-400.

73. Nou X, Skinner B, Braaten B. et al. Regulation of pyelonephritis-associated pili phase-variation in *Escherichia coli*: binding of the PapI and the Lrp regulatory proteins is controlled by DNA methylation. Mol Microbiol 1993; 7:545-53.

74. Quay SC, Kline EL, Oxender DL. Role of leucyl-tRNA synthetase in regulation of branched-chain amino-acid transport. Proc Natl Acad Sci USA 1975; 72:3921-4.

75. Shao ZQ, Lin RT, Newman EB. Sequencing and characterization of the *sdaC* gene and identification of the *sdaCB* operon in *E. coli* K-12. Eur J Biochem 1994; 222:901-7.

76. Van der Woude MW, Low DA. Leucine-responsive regulatory protein and deoxyadenosine methylase control the phase variation and expression of the *sfa* and *daa* pili operons in *Escherichia coli*. Mol Microbiol 1994; 11:605-18.

77. Gazeau M, Delort F, Dessen P et al. *Escherichia coli* leucine-responsive regulatory protein (Lrp) controls lysyl-tRNA synthetase expression. FEBS 1992; 300:254-8.

78. Haney SA, Plakto JV, Oxender DL et al. Lrp, a leucine-responsive protein, regulates branched-chain amino acid transport genes in *Escherichia coli*. J Bacteriol 1992; 174:108-15.

79. Penrose WR, Nichoalds GE, Piperno JR et al. Purification and properties of a leucine-binding protein from *Escherichia coli*. J Biol Chem 1968; 243:5921-8.

80. Huisman TT, Bakker D, Klaasen P et al. Leucine-responsive regulatory protein, IS1 insertions, and the negative regulator FaeA control the expression of the *fae* (K88) operon in *Escherichia coli*. Mol Microbiol 1994; 11:525-36.

81. Bachmann BJ. Linkage map of *Escherichia coli* K-12, ed 8. Microbiol Rev 1990; 54:130-197.

CHAPTER 21

ADAPTIVE RESPONSES TO OXIDATIVE STRESS: THE *soxRS* AND *oxyR* REGULONS

Elena Hidalgo and Bruce Demple

Along with other environmental changes considered elsewhere in this book, bacteria must cope with changing exposure to oxygen radicals. *Escherichia coli* meets these challenges by activating complex resistance mechanisms against these radicals. This chapter focuses on the regulatory mechanisms and functions of two key response systems, the *soxRS* and the *oxyR* regulons.

I. REACTIVE OXYGEN SPECIES

Molecular oxygen (O_2) is used during aerobic metabolism as the final electron acceptor in the respiratory chain, where it gains 4 electrons to produce H_2O. But O_2 can also undergo univalent reduction to form superoxide radical, $O_2 \cdot^-$. This toxic species, generated mainly by reaction with components of the respiratory chain, can gain another electron spontaneously (with transition metals) or enzymatically (by superoxide dismutases) and generate hydrogen peroxide (H_2O_2), a metastable species. The sequential reduction of H_2O_2 in the presence of transition metals produces hydroxyl radical ($\cdot OH$), which is very reactive. Addition of another electron generates H_2O:

$$O_2 \xrightarrow{\;e-\;} O_2 \cdot^- \xrightarrow{\;e-\;} H_2O_2 \xrightarrow{\;e-\;} \cdot OH \xrightarrow{\;e-\;} H_2O$$

These reactive oxygen species are also produced deliberately by some immune cells using the enzyme NADPH oxidase.[1] Murine macrophages[2] also generate another radical, nitric oxide (NO·). These reactive oxygen species, along with other defense molecules, mediate the cytotoxic attack of immune cells on pathogens.[3,4]

What are the biological targets for these reactive species? Lipids are a major target for H_2O_2 and peroxides.[5,6] Reactive oxygen can also damage proteins by either oxidizing amino acids[7,8] or modifying prosthetic groups or metal clusters.[9-11] DNA is also an oxidative target, particularly for hydroxyl radicals[12] which generate many different lesions.[13] The modifications described above are all deleterious to the cell, since

Regulation of Gene Expression in Escherichia coli, edited by E. C. C. Lin and A. Simon Lynch. © 1996 R.G. Landes Company.

they lead to a loss of function of membranes and proteins, and block DNA replication or cause mutations.

II. ANTIOXIDANT DEFENSES

Oxygen-consuming organisms have developed defense mechanisms that keep the concentration of the O_2-derived radicals at acceptable levels or repair oxidative damages. Some molecules are constitutively present and help maintain an intracellular reducing environment or chemically scavenge reactive oxygen. Examples include the NADPH and NADH pools, thioredoxin or glutathione.[14,15]

Specific enzymes decrease the steady-state levels of reactive oxygen. Two superoxide dismutases (SODs), which convert $O_2 \cdot^-$ to H_2O_2 and O_2, have been described in *E. coli*: an iron-containing enzyme (encoded by *sodB*), whose expression is modulated by intracellular iron levels,[16] and a manganese-containing SOD (encoded by *sodA*), the predominant enzyme during aerobic growth, whose expression is transcriptionally regulated by at least six control systems.[17] A third SOD activity with properties like eukaryotic CuZn-SOD has recently been found in the *E. coli* periplasmic space.[18] H_2O_2 is removed by catalase (yielding H_2O and O_2). Two major catalases have been described for *E. coli*: HPI (the gene product of *katG*) is present during aerobic growth and transcriptionally controlled at different levels (see below), and HPII (encoded by *katE*) is induced during stationary phase.[19] Another source of oxygen-derived toxicity is the formation of peroxyl radicals from organic peroxides; they can be eliminated by peroxidases, like HPII[19] and the NADPH-dependent alkyl hydroperoxide reductase encoded by *ahpCF*.[20]

DNA repair enzymes also constitute part of antioxidant defense of *E. coli*. These include endonuclease IV, which is induced by oxidative stress (see below), and exonuclease III, which is induced in stationary phase or starving cells.[21] Still other DNA repair pathways contribute to the overall defense against oxidative damage.[22]

III. OXIDATIVE STRESS

Different situations can produce imbalances in the pro-oxidant (generating reactive oxygen)/antioxidant ratio in cells: either an increase in the production rate of O_2-derived radicals or a limitation in the defense mechanisms generates a condition called oxidative stress.[23]

Increases in the steady-state concentration of free radicals can result from exposure of cells to diffusible chemical oxidants (e.g., H_2O_2 or *tert*-butyl hydroperoxide) or from ionizing radiation, which instantaneously generates radicals within the cell. As mentioned above, immune cells also generate H_2O_2 and nitric oxide, which also cross membranes. Superoxide generated by these cells may either be converted to H_2O_2 and enter the cell, or cause toxic damage to the cell membrane.

Radicals can also be generated intracellularly by compounds called redox-cycling agents, which mediate one-electron reduction of O_2 and divert electrons from the NADPH or NADH pools.[5] Examples include paraquat (PQ), menadione, plumbagin, phenazine methosulfate[24] and 4-nitroquinoline-*N*-oxide.[25]

A different mechanism for oxidative stress is mediated by impairment of the radical-scavenging capacity of the cell. Thus, mutants deficient in SOD may have a 10^4-fold increased steady-state concentration of $O_2 \cdot^-$ over wild-type cells.[26] Surprisingly, *katG*-deleted strains (lacking HPI catalase) do not seem to show an increase in the steady state concentration of H_2O_2 (González-Flecha and Demple, unpublished results), perhaps because the H_2O_2 generated during aerobic growth can be dissipated by either HPII or the alkyl hydroperoxide reductase encoded by *ahpCF*.

IV. GLOBAL RESPONSES TO OXIDATIVE STRESS

Induction of individual proteins as a consequence of oxidative stress was reported 20 years ago: *sodA* (Mn-superoxide dismutase) and *katG* (catalase) expression was induced after exposure of bacteria to redox-cycling agents[27] or to H_2O_2.[28] A more widespread response that coordinates antioxidant defenses was suggested by an adaptive response of *E. coli* to H_2O_2: exposure to micromolar levels of H_2O_2 induced a protective response that conferred resistance to subsequent exposure to millimolar H_2O_2.[29] An analogous adaptive response was later reported for the redox-cycling agent plumbagin.[30] Similar adaptive responses that distinguish oxidative stress by H_2O_2 from that caused by redox-cycling agents have since been reported for the yeast *Saccharomyces cerevisiae*.[31,32]

Analysis of protein expression patterns in bacteria treated with oxidants confirmed the global nature of these responses.[33-35] Collectively, these studies indicate that enteric bacteria induce the expression of at least 40 proteins in response to H_2O_2, and another ~40 proteins in response to superoxide-generating agents (Fig. 21.1). The isolation of gene fusions inducible by oxidative stress[36] and of mutants with increased resistance to H_2O_2 and $O_2\cdot^-$[33,37] allowed the identification of two main oxidative stress response regulons in *E. coli*: the *soxRS* regulon and the *oxyR* regulon. The *soxRS* locus controls a response to $O_2\cdot^-$ generating agents. The *soxRS* regulon contains at least ten genes, including those encoding the manganese-containing SOD, endonuclease IV, glucose-6-phosphate dehydrogenase (G6PD), a fumarase, aconitase, ferredoxin reductase and *micF* RNA, which affects expression of a major outer membrane protein (see below). The *oxyR* gene controls the genes encoding the HPI catalase, glutathione reductase, alkyl hydroperoxide reductase and a protective DNA-binding protein, Dps.

V. THE *SOXRS* REGULON

OVERVIEW

The *soxRS* system coordinates the expression of diverse genes involved both in antioxidant defense and in bacterial resistance to multiple antibiotics. Regulation occurs in two stages: SoxR protein is triggered by an intracellular redox signal to activate transcription of the *soxS* gene; elevated expression of the SoxS transcription factor then switches on ten or more promoters that constitute the *soxRS* regulon.

"Adaptive" responses had suggested the existence of coordinate gene induction by superoxide in *E. coli* (see Section IV above), but demonstrated that a superoxide-stress regulon depended on the genetic studies outlined below. The connection to antibiotics followed only after *soxRS* regulatory mutants were studied and came as a complete surprise.

GENETIC IDENTIFICATION OF A TWO-STAGE REGULATORY SYSTEM

The *soxRS* locus was identified by two different approaches. Our strategy was to isolate mutants that constitutively expressed adaptive resistance to menadione.[37] About one-third of the resulting isolates had constitutive mutations at the *soxRS* locus at 92 minutes. Two-dimensional gel analysis showed that these isolates expressed high levels of at least nine proteins that are inducible by superoxide.[37] Mutants deleted for the *soxRS* locus (via Tn*10*-mediated events[38]) failed to induce

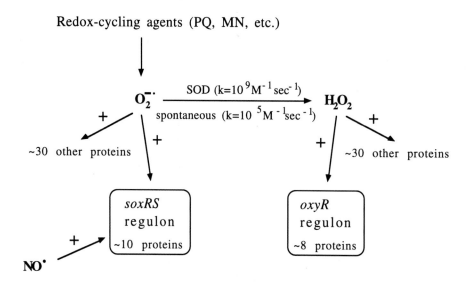

Fig. 21.1. Overall response of E. coli to oxidative stress. Redox-cycling agents such as paraquat (PQ) and menadione (MN) generate intracellular superoxide (O₂·⁻) and trigger the expression of ~40 proteins, of which ~10 are controlled by the soxRS regulon. Superoxide is converted spontaneously or by superoxide dismutase (SOD) to H₂O₂, activating the expression of ~40 additional proteins, of which ~8 are controlled by the oxyR regulon. The regulatory mechanisms governing the expression of the products of genes outside the soxRS and oxyR regulons are unknown.

only these same nine proteins in response to menadione or other redox-cycling agents and were abnormally sensitive to menadione. Thus, *soxRS* exerts overall positive regulation.

An alternative approach[39] employed a *lacZ* operon (transcriptional) fusion to the promoter of the gene encoding endonuclease IV (*nfo*; locus at 47 minutes), which was known to be inducible by PQ and other redox-cycling agents.[40] Strains were isolated with elevated *nfo::lacZ* expression in the absence of PQ. Again, the responsible mutations mapped to a single locus at 92 minutes: the location of *soxRS*. Like the mutations isolated via the menadione resistance phenotype, these mutations also caused constitutively increased expression of other genes besides *nfo* (see below).

Detailed analysis of the cloned *soxRS* locus quickly revealed that two genes, *soxR* and *soxS*, are required for activation of the regulated promoters. The genes are arranged head-to-head in the chromosome (Fig. 21.2). Transcription of the *soxS* gene begins in the intergenic region, but *soxR* transcription initiates within the *soxS* structural gene (Fig. 21.2).[41] The significance of this arrangement remains unknown, although the opposing polarities and overlapping mRNAs invite speculation about regulatory functions.

Several key observations crystallized ideas about the regulatory arrangement in this system. First, both the predicted SoxR and SoxS proteins are related to known DNA-binding transcription activators (Fig. 21.3). Second, the *soxS* gene, but not *soxR*, is strongly induced during activation of the response.[41] Third, controlled expression of SoxS alone activated several regulon genes regardless of the redox status of the cells, while expression of the SoxR protein alone was without effect.[42] These data prompted a hypothesis that gene activation in the *soxRS* system works in two stages:[43] SoxS was predicted to be the direct transcriptional activator of the various regulon genes, with SoxS synthesized under the transcriptional control of the SoxR protein, whose activity in turn would be post-translationally triggered by intracellular signal(s) of oxidative stress (Fig. 21.2).

All the essential elements of this model have been borne out experimentally. Independent studies employed *soxS'::lacZ* transcriptional fusions to show that the *soxS* gene is indeed under the control of *soxR*.[44,45]

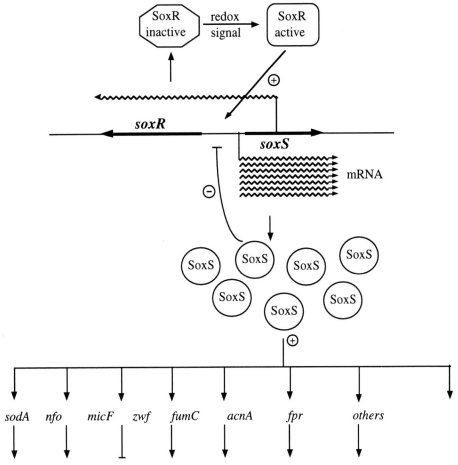

Fig. 21.2. The soxRS regulon. Existing SoxR protein is converted to an active form by a redox stress signal produced by superoxide-generating agents or by nitric oxide. Both inactive and activated SoxR bind the soxS promoter, but only the latter strongly stimulates transcription (up to 100-fold). The increased concentration of SoxS protein activates the target promoters of regulon genes; SoxS protein also represses transcription of its own gene. The soxR and soxS transcripts are shown as wavy lines. "+" symbols and arrowheads indicate positive control; "-" and "⊥" symbols indicate negative regulation. Endo IV, DNA endonuclease IV.

Purified SoxS protein binds tightly and specifically to sites just upstream of the -35 sites of at least four soxRS-regulated genes and recruits RNA polymerase to these promoters in vitro.[46] The stabilization of RNA polymerase binding by soxS at different promoters in vitro paralleled the degree of induction seen for those promoters in vivo.[46] Later work using a MalE-SoxS fusion protein confirmed the specific DNA binding and demonstrated transcriptional activation by the fusion protein in vitro.[47] Thus SoxS seems sufficient to activate the regulon genes.

SoxS protein also binds in vitro to DNA containing its own promoter, although with somewhat lower affinity than found for SoxS binding the micF, sodA, zwf, or nfo promoters. This binding evidently mediates negative autoregulation by the protein. However, this autoregulation does not affect the rate at which soxS transcription was shut down after the removal of PQ.[48] Autoregulation by SoxS could either provide a threshold for triggering the regulon genes or prevent the toxic accumulation of excessive SoxS protein.[46,49]

SoxR: An FeS-Containing Transcription Factor

Initially, the amino acid sequence predicted for SoxR revealed only one clue to its mode of action: homology to the MerR family of one-component transcription activators.[42] SoxR shares with MerR a strongly predicted helix-turn-helix motif near the N-terminus, and a cluster of cysteine residues located near the C-terminus (Fig. 21.3). MerR is converted to a powerful transcription activator by binding Hg^{2+} in a trivalent complex, with cysteines contributed by each subunit of the

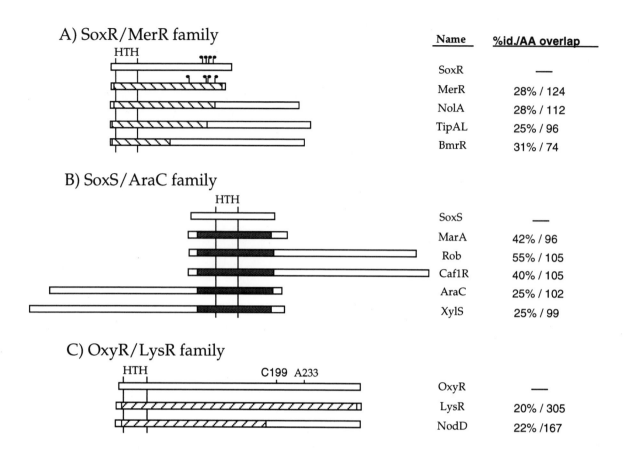

A) SoxR/MerR family

B) SoxS/AraC family

C) OxyR/LysR family

Name	%id./AA overlap
SoxR	—
MerR	28% / 124
NolA	28% / 112
TipAL	25% / 96
BmrR	31% / 74
SoxS	—
MarA	42% / 96
Rob	55% / 105
Caf1R	40% / 105
AraC	25% / 102
XylS	25% / 99
OxyR	—
LysR	20% / 305
NodD	22% /167

Fig. 21.3. Protein families of oxidative stress regulators. The cross-hatched or shaded area in each group indicates the region of homology to the top member of each, with the length of the overlap shown in the column at the right. HTH, predicted helix-turn-helix; %id., percent identity at the amino acid level; AA, amino acid. The HTH motifs for individual proteins are predicted with varying degrees of likelihood; in several cases, the prediction falls below the threshold suggested by Brennan and Matthews.[92] (A) The SoxR/MerR family. The vertical lines and dots above SoxR and MerR indicate the positions of cysteine residues in these proteins. (B) The SoxS/AraC family. (C) OxyR/LysR family. "C199" shows the position of the cysteine whose substitution prevents activation of OxyR; "A233" indicates the alanine that is converted to a valine residue by an oxyR-constitutive mutation. For proteins not mentioned in the text, reports of the individual gene sequences or leading references are as follows: NolA,[93] TipAL,[94] BmrR,[95] Caf1R,[96] AraC and XylS,[97] LysR[98] and NodD.[99]

dimeric protein.[50] Although there is no cross-activation of SoxR by Hg^{2+}, or of MerR by PQ (unpublished data), it was tempting to think that SoxR might employ its cysteine cluster in detecting the signal of superoxide stress. This does seem to be the case.

Purified SoxR, isolated from cells overexpressing the gene product as ~5% of the soluble protein, contains stable iron-sulfur (FeS) centers.[51] Although treatment of the bacteria with PQ prior to extraction has no effect on the Fe content of SoxR, the metal is stripped off if purification is carried out in buffers containing β-mercaptoethanol. Apo-SoxR and the Fe-containing protein (Fe-SoxR) bind with equal affinity to a site encompassing the -35 and -10 elements of the *soxS* promoter. Apo-SoxR has no detectable effect on the subsequent binding of *E. coli* RNA polymerase, and Fe-SoxR stabilizes the polymerase interaction with the *soxS* promoter only 2-fold. However, only Fe-SoxR stimulates in vitro transcription from the *soxS* promoter up to 100-fold, while apo-SoxR was without a significant effect (Fig. 21.4).[51] Thus, in a clear parallel to MerR, only metal-containing SoxR is a powerful transcription activator, while in both systems the apo-protein still binds tightly to the respective target promoter. The stimulatory effect of SoxR is thus downstream of the binding steps, perhaps by causing structural deformation of the promoter DNA, as suggested for MerR.[52a]

Both Fe-SoxR and apo-SoxR are dimers even in dilute solution.[52b] The stoichiometry of 2 Fe and 2 labile sulfides per subunit indicates the presence of either a single Fe$_4$S$_4$ cluster (between the subunits) or two Fe$_2$S$_2$ centers.[52b] Recent physical analysis using electron paramagnetic resonance spectroscopy supports a structure with two binuclear (Fe$_2$S$_2$) centers per (SoxR)$_2$ dimer,[52b,52c] although it remains unknown whether these centers are positioned between the subunits or as a separate entity within each monomer.

NATURE OF THE ACTIVATING SIGNAL FOR SoxR

The *soxRS* system responds to specific signals that are generated only by certain types of oxidative stress. Studies with a *soxS'::lacZ* operon fusion confirmed powerful *soxR*-dependent activation of *soxS* transcription by structurally unrelated redox-cycling agents.[45,25] Conversely, agents that do not generate a flux of superoxide (e.g., H_2O_2, transient heat shocks, UV light, pulses of gamma rays, etc.) failed to induce the system.[45]

Although redox-cycling agents probably also deplete NADPH within the cell, other evidence indicates that superoxide is the likely inducing signal produced under these conditions. Firstly, induction of *soxS* by redox-cycling agents requires oxygen,[45] just as does the generation of superoxide.[5] Secondly, the steady-state level of intracellular superoxide can be increased by eliminating the *E. coli* SOD enzymes (encoded by the *sodA* and *sodB* genes).[53] Under these circumstances, the concentration of superoxide increases >1,000-fold,[26] and this switches on *soxR*-dependent *soxS* expression.[45] Thirdly, superoxide is generated within *E. coli* by expression of *Vibrio harveyi* luciferase, a flavin-containing oxidoreductase, and this metabolic stress activates SoxR.[54]

One important agent unconnected to superoxide triggers SoxR protein to activate the *soxS* gene: nitric oxide (NO·). This free radical gas, which is generated by immune cells as part of their cytotoxic weaponry,[3] permeates membranes, unlike the anion superoxide. Indeed, enzymatic generation of superoxide in the medium *outside E. coli* does not activate *soxRS*.[55] In contrast, pure NO· gas activates *soxS* transcription in a time-, dose- and *soxR*-dependent fashion.[55,56] Induction of *soxS* also occurs over several hours in *E. coli* after phagocytosis by murine macrophages, and the induction is blocked by treatment with an arginine analog that inhibits NO· generation.[56]

The molecular mechanism(s) by which distinct species such as superoxide and NO· can activate the SoxR protein is still unknown. One possibility is that the FeS centers of SoxR are always present, but are converted by direct reaction with these agents from a reduced to an oxidized state (which we know to be transcriptionally active[51]). However, it appears that reduced Fe-SoxR in vitro is just as active in transcription as the oxidized form (E.H. et al, unpublished data). This observation suggests that the "resting" state of SoxR in normal cells could be the apoprotein, but addressing this possibility is a difficult technical challenge.

Perhaps superoxide and NO· generate a common intermediate that activates SoxR. The detailed chemistry of these two agents is rather different, but both agents may react with FeS-containing proteins.[11,3] If apo-SoxR is the inactive form, intracellular superoxide or NO· would somehow mediate the formation of the SoxR Fe_2S_2 centers, while simultaneously destroying the Fe_4S_4 clusters of other proteins, and perhaps other metal centers. The possible biochemistry of such a metal redistribution is virtually unexplored.

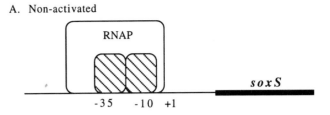

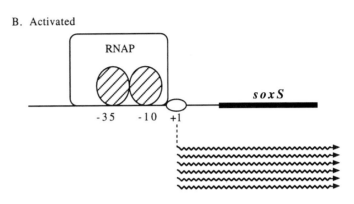

Fig. 21.4. Stimulation of soxS transcription by activated SoxR. RNAP, E. coli RNA polymerase; -35 and -10 indicate the conserved σ^{70} promoter elements, which are spaced 19 base pairs apart in the soxS promoter.[51] The transcription start site is indicated by +1, the soxS transcripts by wavy lines. Nonactivated (A; "resting") and activated (B) SoxR bind the soxS promoter with equal affinity. The likely candidate for the nonactivated state is apo-SoxR, and for the activated state Fe-SoxR. In the presence of RNAP, activated SoxR causes a structural distortion of the promoter DNA, with strong melting of the helix around +1 (shown as a bubble near +1).[52b]

FUNCTIONS OF SOXRS-REGULATED GENES

The multifunctional defense provided by the *soxRS* regulon against oxidative damage and other threats to the cell is remarkable. Several elements of the system function in a seemingly obvious antioxidant role. Thus, increasing the expression of Mn-containing SOD (Mn-SOD), the *sodA* gene product, probably helps keep the steady-state level of superoxide acceptably low. A recent report indicating that Mn-SOD protein in vivo is associated with DNA suggests that protection by SOD may be more effective when it occurs close to the target for damage.[57] Depletion of NADPH by redox cycling and antioxidant reactions (e.g., GSH reductase or alkyl hydroperoxide reductase) would be counteracted by G6PD, encoded by *zwf*.[58] The *nfo*-encoded DNA endonuclease IV acts on a variety of DNA damages that include oxidative lesions.[59-61] Notably, all of the above functions help avert the cytotoxicity to *E. coli* of activated murine macrophages generating nitric oxide,[55,56] which supports their key roles in defending against oxidative damages.

Other *soxRS*-inducible enzymes participate in central metabolism. Thus, the "stable" fumarase encoded by the *fumC* gene is induced to aid or substitute for the *fumAB* gene products, which may be oxidant-sensitive in vivo.[62] Aconitase, a Fe_4S_4 enzyme that is sensitive to superoxide,[11] is also encoded by a *soxRS*-regulated gene.[63] Maintaining the tricarboxyclic acid cycle by inducing aconitase and fumarase helps viability even under oxidative stress. Alternatively, aconitase could function to sequester iron and so prevent oxidative damage to other cellular targets. An NADPH:ferredoxin(oxidized) oxidoreductase (the *fpr* or *mvrA* gene product) under *soxRS* control[64] catalyzes the reduction of PQ (and thus redox cycling), but could help stabilize FeS proteins by maintaining them in the reduced state. Such an enzyme could also influence SoxR by acting on its FeS centers.

Other *soxRS* regulon functions are less clearly related to oxidative stress. These include the *micF*-encoded antisense RNA that down-regulates synthesis of the OmpF outer membrane porin,[65] and an enzyme (probably encoded by the *rimK* gene[66]) that modifies ribosomal protein S6.[37] Both of these functions have been associated with resistance to multiple antibiotics (see Section VII), but not usually with oxidative stress. Perhaps the elimination of OmpF from the outer membrane diminishes the uptake of some environmental redox agents, just as it does an array of classical antibiotics. The possibility that S6 modification is an antioxidant defense has not yet been tested. Despite this limitation and the existence of at least a few unidentified proteins regulated by *soxRS*,[37] the *soxRS* regulon is truly global: not only are multiple promoters of the regulon scattered around the *E. coli* genome, but the functions activated by this system impinge on a very broad array of cellular functions.

VI. THE *OXYR* REGULON

OVERVIEW

Exposure of *E. coli* to 5-200 μM H_2O_2 causes dramatic changes in gene expression: the synthesis of 30-40 proteins is enhanced after the treatment.[33,34] At least 8 of these inducible proteins are controlled under the *oxyR* regulon, including the hydroperoxide scavengers HPI catalase and alkyl hydroperoxide reductase. Other antioxidant activities controlled by *oxyR* are glutathione reductase and Dps protein, which binds and protects DNA from damage.[67] OxyR orchestrates this response by undergoing a post-translational redox activation, which allows the protein to bind and activate target promoters.

IDENTIFICATION OF THE OXYR REGULON

The protein inductions provoked by H_2O_2 confer an adaptive resistance to H_2O_2 or ionizing radiation.[29] In order to identify regulators of this inducible protection, *E. coli* and *Salmonella typhimurium* mutants were isolated with inherently increased resistance to H_2O_2 and organic hydroperoxides.[33] Two-dimensional PAGE showed that eight (*E. coli*[34]) or nine (*S. typhimurium*[33]) H_2O_2-inducible proteins were constitutively expressed in these strains, including the peroxide scavengers and glutathione reductase. The responsible mutations mapped to a single genetic locus in both species, called *oxyR* (89 minutes). Deletions of the *oxyR* locus generated H_2O_2-sensitive strains that could not induce the regulon proteins in response to H_2O_2. Such strains could be complemented by an *E. coli* episome carrying the *oxyR* gene. Therefore, *oxyR* acts as an overall positive regulator, and the *oxyR* constitutive mutations bypass the requirement for H_2O_2-mediated oxidative stress to activate the stress-response genes. Activation of the *oxyR* regulon occurs at the transcriptional level.[68]

OXYR PROTEIN: A REDOX-SENSOR AND TRANSCRIPTIONAL REGULATOR

The OxyR protein is a 34 kDa protein that belongs to the LysR family of transcriptional activators (Fig. 21.3). As found for other members of this family, OxyR acts as a repressor of its own gene under both inducing and noninducing conditions,[69] and also as an activator under inducing conditions from an overlapping promoter on the complementary strand (Fig. 21.5). The gene was cloned independently as a regulator of the phage Mu *mom* gene (see below).

The amount of *oxyR* mRNA[69] and the levels of OxyR protein[70] are not altered after exposure of *E. coli* to H_2O_2. Thus, post-translational regulation by an H_2O_2-generated signal seems to activate pre-existing OxyR protein.

Purified OxyR activates in vitro transcription of the *katG* and *ahpCF* genes by *E. coli* RNA polymerase.[70] Surprisingly, the same level of transcriptional activity was detected for OxyR purified from either H_2O_2-treated or untreated cells. An artifactual activation of OxyR evidently occurs upon cell lysis and exposure of OxyR to oxygen. Storz et al (1990)[70] isolated transcriptionally inactive form of OxyR in vitro by adding high concentrations (100mM) of dithiothreitol (DTT) to the purification buffers; lower DTT concentrations were effective under anaerobic conditions. The process was reversible, since

Fig. 21.5. The oxyR regulon. Existing OxyR protein is activated by an H_2O_2-dependent signal, perhaps hydrogen peroxide itself. Only the activated form binds tightly to its target promoters and stimulates transcription (shown as "+"). Both nonactivated and activated OxyR bind and repress the oxyR promoter. The oxyR and oxyS transcripts are shown as wavy lines.

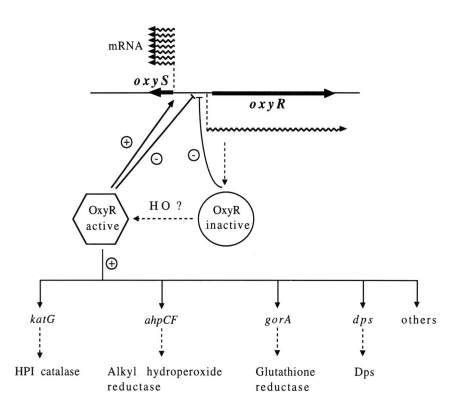

dialysis or dilution of the DTT from the OxyR samples regenerated active OxyR. The authors thus hypothesized that OxyR is activated by an oxidation process in vitro.

The target OxyR of such redox reactions is not yet certain. The strongest candidate is one of the six cysteine residues of OxyR, at position 199 (Fig. 21.3). A noninducible OxyR mutant was engineered by mutating this cysteine to a serine residue.[70,71] The mutant protein in vitro behaves like DTT-treated wild-type OxyR: both are unable to activate the *katG* or *ahp* promoters.[71] The reversibility of the oxidative activation of OxyR suggested that cysteine 199 might react to form sulfenic or sulfinic acid, two reversibly oxidized derivatives.[72] Another residue that may be connected to the activation process is alanine 233 (Fig. 21.3); a mutation that converted this residue to valine generates constitutively active OxyR.[70,71] It has not been determined whether this change increases the oxidant sensitivity of cysteine 199, or causes a conformational change in the protein that mimics the oxidized form. These alternatives might be distinguished by testing the activity of this constitutive mutant protein under anaerobic conditions.

The understanding of OxyR's interaction with promoters it regulates has recently been revised. OxyR behaves as a transcriptional autorepressor under both inducing and noninducing conditions (Fig. 21.5), but activates the different regulon promoters only after H_2O_2 treatment (Fig. 21.5). Initially, transcriptional activation was thought to be controlled at a step after DNA binding, because both the DTT-treated and the oxidized protein formed a complex with the *katG* promoter in band-shift experiments.[70] It is now thought that promoter binding by the DTT-treated OxyR was actually due to reoxidation occurring inside the gels. In footprinting procedures, the cleavage of DNA is performed before loading the samples on gels, which reduces the risk of protein oxidation. Using the latter approach, Toledano et al showed that DTT-treated OxyR and the noninducible cysteine-199-to-serine mutant protein bound only to the *oxyR* promoter (where OxyR acts as a repressor), but not to the *katG* or *ahp* promoters.[71]

The footprinting patterns seen for activated OxyR are rather unusual; the contact extends over 45 bp, and strong sequence similarity among the different binding sites is difficult to find.[73] By selecting OxyR-binding sequences from pools of random oligonucleotides, Toledano et al defined a consensus motif that shows 2-fold symmetry, and in which four sets of four nucleotides each are spaced by three sets of seven random nucleotides.[71] This suggests that an activated OxyR tetramer contacts four consecutive helical turns of DNA in the -35 to -80 region[71] to stimulate RNA polymerase through contact with the α subunit.[74] Nonactivated OxyR would have a different quaternary structure that does not allow stable contact to the DNA of activatable promoters. At the repressible *oxyR* promoter, both the active and the inactive forms bind and repress.

OxyR-Regulated Genes

Although Δ*oxyR* strains can still induce 20-30 *oxyR*-independent proteins after H_2O_2 treatment, the antioxidant defenses of the cells are clearly impaired: these mutants are hyper-sensitive to H_2O_2, hydroperoxides and redox-cycling agents and have a high spontaneous mutation rate in the absence of exogenous H_2O_2.[75,76]

Four enzyme activities were reported to be increased in *oxyR* constitutive mutants: catalase, glutathione reductase, NADPH-dependent hydroperoxide reductase, and Mn-SOD.[33] The report of increased Mn-SOD activity was later shown to be incorrect,[77] but could have been an artifact of protection of SOD by the elevated catalase activity in extracts of *oxyR*-activated cells. The Mn-SOD gene (*sodA*) is instead controlled by *soxRS* (see section IV).

The likely defensive roles of catalase, alkyl hydroperoxide reductase and glutathione reductase are fairly evident, as these activities either eliminate dangerous oxidants or regenerate a critical antioxidant (glutathione). Consistent with this interpretation, the overexpression of HPI catalase, HPII catalase or alkyl hydroperoxide reductase suppressed the H_2O_2 sensitivity of Δ*oxyR* strains.[76]

The *dps* gene is also regulated by OxyR.[67] Dps is a nonspecific DNA-binding protein, originally found to be induced in stationary phase, where it exerts a protective role against hydrogen peroxide damage (see chapter 27 in this volume). Apparently, the *dps* promoter can be recognized by OxyR and σ^{70} subunit of the *E. coli* RNA polymerase during exponential growth and by IHF and σ^S during stationary phase, providing a defense against H_2O_2 all along the growth curve.

Another target of OxyR is the phage Mu *mom* gene, essential for phage development.[78] OxyR-dependent transcription of *mom* is controlled by DNA methylation, but the mechanism of OxyR discrimination between methylation states has not been described. The involvement of OxyR appears to be a case of subversion of a host protein for viral purposes, and seems unrelated to redox regulation.

OxyR also increases transcription from the *oxyS* promoter to generate a small untranslated RNA.[70,71] The *oxyS* gene is arranged head-to-head with *oxyR*, with an overlapping promoter (Fig. 21.5). Recent data suggest that *oxyS* RNA may regulate the translation of several mRNAs (G. Storz, personal communication). The contribution of *oxyS* to antioxidant defenses remains to be determined.

DNA repair functions induced by OxyR have not been described, although the adaptive response to H_2O_2 was reported to include a DNA repair component.[29] Dps protein could play an unexpected repair role, as could an unknown regulon member. Alternatively, some of the 20-30 *oxyR*-independent, H_2O_2-inducible proteins could be involved in repair.

VII. CONTROL OF ANTIBIOTIC RESISTANCE GENES

As mentioned above, activation of the *soxRS* regulon in *E. coli* also switches on genes that mediate cellular resistance to antibiotics. This connection, which was quite unexpected, emerged during genetic manipulation (e.g., transduction of antibiotic resistance markers) of *soxR*-constitutive strains, which consistently produced confluent growth on antibiotic selection plates. Direct measurement of the "minimum inhibitory concentration," first for chloramphenicol and then for other antibiotics, showed that constitutive expression of *soxRS* conferred 3- to 10-fold increased resistance.[37] The resistance applied to a broad array of structurally unrelated antibiotics with different cellular targets: chloramphenicol, tetracycline, quinolones (e.g., nalidixic acid or ciprofloxacin) and β-lactams (e.g., ampicillin).[37,65]

Among the mutagenized isolates that yielded our original set of *soxR*-constitutive mutants,[37] we obtained one menadione-resistant strain whose mutation (which we named *soxQ1*) mapped to 34 minutes,[79] far from the *soxRS* locus.[37,39] Other mutations that confer "multiple antibiotic resistance" in *E. coli* had previously been mapped to a locus called *mar*, also at 34 minutes.[80,81] The *soxQ1* and *mar*-constitutive mutations had other similarities—both increased the expression of Mn-SOD, G6PD, and four other *soxRS*-regulated proteins, including the activity that modifies ribosomal protein S6.[79] More recent work has added the *fumC*-encoded fumarase to this list and demonstrated that *soxQ1* and *mar1* are alleles of the same gene.[82] At least three *soxRS*-regulated genes (e.g., endonuclease IV) are not affected by the *mar* locus, and as many as seven *mar*-regulated proteins are evidently independent of *soxRS*.[79]

Both *soxRS* and *mar* confer antibiotic resistance in part via the *micF* gene. The *micF* transcript is an antisense RNA against the *ompF* message. But the *micF*-dependent down-regulation of OmpF contributes only partially to the overall antibiotic resistance provided by both systems.[81,65] Eliminating OmpF probably diminishes the uptake of antibiotics and other compounds, but may be accompanied by inducible efflux systems to produce a more effective resistance.[83] Active efflux of tetracyline[80] and chloramphenicol[84] has been associated with the *mar* system. Candidates to encode efflux systems controlled by *soxRS* or *mar* are the *emr* genes[85] or *acr* genes.[83]

A physical basis for the regulatory overlap between *soxRS* and *mar* has emerged from a molecular analysis of the *mar* locus. An operon of three genes comprises the site of *mar* mutations that confer antibiotic resistance: *marR*, *marA* and *marB*.[86] The predicted 14 kDa MarA protein is a clear homolog of SoxS (13 kDa), with the two polypeptides sharing 42% amino acid identity (Fig. 21.3).[86] Transcription of the *marRAB* operon is induced by the antibiotics tetracyline or chloramphenicol,[87] or by salicylate,[88] which would provide the means to increase MarA activity in the cell. Indeed, increased expression of MarA alone activates multiple antibiotic resistance in *E. coli*, an approach that was used to clone the *marA* locus.[89]

The *mar* operon is autoregulatory. Introduction of *marR* on a multicopy plasmid down-regulates both *marRAB* transcription and antibiotic resistance.[82] Also, *mar*-constitutive mutations were found either in a putative regulatory region just upstream of the *marRAB* promoter or in the *marR* gene (*marR1*).[86] The *soxQ1* mutation also resides in *marR*, as does a deletion in an allele called *cfxB*; both mutations increase *marRAB* transcription, and are repressible by the cloned *marR* gene.[82] Thus, MarR seems to function as a repressor of *marRAB* transcription, with derepression occurring by an unknown mechanism in cells exposed to tetracycline, chloramphenicol or salicylate. MarR displays limited similarity to the regulatory protein MprA (R. Ariza, personal communication). Recent preliminary data indicate that chloramphenicol and tetracycline may interact directly with MarR to cause derepression (S.B. Levy, personal communication). Increased levels of MarA protein are then hypothesized to activate the *mar* regulon promoters, in a reasonable parallel to the control of SoxS activity. The function of the putative MarB protein is unknown, and it appears unrelated to known protein sequences.[86]

Other proteins closely related to SoxS and MarA have also been reported recently (see Fig. 21.3). Rob protein, which binds the right arm of the *E. coli* replication origin, *oriC*, has an N-terminal segment of ~100 residues that falls clearly within the SoxS/MarA family (51% identity to SoxS).[90] This arrangement contrasts with the XylS/AraC family, in which the SoxS/MarA homology is in the C-terminus, but other proteins with Rob-like structures have now been found (Fig. 21.3).

The close structural relationship among SoxS, MarA and the Rob N-terminus is reflected by the ability of elevated levels of Rob to activate some *soxRS*- or *marRAB*-regulated genes.[49] These include *inaA*,[91] *sodA* and *fumC*, but not the endonuclease IV or G6PD genes. Rob overexpression also activates multiple antibiotic resistance that depends partially on the *micF* gene. Similar antibiotic resistance, as well as

72. Claiborne A, Miller H, Parsonage D et al. Protein-sulfenic acid stabilization and function in enzyme catalysis and gene regulation. FASEB J 1993; 7:1483-1490.

73. Tartaglia LA, Storz G, Ames BN. Identification and molecular analysis of *oxyR*-regulated promoters important for the bacterial adaptation to oxidative stress. J Mol Biol 1989; 210:709-719.

74. Tao K, Fujita N, Ishihama A. Involvement of the RNA polymerase α subunit *C*-terminal region in co-operative interaction and transcriptional activation with OxyR protein. Mol Microbiol 1993; 7:859-864.

75. Storz G, Christman MF, Sies H et al. Spontaneous mutagenesis and oxidative damage to DNA in *Salmonella typhimurium*. Proc Natl Acad Sci USA 1987; 84:8917-8921.

76. Greenberg JT, Demple B. Overproduction of peroxide-scavenging enzymes in *Escherichia coli* suppresses spontaneous mutagenesis and sensitivity to redox-cycling agents in *oxyR*-mutants. EMBO J 1988; 7:2611-2617.

77. Bowen SW, Hassan HM. Induction of the manganese-containing superoxide dismutase in *Escherichia coli* is independent of the oxidative stress (*oxyR*-controlled) regulon. J Biol Chem 1988; 263:14808-14811.

78. Bölker M, Kahmann R. The *Escherichia coli* regulatory protein OxyR discriminates between methylated and unmethylated states of the phage Mu *mom* promoter. EMBO J 1989; 8:2403-2410.

79. Greenberg JT, Chou JH, Monach PA et al. Activation of oxidative stress genes by mutations at the *soxQ-cfxB-marA* locus of *Escherichia coli*. J Bacteriol 1991; 173:4433-4439.

80. George AM, Levy SB. Amplifiable resistance to tetracycline, chloramphenicol, and other antibiotics in *Escherichia coli*: involvement of a non-plasmid-determined efflux of tetracycline. J Bacteriol 1983; 155:531-540.

81. Cohen SP, McMurry LM, Levy SB. *marA* locus causes decreased expression of OmpF porin in multiple-antibiotic-resistant (*mar*) mutants of *Escherichia coli*. J Bacteriol 1988; 170:5416-5422.

82. Ariza RR, Cohen SP, Bachhawat N et al. Repressor mutations in the *marRAB* operon that activate oxidative stress genes and multiple antibiotic-resistance in *Escherichia coli*. J Bacteriol 1994; 176:143-148.

83. Nikaido H. Prevention of drug access to bacterial targets—permeability barriers and active efflux. Science 1994; 264:382-388.

84. McMurry LM, George AM, Levy SB. Active efflux of chloramphenicol in susceptible *Escherichia coli* strains and in multiple-antibiotic-resistant (*mar*) mutants. Antimicrob Agents Chemother 1994; 38:542-546.

85. Lomovskaya O, Lewis K. *emr*, an *Escherichia coli* locus for multidrug resistance. Proc Natl Acad Sci USA 1992; 89:8938-8942.

86. Cohen SP, Hächler H, Levy SB. Genetic and functional analysis of the multiple antibiotic-resistance (*mar*) locus in *Escherichia coli*. J Bacteriol 1993; 175:1484-1492.

87. Hächler H, Cohen SP, Levy SB. *marA*, a regulated locus which controls expression of chromosomal multiple antibiotic resistance in *Escherichia coli*. J Bacteriol 1991; 173:5532-5538.

88. Rosner JL, Slonczewski JL. Dual regulation of *inaA* by the multiple antibiotic resistance (*mar*) and superoxide (*soxRS*) stress response systems of *Escherichia coli*. J Bacteriol 1994; 176:6262-6269.

89. Gambino L, Gracheck SJ, Miller PF. Overexpression of the MarA positive regulator is sufficient to confer multiple antibiotic-resistance in *Escherichia coli*. J Bacteriol 1993; 175:2888-2894.

90. Skarstad K, Thöny B, Hwang DS et al. A novel binding-protein of the origin of the *Escherichia coli* chromosome. J Biol Chem 1993; 268: 5365-5370.

91. White S, Tuttle FE, Blankenhorn D et al. pH-dependence and gene structure of *inaA* in *Escherichia coli*. J Bacteriol 1992; 174:1537-1543.

92. Brennan RG, Matthews BW. The helix-turn-helix DNA binding motif. J Biol Chem 1989; 264:1903-1906.

93. Sadowsky MJ, Cregan PB, Gottfert M, Sharma A et al. The *Bradyrhizobium japonicum nolA* gene and its involvement in the gene type-specific nodulation of soybeans. Proc Natl Acad Sci USA 1991; 88:637-641.

94. Holmes DJ, Caso JL, Thompson CJ. Autogenous transcriptional activation of a thiostrepton-induced gene in *Streptomyces lividans.* EMBO J 1993; 12:3183-3191.

95. Ahmed M, Borsch CM, Taylor SS et al. A protein that activates expression of a multidrug efflux transporter upon binding the transporter substrates. J Biol Chem 1994; 269:28506-28513.

96. Karlyshev AV, Galyov EE, Abramov VM et al. Caf1R gene and its role in the regulation of capsule formation of *Y. pestis.* FEBS Lett 1992; 305:37-40.

97. Gallegos MT, Michan C, Ramos JL. The XylS/AraC family of regulators. Nucleic Acids Res 1993; 21:807-810.

98. Stragier P, Patte JC. Regulation of diaminopimelate decarboxylase synthesis in *Escherichia coli.* III. Nucleotide sequence and regulation of the *lysR* gene. J Mol Biol 1983; 168:333-350.

99. Egelhoff TT, Fisher RF, Jacobs TW et al. Nucleotide sequence of *Rhizobium meliloti* 1021 nodulation genes: *nodD* is read divergently from *nodABC.* DNA 1985; 4:241-248.

CHAPTER 22

THE SOS REGULATORY SYSTEM

John W. Little

A. INTRODUCTION AND CURRENT REGULATORY MODEL

*E*scherichia coli is exposed to a wide variety of stressful environments. Among these are oxidative stress, heat shock, nutrient deprivation, and the condition discussed here, treatments that damage DNA. In each case, *E. coli* has developed responses to these conditions which allow it to counteract the effects of stress. Most of these stress responses involve the increased expression of sets of gene products which, in one way or another, allow the cell to maximize its chances of survival. These responses are also termed "global responses," because they involve a set of functions that are diverse at the molecular and mechanistic level, and because these functions alter the physiology of the cell during the response. Many of these stress responses are discussed elsewhere in this volume.[1-3] In a number of these responses, such as the response to anaerobiosis[3] or oxidative damage,[2] a transcriptional activator controls expression of a group of genes. Others, such as the responses to heat shock[1] or nitrogen limitation,[4] involve the action of alternative σ factors. The SOS response, the subject of this chapter, is controlled instead by the inactivation of a repressor.

The SOS response was among the earliest of the global responses to be recognized and characterized in detail.[5-7] This response is triggered by a diverse set of treatments that damage DNA or inhibit DNA replication. It is controlled by the SOS regulatory network, a system that works to control the expression of about twenty genes at the level of transcription initiation. The SOS regulatory system involves the interplay of two regulatory proteins (Fig. 22.1): the LexA repressor, which represses the SOS genes during normal cell growth (Fig. 22.1, panel A), and the RecA protein, which is present in a quiescent form during this growth state but activated promptly upon inducing treatments to a form that can mediate cleavage of LexA (panel B). Activated RecA is termed a "coprotease", because RecA stimulates a latent self-cleavage activity of LexA, rather than acting itself as a protease. Cleavage inactivates LexA, leading to derepression of the SOS regulon (panel C). Expression of diverse SOS functions then allows the cell to attempt to counteract the effects of DNA damage, both by repair processes and by other means whose role is less obvious. If the damage is not too severe, this effort is successful. Since activation of RecA is

Regulation of Gene Expression in Escherichia coli, edited by E. C. C. Lin and A. Simon Lynch. © 1996 R.G. Landes Company.

reversible, repair of the damage leads to the loss of an inducing signal, and LexA becomes progressively more stable, leading to repression of the SOS genes and re-establishment of the normal growth state (panel D). If the cell is a lysogen of λ or another lambdoid phage, induction of the SOS system can lead to induction of the prophage, leading to lytic growth and cell lysis.[8] This process is termed prophage induction, and it is triggered by RecA-mediated cleavage of the phage repressor in a reaction closely similar to that which cleaves LexA.

The state of the SOS regulatory system is therefore controlled by the level of RecA coprotease activity.[9] When this activity is low or absent, the cells are in the repressed or OFF state (panel A); at high coprotease levels, the cells are in the induced or ON state (panel C); and at intermediate levels of coprotease activity, the cells are in a "subinduced" state (not depicted) in which the SOS genes are partially derepressed. In addition, changes in the levels of coprotease activity control the transitions between the various states of this regulatory system.

Fig. 22.1. Model for the SOS regulatory system. (A) State of the system in exponentially growing cells. (B) Transition to the induced state. (C) Induced state. (D) Transition to the normal growth state. The induced state cannot be perpetuated indefinitely in wild-type E. coli strains, because cell division is blocked (see text). If the cell contains a λ prophage, a prolonged stay in the induced state results in prophage induction. Modified from Little JW et al, Cell 1982; 29:11-22.

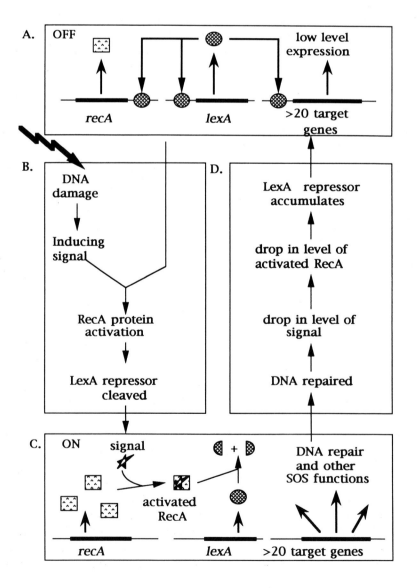

B. DEVELOPMENT OF THE SOS MODEL

The path by which our understanding of this system unfolded is fascinating and reveals how scientists with widely differing strengths and viewpoints take part in scientific progress. A number of historical accounts have documented various stages of this path.[5,7,10-14] A brief recounting of the history is followed by a review of several issues that complicated the course of discovery.

Investigators with highly diverse viewpoints have contributed to our present understanding of the SOS system. Some workers are able to take a large body of phenomenological data, from many different laboratories and often conflicting, and synthesize a general model without needing to understand the molecular details or to reconcile apparently contradictory data.[6,15] Others are skilled at genetic analysis of a complex process such as radiation sensitivity, and their work generally leads to an identification of the important components of a regulatory system that interact to produce the overall phenotype of the cell.[16,17] Finally, many scientists are most adept at testing specific, concrete models at the molecular level. Many workers of this persuasion (including the author) entered the SOS field in the mid-1970s when such models were developed.

1. HISTORY OF THE SOS MODEL

In the SOS field, Evelyn Witkin and Miroslav Radman were the leaders in the synthesis that led to formulations of early SOS models. Witkin offered in 1967 the first proposal that *E. coli* has a coordinated response to DNA damage.[15] She noted a parallel between induction of λ prophage and the radiation sensitivity of *E. coli* strain B. (This strain is sensitive because it cannot divide after irradiation; we now know that strain B is a *lon⁻* strain, and that an SOS gene codes for a cell division inhibitor that is normally degraded by the Lon protease.) Witkin discerned the essential nature of prophage induction, noting that "the primary event in prophage induction is the derepression of an operon. The extraordinary feature of phage induction is the sensitivity of the repressor to relatively slight impairment of the integrity of the DNA. It would be premature, however, to assume that such repressors are found only in association with episomes [such as λ prophage] and never with normal bacterial operons. Repressors that are inactivated when DNA replication stops could, theoretically, play an important part in regulation of cellular activities." She went on to propose that *E. coli* has a repressor that, like λ repressor, is inactivated by DNA-damaging treatments; in strain B, at least, repressor inactivation leads to expression of a cell division inhibitor. At the time, this was an extraordinarily bold hypothesis. Except for λ repressor, which had not yet been characterized biochemically, the molecular players were not yet known; the primary evidence was physiological and from today's vantage point seems vague and indirect. Like her mentor Barbara McClintock,[18] Witkin clearly had a feeling for the organism.

Evidence for the identity of the players in regulation began appearing in 1967 with the discovery that *recA⁻* strains do not support λ induction.[19,20] The pleiotropy of these strains became ever more evident. Another line of evidence linking prophage induction and filamentation was provided by the properties of a thermosensitive mutant called *tif*, later found to carry a *recA* mutation; exposure of this strain to high temperature leads spontaneously to induction of λ lysogens and, in nonlysogenic cells, to filamentation (hence, thermal induction and filamentation).[21] The first involvement of *lexA* came with studies of Defais et al in which two SOS phenomena, Weigle mutagenesis and Weigle reactivation, were shown to be defective in a *lexA* (Ind⁻) strain.[22] Concurrently, it was shown that the *lexA* (Ind⁻) alleles in use at that time were dominant,[17] always a useful clue to the geneticist because it indicates that the mutant allele has an altered function rather than a simple loss of function. Building on these and other clues, Radman developed a more explicit version of the SOS model,[6] although the molecular details by which regulation was proposed to operate were still imprecise.

The development of the SOS model entered a new phase in 1975 with the publication of two landmark papers. Both of these papers provided windows into the early events of the SOS response, a decisive advantage over looking at the distant end-points previously available for study. The first was the demonstration by Roberts and Roberts[23] that λ repressor is cleaved following DNA damage. This study offered a specific mechanism for inactivation of λ repressor during prophage induction. Importantly, the λ *cI ind⁻* mutant, which blocks prophage induction and is dominant to wild type, was resistant to cleavage. The parallel with the dominance of existing *lexA* alleles played an critical role in recognition of the similarities between LexA and λ repressors. Second, Gudas and Pardee[24] studied the expression of an anonymous but abundant protein, termed protein X (now known to be RecA), which was controlled by the SOS system and whose rate of synthesis could readily be analyzed by SDS gel electrophoresis. Again, synthesis of an SOS-regulated protein is an early event in the overall process. Gudas and Pardee examined the effects of mutations in *recA* and *lexA* on expression of protein X, and developed a model based on these studies. This model made explicit and testable predictions about the molecules involved in the regulatory circuitry. In particular, it was proposed that LexA is a repressor, and that DNA damage leads to the production of an anti-repressor, perhaps the RecA protein itself, acting in an unspecified way to antagonize the action of LexA. The studies of Roberts and Roberts with λ repressor, together with the parallel between dominant noninducible mutants in the two repressors, suggested the possibility that LexA was also inactivated by specific cleavage. The stage was set for detailed biochemical and genetic analysis of the regulatory network.

The following year, McEntee identified RecA protein as a 38 kDa protein;[25] several groups soon showed that protein X is RecA protein itself.[26-28] In addition, studies with newly isolated knockout alleles of *lexA* [now termed *lexA* (Def)] showed that protein X is derepressed in such strains, implying that LexA acts to inhibit RecA expression. Derepression was seen even in strains with nonfunctional alleles of *recA* (caused by missense mutations, so that the full-length protein was still made), implying that RecA acts upstream of LexA in the operation of the regulatory circuitry. With these data, the model took shape[26-28] and detailed biochemical analysis could begin.

One surprise soon came when Roberts and colleagues purified the protein responsible for the specific cleavage of λ repressor.[29] This protein they found to be RecA itself, a surprise because RecA plays a central role in the process of genetic recombination. Later work has shown that RecA catalyzes a complex set of DNA strand transfer reactions.[30] Even at the time, it was certain from its involvement in recombination that RecA was a complicated protein. It was hard to understand how it could also be a specific protease; yet the reaction was observed with highly purified protein and, when reaction conditions were optimized, this protein catalyzed multiple cleavage events, so it clearly was acting as an enzyme. Cleavage required ATP, but the poorly hydrolizable analog ATPγS substituted well, indicating that ATP hydrolysis is not required. Later studies showed that cleavage also requires a polynucleotide cofactor, with single-stranded DNA being the most effective.[31] Current evidence indicates that RecA forms a helical filament on single-stranded DNA; ATP, especially ATPγS, stabilizes this filament in a "high-affinity" form that supports cleavage.[30] From the perspective of the SOS system, the requirement for single-stranded DNA was satisfying because it offered a mechanism by which the protein could be activated by inducing treatments (that is, single-stranded DNA was an ideal candidate for an inducing signal). These studies established that RecA acts as an anti-repressor to induce λ, and by inference the SOS system, by catalyzing repressor cleavage.

The role of LexA in SOS regulation was established soon after. We and others took advantage of emerging recombinant DNA technology to clone the *lexA* gene and overproduce the protein.[32-35] Purified protein was shown to be a repressor of several genes; among these was *lexA* itself, which concurrent studies[36,37] showed to be under SOS control. LexA bound in vitro to conserved sites in front of these genes, termed SOS boxes. LexA was also shown to undergo specific cleavage in vitro in a RecA-dependent reaction that closely resembled cleavage of λ repressor; the LexA3 (Ind⁻) mutant protein, encoded by a dominant allele conferring a noninducible phenotype, was shown to be resistant to cleavage. This resistance explained its dominance, since the mutant protein can continue to repress the SOS genes in a merodiploid even though the wild-type protein is inactivated by cleavage.

2. Factors Influencing Development of the SOS Model

Initial recognition and later analysis of this global regulatory system was hampered by several factors. First, most of the physiological manifestations of SOS induction are seen experimentally as distant endpoints of the processes underlying them. For instance, one widely studied SOS function, SOS mutagenesis, is generally only recognized a day after the inducing treatment by the occurrence of mutant colonies or plaques on a plate. Thus, it was difficult to identify proximal events in the induction process. Second, mutations in the two regulatory genes, *recA* and *lexA*, are of several types, and confer highly pleiotropic phenotypes on mutant strains. While this is a clue that these proteins affect multiple cellular functions, in practice it greatly complicated efforts to decipher the underlying mechanism. Third, the *recA* gene was discovered in 1965 by virtue of its defects in genetic recombination,[16] which made most workers at the time think of RecA as a recombination protein. Thus, the first clue that RecA also has a regulatory role, the startling discovery that *recA* mutants are defective in prophage induction,[19,20] was very difficult to relate at the mechanistic level to its role in recombination. In addition, RecA also plays direct mechanistic roles in several SOS functions, such as recombinational repair and mutagenesis.[38] Fourth, *lexA* is an essential gene in normal cells, since the product of an SOS target gene, *sulA*, prevents cell division; knockout *lexA* mutants, now termed *lexA* (Def), could only be isolated after strains lacking SulA became available.[39] In addition, the earliest *lexA* alleles studied, now termed *lexA* (Ind⁻), were not knockout mutations but altered LexA so that it cannot undergo cleavage; hence, such mutations blocked the SOS response.

Finally, it is noteworthy that the combination of genetics and biochemistry played a decisive role at the latter stages of analyzing the SOS system. This combination is powerful in many contexts, because mutants can provide evidence for the existence and nature of relationships among molecules, while biochemistry makes explicit what those relationships are and how the functions of mutant proteins are changed. One example in the SOS case is that of the *tif* mutant described above, now known to be a *recA* allele, *recA441*. The properties of this mutant led to the prediction that RecA must be activated to play its role in the SOS system.[26-28,39] Another example in which biochemistry and genetics complement one another is provided by the analysis of dominant mutants coding for noncleavable repressors, as discussed above. One further powerful feature of this combination is that it is iterative: Biochemistry predicts the existence of certain types of mutants; geneticists can then isolate such mutants (or they can be made by reverse genetics) and their effects on the entire regulatory network can be studied in vivo.

C. RECENT DEVELOPMENTS

As noted above, the overall picture of the SOS regulatory system was in place by 1982. With one surprising exception (section C5 below), ongoing research has largely confirmed this picture, and added to our understanding of the molecular events underlying the response. In addition, a number of new SOS genes have been identified, many physiological consequences of SOS induction have been further explored, and SOS regulation has been studied in other prokaryotes. In this section, some of these more recent developments are reviewed, mostly selected on the basis of their bearing on the operation of the SOS regulatory system. These are discussed more or less in the order in which they occur during the process of SOS induction. An excellent and comprehensive review is given in chapter 10 of ref. 14, which also gives a detailed historical account.

1. LexA as a Repressor

LexA protein has been characterized in detail as a repressor (for a review see ref. 40). LexA differs from many well-characterized repressors (such as Lac and λ repressors) in that most of the operons under LexA control have a substantial basal level of transcription.[41] The strength of LexA binding to various SOS boxes measured in vitro generally correlates with the extent of repression seen in vivo; that is, genes to which LexA binds most tightly are the most completely repressed in normally growing cells. However, this picture is complicated by at least three issues: First, the LexA binding sites at various promoters are located at widely differing positions relative to the -35 and -10 regions, raising the possibility that LexA might interfere with transcription initiation at different stages of this process.[42] Second, it is possible that both LexA and RNA polymerase may occupy certain promoters simultaneously, and that RNA polymerase can initiate at a low rate even when LexA is bound.[42] Finally, at least one LexA-controlled gene, *uvrB*, has two promoters, one controlled by LexA and one that is not, so that even in the presence of LexA a substantial basal level of *uvrB* transcription occurs.[43]

Like most well-characterized repressors, LexA binds to a site (the SOS box) that has dyad symmetry (the consensus sequence is $CTGN_{10}CAG$).[32,34,40] Unlike most repressors, however, LexA dimerizes quite weakly in solution.[44,45] Recent evidence[45] shows that, at least under the in vitro conditions used, LexA binds to DNA in a pathway in which two monomers bind sequentially to the operator; that is, LexA dimerizes on the DNA, as opposed to the binding of preformed dimers. The second monomer binds about 10^6-fold more tightly than the first. We speculate that this enormous degree of cooperativity ensures that relatively small changes in LexA levels will lead to large changes in the occupancy of the various operators, and hence that the system will be very sensitive (see section D2).

2. LexA-Controlled Genes

About twenty *E. coli* genes are known to be directly controlled by LexA (reviewed in refs. 7,14,46). These can be grouped into several functional sets. First, the regulatory genes *recA* and *lexA* are themselves under SOS control. SOS genes involved directly in DNA repair include the excision repair genes *uvrA* and *uvrB*, the DNA helicase *uvrD*, and several genes involved in recombinational repair, including *ruvAB*, *recN*, and *recA* itself. Genes involved in SOS or Weigle mutagenesis include the *umuDC* operon. Other SOS-controlled genes include the DNA polymerase II gene *polB* and the *sulA* gene, coding for a cell division inhibitor. Additionally, a number of SOS-regulated genes of unknown function (*dinB*, *dinD*, *dinF*, *dinG*, *dinH*, and *dinI*—where *din* stands for damage-inducible) have been identified genetically by the use of operon fusions to reporter genes such as *lacZ* (reviewed in ref. 47). Several additional LexA binding sites have been identified by database searches and shown to bind LexA, and it is unclear whether binding of LexA to these sites regulates transcription in vivo.[47]

Several genes, including *dinY*, *dnaA*, *dnaN*, *dnaQ*, *nrdAB*, *phr* and *recQ*, appear to be expressed at higher levels following SOS-inducing conditions but may not be under the direct negative control of LexA.[47,48] The mechanism by which these genes are induced remains unclear; very recent evidence suggests that *dnaN* regulation is post-transcriptional.[49] As a matter of nomenclature, at least one of these genes (*dinY*) has been described as being a part of the SOS regulon,[48] on the grounds that it is induced by treatments that induce the SOS system; it is my view, however, that this term should be reserved for those genes that are under direct control of the SOS regulatory system, and I propose that this usage be restricted to LexA-controlled genes.

A number of genes borne by plasmids or prophages are also under direct LexA control.[14] These include the *mucAB* operon, homologous to the mutagenesis genes *umuD* and *umuC*, carried on the plasmid pKM101; plasmid-borne genes for a number of colicins, such as *cea* and *caa*, coding for colicins E1 and A, respectively; and the phage 186 gene *tum*, which apparently acts as an anti-repressor of 186 repressor and leads to induction of 186 prophage upon SOS induction.[50]

Clearly, these genes control a wide variety of functions, and their actions can be expected to have diverse impacts on cell physiology and behavior of bacterial populations. Discussion of these issues lie beyond the scope of this review, but major advances have resulted in several areas, particularly in our understanding of SOS mutagenesis (see ref. 14, chapters 10 and 12, for a comprehensive review).

3. Identity of the SOS-Inducing Signal

Still the least understood aspect of the SOS-regulatory system is the nature and possible diversity of the inducing signals which trigger this response. This continues to be a difficult experimental problem, in part because induction is a rapid process (see below); to examine signal molecules directly, one would have to study changes occurring within one minute after an inducing treatment. In addition, a diverse set of treatments induces the SOS system; it is not known whether these all lead to the production of a common signal molecule, or whether different effectors arise under different circumstances. Moreover, as pointed out previously,[7] in principle an effective inducing signal could represent any kind of change in a cell, but not necessarily the appearance of a new component; conceivably it could represent a change in some cellular process such as DNA supercoiling or nucleotide ratios.

Biochemical studies by Roberts et al clearly established that RecA protein can be activated to serve its role in cleavage by forming a ternary complex with ATP or ATPγS and single-stranded DNA.[31] These cofactor requirements suggest the likelihood that RecA is activated in vivo by similar types of molecular signals. However, diverse types of treatments activate the SOS system, and it is plausible that a range of effector molecules could serve to activate RecA as well.

Recent studies strongly support the idea that inducing signals can be generated by the process of DNA replication.[51] DNA replication was blocked by treating *dnaC*ts or *dnaE*ts mutants at a nonpermissive temperature; under these conditions, UV irradiation did not result in LexA cleavage. This study provides strong support for the idea that replication past the site of damage, an event that should occur within a few seconds at the UV doses used, generates daughter-strand gaps that serve to activate RecA. The possibility remains, of course, that other effectors would operate under different circumstances. Moreover, as pointed out by these workers,[51] other inducing agents (such as bleomycin or nalidixic acid) may lead to the production of single-stranded DNA by pathways different from that described above for production of daughter-strand gaps.

Another approach to the identification of inducing signals has been the study of *recA* alleles coding for mutant RecA proteins that are constitutively activated for cleavage. Such alleles are called Prt^c for protease-constitutive,[52] or preferably Cpt^c for coprotease-constitutive.[13] Some Cpt^c mutant proteins appear to be activated by an expanded range of effectors in vitro; for instance, RecA1202 protein can be activated by additional nucleotide cofactors such as UTP and CTP, and by polynucleotides such as rRNA and tRNA that are normal constituents of the cell.[53,54] A different type of mechanism has been proposed for other Cpt^c proteins such as RecA730 and RecA441 proteins.[55] Under certain in vitro conditions, RecA competes for limiting amounts of single-stranded DNA with SSB, a single-stranded binding protein involved in DNA replication; the mutant proteins compete more successfully with SSB than wild-type RecA does. By this model, these mutant proteins are activated in vivo by binding to a small amount of single-stranded DNA, which might be present, for example, at the replication fork. Perhaps certain Cpt^c mutant RecA proteins can be activated by both of these mechanisms; to our knowledge, both types of assays have not been carried out on the same mutant protein. Finally, various lines of evidence show that there is a slow but detectable rate of LexA cleavage in normally growing wild-type cells, suggesting that such cells contain a small amount of activated RecA.[9,51,56] Either they contain a pool of effectors for which SSB cannot compete, or wild-type RecA competes to some small extent for endogenous single-stranded DNA. In summary, all these studies are consistent with the proposal that single-stranded DNA is a normal SOS-inducing signal, but again they do not exclude the possibility that other effectors can fulfill this role.

Cleavage of the phage φ80 repressor appears to be more complicated than that of other lambdoid phage repressors or of LexA.[57] In vitro, RecA-mediated cleavage of φ80 repressor is stimulated greatly by the addition of the deoxydinucleotides d(GpG), d(ApG), or d(pGpG). These dinucleotides have no effect on cleavage of LexA or λ repressors. The dinucleotides act by binding to φ80 repressor, probably increasing its affinity for RecA. These findings provide a satisfying explanation for older studies[58] aimed at identifying inducing signals. These studies used an assay based on φ80 repressor cleavage, and used permeabilized cells to ask whether particular exogenous molecules could yield repressor inactivation. The most active effectors in this assay were dinucleotides like d(GpG). However, these findings were largely ignored in subsequent years because these compounds do not affect cleavage of the more widely studied λ and LexA repressors.[57] The need for the diverse cofactor requirements is not understood, but presumably reflects some distinctive feature of φ80 biology that is not shared by λ and other phages. These findings raise the possibility that the rates of RecA-mediated cleavage for other substrates might be modulated by unknown small-molecule effectors.

4. STRUCTURAL ANALYSIS OF RECA, AND ITS INTERACTION WITH LEXA

RecA protein forms right-handed helical filaments on single-stranded and double-stranded DNA.[59-61] These filaments can form under various conditions, and can be observed by electron microscopy. In these filaments, a deep groove spiraling about the outside of the filament is evident. Filaments formed on duplex DNA are particularly regular in their structure, allowing detailed image reconstruction. These filaments appear to be of two types: A so-called low-affinity form, in which RecA binds ADP and single-stranded DNA,[59,61] and a "high-affinity" form containing single- or double-stranded DNA and ATP or ATPγS.[59,60] These forms differ considerably in their helical pitch (76 and 95 Å, respectively). The high-affinity form containing single-stranded DNA is believed to catalyze repressor cleavage, and to be responsible for most of the strand-transfer reactions involved in genetic recombination. In the absence of DNA, RecA alone also forms helical filaments under certain conditions, which have the same appearance as the low-affinity form.

The crystal structure of RecA protein has recently been solved.[62] In the crystal, RecA forms a right-handed helical filament with 6-fold rotational symmetry. The structure does not contain single-stranded DNA or bound nucleotide, although the position of bound ADP was determined in a separate study. The helical filament also exhibits a deep groove, and side chains protrude into this groove from several residues implicated in repressor cleavage on the basis of genetic evidence. It has been proposed that these side chains directly contact LexA and other repressors, perhaps altering their structure so as to promote cleavage.[62] However, this conclusion must remain tentative, because the crystal structure is probably that of the low-affinity form.[59] Hence, the conformation of amino acid side chains in the helical groove might differ in critical ways in the two forms. Low-resolution image reconstruction from electron microscopy of a complex between the high-affinity form of RecA and a noncleavable LexA also suggests that LexA binds in the helical groove, perhaps touching two adjacent subunits of RecA.[63] This conclusion is also tentative because LexA is not bound to every subunit in the filament, and hence the electron density is relatively weak; moreover, these complexes were formed on double-stranded DNA, and LexA may not interact with this complex in the same way as with one containing single-stranded DNA.

Several lines of evidence, some biochemical but most involving in vivo assays for cleavage, have been interpreted to mean that the site at which RecA interacts with cleavable substrates is not the same for all substrates. Several *recA* alleles have been shown to confer an altered pattern of repressor cleavage. RecA1730 protein was selectively deficient in vivo for cleavage of LexA, but normal for cleavage of λ repressor; however, the defect in LexA cleavage was later found to result from a weakened interaction of RecA1730 with single-stranded DNA, since LexA cleavage is at a rate similar to wild-type RecA at elevated RecA levels; it is hypothesized that λ repressor cleavage is less affected because this process is catalyzed by shorter RecA-DNA filaments.[64] In contrast, RecA430 protein cleaves LexA and φ80 repressor slowly, but has no detectable activity against λ repressor.[65-67] In vitro, RecA430 protein is far more deficient at supporting cleavage of λ repressor than of LexA or φ80 repressor; however, it is conceivable that this protein interacts weakly with all substrates, and that this affects λ CI cleavage more than that of LexA or φ80 CI because of differences in K_m for these proteins. The general conclusion from these and similar studies is that, while it may be true that different repressors bind to different sites in RecA, the extreme complexity of RecA makes it difficult to establish this point on the basis of in vivo results. A firm conclusion awaits detailed biochemical analysis.

5. MECHANISM OF LEXA CLEAVAGE

Although the mechanism of RecA-mediated cleavage seemed clear in 1982, one more surprise was in store. It was found during the course of an unrelated experiment (see ref. 13) that purified LexA repressor can undergo a self-cleavage reaction. In this experiment, LexA was treated with trypsin to examine its domain structure; in the control incubation lacking trypsin, LexA was seen to break down to two discrete cleavage products of a size roughly half that of the intact protein. This fortunate chance observation led to an entirely new view of LexA cleavage.

It is interesting to speculate in passing about the likelihood that this discovery would be made. Although the rate of LexA autodigestion is relatively rapid, if this reaction had been 100-fold slower, it is harder to be confident that LexA would be treated in such a way that someone would see cleavage and interpret it correctly. Phage λ repressor autodigests about 50-fold more slowly; this protein had been studied intensively as a repressor for fifteen years, and its self-cleavage had not been seen. In addition, there is no obvious reason that LexA must autodigest at the rate it does. We have been characterizing mutant LexA proteins that appear to have a normal rate of RecA-mediated cleavage but slower autodigestion; the most extreme of these, VS82, autodigests about 1% the rate of wild-type LexA but shows at least 50% the wild-type rate of RecA-mediated cleavage (D. Shepley and J.W. Little, unpublished data). Although it is unclear as yet whether a *lexA* VS82 mutant would show a normal SOS response, had LexA carried this substitution we might not yet be aware of the mechanism by which LexA cleavage operates.

What have we learned about this autodigestion reaction? It is intramolecular; it cuts the same bond as is cleaved in the RecA-dependent reaction; and it is blocked by repressor mutations that also interfere with RecA-mediated cleavage.[13,68-71] Purified λ repressor and all other cleavable substrates tested also can autodigest.[67-69,72-74] These findings suggest that RecA, rather than being a protease, acts as a "coprotease" to stimulate a latent self-cleavage activity of LexA and other cleavable proteins.[13,71] In this view, RecA is still a catalyst, in that it supports multiple cleavage events, but the actual chemistry of cleavage is carried out by groups in LexA, not in RecA. The active site lies in the repressor, and activated RecA acts to increase the rate of this reaction, presumably by making protein-protein contacts with its repressor substrates. Current evidence suggests that RecA favors a particular conformation of LexA that is competent to undergo cleavage.[71,75]

This revised view of specific cleavage is satisfying in that it simplifies the job of RecA: the catalytic activity and most of the specificity lies in the repressors, and RecA need only make some specific protein-protein contacts. It is important to note that this new view of LexA cleavage does not alter in any essential way our understanding of the SOS regulatory circuitry. Indeed, it illustrates a general principle applicable to many regulatory systems: What is important at the level of regulation is that the regulatory interaction has the desired effect, not how it works at the lower, mechanistic level; the mechanistic details of a reaction do not greatly matter as long as the job gets done.

The mechanism of LexA autodigestion involves a serine nucleophile, Ser-119, which probably attacks the peptide bond in a manner similar to other serine proteases such as trypsin; however, unlike trypsin, Ser-119 appears to be activated by a lysine residue, Lys-156 in LexA, which must be deprotonated for cleavage to proceed.[76,77] Hence, autodigestion is rapid only at elevated pH. We postulate that in the reactive conformation favored by RecA protein the pK_a of Lys-156 is reduced to 7 or below, so that cleavage can proceed rapidly at physiological pH.[75]

LexA can also act as a specific protease in an intermolecular reaction in which one molecule of LexA or its C-terminal cleavage product attacks other molecules.[78] Development of this enzymatic reaction was facilitated by the availability of Ind^S mutant proteins, which exhibit greatly increased rates of cleavage.[79,80] Although we cannot yet measure k_{cat} for this trans-cleavage reaction, since we are working below the K_m, its value for the best enzyme and substrate is at least 1 sec^{-1}, implying that LexA is intrinsically a rather good catalyst. Strikingly, a comparable enzyme derived from λ repressor is almost as efficient as its LexA counterpart when assayed on LexA substrates, and the λ repressor enzyme recognizes an Ind^S substrate as well. This finding strongly implies that the structure of the active site is largely conserved between LexA and λ repressor.

It is probable that the structures of these cleavable proteins prevent the interaction between the cleavage site and the active site, because the proteins would otherwise be so unstable that their level could not be regulated. This view is supported by the properties of the Ind^S mutant proteins, which also increase the rate of *trans* cleavage. They lie either near the cleavage site (LP89, QW92) or near the active site lysine (EA152), and our evidence suggests that these mutations strengthen the interaction between the cleavage site and the active site in LexA. We surmise that LexA and λ repressors have modulated their rates of cleavage during evolution to be optimal for their respective niches (see also section D5 below), and that they have done so by modulating the strength of the interaction between the cleavage site and the active site.[71,78]

This relationship between a self-processing reaction (LexA cleavage) and an external effector (activated RecA) is at least formally analogous to a number of other similar self-processing reactions that have recently been described (reviewed in ref. 71). The phosphorylation and dephosphorylation reactions carried out by the sensor and response regulator proteins in two-component regulatory systems are all self-processing reactions, and many of these are stimulated by external effectors. For example, the rate of dephosphorylation of CheY-phosphate is enhanced indirectly by interaction with CheZ. This analogy can also be extended to the Ras protein, whose activity is modulated by hydrolysis of a tightly bound GTP, and which may be considered an "honorary" self-processing protein. This intrinsic GTPase activity can be stimulated greatly by interaction with a class of proteins termed GAPs, which accelerate the rate by a factor of as much as 10^5, leading to deactivation of Ras.

6. CLEAVAGE OF UmuD

Induction of the SOS response leads to an increase in mutation rate, due to a process termed SOS mutagenesis or Weigle mutagenesis. One of the LexA-controlled operons, the *umuDC* operon, plays a crucial role in this process. Both UmuD and UmuC are required for mutagenesis. UmuD protein is cleaved in a specific cleavage reaction analogous to LexA cleavage, yielding UmuD' as a product.[73,74,81] However, unlike the case of repressor cleavage, this reaction *activates* UmuD for its role in mutagenesis. Activation of UmuD is blocked in vivo by the mutation SA60, which changes the putative Ser nucleophile (analogous to Ser119 in LexA). This block can be bypassed by removing the N-terminal portion of UmuD genetically, showing that the primary role of Ser60 is in cleavage, rather than in any functional activity of UmuD'.[81]

Weigle mutagenesis requires activated RecA protein for at least three separable roles. RecA promotes cleavage both of LexA and of UmuD. In addition, RecA has a poorly understood direct mechanistic role in mutagenesis, which may involve formation of a complex with UmuD' and UmuC on damaged DNA.[38,81]

7. HOMOLOGIES AMONG CLEAVABLE PROTEINS

A large number of proteins have been identified that are likely to undergo specific cleavage. Nearly half of these proteins have been shown directly to do so; the remainder probably also do, either because they are analogs of LexA or UmuD, or the repressors of UV-inducible prophages from various organisms. In principle, alignment of these proteins might provide clues as to the chemical mechanism of cleavage and the residues important for this reaction, and perhaps to the identity of a site that binds to activated RecA.

A tentative alignment of the carboxy-terminal portions of these proteins is shown in Fig. 22.2. This portion of these proteins was aligned because, at least in those proteins (LexA and phage λ and P22 repressors) for which data are available, this part of the protein is folded into a discrete functional and structural domain that can carry out autodigestion and RecA-mediated cleavage.[68,69] In making this alignment, the presumed cleavage site (Ala-Gly in almost all cases),[35] the presumed active-site Ser,[76] and the presumed active-site Lys were aligned, and gaps were introduced as necessary to maximize other alignments. A consensus is given at the bottom. Several prominent features are apparent.

The first and most striking is the small degree of similarity among these proteins. As more proteins have been added to this alignment,[47,82,83] the extent of similarity has fallen progressively. Indeed, it seems implausible that homology for many of these proteins could be detected by a computer analysis based solely on the sequence, although this has not been tested directly. Undoubtedly, one factor contributing to the lack of similarity is the fact that this domain in many of these proteins also serves other functional roles, which must in part dictate their sequences. In LexA and several phage repressors, this domain of the protein is known to be involved in dimerization;[44,84] in λ and P22 repressors,[85-87] and probably in HK022 repressor (C. Mao and J.W.

Little, unpublished data), it is also involved in cooperative binding to DNA. Hence, surface residues must accommodate the need for these protein-protein interactions. The mutagenesis protein UmuD′ interacts with UmuC protein and with activated RecA (as part of RecA's direct role in mutagenesis), imposing another constraint.[38] In a few regions of the sequence, most proteins align well but one or a few phage repressors appear anomalous. This might reflect, for instance, the need to interact with a small molecular weight cofactor (as in the case of φ80 repressor).

Second, two groups of proteins with related functions—the LexA homologs and the UmuD homologs—are relatively conserved within each group; this would be expected, because these homologs were isolated by virtue of their ability to provide the same function as their *E. coli* counterparts. By contrast, the phage repressors (except for 434 and P22 repressors) appear to be more diverse. This diversity is not due to the need for different DNA-binding specificities, since that function is carried on a different part of these proteins.[84] Perhaps phage repressors are under selective pressure to diverge so that they will not form heterodimers in cells that happen to contain both phages, since this would impair proper gene regulation.[86]

Third, the limited degree of similarity seen is clustered for the most part in three regions, lying around the cleavage site and around the Ser and Lys residues (underlined) that are thought to be critical for cleavage and hence to lie in an active site. Many *lexA* mutants decreasing the rate of cleavage, and a few increasing the rate (depicted by ↓ and ↑, respectively, below the *lexA* sequence in Fig. 22.2) also cluster in these regions.[56,78-80] Taken together, these data strongly suggest that the conserved residues are conserved for their role in the chemical and structural events of cleavage, rather than for some other function such as dimerization. It is also evident that patches of hydrophobic residues (depicted in the consensus by φ) and clusters of charged or polar residues (depicted as "O") are also apparent, suggesting that the proteins may fold in a conserved way.

It is unlikely that a RecA-binding site is conserved among all these proteins. Two lines of genetic evidence argue that this interaction is not conserved. First, as noted above, *recA* alleles exist which may have selective defects in their ability to support cleavage of certain proteins. Second, in a collection of λ repressor mutants that could not undergo RecA-mediated cleavage, nine of 15 were specifically deficient in RecA-mediated cleavage but normal for autodigestion.[72] The simplest interpretation of these findings is that the mutant repressors are impaired in their interaction with RecA. Strikingly, except for Phe189, 3 residues before the conserved Lys residue, none of these mutations affect residues that are conserved among even a subset of the proteins depicted in Figure 22.2. These findings imply that the RecA binding site on λ repressor is not conserved in other cleavable proteins. It therefore appears likely that different cleavable proteins will be found to interact in different ways with RecA.

8. Kinetics of LexA Cleavage During the SOS Regulatory Cycle

As outlined above, one of the major advances in dissection of the SOS regulatory system was the development of assays to examine relatively proximal events in the induction process. The most proximal event detectable to date is LexA cleavage. Two studies have examined the kinetics of LexA cleavage following inducing treatments, using either pulse-chase methods to detect the rate of cleavage[9] or Western

```
                    cleavage site
                         ↓
ImpA          18  VRPLFADRCQ  AGFPSPATDY  AEQELDLNSY  CISRPAAT..  ........  ...FFL  RASGESMNQA  G....VQNG
SamA          15  TAPLFTERCP  AGFPSPAADY  TEEELDLNAY  CIRRPAAT..  ........  ...FFV  RAIGDSMKEM  G....LHSG
MucA          15  SIPFYLQRIS  AGFPSPAQGY  EKQELNLHEY  CVRHPSAT..  ........  ...YFL  RVSGSSMEDG  R....IHDG
StyumuD       14  PLPFFSYLVP  CGFPSPAADY  IEQRIDLNEL  LVSHPSST..  ........  ...YFV  KASGDSMIEA  G....ISDG
EcoumuD       14  TFPLFSDLVQ  CGFPSPAADY  VEQRIDLNQL  LIQHPSAT..  ........  ...YFV  KASGDSMIDG  G....ISDG
Tuc2009cI    161  DVPI.LGRIA  AGLPLDAVEN  FDGTRPVPAH  FLSSARDY.   ........  ..YWL   MVDGHSMEPK  ......IPYG
D3cI          85  LIPQYTARGE  CGDGYFNDHV  ETTEGLVFKR  DWLKRVNSKP  ...ENLFVI  YADGDSMEPY  ......IFEG
D3112cI      107  YIPLYDGQVS  AGHGSWTDGA  TVLVNLAFTR  YSLRKKGLDP  S...SISAI  RIGGDSMEPL  ......LCDG
HK022cI       93  RVDVLDVQAS  AGPGTMVSNE  FIEKIRAIEY  TTEQARILFN  GRPQESVKVI  TVRGDSMEGT  ......INPG
Phi80cI      101  VPFLKDIEFA  CGDGRVHDED  HNGFKLRFSK  ATLRRVGANS  DGSG..VLCF  PASGDSMEPV  ......IPDG
434cI         80  KYPL.ISMVR  AGSWCEACEP  YDIKDIDEVH  DSDVNLLGNG  ...FWL  KVEGDSMTSP  VGQ..SIPEG
P22c2         85  SYPL.ISWVS  AGQWMEAVEP  YHKRAIENWH  DTTVDCSEDS  ...FWL  DVQGDSMTAP  AGL..SIPEG
Lambda cI    102  EYPV.FSHVQ  AGMFSPELRT  IEEYFPLPDR  MVPPDEHV..  ...FML  EVEGNSMTAP  TGSKPSFPDG
Bsubt_DinR    82  NVPV.IGKVT  AGSPIFAEGT  VEDIFPLPRE  LVGEGTL...  ...FML  EIMGDSMIDA  G....ICDG
LexA_Mleprae 115  FVPI.LGRIA  AGSPITAVEN  IEEYFPLPDR  ...........  ...FLL  KVTGDSMVEA  A....IMDG
LexA_Prett    78  GLPL.IGRVA  AGEPLLAQEH  IESHY.QVDP  ELFKPHAD..  ...FLL  RVNGMSMKDI  G....IMDG
LexA_Ecar     75  GIPL.IGRVA  AGEPLLAQEH  IECRY.QVDP  AMFKPSAD..  ...FLL  RVSGMSMKNI  G....IMDG
LexA_Pputida  78  GLPI.IGRVA  AGAPILAEQH  IEQSC.NINP  AFFHPQAD..  ...YLL  RVHGMSMKDV  G....IFDG
LexA_Paerug   81  GLPI.IGRVA  AGAPILAEQN  IEESC.RINP  AFFNPRAD..  ...YLL  RVRGMSMKDI  G....ILDG
LexA_Ecoli    75  GLPL.VGRVA  AGEPLLAQQH  IEGHY.QVDP  SLFKPNAD..  ...FLL  RVSGMSMKDI  G....IMDG
                                                                                           y             e
Consensus         .φPφ....Ov.  AG.p...A.O.  φe.O......  ...........  .φφ  OφOGdSMO.  ......I.dG


ImpA          DLLVVDRAEK  PQHGDIVIAE  I..DGEFTVK  RLLLRP.RPA  LEPVSDSPEF  RTLYPEN....  .....ICIFGV  VTHVIHRTRE  LR*
SamA          DLMVVDKAEK  PMQGDIVIAE  T..DGEFTVK  RLQLKP.RIA  LLPIN..PAY  PTLYPEE....  .....LQIFGV  VTAFIHKTRS  TD*
MucA          DVLVVDRSLT  ASHGSIVVAC  I..HNEFTVK  RLLLRP.RPC  LMPMNKDFPV  YYIDPDN....  .ESVEIWGV  VTHSLIEHPV  CLR*
StyumuD       DLLVVDSSRN  ADHGDIVIAA  I..EGEFTVK  RLQLRP.TVQ  LIPMNGAYRP  IPVGSED....  ...TLDIFGV  VTFIIKAVS*
EcoumuD       DLLIVDSAIT  ASHGDIVIAA  V..DGEFTVK  RLQLRP.TVQ  LIPMNSAYSP  ITISSED....  ...TLDVFGV  VIHVKAMR*
Tuc2009cI     AYVLIEAVPD  VSDGTIGAVL  FQDDCQATLK  KVYHEIDCLR  LVSINKEFKD  QFATQDNPA.  .....AVIGQ  AVKVEIDL*
D3cI          DVVLFDTSKT  DPQDKQYVYI  RRPDGGVSIK  RLNQQLTGAW  LIRSDNPDKS  AYPDEMASES  SVHELPIIGR  VIWRGGGIG*
D3112cI       DTVLVDHTKS  TVQDAAVYVV  RL.DDHLYAK  RLQRRFDGSV  SIISENKAYT  EMIVPKAKLS  DLE...IIGR  VVWASRWMV*
HK022cI       DEIFVDVSIT  CFDGDGIYVF  V.YGKTMHVK  RLQMQKNRLA  VISDNAAYDR  WYIEEGEEEQ  .....LHILAK  VLIRQSIDYK  RFG*
Phi80cI       ATVAVDTGNK  RVIDGELYAI  ...NQGDLK  RIKQLYRKPG  GKILIRSINR  DYDDEEADE.  .ADVEIIGF  VFWYSVLRYR  R*
434cI         HMVLVDTGRE  PVNGSLVVAK  LTDANEATFK  KLVIDGGQKY  LKGLNPSWPM  TPINGN....  .CKIIGV  VVEARVKFV*
P22c2         MIILVDPEVE  PRNGKLVVAK  LEGENEATFK  KLVMDAGRKF  LKPLNPQYPM  IEINGN....  .CKIIGV  VVDAKLANLP  *
Lambda_cI     MLILVDPEQA  VEPGDFCIAR  L.GGDEFTFK  KLIRDSGQVF  LQPLNPQYPM  IPCNES....  .CSVGK  VIASQWPEET  FG*
Bsubt_DinR    DYVIVKQQNT  ANNGEIVVAM  I..EDDEATVK  RFYKEDTHIR  LQPENPTMEP  IILQN.....  .VSILGK  VIGVFRTVH  Y*
LexA_Mleprae  DWVVVRQQKV  ADNGDIVAAM  I..DGEATVK  TFKRAGGQVN  LIPHNPAFDP  IPGNDA....  .VSILGK  VVTVIRKI*
LexA_Prett    DLLAVHKTQN  VHNGQVVVAR  I..EDEVTVK  RFKQQGNRVE  LIAENPEFEP  IVVDLRQQNF  .TIEGL  AVGVIRNSDW  Y*
LexA_Ecar     DLLAVHKTED  VRNGQIVVAR  I..DDEVTVK  RLKKQGNTVH  LLAENEEFAP  IVVDLRQQSF  .SIEGL  AVGVIRNSDW  S*
LexA_Pputida  DLLAVHTCRE  ARNGQIVVAR  I..GDEVTVK  RFKREGSKVW  LLAENPEFAP  IEVDLKEQEL  .VIEGL  SVGVIRR*
LexA_Paerug   DLLAVHVTRE  ARNGQVVVAR  I..GEEVTVK  RFKREGSKVW  LLAENPEFAP  IEVDLKEQEL  .IEGL  SVGVIRR*
LexA_Ecoli    DLLAVHKTQD  VRNGQVVVAR  I..DDEVTVK  RLKKQGNKVE  LLPENSEFKP  IVVDLRQQSF  .TIEGL  AVGVIRNGDW  L*
                                                        k
Consensus     DφφφVOO.OO  ..OGOφVφA.  φ..O.EφTvK  rφ.O.O...  L.p.N..φ.  i.φO.O...  ..iφG.  vφ.φ.....
```

Fig. 22.2. Homology among cleavable proteins. Amino acid sequences were aligned by hand; sequences were constrained to align at the Ala-Gly cleavage site (marked at the top by ↓), and at the active site residues (Ser-119 and Lys-156 in E. coli LexA), marked with bold type and doubly underlined in the consensus. Similar but not identical alignments, involving subsets of these proteins, have been published.[47,82,83] The numbers at the beginning of each sequence indicate the number of the first amino acid listed for that sequence. The dots below the E. coli LexA sequence denote the positions of every tenth amino acid in LexA. Below the sequences is listed a consensus. In this consensus, amino acids conserved among all 20 proteins are given in large bold letters; those conserved among 15-19 proteins are given by capital letters; those conserved among 10-14 proteins by small letters; residues for which two chemically similar residues are conserved (≥15/20) are given with small letters and the two residues; residues for which hydrophobic residues are conserved (≥ 15/20) are denoted by φ or (20/20) by a bold φ; and residues for which polar or charged groups are conserved (≥ 15/ 20) are denoted by O or (20/20) by a bold O. Amino acids are denoted by the one-letter code.

Sequences are grouped in the following way: At the bottom are five functional lexA genes, followed by a sixth putative lexA gene from Mycobacterium leprae that was identified by random sequencing of a cosmid, and by the B. subtilis dinR gene that may be a LexA homolog. The lexA gene from S. typhimurium, which differs from that of E. coli in only four positions in this interval, was not included. In the middle are eight CI repressors from UV-inducible temperate phages. At the top are five UmuD-like proteins; the recently identified humD gene of phage P1 was not included because it does not appear to be under selective pressure to maintain cleavage activity.[47] The proteins, starting from the bottom, are as follows (name, gene and organism, GenBank accession number; references are in the GenBank files): LexA_Ecoli, E. coli LexA, J01643; LexA_Paerug, Pseudomonas aeruginosa LexA, X63018; LexA_Pputida, P. putida LexA, X63017; LexA_Ecar, Erwinia carotovora LexA, X63189; LexA_Prett, Providencia rettgeri LexA, X70965; LexA_Mleprae, putative LexA from Mycobacterium leprae, U00019; Bsubt_DinR, Bacillus subtilis DinR, M64684; Lambda_cI, CI repressor of coliphage λ;[102] P22c2, C2 repressor of Salmonella phage P22, J02470; 434cI, CI repressor of coliphage 434, M12904; Phi80cI, CI repressor of coliphage φ80, X13065; HK022cI, CI repressor of coliphage HK022, X16093 (note that the amino acid sequence has been corrected from ref. 103); D3112cI, CI repressor of P. aeruginosa phage D3112, X52258; D3cI, CI repressor of P. aeruginosa phage D3, L22692; Tuc2009cI, CI repressor of Lactococcus lactis phage Tuc2009, L26219; EcoumuD, E. coli UmuD, M13387; StyumuD, S. typhimurium UmuD, M57431; MucA, plasmid pKM101 MucA, M13388; SamA, S. typhimurium SamA, a plasmid-borne UmuD analog, D90202; ImpA, another plasmid-borne UmuD analog, X53528. Since the N-terminal Met is removed after synthesis of HK022 CI (I.W. Little, unpublished data) and λ CI repressors,[102] numbering of the amino acids begins at the second residues for these proteins. The following proteins, starting from the bottom, have been shown to undergo in vitro cleavage: E. coli LexA, phage λ, P22, 434, φ80, and HK022 repressors; E. coli UmuD; and MucA. Cleavage of the remaining LexAs is inferred, except for M. leprae, by the fact that the M. leprae gene has not been studied directly. Cleavage of the remaining phage repressors is inferred by the fact that the proteins are functional in E. coli; the function of the M. leprae gene is UV-inducible. Cleavage of the remaining UmuD analogs is inferred from their relationship to E. coli UmuD.

analysis to measure LexA levels.[51] These studies led to several important conclusions about the SOS regulatory system. First, LexA cleavage begins about one minute after UV irradiation (a treatment that quickly creates DNA damage, in comparison to treatment with drugs such as Mitomycin C). Cleavage is not seen in a *recA*⁻ host. Second, cleavage does not require protein synthesis, and cleavage of wild-type LexA can occur in a merodiploid strain also containing a noncleavable LexA mutant protein. These observations confirm that cleavage is an early event in SOS induction and that it is catalyzed by pre-existing RecA molecules. Third, cleavage activity is at a maximum soon after an inducing treatment but, in a cell that can repair DNA damage, the rate of cleavage declines after 30-60 minutes (the time depending on the UV dose) as the cells recover. Hence these approaches allowed an examination of the recovery phase, which has been largely inaccessible to genetic analysis. The power of these approaches in examining proximal events argues for their wider use in studies of SOS regulation.

Recently, steady-state levels of LexA have directly been measured under diverse physiological conditions using Westerns.[88,89] It was found that the stability of LexA varies in a RecA-dependent fashion according to the stage of growth cycle, being less stable in exponentially growing cells than in stationary-phase cells, and particularly unstable when stationary-phase cells are diluted. It will be of interest to investigate further the role of bacterial physiology in modulating the stability of LexA under various conditions.

9. SOS REGULATION IN OTHER PROKARYOTES

The *recA* gene has been found in a wide variety of prokaryotes.[90] This protein evidently plays a crucial role in DNA metabolism throughout the prokaryotic world. Indeed, similar proteins that can form filaments like those formed by RecA have recently been isolated in yeast.[91] However, the role of RecA in potential SOS responses in other organisms is less certain. The existence of SOS regulation in most of the species known to contain RecA homologs has not yet been established. Functional *lexA* genes have also been isolated from several bacteria relatively closely related to *E. coli*, using a functional assay for binding to *E. coli* SOS boxes (see Fig. 22.2). Such an assay would not work in more divergent organisms with a different DNA-binding site specificity.

The best-studied SOS system from an organism other than *E. coli* is that of *Bacillus subtilis*, a gram-positive organism widely diverged from *E. coli*. This system is of interest for several reasons. First, at least some of the SOS genes (of which *recA* is the best studied) appear to be under dual control. These genes are also induced in a RecA-independent manner when these cells enter the competent state.[92] In this state, they are able to take up exogenous DNA; if this DNA is homologous to that of the host, it can then undergo recombination with the genome. A complex regulatory network is involved in establishing the competent state. Future work should reveal how this network interacts with the SOS regulatory system.

A second interesting feature of the *B. subtilis* SOS system is that it offers a striking example of evolutionary conservation. Purified *B. subtilis* RecA protein cleaves *E. coli* LexA protein efficiently.[93] Conversely, the RecA protein of *E. coli* can support SOS induction in *B. subtilis* following DNA damage, implying that it can interact with the *B. subtilis* analog of LexA.[94] Accordingly, the contacts between these two proteins have evidently been conserved over the course of evolution, even though the likely *B. subtilis* analog of LexA has a different DNA-binding site specificity. Although the LexA analog of *B. subtilis* has not yet been analyzed in detail, several different approaches have suggested the existence of one and possibly two LexA analogs.[95,96] Biochemical analysis of these proteins is underway.

D. BEHAVIOR OF THE SOS GENE REGULATORY CIRCUITRY

In this section, I consider several issues related to the operation of the SOS system as a regulatory circuit. Many of these issues are relevant to the behavior of other regulatory systems as well.

1. CONSEQUENCES OF WEAK LEXA DIMERIZATION

As noted above, LexA dimerizes weakly under the in vitro conditions tested to date.[44,45] Although the in vivo dimer dissociation constant is not known, the total in vivo LexA concentration, about 1 μM, is far lower than the value for the dimer dissociation constant measured in vitro (in the range of 15-50 μM). Moreover, most of the cellular LexA is not free in solution, as judged by the contents of minicells,[51] as previously found for Lac repressor;

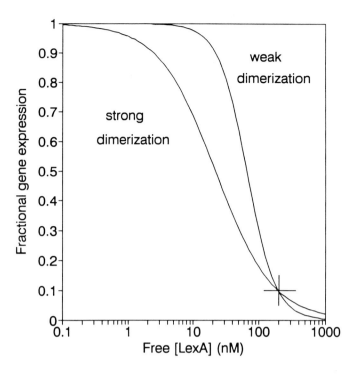

Fig. 22.3. Weak LexA dimerization results in a steep binding curve. The curves were calculated using the following assumptions: First, for the curve marked "strong dimerization" the repressor is assumed to be in the form of dimers over the entire concentration range, and the fraction of bound operator is given by $[LexA]/(K_d + [LexA])$, where K_d is the dissociation constant for the dimer binding to the operator and $[LexA]$ is the concentration of free LexA; for the curve marked "weak dimerization" the repressor is assumed to be in the form of monomers, which dimerize on the DNA with a cooperativity parameter of 10^6; occupancy is calculated as in ref. 105. Although some dimers would be present, this will not alter the proportion of occupied operator, because the concentration of dimers is proportional to that of monomers squared at values well below the dimer dissociation constant, and the equilibria in such a system are coupled (see ref. 45 for detailed discussion). Second, the dissociation constants are adjusted such that both curves give 90% occupancy at 200 nM LexA (marked by the cross) in order to simulate the behavior of a typical LexA-controlled gene with a substantial basal level of expression (10% in this case). Third, the target gene is not expressed when the operator is occupied by repressor. A similar curve was originally described for the case of λ repressor binding cooperatively to two adjacent operators.[97] A value of 200 nM LexA is assumed on the basis of minicell experiments[45] that showed this value in the cytoplasm.

presumably it is bound to operators and nonspecifically to the DNA. If we assume for the sake of discussion that most free LexA is in the monomer form in vivo, the high degree of cooperativity with which LexA binds to its operators has the consequence that small changes in LexA levels will yield large changes in operator occupancy (Fig. 22.3) when compared with a system with stable dimers. This regulatory difference was first noted for cooperative binding of λ repressor to two adjacent operators.[97] This means that the system will be more responsive to relatively minor perturbations in LexA levels. Since LexA cleavage ensues promptly after even relatively mild inducing treatments, this sensitivity ensures a rapid response to such conditions. To my knowledge, the relationship between LexA levels and reporter gene expression has not been explored systematically.

2. AUTOREGULATION AND WEAK REPRESSION OF *LEXA*

The *lexA* gene is under negative control by LexA itself.[36,37] We have speculated[36] that this feature of the regulatory system is important for two reasons: It can damp out small fluctuations in the levels of LexA, arising by chance or by replication of the *lexA* gene; more importantly, it can allow the cell to re-establish the repressed state more rapidly during the recovery phase from an inducing treatment. This situation can be simulated by computer modeling (Fig. 22.4). In this simulation, three cases are modeled: In the first (autoregulated: 2 operators), *lexA* is autoregulated with an induction ratio of 5, as seen in vivo, and LexA binds to 2 operators with no cooperative binding between the two (in the natural *lexA* gene there are two operators, but LexA binds with little or no cooperativity[98]). In the second (no autoregulation), no autoregulation occurs, and gene expression is at a constant rate. In the third case (autoregulated: 1 operator), the system has autoregulation and an induction ratio of five as before, but only a single operator, so as to assess the effects of having two operators. Promoter strengths and binding constants are adjusted to give the same steady-state level of LexA in the absence of inducing treatments. It is also assumed that LexA dimerizes weakly, as in the previous section. A pulse of activated RecA is applied for a short period of time, resulting in decreases in LexA levels. As seen, after RecA is no longer active, the autoregulated systems approach the uninduced steady-state level far more rapidly. Little difference is seen between the curves with one and two operators, implying that the presence of two operators is not crucial for a rapid return to repression. Qualitatively the same behavior is seen (not shown) for a range of RecA levels, and for a system in which LexA is assumed to form stable dimers, although in the latter case the time to return to steady-state levels is longer than in the case with weak dimerization depicted in Figure 22.4.

LexA binds less tightly to its own operators than to most other LexA-controlled operators. This is expected, since a site to which LexA binds more weakly than its own operators would often not be occupied at the LexA levels present in uninduced cells, and the gene would have a small induction ratio.

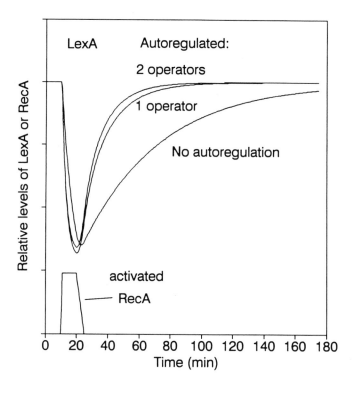

Fig. 22.4. Kinetics of recovery from SOS induction. Results of a computer simulation of the SOS regulatory system are shown. At the outset of this simulation, RecA is inactive, and the levels of LexA are at a steady-state concentration of 200 nM. A pulse of active RecA is applied, and the levels of LexA drop and then recover after RecA is no longer active. For the curve marked "no autoregulation", the rate of gene expression is assumed to be constant. For the curve marked "autoregulated: 1 operator", the gene is controlled by LexA binding to a single operator site; for this curve, the affinity of the repressor and the promoter strength are adjusted to give the same steady-state level of expression as the first curve, and to give an induction ratio of 5 (that is, at steady-state, the rate of gene expression is 20% of the derepressed rate). LexA is assumed to bind as monomers (as in Fig. 22.3), so that occupancy is $\propto$ [LexA]^2. For the curve marked "autoregulated: 2 operators", the gene is controlled by its product binding to two adjacent operators, and it is assumed that occupancy of either operator suffices to repress the gene; the affinities of repressor for these two sites are set to be the same, and binding to the two sites is not cooperative;[98] for this curve, the promoter strength is set to be the same as the previous one, LexA binds as monomers, and the dissociation constants are set to give an induction ratio of 5. The curve marked "RecA" depicts an arbitrary time course for activating RecA; in this curve, RecA is activated quickly (as it would be after UV treatment), it remains maximal for 9 minutes, then it declines linearly to zero over the next 5 minutes. Levels of activated RecA are not intended to be compared directly to the LexA levels. Although not shown for the sake of clarity, levels of activated RecA are 2-fold higher for the two curves depicting autoregulation so as to give a decrease in LexA to the same level as seen with the no-autoregulation case; the time course of RecA activation is the same for all cases. In this simulation, it is assumed that the K_m for the RecA:LexA interaction is 1 μM, that is, somewhat above the steady-state LexA level, so that the rate of LexA cleavage at any given time is given by [RecA] * V_{max} * [LexA]/([LexA] + K_m). A term is included for dilution of cellular contents during cell growth, assuming a doubling time of 35 minutes; it is assumed that this term is constant even while the SOS system is induced. Experimental curves qualitatively similar to those shown here are observed after low doses of UV irradiation,[51] although in such experiments the levels of activated RecA are not likely to match the arbitrary time-course used here. The computer program generating these curves is a numerical simulation.[105] In this type of simulation, the concentrations of all components are set at an initial value; rates of expression are calculated based on these values, and the system is allowed to run for a short period of time (Δt); the concentrations of all components are recalculated, and are input into equations for rates of expression during the next Δt period. This process is repeated in a loop for as many iterations as desired. To guard against nonlinear behavior, the system should show the same behavior when Δt is made shorter. Annotated BASIC programs for this and the curves in Figs. 22.5 and 22.6 are available by anonymous FTP at biosci.arizona.edu.

3. THE SUBINDUCED STATE

Under many conditions, the rate of LexA cleavage suffices to give low but substantial levels of LexA. This "subinduced" state[99] results in partial derepression of SOS genes. Since different SOS genes have different induction ratios, the pattern of derepression will differ at different levels of LexA. The most weakly repressed genes will be completely derepressed first, but larger relative changes should occur for the most strongly repressed genes, at least at simple promoters for which occupancy gives complete repression (Fig. 22.5). In panel A, hypothetical curves like that in Figure 22.3 ("weak dimerization") are shown for genes with induction ratios of 5, 20, 100 and 500, relative to their values at 200 nM LexA (shown by the dashed line). In panel B, the data are replotted to show the factor by which each gene is derepressed at various LexA levels. If the level of LexA is reduced 10-fold, for example (dashed line), the most weakly repressed gene (bottom curve) is almost fully derepressed, but the extent of induction for the other genes is progressively greater for those genes that are most strongly repressed. This conclusion does not depend on the assumption of weak dimerization; if LexA were present at stable dimers, the same conclusion would obtain, but over a wider range of LexA levels.

The most thorough study to date of this state examined the expression of several LexA-controlled genes in a wild-type strain in the absence or presence of Mitomycin C.[41] Behavior generally consistent with Figure 22.5 was observed, although only one level of Mitomycin C, and hence one level of subinduction, was examined.

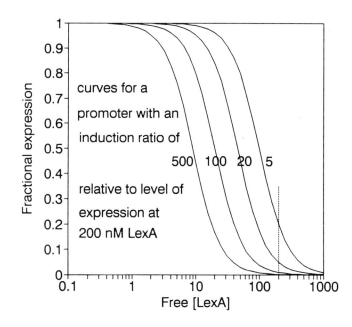

A

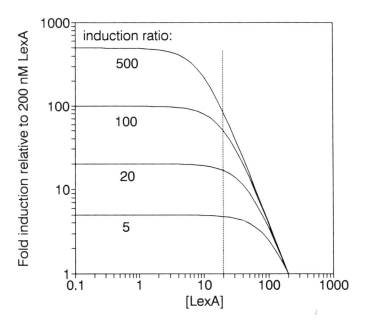

B

Fig. 22.5. The sub-induced state: Induction ratios for several LexA-controlled genes. In panel A, curves similar to the one in Fig. 22.3 (for the case of weak dimerization) are plotted for four hypothetical genes. These genes have induction ratios of 5, 20, 100 and 500, as indicated; these ratios are calculated by the ratio for full derepression over the fractional expression at 200 nM LexA (indicated by the dashed vertical line). Assumptions for calculating these curves are the same as in Fig. 22.3. In panel B, the same data are replotted. The value plotted is the ratio of gene expression at a given LexA level to that at 200 nM LexA, which is taken to represent the value of free LexA in vivo (see text). The dashed vertical line represents the case discussed in the text for which the value of free LexA has fallen by a factor of ten.

4. RELATIONSHIP OF CLEAVAGE RATES TO SOS GENE EXPRESSION

We can extend the models developed above to ask how rates of LexA cleavage affect the rates of SOS gene expression. Because there are several steps between these two processes, it is not obvious that there will be a simple relationship between the two. First, it is not known whether the weak dimerization discussed above in fact gives the steep binding curves shown in Figure 22.3; for the simulation, this will be assumed to be the case. Second, the fact that *lexA* is autoregulated means that LexA levels respond in a complicated way to cleavage (as in Fig. 22.4). Third, it is unclear whether the distribution of LexA molecules in the cell changes with differing LexA levels; for example, it is not known whether the fraction of free protein (not bound to DNA) is constant. For the simulation, this is assumed to be the case. Finally, it is not known what the in vivo K_m for LexA (treated as substrate) is in RecA-mediated cleavage. In consequence, it is unclear how the rate of LexA cleavage depends on LexA concentration.

The effects of varying K_m are seen in Figure 22.6, in which levels of gene expression are plotted as a function of the level of activated RecA in arbitrary units. In this simulation, the induction ratio of the gene was assumed to be 100 (as in Fig. 22.5 above). So that the shapes of all the curves can be compared on the same plot, the V_{max} for cleavage was adjusted so that a RecA level of 0.1 gave a 4-fold drop in LexA levels for all K_m values. It can be seen that, if the value of K_m is high, the levels of expression are roughly proportional to levels of activated RecA, but if the value of K_m is low (tight binding of LexA to activated RecA) the response is markedly nonlinear, and the system exhibits threshold behavior. I am not aware of any evidence for such threshold behavior in the SOS system, and one would not expect it to occur, since the system should be responsive to DNA damage in a graded fashion, and subinduced states can be observed.[41]

Two tentative conclusions may be reached from this analysis. First, if it approximates the real situation, it suggests that rates of reporter gene expression are at least a rough measure of cleavage rates. However, the simulation involves so many assumptions that this conclusion is not firm. Second, it suggests that the K_m for the RecA:LexA interaction is fairly high (> 100 nM), a conclusion that also makes sense in terms of the needs of the system.

Two lines of experimental evidence also suggest that the K_m in the cell is relatively high. First, a rough value for K_m measured in vitro was 500 nM.[70] Second, when the kinetics of LexA cleavage was measured after UV irradiation (in the presence of chloramphenicol to block resynthesis) using Westerns, first-order kinetics was observed,[51] as would be expected if the reaction is below the K_m. This interpretation does assume that the level of activated RecA is constant over the measured time course.

Fig. 22.6. Response of gene expression to levels of activated RecA. For each curve, steady-state levels of LexA were calculated for various levels of activated RecA (present at constant levels) by simulations like that in Fig. 22.4, using the two-operator model and the value of K_m given, and a LexA level of 200 nM in the absence of RecA. The V_{max} value for the RecA-mediated cleavage reaction is adjusted for each curve so that all values for K_m give a 4-fold drop in LexA levels at the same arbitrary level of RecA (at the point where all the curves cross). LexA levels were then converted into fractional levels of gene expression for a promoter with an induction ratio of 100, as shown in Fig. 22.6. Values for K_m are in units of nM.

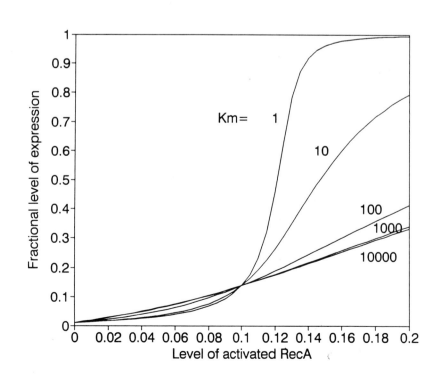

One last complication deserves mention. Workers generally measure rates of SOS gene expression by assaying levels of reporter gene product, rather than rates of gene expression; in many experiments, the rate of expression varies with time (since the level of activated RecA changes with time), further complicating analysis of rates. In summary, although it is possible that rates of gene expression are roughly proportional to levels of activated RecA, this assumption has not been validated. At present, the only dependable measure of cleavage rates is a direct examination, using Westerns or preferably pulse-chase methods.

5. EVOLUTION OF RELATIVE CLEAVAGE RATES

In vivo and in vitro data show that cleavage of LexA is about 50-fold faster than that of λ CI repressor.[7,9,23,41] This difference is due in part to intrinsic differences in the maximal rate of RecA-mediated cleavage, and in part to the fact that dimers of λ repressor are poor substrates for this reaction; at the λ repressor levels in a lysogen, most of the repressor is probably in the form of dimers. In any case, the relatively slow rate of λ repressor cleavage is consistent with the expectation that LexA cleavage should ensue promptly upon inducing treatments, since it is in the cell's interest to respond quickly even to small amounts of damage. By contrast, λ is induced efficiently only upon treatments that would kill a substantial fraction of nonlysogenic cells by lethal DNA damage,[99] suggesting that the rate of λ repressor cleavage has evolved such that it leads to induction only when the cell is in trouble and may not survive. In this view, λ has evolved to take advantage of a cellular regulatory mechanism, as it has in many other aspects of λ biology.[100]

Similarly, the rate of UmuD cleavage is far lower than that of LexA.[73,74] It is plausible that mutagenesis is a desperation measure taken by the cell if the damage cannot be repaired; it is a means of last resort to avoid cell death, even though the surviving cell suffers mutational change. One may surmise, once again, that UmuD cleavage has evolved to an optimum rate.

Speculations about the driving forces behind evolution are inevitably linked to an anthropocentric view of nature, and may be regarded as suspect for that reason. Nonetheless, the two situations just described have often cited parallels in human affairs. Prophage induction is frequently likened to rats deserting a sinking ship. Similarly, SOS mutagenesis is aptly summarized by an aphorism more widely understood in the past than now; namely, the Cold War phrase of the 1950s, "Better Red than dead."

6. REVERSIBLE STATES AND GENETIC SWITCHES

Despite being triggered by the same biochemical mechanism, SOS induction and prophage induction differ in one decisive way. SOS induction is designed to be reversible, and to allow a return to the normal growth state when the inducing treatment has been overcome. Such reversibility is the hallmark of prokaryotic regulation, since bacteria are exposed to unpredictably changing environments, to which they must respond. By contrast, prophage induction results from triggering a genetic switch,[84] which changes irreversibly the regulatory state of the viral genome and sets λ on the path to lytic growth.

An interesting recent instance has been described in which the SOS system has been converted by mutation into a genetic switch.[101] The *recA432* allele changes the properties of RecA such that, when present at high RecA432 concentrations, the protein becomes constitutively activated. Presumably, normal constituents of the cell can serve as inducing signals. At repressed levels of RecA432, as would be present in normally growing cells, the protein is not spontaneously activated, but it can be activated by DNA damaging treatments. Accordingly, when the SOS response is triggered in *recA432* hosts by DNA damage and high levels of RecA432 are produced, the system becomes locked into the induced state, and no longer requires DNA damage to perpetuate this state. Induction persists for many hours, and *recA432 sulA* strains can continue to divide. It is not known whether these cells can eventually return to the repressed state.

E. FUTURE PROSPECTS

Although the SOS regulatory system is relatively well understood, it still offers opportunities to investigate several areas of wide general interest. First, since the SOS boxes in different LexA-controlled genes are located in various places relative to the -35 and -10 regions, it is likely that the same repressor, LexA, will operate at different stages in the process of transcription initiation.[40] This diversity may lead to insights into the steps in this complex process. Moreover, LexA shows cooperative binding to certain operators, in a way that depends strongly on the spacing between operators; it can also promote bending at other operators.[40] The diversity of its actions at different sites is of general interest in the study of DNA-binding proteins. Second, the mechanism of LexA cleavage is unusual both in its chemical mechanism and in its stimulation by an external effector.[71] Further exploration of these factors should provide insights into enzyme mechanisms and self-processing reactions. Third, the physiological changes provoked by the SOS response are still poorly understood. These should provide inroads into many aspects of bacterial physiology.[14] The functions of many LexA-regulated genes are not yet known;[47] identifying these functions should broaden this picture. In addition, a number of genes are also regulated during the SOS response in ways that may be independent of LexA; it will be of interest to identify how they are regulated and how their products contribute to this response. Moreover, the physiological state of the cell apparently affects the stability of LexA;[88,89] it will be of interest to learn the factors involved in such control. Finally, the SOS system is well suited to exploring the behavior of gene regulatory systems, as typified by the examples in the previous section. An increased understanding of such systems should be applicable to a wide range of regulatory phenomena in biology.

ACKNOWLEDGMENTS

I am grateful to Carol Dieckmann for helpful comments. Work from my laboratory is supported by grants from the National Institutes of Health and the National Science Foundation.

REFERENCES

1. Georgopoulos C. Heat shock regulation. In: Lin EC, ed. This volume. 1996.
2. Demple B. Adaptive responses to oxidative stress: The soxRS and oxyR regulons. In: Lin EC, ed. This volume. 1996.
3. Guest JR. The FNR modulon and FNR-regulated gene expression. In: Lin EC, ed. This volume. 1996.
4. Kustu S, North AK, Weiss DS. Prokaryotic transcriptional enhancers and enhancer-binding proteins. Trends Biochem Sci 1991; 16:397-402.
5. Witkin EM. Ultraviolet mutagenesis and inducible DNA repair in *Escherichia coli*. Bacteriol Rev 1976; 40:869-907.

6. Radman M. SOS repair hypothesis: phenomenology of an inducible DNA repair which is accompanied by mutagenesis. In: Hanawalt P, Setlow RB, eds. Molecular Mechanisms for Repair of DNA. NY: Plenum Press, 1975:355-367.

7. Little JW, Mount DW. The SOS regulatory system of Escherichia coli. Cell 1982; 29:11-22.

8. Roberts JW, Devoret R. Lysogenic induction. In: Hendrix RW, Roberts JW, Stahl FW, Weisberg RA, eds. Lambda II. Cold Spring Harbor, NY: Cold Spring Harbor Laboratory, 1983:123-144.

9. Little JW. The SOS regulatory system: control of its state by the level of RecA protease. J Mol Biol 1983; 167:791-808.

10. Witkin EM. From Gainesville to Toulouse: The evolution of a model. Biochimie 1982; 64:549-555.

11. Walker GC. Mutagenesis and inducible responses to deoxyribonucleic acid damage in *Escherichia coli*. Microbiol Rev 1984; 48:60-93.

12. Witkin EM. RecA protein in the SOS response: Milestones and mysteries. Biochimie 1991; 73:133-141.

13. Little JW. Mechanism of specific LexA cleavage: Autodigestion and the role of RecA coprotease. Biochimie 1991; 73:411-422.

14. Friedberg EC, Walker GC, Siede W. DNA Repair and Mutagenesis. Amer. Soc. Microbiol. Press, 1995.

15. Witkin EM. The radiation sensitivity of *Escherichia coli* B: A hypothesis relating filament formation and prophage induction. Proc Natl Acad Sci USA 1967; 57:1275-1279.

16. Clark AJ, Margulies AD. Isolation and characterization of recombination-deficient mutants of *Escherichia coli* K-12. Proc Natl Acad Sci USA 1965; 53:451-459.

17. Mount DW, Low KB, Edmiston SJ. Dominant mutations (*lex*) in *Escherichia coli* K-12 which affect radiation sensitivity and frequency of ultraviolet-induced mutations. J Bacteriol 1972; 112:886-893.

18. Keller EF. A feeling for the organism: The life and work of Barbara McClintock. San Francisco:W.H.Freeman, 1983.

19. Brooks K, Clark AJ. Behavior of λ bacteriophage in a recombination deficient strain of *E. coli*. Virology 1967; 1:283-293.

20. Hertman I, Luria SE. Transduction studies on the role of a *rec*⁺ gene in ultraviolet induction of prophage lambda. J Mol Biol 1967; 23:117-133.

21. Kirby EP, Jacob F, Goldthwait DA. Prophage induction and filament formation in a mutant strain of *Escherichia coli*. Proc Natl Acad Sci USA 1967; 58:1903-1910.

22. Defais M, Fauquet P, Radman M, Errera M. Ultraviolet reactivation and ultraviolet mutagenesis of λ in different genetic systems. Virology 1971; 43:495-503.

23. Roberts JW, Roberts CW. Proteolytic cleavage of bacteriophage lambda repressor in induction. Proc Natl Acad Sci USA 1975; 72:147-151.

24. Gudas LJ, Pardee AB. Model for regulation of *Escherichia coli* DNA repair functions. Proc Natl Acad Sci USA 1975; 72:2330-2334.

25. McEntee K, Hesse JE, Epstein W. Identification and radiochemical purification of the *recA* protein of *Escherichia coli* K-12. Proc Natl Acad Sci USA 1976; 73:3979-3983.

26. Emmerson PT, West SC. Identification of protein X of *Escherichia coli* as the *recA*⁺/*tif* gene product. Mol Gen Genet 1977; 155:77-85.

27. Gudas LJ, Mount DW. Identification of the *recA* (*tif*) gene product of *Escherichia coli*. Proc Natl Acad Sci USA 1977; 74:5280-5284.

28. McEntee K. Protein X is the product of the *recA* gene of *Escherichia coli*. Proc Natl Acad Sci USA 1977; 74:5275-5279.

29. Roberts JW, Roberts CW, Craig NL. *Escherichia coli recA* gene product inactivates phage λ repressor. Proc Natl Acad Sci USA 1978; 75: 4714-4718.

30. Kowalczykowski SC, Dixon DA, Eggleston AK, Lauder SD, Rehrauer WM. Biochemistry of homologous recombination in *Escherichia coli*. Microbiol Rev 1994; 58:401-465.

31. Craig NL, Roberts JW. *E. coli* recA protein-directed cleavage of phage λ repressor requires polynucleotide. Nature 1980; 283:26-30.

32. Little JW, Mount DW, Yanisch-Perron CR. Purified lexA protein is a repressor of the *recA* and *lexA* genes. Proc Natl Acad Sci USA 1981; 78:4199-4203.

33. Little JW, Edmiston SH, Pacelli LZ, Mount DW. Cleavage of the *Escherichia coli lexA* protein by the *recA* protease. Proc Natl Acad Sci USA 1980; 77:3225-3229.

34. Brent R, Ptashne M. Mechanism of action of the *lexA* gene product. Proc Natl Acad Sci USA 1981; 78:4204-4208.

35. Horii T, Ogawa T, Nakatani T, Hase T, Matsubara H, Ogawa H. Regulation of SOS functions: purification of E. coli LexA protein and determination of its specific site cleaved by the RecA protein. Cell 1981; 27:515-522.

36. Little JW, Harper JE. Identification of the *lexA* gene product of *Escherichia coli* K-12. Proc Natl Acad Sci USA 1979; 76:6147-6151.

37. Brent R, Ptashne M. The *lexA* gene product represses its own promoter. Proc Natl Acad Sci USA 1980; 77:1932-1936.

38. Frank EG, Hauser J, Levine AS, Woodgate R. Targeting of the UmuD, UmuD', and MucA' mutagenesis proteins to DNA by RecA protein. Proc Natl Acad Sci USA 1993; 90:8169-8173.

39. Mount DW. A mutant of *Escherichia coli* showing constitutive expression of the lysogenic induction and error-prone DNA repair pathways. Proc Natl Acad Sci USA 1977; 74:300-304.

40. Schnarr M, Oertel-Buchheit P, Kazmaier M, Granger-Schnarr M. DNA binding properties of the LexA repressor. Biochimie 1991; 73:423-431.

41. Peterson KR, Mount DW. Differential repression of SOS genes by unstable LexA41 (Tsl-1) protein causes a "split-phenotype" in *Escherichia coli* K-12. J Mol Biol 1987; 193:27-40.

42. Weisemann JM, Weinstock GM. The promoter of the *recA* gene of *Escherichia coli*. Biochimie 1991; 73:457-470.

43. Sancar GB, Sancar A, Little JW, Rupp WD. The *uvrB* gene of *Escherichia coli* has both lexA-repressed and lexA-independent promoters. Cell 1982; 28:523-530.

44. Schnarr M, Pouyet J, Granger-Schnarr M, Daune M. Large-Scale purification, oligomerization equilibria, and specific interaction of the LexA repressor of *Escherichia coli*. Biochem 1985; 24:2812-2818.

45. Kim B, Little JW. Dimerization of a specific DNA-binding protein on the DNA. Science 1992; 255:203-206.

46. Peterson KR, Ossanna N, Thliveris AT, Ennis DG, Mount DW. Derepression of specific genes promotes DNA repair and mutagenesis in *Escherichia coli*. J Bacteriol 1988; 170:1-4.

47. Lewis LK, Harlow GR, Gregg-Jolly LA, Mount DW. Identification of high affinity binding sites for LexA which define new DNA damage-inducible genes in *Escherichia coli*. J Mol Biol 1994; 241:507-523.

48. Petit C, Cayrol C, Lesca C, Kaiser P, Thompson C, Defais M. Characterization of *dinY*, a new *Escherichia coli* DNA repair gene whose products are damage inducible even in a *lexA*(Def) background. J Bacteriol 1993; 175:642-646.

49. Tadmor Y, Bergstein M, Skaliter R, Shwartz H, Livneh Z. β subunit of DNA polymerase III holoenzyme is induced upon ultraviolet irradiation or nalidixic acid treatment of *Escherichia coli*. Mutat Res Fundam Mol Mech Mutagen 1994; 308:53-64.

50. Lamont I, Brumby AM, Egan JB. UV induction of coliphage 186: Prophage induction as an SOS function. Proc Natl Acad Sci USA 1989; 86:5492-5496.

51. Sassanfar M, Roberts JW. Nature of the SOS-inducing signal in *Escherichia coli*. The involvement of DNA replication. J Mol Biol 1990; 212:79-96.

52. Tessman ES, Peterson P. Plaque color method for rapid isolation of novel *recA* mutants of *Escherichia coli* K-12: new classes of protease-constitutive *recA* mutants. J Bacteriol 1985; 163:677-687.

53. Wang W-B, Sassanfar M, Tessman I, Roberts JW, Tessman ES. Activation of protease-constitutive RecA proteins of *Escherichia coli* by all of the common nucleoside triphosphates. J Bacteriol 1988; 170:4816-4822.

54. Wang W-B, Tessman ES, Tessman I. Activation of protease-constitutive RecA proteins of *Escherichia coli* by rRNA and tRNA. J Bacteriol 1988; 170:4823-4827.

55. Lavery PE, Kowalczykowski SC. Biochemical basis of the constitutive repressor cleavage activity of recA730 protein. A comparison to recA441 and recA803 proteins. J Biol Chem 1992; 267:20648-20658.

56. Lin LL, Little JW. Isolation and characterization of noncleavable (Ind⁻) mutants of the LexA repressor of *Escherichia coli* K-12. J Bacteriol 1988; 170:2163-2173.

57. Eguchi Y, Ogawa T, Ogawa H. Stimulation of RecA-mediated cleavage of phage ϕ80 cI repressor by deoxydinucleotides. J Mol Biol 1988; 204:69-77.

58. Irbe RM, Morin LME, Oishi M. Prophage (ϕ80) induction in *Escherichia coli* K-12 by specific deoxyoligonucleotides. Proc Natl Acad Sci USA 1981; 78:138-142.

59. Yu X, Egelman EH. Structural data suggest that the active and inactive forms of the RecA filament are not simply interconvertible. J Mol Biol 1992; 227:334-346.

60. DiCapua E, Cuillel M, Hewat E, Schnarr M, Timmins PA, Ruigrok RWH. Activation of RecA protein. The open helix model for LexA cleavage. J Mol Biol 1992; 226:707-719.

61. Ruigrok RWH, Bohrmann B, Hewat E, Engel A, Kellenberger E, DiCapua E. The inactive form of recA protein: The 'compact' structure. EMBO J 1993; 12:9-16.

62. Story RM, Weber IT, Steitz TA. The structure of the *E. coli recA* protein monomer and polymer. Nature 1992; 355:318-325.

63. Yu X, Egelman EH. The LexA repressor binds within the deep helical groove of the activated RecA filament. J Mol Biol 1993; 231:29-40.

64. Dutreix M, Burnett B, Bailone A, Radding CM, Devoret R. A partially deficient mutant, *recA1730*, that fails to form normal nucleoprotein filaments. Mol Gen Genet 1992; 232:489-497.

65. Menetski JP, Kowalczykowski SC. Biochemical properties of the *Escherichia coli* RecA430 protein. Analysis of a mutation that affects the interaction of the ATP-recA protein complex with single-stranded DNA. J Mol Biol 1990; 211:845-855.

66. Roberts JW, Roberts CW. Two mutations that alter the regulatory activity of *E. coli* recA protein. Nature 1981; 290:422-424.

67. Eguchi Y, Ogawa T, Ogawa H. Cleavage of bacteriophage ϕ80 CI repressor by RecA protein. J Mol Biol 1988; 202:565-573.

68. Little JW. Autodigestion of lexA and phage lambda repressors. Proc Natl Acad Sci USA 1984; 81:1375-1379.

69. Slilaty SN, Rupley JA, Little JW. Intramolecular cleavage of LexA and phage lambda repressors: dependence of kinetics on repressor concentration, pH, temperature, and solvent. Biochemistry 1986; 25:6866-6875.

70. Lin LL, Little JW. Autodigestion and RecA-dependent cleavage of Ind⁻ mutant LexA proteins. J Mol Biol 1989; 210:439-452.

71. Little JW. LexA cleavage and other self-processing reactions. J Bacteriol 1993; 175:4943-4950.

72. Gimble FS, Sauer RT. λ repressor inactivation: properties of purified Ind⁻ proteins in the autodigestion and RecA-mediated cleavage reactions. J Mol Biol 1986; 192:39-47.

73. Burckhardt SE, Woodgate R, Scheuermann RH, Echols H. UmuD mutagenesis protein of *Escherichia coli*: Overproduction, purification, and cleavage by RecA. Proc Natl Acad Sci USA 1988; 85:1811-1815.

74. Shinagawa H, Iwasaki H, Kato T, Nakata A. RecA protein-dependent cleavage of UmuD protein and SOS mutagenesis. Proc Natl Acad Sci USA 1988; 85:1806-1810.

75. Roland KL, Smith MH, Rupley JA, Little JW. In vitro analysis of mutant LexA proteins with an increased rate of specific cleavage. J Mol Biol 1992; 228:395-408.

76. Slilaty SN, Little JW. Lysine-156 and serine-119 are required for LexA repressor cleavage: a possible mechanism. Proc Natl Acad Sci USA 1987; 84:3987-3991.

77. Roland KL, Little JW. Reaction of LexA repressor with diisopropyl fluorophosphate. A test of the serine protease model. J Biol Chem 1990; 265:12828-12835.

78. Kim B, Little JW. LexA and λ CI repressors as enzymes: Specific cleavage in an intermolecular reaction. Cell 1993; 73:1165-1173.

79. Smith MH, Cavenagh MM, Little JW. Mutant LexA proteins with an increased rate of in vivo cleavage. Proc Natl Acad Sci USA 1991; 88:7356-7360.

80. Slilaty SN, Vu HK. The role of electrostatic interactions in the mechanism of peptide bond hydrolysis by a Ser-Lys catalytic dyad. Protein Eng 1991; 4:919-922.

81. Nohmi T, Battista JR, Dodson LA, Walker GC. RecA-mediated cleavage activates UmuD for mutagenesis: mechanistic relationship between transcriptional derepression and posttranslational activation. Proc Natl Acad Sci USA 1988; 85:1816-1820.

82. Sauer RT, Yocum RR, Doolittle RF, Lewis M, Pabo CO. Homology among DNA-binding proteins suggests use of a conserved super-secondary structure. Nature 1982; 298:447-451.

83. Battista JR, Ohta T, Nohmi T, Sun W, Walker GC. Dominant negative *umuD* mutations decreasing RecA-mediated cleavage suggest roles for intact UmuD in modulation of SOS mutagenesis. Proc Natl Acad Sci USA 1990; 87:7190-7194.

84. Ptashne M. A Genetic Switch: Phage λ and Higher Organisms. Cambridge, MA:Cell Press and Blackwell Scientific Publications, 1992. 2nd Ed.

85. Valenzuela D, Ptashne M. P22 repressor mutants deficient in co-operative binding and DNA loop formation. EMBO J 1989; 8:4345-4350.

86. Whipple FW, Kuldell NH, Cheatham LA, Hochschild A. Specificity determinants for the interaction of λ repressor and P22 repressor dimers. Genes Dev 1994; 8:1212-1223.

87. Benson N, Adams C, Youderian P. Genetic selection for mutations that impair the co-operative binding of lambda repressor. Mol Microbiol 1994; 11:567-579.

88. Dri A-M, Moreau PL. Phosphate starvation and low temperature as well as ultraviolet irradiation transcriptionally induce the *Escherichia coli* LexA-controlled gene *sfiA*. Mol Microbiol 1993; 8:697-706.

89. Dri A-M, Moreau PL. Control of the LexA regulon by pH: evidence for a reversible inactivation of the LexA repressor during the growth cycle of *Escherichia coli*. Mol Microbiol 1994; 12:621-629.

90. Roca AI, Cox MM. The RecA protein: structure and function. Crit Rev Biochem Mol Biol 1990; 25:415-456.

91. Ogawa T, Yu X, Shinohara A, Egelman EH. Similarity of the yeast RAD51 filament to the bacterial RecA filament. Science 1993; 259:1896-1899.

92. Cheo DL, Bayles KW, Yasbin RE. Elucidation of regulatory elements that control damage induction and competence induction of the *Bacillus subtilis* SOS system. J Bacteriol 1993; 175:5907-5915.

93. Lovett CM,Jr., Roberts JW. Purification of a RecA protein analog from *Bacillus subtilis*. J Biol Chem 1985; 260:3305-3313.

94. Love PE, Yasbin RE. Induction of the *Bacillus subtilis* SOS-like response by *Escherichia coli* RecA protein. Proc Natl Acad Sci USA 1986; 83:5204-5208.

95. Lovett CM,Jr., Cho KC, O'Gara TM. Purification of an SOS repressor from *Bacillus subtilis*. J Bacteriol 1993; 175:6842-6849.

96. Raymond-Denise A, Guillen N. Identification of *dinR*, a DNA damage-inducible regulator gene of *Bacillus subtilis*. J Bacteriol 1991; 173: 7084-7091.

97. Johnson AD, Poteete AR, Lauer G, Sauer RT, Ackers GK, Ptashne M. λ Repressor and cro—components of an efficient molecular switch. Nature 1981; 294:217-223.

98. Brent R. Regulation and autoregulation by *lexA* protein. Biochimie 1982; 64:565-569.

99. Bailone A, Levine A, Devoret R. Inactivation of prophage λ repressor in vivo. J Mol Biol 1979; 131:553-572.

100. Friedman DI. Interaction between bacteriophage lambda and its Escherichia coli host. Curr Opin Genet Dev 1992; 2:727-738.

101. Ennis DG, Little JW, Mount DW. Novel mechanism for UV sensitivity and apparent UV nonmutability of *recA432* mutants: Persistent LexA cleavage following SOS induction. J Bacteriol 1993; 175:7373-7382.

102. Sauer RT. DNA sequence of the bacteriophage λ cI gene. Nature 1978; 276:301-302.

103. Oberto J, Weisberg RA, Gottesman ME. Structure and function of the *nun* gene and the immunity region of the lambdoid phage HK022. J Mol Biol 1989; 207:675-693.

104. Carlson NG, Little JW. Highly cooperative DNA binding by the coliphage HK022 repressor. J Mol Biol 1993; 230:1108-1130.

105. Keen RE, Spain JD. Computer Simulation in Biology: A BASIC Introduction. New York:Wiley-Liss, 1992.

CHAPTER 23

HEAT SHOCK REGULATION

Dominique Missiakas, Satish Raina and Costa Georgopoulos

I. INTRODUCTION

A. OVERVIEW

The heat shock or stress response of *Escherichia coli* has evolved in order to detect and deal with the presence of unfolded, misfolded, damaged or aggregated polypeptide chains. At the present, two major regulons are known to control this response. The "classical" heat shock regulon has evolved to deal with intracellular protein perturbations and is under the positive control of the σ^{32} transcription factor (the *rpoH* gene product) and the negative control of some of the heat shock proteins themselves. The newly discovered second heat shock regulon has evolved to deal with protein misfolding/aggregation/imbalance in the outer cellular compartments. It is under the positive control of the σ^E transcription factor (the *rpoE* gene product). The two heat shock regulons appear to be interconnected, inasmuch as the σ^E factor participates in the transcriptional regulation of the σ^{32}-encoding gene, especially at very high temperatures.

B. THE PROBLEM OF PROTEIN FOLDING, UNFOLDING AND AGGREGATION

The pioneering work of Anfinsen[1] with the in vitro refolding of purified ribonuclease A left the long-lasting impression that the folding of a newly synthesized polypeptide chain is an intrinsic feature of its primary structure, independent of other factors (reviewed in ref. 2). Most of the in vitro protein refolding experiments are carried out by first denaturing a given purified polypeptide, and then removing the denaturant. Under these conditions, most polypeptides quickly collapse into a compact structure, usually referred to as "molten globule," which possesses extensive secondary structure, yet exposes hydrophobic groups, making the improperly folded molecule prone to aggregation. The probability that a given unfolded polypeptide will fold properly increases at relatively low protein concentrations (which limit interpolypeptide aggregation) and low temperature (which attenuates hydrophobic interactions). However, in vivo there is an acute problem of protein aggregation primarily because of the extremely high intracellular protein concentration (100-150 mg/ml) and the relatively high temperature (37°C in humans and *E. coli*) which favor hydrophobic interactions.

Regulation of Gene Expression in Escherichia coli, edited by E. C. C. Lin and A. Simon Lynch. © 1996 R.G. Landes Company.

As a first approximation, it appears as if all organisms have evolved to live in a particular niche so that the overall structure and stability of their proteins are maintained within the normal temperature range encountered by them. Outside this temperature range, and especially at the high end, significant protein unfolding and aggregation of both newly synthesized and pre-existing folded proteins occur. In the case of *E. coli,* which can grow at temperatures up to 45-46°C, the protein unfolding/aggregation problem becomes particularly acute at temperatures above 43°C (see below).

To deal with the problem of potential protein unfolding/aggregation, a set of universally conserved proteins has evolved, collectively referred to as molecular chaperones.[3-5] These chaperone proteins act primarily by binding to the reactive surfaces of polypeptides (such as the hydrophobic surfaces exposed on "molten globule" intermediate structures). In doing so, chaperones sequester polypeptides from the rest of the reactive surfaces present in their vicinity, thus effectively preventing aggregation and favoring the proper folding pathway. The chaperone proteins act without covalently modifying their polypeptide substrates and without being part of the finished product.[3] Because high temperatures tend to favor both protein unfolding on the one hand and hydrophobic interactions on the other, there is an extra need for chaperones to prevent drastic protein unfolding/aggregation, to enhance disaggregation and even to promote proteolysis of terminally damaged proteins (see below). This is most likely the reason why many chaperones and proteases are expressed at higher levels in cells at elevated temperatures, especially following a heat shock. In the case of *E. coli* at 46°C, a temperature at which bacterial growth almost ceases, at least 30% of all polypeptides synthesized are either chaperones or proteases (see below).[6]

Although the bulk of the so-called heat shock or stress proteins, defined as those whose rate of synthesis preferentially increases with an increase in temperature, are either chaperones or proteases, not all molecular chaperones or proteases belong to the heat shock class of proteins (Table 23.1). What is clear, however, is that without adequate levels of chaperones, many of the intracellular proteins would aggregate in vivo. This was clearly shown in *E. coli* by Gragerov et al,[7] who demonstrated wholesale protein aggregation in the absence of sufficient "chaperone power" and the ability of either the DnaK or GroEL chaperone machine (see below) to deal successfully with such aggregation when overproduced.

II. PROPERTIES OF IMPORTANT HEAT SHOCK PROTEINS

A. THE DNAK/DNAJ/GRPE CHAPERONE MACHINE

The *dnaK, dnaJ* or *grpE* genes were originally discovered because mutations in them blocked bacteriophage λ DNA replication (reviewed in refs. 8-11). The genes *dnaK* and *dnaJ* were so named because mutations in either of them blocked *E. coli* DNA replication at nonpermissive temperature. The *grpE* gene was so named because mutations in it blocked bacteriophage λ DNA replication and compensatory mutations mapped in the bacteriophage P gene (*grpE, groP*-like complementation group E). The reason why the DnaK, DnaJ and GrpE proteins are referred to as a chaperone "machine"[9] is because they work together to carry out an interesting array of biological functions. These functions include the protection of various unfolded proteins from premature aggregation, the protection of proteins from thermally induced

Table 23.1. Heat shock proteins in E. coli

	Gene position min	MW (kDa) of monomer	Comments	Reference
σ³² Controlled				
Chaperones				
DnaK	0.3	70	Hsp70 homolog	Georgopoulos et al[11]
DnaJ	0.3	39	co-transcribed with *dnaK*	Georgopoulos et al[11]
ClpX	10	46	also a protease activator	Wawrzynow et al[48]
HtpG	10.8	70	suppressor of *secY*	Ueguchi and Ito[105]
GrpE	56.8	26	nucleotide exchange factor	Liberek et al[21]
IbpB (HtpE, HslS)	83.0	16.3		Allen et al[106]
IbpA(HtpN, HslT)	83.0	15.8	co-transcribed with IbpB	Allen et al,[106] Chuang et al[107]
GroEL	94.2	60	essential	Fayet et al,[26] Georgopoulos et al[11]
GroES	94.2	16	essential co-transcribed with *GroEL*	Fayet et al,[26] Georgopoulos et al[11]
Proteases				
Lon/La	10.0	89		Gottesman and Maurizi[40]
ClpP	10.0	24 (22)		Gottesman and Maurizi[40]
ClpX	10.0	46	protease/chaperone co-transcribed with *clpP*	Wojkowiak et al,[46] Gottesman et al[44]
ClpA	19.0	87	protease/chaperone	Gottesman and Maurizi[40]
ClpB	56.0	84	putative protease	Gottesman and Maurizi[40]
HflB (FtsH)	69.2	70	essential	Herman et al[61,62]
HslV (HtpO)	89.0	19	co-transcribed with *hslU* ATP(HslU)-dependent Thr protease	Chuang et al[107] Missiakas and Raina (in preparation)
HslU (HtpI)	89.0	49.5	highly homologous to ClpX	Chuang et al[107]
Regulators				
HtpY	0.3	21	exact function unknown	Missiakas et al[92]
σ⁷⁰	67.0	70	housekeeping sigma factor	Burton et al[108]
HtrC	90.0	21	exact function unkown	Raina and Georgopoulos[109]
Metabolic Enzymes				
GAPDH	22.6	35	biosynthetic enzyme	Charpentier and Branlant[110]
HtrM (RfaD)	81.2	34	epimerase	Raina and Georgopoulos[87]
Unknown Functions				
HslA	10.0	65		Chuang and Blattner[111]
HslC	80.0	80		Chuang and Blattner[111]
HslK	39.8	49		Kornitzer et al[112]
HtpX	40.3	32		Chuang and Blattner[111]
FtsJ	69.2	26	co-transcribed with *hflB* (*ftsH*)	Tomoyasu et al[113]
HslO	75.0	33		Chuang and Blattner[111]
HslP	75.0	30		Chuang and Blattner[111]
HslW	94.2	22		Chuang and Blattner[111]
HslX	94.8			Chuang and Blattner[111]
HslY	94.8	45		Chuang and Blattner[111]
HslZ	94.8	37		Chuang and Blattner[111]
σ²⁴ Controlled				
Protease				
HtrA (DegP)	3.9	50	periplasmic protein	Lipinska et al[85]
Regulators				
σ²⁴/σ^E	55.5	21.6		Raina et al,[75] Rouvière et al[77]
RseA	55.5	24.3	co-transcribed with σ^E negative regulator of σ^E	Raina et al[75] Missiakas and Raina (in preparation)
RseB	55.5	35.8	co-transcribed with σ^E negative regulator of σ^E	Missiakas and Raina (in preparation)
RseC	55.5	17	co-transcribed with σ^E positive regulator of σ^E	Missiakas and Raina (in preparation)
σ³²	77.5	32		Yura et al[53]
σ⁵⁴ Controlled				
PspABCD	29.2	28 (PspA)	stationary phase survival protein transport	Weiner et al[100] Kleerebem et al[99]

aggregation, the disaggregation of certain protein aggregates (such as heat-inactivated RNA polymerase of *E. coli*),[12,13] as well as the promotion of protein export.[14,15]

Liberek et al[16] showed that part of the synergistic action of the DnaK/DnaJ/GrpE chaperone machine is due to the fact that DnaK's weak ATPase activity is stimulated up to 50-fold in the joint presence of DnaJ and GrpE. The DnaJ protein can act catalytically to specifically accelerate the rate of hydrolysis of the DnaK-bound ATP,[16] whereas GrpE induces the release of all DnaK-bound nucleotides. This synergistic action of DnaJ and GrpE allows DnaK to efficiently cycle through its various forms, i.e., free DnaK, DnaK-ADP, and DnaK-ATP. Various studies, summarized by Hendrick and Hartl,[5] have shown that the DnaK chaperone appears to bind polypeptides in the "stretched" conformation,[17] whereas the DnaJ chaperone binds to polypeptides in their "molten globule" state.[18]

Schmid et al[19] recently showed, using fast-flow kinetics, that whereas ATP binding to DnaK causes a 47-fold increase in the rate of DnaK-substrate peptide complex formation, it causes an even greater, 440-fold increase in the rate of peptide dissociation. As a net result, the release of substrate from DnaK is favored in the presence of ATP, in agreement with the results of Palleros et al[20] and Liberek et al.[21] In contrast, in the presence of DnaJ, the DnaK chaperone can bind even more tightly to its polypeptide substrates, but in this case ATP hydrolysis is required (summarized in ref. 5). Although the resulting DnaJ/substrate/DnaK three-body complex is very stable, certain complexes can be destabilized in the presence of GrpE, an action thought to be mediated through the release of DnaK-bound ADP.[18] However, not all such complexes respond to GrpE to the same degree, as evidenced by the fact that GrpE does not play a major role in either the formation or destabilization of the DnaK/σ^{32}/DnaJ complex.[22,23]

DnaJ does not need to be part of the DnaK/substrate/DnaJ complex, but can also act catalytically to accelerate DnaK's ATPase activity and enable it to bind effectively to its substrate in the presence of ATP.[23] This "activation" model of DnaK by DnaJ may have physiological relevance since there is approximately 10-fold more DnaK than DnaJ in an *E. coli* cell.[24] Thus, when DnaJ is in excess, a stable DnaK/substrate/DnaJ complex will form. When DnaJ is limiting, it can still act catalytically to allow the formation of a DnaK/substrate complex in the presence of ATP.

B. The GroES/GroEL Chaperonin Machine

The *groE* locus of *E. coli* was originally discovered in the early 1970s because mutations in it blocked the morphogenesis of many bacteriophages (reviewed in refs. 8, 25). Subsequently, the *groE* locus was shown to comprise an operon of two genes, *groES* and *groEL* ("S" indicates the smaller and "L" the larger of the two gene products). The name *groE* was assigned because bacteriophage growth (*gro*) was blocked and because the first isolated bacteriophage λ compensatory mutations mapped in the λ*E* gene (hence, *groE*).

From the early observations made with bacteriophages, it was thought that GroES/GroEL were uniquely involved in the oligomerization of morphogenetic proteins. However, subsequent genetic and biochemical studies showed that the *groES* and *groEL* genes were necessary for host macromolecular synthesis as well (summarized in ref. 25). Of particular importance were the genetic discoveries that neither of the two genes could be deleted under all conditions studied[26] and that their overproduction suppressed the temperature-sensitive (Ts⁻) phenotype of various missense mutations.[27] These studies, coupled with the observation that *groES* or *groEL* mutations often result in a decrease in overall proteolysis,[28] strongly suggested a potential role in protein stability (see below).

The GroES and GroEL proteins have been purified to homogeneity and their biochemical properties studied in detail in many different laboratories.[10,29] The GroEL protein has been recently crystallized and its structure solved.[30] Briefly, the native GroEL protein is a large cylindrical structure consisting of two stacked heptameric rings with a central hole, the 14 subunits being 60 kDa each. The GroES protein is also a ring of 7 subunits, each subunit being 10.5 kDa. The GroES and GroEL structures interact physically, the GroES moiety binding to either one or both ends of the GroEL cylinder to give rise to either a bullet-shaped or American football-shaped particle.[31,32] GroES both inhibits and synchronizes the weak ATPase activity of GroEL.[25] Interestingly, the bacteriophage T4-encoded gp31 protein interacts with and modulates GroEL's activities in a manner identical to that of GroES despite extremely limited amino acid sequence homology.[33]

Various laboratories have shown that GroEL can bind promiscuously to many unfolded polypeptides, but not their corresponding folded forms (pioneering this work were Bochkareva et al[34] and Goloubinoff et al[35]). The unfolded polypeptide form that GroEL recognizes appears to be that of a "molten globule," a collapsed state that possesses a great deal of secondary structure, but still exhibits hydrophobic groups. Most likely the polypeptide substrate binds in the central hole of the GroEL structure, shielding reactive groups and minimizing possible inadvertent interpolypeptide aggregation. Most studies suggest that either one or at the most two substrate molecules are bound by GroEL, its central cavity being large enough to accommodate a protein perhaps as large as the 75 kDa peroxisomal alcohol oxidase.[32,36] Hartl and co-workers[5,18,29] proposed a sequential protein folding pathway, with the DnaK chaperone machine first binding to the nascent polypeptide chain as it emerges from the ribosome thus preventing its early aggregation. The participation of GrpE accelerates the release of the DnaK/DnaJ/nascent polypeptide, whose semifolded, molten-globule form can be bound by GroEL. The GroEL-bound polypeptide will be released with the aid of GroES. The released polypeptide will quickly rebind to GroEL if it still exhibits sufficient hydrophobic surfaces, but will not if it is on its way to proper folding. Thus, although it may take a few rounds of binding/release, a polypeptide chain will eventually fold properly.

An interesting observation made with the GroES/GroEL chaperone machine is that it may possess the ability to "unfold" certain polypeptide chains, such as refolded pre-β-lactamase,[37] or cyclophilin.[38] Perhaps the GroEL-bound polypeptide chain can undergo many reversible conformational changes, which enable it to follow a productive folding pathway. The apparent participation of the GroES/GroEL chaperone machine in generalized proteolysis could be mediated by its ability to induce such net unfolding in proteins, or to disaggregate certain protein aggregates, such as the RNAP-σ^{70} holoenzyme heat-induced aggregates (see below).[13,23]

C. THE LON HEAT SHOCK PROTEASE

The Lon (La) protease was one of the first heat shock proteins to be identified, purified and extensively characterized. The gene was named *lon* because mutations in it resulted in long (lon) filament formation following UV treatment.[39] The Lon protein serves as the prototype of the ATP-dependent proteases. An extensive summary of its biology and biochemistry has been presented by Gottesman and Maurizi (see also chapter 24).[40] An interesting observation is the discovery that the Lon protease can be found as part of a large complex, which contains in addition to an unfolded polypeptide, such as PhoA61, the DnaK and GrpE chaperones.[41] The association of the PhoA61 protein with DnaK and Lon correlates with its rate of degradation, because in a *dnaK* deletion background PhoA61 is degraded half as rapidly. In a *dnaK756* mutant background, the rate of proteolysis of PhoA61 is even further enhanced. The reason for this acceleration could be due to the fact that DnaK756 is defective in the release of its substrates,[21] thus enhancing the accessibility of the polypeptide to proteases. The recent mitochondrial studies of Wagner et al[42] are in full agreement with the conclusion that the Lon-equivalent protease and the DnaK-equivalent chaperone machine play crucial roles in mitochondrial protein degradation.

D. The Clp Proteases

One of the most interesting developments in the fields of both ATP-dependent proteases and chaperone systems has been the discovery and extensive characterization of the Clp family of ATP-dependent proteases. The genes are named *clp* because they code for caseinolytic proteases.[43] This family of proteases has a bipartite nature, being composed of two types of subunits (reviewed in refs. 40, 44). Briefly, the ClpP component is made up of two hexameric rings and possesses an ATP-independent protease activity for very short peptides but not for longer polypeptides. However, the $ClpP_{12}/ClpA_7$ bipartite system becomes an ATP-dependent protease capable of degrading specific polypeptides such as casein and the P1 RepA protein, but not others, such as λO.[10,40,45] In contrast, ClpP can also function in conjunction with the ClpX protein as an ATP-dependent protease to hydrolyze λO, but not casein.[46] Most likely, the ClpA and ClpX proteins govern substrate specificity by somehow "presenting" to ClpP the appropriate polypeptide substrate. In this respect, it is of interest that both the ClpA and ClpX proteins have been recently shown to behave as bona fide chaperone proteins.

Wickner et al[45] showed that purified ClpA protein (a) dissociates P1 RepA dimers into monomers, with the help of ATP, thus functioning in a fashion analogous to that of the DnaK/DnaJ chaperone system,[47] (b) ClpA forms a stable complex with RepA in the presence of a nonhydrolyzable ATP analog, (c) ClpA's presence is required for the proteolysis of RepA protein by ClpP, even when RepA is first monomerized by the DnaK/DnaJ system, and (d) ClpA protects firefly luciferase from irreversible aggregation at high temperature in the absence of ATP, but cannot rejuvenate such aggregates once formed, either with or without ATP. Similarly, Wawrzynow et al[48] showed that purified ClpX protein can also act as a bona fide chaperone, since (a) ClpX protects λO protein from heat-induced aggregation, (b) dissociates heat-induced λO aggregates with the help of ATP, (c) ClpX's ATPase activity is stimulated in the presence of λO protein, and (d) ClpX binds better to denatured firefly luciferase than to the native form.

The ClpA, ClpB and ClpX proteins of *E. coli* belong to a very large, highly conserved superfamily of proteins that include the Hsp100 proteins of eukaryotes.[44,49] Recently, Parsell et al[50] have demonstrated that the Hsp104 family member of yeast behaves like a bona fide chaperone in vivo since it mediates the resolubilization of heat-induced aggregates of the *Vibrio harveyi* luciferase, as well as the resolubilization of wholesale, large electron-dense aggregates in the cytoplasm and nucleus. Surprisingly however, it does not protect yeast from the formation of such aggregates at high temperature. All of these recent observations demonstrate that this superfamily of proteins plays interesting and diverse roles in maintaining cellular proteins in a competent state and helps in the disposal of damaged proteins.

III. REGULATION OF THE σ³²-PROMOTED HEAT SHOCK RESPONSE

From pioneering work in the laboratories of T. Yura, F. C. Neidhardt and C. Gross, it is known that the "classical" heat shock response is under the transcriptional regulation of the *rpoH* (*htpR*; *hin*) gene product, σ³², which complexes with the RNAP core (E) to constitute the Eσ³² holoenzyme. Eσ³² recognizes the classical heat shock promoters located upstream of heat shock genes such as *dnaK*, *groES* and *lon*.[51,52] Because there are considerable differences in the nucleotide sequences of the Eσ³²- and Eσ⁷⁰-directed promoters (Table 23.2), each holoenzyme transcribes exclusively their corresponding promoters.

Gross et al,[51] Neidhardt and van Bogelen,[52] Yura et al[53] and Bukau[54] have elegantly summarized the early history and more recent studies on the regulation and function of the *rpoH* gene and its product. Hence, we will highlight briefly the transcriptional and translational regulation of the *rpoH* gene, with an emphasis on the various mechanisms by which the DnaK chaperone machine can negatively regulate both the intracellular levels and activity of σ³² (Table 23.3).

A. TRANSCRIPTIONAL REGULATION OF THE *RPOH* GENE

Figure 23.1A highlights many interesting features of *rpoH* gene regulation that operate either at the transcriptional or translational level. Under normal physiological conditions, the P1 promoter contributes the bulk of *rpoH* transcription while the rest of its promoters contribute varying minor amounts.[51,53] The P3 promoter is under the control of the Eσᴱ holoenzyme (see below) and its usage pattern increases with a corresponding increase in temperature, so that at 51°C, P3 constitutes the sole operating transcription system of the *rpoH* gene.[55] At such extreme temperatures, Eσ⁷⁰-directed transcription ceases, whereas Eσᴱ transcription continues unabatedly. Continuous transcription of the *rpoH* gene at high temperatures is very important because it ensures the replenishment of σ³², which has an extremely short half-life (≈ 1 minute; see below). The continuous presence of σ³² at high temperatures assures the unabated transcription of the classical heat shock genes and the corresponding accumulation of heat shock proteins, such as the GroEL and DnaK chaperone machines and the Lon and Clp protease systems, whose protective, disaggregating, and proteolytic activities are especially needed under these conditions (see below).

The only negative control known to be exerted on *rpoH* gene transcription is by DnaA, the DNA replication and master regulator protein. Wang and Kaguni[56] showed in vitro that purified DnaA protein can bind to the two DnaA boxes present in the *rpoH* promoter region (Fig. 23.1A), and attenuate transcription from the P3 and P4 promoters. An analogous result was found in vivo following overproduction of the DnaA protein. Since transcription from the major *rpoH* promoter, P1, is not affected by DnaA, most likely this negative regulation serves to "fine-tune" *rpoH* gene transcription under various physiological conditions.

Table 23.2. Consensus sequences for selected RNAP promoters in E. coli

Holoenzyme	Consensus Promoter		
	-35 nt position	nt spacing	-10 nt position
Eσ⁷⁰	TTGACA	16-18 bp	TATAAT
Eσ³²	CTTGAA	13-15 bp	CCCCATxT
Eσᴱ	GAACTT	16 bp	TCTGA

B. POST-TRANSCRIPTIONAL REGULATION OF RPOH MRNA

Following a shift from 30 to 42°C, there is a large, transient increase in intracellular σ^{32} levels. Two factors contribute significantly to this: an increased rate of translation of the *rpoH* mRNA, and a transient stabilization in the half-life of σ^{32}.[51,53,57]

Through the extensive use of *rpoH-lacZ* protein fusions and their various deletion derivatives, three important *cis*-acting regulatory regions were defined (Yura et al;[53] Fig. 23.1A). The first region, designated as A, corresponds to nucleotides 6-20 of the coding sequence. It plays a positive role, since its deletion results in a 15-fold decrease in the rate of *rpoH* mRNA translation.[53]

The second region, designated as B, has been localized to *rpoH* nucleotides 153-247. It behaves as a negative element, since its deletion leads to a substantial increase in the rate of translation of the *rpoH* mRNA. Evidence exists to suggest that the base pairing of certain nucleotides in region B with those of region A is important in this thermoregulation of the translation of *rpoH* mRNA. One of the possibilities envisioned is that a thermosensitive protein factor may bind to and stabilize this inhibitory double-stranded mRNA structure (in analogy with the Rom/Rop protein encoded by plasmid ColE1).[53]

Table 23.3. Various factors modulating the σ^{32}***- and*** σ^E***-promoted heat shock responses***

A. Eσ^{32}-promoted heat shock response

1. Positive modulation
   - (a) *Trans*-acting factors
     - (i) Eσ^{70} holoenzyme transcription of P1, P4 and P5 promoters of *rpoH* gene
     - (ii) cAMP·CAP-dependent transcription of P5 promoter of *rpoH* gene
     - (iii) unfolded/misfolded intracellular polypeptides
     - (iv) HtpY protein
   - (b) *Cis*-acting regions
     - (i) P1, P3, P4, P5 promoters
     - (ii) nucleotides 6-20 of coding sequence enhance rate of translation

2. Negative modulation
   - (a) *Trans*-acting factors
     - (i) DnaK chaperone machine sequesters σ^{32} from RNAP core, interferes with formation of "open" complex at σ^{32}-directed promoters, enhances σ^{32} proteolysis, participates in translational repression of *rpoH* mRNA (?)
     - (ii) HflB heat shock protein (protease?) directly or indirectly participates in σ^{32} proteolysis
     - (iii) HtrC protein
   - (b) *Cis*-acting regions
     - (i) the DnaA binding sites located in the *rpoH* promoter region
     - (ii) nucleotides 153-247 of the *rpoH* structural gene lower the rate of mRNA translation
     - (iii) nucleotides 364-433 of the *rpoH* structural gene code for a polypeptide segment responsible for mRNA translational repression

B. Eσ^E-directed heat shock response

1. Positive modulation
   - (a) *Trans*-acting factors
     - (i) Eσ^E holoenzyme directs transcription from the P2 promoter of the *rpoE* gene
     - (ii) unknown RNAP holoenzyme form and/or positive factor(s) directs transcription from the P1 promoter of the *rpoE* gene
     - (iii) unfolded/misfolded polypeptides in periplasmic space and/or changes in outer membrane composition
   - (b) *Cis*-acting regions
     - (i) P1 and P2 promoters of the *rpoE* gene

2. Negative modulation
   - (a) *Trans*-acting factors
     - (i) σ^E negative regulators encoded by the *rseA* and *rseB* genes co-transcribed with *rpoE*

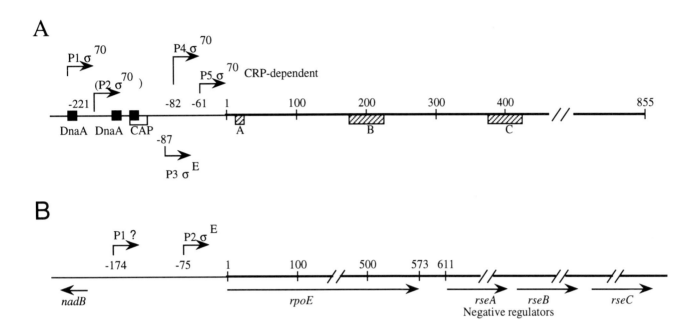

The third *cis*-acting region, designated as C, has been localized to *rpoH* nucleotides 364-433. Nagai et al,[58] utilizing a set of in-frame deletion derivatives, were able to pinpoint region C as being responsible for a translational repression control. The use of a frameshift mutation that altered the amino acid sequence (residues 122-144) abolished translational repression, suggesting that most likely it is the specific amino acid sequence that mediates translational repression. It could be that one or more members of the DnaK chaperone machine binds to this σ^{32} segment as it emerges from the ribosome and participates in the arrest of σ^{32} synthesis directly (by acting as a signal recognition particle) or indirectly (perhaps by "presenting" the unfinished σ^{32} polypeptide chain to proteases; see below).

The σ^{32} polypeptide has been shown to be one of the most unstable proteins of *E. coli*, its half-life being 45-60 seconds between 30 and 42°C, but substantially longer at 22°C.[57,59] A variety of factors must contribute to this extreme instability of σ^{32} since its half-life is known to be modulated under a variety of experimental conditions; namely, (i) following a shift from 30 to 42°C, σ^{32} is stabilized for a very short time, after which its extreme instability is restored,[57,60] (ii) in a *dnaK*, *dnaJ*, or *grpE* mutant background, the σ^{32} half-life is substantially increased at all temperatures, with the curious lack of participation of *dnaJ* specifically at 42°C.[59,60] This result strongly suggests that the DnaK chaperone machine directly or indirectly participates in σ^{32} degradation, and (iii) Herman et al[61] have recently shown that under conditions of low FtsH activity, σ^{32} is dramatically stabilized, suggesting that the FtsH protein directly or indirectly participates in σ^{32} degradation. Interestingly, the *ftsH* gene is under σ^{32}-dependent transcriptional control[61] and is identical to *hflB* (<u>h</u>igh <u>f</u>requency of <u>l</u>ysogenization; Herman et al[62]), a locus previously identified as regulating bacteriophage λcII proteolysis.[63] Because of its demonstrated role as a protease, we will refer to it as *hflB*. Thus, at least one of the σ^{32}-promoted heat shock protease systems, and perhaps more, can degrade full-length σ^{32}. The in vivo folded state of the degradable σ^{32} is not known.

Fig. 23.1. Control of expression of the rpoH *and* rpoE *genes. (A) Relevant features of the* rpoH *gene are shown. Nucleotide positions 1 to 855 refer to the first and last nucleotides of the* rpoH *structural gene, respectively. The transcriptional start sites are shown by arrows, along with the positive factors needed for initiation. The position of two DnaA DNA binding sites are shown which inhibit transcription from the P3 and P4 promoters. The positions of the cAMP·CAP binding site which activates transcription from the P5 promoter is also shown. The P2 promoter is in parenthesis because it appears to be strain-specific. The cis-acting elements A, B, C are involved in translational control (see text for details). (B) Relevant features of the* rpoE *gene are shown. Nucleotide positions 1 to 573 refer to the first and last nucleotide of the* rpoE *structural gene, respectively. The two known transcriptional start sites and their positions are shown. Also shown are the positions of the* nadB *divergent gene, and* rseA rseB *which code for two polypeptides that inhibit* σ^E *action and* rseC.

C. Molecular Mechanisms of Autoregulation of the σ^{32}-Dependent Heat Shock Response

As stated above, the products of four σ^{32}-dependent heat shock genes, namely, *dnaK, dnaJ, grpE,* and *hflB,* negatively modulate the rate of their own gene expression. All of the accumulated data are consistent with the interpretation that the intracellular levels of unfolded/misfolded proteins in *E. coli* control the extent of heat shock gene expression.[64-66] Most likely, these unfolded proteins exert their control through the titration of the DnaK chaperone machine and the HflB putative protease (for a discussion see refs. 53, 54, 61, 66, 67). These regulatory checkpoints include: (a) sequestration of the σ^{32} polypeptide from the RNAP core by the DnaK chaperone machine, thus interfering with the formation of the Eσ^{32} holoenzyme.[22] It turns out that by itself DnaK binds weakly to σ^{32} in an ATP-sensitive manner[16,68] whereas DnaJ binds to σ^{32} in an ATP-independent manner. Yet the DnaJ and DnaK proteins together make an effective, three-body complex with σ^{32} in an ATP-dependent reaction.[22] DnaJ can either be part of this DnaK/σ^{32}/DnaJ complex or act catalytically to enable DnaK to bind σ^{32} effectively in an ATP-dependent reaction.[22] Furthermore, although the DnaJ12 truncated protein (composed of only the first 108 amino acid residues of DnaJ) itself does not bind σ^{32}, nevertheless it enables DnaK to bind σ^{32} in an ATP-dependent reaction, thus leading to effective autoregulation of the heat shock response at least at 30°C.[23,69] The specific role that GrpE plays in the autoregulation of the heat shock response is not known, since it does not appear to play a role in either formation or stability of the DnaK/σ^{32}/DnaJ complex.[22] Very likely, GrpE's role in vivo is to enable the efficient "recycling" of the DnaK chaperone; (b) the specific blockage of Eσ^{32} function following binding to its heat shock promoters. Specifically, DnaJ with DnaK's help can effectively block the formation of the Eσ^{32} "open" transcriptional complex, but not that of Eσ^{70} (B. Bukau, personal communication); (c) the arrest of σ^{32} translation, normally seen following an initial burst at 42°C,[53,58] does not take place in either *dnaK, dnaJ,* or *grpE* mutant bacteria. A likely possibility for the molecular basis of this control is the following. When the intracellular levels of the DnaK chaperone machine are in excess over those of unfolded/misfolded proteins, the DnaK and DnaJ chaperones can effectively "capture" portions of the σ^{32} polypeptide as they emerge from the ribosome, most likely the 122-144 amino acid region C, and either block translation *per se* (by acting as a signal recognition particle) or "present" the growing σ^{32} polypeptide chain to various proteases, including HflB.[53,58,61] In the latter case, translational arrest would be synonymous to σ^{32} breakdown, even before a completed σ^{32} polypeptide chain is finished.

D. The Role of Ambient Temperature in RNAP Holoenzyme Function

Recent studies carried out by K. Liberek's group have shed more light on the role of ambient temperature on the formation and function of Eσ^{32} versus Eσ^{70} holoenzymes. It turns out that ambient temperature potentially plays two roles in regulating Eσ^{32}- versus Eσ^{70}-promoted transcription; namely, (a) the Eσ^{70} holoenzyme is more prone to heat inactivation compared to Eσ^{32}. This heat sensitivity has been traced to a specific aggregation of the σ^{70} transcription factor at temperatures above 50°C.[70] The DnaK chaperone machine can disaggregate Eσ^{70} heat-induced aggregates, formed at 51°C, thus restoring enzymatic activity.[12,13,70] The RNAP core E itself is resistant to heat inactivation at 51°C, but is irreversibly inactivated at 57°C.[70] Purified σ^{70} factor aggregates at 51°C or 57°C, but such aggregates can be disaggregated by the DnaK chaperone machine.[70] In vitro σ^{32} is relatively stable at 51°C, aggregates at 57°C but some of it can be disaggregated in the presence of the RNAP core,[70] and (b) $E\sigma^{32}$ holoenzyme formation is favored over that of $E\sigma^{70}$ with a gradual rise in temperature, the preference becoming very pronounced at 51°C.[70] However, it is not clear whether this apparent preference of the RNAP core for σ^{32} at higher temperatures simply reflects σ^{70}'s differential aggregation at high temperatures, or whether it is also due to a higher affinity per se for σ^{32} by the RNAP core.

IV. A SECOND HEAT SHOCK REGULON

A. INTRODUCTION

As stated above, earlier studies showed that neither $E\sigma^{32}$ nor $E\sigma^{70}$ could recognize the P3 promoter of the *rpoH* gene, whereas crude RNA polymerase preparations were found to carry such an activity. Indeed, starting with gel-fractionated RNA polymerase preparations, two groups independently showed that a 24 kDa species transcribes the P3 promoter when supplemented with core RNA polymerase. This 24 kDa species was named σ^{24} by Wang and Kaguni[71] and σ^E by Erickson and Gross.[55] In this review, we will refer to this activity as σ^E. It was also demonstrated that σ^E is required for the transcription of another gene, *htrA* (high temperature requirement complementation group A; also identified as *degP* by Strauch et al[72]). We will refer to it as *htrA*. Transcriptional mapping of the *htrA* gene revealed a promoter with remarkable resemblance to that of the *rpoH*P3 promoter.[55,73] Subsequently, this promoter was found to be transcribed by the gel-purified σ^E factor when complexed to core RNA polymerase.[55] These preliminary results suggested that there might be a second heat shock regulon in *E. coli*, since transcription starting from either the *rpoH*P3 or the *htrA* promoters was shown to be heat-inducible.[55,73,74]

B. IDENTIFICATION AND CHARACTERIZATION OF *RPOE*, THE σ^E ENCODING GENE

By constructing *lacZ* transcriptional fusions to either the *rpoH*P3 or *htrA* promoters, *trans*-acting mutations were sought which specifically abolished transcription from these two promoters. Such a search led to the isolation of mutations in a few genes, including the *bona fide* σ^E-encoding gene, designated as *rpoE*.[75] In a complementary approach, wild-type *E. coli* DNA multicopy libraries were introduced into these two fusion strains and clones yielding increased σ^E-dependent LacZ activity were selected.[75,76] Using this screen, Raina et al[75] again identified the σ^E-encoding gene as well as five other genes. In parallel studies, Mecsas et al[76] found that overexpression of outer membrane proteins (OMPs) specifically led to enhanced σ^E activity. In parallel studies, Rouvière et al[77] also identified the σ^E-encoding gene by noticing that an unidentified open reading frame in the database contained a stretch of 24 amino acids whose sequence matched perfectly that of one of the peptides resulting from the tryptic digestion of gel-purified σ^E. Sequencing of the *rpoE* gene revealed that most of the features common to sigma factors are conserved in its protein product.[75,78,79] Interestingly, two of the point mutations in the *rpoE* gene leading to a lower *lacZ* activity alter amino acid residues located in the helix-turn-helix motif of region 4.2. Another mutation was found in region 2.1, which is involved in binding to the β subunit of the RNA polymerase core.[75] The σ[E] protein was purified following overexpression from the cloned *rpoE* gene and shown to selectively transcribe the *htrA* and *rpoH*P3 promoters when supplemented with core RNA polymerase.[75,77]

Due to the nature of the applied screen, all *rpoE* point mutations resulted in a severely reduced expression of all heat shock proteins at 50°C.[75] However, no strong bacterial growth defect was noticed at lower temperatures, such as 30°C. Subsequent experiments showed that bacteria carrying a null allele of *rpoE* are viable at temperatures up to 40°C.[75,77] This finding suggests that the σ[E] regulon possesses the specific function of insuring viability under extreme growth conditions, whereas σ[32], which is required even at 20°C,[53] controls the major heat shock regulon, whose primary function is to cope with intracellular protein damage at all temperatures. This is consistent with the earlier findings that *rpoH* gene transcription is mostly regulated through its σ[70]-dependent promoters under a wide range of physiological growth conditions.[53]

C. Stimuli Inducing the σ[E] Regulon

Since overexpression of OMPs induced a σ[E]-dependent stress response, Mecsas et al[76] proposed that the stress signal was generated by the accumulation of misfolded or immature OMPs in the extracytoplasmic space. Additional support for this proposal comes from the close relatedness of σ[E] to a subgroup of sigma factors whose members regulate extracytoplasmic functions (ECF).[79] Actually, the σ[E] protein sequence is 66% identical to the putative ECF transcription factor AlgU (AlgT) from *Pseudomonas aeruginosa*,[80] known to regulate alginate biosynthesis via AlgD.[81] CarQ from *Myxococcus xanthus*, another ECF factor, regulates the synthesis of extracytoplasmic localized carotenoids.[82]

Raina et al[75] found that mutations in additional genes, such as those of the *dsb* family, whose products affect proper folding of periplasmically located or membrane bound disulfide-containing proteins,[83,84] induced a σ[E]-dependent response (2-fold). A double *dsbA htrA* null mutant induced an even higher increase (6-fold) in σ[E]-dependent transcription, whereas an *htrA* null mutant (which codes for a periplasmic protease[72,85,86]) did not show any noticeable effect by itself. Based on this lack of an effect of the *htrA* null mutant, Mecsas et al[76] concluded that HtrA substrates may not be part of the σ[E] signal transduction system. However, this conclusion is not in accordance with the phenotype exhibited by the double *dsbA htrA* mutant.[75] Also, overexpression of the periplasmically located asparaginase B protein, which contains one disulfide bond, also induced the σ[E]-dependent response (D. Missiakas and S. Raina, unpublished results). Furthermore, mutations leading to altered lipopolysaccharide (LPS), such as those in the *htrM/rfaD* gene,[87] result in constitutive expression of the σ[E] protein, at levels high enough to be detectable by two-dimensional polyacrylamide gel electrophoresis.[75] It is known that altered LPS greatly affects the ratio of OMPs,[88,89] thus linking the HtrM/RfaD phenotype with the observations made by Mecsas et al.[76]

An additional interesting finding is that *rpoE* transcription is also induced in strains carrying a mutant MalE protein which aggregates in the periplasmic space of *E. coli* (J-M. Betton and D. Missiakas, unpublished results). This could be analogous to the mechanism of inducing a σ[32]-dependent response in the presence of unstable cytoplasmic proteins.[65] Taken together, these data suggest that not only misfolding of OMPs, but in general, misfolding of many secretory proteins (particularly those which are substrates of the HtrA and Dsb proteins) may result in an increased σ^E-dependent response.

In all cases mentioned above involving stimulation of the σ^E response, normal levels of *rpoE* transcripts were restored following the overexpression of the *surA* gene.[90] The *surA* gene was cloned independently as a multicopy suppressor of the lethality exhibited by an *htrM/rfaD* null mutant on MacConkey medium (D. Missiakas and S. Raina, unpublished results). Furthermore, we found that purified SurA protein crossreacts with anti-parvulin antibodies (D. Missiakas, Rahfeld J.-U., Fischer G. and S. Raina, unpublished results), a recently identified peptidyl-prolyl isomerase activity in the cytoplasm.[91] This result is consistent with the fact that SurA contains two parvulin-like conserved blocks of sequences.[91] Hence, the SurA protein may act as a chaperone and/or peptidyl-prolyl isomerase in the periplasm, and its overproduction may accelerate protein folding, thus dampening the extracellular signal due to misfolded proteins.

D. REGULATION OF σ^E

1. Positive autoregulation

Promoter analyses of the *rpoE* gene revealed two transcriptional start sites (Fig. 23.1B). The most proximal to the initiation start site, designated as *rpoE*P2, was found to contain -35 and -10 boxes with very high sequence homology to those of the *rpoH*P3 and *htrA* promoters.[75,77] Moreover, the transcriptional activity from the *rpoE*P2 promoter was found to be the only one sustained upon a temperature shift-up from 30 to 50°C. Usually at 50°C, transcription of housekeeping genes is turned off (presumably due to inactivation of the primary σ^{70} factor), whereas transcription of classical σ^{32}-dependent heat shock genes is highly induced. It was therefore not surprising to find that in run-off transcription assays *rpoE*P2 is transcribed by the σ^E-containing RNA polymerase holoenzyme,[75,77] a result that explains as to how the σ^{32} regulon remains active at such extreme temperatures.

Regarding the most distal promoter *rpoE*P1, it does not show any clear homology to other promoters known to be transcribed either by σ^{70}, σ^{32} or σ^E. Consistent with this, the *rpoE*P1 promoter is not transcribed by the Eσ^{70}, Eσ^{32} or Eσ^E holoenzymes.[75,77] Nevertheless, mRNA synthesis initiating at the *rpoE*P1 promoter declines with increasing temperature, in contrast to that emanating from the *rpoE*P2 promoter. Perhaps an additional positive-acting factor(s) is needed for transcription from the *rpoE*P1 promoter.

2. Negative regulation

Like the other ECF sigma factors, such as AlgU (AlgT) of *P. aeruginosa* or CarQ of *M. xanthus*, σ^E is subject to a negative control involving putative anti-sigma factor(s). The reasoning for this conclusion is based on the following two observations: clones which overexpress the *rpoE* gene alone exhibit a transcriptional activity of the *htrA*, *rpoH*P3 and *rpoE*P2 promoters that is in close stoichiometry with the dosage of the *rpoE* gene. However, when *rpoE* is co-expressed with additional adjacent, downstream DNA sequences, its activity exhibits a 4-fold decline (D. Missiakas and S. Raina, unpublished results). Sequence and transcription analyses show that this downstream region carries three additional genes which are co-expressed with *rpoE*

(Fig. 23.1B). The second and third genes of this operon are those encoding the negative regulatory elements. We designate them as *iseA* and *iseB* (regulator of sigma E). The last gene of this operon is probably encoding a positive regulator of σ^E activity and is called *rseC* (accession number U37089; D. Missiakas and S. Raina, unpublished results). RseA and RseB proteins exhibit 30% homology with MucA/MucB, two presumed negative regulators of AlgU (AlgT) protein.[81] However, the highest degree of homology (50 to 70%) for the *E. coli rpoE* operon was found to match another sequence in the databank. This sequence corresponds to an identically arranged four gene operon from photobacterium species (accession number L41667).

E. SENSING EXTRACYTOPLASMIC STRESSES

Recent studies, conducted mainly with the *htrA* promoter region, have led to the conclusion that there are at least two different systems used by *E. coli* to sense extracytoplasmic stresses. One system transmits the signal to σ^E by a mechanism involving the transmembrane protein RseA, whereas the other specifically induces *htrA* transcription without affecting σ^E. The various lines of evidence which led to these conclusions can be summarized as follows: (i) among the mutants isolated as reducing transcriptional activity using the *htrA-lacZ* or *rpoH*P3-*lacZ* fusions (which yielded the *rpoE* gene), one class mapped to the *cpxA/cpxR* operon.[75] Closer examination of the phenotypes exhibited by this class of mutations revealed that only the *htrA-lacZ* transcriptional activity was affected; (ii) this result was consistent with our previous findings of differences in the usage of the *rpoH*P3 and *htrA* promoters following overexpression of HtpY;[92] (iii) all such differences turned out to be due to the presence of an additional promoter in the *htrA* gene, located a few bases downstream from the σ^E-dependent promoter, and whose expression is modulated by *cpxA/cpxR*;[93] and finally (iv) Danese et al[93] noticed that overexpression of the NlpE lipoprotein induces only *htrA* transcription and not *rpoH*P3 transcription.

The CpxA protein possesses the sequence signature of a histidine kinase,[94] whereas CpxR resembles proteins of the two-component regulatory systems such as OmpR.[95] Interestingly, two new *E. coli* genes *prpA* and *prpB*, whose products in multicopy enhance *htrA* transcription without affecting *rpoH*P3, were found to encode for phosphatase activities (D. Missiakas and S. Raina, unpublished results). Interestingly, both genes share about 50% identity at the amino acid level. The sequences as well as subsequent biochemical characterizations suggest that both proteins belong to the type I class of phosphatases. In vivo, these phosphatases probably modulate the activity of the CpxA/CpxR signaling pathway by changing the half-life of the phosphorylated proteins through a regulatory mechanism as yet unknown and possibly also used by the other two-component signaling systems in *E. coli* (D. Missiakas and S. Raina, unpublished observations).

Obviously, a great deal more remains to be learned about the extracellular signals, the relay systems and the, as yet, undiscovered genes involved in the induction and regulation of the heat shock response. As shown in Figure 23.2, one can notice that in addition to induction of σ^{32} regulon, many new additional proteins are accumulated when σ^E is induced in a controlled fashion. Thus, in addition to the known six genes (Table 23.1) at least four more unidentified genes seem to be positively regulated by σ^E. If misfolding in the periplasm and the outer membrane indeed induces two additive signal pathways which affect σ^E-dependent and *cpxA/cpxR*-mediated transcription, it may not be surprising to find that the σ^{32}-dependent regulon is also under the control of separable pathway systems. Clearly, many interesting new insights will be discovered in the regulation of both heat shock regulons during the next few years.

V. THE σ^{54}-PROMOTED STRESS RESPONSE

The σ^{54} transcription factor is the product of the *rpoN* gene. It helps RNA polymerase transcribe a limited number of genes, including those involved in nitrogen metabolism.[96] P. Model's laboratory reported a few years ago that the expression of the *psp* operon of *E. coli* is induced following infection by filamentous bacteriophage, heat, ethanol or osmotic shock.[97] This stress-induced induction of *psp* was shown to be under the positive control of σ^{54}.[98] Subsequent studies established that *psp* can be also induced under intracellular conditions that block the protein exporting pathway of *E. coli*[99] or by following treatment by agents that interfere with oxidative phosphorylation.[100] Although the *psp* operon is not essential for *E. coli* growth, it does appear to play some role in its stationary-phase survival.[100]

One of the five *groE* operons of *Bradyrhizobium japonicum* was shown to be coregulated with symbiotic nitrogen fixation genes, under the control of the σ^{54} transcription factor.[101] Perhaps this coregulation reflects a more specialized chaperone role in nitrogen fixation for this particular GroES/GroEL chaperone system and/or a need to fine tune the chaperone levels in response to specific extracellular conditions and physiological needs.

VI. HEAT SHOCK OR STRESS RESPONSES IN OTHER EUBACTERIA

Recent studies suggest that the σ^{32}-like transcription regulation mechanism, used by *E. coli* to regulate the expression of its major chaperone and protease systems, may not be universal in all eubacteria. For example, in gram-positive bacteria there is overwhelming evidence that the heat-inducible transcription of the major *groE* and *dnaK*-like operons is promoted by the vegetative, σ^{70}-like transcription factor. A unifying feature of such heat shock gene induction is the presence of a 9-nucleotide inverted repeat, separated by a spacer of nine nucleotides, the so-called CIRCE element.[102] While the nucleotide sequences of the inverted repeat of the CIRCE element has been extremely conserved across evolution, that of the spacer region has not.[102] Very likely, a repressor-like molecule, that is somehow inactivated following heat shock binds to the CIRCE element to down-regulate the levels of transcription. In addition to the CIRCE element regulation in *B. subtilis*, the σ^{B} transcription factor also plays a role in heat- or stress-regulation of a limited number of additional genes.[103] These findings, coupled with the fact that at least the *groE* operon of the gram-negative bacteria *Agrobacterium tumefaciens* is also under CIRCE regulation,[104] underline the lack of universality of the σ^{32}-promoted response of *E. coli* among eubacteria.

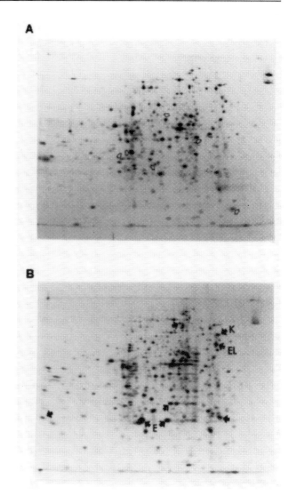

Fig. 23.2. Effects of σ^E overexpression on the protein pattern of E. coli. Bacteria carrying the rpoE gene under the control of the lacI repressor were induced with 1 mM IPTG for 30 minutes at 30°C. Following this, the cultures were labeled with ^{35}S-methionine for 10 minutes, concentrated, and their proteins analyzed by standard two-dimensional gel electrophoresis and autoradiography. (A) Control, uninduced culture, (B) induced culture. The open arrows indicate the positions of those proteins whose synthesis is decreased following σ^E overproduction. The closed arrows indicate the positions of those proteins whose synthesis is markedly enhanced following the σ^E overproduction. The arrows marked by the letter "E" identify the positions of two putative σ^E isoforms, "K" the position of DnaK and "EL" the position of GroEL.

REFERENCES

1. Anfinsen CB. Principles that govern the folding of protein chains. Science 1973; 181:223-230.
2. Jaenicke R. Folding and association of proteins. Prog Biophys Mol Biol 1987; 49:117-237.
3. Ellis RJ, van der Vies SM. Molecular chaperones. Ann Rev Biochem 1991; 60:321-347.

4. Georgopoulos C, Welch WJ. Role of major heat shock proteins as molecular chaperones. Annu Rev Cell Biol 1993; 9:601-635.

5. Hendrick JP, Hartl F-U. Molecular chaperone functions of heat-shock proteins. Annu Rev Biochem 1993; 62:349-384.

6. Herendeen SL, VanBogelen RA, Neidhardt FC. Levels of major proteins of *Escherichia coli* during growth at different temperatures. J Bacteriol 1979; 139:185-194.

7. Gragerov A, Nudler E, Komissarova N et al. Cooperation of GroEL/GroES and DnaK/DnaJ heat shock proteins in preventing protein misfolding in *Escherichia coli*. Proc Natl Acad Sci USA 1992; 89:10341-10344.

8. Friedman DI, Olson ER, Tilly K et al. Interactions of bacteriophage λ and host macromolecules in the growth of bacteriophage λ. Microbiol Rev 1984; 48:299-325.

9. Georgopoulos C. The emergence of the chaperone machines. Trends Biochem Sci 1992; 17:295-299.

10. Georgopoulos C, Linder CH. Molecular chaperones in T4 assembly. In: Karam J, ed. Bacteriophage T4 II. Washington D.C.: American Society for Microbiology, 1993.

11. Georgopoulos C, Liberek K, Zylicz M et al. Properties of the heat shock proteins of *Escherichia coli* and the autoregulation of the heat shock response. In: Morimoto R, Tissières A, Georgopoulos C, eds. The biology of heat shock proteins and molecular chaperones. Cold Spring Harbor: Cold Spring Harbor Laboratory Press, 1994:209-249.

12. Skowyra D, Georgopoulos C, Zylicz M. The *Escherichia coli* dnaK protein, the hsp70 homolog, can reactivate heat-inactivated RNA polymerase in an ATP hydrolysis-dependent reaction. Cell 1990; 62:939-944.

13. Ziemienowicz A, Skowyra D, Zeilstra-Ryalls J et al. Either of the *Escherichia coli* GroEL/GroES and DnaK/DnaJ/GrpE chaperone machines can reactivate heat-treated RNA polymerase: different mechanisms for the same activity. J Biol Chem 1993; 268:25425-25431.

14. Phillips GJ, Silhavy TJ. Heat-shock proteins DnaK and GroEL facilitate export of LacZ hybrid proteins in *E. coli*. Nature 1990; 344:882-884.

15. Wild J, Altman E, Yura T et al. DnaK and dnaJ heat shock proteins participate in protein export in *Escherichia coli*. Genes Dev 1992; 6:1165-1172.

16. Liberek K, Marszalek J, Ang D et al. The *Escherichia coli* DnaJ and GrpE heat shock proteins jointly stimulate DnaK's ATPase activity. Proc Natl Acad Sci USA 1991; 88:2874-2878.

17. Landry SJ, Jordan R, McMacken R et al. Different conformations of the same polypeptide bound to chaperone DnaK and GroEL. Nature 1992; 355:455-457.

18. Langer T, Lu C, Echols H et al. Successive action of DnaK, DnaJ and GroEL along the pathway of chaperone-mediated protein folding. Nature 1992; 356:683-689.

19. Schmid D, Baici A, Gehring H et al. Kinetics of molecular chaperone action. Science 1994; 263:971-973.

20. Palleros DR, Reid KL, Shi L et al. ATP-induced protein-Hsp70 complex dissociation requires K⁺ but not ATP hydrolysis. Nature 1993; 365: 664-666.

21. Liberek K, Skowyra D, Zylicz M et al. The *Escherichia coli* DnaK chaperone protein, the Hsp70 eukaryotic equivalent, changes its conformation upon ATP hydrolysis, thus triggering its dissociation from a bound target protein. J Biol Chem 1991; 266:14491-14496.

22. Liberek K, Georgopoulos C. Autoregulation of the *Escherichia coli* heat shock response by the DnaK and DnaJ heat shock proteins. Proc Natl Acad Sci USA 1993; 90:11019-11023.

23. Liberek K, Wall D, Georgopoulos C. The DnaJ chaperone catalytically activates the DnaK chaperone to specifically bind the σ^{32} heat shock transcriptional regulator. Proc Nat Acad Sci (USA) 1995; 92:6224-6228.

24. Bardwell JCA, Tilly K, Craig E et al. The nucleotide sequence of the *Escherichia coli* K12 *dnaJ* gene: a gene that encodes a heat shock protein. J Biol Chem 1986; 261:1782-1785.

25. Zeilstra-Ryalls J, Fayet O, Georgopoulos C. The universally conserved GroE chaperonins. Annu Rev Microbiol 1991; 45:301-325.

26. Fayet O, Ziegelhoffer T, Georgopoulos C. The *groES* and *groEL* heat shock gene products of *Escherichia coli* are essential for bacterial growth at all temperatures. J Bacteriol 1989; 171:1379-1385.

27. Van Dyk TK, Gatenby AA, LaRossa RA. Demonstration by genetic suppression of interaction of GroE products with many proteins. Nature 1989; 342:451-453.

28. Straus DB, Walter WA, Gross CA. *Escherichia coli* heat shock gene mutants are defective in proteolysis. Genes Dev 1988; 2:1851-1858.

29. Hendrick JP, Langer T, Davis TA et al. Control of folding and membrane translocation by binding of the chaperone DnaJ to nascent polypeptides. Proc Natl Acad Sci USA 1993; 90:10216-10220.

30. Braig K, Otwinowski Z, Hegde R et al. The crystal structure of the bacterial chaperonin GroEL at 2.8 Å. Nature 1994; 371:578-586.

31. Azem A, Kessel M, Goloubinoff P. Characterization of two functional GroEL14 (GroES7)2 chaperonin hetero-oligomers. Science 1994; 265:653-656.

32. Langer T, Pfeifer G, Martin J et al. Chaperonin-mediated protein folding: GroES binds to one end of the GroEL cylinder, which accommodates the protein substrate within its central cavity. EMBO J 1992; 11:4757-4765.

33. van der Vies SM, Gatenby AA, Georgopoulos C. Bacteriophage T4 encodes a co-chaperonin that can substitute for *Escherichia coli* GroES in protein folding. Nature 1994; 368:654-656.

34. Bochkareva ES, Lissin NM, Girshovich AS. Transient association of newly synthesized unfolded proteins with the heat-shock GroEL protein. Nature 1988; 336:254-257.

35. Goloubinoff P, Christeller JT, Gatenby AA et al. Reconstitution of active dimeric ribulose bisphosphate carboxylase from an unfolded state depends on two chaperonin proteins and Mg-ATP. Nature 1989; 342:884-889.

36. Braig K, Simon M, Furuya F et al. A polypeptide bound by the chaperonin groEL is localized within a central cavity. Proc Natl Acad Sci USA 1993; 90:3978-3982.

37. Laminet AA, Ziegelhoffer T, Georgopoulos C et al. The *E.coli* heat shock proteins GroEL and GroES modulate the folding of the β-lactamase precursor. EMBO J 1990; 9:2315-2319.

38. Zahn R, Spitzfaden C, Ottiger M et al. Destabilization of the complete protein, secondary structure on binding to the chaperone GroEL. Nature 1994; 368:261-265.

39. Howard-Flanders P, Simson E, Theriot L. A locus that controls filament formation and sensitivity to radiation in *Escherichia coli* K12. Genetics 1964; 49:237-241.

40. Gottesman S, Maurizi MR. Regulation by proteolysis: Energy-dependent proteases and their targets. Microbiol Rev 1992; 56:592-621.

41. Sherman MY, Goldberg AL. Involvement of the chaperonin DnaK in the rapid degradation of a mutant protein in *Escherichia coli*. EMBO J 1992; 11:71-77.

42. Wagner I, Arlt H, van Dyck L et al. Molecular chaperones cooperate with PIM1 protease in the degradation of misfolded proteins in mitochondria. EMBO J 1994; 13:5135-5145.

43. Katayama Y, Gottesman S, Pumphrey J et al. The two-component ATP-dependent Clp protease of Escherichia coli: purification, cloning, and mutational analysis of the ATP-binding component. J Biol Chem 1988; 263:15226-15236.

44. Gottesman S, Clark WP, de Crecy-Lagard V et al. ClpX, an alternative subunit for the ATP-dependent Clp protease of *Escherichia coli*. J Biol Chem 1993; 268:22618-22626.

45. Wickner S, Gottesman S, Skowyra D et al. A molecular chaperone, ClpA, functions like DnaK and DnaJ. Proc Natl Acad Sci USA 1994; 91:12218-12222.

46. Wojkowiak D, Georgopoulos C, Zylicz M. Isolation and characterization of ClpX, a new ATP-dependent specificity component of the Clp protease of *Escherichia coli*. J Biol Chem 1993; 268:22609-22617.

47. Wickner S, Hoskins J, McKenney K. Monomerization of RepA dimers by heat shock proteins activates binding to DNA replication origin. Proc Nat Acad Sci USA 1991; 88:7903-7907.

48. Wawrzynow A, Wojtkowiak D, Marszalek J et al. The ClpX heat shock protein of *Escherichia coli*, the ATP-dependent substrate specificity component of the ClpP/ClpX protease, is a novel molecular chaperone. EMBO J 1995; 14:1867-1877.

49. Squires C, Squires CL. The Clp proteins: proteolysis regulators or molecular chaperones? J Bacteriol 1992; 174:1081-1085.

50. Parsell DA, Kowal AS, Singer MA et al. Protein disaggregation mediated by heat-shock protein Hsp104. Nature 1994; 372:475-477.

51. Gross CA, Straus DB, Erickson JW et al. The function and regulation of heat shock proteins in *Escherichia coli*. In: Morimoto R, Tissières A, Georgopoulos C, eds. Stress proteins in biology and medicine. Cold Spring Harbor: Cold Spring Harbor Laboratory Press, 1990:167-189.

52. Neidhardt FC, VanBogelen RA. Heat shock response. In: Neidhardt FC, Ingraham JL, Low KB, Magasanik B, Schaechter M, Umbarger HE, eds. *Escherichia coli* and *Salmonella typhimurium*. Washington D.C.: American Society for Microbiology, 1987:1334-1345.

53. Yura T, Nagai H, Mori H. Regulation of the heat-shock response in bacteria. Annu Rev Microbiol 1993; 47:321-350.

54. Bukau B. Regulation of the *Escherichia coli* heat-shock response. Mol Microbiol 1993; 9:671-680.

55. Erickson JW, Gross CA. Identification of the σ^E subunit of *Escherichia coli* RNA polymerase; a second alternate σ factor involved in high temperature gene expression. Genes Dev 1989: 3:1462-1471.

56. Wang Q, Kaguni JM. DnaA protein regulates transcription of the *rpoH* gene of *Escherichia coli*. J Bacteriol 1989; 264:7338-7344.

57. Straus DB, Walter WA, Gross CA. The heat shock response of E.coli is regulated by changes in the concentration of σ^{32}. Nature 1987; 329:348-351.

58. Nagai H, Yuzawa H, Kanemori M et al. A distinct segment of the σ^{32} polypeptide is involved in DnaK-mediated negative control of the heat shock reponse in *Escherichia coli*. Proc Natl Acad Sci USA 1994; 91:10280-10284.

59. Tilly K, Spence J, Georgopoulos C. Modulation of the stability of *Escherichia coli* heat shock regulatory factor σ^{32}. J Bacteriol 1989; 171:1585-1589.

60. Straus DB, Walter W, Gross CA. DnaK, DnaJ, and GrpE heat shock proteins negatively regulate heat shock gene expression by controlling the synthesis and stability of σ^{32}. Genes Dev 1990; 4:2202-2209.

61. Herman C, Thévenet D, D'Ari R et al. Degradation of σ^{32}, the heat shock regulator in *Escherichia coli*, is governed by HflB. Proc Natl Acad Sci USA 1995; 92:3516-3520.

62. Herman C, Ogura T, Tomoyasu T et al. Cell growth and λ phage development controlled by the same essential *Escherichia coli* gene, *ftsH/hflB*. Proc Natl Acad Sci USA 1993; 90:10861-10865.

63. Banuett F, Hoyt MA, McFarlane L et al. *hflB*, a new *Escherichia coli* locus regulating lysogeny and the level of bacteriophage lambda cII protein. J Mol Biol 1986; 187:213-224.

64. Kanemori M, Hirotada M, Yura T. Induction of heat shock proteins by abnormal proteins results from stabilization and not increased synthesis of s^{32} in *Escherichia coli*. J Bacteriol 1994; 176:5648-6553.

65. Parsell DA, Sauer RT. Induction of a heat shock-like response by unfolded protein in *Escherichia coli*: dependence on protein level not protein degradation. Genes Dev 1989; 3:1226-1232.

66. Wild J, Walter WA, Gross CA et al. Accumulation of secretory protein precursors in *Escherichia coli* induces the heat shock response. J Bacteriol 1993; 175:3992-3997.

67. Craig EA, Gross CA. Is hsp70 the cellular thermometer? Trends Biochem Sci 1991; 16:135-140.

68. Gamer J, Bujard H, Bukau B. Physical interaction between heat shock proteins DnaK, DnaJ, and GrpE and the bacterial heat shock transcription factor σ^{32}. Cell 1992; 69:833-842.

69. Wall D, Zylicz M, Georgopoulos C. The conserved G/F motif of the DnaJ chaperone is necessary for the activation of the substrate binding properties of the DnaK chaperone. J Biol Chem 1995; 270:2139-2144.

70. Blaszczak A, Zylicz M, Georgopoulos C et al. Both ambient temperature and the DnaK chaperone machine modulate the heat shock response in *Escherichia coli* by regulating the switch between σ^{70} and σ^{32} factors assembled with RNA polymerase. EMBO J 1995; 14:5085-5093.

71. Wang Q, Kaguni JM. A novel sigma factor is involved in expression of the *rpoH* gene of *Escherichia coli*. J Bacteriol 1989; 171:4248-4253.

72. Strauch KL, Beckwith. *Escherichia coli* mutation preventing degradation of abnormal periplasmic proteins. Proc Natl Acad Sci USA 1988; 85:1576-1580.

73. Lipinska B, Sharma S, Georgopoulos C. Sequence analysis and transcriptional regulation of the *htrA* gene of *Escherichia coli*: a σ^{32}-independent mechanism of heat-inducible transcription. Nucleic Acids Res 1988; 16:10053-10067.

74. Erickson JW, Vaughn V, Walter WA et al. Regulation of the promoters and transcripts of *rpoH*, the *Escherichia coli* heat shock regulatory gene. Genes Dev 1987; 1:419-432.

75. Raina S, Missiakas D, Georgopoulos C. The *rpoE* gene encoding the σ^E (σ^{24}) heat shock sigma factor of *Escherichia coli*. EMBO J 1995; 14:1043-1055.

76. Mecsas J, Rouvière PE, Erickson JW et al. The activity of σ^E, an *Escherichia coli* heat-inducible sigma factor, is modulated by expression of outer membrane proteins. Genes Dev 1993; 7:2618-2628.

77. Rouvière P, de las Penas A, Mecsas J et al. *rpoE*, the gene encoding the second heat-shock sigma factor, σ^E, in *Escherichia coli*. EMBO J 1995; 14:1032-1042.

78. Lonetto M, Gribskov M, Gross CA. The σ^{70} family: sequence conservation and evolutionary relationships. J Bacteriol 1992; 174:3843-3849.

79. Lonetto M, Brown KL, Rudd KE et al. Analysis of the *Streptomyces coelicolor* sigmaE gene reveals the existence of a subfamily of eubacterial RNA polymerase sigma factors involved in the regulation of extracytoplasmic functions. Proc Natl Acad Sci USA 1994; 91:7573-7577.

80. Martin DW, Holloway BW, Deretic V. Characterization of a locus determining the mucoid status of *Pseudomonas aeruginosa*: AlgU shows sequence similarities with a *Bacillus* sigma factor. J Bacteriol 1993; 175:1153-1164.

81. Deretic V, Schurr MJ, Boucher JC et al. Conversion of *Pseudomonas aeruginosa* to mucoidy in cystic fibrosis: environmental stress and regulation of bacterial virulence by alternative sigma factors. J Bacteriol 1994; 176:2773-2780.

82. McGowan SJ, Gorham HC, Hodgson DA. Light-induced carotenogenesis in *Myxococcus xanthus*: DNA sequence analysis of the *carR* region. Mol Microbiol 1993; 10:713-735.

83. Bardwell JCA, McGovern K, Beckwith J. Identification of a protein required for disulfide bond formation in vivo. Cell 1991; 67:581-589.

84. Missiakas D, Georgopoulos C, Raina S. The *Escherichia coli dsbC* (*xprA*) gene encodes a periplasmic protein involved in disulfide bond formation. EMBO J 1994; 13:2013-2020.

85. Lipinska B, Zylicz M, Georgopoulos C. The HtrA (DegP) protein essential for *Escherichia coli* growth at high temperatures, is an endopeptidase. J Bacteriol 1990; 172:1791-1797.

86. Strauch KL, Johnson K, Beckwith J. Characterization of *degP*, a gene required for proteolysis in the cell envelope and essential for growth of *Escherichia coli* at high temperature. J Bacteriol 1989; 171:2689-2696.

87. Raina S, Georgopoulos C. The *htrM* gene, whose product is essential for *Escherichia coli* viability only at elevated temperatures, is identical to the *rfaD* gene. Nucleic Acids Res 1991; 19:3811-3819.

88. Nikaido H, Vaara M. Molecular basis of bacterial outer membrane permeability. Microbiol Rev 1985; 49:1-32.

89. Schnaitman CA, Klena JD. Genetics of lipopolysaccharide biosynthesis in enteric bacteria. Microbiol Rev 1993; 57:655-682.

90. Tormo A, Almiron M, Kolter R. *surA*, an *Escherichia coli* gene essential for survival in stationary phase. J Bacteriol 1990; 172:4339.

91. Rahfeld J-U, Rücknagel KP, Schelbert B et al. Confirmation of the existence of a third family among peptidyl-prolyl *cis/trans* isomerases: Amino acid sequence and recombinant production of parvulin. FEBS Lett 1994; 352:180-184.

92. Missiakas D, Georgopoulos C, Raina S. The *Escherichia coli* heat shock gene *htpY*: mutational analysis, cloning, sequencing and transcriptional regulation. J Bacteriol 1993; 175:2613-2624.

93. Danese P, Snyder WB, Cosma C et al. The Cpx two-component signal transduction pathway of *Escherichia coli* regulates transcription of the gene specifying the stress-inducible periplasmic protease, DegP. Genes Dev 1995; 9:387-398.

94. Weber RF, Silverman PJ. The Cpx proteins of *Escherichia coli* K12. Structure of the CpxA polypeptide as an inner membrane component. J Mol Biol 1988; 203:467-476.

95. Dong J, Iuchi S, Kwan SH et al. The deduced amino-acid sequence of the cloned *cpxR* gene suggests the protein is the cognate regulator for the membrane sensor, CpxA, in a two-component signal transduction system of *Escherichia coli*. Gene 1993; 136:227-230.

96. Kustu S, Santero E, Keener J et al. Expression of σ^{54} (*ntrA-*) dependent genes is probably united by a common mechanism. Microbiol Rev 1989; 53:367-376.

97. Brissette JL, Weiner L, Ripmaster TL et al.. Characterization and sequence of the *Escherichia coli* stress-induced *psp* operon. J Mol Biol 1991; 220:35-48.

98. Weiner L, Brissette JL, Model P. Stress-induced expression of the *Escherichia coli* phage shock protein operon is dependent on σ^{54} and modulated by positive and negative feedback mechanisms. Genes Dev 1991; 5:1912-1923.

99. Kleerebezem M, Tommassen J. Expression of the *pspA* gene stimulates efficient protein export in *Escherichia coli*. Mol Microbiol 1993; 7:947-956.

100. Weiner L, Model P. Role of an *Escherichia coli* stress-response operon in stationary-phase survival. Proc Nat Acad Sci USA 1994; 91:2191-2195.

101. Fischer HM, Babst M, Kaspar T et al. One member of a *groESL*-like chaperonin multigene family in *Bradyrhizobium japonicum* is co-regulated with symbiotic nitrogen fixation genes. EMBO J 1993; 12:2901-2912.

102. Zuber U, Schumann W. CIRCE, a novel heat shock element involved in regulation of heat shock operon *dnaK* of *Bacillus subtilis*. J Bacteriol 1994; 176:1359-1363.

103. Boylan SA, Redfield AR, Brody MS et al.. Stress-induced activation of the sigma B transcription factor of *Bacillus subtilis*. J Bacteriol 1993; 175:7931-7937.

104. Segal G, Ron EZ. Heat shock transcription of the *groESL* operon of *Agrobacterium tumefaciens* may involve a hairpin-loop structure. J Bacteriol 1993; 175:3083-3088.

105. Ueguchi C, Ito K. Multicopy suppression: an approach to understanding intracellular functioning of the protein export system. J Bacteriol 1992; 174:1454-1461.

106. Allen SP, Polazzi JO, Gierse JK et al. Two novel heat shock genes encoding proteins produced in response to heterologous protein expression in *Escherichia coli*. J Bacteriol 1992; 174:6938-6947.

107. Chuang S-E, Burland V, Plunkett G et al. Sequence analysis of four new heat shock genes constituting the *hslTS/ibpAB* and *hslVU* operons in *Escherichia coli*. Gene 1993; 175:2026-2036.

108. Burton Z, Burgess RR, Lin J et al. The nucleotide sequence of the cloned *rpoD* gene for the RNA polymerase sigma subunit from *E. coli* K12. Nucl Acids Res 1981; 9:2889-2903.

109. Raina S, Georgopoulos C. A new *Escherichia coli* heat shock gene, *htrC*, whose product is essential for viability only at high temperatures. J Bacteriol 1990; 172:3417-3426.

110. Charpentier B, Branlant C. The *Escherichia coli gapA* gene is transcribed by the vegetative RNA polymerase holoenzyme $E\sigma^{70}$ and by the heat shock RNA polymerase $E\sigma^{32}$. J Bacteriol 1994; 176:830-839.

111. Chuang SE, Blattner FR. Characterization of twenty-six new heat shock genes of *Escherichia coli*. J Bacteriol 1993; 175:5242-5252.

112. Kornitzer RD, Teff D, Altuvia S et al. Isolation, characterization and sequence of an *Escherichia coli* heat shock gene, *htpX*. J Bacteriol 1991; 173:2944-2953.

113. Tomoyasu T, Yuki T, Morimura S et al. The *Escherichia coli* FtsH protein is a prokaryotic member of a protein family of putative ATPases involved in membrane functions, cell cycle control, and gene expression. J Bacteriol 1993; 175:1344-1351.

ROLES FOR ENERGY-DEPENDENT PROTEASES IN REGULATORY CASCADES

Susan Gottesman

INTRODUCTION

In theory, the rapid degradation of a protein may play as critical a role in regulating protein activity as controls on transcription and translation. However, the role of protein turnover in regulation received relatively little attention during the years when the elegant systems for regulation of transcription initiation were first being investigated. The first dramatic example of a proteolytic event with clear regulatory consequences was provided by the demonstration, in 1975, that induction of bacteriophage lambda after DNA damage was due to cleavage of the lambda repressor.[1] Since those experiments, an increasing number of proteases and interesting protein targets have been identified, both in prokaryotes and eukaryotes.

In many of these cases, rapid degradation of the target protein occurs under all circumstances, suggesting that proteolysis per se does not play a regulatory role. Even in these cases, however, rapid degradation dramatically affects the consequences of changing the rate of synthesis of the unstable protein. Therefore, degradation must be considered a critical component of the overall regulatory cascade.

One of the threads of investigation that led to an understanding of the nature of this regulatory proteolysis in *E. coli* was the observation that abnormal proteins are degraded by an energy-dependent process.[2,3] In both prokaryotes and eukaryotes, the incorporation of amino acid analogs into proteins, the production of truncated or otherwise mutant protein products, or the overproduction of specific proteins in foreign hosts leads to either the rapid turnover of these abnormal proteins or their accumulation in inclusion bodies.[3] The rapid turnover of these proteins is blocked when energy sources are removed, suggesting that degradation is carried out by energy-dependent proteases.

It has become clear that the degradation of specific unstable proteins is also energy-dependent. Therefore, the investigation of proteases responsible for destruction of both abnormal proteins and naturally

Regulation of Gene Expression in Escherichia coli, edited by E. C. C. Lin and A. Simon Lynch. © 1996 R.G. Landes Company.

unstable proteins has converged on a set of interesting, intracellular enzymes that use energy to recognize and destroy specific targets.[4-6] Here, I will review the strategies that led to recognition of both the proteases and their targets, and our current understanding of how the rapid degradation of the target plays a role in regulation.

THE PROTEASES AND THEIR TARGETS

IDENTIFYING UNSTABLE REGULATORY PROTEINS: LAMBDA LEADS THE WAY

In prokaryotes, bacteriophage lambda not only provided the first example of a key regulatory role for protein degradation, but has continued to provide an easily accessible set of unstable proteins with important biological roles. Surprisingly, each one of these unstable proteins appears to be degraded by a distinct cellular protease.

Establishment of the role of turnover and the requirement for energy was extended from abnormal proteins to naturally unstable proteins by observations on the functional instability of a variety of lambda proteins. N protein, the antitermination regulatory protein necessary for lytic gene expression, is functionally unstable, requiring continuing synthesis for continued activity.[7,8] Differential stability of the site-specific recombination proteins Xis and Int was hypothesized by Weisberg and Gottesman[9] as the explanation for some asymmetries of excision and integration. Int is required for both the integration and excision of lambda from the bacterial chromosome, while Xis is only required for excision. Weisberg and Gottesman showed that Xis was in fact functionally unstable while Int was much more stable. If Xis decays while Int is still available, after immunity is established, this differential stability may contribute to stable lysogenization. Xis instability is energy-dependent, as the turnover of abnormal proteins is.[9] Replication of lambda is dependent on the two lambda-encoded proteins O and P; O was shown to be functionally unstable.[10] A central regulator for the decision between λ lytic and lysogenic growth is the cII protein; cII was also found to be functionally unstable.[11-13] N protein degradation proved to be due to the Lon protease, O is degraded by ClpXP, while cII degradation depends on the HflA and HflB proteases (see below); the protease responsible for Xis degradation has not yet been identified.

The first biochemical evidence for a regulatory role of protein degradation came from the demonstration that the basis of lambda prophage induction after exposure of cells to ultraviolet irradiation was the rapid, energy-dependent degradation of the lambda repressor.[1,14] The dependence of proteolysis on the RecA protein, both in vivo and in vitro[15] was unexpected. RecA had been identified as an essential recombination protein; the results of Roberts and co-workers suggested a completely different role, although it has become clear that the ability of RecA to recognize and bind to single-stranded DNA is probably critical for both activities.[16]

The regulatory cascade leading to lambda induction is directly paralleled by the induction of the general DNA-damage response in *E. coli*, the SOS response. The cellular repressor of DNA repair genes, LexA, is also cleaved in a RecA-dependent process after DNA damage.[17] Presumably lambda, detecting levels of DNA damage which might interfere with its survival and growth, chooses to move on to a new host under these conditions. A third target is UmuD, necessary for error prone repair; in this case the RecA-stimulated cleavage leads to activation of the protein.[18]

Some aspects of the RecA-dependent, DNA damage-induced proteolysis have proven to be relatively unique. The active sites for cleavage of both cI and LexA lie not in RecA but in the target proteins themselves.[19,20] Under appropriate (but nonphysiological) conditions, LexA and cI can undergo self-cleavage at the same site found in the RecA-dependent reaction without the participation of RecA.[21] In vivo, the requirement for RecA and the necessity for RecA activation by DNA damage provides a regulatory step for the initiating induction of the SOS response, including lambda induction. In vitro, RecA protease activity depends on ATP and on single-stranded DNA or oligonucleotides.[22] It is believed that the interaction with these oligonucleotides is the in vitro analog of the in vivo inducing signal, presumably provided by single-stranded DNA stretches present after DNA damage and known to interact with RecA.[16] Therefore, for this protease, the degradation is quite specific, both in terms of the target proteins and in terms of the cleavage site; activation of proteolysis depends on RecA interaction with DNA in a manner that has not been fully defined. While LexA cleavage is clearly an irreversible induction step, induction is made rapidly reversible because LexA represses its own synthesis. Therefore, LexA synthesis also increases after SOS induction. As soon as DNA damage is repaired, RecA is no longer as active in promoting proteolysis, the rate of LexA cleavage decreases, and LexA rapidly accumulates and returns the system to equilibrium by shutting down transcription of the SOS genes.[20,23]

ABNORMAL PROTEIN DEGRADATION LEADS TO IDENTIFICATION OF THE LON (LA) PROTEASE

In a separate line of experiments, cloning and purification of the product of the *lon* (<u>lon</u>g cells; see next section) gene of E. coli led to the demonstration that *lon* encoded an ATP-dependent protease capable of degrading casein in vitro, and, subsequently, to the identification of a number of naturally unstable target substrates.

lon mutants were first identified because they gave rise to UV-sensitive cells, which grew into *lon*g filaments after DNA damage.[24] Mutations in the same locus were isolated by Markovitz, who called them *capR*, for regulator of capsular polysaccharide synthesis, based on the mucoid appearance of the mutant colonies. Markovitz demonstrated the increased synthesis of colanic acid capsular polysaccharide, a polymer of glucose, galactose, glucuronic acid and fucose and increases in activity of some of the enzymes in the capsule synthesis pathway in *lon* mutants, suggesting that Lon might act as a repressor of genes for capsule synthesis.[25] The relationship of capsule overproduction to filamentation remained unclear, however.

In 1973, Bukhari and Zipser again came upon the same locus, in this case in a search for mutations that would prolong the half-life of fragments of β-galactosidase, thereby allowing intramolecular complementation between these normally rapidly degraded fragments.[26] They called the mutated gene *degR*. We later demonstrated that *degR* and a second mutation, *degT*, both contained mutations at the *lon* locus and that *lon* mutations were sufficient to confer the Deg phenotype.[27] The purification and biochemical analysis of the Lon protein demonstrated that it was an ATP-dependent protease.[28,29]

Our work with the *lon* mutation began as a side product of in vitro studies of lambda site-specific recombination. As noted above, Xis, necessary for excision of lambda from the host chromosome, is functionally unstable in vivo.[9] If one assumes degradation of the protein is the cause for this instability, stabilization in a protease mutant host would be useful for isolation of more active protein. (An interest in overproducing unstable proteins continues to be the major motivation for requests for *lon*-deficient bacterial strains.)

I began to work with the *deg* mutants of Bukhari and Zipser[26] with the hope that *deg/lon* mutants would lead to stabilization of Xis. The first aim was to move these mutations to a clean genetic background in which their effect on Xis functional decay could be easily assayed. To make the genetic screening independent of the partially diploid *lac* mutant strains originally used by Zipser and co-workers (and avoid the necessity for running protein gels to confirm the phenotypes), I considered other abnormal proteins which might be subject to degradation and might lend themselves to simple screening. Temperature-sensitive mutations seemed likely candidates. If degradation was responsible for the loss of function of a particular temperature-sensitive mutant protein at high temperature, and the Lon protease was responsible for this degradation, one might predict that *lon* mutations would suppress the temperature-sensitive phenotype. Temperature-sensitive phage mutations promised to be particularly convenient, since the growth of the mutant phage could be easily assayed on a variety of host strains at different temperatures. In fact, of the temperature-sensitive phage mutants I was able to collect among groups at MIT, many proved to be at least partially suppressible in *lon* or *deg* mutant hosts and provided a convenient marker for further manipulation of the strains.[27] Therefore, many mutations leading to in vivo temperature-sensitive protein activity render the proteins more sensitive to degradation at high temperatures; the Lon protease, encoded by *lon/degR/capR*, is at least in part responsible for this degradation. In at least one case, a *lon* mutant was shown to lead to both slower degradation of a temperature sensitive protein (in this case, a mutant σ70) and suppression of the temperature sensitive growth of the mutant cell.[30]

In fact, Xis is not stabilized in *lon* mutants, but a screen of the other known unstable lambda proteins demonstrated a profound effect of Lon on both the functional instability and chemical decay of lambda N protein in vivo.[31] In addition, N protein could be degraded in vitro by purified Lon protein in an energy-dependent process, providing the first direct evidence that the dependence on Lon for in vivo degradation reflected in vitro reality, without the necessary participation of other cellular functions.[32]

The demonstration by the Goldberg and Markovitz laboratories that Lon itself encoded an ATP-dependent protease[28,29] immediately suggested the possibility that the phenotypes associated with *lon* mutants all derived from loss of degradation of particular substrates. The obvious approach to identifying the genes encoding such substrates would be to look for suppressors of the *lon* phenotypes. In addition, if the pleiotropic phenotype of *lon* mutants were due to stabilization of multiple substrates, suppressor mutations that inactivated the gene for a given substrate should suppress only those phenotypes associated with overproduction of that substrate. We and others undertook this approach, and the expected predictions were borne out and are summarized below.

THE BASIS FOR UV SENSITIVITY OF *LON* MUTANTS: STABILIZATION OF THE DIVISION INHIBITOR SULA (SFIA)

George and co-workers first suggested that the loss of proteolytic activity in *lon* mutants might be responsible for the UV sensitivity of these strains.[33] After UV treatment of *lon* strains, cells fail to septate but continue to grow, forming long filaments.[24] George et al found that a strain carrying a *lon* mutation in combination with *tif* (thermal induction of λ and thermal filamentation),[34] a mutation allowing high temperature induction of the SOS response, led to filamentation and death of cells when the temperature was raised. This result suggested that filamentation itself, rather than DNA damage, might be lethal in *lon* mutants. Revertants of the *lon tif* strain that could grow at high temperature included, in addition to cells no longer able to induce the SOS response, two classes of second site suppressors that no longer filamented but did not impair the induction of lambda or other expression of the DNA damage response.[33] The authors suggested that either *sfiA* or *sfiB* (for suppressor of filamentation), the two genes identified in this selection, might encode a division inhibitor sensitive to the Lon protease.[33] Mutations at the same two loci were identified independently as suppressors of the UV sensitivity of *lon* strains and called *sulA* and *sulB* (for suppressor of lon).[35-39] Because mutations at the *sfiA/sulA* locus arise about ten times more frequently than those at *sfiB/sulB* and are fully recessive to wild type, whereas mutations at *sfiB/sulB* are partially dominant,[35,36] the product of the *SulA* gene seemed to us a more likely candidate for the postulated Lon target and division inhibitor. This proved to be the case; SulA is a highly unstable protein (half-life of 1-2 minutes) that is stabilized in *lon* mutants, and SulA overproduction is sufficient to inhibit cell division.[40-42] As predicted by the dependence of filamentation on either UV damage or induction of the DNA damage SOS response by a *tif* mutation, SulA is expressed as part of the SOS DNA damage response.[43,44] SulA apparently interacts with FtsZ, now known to be the protein primarily responsible for initiating septum formation, and blocks its action, in a reversible reaction.[45-47]

The assumption, not yet proven, is that SulA contributes to repair of DNA damage by transiently inhibiting septation and the consequent separation of nucleoids into separate compartments. Once the DNA is repaired, new SulA synthesis should cease as LexA repressor accumulates. The SulA that is present will be rapidly depleted and septation will be restored. Therefore, SulA degradation provides a mechanism for rapidly reversing cell division inhibition when cellular conditions (i.e., the amount of DNA damage) change. In *lon* mutants, SulA degradation does not occur, and cells die.

The availability of *sulA* mutations allowed us to ask if other phenotypes associated with *lon* mutations were also dependent on the increased accumulation of SulA or were due to other possible Lon targets. SulA-dependent filamentation is responsible for the poor lysogenization of Pl phage in *lon* mutants, and possibly instability of plasmids, but does not account for the poor lysogenization by bacteriophage lambda.[31,35] Furthermore, the *sul/sfi* suppressors of UV sensitivity do not block capsule overproduction, suggesting the existence of at least one other naturally unstable cellular target for Lon.

THE BASIS FOR CAPSULE OVERPRODUCTION IN *LON* MUTANTS: STABILIZATION OF THE LIMITING POSITIVE REGULATOR, RcsA

If one makes the assumption that accumulation of a Lon substrate is responsible for the overproduction of capsular polysaccharide in *lon* mutants, the substrate protein would be expected to be a limiting positive regulator for capsule gene expression. The advent of relatively simple methods for the isolation and characterization of *lac* fusions to random *E. coli* promoters[48] facilitated the task of looking for this positive regulator. Mudlac transcriptional fusions which blocked capsule synthesis were isolated; many were found to respond to Lon, giving high expression in *lon⁻* cells and low expression in *lon⁺* cells. Therefore, the previously observed increase in enzymes in the capsule synthesis pathway in *lon* mutants[25,49] was reflected at the level of transcription of genes necessary for capsule synthesis (*cps* genes).[50] Using these *lac* fusions, regulatory mutations which decreased or increased *lac* expression were isolated and subsequently characterized.[51] Mutations defined two genes, *rcsA* and *rcsB* (regulators of capsule synthesis), necessary for expression of the *cps-lac* fusions in *lon* mutants. *rcsA* had a characteristic expected of a limiting positive regulator—*cps* expression was highly sensitive to the gene dosage of *rcsA*.[52] Identification of the protein product, RcsA, confirmed that it was rapidly degraded in *lon⁺* hosts but was relatively stable in *lon* mutants.[52,53]

The continued investigation of the role of RcsA in capsule synthesis has demonstrated that not only does RcsA degradation keep levels of this protein low, but that RcsA synthesis[54] and RcsA activity (Jubete, Maurizi and Gottesman, in preparation) are also subject to stringent regulation. In addition, the stability of RcsA is modulated by interactions with RcsB.[53] Therefore, in a pattern which is becoming common for many unstable proteins, the activity of RcsA is regulated at every level (see ref. 55 for a recent review).

ADDITIONAL TARGETS OF LON PROTEOLYSIS

lon mutants continue to be isolated in a variety of guises. In cases where a mutant protein retains some activity but is depleted by Lon-dependent degradation, stabilization in *lon* mutants may be sufficient to allow suppression of the original mutant phenotype. Mutations in *lon* have also been found as suppressors of *prlF*1, a mutation in an export pathway.[56] The original export mutation causes an increase in Lon activity, possibly as a consequence of the increased accumulation of cytoplasmically localized precursors. Apparently at least some of the phenotypes of the export mutation are due to rapid degradation of precursors in the cytoplasm; when degradation is blocked in a *lon* mutation, many of the phenotypes are reversed.[56] In other cases, the reason why mutations in *lon* are isolated is not clear.[57]

Low copy number plasmids like F and R100 encode functions leading to post-segregational killing. In general, one of these is a protein which kills the cell after plasmid loss (toxin), and a second is an unstable function which protects from killing while the plasmid is still present and synthesis continues (antidote).[58] For F and R100, the unstable functions are proteins which block killing and are stabilized in *lon* mutants.[59,60]

One can consider degradation of the antidote after all possibility of new synthesis has been lost (because the gene encoding it is no longer present) as a commitment to a developmental pathway, in this case leading to cell death. The identification of proteolysis of particular substrates as part of such developmental pathways is becoming increasingly common.

Lon has now been identified in *B. subtilis* and *Myxococcus*, in both cases functioning in a developmental pathway,[61-65] and has been implicated in such a pathway in *Caulobacter crescentus*.[66] In *Bacillus subtilis*, Lon participates in keeping the basal level of developmentally regulated genes low.[65] Finally, Lon analogs have been identified in *S. cerevisiae* and humans, in both cases as mitochondrial enzymes. In *S. cerevisiae*, mutations defective in Lon expression have defective mitochondria, although the Lon targets have not yet been identified.[67-69]

OTHER ENERGY-DEPENDENT PROTEASES IN *E. COLI*

Even in null *lon* mutants, it is clear that there is residual turnover, both of abnormal proteins and of specific Lon substrates.[70] This observation, and the finding that *lon* mutations do not perturb the degradation of unstable proteins such as lambda Xis, lambda O protein, and the heat shock σ factor σ[32], all made it highly likely that other energy-dependent proteases were present in *E. coli*. We began a search for these, pursuing both a biochemical approach (identification and purification of energy-dependent protease activities from *lon* mutant extracts) and a genetic approach (screening for mutants or plasmids leading to suppression of *lon* phenotypes attributable to more than one Lon target). The first of these approaches led more rapidly to identification of a second major protease activity in *E. coli*, the ClpAP protease (for caseinolytic protease); the second approach has uncovered evidence for yet another activity, which we have called Alp, although the protease itself remains elusive.[71-73]

THE CLP FAMILY OF PROTEASES AND THEIR ROLE IN REGULATION

Biochemical identification of a second ATP-dependent protease in *E. coli* was based on the ATP-dependent degradation of casein in extracts, and thus did not assume any ability to degrade a specific naturally unstable protein. The protease which was identified is composed of two subunits: ClpA, an ATPase with the ability to recognize substrates, and ClpP, a serine peptidase able to handle proteins when ClpA is present.[74-80] The genes encoding these subunits are widely separated on the *E. coli* genome (Table 24.1); *clpP* but not *clpA* has a heat shock promoter.[79,81] Thus far, no specific, naturally unstable substrate has in

Table 24.1. Energy-dependent proteases in E. coli

Protease	Gene	Gene Location	MW (daltons)	Activities	Known Substrates	References
Lon(La)	lon	10'	94,000 x 4	ATPase peptidase	RcsA	52,53
					SulA	41
					λN	31,32
					CcdA, Peml	59, 60
ClpAP (Ti)					ClpA, some lac fusions	75,78,79
	clpA	19'	83,000 x 6	ATPase	N-end substrates	82
	clpP	10'	21,000 x 14	peptidase	RepA (chaperone)	87
ClpXP					O protein	83, 84
	clpX	10'	46,300 x ?	ATPase	Phd	86
	clpP				Muvir	85
					Mu A (chaperone)	88
HflA					cII, cIII	96
	hflX,	95'	50,000 x ?	GTPase		
	hflK	95'	46,000 x ? (membrane)			
	hflC	95'	37,000 x ? (membrane) (peptidase?)			
HflB/FtsH	ftsH	69'	70,700 x ?	ATPase Zn protease	cII, cIII	96
					sigma 32	103, 104

The text gives additional details and references for these proteases. With the exception of ClpA action as a chaperone with RepA, all the other substrates are those defined in vivo; the references shown are those for the specific substrate.

fact been identified for ClpAP. ClpAP appears to be partially responsible for degradation of abnormal proteins. In *lon* mutants, some of the residual turnover of abnormal proteins is decreased in *clpA* or *clpP* mutants.[75,78] In addition, some *lac* fusion proteins are subject to degradation by ClpAP.[75,78,79] (Gottesman and Maurizi, unpublished results) Varshavsky and co-workers have shown that the "N-end rule", targeting proteins with specific, unusual N-terminal amino acids for degradation, operates in *E. coli*, and is due to the ClpAP protease.[82] Whether this reflects a pathway for degradation of abnormal proteins with unusual N-termini (for instance, improperly processed or improperly exported processed proteins) or a control mechanism for some specific unstable proteins is still unclear.

The resolution of the ATPase activity and the protease active site of ClpAP into separate subunits immediately raised the possibility that either ClpP or ClpA might act alone or in concert with other cellular proteins to promote other activities. This has in fact proven to be the case. M. Zylicz and co-workers purified activities capable of degrading the lambda replication protein, lambda O, in vitro, and found an ATP-dependent protease, consisting of two subunits, ClpP and a unique ATPase protein which was clearly not the size of ClpA.[83] In a collaborative study, the N-terminal sequence of the new ATPase subunit was sent to us, and matched precisely the open reading frame downstream of *clpP*, which we had provisionally named *clpX*.[84] ClpXP is responsible for the rapid degradation of lambda O protein, for the turnover of certain mutant Mu repressors,[85] and for the turnover of the antidote for plasmid P1.[86] Since the substrate specificity of ClpP-dependent degradation is different between ClpAP and ClpXP, the substrate specificity of this class of proteases must reside in the ATPase subunit.

ClpAP- and ClpXP-dependent proteolysis may represent only a small portion of the activities these classes of proteins are able to carry out. In the absence of ClpP, ClpA has been shown to act as a chaperone in vitro, activating proteins which it can also target for degradation when ClpP is present,[87] and *clpA* mutants show a few properties not shared by *clpP* mutants (S. Gottesman, unpublished observations). *clpX* mutants are defective for Mu replication, a property not shared by *clpP* mutants, probably a reflection of a chaperone activity of ClpX.[88,88a] Finally, some properties of *clpP* mutants are not shared by either *clpA*, *clpX*, or appropriate double mutants, suggesting that there may be additional ATPase components for ClpP, or that in some cases ClpP can act alone (S. Gottesman, unpublished data).[89]

Based on the sequence analysis of *clpA*, another candidate ATPase with high homology to ClpA has been known for some years. Called ClpB, this protein is part of the *E. coli* heat shock response and is necessary for full thermotolerance in *E. coli*.[90,91] A homolog in yeast, Hsp104, plays a major role in thermotolerance.[92,93] In neither case is there evidence that these proteins participate in energy-dependent proteolysis; however, there is evidence that Hsp104 can itself act as a chaperone in the disaggregation of proteins.[94]

THE HFL PROTEASES

The identification of unstable proteins with important roles in lambda development led not only to the identification of the RecA-dependent pathway of cI and LexA degradation, the identification of the ClpXP protease, and our original interest in Lon protease, but also to the identification of two other energy-dependent systems, one of which may prove to be the first essential *E. coli* protease. Lambda cII protein plays a central role in determining the balance between the lytic and lysogenic response for bacteriophage lambda. *c*II mutants were identified among the first classes of lambda plaque morphology mutants as clear plaques (c), of a different sort than *c*I mutations (which were eventually found to encode the repressor) because establishment but not maintenance of repression was defective.[95] λ cIII protein also participates in establishment of repression, by modulating the availability of cII. Since cII is necessary for the initial burst of repressor synthesis during establishment of a lysogen, and for synthesis of integrase, levels of cII activity are apparently critical in the choice between lysogen formation and lytic growth. cII is very unstable,[12] and its stability can be affected either by the presence of cIII, which stabilizes cII, or by mutations in two *E. coli* loci, *hflA* and *hflB* (see ref. 96 for review). *hfl* stands for <u>h</u>igh <u>f</u>requency <u>l</u>ysogenization; as expected, stabilizing cII leads to an increase in the frequency of lysogenization. Recently, evidence has accumulated that both the *hflA* and *hflB* loci encode energy-dependent proteases. Mutations in both loci lead to cII stabilization in vivo.[97] *hflA* consists of three genes, *hflX*, *hflK*, and *hflC*;[98] the protein encoded by *hflX* has homology to GTP-binding proteins.[99] In vitro studies of cII degradation showed a low level degradation by HflD and HflC;[100] Noble et al suggest that the GTP-binding component, HflX, may allow more efficient activity. *hflB* has recently been found to be identical to an essential gene, *ftsH*, with homology to the ATPase components of eukaryotic energy-dependent proteases.[101,102] Mutations in *ftsH* lead to stabilization of both cII and an important cellular regulatory protein, the heat shock σ factor $σ^{32}$.[103,104] Finally, in vitro the FtsH protein carries out energy-dependent degradation of a $σ^{32}$ derivative.[104] Both of these proteases differ from Lon and Clp in that they are localized to the membrane, but since the substrates are cytoplasmic, and presumably the active sites are within the cytoplasm, it is unclear what the significance of this location is. The transient stabilization and consequent accumulation of $σ^{32}$ is an important part of the induction of heat shock,[105] and lambda lysogeny is known to respond to a variety of environmental signals in a way which is still unclear.[96] Possibly the membrane domains help to sense the physiological status of the cell and modulate protease activity.

Degradation of $σ^{32}$ is rapid in cells growing under normal, nonstressed conditions, and slows transiently immediately after a heat shock.[105] The consequent accumulation of $σ^{32}$ is a primary contributor to increased synthesis of the $σ^{32}$-dependent heat shock genes. Accumulation of heat shock genes in turn leads to a return to rapid degradation of $σ^{32}$. It is unclear if the transient $σ^{32}$ stabilization reflects a change in availability of $σ^{32}$ in a protease-sensitive form (for instance, unassociated with RNA polymerase core or associated with chaperones such as DnaJ/DnaK) and/or transient titration of the protease itself.[106-108]

REGULATING PROTEASE ACTIVITY IN VIVO

There is relatively little evidence that the turnover of the highly unstable proteins discussed in this review is regulated by regulation of the protease activity itself. With the exception of RecA-dependent protease activity, which is clearly regulated by the availability of the inducing signal from damaged DNA, Lon and Clp, and probably HflA and HflB, are relatively abundant proteins. While *lon*, *clpXP* and *hflB* are heat shock operons, even at low temperatures there is sufficient protease activity to rapidly turn over the substrates identified thus far. Whether limitation of energy (ATP) is ever sufficient in the cell to block energy-dependent degradation is unclear, but probably unlikely. It is known that even under starvation conditions, energy-dependent degradation continues and even increases,[3,109] although it is unclear what proteases are responsible for this increased degradation. It seems more likely that the use of ATP by these proteases is for appropriate substrate selectivity and processivity rather than for regulation of overall protease activity per se (see ref. 5).

Two other approaches to regulating protein degradation also occur in *E. coli*. The first is the existence of specific protease inhibitors; those identified thus far are all encoded by bacteriophages. cIII stabilizes the targets of the Hfl proteases, although it is unclear yet whether this is by interaction with the protease or with the substrates.[96,97,103] Lambda RexB protein protects lambda O protein from degradation,[110] and apparently does so by a general inhibition of the ClpXP protease (H. Engelberg-Kulka, personal communication). Phage T4 shuts off most host proteolysis on infection,[111] presumably a useful feature for the demonstration of a correlation between position of nonsense mutations and protein size using T4 proteins,[112] since normally amber fragments of proteins are rapidly degraded in *E. coli*. A protein responsible for inhibiting Lon-dependent proteolysis has been identified as the product of the *pinA* (proteolysis inhibition) gene. In vivo, a lambda phage carrying *pinA* turns off Lon-dependent protein degradation.[113] In vitro, PinA interacts with Lon to block the use of ATP for protein turnover (J. Hilliard, L.D. Simon and M. Maurizi, in preparation). Since the shutdown of protein degradation by infection by T4 phage is more drastic than that seen with *pinA* expression alone, it seems likely that T4 carries analogous functions which act as specific inhibitors of other cellular proteases. The function of this protein degradation shut-off for T4 biogenesis remains unclear. It remains possible that *E. coli* also encodes such inhibitors.

Another strategy for modulating specific protein degradation is the interaction of substrates with other components—proteins or DNA. SulA is partially stabilized by the presence of FtsZ,[47] RcsA is stabilized by interaction with RcsB,[53] and CcdA is stabilized by CcdB.[60] These results suggest that degradation by Lon, at least, requires access to regions of the substrates which may also participate in other protein-protein interactions. In vitro degradation of RepA by ClpAP is blocked by interaction of RepA with its DNA target.[87] Interactions are not always protective, however. Mu repressor can be destabilized by interaction with an unstable mutant repressor, suggesting that ClpXP recognition of a substrate and degradation do not have to proceed on the same polypeptide.[85] Models for regulation of the heat shock response postulate that changes in the interaction of σ^{32} with chaperones dramatically affect the rate of σ^{32} degradation and therefore the regulation of synthesis of heat shock genes; in this case, the assumption is that rapid degradation requires interaction with the chaperones.[106-108] σ^S, the stationary phase σ factor encoded by *rpoS*, also appears to be an unstable protein whose degradation varies with cell growth conditions. In this case, stabilization is apparent during stationary phase; the basis for the growth-dependent stabilization is not yet known.[114] It has recently been reported that ClpXP is responsible for the degradation of σ^S during exponential growth.[115] The ability to modulate protein degradation via interactions of the substrate protein with other cellular components means that a full understanding of the role of degradation in regulation requires an understanding of the availability and activity of all the other possible participants as well.

SUMMARY AND GENERAL CONCLUSIONS

E. coli contains at least six to ten energy-dependent proteases, and for most of the examples already studied, degradation of specific substrates is generally due to a single one of these proteases. The basis for a given protease degrading a given target is not yet clear. The addiction protein CcdA for one system (F) is degraded by Lon, and for another, the Phd protein of phage P1, by ClpXP. There is relatively little evidence for regulation of the proteases themselves; in general, rapid proteolysis is a way of keeping basal levels of proteins low when they are not needed and returning levels rapidly to a low basal level after a transient increase.

A hallmark of the proteins subject to rapid degradation is that proteolysis is likely to be only one of multiple controls on synthesis and activity.[55,114,116] This implies that degradation is usually a strategy used only in those cases where small variations in accumulation of active protein have significant consequences for the cell. Therefore, we can expect rapidly degraded proteins to occupy critical regulatory niches, and/or to have activities that are advantageous to the cell only under a limited set of conditions, and that may be quite detrimental in other cases.

REFERENCES

1. Roberts JW, Roberts CW. Proteolytic cleavage of bacteriophage lambda repressor in induction. Proc Natl Acad Sci USA 1975; 72:147-151.
2. Goldberg AL. Degradation of abnormal proteins in *Escherichia coli*. Proc Natl Acad Sci 1972; 69:422-426.
3. Goldberg AL, John ACS. Intracellular protein degradation in mammalian and bacterial cells. Annu Rev Biochem 1976; 45:747-803.
4. Hershko A, Ciechanover A. The Ubiquitin system for protein degradation. Annu Rev Biochem 1992; 61:761-807.
5. Gottesman S, Maurizi MR. Regulation by proteolysis: Energy-dependent proteases and their targets. Microbiol Rev 1992; 56:592-621.
6. Maurizi MR. Proteases and protein degradation in *Escherichia coli*. Experientia 1992; 48:178-201.
7. Konrad MW. Dependence of 'early' λ bacteriophage RNA synthesis on bacteriophage-directed protein synthesis. Proc Natl Acad Sci USA 1968; 59:171-178.
8. Schwartz M. On the function of the N cistron in phage λ. Virology 1970; 40:23-33.
9. Weisberg RA, Gottesman ME. The stability of Int and Xis functions. In: Hershey AD, ed. The Bacteriophage Lambda. Cold Spring Harbor, NY: Cold Spring Harbor Laboratory, 1971:489-500
10. Wyatt WM, Inokuchi H. Stability of lambda O and P replication functions. Virology 1974; 58:313-315.
11. Belfort M, Wulff D. The roles of the lambda cIII gene and the *Escherichia coli* catabolite gene activation system in the establishment of lysogeny by bacteriophage lambda. Proc Natl Acad Sci USA 1974; 71:779-782.
12. Reichardt LF. Control of bacteriophage lambda repressor synthesis after phage infection: the role of the N, cII, cIII and cro products. J Mol Biol 1975; 93:267-288.
13. Jones MO, Herskowitz I. Mutants of bacteriophage λ which do not require the cIII gene for efficient lysogenization. Virology 1978; 88:199-212.
14. Roberts JW, Roberts CW, Mount DW. Inactivation and proteolytic cleavage of phage λ repressor in vitro in an ATP-dependent reaction. Proc Natl Acad Sci USA 1977; 74:2283-2287.

15. Roberts JW, Roberts CW, Craig NL. *Escherichia coli* recA gene product inactivates phage λ repressor. Proc Natl Acad Sci USA 1978; 75:4714-4718.

16. Roberts JW, Devoret R. Lysogenic Induction. In: Hendrix RW, Roberts JW, Stahl FW, Weisberg RA, eds. Lambda II. Cold Spring Harbor, NY: Cold Spring Harbor Laboratory, 1983:123-144

17. Little JW, Edmiston SH, Pacelli LZ et al. Cleavage of the *Escherichia coli* lexA protein by the *recA* protease. Proc Natl Acad Sci USA 1980; 77:3225-3229.

18. Nohmi T, Battista JR, Dodson LA et al. RecA-mediated cleavage activates UmuD for mutagenesis: Mechanistic relationship between transcriptional derepression and posttranslational activation. Proc Natl Acad Sci USA 1988; 85:1816-1820.

19. Slilaty SN, Little JW. Lysine-156 and serine-119 are required for LexA repressor cleavage: a possible mechanism. Proc Natl Acad Sci USA 1987; 84:3987-3991.

20. Little JW. Mechanism of specific LexA cleavage: autodigestion and the role of RecA coprotease. Biochimie 1991; 73:411-422.

21. Little JW. Autodigestion of LexA and phage lambda repressors. Proc Natl Acad Sci USA 1984; 81:1375-1379.

22. Craig NL, Roberts JW. *E. coli recA* protein-directed cleavage of phage λ repressor requires polynucleotide. Nature 1980; 283:26-30.

23. Little JW. LexA cleavage and other self-processing reactions. J Bacteriol 1993; 175:4943-4950.

24. Howard-Flanders P, Simson E, Theriot L. A locus that controls filament formation and sensitivity to radiation in *Escherichia coli* K12. Genetics 1964; 49:237-246.

25. Markovitz A. Regulatory mechanisms for synthesis of capsular polysaccharide in mucoid mutants of *Escherichia coli* K12. Proc Natl Acad Sci USA 1964; 51:239-246.

26. Bukhari AI, Zipser D. Mutants of *Escherichia coli* with a defect in the degradation of nonsense fragments. Nature 1973; 243:238-241.

27. Gottesman S, Zipser D. Deg phenotype of *Escherichia coli lon* mutants. J Bacteriol 1978; 113:844-851.

28. Charette M, Henderson GW, Markovitz A. ATP hydrolysis- dependent activity of the *lon(capR)* protein of *E. coli* K12. Proc Natl Acad Sci USA 1981; 78:4728-4732.

29. Chung CH, Goldberg AL. The product of the *lon (capR)* gene in *Escherichia coli* is the ATP-dependent protease, protease La. Proc Natl Acad Sci USA 1981; 78:4931-4935.

30. Grossman AD, Burgess R, Walter W et al. Mutations in the *lon* gene of *E. coli* K12 phenotypically suppress a mutation in the σ subunit of RNA polymerase. Cell 1983; 32:151-159.

31. Gottesman S, Gottesman ME, Shaw JE et al. Protein degradation in *E. coli*: the *lon* mutation and bacteriophage lambda N and cII protein stability. Cell 1981; 24:225-233.

32. Maurizi MR. Degradation *in vitro* of bacteriophage lambda N protein by Lon protease from *Escherichia coli*. J Biol Chem 1987; 262:2696-2703.

33. George J, Castellazzi M, Buttin G. Prophage induction and cell division in *E. coli*. III. Mutations *sfiA* and *sfiB* restore division in *tif* and *lon* strains and permit the mutator properties of *tif*. Mol Gen Genet 1975; 140:309-332.

34. Castellazi M, George J, Buttin G. Prophage induction and cell division in *E. coli*. I. Further characterization of the thermosensitive mutation tif-1 whose expression mimics the effect of UV irradiation. Mol Gen Genet 1972; 119:139-152.

35. Gottesman S, Halpern E, Trisler P. Role of *sulA* and *sulB* in filamentation by *lon* mutants of *Escherichia coli* K-12. J Bacteriol 1981; 148:265-273.

36. Huisman O, D'Ari R, George J. Further characterization of *sfiA* and *sfiB* mutations in *Escherichia coli*. J Bacteriol 1980; 144:185- 191.

37. Gayda RC, Yamamoto LT, Markovitz A. Second-site mutations in *capR* (*lon*) strains of *Escherichia coli* K-12 that prevent radiation sensitivity and allow bacteriophage lambda to lysogenize. J Bacteriol 1976; 127:1208-1216.

38. Johnson BF. Fine structure mapping and properties of mutations suppressing the *lon* mutation in *Escherichia coli* K12 and B strains. Genet Res 1977; 30:273-286.

39. Johnson BF, Greenberg J. Mapping of *sul*, the suppressor of *lon* in *Escherichia coli*. J Bacteriol 1975; 122:570-574.

40. Huisman O, D'Ari R, Gottesman S. Cell division control in *Escherichia coli*: specific induction of the SOS SfiA protein is sufficient to block septation. Proc Natl Acad Sci USA 1984; 81:4490-4494.

41. Mizusawa S, Gottesman S. Protein degradation in *Escherichia coli*: the *lon* gene controls the stability of the SulA protein. Proc. Natl Acad Sci USA 1983; 80:358-362.

42. Schoemaker JM, Gayda RC, Markovitz A. Regulation of cell division in *Escherichia coli*: SOS induction and cellular location of the SulA protein, a key to *lon*-associated filamentation and death. J Bacteriol 1984; 158:551-561.

43. Mizusawa S, Court D, Gottesman S. Transcription of the *sulA* gene and repression by LexA. J Mol Biol 1983; 171:337-343.

44. Huisman O, D'Ari R. An inducible DNA-replication-cell division coupling mechanism in *E. coli*. Nature 1981; 290:797-799.

45. Lutkenhaus JF. Coupling of DNA replication and cell divison: *sulB* is an allele of *ftsZ*. J Bacteriol 1983; 154:1339-1346.

46. Maguin E, Lutkenhaus J, D'Ari R. Reversibility of SOS-associated division inhibition in *Escherichia coli*. J Bacteriol 1986; 166:733-738.

47. Jones C, Holland IB. Role of the SulB (FtsZ) protein in division inhibition during the SOS response in *Escherichia coli*: FtsZ stabilizes the inhibitor SulA in maxicells. Proc Natl Acad Sci USA 1985; 82:6045-6049.

48. Casadaban MJ, Cohen SN. Lactose genes fused to exogenous promoters in one step using a Mu-lac bacteriophage: in vivo probe for transcriptional control sequences. Proc Natl Acad Sci USA 1979; 76:4530-4533.

49. Markovitz A. Genetics and regulation of bacterial capsular polysaccharide biosynthesis and radiation sensitivity. In: Sutherland I, ed. Surface Carbohydrates of the Prokaryotic Cell. London: Academic Press, 1977:415-462

50. Trisler P, Gottesman S. *lon* transcriptional regulation of genes necessary for capsular polysaccharide synthesis in *Escherichia coli* K-12. J Bacteriol 1984; 160:184-191.

51. Gottesman S, Trisler P, Torres-Cabassa AS. Regulation of capsular polysaccharide synthesis in *Escherichia coli* K12: characterization of three regulatory genes. J Bacteriol 1985; 62:1111-1119.

52. Torres-Cabassa AS, Gottesman S. Capsule synthesis in *Escherichia coli* K-12 is regulated by proteolysis. J Bacteriol 1987; 169:981-989.

53. Stout V, Torres-Cabassa A, Maurizi MR et al. RcsA, an unstable positive regulator of capsular polysaccharide synthesis. J Bacteriol 1991; 173:1738-1747.

54. Sledjeski D, Gottesman S. A small RNA acts as an antisilenceer of the H-NS-silenced *rcsA* gene of *Escherichia coli*. Proc Natl Acad Sci USA 1995; 92:2003-2007.

55. Gottesman S. Regulation of Capsule Synthesis: Modification of the two-component paradigm by an accessory unstable regulator. In: Hoch JA,

Silhavy TJ, eds. Signal Transduction in Bacteria. Washington, DC: American Society for Microbiology, 1995:253-262

56. Snyder WB, Silhavy TJ. Enchanced export of β-galactosidase fusion proteins in *prlF* mutants is Lon dependent. J Bacteriol 1992; 174:5661.

57. Lam H-M, Tancula E, Dempsey WB et al. Suppression of insertions in the complex *pdxJ* operon of *Escherichia coli* K-12 by *lon* and other mutations. J Bacteriol 1992; 174:1554-1567.

58. Yarmolinsky M. Programmed Cell Death in Bacterial Populations. Science 1995; 267:836-837.

59. Tsuchimoto S, Nishimura Y, Ohtsubo E. The stable maintenance system *pem* of plasmid R100: Degradation of PemI protein may allow PemK protein to inhibit cell growth. J Bacteriol 1992; 174:4205-4211.

60. Van Melderen L, Bernard P, Couturier M. Lon-dependent proteolysis of CcdA is the key control for activation of CcdB in plasmid-free segregant bacteria. Molec Microbiol 1994; 11:1151-1157.

61. Tojo N, Inouye S, Komano T. Cloning and nucleotide sequence of the *Myxococcus xanthus lon* gene: indispensability of *lon* for vegetative growth. J Bacteriol 1993; 175:2271-2277.

62. Tojo N, Inouye S, Komano T. The *lonD* gene is homologous to the *lon* gene encoding an ATP-dependent protease and is essential for the development of *Myxococcus xanthus*. J Bacteriol 1993; 175:4545-4549.

63. Gill RE, Karlok M, Benton D. *Myxococcus xanthus* encodes an ATP-dependent protease which is required for developmental gene transcription and intercellular signaling. J Bacteriol 1993; 175:4538-4544.

64. Riethdorf S, Volder U, Gerth U et al. Cloning, nucleotide sequence, and expression of the *Bacillus subtilis lon* gene. J Bacteriol 1994; 176:6518-6527.

65. Schmidt R, Decatur AL, Rather PN et al. *Bacillus subtilis* Lon protease prevents inappropriate transcription of genes under the control of the sporulation transcription factor σ^G. J Bacteriol 1994; 176:6528-6537.

66. Alley MRK, Maddock JR, Shapiro L. Requirement for the carboxyl terminus of a bacterial chemoreceptor for its targeted proteolysis. Science 1993; 239:1754-1757.

67. Wang N, Gottesman S,.Willingham MC et al. A human mitochondrial ATP-dependent protease that is highly homologous to bacterial Lon protease. Proc Natl Acad Sci USA 1993; 90:11247-11251.

68. Suzuki CK, Suda K, Wang N et al. Requirement for the yeast gene *LON* in intramitochondrial proteolysis and maintenance of respiration. Science 1994; 264:273-276.

69. Van Dyck L, Pearce DA, Sherman F. PIM1 encodes a mitochondrial ATP-dependent protease that is required for mitochondrial function in the yeast Saccharomyces cerevisiae. J Biol Chem 1994; 269:238-242.

70. Maurizi MR, Trisler P, Gottesman S. Insertional mutagenesis of the *lon* gene in *Escherichia coli*: *lon* is dispensable. J Bacteriol 1985; 164:1124-1135.

71. Kirby JE, Trempy JE, Gottesman S. Excision of a P4-like cryptic prophage leads to Alp protease expression in *Escherichia coli*. J Bacteriol 1994; 176:2068-2081.

72. Trempy JE, Kirby JE, Gottesman S. Alp suppression of Lon: Dependence on the *slpA* gene. J Bacteriol 1994; 176:2061- 2067.

73. Trempy JE, Gottesman S. Alp, a suppressor of *lon* protease mutants in *Escherichia coli*. J Bacteriol 1989; 171:3348-3353.

74. Hwang BJ, Woo KM, Goldberg AL et al. Protease Ti, a new ATP-dependent protease in *Escherichia coli* contains protein- activated ATPase and proteolytic functions in distinct subunits. J Biol Chem 1988; 263:8727-8734.

75. Katayama Y, Gottesman S, Pumphrey J et al. The two-component ATP-dependent Clp protease of *Escherichia coli*: purification, cloning, and mutational analysis of the ATP-binding component. J Biol Chem 1988; 263:15226-15236.

75a. Katayama-Fujimura Y, Gottesman S, Maurizi MR. a multiple component ATP-dependent protease from *Escherichia coli*. J Biol Chem 1987; 262:4477-4485.

76. Woo KM, Chung WJ, Ha DB et al. Protease Ti from *Escherichia coli* requires ATP hydrolysis for protein breakdown but not for hydrolysis of small peptides. J Biol Chem 1989; 264:2088- 2091.

77. Maurizi MR, Clark WP, Kim S-H et al. ClpP represents a unique family of serine proteases. J Biol Chem 1990; 265:12546- 12552.

78. Maurizi MR, Clark WP, Katayama Y et al. Sequence and structure of ClpP, the proteolytic component of the ATP-dependent Clp protease of *Escherichia coli*. J Biol Chem 1990; 265:12536- 12445.

79. Gottesman S, Clark WP, Maurizi MR. The ATP-dependent Clp protease of *Escherichia coli*: sequence of *clpA* and identification of a Clp-specific substrate. J Biol Chem 1990; 265:7886-7893.

80. Gottesman S, Squires C, Pichersky E et al. Conservation of the regulatory subunit for the Clp ATP-dependent protease in prokaryotes and eukaryotes. Proc Natl Acad Sci USA 1990; 87:3513-3517.

81. Kroh HE, Simon LE. The ClpP component of Clp protease is the σ^{32}-dependent heat shock protein F21.5. J Bacteriol 1990; 172:6026-6034.

82. Tobias JW, Shrader TE, Rocap G et al. The N-End rule in bacteria. Science 1991; 254:1374-1376.

83. Wojtkowiak D, Georgopoulos C, Zylicz M. ClpX, a new specificity component of the ATP-dependent Escherichia coli Clp protease, is potentially involved in λ DNA replication. J Biol Chem 1993; 268:22609-22617.

84. Gottesman S, Clark WP, de Crecy-Lagard V et al. ClpX, an alternative subunit for the ATP-dependent Clp protease of *Escherichia coli*. Sequence and *in vivo* activities. J Biol Chem 1993; 268:22618-22626.

85. Geuskens V, Mhammedi-Alaoui A, Desmet L et al. Virulence in bacteriophage Mu: a case of trans-dominant proteolysis by the *Escherichia coli* Clp serine protease. EMBO J 1992; 11:5121- 5127.

86. Lehnherr H, Yarmolinsky MB. Addiction protein Phd of plasmid prophage P1 is a substrate of the ClpXP serine protease of *Escherichia coli*. Proc Natl Acad Sci USA 1995; 92:3274-3277.

87. Wickner S, Gottesman S, Skowyra D et al. A molecular chaperone, ClpA, functions like DnaK and DnaJ. Proc Natl Acad Sci USA 1994; 91:12218-12222.

88. Mhammedi-Alaoui A, Pato M, Gamma M-J et al. A new component of bacteriophage Mu replicative transposition machinery: the *Escherichia coli* ClpX protein. Molec Microbiol 1994; 11:1109-1116.

88a. Leuchenko I, Luo L, Baker TA. Disassembly of the Mu transposase tetramer by the ClpX chaperone. Genes and Dev 1995; 9:2399-2408.

89. Damerau K, St. John AC. Role of Clp protease subunits in degradation of carbon starvation proteins in *Escherichia coli*. J Bacteriol 1993; 175:53-63.

90. Kitagawa M, Wada C, Yoshioka S et al. Expression of ClpB, an analog of the ATP-dependent protease regulatory subunit in *Escherichia coli*, is controlled by a heat shock σ facter (σ32). J Bacteriol 1991; 173:4247-4253.

91. Squires CL, Pedersen S, Ross BM et al. ClpB is the *Escherichia coli* heat shock protein F84.1. J Bacteriol 1991; 173:4254-4262.

92. Sanchez Y, Taulien J, Borkovich KA et al. Hsp104 is required for tolerance to many forms of stress. EMBO J 1992; 11:2357- 2364.

93. Parsell DA, Sanchez Y, Stitzel JD et al. Hsp104 is a highly conserved protein with two essential nucleotide-binding sites. Nature 1991; 353:270-273.

94. Parsell DA, Kowal AS, Singer MA et al. Protein disaggregation mediated by heat-shock protein Hsp104. Nature 1994; 372:475-478.

95. Kaiser AD. Mutations in a temperate bacteriophage affecting its ability to lysogenize. *Escherichia coli.* Virology 1957; 3:42-.

96. Wulff DL, Rosenberg M. Establishment of repressor synthesis. In: Hendrix RW, Roberts JW, Stahl FW, Weisberg RA, eds. Lambda II. Cold Spring Harbor, NY: Cold Spring Harbor Laboratory, 1983:53-73

97. Hoyt MA, Knight DM, Das A et al. Control of phage lambda development by stability and synthesis of cII protein: Role of the viral cIII and host *hflA, himA* and *himD* genes. Cell 1982; 31:565-573.

98. Banuett F, Herskowitz I. Identification of polypeptides encoded by an *Escherichia coli* locus (*hflA*) that governs the lysis-lysogeny decision of bacteriophage λ. J Bacteriol 1987; 169:4076-4085.

99. Noble JA, Innis MA, Koonin EV et al. The *Escherichia coli hflA* locus encodes a putative GTP-binding protein and two membrane proteins, one of which contains a protease-like domain. Proc Natl Acad Sci USA 1993; 90:10866-10870.

100. Cheng HH, Muhlrad PJ, Hoyt A et al. Cleavage of the cII protein between lysis and lysogeny. Proc Natl Acad Sci USA 1988; 85:7882-7886.

101. Tomoyasu T, Yuki T, Morimura S et al. The *Escherichia coli* FtsH protein is a prokaryotic member of a protein family of putative ATPases involved in membrane functions, cell cycle control, and gene expression. J Bacteriol 1993; 175:1344-1351.

102. Herman C, Ogura T, Tomoyasu T et al. Cell growth and λ phage development controlled by the same essential *Escherichia coli* gene, *ftsH/hflB.* Proc Natl Acad Sci USA 1993; 90:10861- 10865.

103. Herman C, Thévenet D, D'Ari R et al. Degradation of σ32, the heat shock regulator in *Escherichia coli*, is governed by HflB. Proc Natl Acad Sci USA 1995; 92:3516-3520.

104. Tomoyasu T, Gamer J, Bukau B et al. *Escherichia coli* FtsH is a membrane-bound ATP-dependent protease which degrades the heat-shock transcription factor σ^{32}. EMBO J 1995; 14:2551-2560.

105. Straus DB, Walter WA, Gross CA. The heat shock response of *E. coli* is regulated by changes in the concentration of σ32. Nature 1987; 329:348-391.

106. Bukau B. Regulation of the *Escherichia coli* heat-shock response. Molec Microbiol 1993; 9:671-680.

107. Georgopoulos C, Ang D, Liberek K et al. Properties of the *Escherichia coli* heat shock proteins and their role in bacteriophage λ growth. In: Morimoto RI, Tissieres A, Georgopoulos C, eds. Stress Proteins in Biology and Medicine. New York: Cold Spring Harbor Laboratory Press, 1990:191-221

108. Craig EA, Gross CA. Is hsp70 the cellular thermometer? Trends Biochem Sci 1991; 16:135-140.

109. St. John AC, Goldberg AL. Effects of reduced energy production on protein degradation, guanosine tetraphosphate, and RNA synthesis in *Escherichia coli.* J Biol Chem 1978; 253:2705- 2711.

110. Schoulaker-Schwarz R, Dekel-Gorodetsky L, Engelberg-Kulka H. An additional function for bacteriophage λ *rex*: The rexB product prevents degradation of the λ O protein. Proc Natl Acad Sci USA 1991; 88:4996-5000.

111. Simon LD, Tomczak K, John ACS. Bacteriophages inhibit degradation of abnormal proteins in *E. coli*. Nature 1978; 275:424-428.

112. Sarabhai AS, Stretton AOW, Brenner S et al. Colinearity of the gene with the polypeptide chain. Nature 1964; 201:13-17.

113. Skorupski K, Tomaschewski J, Ruger W et al. A bacteriophage T4 gene which functions to inhibit Lon protease. J Bacteriol 1988; 170:3016-3024.

114. Lange R, Hengge-Aronis R. The cellular concentration of the σ^s subunit of RNA polymerase in *Escherichia coli* is controlled at the levels of transcription, translation and protein stability. Genes & Develop 1994; 1994:1600-1612.

115. Lee K-H, Schweder T, Lomovskaya O et al. ClpPX proteolytic activity plays a major role in lowering $\sigma 38$ levels in exponential phase *Escherichia coli*. ASM General Meeting Abstracts 1995:509; H-100.

116. Yura T, Nagai H, Mori H. Regulation of the heat-shock response in bacteria. Annu Rev Microbiol 1993; 47:321-350.

CHAPTER 25

CONTROL OF rRNA AND RIBOSOME SYNTHESIS

Richard L. Gourse and Wilma Ross

1. INTRODUCTION

In rapidly growing bacteria, the synthesis of ribosomes accounts for the cell's single largest expenditure of biosynthetic energy. Under these conditions, the cell contains more than 70,000 ribosomes, each of which is constructed from more than 50 ribosomal proteins and 3 ribosomal RNAs.

RIBOSOMAL PROTEIN SYNTHESIS

The coordination of ribosomal RNA (rRNA) synthesis with ribosomal protein (r-protein) synthesis results from the mechanism of regulation of r-protein synthesis, since the rate of r-protein synthesis is ultimately determined by the level of the rRNAs. Ribosomal protein expression is determined by a feedback mechanism in which rRNAs and r-protein mRNAs compete for so-called repressor r-proteins (e.g., S4, S7, S8, S20, L1, L4, L10). If the r-proteins are in excess of the rRNAs to which they normally bind during the process of ribosome assembly, then these r-proteins bind to their own mRNAs, in most cases directly repressing translation (reviewed in refs. 1-3).

In a few instances, feedback regulation of r-protein synthesis is achieved by other mechanisms in addition to, or instead of, direct repression of translation. For example, in the case of the S10 operon, excess L4 protein causes premature transcription termination in addition to acting as a translational repressor.[4-6] In the case of the *spc* and *str* operons, repressor r-proteins S8 and S7, respectively, act at a single site in their respective mRNAs, regulate the expression of the translationally coupled downstream genes directly by translational repression and regulate the expression of the first one or two genes in the operon indirectly by causing mRNA degradation.[7-11] These feedback mechanisms coordinate the synthesis rates of most r-proteins with rRNA, and are likely to be sufficient to explain the regulation of r-protein synthesis.[12]

RIBOSOMAL RNA SYNTHESIS

Since the rate-limiting step in ribosome synthesis is the synthesis of rRNA, we will concentrate in this chapter almost entirely on rRNA

Regulation of Gene Expression in Escherichia coli, edited by E. C. C. Lin and A. Simon Lynch. © 1996 R.G. Landes Company.

synthesis. In order to make the large number of ribosomes present at high growth rates, the seven *E. coli* rRNA operons transcribe well over half the cell's total RNA under these conditions, even though there are approximately 2000 other operons in the cell. There are special mechanisms which make rRNA promoters so active, and others which carefully regulate the rate of rRNA transcription in order to prevent an over-investment in ribosome synthesis in slower growing or amino acid starved cells. Two control systems appear to account for the reduction in rRNA transcription under conditions of less than maximal growth: growth rate dependent control and stringent control.[13-15] Growth rate dependent control refers to the system that increases rRNA synthesis with approximately the square of the steady state growth rate, while stringent control refers to the virtual shut-off of rRNA transcription after amino acid starvation.

In this review, we first describe the mechanisms contributing to the extraordinary rate of rRNA transcription and then discuss the mechanisms contributing to rRNA regulation. A recurring theme is that the cell has multiple systems that contribute to overall rRNA transcription, and that rRNA output is determined by the interplay of these various systems. We have concentrated on new information available on rRNA transcription and regulation. Earlier reviews should be consulted for information on subjects not covered in this chapter.[1,16] A comprehensive review on the control of r-protein synthesis appeared recently.[3] We have made no attempt to be all inclusive in the information covered, and we apologize to those investigators whose work has been omitted.

2. rRNA GENE ORGANIZATION

In *E. coli*, there are seven rRNA operons (*rrnA, rrnB, rrnC, rrnD, rrnE, rrnG, rrnH*), each of which contains 16S, 23S, and 5S genes in that order. Each operon has two promoters, *rrn* P1 and *rrn* P2, which produce transcripts that are then processed to make the three mature rRNAs. Since the three rRNAs are essentially stable when incorporated into ribosomes, the amounts of the three rRNAs are therefore coordinated. The general structure of the *rrnB* operon is shown in Figure 25.1.

The complete DNA sequences of the rRNA genes from five of the seven operons have been published (all but *rrnD* and *rrnG*).[17,18] The positions of bases in 16S rRNA that are not conserved between operons has been inferred from rRNA sequence information.[19] The sequences of the promoter regions are available for all seven operons.[16] Genetic fusions have been used to compare the transcription of each of the seven operons, and it appears that the transcriptional activities and regulation of the different operons differ in only minor respects, if at all.[20]

The *rrn* P1 promoters are subject to regulation and account for most rRNA transcription at all but the slowest growth rates.[21-23] It was proposed that the *rrn* P2 promoters are low-level constitutive promoters, important for rRNA expression only when the *rrn* P1 promoters are turned off.[24] However, the *rrn* P2 promoters are actually quite strong when separated from *rrn* P1.[25-27] It has been suggested that P2 promoters are occluded by transcription from P1.[28] However, occlusion cannot explain why the P2 promoters are relatively inactive at low steady state growth rates, when P1 promoter activity is significantly reduced (Gourse et al, 1986).

Transfer RNA genes are located between the 16S and 23S rRNA genes of all seven *E. coli* rRNA operons, and in some operons tRNA

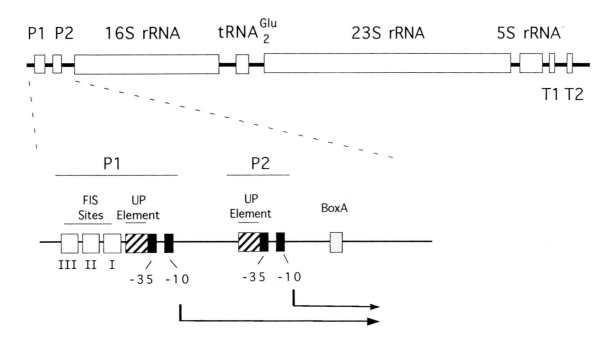

Fig. 25.1. *The general structure of the E. coli rrnB operon. The regions coding for the mature rRNAs and the spacer tRNA are also shown, as is the transcription termination site T1 T2. The promoter region is expanded to show the Fis binding sites I, II, and III upstream of rrnB P1 and the UP Elements (shaded rectangles) in both P1 and P2. The -10 and -35 consensus hexamers and the boxA sequence required for antitermination are also indicated. The lines with arrows represent the transcription initiation sites.*

genes are found distal to the rRNA genes as well. Although these tRNAs are cotranscribed with the rRNAs, most tRNA genes are not in rRNA operons. tRNA gene organization has recently been reviewed.[29]

3. HIGH ACTIVITY OF rRNA SYNTHESIS RATES

As is apparent from the Introduction, each operon must have promoters which are significantly stronger than typical *E. coli* promoters in order to account for so much of the cell's total RNA synthesis. Our estimates indicate that a typical *rrn* P1 promoter, *rrnB* P1, for example, is 10- to 100-fold more active in rich medium than the *lacUV5* promoter. It is only very recently that the *cis* and *trans*-acting elements responsible for the remarkable efficiency of the rRNA transcription machinery have begun to be identified. Molecular mechanisms that increase both the frequency of transcription initiation by RNA polymerase and the efficiency with which the initiated transcripts are completed contribute to the high rate of rRNA synthesis. Special features of the promoters themselves, the UP Element, and the action of a transcription factor, Fis (factor for inversion stimulation), contribute to the rapid rate of initiation, while an rRNA antitermination system ensures that RNA polymerases that have initiated transcription actually make complete products. The features of the P1 and P2 promoters are discussed in detail below (Fig. 25.1).

I. CORE PROMOTER STRUCTURE

The activities of *E. coli* promoters transcribed by Eσ[70] RNAP correlate quite closely with the degree of the promoters' similarity to the core promoter (-10 and -35 recognition region) consensus sequences.[30] It is not surprising, therefore, that the -10 and -35 consensus sequences of the *rrn* P1 promoters match the *E. coli* consensus very closely. The -10 hexamers of all seven operons are exact matches to consensus (TATAAT), while the -35 hexamers vary slightly, but all provide good matches to consensus. The spacing between the two hexamers, however, is only 16 bp, rather than the 17 bp found most frequently in *E. coli* promoters.

In order to determine the positions most important to rRNA transcriptional strength and regulation, mutations were constructed within *rrnB* P1, and promoter activities of more than 50 mutants were examined using *lacZ* fusions.[31] As expected, mutations in the -10 hexamer or in the TTG of the -35 hexamer were detrimental to promoter activity. On the other hand, a mutation that changed the one nonconsensus position (position -33) in the -35 hexamer to consensus (TTGTCA to TTGACA) or a 1 bp insertion mutation that increased the -10, -35 spacer length to 17 bp, dramatically increased promoter activity in vivo. Interestingly, neither 17 bp spacer mutants nor the -33 mutant were growth rate regulated, implying that rRNA promoters have evolved not for maximal promoter activity, but to allow for their regulation as well.[32]

The *rrn* P1 promoters have unusual mechanistic characteristics in vitro that may be relevant to their unusual strength or regulatory capacities. For example, *rrn* P1 promoters are unusually salt sensitive and are affected by template supercoiling.[33-35] The open complex at *rrnB* P1 is unusually short-lived, but it can be stabilized by the addition of the two initiating nucleotides, ATP and CTP.[33,36,37] The addition of the initiating nucleotides, in the absence of the other two nucleotides UTP and GTP, creates an unusual situation in which RNAP starts transcription at position -3 rather than at +1.[37] It is unclear whether this phenomenon ever occurs in vivo and, if so, whether it is of any physiological significance. However, the instability of the open complex initiating at +1 might contribute to the rapid promoter clearance rate characteristic of the *rrn* P1 promoter.[38] In addition, the strong, growth rate regulated, stringently controlled tRNA promoters *tyrT* and *leuV* share some of the same characteristics (W. Ross and R. Gourse, unpublished and ref. 39). Further studies are needed to relate the unusual mechanistic aspects of rRNA promoter function to biological activity and regulation in vivo.

II. THE UP ELEMENT

rrn P1 promoters contain sequences upstream of the -35 hexamer that make a major contribution to promoter activity both in vivo and in vitro.[26,34,40-43] *rrnB* P1 promoter derivatives with different upstream endpoints were examined in vivo as *lacZ* fusions and in vitro using transcription and RNAP binding assays. The -40 to -60 region stimulates *rrnB* P1 promoter activity dramatically in vitro in the absence of proteins other than RNAP and increases *rrnB* P1 transcription about 30-fold in vivo.[43] This region, the UP Element (for promoter Upstream Element,[44] constitutes a third recognition element for RNAP (besides the -10 and -35 hexamers).

The effect of the UP Element on transcription was examined in detail in vitro.[34,43] Sequences between -60 and -50 increase primarily the initial equilibrium constant (K_B), while the -50 to -40 region may increase the rate of RNAP isomerization as well. The *rrnB* P1 UP Element is a separable promoter element in that it also can increase transcription from non-*rrn* promoters in hybrid promoter constructs. For example, the UP Element has virtually the same effect on transcription of a *lac* core promoter as it does on the *rrnB* P1 core promoter.[43,44]

Overlapping 3 bp substitutions were constructed throughout the -40 to -60 region and defined two segments of functional importance in the UP Element centered at approximately -40 and -52 (S. Estrem,

T. Gaal, W. Ross, R. Gourse, unpublished). Substitutions in these segments reduced promoter activity roughly 6-fold each, and a promoter containing mutations in both regions had activity equivalent to a promoter entirely lacking an UP Element. The two regions defined by the three-bp substitutions are protected by RNAP from hydroxyl radical attack in footprinting experiments.[45] Thus, it appears that the two DNA regions centered at approximately -40 and -52 define at least a portion of the DNA surface that interacts with RNAP.

In order to determine the region in RNAP responsible for UP Element function in *rrnB* P1, we examined the effects of mutations in specific RNAP subunits using transcription and footprinting assays. The sigma subunit of RNAP is responsible for interactions with the -10 and -35 hexamers of typical *E. coli* promoters,[46] and there is every reason to believe this holds true for rRNA promoters as well. However, we suspected that the α subunit might be involved in UP Element interactions, since the C-terminal region of the α subunit was known to affect the function of transcription factors which bind upstream of certain promoters.[47]

We tested mutations in the α subunit for their effects on UP Element-dependent stimulation of *rrnB* P1. We found that the activity of the *rrnB* P1 core promoter was unaffected by deletion of the C-terminal region of the α subunit, but that RNAP holoenzymes lacking the C-terminal region of α fail to make contact with the UP Element and fail to utilize the UP Element for stimulating transcription.[44] The effect of the C-terminal region of α on transcription is direct: the C-terminal region of α is a DNA binding domain that binds to the -40 to -60 region of *rrnB* P1 in RNAP holoenzyme, in purified intact α, or as a C-terminal purified peptide of 85 amino acids.[44,48] Thus, the *rrnB* P1 promoter and probably all *rrn* P1 promoters are "extended" by interactions with the α subunit of RNAP, and these interactions are in large part responsible for the high activity of these promoters.

Residues in the α subunit responsible for interactions with the UP Element have been defined.[48a] Two segments of the C-terminal domain of α, including residues 265-269 and residues 296-299, appear to be responsible for interactions with the UP Element. Preliminary results suggest a model in which each of the DNA segments at -40 and -52 protected by holoenzyme interacts with one α monomer, and in which the 265-269 and 296-299 regions of α each contribute to the DNA recognition domain. However, additional information will be required to confirm this model.

A section of the UP Element was displaced from the core promoter region by one turn of the DNA helix by inserting 11 bp between positions -46 and -47. Contacts with RNAP were still maintained, and promoter activity was as high as in *rrnB* P1 promoters with uninterrupted UP Elements. Thus, the α-UP Element DNA interaction can be displaced further upstream by a full turn of the DNA helix and retain *rrnB* P1 promoter activation function, supporting the model that the UP Element constitutes an independent recognition element for RNAP.[49] The residues in α responsible for interacting with the UP Element are close to but distinct from the amino acids in α primarily responsible for interactions with the transcription factor CAP, the activator of the *lac* promoter.[50] A number of other activator proteins also interact with the C-terminal region of α.[47] Therefore, it appears that α contains a flexibly tethered C-terminal domain which can mediate promoter activation either by contacting DNA or transcription factors or both.[48]

UP Elements are not restricted to *rrnB* P1. We found several other naturally occurring promoters that contain upstream sequences contributing to promoter activity, extending conclusions from previous investigators that DNA upstream of the -35 hexamer can affect transcription in synthetic promoter constructs.[51] For example, we found that the *rrnB* P2 promoter also contains an UP element, but the P2 UP Element does not have as large an effect on promoter activity as that in the *rrnB* P1 promoter.[44] Sequences upstream of the *rrnB* P2, *leuV*, *merT*, and *RNA II* promoters all contribute to the activity of their natural core promoters and increase transcription from a *lac* core promoter in synthetic hybrid constructs (ref. 44 and W. Ross, J. Salomon, S. Aiyar and R. Gourse, unpublished). Promoter upstream elements are found in other bacteria, as well. For example, an UP Element was recently identified in the *B. subtilis* flagellin promoter.[52]

III. FIS-DEPENDENT UPSTREAM ACTIVATION

Fis is a small (11.2 kDa) DNA binding protein that increases transcription initiation of *rrn* P1 and several tRNA promoters.[20,42,53-55] In *rrnB* P1, Fis binds in vitro to three sites centered at -71 (Site I), -102 (Site II), and -143 (Site III).[53] Binding of Fis to the three sites in vivo was also assayed directly by in vivo footprinting (W. Ross, J. Salomon, R. Gourse, unpublished). Purified Fis protein activates *rrnB* P1 transcription in vitro, and mutations in the Fis binding sites reduce transcription from the promoter in vivo and in vitro.[53] Primarily as a result of Fis, the region between -60 and -150 stimulates transcription as much as 10-fold. Together with the effect of the UP Element, sequences upstream of the core promoter stimulate transcription more than 300-fold in vivo.[43]

Fis was originally identified as a host factor in several site-specific recombination systems,[56-58] and it plays a number of other roles in *E. coli* as well. For example, in addition to its role as a positively acting transcription factor, Fis is required for DNA replication under certain conditions,[59,60] and it functions as a transcriptional repressor.[61] Fis expression is maximal early in exponential phase and disappears in stationary phase.[62,63] Since Fis has multiple host functions, a priori it was not clear whether the timing of Fis expression would be indicative of when it participates in rRNA transcription. In fact, we have found that activation of rRNA transcription by Fis is limited to cells growing exponentially (J. Appleman, W. Ross, J. Salomon, and R. Gourse, unpublished) and parallels the time of Fis expression.

Although Fis clearly activates rRNA transcription about 5- to 10-fold during steady state growth, rRNA is still transcribed efficiently when the *fis* gene is eliminated. Furthermore, cells lacking Fis have almost normal growth rates.[53] We attribute this apparent paradox to the presence of compensating effects of other systems mediating rRNA and tRNA transcription. One of these compensating mechanisms is a negative feedback system that is proposed to monitor the level of protein synthesis and maintain the appropriate rRNA synthesis rate for the growth (protein synthesis) rate (see "growth rate dependent regulation" section, below). Under normal growth conditions, even in rich media, this feedback system partially represses *rrn* P1 promoters by an unknown mechanism involving only the core promoter sequences. This repression becomes relaxed in the *fis* deletion strain to make up for the lack of upstream activation by Fis. *rrnB* P1-*lacZ* fusions lacking Fis sites have been employed as reporters of the compensatory system in *fis⁻* strains: β-galactosidase activities from such fusions increase in *fis⁻* strains compared to the activities of the same fusions in *fis⁺* strains.[43,53,64]

The mechanism by which Fis activates rRNA transcription has been investigated extensively. Several lines of evidence suggest that Fis activates rRNA transcription by interacting directly with RNA polymerase rather than by altering DNA structure and affecting transcription indirectly: (1) Fis binds to the same face of the helix as RNA polymerase in *rrnB* P1.[65] (2) Activation by Fis is face-of-the-helix dependent.[49] (3) The binding of Fis and RNAP is mutually cooperative.[65] (4) Fis activates transcription in vitro even when no DNA is present upstream of the Fis binding site, indicating that Fis does not function by bringing DNA further upstream into the proximity of RNAP bound at the promoter.[65] (5) Mutant Fis proteins have been identified that reduce transcription activation in vitro under conditions where Fis fully occupies its binding sites in the *rrnB* P1 upstream activating region (UAR).[64]

Although Fis almost certainly interacts with RNAP and its binding site is adjacent to the UP Element, activation by Fis does not require the C-terminal domain of the α subunit,[44] or depend on the presence of a functional UP Element in the DNA.[43] Thus, Fis does not appear to activate transcription by contacting the C-terminal domain of α or by facilitating UP Element-α interactions. However, the site that Fis contacts on RNAP has not been identified.

IV. DNA CURVATURE AND rRNA PROMOTER FUNCTION.

The region upstream of the *rrnB* P1 promoter contains A+T-rich sequences characteristic of bent DNA,[66] and early reports indicated that restriction fragments containing these sequences appeared to display curvature in vitro.[26,41] Therefore, it was proposed that curvature might be related to the ability of this region to increase transcription rates. However, further examination indicated that the region responsible for most of the curvature detected by gel electrophoresis is located at about -100, well upstream of the UP Element and Fis Site I which are responsible for almost all of the effect of the upstream region on transcription.[67] Therefore, intrinsic bending of the DNA does not appear to play a major role in *rrnB* P1 transcription. Whether the A+T-rich *rrnB* P1 UP Element displays any unusual structural characteristics remains to be determined. Since the RNAP α subunit interacts with A+T-rich sequences, and similar sequences sometimes display curvature, it is possible that DNA bending might play an indirect role in transcription by facilitating α subunit interactions at some promoters.

V. ANTITERMINATION/TERMINATION

Cis-acting sequences *downstream* of the rRNA promoters affect rRNA synthesis rates. These effects result from an *rrn* antitermination system (last reviewed in ref. 68), which has DNA determinants that include some of those found in bacteriophage lambda *nut* sites.[69] In lambda, Box A (consensus sequence 5'-CGCTCTTTA-3'[70]) is a conserved sequence that is important for NusB function, while Box B is a region of hyphenated dyad symmetry downstream of BoxA that is responsible for interacting with lambda N.[71] Box A sequences occur in all seven rRNA operons, approximately 30 bp downstream of the P2 start site.[16] The full *cis*- and *trans*-acting determinants of rRNA antitermination have yet to be defined, although requirements for Box A[26,72] and for NusB[73] have been demonstrated. Roles for NusA, NusE (r-protein S10), and NusG have also been proposed.[74,75]

Although the similarities between the lambda and *rrn* antitermination systems are striking, it is clear that the two systems are not identical. A lambda N-like protein has yet to be identified from *E. coli*. The cis-acting requirements for *rrn* antitermination are also different from those found in lambda. In *rrn*, a Box B-like hairpin precedes rather than follows Box A,[69] but Box A sequences are both necessary and sufficient for terminator readthrough.[26,72,74] Although the *rrn* Box B-like sequence is not essential, some mutations in this region decrease the ability of some *rrn* Box A mutations to promote readthrough.[72] Box A mutations are by far the largest class of *cis*-acting negative mutations affecting *rrn* antitermination, while Box B mutations are most abundant in the lambda system.[72] A role for the *rrnB* Box B-like hairpin in *negative* regulation has also been proposed.[76,77]

An *rrn* Box A site has also been identified in the 16S-23S spacer in all 7 rRNA operons, following the spacer tRNA(s) and preceding the 23S rRNA gene.[72,78] Mutations in this spacer Box A affect the synthesis of 23S rRNA.[78] Like the leader Box A sequence, the spacer Box A sequence is found in the spacers of rRNA operons of a variety of other species at the same position as in *E. coli*.[72] It may be that the spacer Box A sequence is a "re-loading" site for antitermination factor(s) in order to prevent premature transcription termination within the long, untranslated 23S rRNA gene.[72,78,79] A role for this region in ribosome assembly or rRNA processing cannot be ruled out, however.[80]

The *rrn* leader region downstream of Box A but upstream of the mature 16S rRNA 5' end contains sites where RNA polymerase pauses in vitro.[81,82] This region is important for the stability and processing of 16S rRNA and the assembly of 30S subunits,[83-86] but the duration of pauses in this region could also potentially affect the rate of rRNA synthesis in vivo.[82,87]

Much of the evidence for antitermination was based on studies of artificial constructs; i.e., involving situations where known terminators were placed downstream of an *rrn* "nut" site, and readthrough was assayed by measurement of downstream reporter gene activity. In order to quantify the biological effect of the *rrn* antitermination system in vivo, it would be necessary to determine how much premature termination would occur in intact rRNA operons in the absence of antitermination function.

More efficient transcriptional readthrough of an *rrnB* 16S fragment was found when transcription was initiated from the *rrnG* promoter/nut site region, compared to transcription initiated from a promoter without an *rrn* nut site.[88] When rRNA synthesis rates were measured directly from proximal and distal sections of rRNA operons in a *nusB* mutant, it was found that *nusB* mutants do in fact exhibit polarity in rRNA transcription, further implying that the antitermination machinery has functional significance to rRNA elongation.[73]

The *rrn* antitermination system allows readthrough primarily of rho-dependent termination sites,[89,90] consistent with the results of earlier studies in which it was shown that a mutation in *rho* restored readthrough of a 16S fragment initiated from promoters lacking "nut" sites.[88] The *rrn* antitermination machinery is able to transcribe through even multiple rho-dependent termination sites efficiently, but reads through rho-independent termination sites only very poorly and with some variation depending on the specific terminator.[89] These authors concluded that rho-dependent terminators that may exist within the 16S gene should be easily overcome by the *rrn* antitermination system.

rRNA transcripts are terminated efficiently at the end of rRNA operons, implying either that these terminators are insensitive to the *rrn* antitermination machinery (Model I) or that the antitermination complex gradually loses its resistance to terminators as it moves through the long rRNA operon (Model II). The terminators at the end of *rrnB* and *rrnG* have been examined in detail and were found to cause efficient termination of antitermination proficient rRNA complexes, supporting Model I.[89-91]

It was proposed that the terminators at the ends of *rrn* operons might possess special qualities, since they terminate antitermination proficient rRNA complexes. This now appears not to be the case, since *rrn* terminators fail to terminate transcripts modified by λN.[92] Apparently, the rho-dependent terminators within rRNA operons are not very efficient, or rRNA antitermination complexes need not be as efficient at reading through terminators as the antitermination complexes that read through the early terminators in phage λ.

The DNA sequences of the *rrn* termination regions have been published for all *rrn* operons but *rrnA*.[17,93-97] In cases where termination has been examined in vivo, it appears that the termination regions are quite complex. *rrnG* contains both rho-dependent and rho-independent sites, but the rho-dependent site terminates transcription as efficiently as both together in vitro (98% termination) even in the absence of the rho-independent site.[90] *rrnB* contains two rho-independent terminators, *rrnB t1* and *rrnB t2*. *rrnB t2*, which displays near sequence identity with the *rrnG* rho-independent terminator, is sufficient to terminate more than 96% of transcripts. *rrnB t1* is somewhat less efficient, however, even when the sequences just downstream of the *t1* stem-loop are included.[90,91] The *rrnD* termination region looks similar to *rrnB*. *rrnC* and *rrnH* have rho-independent terminators, and there is evidence that the termination region of *rrnC*, like those of *rrnB* and *rrnG*, is complex.[90]

VI. RRNA CHAIN ELONGATION

Early estimates of the rate of rRNA chain elongation in *E. coli* ranged from 10 to 100 nucleotides per second (cited in refs. 98,99), although the rate of mRNA chain elongation is usually thought to be approximately 40-50 nucleotides per second in vivo. Electron microscopic analyses concluded the rRNA chain elongation rate is 42 nucleotides per second at 37°C. in rich medium,[99] but recent estimates using other techniques suggest a value roughly double this rate.[100,101] It has been proposed that the putative rapid chain elongation rate of rRNA operons results from the function of the rRNA antitermination system.[101]

Previously, the average RNA (mRNA, tRNA, rRNA) chain elongation rate was considered to be invariant under different nutritional conditions. This assumption was challenged by Jensen and Pedersen[102] who reevaluated the published information and concluded that the data are in best agreement with a 20 to 30% variation in elongation rate between the slowest and the fastest growth rates. Jensen and colleagues then measured the general RNA chain elongation rate and found it decreases substantially after amino acid starvation but varies only very slightly (much less than 20%) in exponential cells growing with different doubling times.[101,103-105] Furthermore, the chain elongation rate in rRNA operons does not vary at all.[104] These results will be discussed below in relation to "passive" models for the control of rRNA transcription at different steady state growth rates and after amino acid starvation.

It was shown previously that (p)ppGpp (ppGpp and pppGpp; see next section) inhibits the general RNA chain elongation reaction in vitro.[106] Questions remain to be answered on the mechanism of this inhibition, on the effects of (p)ppGpp on mRNA and rRNA chain elongation rates in vivo, and on the effect of antitermination on these rates.

4. STRINGENT CONTROL

Amino acid starvation leads to a variety of cellular responses, collectively called the stringent response. In this review, we will discuss only the 10- to 20-fold decrease in rRNA transcription that follows within a few minutes of amino acyl tRNA limitation. Concurrently, (p)ppGpp accumulates to high levels, and there is a large body of evidence indicating that the inhibition of rRNA transcription and the other regulatory effects observed during the stringent response are directly or indirectly caused by (p)ppGpp. It was shown that changes in transcription are not a secondary effect of starvation, since (p)ppGpp, produced conditionally by induction of a *lac* promoter-*relA* fusion, mimics the effects of amino acid starvation.[107] Whether or not (p)ppGpp itself interacts directly with the transcription apparatus is still, however, a primary question in rRNA control. The possible role of basal levels of (p)ppGpp will be discussed in the section on growth rate dependent control (below). We refer the reader to refs. 108 and 109 for aspects of the stringent response not directly related to rRNA control, e.g., the cellular events that signal the ribosome to generate (p)ppGpp and the pleiotropic effects of starvation on biosynthetic pathways other than rRNA.

Both the *rrnB* P1 and P2 promoters are subject to stringent control.[28,110-112] Transcription of many tRNA promoters is also inhibited in response to amino acid starvation.[113] In *rrnB* P1, sequences between -41 and +1 with respect to the transcription start site are sufficient to respond to starvation.[112] Thus, other features of this promoter (the UP Element, Fis sites, and the antitermination sites) are all dispensable for stringent regulation. There is a GC-rich DNA region between the -10 hexamer and the transcription start site (the "discriminator") which shows sequence conservation between many rRNA and tRNA promoters[114] and appears to be essential for the response to starvation.[112,115] Whether the discriminator is a direct regulatory target or whether mutations in this sequence eliminate control because they change the kinetic characteristics of the promoter remains to be determined.

Since the increase in (p)ppGpp concentration following amino acid starvation correlates with the shut-off of stable RNA (rRNA and tRNA) synthesis in vivo, it has long been assumed that (p)ppGpp is a direct inhibitor of rRNA transcription. Although there have been numerous published reports of rRNA transcription inhibition by (p)ppGpp in vitro (see ref. 1 for review), the reproducibility and/or specificity of such results have been questioned. Many investigators have failed to obtain specific inhibition upon adding (p)ppGpp to transcription systems containing only DNA, buffer components, nucleotides, and RNAP. However, there are always alternative explanations for negative results, and it is not possible here to review each individually. Suffice it to say that the question will not be answered until a target and mechanism for the action of (p)ppGpp have been defined in molecular detail.

Several groups have attempted to identify sites in RNA polymerase that are responsible for interactions with (p)ppGpp. For example, Glass and collaborators used a genetic approach to look for mutants in RNAP that failed to respond to isoleucine starvation.[116-118] However, further studies on the mutant RNA polymerases indicated that the effects on rRNA synthesis in these strains were indirect; i.e., that the mutant strains produced reduced amounts of (p)ppGpp but the RNA polymerases still responded to (p)ppGpp concentration changes in vivo.[119]

It was claimed that the 10 kDa omega factor, which has long been known to co-purify with RNA polymerase but is without known function, is necessary for inhibition of RNA polymerase by (p)ppGpp in vitro.[120] Based on these results, it was suggested that only RNAP containing omega, which is encoded by the *rpoZ* gene mapping next to *spoT*,[121] is subject to stringent control of rRNA transcription in vivo. However, this interpretation was questioned when it was found that strains deleted for *rpoZ* are still capable of stringent control.[122]

Another report proposed a nucleotide binding site for (p)ppGpp on RNA polymerase distinct from substrate binding sites based on photoaffinity labeling with a (p)ppGpp photoanalog (8-azidoguanosine -3'-phosphate-5'-[5'-^{32}P]phosphate).[123] The analog labeled the β, β', and s subunits of RNAP. It is not clear whether the proposed binding domain results from a pocket formed from all three subunits, whether there is a lack of specificity in the binding of the analog, or whether the analog is sufficiently similar to (p)ppGpp to be biologically relevant.

Most models for stringent control have concentrated on a direct role for (p)ppGpp in transcription *initiation* specific to rRNA promoters. However, Jensen and Pedersen[102] suggested that a nonspecific inhibitory effect of (p)ppGpp on transcription *elongation* was indirectly responsible for stringent control of stable RNA transcription by sequestering RNAP in elongating transcription complexes, reducing the free RNAP concentration, and thereby reducing transcription of promoters limited at the RNAP concentration-dependent transcription step. Consistent with the model, (p)ppGpp concentrations in the 50 to 200 μM range (i.e., within the range of concentrations present after amino acid starvation) inhibit RNA chain elongation in vitro,[106] and RNA chain elongation rates do in fact decrease dramatically after amino acid starvation.[103,104]

There are at least two prerequisites for this "passive model" proposed by Jensen and colleagues: (1) rRNA promoters would have to be especially sensitive to changes in the concentration of RNAP; and (2) after amino acid starvation the free RNAP concentration would have to be low enough to limit binding and initiation at rRNA promoters but not at nonstringent promoters. Neither of these have been proven. In support of the model (and in contrast to the situation in balanced growth, where continuing RNAP synthesis makes new RNAP available for rRNA promoter binding and where RNAP from the existing pool can be recruited to rRNA promoters after dissociation from low affinity specific or nonspecific sites), it is reasonable to suspect that RNAP concentrations could in fact be limiting for rRNA transcription in the first few minutes after amino acid starvation. However, intuitively it would seem that many or most mRNA promoters should also be sensitive to a drop in the RNAP concentration, since most other promoters have a lower binding constant for RNAP than rRNA promoters, and the RNAP concentration-dependent step in transcription should be limiting at least in a subset of these promoters. In summary, the passive model is highly speculative as an explanation for the stringent control of rRNA but not mRNA promoters, and there is little supporting data at present. However, it remains as an interesting alternative to most direct models for the action of (p)ppGpp on rRNA transcription.

5. GROWTH RATE DEPENDENT CONTROL

As cells grow faster and faster (i.e., have an increased rate of protein synthesis), they must manufacture more and more ribosomes to accommodate the higher rate of protein synthesis. We have already mentioned that the *rrn* P1 promoters are regulated such that rRNA synthesis increases with the square of the cellular growth rate and ribosomes therefore accumulate in proportion to the growth rate. Genetic studies indicated that *rrnB* P1 core promoter sequences (-41 to +1) are sufficient for growth rate dependent control, and certain mutations within this region eliminate regulation.[32,112,124,125] Nevertheless, these studies did not reveal an obvious "target sequence" for a growth rate regulator. Since the RNAP binding region and the region necessary for growth rate control overlap, it seems likely that the way that RNAP recognizes the -10 and -35 recognition region must be regulated in some manner leading to growth rate dependent regulation of *rrn* P1 promoters. In support of this idea, a mutation in the region of *rpoD* which recognizes the -35 hexamer was isolated in a screen for mutations affecting rRNA regulation.[126]

The proportionality between ribosome concentrations and growth rate holds only in the medium to fast growth rate range but not in slow growth conditions. *rrn* P2 promoters are transcribed the same at all steady state growth rates.[26] It is therefore likely that P2 is responsible for making the excess rRNA and hence excess ribosomes found in slowly growing cells. Overproduction of ribosomes at low growth rates may be a great advantage for *E. coli* cells in order to adapt promptly to a nutritional upshift, as pointed out by Koch.[127]

Although progress has been made in the definition of the *cis*-acting sites in *rrn* P1 promoters responsible for growth rate regulation, the identification of the *trans*-acting factors involved in altering the manner in which RNAP interacts with the target sequences remains elusive. Until the system has been defined in molecular detail, all that can be done is to ask whether the available evidence is consistent with the hypotheses currently being entertained. In this review, three hypotheses will be discussed: (1) active models involving an unidentified feedback effector; (2) active models based on a role for (p)ppGpp; and (3) passive control models incorporating or not incorporating a role for (p)ppGpp.

I. Feedback Mechanism

Based on the results of rRNA gene dosage experiments, it was proposed that rRNA synthesis is under the control of a feedback system that somehow signals the presence of excess functional ribosomes and regulates rRNA transcription accordingly.[26,128-130] This model attempts to explain the balance between energy devoted to the various biosynthetic pathways and to the protein synthesizing system itself: in steady state growth, the level of ribosomes never exceeds for long that needed for a specific growth rate. The "signal" of the excess ribosomes is the product of the system, excess protein synthesis, while the "target" of the regulation system is the rRNA promoter.

The original model proposed a feedback mechanism that senses the level of free, nontranslating ribosomes. Subsequent work indicated, however, that inhibition of the initiation of translation (caused by limitation for IF2 or by defects in mRNA binding) leads to a breakdown of the feedback system and consequently to overproduction of rRNA.[131,132] It was concluded that cells probably sense the presence of excess ribosomes from their excess translation activity and not directly from the level of free ribosomes. Therefore, it was proposed that excess rRNA (and therefore ribosome) synthesis causes a small, temporary increase in protein synthesis over that appropriate for the nutritional conditions, and this excess translation generates signals to decrease rRNA synthesis. The feedback system is responsible for balancing the ribosome synthesis rate with the need for protein synthesis: during conditions of slow steady state growth, the feedback system sets up an equilibrium between active/inactive rrn P1 promoters that is shifted toward the repressed state, while the equilibrium is shifted more toward the derepressed state at faster growth rates.

While the identity of the feedback effector generated by excess translation (for a specific growth rate) remains unknown at this time, the model has considerable predictive power. For example, the model predicts that all but the most severe trans-acting mutations that reduce the efficiency of the rRNA transcription apparatus should be phenotypically silent under most conditions, since the cell should derepress the feedback system to keep rRNA transcription constant; i.e., that the cell should be able to compensate for defects in rRNA transcription. In agreement with this prediction, nus mutations (leading to defects in antitermination and a decrease in the overall rate of rRNA transcription elongation) were found to increase rRNA and tRNA transcription initiation.[73] Likewise, deletion of the fis gene, which decreases the rate of RNAP binding to rRNA promoters,[65] results in an increase in rrn P1 core promoter activity.[43,53] Similarly, mutations in the α subunit of RNAP, which decrease the efficiency of UP Element utilization at rrn P1 promoters, increase the activity of rrn P1 core promoters.[44] The feedback model also predicts that the cell should be able to survive with a reduced rRNA gene dose by derepressing transcription of the remaining operons. In fact, E. coli cells survive the deletion of several rRNA operons without a major decrease in rRNA synthesis rate (or growth rate).[20,100,133] These results are consistent with the predictions of the feedback model: any situation that would reduce the output of rRNA leads to derepression of the core rRNA promoters in order to keep rRNA synthesis at the appropriate rate for the growth rate of the cell.

II. (p)ppGpp

Initially, it was reasonable to suspect that the likely effector of stringent control, (p)ppGpp, might be responsible for growth rate dependent control as well.[134] We summarize the evidence supporting this proposition briefly and then discuss evidence that is in conflict with the proposed role for (p)ppGpp in growth rate control and argues for separate mechanisms of growth rate and stringent control.

Bremer and colleagues have argued that the distinction between growth rate control and stringent control is arbitrary because the concentration of (p)ppGpp always varies inversely with the fraction of rRNA transcription to total transcription, whether the (p)ppGpp concentration is high as after amino acid starvation or low as in exponential growth.[135,136] The low (p)ppGpp concentrations found during exponential growth in relA+ or relA− cells correlate inversely with the differential rRNA synthesis rate (rRNA synthesis rate/total RNA synthesis rate) under most steady state conditions.[135,137,138] Because of this correlation, because it has been reported that (p)ppGpp affects transcription in vitro, and because a unified mechanism for stringent control and growth rate control has aesthetic appeal, many investigators have favored the idea that (p)ppGpp is the effector of growth rate dependent control.

If (p)ppGpp were the sole cause of growth rate dependent control, its concentration should always vary inversely with rRNA transcription rates. In fact, there have been several claims in the past that deviations from the normal inverse correlation occur (reviewed in ref. 1).

More recently, it was found that cells undergoing partial pyrimidine limitation display *increasing* (p)ppGpp concentrations with increasing growth rates and increasing differential rRNA synthesis rates.[139] It was also reported that reduced rRNA gene dose leads to an increase in rRNA promoter activity from the remaining rRNA operons without a corresponding change in (p)ppGpp levels.[100] Furthermore, it was reported that the same (p)ppGpp concentration can correspond to more than one growth rate (and presumably rRNA synthesis rate).[140] Thus, the inverse correlation between (p)ppGpp concentration and rRNA transcription may not be as perfect as previously proposed.

The construction of a strain devoid of (p)ppGpp made possible a crucial test of the proposed essential role of (p)ppGpp in growth rate dependent control. Cashel and colleagues proposed that the *spoT* gene, which previously was thought to play a role only in the degradation of (p)ppGpp, is responsible for the synthesis of the residual (p)ppGpp observed in *relA⁻* strains, and they demonstrated that a *relA⁻spoT⁻* strain makes no detectable (p)ppGpp.[141]

The double mutant strain (ppGpp°) requires multiple amino acid supplements and has other pleiotropic effects, supporting the idea that basal levels of (p)ppGpp are necessary for many cellular processes including amino acid biosynthesis. However, when Gourse and co-workers measured growth rate dependent regulation of rRNA synthesis in the double mutant (ppGpp°) strain, both by examination of total RNA/protein ratios and by monitoring *rrn* P1 promoter activities at different growth rates using *rrnB* P1-*lacZ* fusions, the double mutant strain was found to exhibit normal growth rate dependent regulation.[125,142] Bremer and colleagues measured the differential rRNA synthesis rate in the ppGpp° strain and found it was subject to growth rate dependent regulation.[143] These authors also tested the regulation of an *rrnB* P1-*lacZ* fusion (different from that used by Gourse and co-workers), but they found that this fusion appeared to exhibit the same activity at all growth rates, in contrast to the results of the direct rRNA synthesis rate measurements. An explanation for the discrepancy has been proposed.[125]

An rRNA promoter lacking Fis sites, an UP Element, and antitermination determinants still retains growth rate dependent control in the ppGpp° strain,[125] indicating that there must be a regulatory mechanism for rRNA transcription that does not involve Fis, (p)ppGpp, α-UP Element DNA interactions, or Nus factors. Consistent with the idea that feedback regulation of rRNA transcription is responsible for growth rate regulation, *relA⁻spoT⁻* cells are still susceptible to feedback inhibition by increased *rrn* gene dose.[142]

If stringent control and growth rate control work through the same mechanism, the promoter targets of the two responses should be the same. However, several promoters are stringently controlled but not growth rate regulated.[112] This result is not easily accounted for by the (p)ppGpp model for growth rate dependent control.

III. PASSIVE MODELS FOR GROWTH RATE DEPENDENT CONTROL

Passive control models, originally proposed by Maaloe,[144,145] were summarized in an earlier review.[1] These models are based on the idea that the activities of non-rRNA promoters are actively regulated and change with differences in environmental or nutritional conditions, but that the rate of rRNA synthesis is determined only by the RNA polymerase concentration. According to the passive model, the RNA polymerase concentration available for rRNA promoters depends on the amount that is used up by non-rRNA promoters. As discussed above with respect to the Jensen and Pedersen model for stringent control, two corollaries of these models are that RNA polymerase is limiting, and that rRNA promoters compete poorly for limiting RNA polymerase.

What evidence we have regarding the availability of RNA polymerase for rRNA transcription at different steady state growth rates suggests that RNA polymerase concentration is not limiting. Cells carrying plasmids encoding mutant *rrn* operons making nonfunctional rRNAs nevertheless had a rate of rRNA synthesis (defective plus wild type) that was increased approximately 2-fold.[128] This shows that there was enough RNAP available in the cell to transcribe twice the amount of rRNA. Conversely, reducing the total RNAP concentration 2-fold by limiting transcription of *rpoB* and *rpoC* did not substantially reduce rRNA transcription.[146] Similarly, when a temperature-sensitive *rpoC* allele was used to manipulate the amount of active RNAP, it was found that rRNA genes continued to be actively transcribed at the expense of nonribosomal genes when RNAP concentrations were reduced.[147]

Therefore, it would appear that RNA polymerase is not limiting for rRNA transcription during exponential growth. There are at least two possible explanations. First, most RNAP in the cell is bound nonspecifically to DNA, and it could be that rRNA promoters can recruit RNAP from this pool. In addition, although the mechanisms responsible for the regulation of RNA polymerase synthesis are not well understood, it is likely there are mechanisms that can adjust RNAP synthesis to an increased demand.

Furthermore, measurements of the relative strength of rRNA promoters in vitro do not support the idea that rRNA promoters are unable to compete for RNAP with non-rRNA promoters in vivo.[33,34,43,53] In fact, we have not observed other promoters that can compete successfully for limiting RNAP in vitro with rRNA promoters on supercoiled templates in the presence of Fis (unpublished observations). The ability of rRNA promoters to compete with non-rRNA promoters for RNAP in vivo would conflict with the central assumption of all passive models for growth rate dependent control.

The Jensen and Pedersen model for growth rate dependent control[102] proposes that (p)ppGpp increases at lower growth rates; this reduces RNA chain elongation rates, which in turn lowers the available RNAP, which is proposed to reduce transcription initiation of rRNA promoters as the growth rate decreases. This model is subject to the same limitations as other passive models, as just discussed. In addition, the variation in mRNA chain elongation rates at different steady state growth rates is quite small, and the rRNA chain elongation rate appears to be invariant (see rRNA chain elongation section, above). Therefore the variation in the amount of RNAP sequestered in transcribing complexes at different growth rates might not be as large as that proposed in the original formulation of the model. Furthermore, something besides (p)ppGpp would have to be responsible for slowing the transcription elongation rate, since (p)ppGpp is not required for growth rate dependent control.[125,142]

6. ADDITIONAL CONSIDERATIONS

I. tRNA Synthesis

The transcription characteristics of many tRNA genes are similar to rRNA operons in terms of level of expression, stringent control, and growth rate dependent control. In some cases tRNA genes are found within rRNA transcription units, thus explaining the coordination, but in other cases similar or the same mechanisms must work independently on tRNA promoters. Many tRNAs have been shown to be stringently controlled,[114,148] subject to the same system that feedback regulates rRNA transcription,[128,129] or activated by Fis.[25,149,150] For example, the *leuV* promoter, which transcribes the highly expressed tRNA$_1^{leu}$, is activated by both Fis and by an UP Element (W. Ross, W.M. Holmes, and R.L. Gourse, unpublished), and it is stringently controlled and growth rate dependent.[151] However, many tRNA promoters, especially those for tRNAs recognizing rare codons, apparently are not highly expressed,[152] subject to growth rate control,[152] or activated by Fis.[25] Therefore, we should not expect the same mechanisms to be responsible for expression of all tRNAs.

II. Carbon Metabolism

Liebke and Speyer[153] isolated a temperature-sensitive mutation, originally termed *ts8*, that preferentially affects rRNA synthesis with only weak effects on protein synthesis. This mutation is an allele of *fda*, the gene that encodes the glycolytic enzyme fructose-1,6-diphosphate aldolase.[154] The *ts8* mutation shuts down rRNA transcription at the restrictive temperature, and it affects primarily promoters that are competent for growth rate control.[155] A different mutation in *fda*, isolated almost 30 years ago, has some of the same characteristics.[156,157]

Since the *fda* mutation affects energy production from glucose, growth of the mutant strain at the nonpermissive temperature in glucose medium might correspond simply to nutritional shift-down; the observed specific inhibition of rRNA synthesis might then result simply from the feedback system involved in growth rate dependent control. However, the observed inhibition of rRNA synthesis in these mutants might be related to increased accumulation of the fructose-1,6-diphosphate resulting from the mutational block of the aldolase.[155,158] It is possible that the *fda* effect on rRNA synthesis might reflect a role for *fda* in growth rate dependent regulation or in an additional system regulating rRNA synthesis, distinct from the growth rate control system. For example, the function of such a system could be to reduce ribosome synthesis in response to potentially toxic metabolic intermediates (see discussion in ref. 155). This possibility deserves further study.

III. Heat Shock Response

Transcription of heat shock genes in *E. coli* is accomplished by an RNA polymerase holoenzyme containing an alternative sigma factor, σ^{32}, rather than the major sigma factor, σ^{70}. The core regions of the *rrn* P1 promoters contain consensus hexamers for recognition by $E\sigma^{32}$ that overlap those for $E\sigma^{70}$. It was shown that $E\sigma^{32}$ binds and transcribes from *rrnB* P1 in vitro.[159] Since there is evidence that $E\sigma^{70}$ and ribosomes are labile at high temperatures (48-55°C; see discussion in ref. 159), it seems reasonable that transcription of rRNA by $E\sigma^{32}$ might play a role in ribosome synthesis at high temperatures in order to provide the temporary protein synthesis capacity necessary for survival should conditions improve.

7. CONCLUSION AND FUTURE PROSPECTS

In this review, we have attempted to convey the current status of studies on the control of rRNA synthesis in *E. coli*. It should not be surprising that for something like rRNA that is so central to cell growth and survival, there would be multiple contributing mechanisms for expression. The multiple regulatory components, feedback system(s), and pleiotropic effects of mutations that affect rRNA synthesis have made rRNA synthesis a challenging subject to address by the genetic methodologies used for studying regulation questions in other systems. It is clear that multiple mechanisms contribute to rRNA synthesis and regulation concurrently during steady state growth. The accumulated evidence makes it unlikely that (p)ppGpp, or any of the other identified trans-acting factors known to affect rRNA transcription, is an obligatory participant in the growth rate control mechanism. In steady state growth, the feedback system most likely adjusts the rate of rRNA synthesis to the growth rate after Fis-RNAP interactions, UP Element-RNAP interactions, and Nus factor-RNAP interactions have made their contributions. Nevertheless, there has been substantial progress in recent years in understanding the factors important for high rRNA expression, and we are optimistic that the factors contributing to rRNA regulation will be elucidated next.

REFERENCES

1. Nomura M, Gourse RL, Baughman G. Regulation of the synthesis of ribosomes and ribosomal components. Annu Rev Biochem 1984; 53:75-117.
2. Draper DE. Translational regulation of ribosmal proteins in *Escherichia coli*: Molecular mechanisms. In: Ilan J, ed. Translation of Gene Expression. New York: Plenum Press, 1987:1-26.
3. Zengel JM, Lindahl L. Diverse mechanisms for regulating ribosomal protein synthesis in *Escherichia coli*. Prog. Nucl Acid Res Molec Biol 1994; 47:331-370.
4. Lindahl L, Zengel J. Ribosomal Genes in *Escherichia coli*. Annu Rev Genet 1986; 20:297-326.
5. Yates JL, Nomura M. *E. coli* ribosomal protein L4 is a feedback regulatory protein. Cell 1980; 21:517-522.
6. Freedman LP, Zengel JM, Archer RH, Lindahl L. Autogenous control of the S10 ribosomal protein operon of *Escherichia coli*: genetic dissection of transcriptional and post-transcriptional regulation. Proc Natl Acad Sci USA 1987; 84:6516-6520.
7. Gregory RJ, Cahill PBF, Thurlow DL, Zimmermann RA. Interaction of *Escherichia coli* ribosomal protein S8 with its binding sites in ribosomal RNA and messenger RNA. J Mol Biol 1988; 204:295-307.
8. Cerretti DP, Mattheakis LC, Kearney KR, Vu L, Nomura M. Translational regulation of the *spc* operon in *Escherichia coli*. Identification and structural analysis of the target site for S8 repressor protein. J Mol Biol 1988; 204:309-329.
9. Mattheakis LC, Nomura M. Feedback regulation of the *spc* operon in *Escherichia coli*: Translational coupling and mRNA processing. J Bacteriol 1988; 170:4484-4492.
10. Mattheakis LC, Vu L, Sor F, Nomura M. Retroregulation of the synthesis of ribosomal proteins L14 and L24 by feedback repressor S8 in *Escherichia coli*. Proc Natl Acad Sci. USA 1989; 86:448-452.
11. Saito Y, Mattheakis LC, Nomura M. Post-transcriptional regulation of the *str* operon in *Escherichia coli*. Ribosomal protein S7 inhibits coupled translation of S7 but not its independent translation. J Mol Biol 1994; 235:111-124.
12. Cole JR, Nomura M. Translational regulation is responsible for growth-rate-dependent and stringent control of the synthesis of ribosomal proteins L11 and L1 in *Escherichia coli*. Proc Natl Acad Sci USA 1986; 83:4129-4133.
13. Maaloe O, Kjeldgaard NO. Control of macromolecular synthesis: a study of DNA, RNA, and protein synthesis in bacteria. New York: Benjamin, 1966.

14. Gausing K. Regulation of ribosome biosynthesis in *E. coli*. In: Chambliss G et al, ed. Ribosomes: Structure, Function, and Genetics. Baltimore: University Park Press, 1980:693-718.

15. Stent GS, Brenner S. A genetic locus for the regulation of ribonucleic acid synthesis. Proc Natl Acad Sci USA 1961; 47:2005-2014.

16. Jinks-Robertson S, Nomura M. Ribosomes and tRNA. In: Neidhardt FC, ed. *Escherichia coli* and *Salmonella typhimurium*: cellular and molecular biology. Washington, DC: American Society for Microbiology, 1987:1358-1385.

17. Brosius J, Dull TJ, Sleeter DD, Noller HF. Gene organization and primary structure of of a ribosomal RNA operon from *E. coli*. J Mol. Biol 1980; 148:107-127.

18. Blattner FR, Burland V, Plunkett G, Sofia HJ, Daniels DL. Analysis of the *Escherichia coli* genome. IV. DNA sequence of the region from 89.2 to 92.8 minutes. Nucl Acids Res 1993; 21:5408-5417.

19. Carbon P, Ehresmann C, Ehresmann B, Ebel J-P. The complete nucleotide sequence of 16-S RNA from *Escherichia coli*. Eur. J Biochem 1978; 100:399-410.

20. Condon C, Philips J, Fu Z-Y, Squires C, Squires CL. Comparison of the expression of the seven ribosomal RNA operons in *Escherichia coli*. EMBO J 1992; 11:4175-4185.

21. deBoer HA, Nomura M. In vivo transcription of rRNA operons in *Escherichia coli* initiates with purine nucleotide trophosphates at the first promoter and with CTP at the second promoter. J Biol Chem 1979; 254:5609-5612.

22. Lund E, Dahlberg JE. Initiation of *Escherichia coli* ribosomal RNA synthesis in vivo. Proc Natl Acad Sci USA 1979; 76:5480-5484.

23. Sarmientos P, Cashel M. Carbon starvation and growth rate-dependent regulation of the *Echerichia coli* ribosomal RNA promoter: differential control of dual promoters. Proc Natl Acad Sci USA 1983; 80:7010-7013.

24. Sarmientos P, Contente S, Chinali G, Cashel M. Ribosomal RNA operon promoters P1 and P2 show different regulatory responses. In: Hamer DH, Rosenberg M. eds. Gene Expression. New York: Alan R. Liss, 1983:65-74.

25. Thayer G, Brosius J. In vivo transcription from deletion mutations introduced near *Escherichia coli* ribosomal RNA promoter P_2. Mol Gen Genet 1985; 199:55-58.

26. Gourse RL, deBoer HA, Nomura M. DNA determinants of rRNA synthesis in *E. coli*: growth rate dependent regulation, feedback inhibition, upstream activation, and anti-termination. Cell 1986; 44:197-205.

27. Lukacsovich T, Gaal T, Venetianer P. The structural basis of the in vivo strength of the rRNA P_2 promoter of *Escherichia coli*. Gene 1989; 78:189-194.

28. Gafny R, Cohen S, Nachaliel N, Glaser G. Isolated P2 rRNA promoters of *Escherichia coli* are strong promoters that are subject to stringent control. J Mol Biol 1994; 243:152-156.

29. Komine Y, Adachi T, Inokuchi H, Ozeki H. Genomic organization and physical mapping of the transfer RNA geens in *Escherichia coli* K12. J Mol Biol 1990; 212:579-598.

30. McClure WR. Mechanism and control of transcription initiation in prokaryotes. Annu. Rev. Biochem 1985; 54:171-204.

31. Gaal T, Barkei J, Dickson RR, deBoer HA, deHaseth PL, Alavi H, Gourse RL. Saturation mutagenesis of an *E. coli* rRNA promoter and initial characterization of promoter variants. J Bacteriol 1989; 171:4852-4861.

32. Dickson RR, Gaal T, deBoer HA, deHaseth PL, Gourse, R.L. Identification of promoter mutants defective in growth rate dependent regulation of rRNA transcription in *Escherichia coli*. J Bacteriol 1989; 171:4862-4870.

33. Gourse RL. Visualization and quantitative analysis of complex formation between *E. coli* RNA polymerase and an rRNA promoter in vitro. Nucl Acids Res 1988; 16:9789-9809.

34. Leirmo S, Gourse RL. Factor-independent activation of rRNA transcription. I. Kinetic analysis of the roles of the upstream activator region and supercoiling on the *rrnB* P1 promoter in vitro. J Mol Biol 1991; 220:555-568.

35. Ohlsen K, Gralla JD. Interrelated effects of DNA supercoiling, ppGpp, and low salt on melting within the *Escherichia coli* ribosomal RNA *rrnB* P1 promoter. Mol Microbiol 1992; 6:2243-2251.

36. Ohlsen KL, Gralla JD. DNA melting within stable closed complexes at the *Escherichia coli rrnB* P1 promoter. J Biol Chem 1992; 267: 19813-19818.

37. Borukhov S, Sagitov V, Josaitis CA, Gourse RL, Goldfarb A. Two modes of transcription initiation in vitro at the *rrnB* P1 promoter of *Escherichia coli*. J Biol Chem 1993; 268:23477-23482.

38. Langert W, Meuthen M, Mueller K. Functional characteristics of the *rrnD* promoters of *Escherichia coli*. J Biol Chem 1991; 266:21608-21615.

39. Kupper H, Contreras R, Khorana HG, Landy A. In: Losick R, Chamberlin M. ed. RNA polymerase. Cold Spring Harbor, NY: Cold Spring Harbor Laboratory Press; 1976:473-484.

40. Petho A, Belter J, Boros I, Venetianer P. The role of the upstream sequences in determining the strength of an rRNA promoter of *E. coli*. Biochim Biophys Acta 1986; 866:37-43.

41. Zacharias M, Goringer HU, Wagner R. Analysis of the Fis-dependent and Fis-independent transcription activation mechanisms of the *Escherichia coli* ribosomal RNA P1 promoter. Biochem 1992; 31:2621-2268.

42. Sander P, Langert W, Mueller K. Mechanisms of upstream activation of the *rrnD* promoter P_1 of *Escherichia coli*. J Biol Chem 1993; 268:16907-16916.

43. Rao L, Ross W, Leirmo S, Schlax PJ, Gourse RL. Factor-independent activation of *rrnB* P1: An "extended" promoter with an upstream element that dramatically increases promoter strength. J Mol Biol 1994; 235:1421-1435.

44. Ross W, Gosink KK, Salomon J, Igarashi K, Zou C, Ishihama A, Severinov K, Gourse RL. A third recognition element in bacterial promoters: DNA binding by the a subunit of RNA polymerase. Science 1993; 262: 1407-1413.

45. Newlands JT, Ross W, Gosink K, Gourse RL. Factor-independent activation of rRNA transcription. II. Characterization of complexes of *rrnB* P1 promoters containing or lacking the upstream activator region with *E. coli* RNA polymerase. J Mol Biol 1991; 220:569-583.

46. Dombroski AJ, Walter WA, Record MT Jr, Siegele D, Gross CA. Polypeptides Containing Highly Conserved Regions of Transcription Initiation Factor σ70 Exhibit Specificity of Binding to Promoter DNA. Cell 1992; 70:501-512.

47. Russo F, Silhavy T. Alpha: the Cinderella subunit of RNA polymerase. J Biol Chem 1992; 267:14515-14518.

48. Blatter EE, Ross W, Tang H, Gourse RL, Ebright RH. Domain organization of RNA polymerase a subunit: C-terminal 85 amino acids constitute an independently folded domain capable of dimerization and DNA binding. Cell 1994; 78:889-896.

48a. Gaal T, Ross W, Blatter EE et al. DNA binding determinants of the α subunit of RNA polymerase: a novel DNA binding domain architecture. Genes Dev 1995; (in press).

49. Newlands JT, Josaitis CA, Ross W, Gourse RL. Both fis-dependent and factor-independent upstream activation of the *rrnB* P1 promoter are face

of the helix dependent. Nucl Acids Res 1992; 29:719-726.

50. Tang H, Severinov K, Goldfarb A, Fenyo D, Chait B, Ebright RH. Location, structure, and function of the target of a transcriptional activator protein. Genes Dev 1994; 8:3058-3067.

51. Bujard H, Brenner M, Deuschle U, Kammerer W, Knaus R. Structure-Function Relationship of *Escherichia coli* Promoters. In: Reznikoff WS et al, eds. RNA polymerase and the regulation of transcription. New York: Elsevier, 1987:95-103.

52. Fredrick K, Caramori T, Chen Y-C, Galizzi A, Helmann JD. Promoter architecture in the flagellar regulon of *Bacillus subtilis*: high level expression of flagellin by the s^D RNA polymerase requires an upstream promoter element. Proc Natl Acad Sci USA 1995; 92:2582-2586.

53. Ross W, Thompson JF, Newlands JT, Gourse RL. *E. coli* Fis protein activates rRNA transcription in vitro and in vivo. EMBO J 1990; 9:3733-3742.

54. Nilsson L, Vanet A, Vijgenboom E, Bosch L. The role of Fis in *trans*-activation of stable RNA operons of *E. coli*. EMBO J 1990; 9:727-734.

55. Nilsson L, Emilsson V. Factor for Inversion stimulation-dependent growth rate regulation of individual tRNA species in *Escherichia coli*. J Biol Chem 1994; 269:9460-9465.

56. Johnson RC, Bruist MF, Simon MI. Host protein requirements for in vitro site-specific inversion. Cell 1986; 46:531-539.

57. Koch C, Kahmann R. Purification and properties of the *Escherichia coli* host factor required for inversion of the G segment in bacteriphage Mu. J Biol Chem 1986; 261:15673-15678.

58. Thompson JF, Moitosa de Vargas L, Koch C, Kahmann R, Landy A. Cellular factors couple recombination with growth phase: characterization of a new component in the λ site-specific recombination pathway. Cell 1987; 50:901-908.

59. Filutowicz M, Ross W, Wild J, Gourse RL. Involvement of Fis protein in replication of the *E. coli* chromosome. J Bacteriol 1992; 174:398-407.

60. Gille H, Egan JB, Roth A, Messer W. The Fis protein binds and bends the origin of chromosomal DNA replication, *oriC*, of *Escherichia coli*. Nucl Acids Res 1991; 19:4167-4172.

61. Xu J, Johnson RC. Isolation of genes repressed by Fis: Fis and RpoS co-modulate growth phase dependent gene expression in *Escherichia coli*. J Bacteriol 1995; 177:938-947.

62. Ball CA, Osuna R, Ferguson KC, Johnson RC. Dramatic changes in Fis levels upon nutrient upshift in *Escherichia coli*. J Bacteriol 1992; 174:8043-8056.

63. Nilsson L, Verbeek H, Vijgenboom E, van Drunen C, Vanet A, Bosch L. Fis-dependent *trans*-activation of stable RNA operons of *E. coli* under varying growth conditions. J Bacteriol 1992; 174:921-929.

64. Gosink KK, Ross W, Leirmo S, Osuna R, Finkel SE, Johnson RC, Gourse RL. DNA binding and bending are necessary but not sufficient for Fis-dependent activation of *rrnB* P1. J Bacteriol 1993; 175:1580-1589.

65. Bokal AJ IV, Ross W, Gourse RL. The transcriptional activator protein Fis: DNA interactions and cooperative interactions with RNA polymerase at the *Escherichia coli rrnB* P1 promoter. J Mol Biol 1995; 245:197-207.

66. Plaskon RR, Wartell RM. Sequence distributions associated with DNA curvature are found upstream of strong *E. coli* promoters. Nucl Acids Res 1987; 15:785-796.

67. Gaal T, Rao L, Estrem ST, Yang J, Wartell RM, Gourse RL. Localization of the intrinsically bent DNA region upstream of the *E. coli rrnB* P1 promoter. Nucl Acids Res 1994; 22:2344-2350.

68. Morgan EA. Antitermination mechanisms in rRNA operons of *Escherichia coli*. J Bacteriol 1986; 168:1-5.

69. Li S, Squires C, Squires CL. Antitermination of *Escherichia coli* ribosomal RNA transcription is caused by a control region segment containing lambda *nut*-like sequences. Cell 1984; 38:851-860.

70. Friedman DI, Olson E, Johnson LL, Alessi D, Craven MG. Transcription-dependent competition for a host factor: the function and optimal sequence of the λ*boxA* transcription antitermination signal. Genes Dev 1990; 4:2210-2222.

71. Lazinski D, Grzadzielska E, Das A. Sequence-specific recognition of RNA hairpins by bacteriophage antiterminators requires a conserved arginine-rich motif. Cell 1989; 59:207-218.

72. Berg K, Squires C, Squires CL. Ribosomal RNA operon anti-termination. Function of leader and spacer region Box B-Box A sequences and their conservation in diverse micro-organisms. J Mol Biol 1989; 209:345-358.

73. Sharrock RA, Gourse RL, Nomura M. Defective antitermination of rRNA transcription and derepression of rRNA and tRNA synthesis in the *nus* B5 mutant of *Escherichia coli*. Proc Natl Acad Sci USA 1985; 82:5275-5279.

74. Theissen G, Behrens SE, Wagner R. Functional importance of the *Escherichia coli* ribosomal RNA leader box A sequence for post-transcriptional events. Mol Microbiol 1990; 4:1667-1678.

75. Squires CL, Greenblatt J, Li J, Condon C, Squires CL. Ribosomal RNA antitermination in vitro: requirement for Nus factors and one or more unidentified cellular components. Proc Natl Acad Sci USA 1993; 90:970-974.

76. Lukacsovich T, Boros I, Venetianer P. New regulatory features of the promoters of an *Escherichia coli* rRNA gene. J Bacteriol 1987; 169:272-277.

77. Csiszar K, Lukacsovich T, Venetianer P. Regulatory elements of the promoter of an rRNA gene of *E. coli*. Biochim Biophys Acta 1990; 1050:312-316.

78. Stark MJR, Gourse RL, Jemiolo DK, Dahlberg AE. A mutation in an *Escherichia coli* ribosomal RNA operon which blocks the production of precursor 23S ribosomal RNA by RNase III in vivo and in vitro. J Mol Biol 1985; 182:205-216.

79. Zacharias M, Wagner R. Functional characterization of a putative internal promoter sequence between the 16S and the 23S RNA genes within the *Escherichia coli rrnB* operon. Mol Microbiol 1989; 3:405-410.

80. Srivastava AK, Schlessinger D. Mechanism and regulation of bacterial ribosomal RNA processing. Annu Rev Microbiol 1990; 44:105-129.

81. Kingston RE, Chamberlin MJ. Pausing and attenuation of in vitro transcription in the *rrnB* operon of *E. coli*. Cell 1981; 27:523-531.

82. Krohn M, Pardon B, Wagner R. Effects of template topology on RNA polymerase pausing during in vitro transcription of the *Escherichia coli rrnB* leader region. Mol Microbiol 1992; 6:581-589.

83. Krych M, Sirdeshmukh R, Gourse R, Schlessinger D. Processing of *Escherichia coli* 16S rRNA with phage λ leader sequences. J Bacteriol 1987; 169:5523-5529.

84. Theissen G, Eberle J, Zacharias M, Tobias L, Wagner R. The T_L structure within the leader region of *Escherichia coli* ribosomal RNA operons has post-transcriptional functions. Nucl Acids Res 1990; 18:3893-3901.

85. Mori H, Dammel C, Becker E, Triman K, Noller HF. Single base alterations upstream of the *E. coli* 16S rRNA coding region result in temperature-sensitive 16S rRNA expression. Biochim Biophys Acta 1990; 1050:323-327.

86. Theissen G, Thelen L, Wagner R. Some base substitutions in the leader of an *Escherichia coli* ribosomal RNA operon affect the structure and function of ribosomes. J Mol Biol 1993; 233:203-218.

87. Zacharias M, Wagner R. Deletions in the t_L structure upstream to the rRNA genes in the *E. coli* rrnB operon cause transcription polarity. Nucl Acid Res 1987; 15:8235-8248.

88. Aksoy S, Squires CL, Squires C. Evidence for antitermination in *Escherichia coli* rRNA transcription. J Bacteriol 1984; 159:260-264.

89. Albrechtsen B, Squires CL, Li S, Squires C. Antitermination of characterized transcriptional terminators by the *Escherichia coli rrnG* leader region. J Mol Biol 1990; 213:123-134.

90. Albrechtsen B, Ross BM, Squires C, Squires CL. Transcriptional termination sequence at the end of the *Escherichia coli* ribosomal RNA *G* operon: complex terminators and antitermination. Nucl Acids Res 1991; 19:1845-1852.

91. Orosz A, Boros I, Venetianer P. Analysis of the complex transcription termination region of the *Escherichia coli rrnB* gene. Eur J Biochem 1991; 201:653-659.

92. Ghosh B, Grzadzielska E, Bhattacharya P, Peralta E, DeVito J, Das A. Specificity of antitermination mechanisms: suppression of the terminator cluster T1-T2 of *Escherichia coli* ribosomal RNA operon, *rrnB*, by phage λ antiterminators. J Mol Biol 1991; 222:59-66.

93. Young RA. Transcription termination in the *Escherichia coli* ribosomal RNA operon *rrnC*. J Biol Chem 1979; 254:12725-12731.

94. Duester G, Holmes WM. The distal end of the ribosomal RNA operon *rrnD* of *Escherichia coli* contains a tRNA^{Thr} gene, two 5S genes and a transcription terminator. Nucl Acids Res 1980; 8:3793-3807.

95. Sakiya T, Mori M, Takahashi N, Nishimura S. Sequence of the distal tRNA^{Asp} gene and the transcription termination signal in the *Escherichia coli* ribosomal RNA operon rrnF (or G). Nucl Acids Res 1980; 8:3809-3827.

96. Liebke H, Hatfull G. The seqeunce of the distal end of the *E. coli* ribosomal RNA operon indicates conserved features are shared by *rrn* operons. Nucl Acids Res 1985; 13:5515-5525.

97. Seol W, Shatkin AJ. Sequence of the distal end of the *E. coli* ribosomal RNA *rrnG* operon. Nucl Acids Res 1990; 18:3056.

98. Bremer H, Dennis PP. Modulation of chemical composition and other parameters of the cell by growth rate. In: Neidhardt FC et al, ed. *Escherichia coli* and *Salmonella typhimurium*: cellular and molecular biology. Washington, DC: American Society for Microbiology, 1987: 1527-1542.

99. Gotta SL, Miller OL, French SL. rRNA Transcription rate in *Escherichia coli*. J Bacteriol 1991; 173:6647-6649.

100. Condon C, French S, Squires C, Squires CL. Depletion of functional ribosomal RNA operons in *Escherichia coli* causes increased expression of the remaining intact copies. EMBO J 1993; 12:4305-4315.

101. Vogel U, Jensen KF. The RNA chain elongation rate in *Escherichia coli* depends on the growth rate. J Bacteriol 1994; 176:2807-2813.

102. Jensen KF, Pedersen S. Metabolic growth rate control in *Escherichia coli* may be a consequence of subsaturation of the macromolecular biosynthetic apparatus with substrates and catalytic components. Microbiol Rev 1990; 54:89-100.

103. Vogel U, Sorensen M, Pedersen S, Jensen KF, Kilstrup M. Decreasing transcription elongation rate in *Escherichia coli* exposed to amino starvation. Mol Microbiol 1992; 6:2191-2200.

104. Vogel U, Jensen KF. Effects of guanosine 3',5'-bisdiphosphate (ppGpp) on rate of transcription elongation in isoleucine-starved *Escherichia coli*. J Biol Chem 1994; 269:16236-16241.

105. Sorensen MA, Jensen KF, Pedersen S. High concentrations of ppGpp decrease the RNA chain growth rate. J Mol Biol 1994; 236:441-454.

106. Kingston RE, Nierman WC, Chamberlin MJ. A direct effect of guanosine tetraphosphate on pausing of *Escherichia coli* RNA polymerase during RNA chain elongation. J Biol Chem 1981; 256:2787-2797.

107. Schreiber G, Metzger SG, Aizenman E, Roza S, Cashel M, Glaser G. Overexpression of the *relA* gene in *Escherichia coli*. J Biol Chem 1991; 266:3760-3767.

108. Cashel M, Rudd KE. The stringent response. In: *Escherichia coli* and *Salmonella typhimurium*: cellular and molecular biology. Neidhardt FC et al, ed. Washington, DC: American Society for Microbiology, 1987:1410-1438.

109. Goldman E, Jakubowski H. Uncharged tRNA, protein synthesis, and the bacterial stringent response. Mol Microbiol 1990; 4:2035-2040.

110. Gourse RL, Stark MJR, Dahlberg AE. Regions of DNA involved in the stringent control of plasmid-encoded rRNA in vivo. Cell 1983; 32:1347-1354.

111. Sarmientos P, Sylvester JE, Contente S, Cashel M. Differential stringent control of the tandem *E. coli* ribosomal RNA promoters from the *rrnA* operon expressed in vivo in multicopy plasmids. Cell 1983; 32:1337-1346.

112. Josaitis CA, Gaal, T, Gourse RL. Stringent control and growth rate dependent control have nonidentical promoter sequence determinants. Proc Natl Acad Sci USA 1995; 92:1117-1121.

113. Ikemura T, Dahlberg JE. Small ribonucleic acids in *Escherichia coli*. II. Noncoordinate accumulation during stringent control. J Biol Chem 1973; 248:5033-5041.

114. Travers AA. Conserved features of coordinately regulated *E. coli* promoters. Nucl Acids Res 1984; 12:2605-2618.

115. Travers AA. Promoter sequence for stringent control of bacterial ribonucleic acid synthesis. J Bacteriol 1980; 141:973-976.

116. Nene V, Glass RE. Relaxed mutants of *Escherichia coli* RNA polymerase. FEBS Lett 1983; 153:307-310.

117. Glass, RE, Jones ST, Ishihama A. Genetic studies on the β subunit of *Escherichia coli* RNA polymerase. VII. RNA polymerase *is* a target for ppGpp. Mol Gen Genet 1986; 203:265-268.

118. Glass RE, Jones ST, Nomura T, Ishihama A. Hierarchy of the strength of *Escherichia coli* stringent control signals. Mol Gen. Genet 1987; 210:1-4

119. Baracchini E, Glass R, Bremer H. Studies in vivo on *Escherichia coli* RNA polymerase mutants altered in the stringent response. Mol Gen. Genet 1988; 213:379-387.

120. Igarashi K, Fujita N, Ishihama A. Promoter selectivity of *Escherichia coli* RNA polymerase: omega factor is responsible for ppGpp sensitivity. Nucl Acids Res 1989; 17:8755-8765.

121. Gentry DR, Burgess RR. *rpoZ*, encoding the omega subunit of *Escherichia coli* RNA polymerase, is in the same operon as *spoT*. J Bacteriol 1989; 171:1271-1277.

122. Gentry D, Xiao H, Burgess RR, Cashel M. The omega subunit of *Escherichia coli* K-12 RNA polymerase is not required for stringent RNA control in vivo. J Bacteriol 1991; 173:3901-3903.

123. Owens JR, Young A, Woody M, Haley BE. Characterization of the guanosine-3'-diphosphate-5'-diphosphate binding site on *E. coli* RNA polymerase using a photoprobe, 8-azidoguanosine-3'-5'-bisphosphate. Biochem Biophys Res Comm 1987; 142:964-971.

124. Zacharias M, Goringer HU, Wagner R. The signal for growth rate control and stringent sensitivity in *E. coli* is not restricted to a particular sequence motif within the promoter region. Nucl Acids Res 1990; 18:6271-6275.

125. Bartlett MS, Gourse RL. Growth rate dependent control of the *rrnB* P1 core promoter in *Escherichia coli*. J Bacteriol 1994; 176:5560-5564.

126. Keener J, Nomura M. Dominant lethal phenotype of a mutation in the -35 recognition region of *Escherichia coli*[70]. Proc Natl Acad Sci USA 1993; 90:1751-1755.

127. Koch AL. Overall controls on the biosynthesis of ribosomes in growing bacteria. J Theor Biol 1970; 28:203-231.

128. Jinks-Robertson S, Gourse RL, Nomura M. Expression of rRNA and tRNA genes in *Escherichia coli*: evidence for feedback regulation by products of rRNA operons. Cell 1983; 33:865-876.

129. Gourse RL, Nomura M. The level of rRNA, not tRNA, synthesis controls transcription of rRNA operons in *E. coli*. J Bacteriol 1984; 160:1022-1026.

130. Gourse RL, Takebe Y, Sharrock RA, Nomura M. Feedback regulation of rRNA and tRNA synthesis and accumulation of free ribosomes after conditional expression of rRNA. Proc Natl Acad Sci USA 1985; 82:1069-1073.

131. Cole JR, Olsson CL, Hershey JWB, Grunberg-Manago M, Nomura M. Feedback regulation of rRNA synthesis in *Escherichia coli*. Requirement for initiation factor IF2. J Mol Biol 1987; 198:383-392.

132. Yamagishi M, deBoer HA, Nomura M. Feedback regulation of rRNA synthesis: a mutational alteration in the anti-Shine-Dalgarno of the 16S rRNA gene abolishes regulation. J Mol Biol 1987; 198:547-550.

133. Ellwood M, Nomura M. Deletion of a ribosomal ribonucleic acid operon in *Escherichia coli*. J Bacteriol 1980; 143:1077-1080.

134. Travers AA. RNA polymerase and the control of growth. Nature 1976; 263:641-646.

135. Ryals J, Little R, Bremer H. Control of rRNA and tRNA synthesis in *Escherichia coli* by guanosine tetraphosphate. J Bacteriol 1982; 151:1261-1268.

136. Baracchini E, Bremer H. Stringent and growth control of rRNA synthesis in *Escherichia coli* are both mediated by ppGpp. J Biol Chem 1988; 263:2597-2602.

137. Hernandez VJ, Bremer H. Guanosine tetraphosphate (ppGpp) dependence of the growth rate control of *rrnB* P1 promoter activity in *Escherichia coli*. J Biol Chem 1990; 265:11605-11614.

138. Hernandez VJ, Bremer H. *Escherichia coli* ppGpp synthetase II activity requires *spoT*. J Biol Chem 1991; 266:5991-5999.

139. Vogel U, Pedersen S, Jensen KF. An unusual correlation between ppGpp pool size and rate of ribosome synthesis during partial pyrimidine starvation of *Escherichia coli*. J Bacteriol 1991; 173:1168-1174.

140. Joseleau-Petit D, Thevenet D, D'Ari R. ppGpp concentration, growth without PBP2 activity, and growth-rate control in *Escherichia coli*. Mol Microbiol 1994; 13:911-917.

141. Xiao H, Kalman M, Ikehara K, Zemel S, Glaser G, Cashel M. Residual guanosine 3',5'-Bispyrophosphate synthetic activity of *relA* null mutants can be eliminated by *spoT* null mutations. J Biol Chem 1991; 266:5980-5990.

142. Gaal T, Gourse RL. Guanosine 3'-diphosphate 5'-diphosphate is not required for growth rate-dependent regulation of rRNA transcription in *Escherichia coli*. Proc Natl Acad Sci USA 1990; 87:5533-5537.

143. Hernandez VJ, Bremer H. Characterization of RNA and DNA synthesis in *Escherichia coli* strains devoid of ppGpp. J Biol Chem 1993; 268:10851-10862.

144. Maaloe O. An analysis of bacterial growth. Dev Biol Supp 1969; 3:33-58.

145. Maaloe O. Regulation of the protein synthesizing machinery-ribosomes, tRNA, factors, and so on. In: Goldberger R. ed. Biological regulation and development. New York: Plenum, 1979:487-542.

146. Nomura M, Bedwell D, Yamagishi M, Cole JR, Kolb JM. RNA polymerase and the regulation of RNA synthesis in *Escherichia coli*. RNA polymerase concentration, stringent control, and ribosome feedback regulation. In: Reznikoff WS et al, eds. RNA polymerase and the regulation of transcription. New York: Elsevier Science Publishing, 1986:137-149.

147. Downing W, Dennis PP. RNA polymerase activity may regulate transcription initiation and attenuation in the *rpl*KAJL*rpo*BC operon in *Escherichia coli*. J Biol Chem 1991; 266:1304-1311.

148. Lamond AI, Travers AA. Stringent control of bacterial transcription. Cell 1985; 41:6-8.

149. Josaitis C, Gaal T, Ross W, Gourse RL. Sequences upstream of the -35 hexamer of *rrnB* P1 affect promoter strength and upstream activation. Biochim Biophys Acta 1990; 1050:307-311.

150. Verbeek H, Nilsson L, Baliko G, Bosch L. Potential binding sites of the *trans*-activator Fis are present upstream of all rRNA operons and of many but not all tRNA operons. Biochim Biophys Acta 1990; 1050:302-306.

151. Rowley KB, Elford RM, Roberts I, Holmes WM. In vivo regulatory responses of four *Escherichia coli* operons which encode leucyl-tRNAs. J Bacteriol 1993; 175:1309-1315.

152. Emilsson V, Kurland CG. Growth rate dependence of transfer RNA abundance in *Escherichia coli*. EMBO J 1990; 8:4359-4366.

153. Liebke HH, Speyer JF. A new gene in *E. coli* rRNA synthesis. Mol Gen. Genet 1983; 189:314-320.

154. Singer M, Rossmeissl P, Cali BM, Liebke H, Gross CA. The *Escherichia coli ts8* mutation is an allele of *fda*, the gene encoding fructose-1,6-diphosphate aldolase. J Bacteriol 1991; 173:6242-6248.

155. Singer M, Walter WA, Cali BM, Liebke H, Gourse RL, Gross CA. Physiological effects of the fructose-1,6-diphosphate aldolase ts8 mutation on stable RNA synthesis in *Escherichia coli*. J Bacteriol 1991; 173:6249-6257.

156. Bock A, Neidhardt FC. Isolation of a mutant of *Escherichia coli* with a temperature sensitive fructose-1,6-diphosphate aldolase activity. J Bacteriol 1966; 92:464-469.

157. Bock A, Neidhardt FC. Properties of a mutant of *Escherichia coli* with a temperature sensitive fructose-1,6-diphosphate aldolase activity. J Bacteriol 1966; 92:470-476.

158. Schreyer R, Bock KA. Phenotypic suppression of a fructose-1,6-diphosphate aldolase mutation in *Escherichia coli*. J Bacteriol 1973; 115:268-276.

159. Newlands JT, Gaal T, Mecsas J, Gourse RL. Transcription of the *E. coli rrnB* P1 promoter by the heat shock RNA polymerase ($E\sigma^{32}$) in vitro. J Bacteriol 1993; 175:661-668.

CELL DIVISION

Lawrence I. Rothfield and Jorge Garcia-Lara

INTRODUCTION

The *E. coli* division cycle must be regulated both temporally and topologically. Temporal regulation is required to assure that septation does not occur until chromosome replication has been completed and until the two daughter chromosomes have been segregated to opposite ends of the cell. Topological regulation is required to assure that the septum is formed at its proper midcell location and not at eccentric sites that would lead to formation of anucleate cells. A considerable amount is known about genes involved in these aspects of cell division. However, surprisingly little is understood about the regulation of these processes and studies of this important problem are still in their infancy. Although there is reason to believe that the expression of important cell division genes is subject to regulatory control, it is not known whether orderly progression through the cell cycle is mediated by changes in gene expression. On the other hand, it is clear that regulation of gene expression does play an important role in modulating the response of the division process to physiological aberrations such as interference with DNA replication.

Modern genetic studies of the division process began more than a quarter century ago with the pioneering work of Hirota and his collaborators, who isolated a collection of *E. coli* mutants that failed to divide when grown at elevated temperature.[1] The mutants were called *fts* mutants, for filamentation thermosensitive. This work, and that of other investigators in succeeding years, led to the identification of a large number of *fts* mutants. Only a subset of these are thought to code for proteins that play a direct role in the division process, as defined by the failure of null mutants to form the division septum and the failure to identify any other metabolic or physiologic defects in the mutant cells. In some cases there is more direct evidence for a direct role for specific Fts proteins in the regulation or catalysis of the division process, as described more fully in the succeeding sections of this chapter.

Many of the original and many of the subsequently identified *fts* mutants were later shown to code for proteins required for important cellular processes that had no obvious relation to cell division. In addition, mutations in other genes that are required for other cellular functions were shown to also affect the ability of cells to divide. They are discussed further in a later section of this chapter.

Regulation of Gene Expression in Escherichia coli, edited by E. C. C. Lin and A. Simon Lynch. © 1996 R.G. Landes Company.

THE CELL DIVISION PROCESS

Cell division in *E. coli* and most other bacteria occurs by formation of a division septum at the midpoint of the cell. Since the cell envelope of *E. coli* and other gram-negative bacteria consists of three layers—the inner membrane, murein and outer membrane layers—septum formation requires the coordinate invagination of three distinct structures. The ingrowth of the three layers is not obligatorily coupled since mutants of *E. coli* and its closely related cousin *Salmonella typhimurium* have been described in which the ingrowth of outer membrane is uncoupled from ingrowth of the murein and inner membrane layers.[2,3] In these mutants, the inner membrane and murein layers appear to invaginate normally whereas the outer membrane fails to invaginate. As a result, the cells exist as chains in which discrete cytoplasmic units are separated by inner membrane-murein septa, with the units being held together by outer membrane bridges. The products of the two genes, *cha* and *lkyD*, in *E. coli* and *S. typhimurium*, respectively, are therefore thought to be required for the attachment of outer membrane to the inner layers of the septum during the invagination process. It is striking that little has been done to further explore this portion of the division process.

Septal invagination normally begins at a discrete point in the cell cycle, approximately 10 minutes after completion of chromosome replication. However, septation does not require that chromosomes be completely replicated. Thus, septation occurs in cells in which chromosome replication is blocked prior to termination if the SOS response is prevented.[4-6a] When cells are blocked at the stage of initiation of chromosome replication, septation can also continue, giving rise to a population of anucleate cells.[5,6,6b] It has been suggested that the signal for septation to begin may be related to the achievement of a certain critical cell length,[7] or to the presence of the DNA-free region that appears at the midpoint of the rod-shaped cell after the daughter chromosomes have moved apart toward the cell poles.[8]

ESSENTIAL CELL DIVISION GENES

THE "TWO-MINUTE CLUSTER" OF CELL DIVISION GENES

It is striking that most genes that are known to play a direct role in the division process are located in a small region of the *E. coli* chromosome, near 2 minutes on the genetic map (Fig. 26.1). The same region also contains at least seven genes whose protein products are required for synthesis of the cell wall peptidoglycan (murein). The genes in the two-minute region that are specifically required for cell division include *ftsL, ftsI, ftsW, ftsQ, ftsA* and *ftsZ*.[7]

Although all of the gene products of the *fts* genes in the two minute region are required for septation, their cellular concentrations differ. FtsZ is a high abundance protein, present at approximately 20,000 copies per cell.[9] In contrast, FtsI, FtsL and FtsQ are present at very low concentrations, probably less than 50 molecules per cell.[10-12] FtsA is present at a slightly higher concentration of 50-200 molecules per cell.[13]

FtsI, FtsL and FtsQ appear to be transmembrane proteins (Fig. 26.2).[11,12,14] A membrane location for *ftsW* has also been suggested on the basis of its hydropathicity profile.[15] FtsA is likely to be a peripheral inner membrane protein whose membrane-binding capacity may be modulated by its state of phosphorylation[16,17] whereas FtsZ (see below) changes its cellular location from cytoplasm to membrane as a function of the division cycle.

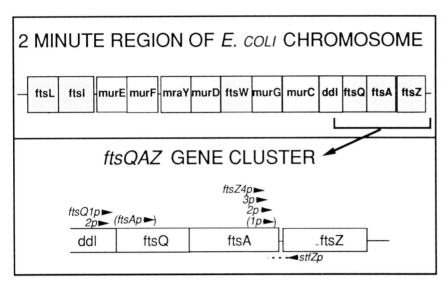

Fig. 26.1. The "two-minute cluster" of genes involved in cell division and in cell wall biosynthesis. Cell division genes are shaded. The lower panel represents the ftsQAZ region of the cluster (indicated by the brackets in the upper panel).

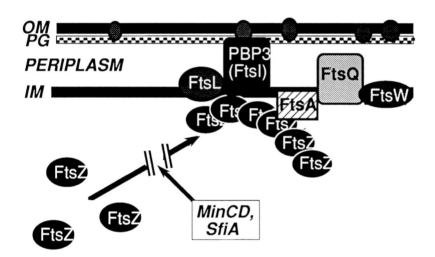

Fig. 26.2. E. coli cell division proteins. The cartoon includes the Fts proteins that are coded for by genes in the two-minute region, and the MinCD and SfiA division inhibitors. Although it is likely that the Fts proteins function at the division site, as indicated in the figure, the only one that has been directly shown to be present at the division site is FtsZ. FtsZ moves from cytosol to the division site and formation of the FtsZ ring at that site is believed to be the initial event in the septation process. Division inhibitors MinCD and SfiA appear to act by blocking formation of the FtsZ ring. OM, outer membrane; IM, inner membrane.

The *ftsQ*, *ftsA* and *ftsZ* genes form a contiguous subcluster within the two minute region and have been the most intensively studied of the *fts* genes.

FtsZ and FtsW appear to act before any of the other known cell division gene products. This conclusion is largely based on studies of the nonseptate cells that are formed when temperature-sensitive *fts* mutants or double mutants containing a *rodA* mutation in combination with one of the temperature-sensitive *fts* mutations are grown at elevated temperature.[18-19a] In *ftsQ*, *ftsA* and *ftsI* mutants the nonseptate cells show broad constrictions ("abortive constrictions") at regular intervals along the length of the cells. In contrast, the abortive constrictions are not seen in *ftsZ* or *ftsW* filaments or in *rodAftsZ* or *rodAftsW* cells, implying that FtsZ and FtsW act at an earlier stage in the division pathway than FtsQ, FtsA or FtsI.

FTSZ

Immunoelectronmicroscopic studies from the laboratory of J. Lutkenhaus have shown that FtsZ changes its cellular location as cells progress through the division cycle (Figs. 26.2 and 26.3).[9] FtsZ

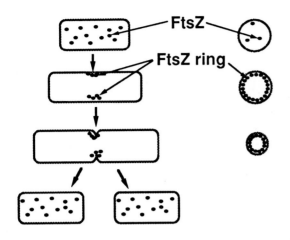

Fig. 26.3. Cyclic change in the distribution of FtsZ during the division cycle. Cross-sections through the division site at midcell are shown on the right. The corresponding longitudinal sections are shown on the left. Adapted from Bi E et al, Nature 1991; 354:161-164.

is present almost exclusively in the cytoplasm throughout most of the cell cycle. At or immediately before the onset of septal invagination, FtsZ moves from the cytoplasm to the inner surface of the cytoplasmic membrane at midcell where it forms a ring that extends around the circumference of the cell. Since these conclusions are based only on immunomicroscopy, it is not certain that the ring consists of a continuous FtsZ polymer. However, this appears likely since there are more than enough molecules of FtsZ in the cell to form such a continuous structure and FtsZ has been shown to polymerize in vitro.[20,21] FtsZ remains localized at the leading edge of the ingrowing septum until cell separation occurs. At that time, FtsZ disappears from the division site and reappears in the cytoplasm. The same sequence of events occurs when septa are formed at polar sites in minicell-forming mutants, where the ring is formed adjacent to the cell pole instead of midcell.

These observations imply that the initial step in septum formation is formation of the FtsZ ring. FtsZ is a GTP-binding protein with GTPase activity,[22-24] and FtsZ has been shown to polymerize in vitro in the presence of GTP.[20,21] Mutations that lead to changes within the putative GTP-binding site are associated with a marked reduction in GTPase activity that is associated with a temperature-sensitive filamentation phenotype.[22,23] Based on these observations, the following model can be proposed. In this model, an event occurs at a specific time in the cell cycle that permits FtsZ to recognize and bind to a membrane receptor that is located at midcell (Fig. 26.2). The event could be a change in concentration or a post-translational modification of the structure of FtsZ or another cytosolic factor, or a change in the putative membrane binding site for FtsZ. This event would be followed by the GTP-dependent polymerization of FtsZ to form the FtsZ ring. This is followed by septal ingrowth, catalyzed by other components of the division machinery, leading finally to the formation of the two daughter cells.

FtsZ not only acts prior to the other division proteins, but appears to be the rate-limiting component in the division process. Thus, when FtsZ is moderately overexpressed, cells increase the number of division events per unit increase in cell mass.[26] This is not true of any other known division protein. The additional septa are formed adjacent to the cell poles, resulting in the formation of anucleate minicells.

This type of evidence is consistent with the view that division cannot occur unless the cellular concentration of FtsZ achieves a threshold level at the time that septation is initiated, although this has never been rigorously tested. On the other hand, the present evidence does not speak to the question of whether division is automatically triggered when the concentration of FtsZ reaches the threshold level, or whether another event is responsible for the timing of septation.

It has also been suggested that division requires that the cell contain a fixed number of molecules of FtsZ.[25] This could, for example, represent the number of molecules required to form a complete FtsZ ring at the division site. In this view, transcriptional regulatory mechanisms exist to ensure that the critical concentration of FtsZ is maintained. This is discussed further below (Transcriptional Regulation of Cell Division Genes: Regulation of *ftsQ1p* by growth rate).

A decrease in FtsZ concentration to approximately 20% of normal leads to a partial or complete arrest of division.[26,27] This is ascribable to a postponement of the onset of septation relative to the termination of chromosome replication.[28] A 5- to 6-fold *increase* in FtsZ concentration also causes a division block. Therefore it is critical that the cell maintains the concentration of FtsZ within narrow limits under a wide range of growth conditions.

The normal division pattern also requires that the FtsA/FtsZ ratio be maintained within narrow limits. When *ftsA* or *ftsZ* are expressed at abnormally high levels, a division block occurs. In each case the division block can be suppressed if the other protein is concomitantly overexpressed.[29,30] This suggests that the FtsA and FtsZ proteins interact during the division process. The cell contains several hundred molecules of FtsZ for every molecule of FtsA. If an FtsA-FtsZ interaction plays a role in the division process, it could operate at the initial stage of interaction of FtsZ with the membrane, providing a nucleation site for the subsequent FtsZ polymerization that leads to formation of the FtsZ ring. Alternatively, FtsA could function after formation of the FtsZ ring, acting to transduce a signal from the ring structure to other division proteins that facilitate septal ingrowth.

FTSI

The FtsI protein [also called penicillin-binding protein 3 (PBP3)] is believed to play a direct role in the synthesis of septal murein. In vitro studies have shown that FtsI catalyzes enzyme reactions (murein transglycosylase and murein transpeptidase) that are likely to play a role in murein biosynthesis.[31] In vivo studies support the view that the activity of FtsI is limited to the synthesis of septal murein. Cell division ceases when cells are exposed to antibiotics, such as furazlocillin or cephalexin, that bind to PBP3 but not to other penicillin-binding proteins. This leads to the formation of long, nonseptate filaments since cell elongation continues in the presence of the PBP3-specific antibiotics.[10,32] Antibiotics that bind to the other penicillin-binding proteins do not lead to this phenotype. Filamentation also occurs when FtsI is inactivated by growth of *ftsIts* strains at elevated temperature. Taken together, this is strong presumptive evidence that FtsI, either alone or in cooperation with other proteins, is required for synthesis of the murein component of the division septum.

FTS GENES LOCATED OUTSIDE THE TWO MINUTE CLUSTER

The *ftsN* gene is essential for cell division but maps outside of the two-minute cluster, at 88.5 units on the *E. coli* genetic map. *ftsN* was identified during a search for genes which, when present in multiple copies, would suppress the temperature-dependent division block in an *ftsAts* strain.[32a] Since insertional inactivation of the chromosomal *ftsN* gene leads to filamentation without apparently interfering with other aspects of cell growth, *ftsN* is classified as an essential cell division gene. Most interestingly, overexpression of *ftsN* not only suppresses the *ftsA* division block, but also partially suppresses the temperature-sensitive filamentation phenotype of certain *ftsQts* and *ftsIts* mutants. The mechanism of suppression is not known. The possibility that FtsN may act by up-regulating transcription or translation of these genes or gene products has not been excluded.

A second gene cluster, located at 76 units on the *E. coli* genetic map, includes several other genes that may be involved in the division process.[33] These include *ftsE*, whose product shows regions of strong homology to a specialized group of membrane transport proteins. The role of these genes is not clear.

TRANSCRIPTIONAL REGULATION OF CELL DIVISION GENES

POSSIBLE ROLES FOR TRANSCRIPTIONAL REGULATION OF CELL DIVISION GENES

One can imagine two possible roles for regulation of expression of cell division genes. The two roles are not mutually exclusive.

First, changes in the expression of essential cell division genes could play a role in the cyclic nature of the division process. For example, one or more of the *fts* genes could be upregulated at a specific point in the cell cycle to increase the concentration of a protein such as FtsZ above a threshold level that is needed to initiate the septation process. Following division, the protein concentration would fall below the threshold level due to protein modification or degradation, or because of simple dilution due to cell growth at a rate that exceeded the rate of new synthesis. In this scenario, the cyclic nature of the division process would reflect a cyclic oscillation in gene expression.

Second, the expression of one or more of the essential division genes could be altered as part of the cellular response to physiological stimuli such as changes in nutritional status or alterations in DNA structure or replication.

Both roles presume that transcriptional regulatory factors exist that can modulate the expression of one or more *fts* genes. Evidence is beginning to appear that supports the existence of such factors. Most attention has focused on the *ftsQAZ* gene cluster because of the primary role of FtsZ in initiating the septation process and because of the large number of promoters that can affect expression of the genes within the cluster.

PROMOTER ORGANIZATION OF THE *FTSQAZ* CLUSTER

The *ftsQAZ* gene cluster has a complex promoter organization (Fig. 26.1). Two promoters (*ftsQ1*p and *ftsQ2*p) are located upstream of *ftsQAZ*.[25] Since there are no strong transcriptional terminators within the locus, transcription from the two upstream promoters leads to expression of *ftsQ*, *ftsA* and *ftsZ*.

Several additional promoters (*ftsZ2*p *ftsZ3*p and *ftsZ4*p) are located within the *ftsA* coding region. Expression from these promoters leads to transcription only of *ftsZ*.[25,34,35] An additional promoter (*ftsZ1p*) that had been thought to exist based on RNA analysis probably does not exist. RNaseE cleaves at the position of "*ftsZ1*p", and the 5' end of the RNA species that had been ascribed to *ftsZ1*p appears to come from this cleavage event (J.P. Bouché, personal communication). A weak promoter (*ftsA*p) may also exist within *ftsQ* that initiates transcription into *ftsA* and *ftsZ*[36] although a corresponding RNA transcript has not been identified.

There is disagreement concerning the contributions of the different promoters to transcription of *ftsZ*. In the most direct measurements, S1 nuclease protection assays using probes that overlapped the transcriptional start sites indicated that a large proportion of the mRNA that entered *ftsZ* originated from the upstream promoters *ftsQ1,2*p.[35] In another study, a more indirect transcription titration assay using reverse transcriptase and PCR amplification suggested that only 10% of *ftsZ* transcripts originated from *ftsQ1,2*p.[34] In all of these studies it should be noted that the quantity of each RNA species was measured. If there were differences in the rates of degradation of transcripts from the different promoters, the results might not represent their relative rates of transcription. Of course, from the point of view of the cell, it is presumably the concentration of the message that is most important. Recent studies using *lacZ* transcriptional fusions suggested that the rate of transcription from *ftsQ1,2*p was approximately 65% that of *ftsZ2-4*p[37] although here, too, there is disagreement between different studies.[38-40]

The question of the relative contributions of the two groups of promoters is of more than academic interest since, at this time, it is only the upstream *ftsQ1,2*p promoters that have been shown to be subject to regulation (discussed below). Some of the differences between studies could reflect differences in the physiological state of the cultures. Further studies in cells growing at different rates and sampled at different stages of the growth curve are needed to resolve this question.

Recently, an "antisense" promoter (*stfZ*) has been identified within the *ftsZ* coding sequence.[41] This is further discussed below.

REGULATION OF *FTSQ2*P BY THE *SDIA* GENE PRODUCT

The only protein that has been shown to regulate transcription of the *ftsQAZ* gene cluster is the product of the *sdiA* gene.[37] Overexpression of *sdiA* leads to a 5- to 10-fold increase in transcription from the *ftsQ2*p promoter, the major upstream promoter that initiates transcription into *ftsQ, ftsA* and *ftsZ*. This leads to a corresponding increase in the cellular concentration of the FtsZ protein. Conversely, an *sdiA::kan* insertion mutation leads to a 40% decrease in *ftsQ2*p expression. Thus, SdiA acts as a positive transcriptional regulator of *ftsQ2*p (Fig. 26.1). The carboxy-terminal end of SdiA contains a classical helix-turn-helix DNA-binding motif and shows significant sequence similarity to a number of bacterial transcriptional regulatory proteins.[37] Although this is consistent with SdiA acting directly on *ftsQ2*p, this has not been directly shown and it is possible that the effect of SdiA on *ftsQ2*p expression is mediated by another protein.

It is interesting to note that the *sdiA* gene was first identified not as part of a search for regulatory genes, but rather as a gene which appeared to function as an *ftsZ* homolog. Thus, overexpression of *sdiA* led to the suppression of the action of division inhibitors MinCD and SfiA, whose target appears to be FtsZ (discussed later in this chapter).[42,43] Overexpression of *sdiA* also resulted in an increase in the frequency of division events, as shown by the formation of minicells and of rod-shaped cells that were shorter than normal. Further, overexpression of *sdiA* suppressed the division defect of an *ftsZts* mutant. All of these effects reflect the increase in FtsZ concentration that results from increased transcription from *ftsQ2*p.[37] The suppression of the temperature-sensitive division defect of the *ftsZts* mutant presumably reflected the increase in concentration of the mutant protein to a level that could support the division process.

As described above, the division pattern of the cell is altered when *sdiA* is overexpressed, presumably due to the resulting increase in cellular FtsZ. However, the normal physiological role of SdiA is unclear. Loss of SdiA function due to insertional mutation at a position corresponding to amino acid 50 of the 241 amino acid protein did not significantly affect the division pattern of the cell although it did cause a decrease in transcription from *ftsQ2*p and a corresponding decrease in the cellular concentration of FtsZ.[37] This suggests that SdiA is not required for the normal division process. However, it remains possible that SdiA is required for normal division and that the amino-terminal 21% of the protein is sufficient to perform this role. Two other possibilities should also be considered. First, SdiA may be one of several proteins that carry out analogous functions in the regulation of expression of essential cell division genes. This type of redundancy is not unknown where critical cellular functions such as cell cycle control are concerned.[44] If this were correct, one or more other regulatory proteins would exist that could substitute for SdiA and a knockout of only one of the redundant proteins would not significantly affect the division cycle. Second, SdiA may play a role in the ability of the cell to modulate the division process in response to special physiological events. At this point it is not possible to choose between these several possibilities. The possibility that SdiA may also regulate transcription of other cellular genes in addition to *ftsQAZ* should also be kept in mind.

If SdiA played a role in modulating cell division during the normal cell cycle or in response to certain physiological stimuli, it would be reasonable to expect that the *sdiA* gene itself would be subject to regulation. The presence of *cis*-acting upstream sequences that affect the level of *sdiA* expression (unpublished data) and the demonstration of an extracellular factor that affects *sdiA* expression (see below) lend support to this idea.

REGULATION OF SDIA EXPRESSION BY AN EXTRACELLULAR FACTOR

It has recently been shown that *E. coli* K-12 cells release a factor into the culture medium that down-regulates *sdiA* expression in target cells (Figs. 26.4 and 26.5a).[45] This leads to a moderate decrease in expression from *ftsQ2*p, the target of the SdiA transcriptional regulator (Fig. 26.5b). The effect on *sdiA* is promoter-specific since the extracellular factor does not affect expression from the *ftsQAZ* promoters that are not regulated by SdiA (*ftsQ1*p and *ftsZ2-4*p) (Fig. 26.5c, d).

Extracellular factors that are released by bacteria and that have specific transcriptional effects on recipient cells have been described in a number of bacterial species.[46] The most extensively studied of these *trans*-acting factors are acylated homoserine lactones, but short peptides and other small molecules may play a similar role in some species.[47] The extracellular factor that regulates the expression of *sdiA* has properties suggesting that it may belong to the class of homoserine lactones (unpublished data). Other systems that utilize homoserine lactone derivatives as extracellular signals include the *lux* system that is responsible for bioluminescence in several *Vibrio* species,[48] the *tra* system that is required for conjugal plasmid transfer in *Agrobacterium tumefaciens*,[49,50] the *exp* system that is required for exoenzyme production in *Erwinia carotovora*,[51] and the *las* system that is responsible for elastase production in *Pseudomonas aeruginosa*.[52,53] In most of these cases, the extracellular signal has been shown to lead to a specific transcriptional response in the target cell (illustrated in Fig. 26.6 for the Vibrio *fischeri lux* system). In each case the transcriptional response appears to be mediated by a cellular protein (R-protein) that recognizes the cognate

Fig. 26.4. Regulation of the ftsQAZ gene cluster. Expression of ftsQ1p is positively regulated in response to decreases in growth rate by a yet-to-be-identified factor (Y). Expression of ftsQ2p is positively regulated by SdiA. Expression of sdiAp is negatively regulated by an extracellular factor (ECF) present in the medium of stationary phase cells. The effect of ECF on expression of sdiAp is presumed to be mediated by another protein (X) that has not yet been identified. The filled boxes within ftsQ and ftsA represent sequences with homology to DnaA-binding sites.

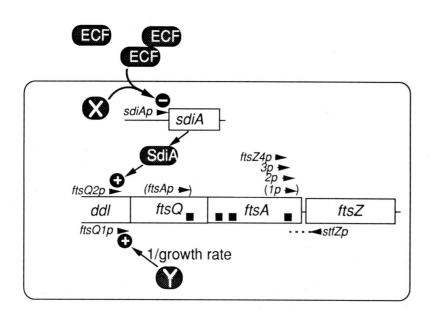

signaling molecule.[48] In the case of the V. *fischeri* LuxR protein, it has been shown that the putative transcriptional activator domain of the R-protein binds to its target promoter at a specific site, synergistically with RNA polymerase.[54]

SdiA shows significant sequence similarity to the group of R-proteins. It was this similarity that prompted the search for an extracellular signaling molecule that affected *sdiA* and its transcriptional target. In the *lux* system of V. *fischeri*, the LuxR protein acts to regulate its own expression when the cognate homoserine lactone derivative (OHHL) is present (Fig. 26.6). However, the effect of the *E. coli* extracellular factor on expression of *sdiA*p does not appear to require the SdiA protein since *sdiA*p expression, as monitored by a *sdiA*p-*lacZ* transcriptional fusion, is down-regulated to a similar extent in *sdiA*⁺ and *sdiA::kan* cells. This implies that another cellular protein (indicated as **X** in Fig. 26.4) is responsible for down-regulating *sdiA*p in the presence of the extracellular factor.

We can only speculate about the possible physiological role of the extracellular signaling factor. The secondary effect on *ftsQ2*p expression is insufficient to block cell division. It may be that the accumulation of the factor to a high concentration in the medium as cell density increases plays a role in mediating the down-regulation of division that is required when cells slow their growth rate as they enter into stationary phase. The possibility should also be considered that the *E. coli* factor may modulate the expression of other genes, with the changes in expression from *sdiA*p and *ftsQ2*p being downstream effects of the primary response. Further study will be needed to distinguish between these and other possibilities.

REGULATION OF *FTSQ1*P BY GROWTH RATE

Expression of the second upstream promoter of the *ftsQAZ* cluster, *ftsQ1*p, varies inversely with growth rate.[55,56] As a result, when cells enter stationary phase, *ftsQ1*p expression and FtsZ concentration are increased (Fig. 26.4). The transcriptional regulatory protein that mediates this effect has not yet been identified.

A similar pattern of expression was also found for the "morphogene" *bolA* and for the *mcb* operon, responsible for production of microcin B17. Aldea and co-workers[56] have proposed the term "gearbox" to designate promoters of this type, that increase transcription as growth rate decreases. From sequence comparison of *ftsQ1*p, *bolA1*p and *mcb*p, Aldea et al have suggested a consensus sequence for the -35 (ctgCAA) and -10 (CGGcaagT) regions of gearbox promoters.[56] A subset of stationary phase-inducible promoters show some homology to this sequence. These include those mentioned above and the stringent starvation protein promoter *ssp*p.[57,58] The relation of the proposed consensus sequences to the regulation of transcription in response to decrease in growth rate is still unclear since a number of other genes whose expression is increased in stationary phase cells do not contain the suggested "gearbox" promoter sequence.

The stationary phase sigma factor, σ^s,[59] would be a likely candidate to regulate expression of *ftsQ1*p and other gearbox promoters whose

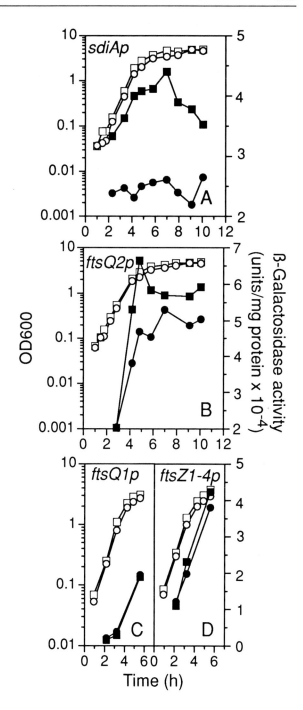

Fig. 26.5. Regulation of fts *and* sdiA *promoters by an extracellular factor released by* E. coli *into the growth medium.* E. coli *UT481containing transcriptional fusions sdiAp-lacZ (A), ftsQ2p-lacZ (B), ftsQ1p-lacZ (C) and ftsZ1-4p-lacZ (D) carried on mini-F plasmids was grown in LB medium (squares) or in LB medium that had previously supported growth of* E. coli *DH5α ("conditioned medium", circles). β-galactosidase activity (filled symbols) was measured[96] at intervals to monitor transcription from each of the promoters (1 unit = 1 nm ONPG hydrolyzed per minute). Growth was followed by measuring OD600 (empty symbols). 0 time is the time the cells were suspended in either LB or in conditioned medium.*

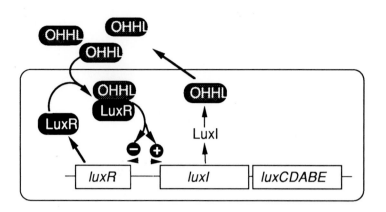

Fig. 26.6. Regulation of the V. fischeri *lux system. An extracellular factor [N-3(oxohexanoyl)homoserine lactone, OHHL] accumulates in the medium of stationary phase cells. OHHL acts together with LuxR to up-regulate transcription into the luxICDABE operon and to down-regulate transcription of luxR. Synthesis of OHHL requires the luxI gene product. Adapted from Swift S et al, Trends in Microbiology 1994; 2:193-198.*

expression is inversely related to growth rate. However, although σ[s] positively regulates *bolA1*p, it negatively regulates *mcb*p, and does not affect transcription from *ssp*p.[58,60,61] There are no reports that σ[s] mediates the growth rate-dependent expression of *ftsQ1*p.

Why would the cell want to increase the concentration of the *ftsQAZ* gene products when growth slows and cells enter the stationary phase? To explain this, it has been proposed that FtsZ (and/or FtsQ and FtsA) must be present at a constant number of molecules per cell for division to occur.[56,56a] One could imagine, for example, that this represents the number of FtsZ molecules required to form the FtsZ ring (see Figs. 26.2, 26.3). It has long been known that cells are smaller in slowly growing or stationary phase cultures than in rapidly growing cultures. Aldea et al pointed out that if each cell must contain a critical number of FtsZ molecules in order to divide, then the FtsZ concentration (expressed as molecules per unit volume) must be higher in the smaller cells that are produced at slow growth rates than in the larger cells that are present in rapidly growing cultures. According to this model, the inverse relation of the concentration of FtsZ (and/or FtsQ and FtsA) to the volume of the cell is maintained by increasing the transcriptional activity of *ftsQ1*p when the growth rate slows.

This view has been challenged by Tétart and Bouché,[27] based on studies in which the cellular concentration of FtsZ was progressively decreased by increasing the expression of *dicF* RNA, an antisense RNA that down-regulates *ftsZ* expression (see below, Regulation of *ftsZ* Expression by Antisense RNA). The resulting decrease in FtsZ concentration led to a progressive increase in cell size. If it were correct that a certain number of FtsZ molecules per cell must be present before septation can occur, then a 50% drop in FtsZ concentration (expressed as molecules per unit volume) should require cells to double their volume before they septate, that is, the increase in cell size should be proportional to the decrease in FtsZ concentration. However, Tétart and Bouché observed that the percent increase in cell size was always proportionately less than the decrease in FtsZ concentration or the decrease in the number of FtsZ molecules per cell. This prompted them to argue that division does not require a fixed number of FtsZ molecules per cell. However, the results do not appear inconsistent with the idea that division does not occur until the number of FtsZ molecules in the cell reaches a certain level, as suggested by Aldea et al.[25] Thus far, all of the relevant experiments on this question[25,27] have measured the average concentration of FtsZ in populations that contained cells at all stages of the cell cycle, whereas the important question is the level of FtsZ at the time that septation is initiated. Until similar experiments are performed on synchronized populations it will not be possible to choose between these or other models.

REGULATION OF *FTSZ2-4*P

The *ftsZ2-4*p promoters seem prime candidates for regulation since they are the only promoters that regulate *ftsZ* expression without also affecting transcription into *ftsQ* and *ftsA*. However, at this time there is no evidence that *ftsZ* expression can be regulated by modulation of expression of the *ftsZ2-4*p promoters.

Several putative DnaA-binding sites (DnaA boxes) are located within *ftsQ* and *ftsA*, upstream of *ftsZ2-4*p (Fig. 26.4). This had suggested that DnaA, which is required for initiation of chromosome replication, might play a role in regulating expression of FtsZ by changing the activity of one or more of the *ftsZ2-4*p promoters. This would have provided the missing link between the DNA replication cycle and the division cycle. However, this does not appear to be the case. Several studies have now shown that the *ftsZ2-4*p promoters do not respond specifically to changes in functional DnaA protein, and overexpression of *dnaA* does not alter expression of *ftsZ2-4*p.[34,62]

REGULATION OF *FTSZ* EXPRESSION BY ANTISENSE RNA

Among the most interesting findings of recent years was the demonstration that *ftsZ* expression can be regulated at the translational level by the action of at least two chromosomally encoded antisense RNAs whose expression leads to a block in cell division.

Dewar and Donachie have shown that an antisense promoter exists within the initial portion of the *ftsZ* coding sequence (*stfZ* in Fig. 26.1).[41] When expressed from a multicopy plasmid the antisense transcript leads to the formation of nonseptate filaments. The division block is likely to reflect an inhibitory effect of *stfZ* RNA on *ftsZ* translation, secondary to the binding of the antisense RNA to the ribosome-binding site of *ftsZ* mRNA. The resulting down-regulation of translation would lower the cellular FtsZ concentration below the level required for division. This model remains speculative since the predicted decrease in FtsZ concentration has not yet been demonstrated. The antisense transcript also includes an open reading frame that could code for a 25 amino acid protein and the division inhibitory effect that is associated with expression of *stfZ* could reflect a direct effect of this protein on the division process.

A second chromosomal gene that inhibits cell division by an antisense mechanism is *dicF*.[63] *dicF* is part of an operon, *dicBF*, which is normally not expressed (discussed later in this chapter). When *dicF* is induced experimentally, cell division ceases. Since the 53 nucleotide transcript does not code for a protein product, *dicF* RNA is responsible for this effect. The division block is caused by a 31 nucleotide region of *dicF* RNA that shows marked complementarity to the translation initiation region of *ftsZ* mRNA.[27,64] Binding of *dicF* RNA to this region of the *ftsZ* message would inhibit translation of *ftsZ*. Consistent with this view, studies of transcriptional and translational *ftsZ-lacZ* fusions indicated that *dicF* expression leads to a significantly greater inhibition of *ftsZ* translation than of *ftsZ* transcription.[27]

It is not yet known whether mechanisms exist to regulate the expression of the *stfZ* or *dicF* antisense RNAs or whether changes in expression of one or both of these RNAs play a role in regulating the cell division process.

REGULATION OF *FTSI* EXPRESSION BY *MREB*

The *mreB* gene is located in a cluster of genes at 71 minutes on the *E. coli* genetic map that are involved in shape determination. Mutational analysis has suggested that the *mreB* gene product plays a role in regulating *ftsI* expression.[65] Thus, a mutation in *mreB* was associated with an increased cellular concentration of penicillin-binding protein 3 (the product of the *ftsI* gene) and of a FtsI-LacZ hybrid protein. Chromosomal deletions that included *mreB* (and other *mre* genes that are required to maintain the rod-shape of *E. coli*) showed a similar phenotype. In addition, cells that overexpressed *mreB* were unable to divide. This presumably reflected a decrease in the cellular concentration of FtsI although the predicted change in concentration of PBP3 has not been directly shown. All of these observations are consistent with the view that the *mreB* gene product negatively regulates the cellular level of FtsI.

The role of MreB as a negative regulator of cell division has not been fully defined. It is not known whether the observed effects on FtsI levels reflect changes in transcription of *ftsI*, or changes in translation or stability of the FtsI protein, nor is it known whether MreB acts directly or indirectly in modulating the cellular concentration of FtsI.

CYCLIC OSCILLATIONS IN *FTSZ* TRANSCRIPTION

After the *ftsQAZ* cluster of cell division genes was identified and its promoter organization defined, several laboratories asked whether the periodic nature of the division process resulted from a periodicity of transcription that affected the cellular level of one or more of the gene products as cells progressed through the cell cycle.

Early studies using transcriptional gene fusions gave conflicting results concerning the periodicity of transcription into *ftsZ* from *ftsZ2-4*p.[66,67] More recent studies, in which RNA levels were monitored directly, showed that the concentration of transcripts extending into *ftsZ* did oscillate during the cell cycle. In one study, the increase occurred approximately 10-15 minutes before septum formation, suggesting a possible link to the timing of division.[34] In another study, the oscillation was ascribed to a brief inhibition of transcription that coincided with the time of replication of *ftsZ*, rather than being related to the timing of septation.[35]

However, one thing is quite clear. The division cycle does not depend on cyclic changes in *ftsZ* transcription. Garrido et al showed[34] and we have confirmed (unpublished studies) that there is no apparent division defect in the presence of an *ftsZnull* mutation of the chromosomal *ftsZ* allele when *ftsZ* is expressed under IPTG control at a steady rate from a copy of P_{lac}-*ftsZ* in an integrated prophage. These results imply that cell division does not require periodic changes in *ftsZ* gene expression.

It should be kept in mind that the observed periodicity of expression from *ftsQ1,2*p also leads to oscillations in *ftsQ* and *ftsA* expression. The possibility that oscillations in expression of division genes other than *ftsZ*, such as *ftsQ* and *ftsA,* may play a role in regulating the periodicity of division has not been investigated.

TRANSLATIONAL CONTROL

It is striking that the molar ratio of the FtsZ protein to FtsA and FtsQ (approximately 100/1 and 400/1, respectively) is much higher than predicted by the concentrations of the mRNAs for the three proteins. This is likely to reflect a significantly lower rate of translation of *ftsA* and *ftsQ* relative to *ftsZ*.[34a]

DIVISION INHIBITORS

DIVISION INHIBITION IN RESPONSE TO DNA PERTURBATIONS

Cell division is inhibited as part of the SOS response to perturbations of DNA structure or replication (see chapter 22). Two mechanisms have been described that lead to the inhibition of division, one being part of the classical response, dependent on RecA and LexA, and the other being part of a RecA-dependent, LexA-independent response.

In the classical SOS response, DNA perturbations lead to activation of RecA protease, leading to cleavage of the LexA repressor protein. Among the large number of genes whose expression is normally repressed by LexA (see chapter 22) and that are derepressed in the SOS response is the *sfiA* gene (also called *sulA*).

The *sfiA* gene product is a division inhibitor, and in most strains is solely responsible for the filamentation that occurs as part of the SOS response. The division block due to derepression of *sfiA* continues until the stimulus to SOS induction is relieved. SfiA is then degraded by the Lon protease, permitting division to resume.[68] The SfiA division inhibitor appears to act by blocking formation of the FtsZ ring (Fig. 26.2).[69]

The SfiA-mediated division block is beneficial to the cell since it prevents the passage of damaged DNA to progeny cells until DNA repair processes have had an opportunity to repair the damage. In addition, in cells in which DNA replication has been temporarily inhibited, the division block prevents the formation of anucleate cells or the fragmentation of the chromosome due to a guillotine effect of septal closure on the incompletely replicated chromosome at midcell. The mechanism of LexA repression of *sfiA* expression is similar to that of other members of the SOS regulon (see chapter 22).

The nomenclature of the *sfiA/sulA* gene illustrates the illogic that sometimes characterizes scientific discourse. The gene had been discovered independently by two groups.[70,71] For nearly 20 years the same gene has been called *sfiA* (for <u>s</u>uppressor of <u>fi</u>lamentation) by about half of the scientific community, and *sulA* (for <u>s</u>uppressor of <u>l</u>on) by the other half, usually based on the country of origin or scientific background of the investigator. Similarly, mutations in *ftsZ* that confer resistance to the *sfiA/sulA* division inhibitor have been called either *sfiB* or *sulB*. It would seem logical for investigators in this field to agree to use one name or the other to avoid the present confusion, but this has not yet been accomplished. Since the mnemonic "<u>s</u>uppressor of <u>fi</u>lamentation" is more suggestive of the division-related role of the gene than "<u>s</u>uppressor of <u>l</u>on", the *sfi* nomenclature seems most logical.

A second mode of division inhibition associated with the SOS response is mediated by the *sfiC* gene. *sfiC* resides in an excisable element, *ε14*, that behaves like a defective prophage.[72,73] *ε14* is present in some, but not all, strains of *E. coli*, so that the SfiC-mediated division block is not a universal part of the SOS response. *ε14* undergoes site-specific integration at a specific chromosomal locus, the *ε14 att* site, in a RecA and RecB-independent manner, thereby resembling several temperate bacteriophages.[72]

ε14 is excised from the chromosome as part of the general RecA-dependent SOS response. At the same time, division ceases in a response that is ascribed to the *sfiC* gene product. *ε14* also contains an invertible DNA sequence (the "P-region") and the *pin* gene that catalyzes its inversion.[74] The *pin* gene product can substitute for the *gin* gene product of bacteriophage Mu, a protein responsible for catalyzing an analogous DNA inversion event within Mu.

Unlike *sfiA*, expression of *sfiC* is not regulated by the LexA repressor. Therefore, the up-regulation of *sfiC* expression during the SOS response is not mediated by cleavage of LexA. Instead, the RecA-dependent expression of *sfiC* is likely to reflect cleavage by the RecA protease of another molecule, either a specific repressor of *sfiC* expression or a repressor that plays a role both in maintaining the integrated state of *ε14* and in preventing *sfiC* expression. The possibility can also be considered that inversion at the invertible P-region in *ε14* may be involved in the regulation of *sfiC* expression. SfiC also differs from SfiA in its stability, so that the SfiC-mediated division block is essentially irreversible. This irreversibility confers no apparent advantage to the cell, and in fact would be deleterious in the event of a transient perturbation of the host chromosome.

The mechanism by which SfiC prevents division is unclear and it has not even been established that *sfiC* codes for a protein product. Surprisingly, the *sfiC*-mediated division block cannot be overcome by overexpression of FtsZ.[75] This distinguishes SfiC from the other known endogenous division inhibitors (SfiA, MinCD, StfZ, DicB and DicF). On the other hand, an *ftsZ*[sfiB] mutation that imparts resistance to the SfiA division inhibitor also imparts resistance to *sfiC*-mediated division inihibition,[73] implying the FtsZ is likely to be the target of the SfiC division inhibitor. Further study of *ε14* and *sfiC* represents a potentially fertile field for future research.

In addition to these two SOS-related responses, there is also an SOS-independent partial division block that results when DNA synthesis is blocked.[6] This RecA-independent response does not require SfiA or SfiC. The mechanism is unknown.

A DIVISION INHIBITOR INVOLVED IN PLACEMENT OF THE DIVISION SEPTUM

The only division inhibitor that functions during the normal division cycle is the product of the *minC* gene. The MinC division inhibitor functions as part of a mechanism that ensures that the division site is placed at midcell instead of at other potential division sites that are located near the cell poles.[76] MinC is capable by itself of blocking division at all potential division sites but only does so at unphysiologically high levels of expression. At normal levels of MinC expression, a second protein is required to activate the MinC division inhibitor.[42] The activator protein is the product of the *minD* gene. A third protein, MinE, then gives topological specificity to the MinCD division inhibitor. As a result, under normal circumstances the division inhibitor prevents septation at the aberrant polar sites but does not block division at the normal site at midcell, permitting division to proceed normally. The *minC*, *minD* and *minE* genes form a gene cluster at approximately 26 minutes on the *E. coli* genetic map.

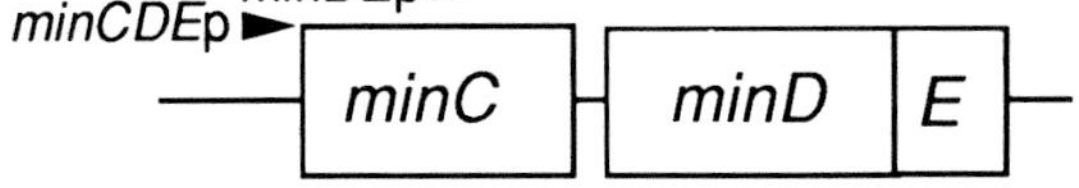

The *minCDE* locus contains two promoters (Fig. 26.7).[76] A stronger upstream promoter (*minCDEp*) leads to expression of all three genes. A second promoter within *minC* (*minDEp*) drives transcription only of *minD* and *minE*. The *minDEp* promoter is subject to regulation by the cAMP·CAP system described in chapter 12. It is not known whether the expression of either promoter is differentially regulated during normal cell growth or in response to specific physiological stimuli.

Fig. 26.7. Promoter organization of the minCDE locus. Adapted from de Boer PA J et al, Cell 1989; 56:641-649.

DICB

A second chromosomal gene whose expression leads to inhibition of division is *dicB*.[77] DicB does not block division by itself, but functions as a second activator of the latent MinC division inhibitor (discussed above). The *dicB* gene is part of a locus that also includes *dicF*, which inhibits division by an antisense effect on *ftsZ* expression (discussed above). The *dicB* and *dicF* genes are normally tightly repressed so that *dicB* and *dicF* are not expressed under any known physiological conditions. The *dicA* and *dicC* genes that are involved in repression of the *dicBF* operon show some similarity to the *c2* and *cro* genes of bacteriophage P22 and it is believed that the *dic* genes represent part of a defective prophage.[77,78] It is not known whether any cellular mechanism exists to derepress the *dicBF* operon or whether these division inhibitor genes are merely vestigial remnants that have no function in the life of the host cell.

PLASMID-ENCODED DIVISION INHIBITORS

A very interesting mechanism of division inhibition is used by low copy number plasmids such as F, P1 and R, to ensure that every cell in the culture contains the plasmid. These genetic elements are usually maintained as single copy plasmids. During the cellular division cycle, plasmid-encoded partition proteins are responsible for the equipartition of the two products of plasmid replication into the two daughter cells (reviewed in ref. 79). If this process fails, one of the daughter cells will find itself plasmid-free.

The further division of these plasmid-free cells is prevented in the following way, as illustrated for the best studied example, the F-plasmid. The plasmid chromosome contains two genes, *ccdA* and *ccdB* (also called *letA* and *letD*), and both gene products are present in plasmid-containing cells. The *ccdB* gene product is a division inhibitor that is capable of blocking cell division and the *ccdA* gene product is a suppressor of the activity of CcdB.[80] Thus, division is not inhibited in plasmid-containing cells because the inhibitor and its antidote are both present. However, the CcdA protein is much more labile than CcdB protein. In plasmid-free cells it is rapidly degraded by the Lon protease, thereby removing the antidote to the more stable CcdB.[81] As a result, in plasmid-free cells division is blocked by CcdB and the plasmid-free segregants fail to divide. This explains the paradox of division inhibition that occurs only in plasmid-free cells despite the fact that the inhibitor protein is a plasmid-encoded gene product. Although CcdB induces the SOS response, division inhibition occurs even in *recA* mutants.[80] Therefore, the SfiA-mediated division inhibition that occurs as part of the SOS response is not solely responsible for the CcdB-induced division block. The CcdB protein interacts with DNA gyrase, leading to an ATP-dependent cleavage of DNA.[80a,80b] It is presumably this activity that is responsible for the lethal effect of CcdB rather than any direct effect on the cell division process. Consistent with this view, mutations in the GyrA subunit of DNA gyrase are associated with resistance to the cytotoxic effect of CcdB.[80c,80d]

CELL DIVISION INHIBITION BY MUTATIONS IN GENES THAT DO NOT CODE FOR CELL DIVISION PROTEINS

It is becoming increasingly clear that cell division can be significantly affected by mutations in many genes whose products are not directly involved in the division process. Among others, the gene products of these genes include: SecA, required for protein export;[82] GroAB, a molecular chaperone;[83] Fic, a protein involved in folic acid biosynthesis;[84] NrdB(FtsB), a subunit of ribonucleoside diphosphate reductase;[85,86] HflB (FtsH), a protein involved in degrading a subset of proteins that includes the cII protein of bacteriophage λ and the heat shock sigma factor σ^{32};[87] ribosomal proteins;[88] RpoB, the β subunit of RNA polymerase;[89] tRNA$_2^{Ser}$;[90] tRNA$_3^{Leu}$;[91] and elongation factor EFTu.[92] In terms of our understanding of cell division, these mutants have usually been considered to be less interesting than the *fts* genes that are directly involved in the division process. However, in most cases the reason for the division defect that is associated with mutations in non-*fts* genes has not been satisfactorily explained and this group could well provide an important entry into poorly understood aspects of regulation of the division process.

It may be significant that a number of these genes code for products involved in protein synthesis. In an interesting series of studies, Vinella et al have observed that, under certain circumstances, a decrease in the cellular pool of ppGpp is associated with inhibition of cell division.[93] The division inhibition is reversed when the ppGpp pool is increased or when the level of FtsZ (and probably also FtsA and FtsQ) is increased by introduction of a plasmid that expresses *ftsQAZ*.[89,93] This led to the suggestion that ppGpp might positively regulate expression of *ftsZ*.

PAST, PRESENT AND FUTURE

After more than three decades of intense study, we still know very little about the regulatory mechanisms used to orchestrate the orderly progression through the bacterial division cycle.

As described in this chapter, most (but probably not all) of the important cell division genes and gene products have been identified and the general outlines of their transcriptional organization are becoming clear. We have just begun to identify factors that regulate the expression of these genes and can anticipate that the list of regulatory factors and an understanding of their molecular mechanisms of action will progress significantly within the next few years.

We are also making good progress in understanding the initial events in initiating septum formation. It now appears that an important point of regulation will be the movement of FtsZ from cytosol to the membrane at the division site and its polymerization to form the FtsZ ring. There is a strong probability that the timing of septation will be determined by a signal that initiates this event. The challenge will now be to identify this signal.

Beyond this, however, we know very little. The question of how the cell knows when to initiate the septation event is a total mystery. As discussed above, it is likely that septation requires that the FtsZ concentration exceed some threshold level at the time of initiation of septal invagination. However, it is not at all clear that this is sufficient to trigger the septation process.

Several models for regulating the timing and frequency of septation can be considered. For example, the periodic nature of the septation event could be mediated by changes in a division inhibitor that acted as a negative regulator of septation during most of the cell cycle. In this scenario, the FtsZ ring would form spontaneously when the division inhibitor was inactivated or perhaps allowed to decay below a threshold concentration by a periodic change in its rate of synthesis. It is clear that the SfiA division inhibitor is not the hypothetical negative regulatory factor since division is unperturbed in *sfiA* mutants. MinCD could be such a factor although the cyclic change in MinCD concentration would require careful regulation since a significant fall in MinCD activity would result in minicell formation. In an alternative model, the periodic nature of the septation event would be orchestrated by a positively acting factor that promoted formation of the FtsZ ring. Such a factor might modify FtsZ itself or facilitate the synthesis or modification of a membrane protein that served as the site of FtsZ interaction with the membrane.

In both of these general models, the hypothetical positive or negative effectors could also operate by changing the cellular concentration of FtsZ or another key element of the division process. The major challenge of the next few years will be to distinguish between these various possibilities, to identify the putative signals and, most important, to understand the molecular basis for the cyclic nature of the process.

In addition to the obvious need to regulate the timing of division during the division cycle, there is ample evidence that cells can regulate the division process in response to external events that supervene during the life of the cell. The most dramatic of these is the division inhibition that accompanies the SOS response to perturbations of DNA synthesis or structure, mediated chiefly by the SOS-associated division inhibitor SfiA. The molecular mechanism that leads to derepression of *sfiA* is well understood (see chapter 22). The next important challenge is to explain how SfiA and other division inhibitors, such as MinCD, act to prevent the formation of the FtsZ division ring and thereby interfere with the first step in the septation process. Based on the rapid rate of recent progress in understanding FtsZ function, it is likely that the molecular mechanisms will be clarified in the near future.

However, there are a number of other situations in which the timing of division is perturbed, about which we know very little. These could be fruitful starting points for future study. One example is the response of cells to nutritional upshifts or downshifts. It has been known for many years that cells growing in rich media, with rapid generation times, are longer than cells that are growing more slowly in minimal media. When cells are shifted from poor to rich medium the increase in length of the up-shifted cells is brought about by a delay in the first division that follows the upshift.[94,95] The delay allows the cells to reach the new length that is characteristic of rapidly growing cells. Subsequent divisions occur at regular intervals to maintain the increased cell length. This implies that the mechanism that determines the timing of division can be rapidly readjusted by factors involved in sensing changes in the nutritional state or rate of growth of the cells. The resetting of the mechanism requires only a short pulse of exposure to rich medium and then continues for several generations after shift back to poor medium. This is an example of one type of system that could provide an entry into the key question of how cells adjust the timing of the division event.

Taken as a whole, it seems fair to conclude that we have arrived at *the end of the beginning* in the search for an understanding of the cell division process and its regulation. We are, however, far from *the beginning of the end* in this search. In the context of this volume, the possible role of gene regulation in cell division is still unknown. The coming years should be especially exciting in moving toward an understanding of this key biological process and there are likely to be many surprises still in store.

ACKNOWLEDGMENTS

We thank numerous colleagues for helpful discussions and for sharing the results of unpublished work. Studies from the authors' laboratory were supported by grants from the National Institute of General Medical Sciences and the National Institute of Allergy and Infectious Diseases, USPHS. J. G-L. was supported by a grant from NATO.

REFERENCES

1. Hirota Y, Ryter A, Jacob F. Thermosensitive mutants of *E. coli* affected in the process of DNA synthesis and cellular division. Cold Spring Harbor Symp Quant Biol 1968; 33:677-693.

2. Weigand RA, Vinci KD, Rothfield LI. Morphogenesis of the bacterial division septum: a new class of septation-defective mutants. Proc Natl Acad Sci USA 1976; 73:1882-1886.

3. Chakraborti AS, Ishidate K, Cook WR, Zrike J, Rothfield LI. Accumulation of a murein-membrane attachment site fraction when cell division is blocked in *lkyD* and *cha* mutants of *Salmonella typhimurium* and *Escherichia coli.* J Bacteriol 1986; 168:1422-1429.

4. Howe WE, Mount DW. Production of cells without deoxyribonucleic acid during thymine starvation of *lexA* cultures of *Escherichia coli.* J Bacteriol 1975; 124:1113-1121.

5. Tang M-S, Helmstetter CE. Coordination between chromosome replication and cell division in *Escherichia coli.* J Bacteriol 1980; 141:1148-1156.

6. Jaffé A, D'Ari R, Norris V. SOS-independent coupling between DNA replication and cell division in *Escherichia coli.* J Bacteriol 1986; 165:66-71.

6a. Inouye M. Unlinking of cell division from DNA replication in a temperature-sensitive DNA synthesis mutant of *Escherichia coli.* J Bacteriol 1969; 99:842-850.

6b. Hirota Y, Jacob F, Ryter A, Buttin G, Nakai T. On the process of cellular division in *E. coli.* I. Asymmetric cell division and production of DNA-less bacteria. J Molec Biol 1968; 35:175-192.

7. Donachie WD. The cell cycle of *Escherichia coli.* Annual Review of Microbiology 1993; 47:199-230.

8. Woldringh CL, Mulder E, Huls PG, Vischer NOE. Toporegulation of bacterial division according to the nucleoid occlusion model. Res Microbiology 1991; 142:309-320.

9. Bi E, Lutkenhaus J. FtsZ ring structure associated with division in *Escherichia coli.* Nature 1991; 354:161-164.

10. Spratt BG. Distinct penicillin binding proteins involved in the division, elongation and shape of *Escherichia coli.* Proc Natl Acad Sci USA 1975; 72:2999-3003.

11. Guzman, L-M, Barondess JJ, Beckwith J. FtsL, an essential cytoplasmic membrane protein involved in cell division in *Escherichia coli.* J Bacteriol 1992; 174:7716-7728.

12. Carson MJ, Barondess J, Beckwith M. The FtsQ protein of *Escherichia coli*: membrane topology, abundance, and cell division phenotypes due to overproduction and insertion mutations. J Bacteriol 1991; 173:2187-2195.

13. Wang H, Gayda RC. Quantitative determination of FtsA at different growth rates in *Escherichia coli* using monoclonal antibodies. Molec Microbiol 1992; 6:2517-2524.

14. Bowler DL, Spratt BG. Membrane topology of penicillin-binding protein 3 of *Escherichia coli.* Molec Microbiol 1989; 3:1277-1286.

15. Ikeda M, Sato T, Wachi M, Jung HK, Ishino F, Kobayashi Y, Matsuhashi M. Structural similarity among *Escherichia coli* FtsW and RodA proteins and *Escherichia coli* SpoVE protein, which function in cell division, cell elongation, and spore formation respectively. J Bacteriol 1989; 171:6375-6378.

16. Pla J, Dopazo A, Vicente M. The native form of FtsA, a septal protein of *Escherichia coli*, is located in the cytoplasmic membrane. J Bacteriol 1990; 172:5097-5102.

17. Sanchez M, Valencia A, Ferrandiz M-J, Sander C, Vicente M. Correlation between the structure and biochemical activities of FtsA, an essential cell division protein of the actin family. EMBO J 1994; 13:4919-4925.

18. Taschner PEM, Huls PG, Pas E, Woldringh CL. Division behavior and shape changes in isogenic *ftsZ, ftsQ, ftsA, pbpB,* and *ftsE* cell division mutants of *Escherichia coli* during temperature shift experiments. J Bacteriol 1988; 170:1533-1540.

19. Begg K, Donachie W. Cell shape and division in *Escherichia coli*: experiments with shape and division mutants. J Bacteriol 1985; 163:615-622.

19a. Khattar M, Begg K, Donachie W. Identification of FtsW and characterization of a new *ftsW* division mutant of *Escherichia coli*. J Bacteriol 1994; 176:7140-7147.

20. Mukherjee A, Lutkenhaus J. Guanine nucleotide-dependent assembly of FtsZ into filaments. J Bacteriol 1994; 176:2754-2758.

21. Bramhill D, Thompson C. GTP-dependent polymerization of *Escherichia coli* FtsZ protein to form tubules. Proc Natl Acad Sci USA 1994; 91:5813-5817.

22. de Boer P, Crossley R, Rothfield L. The essential bacterial cell-division protein FtsZ is a GTPase. Nature 1992; 359:254-256.

23. RayChaudhuri D, Park JT. *Escherichia coli* cell-division gene *ftsZ* encodes a novel GTP-binding protein. Nature 1992; 359:251-254.

24. Mukherjee A, Dai K, Lutkenhaus J. *Escherichia coli* cell-division protein FtsZ is a guanine nucleotide binding protein. Proc Natl Acad Sci USA 1993; 90:1053-1057.

25. Aldea M, Garrido T, Pla J, Vicente M. Division genes in *Escherichia coli* are expressed coordinately to cell septum requirements by gearbox promotors. EMBO J 1990; 9:3787-3794.

26. Ward JE, Lutkenhaus J. Overproduction of FtsZ induces minicell formation in *Escherichia coli*. Cell 1985; 42:941-949.

27. Tétart F, Bouché J-P. Regulation of the expression of the cell cycle gene *ftsZ* by DicF antisense RNA. Division does not require a fixed number of FtsZ molecules. Molec Microbiol 1992; 6:615-620.

28. Tétart F, Albigot R, Conter A, Mulder E, Bouché J-P. Involvement of FtsZ in coupling of nucleoid separation with septation. Molec Microbiol 1992; 6:621-627.

29. Dai K, Lutkenhaus J. The proper ratio of FtsZ to FtsA is required for cell division to occur in *Escherichia coli*. J Bacteriol 1992; 174:6145-6151.

30. Dewar SJ, Begg KJ, Donachie WD. Inhibition of cell division initiation by an imbalance in the ratio of FtsA to FtsZ. J Bacteriol 1992; 174:6314-6316.

31. Ishino F, Matsuhashi M. Peptidoglycan synthetic activities of highly purified penicillin-binding protein 3 in *Escherichia coli*: a septum-forming reaction sequence. Biochem Biophys Res Commun 1981; 101:905-911.

32. Park JT, Burman L. A new penicillin with a unique mode of action. Biochem Biophys Res Commun 1973; 51:863-868.

32a. Dai K, Xu Y, Lutkenhaus J. Cloning and characterization of *ftsN*, an essential cell division gene in *Escherichia coli* isolated as a multicopy suppressor of *ftsA12*(Ts). J Bacteriol 1993; 175:3790-3797.

33. Gibbs TW, Gill DR, Salmond GPC. Localised mutagenesis of the *ftsYEX* operon: conditionally lethal missense substitutions in the FtsE cell division protein of *Escherichia coli* are similar to those found in the cystic fibrosis transmembrane conductance regulator protein (CFTR) of human patients. Mol Gen Genet 1992; 234:121-128.

34. Garrido T, Sànchez M, Placios P, Aldea M, Vicente M. Transcription of *ftsZ* oscillates during the cell cycle of *Escherichia coli*. EMBO J 1993; 12:3957-3965.

34a. Mukherjee A, Donachie W. Differential translation of cell division proteins. J Bacteriol 1990; 172:6106-6111.

35. Zhou P, Helmstetter CE. Relationship between *ftsZ* gene expression and chromosome replication in *Escherichia coli*. J Bacteriol 1994; 176: 6100-6106.

36. Dewar SJ, Donachie WD. Regulation of expression of the *ftsA* cell division gene by sequences in upstream genes. J Bacteriol 1990; 172: 6611-6614.

37. Wang X, de Boer PAJ, Rothfield LI. A factor that positively regulates cell division by activating transcription of the major cluster of essential cell division genes of *Escherichia coli*. EMBO J 1991; 10:3363-3372.

38. Robinson AC, Kenan DJ, Hatfull GF, Sullivan NF, Spiegelberg R, Donachie WD. DNA sequence and transcriptional organization of essential cell division genes *ftsQ* and *ftsA* of *Escherichia coli*: Evidence for overlapping transcriptional units. J Bacteriol 1984; 160:546-555.

39. Robinson AC, Kenan DJ, Sweeney J, Donachie WD. Further evidence for overlapping transcriptional units in an *Escherichia coli* cell envelope-cell division gene cluster: Dna sequence and transcriptional organization of the *ddl ftsQ* region. J Bacteriol 1986; 167:809-817.

40. Yi, Q-M, Rockenbach S, Ward JE, Lutkenhaus J. Structure and expression of the cell division genes *ftsQ, ftsA* and *ftsZ*. J Mol Biol 1985; 184:399-412.

41. Dewar S, Donachie W. Antisense transcription of the *ftsA-ftsZ* gene junction inhibits cell division in *E. coli*. J Bacteriol 1993; 175:7097-7101.

42. de Boer PAJ, Crossley RE, Rothfield LI. Central role for the *Escherichia coli minC* gene product in two different cell division-inhibition systems. Proc Natl Acad Sci USA 1990; 87:1129-1133.

43. Bi E, Lutkenhaus J. Interaction between the *min* locus and *ftsZ*. J Bacteriol 1990; 172:5610-5616.

44. Cross FR. CLN- and CDC28-dependent stimulation of CLN1 and CLN2 RNA levels: implications for regulation by alpha-factor and by cell cycle progression. Cold Spring Harbor Symp on Quantitative Biology 1991; LVI:1-8.

45. García-Lara J, Shang LH, Rothfield LI. An extracellular factor regulates expression of *sdiA*, a transcriptional activator of cell division genes in *E. coli*. 1995; (submitted).

46. Swift S, Bainton NJ, Winson MK. Gram-negative bacterial communication by N-acyl homoserine lactones: A universal language? Trends in Microbiology 1994; 2:193-198.

47. Kaiser D, Losick R. How and why bacteria talk to each other. Cell 1993; 73:873-885.

48. Fuqua WC, Winans SC, Greenberg EP. Quorum sensing in bacteria:the LuxR-LuxI family of density-responsive transcriptional regulators. J Bacteriol 1994; 176:269-275.

49. Piper KR, Beck vov Bodman S, Farrand SK. Conjugation factor of *Agrobacterium tumefaciens* regulates Ti plasmid transfer by autoinduction. Nature (London) 1993; 362:448-450.

50. Zhang L, Murphy PJ, Kerr A, Tate ME. *Agrobacterium* conjugation and gene regulation by N-acyl-homoserine lactones. Nature (London) 1993; 362:446-448.

51. Pirhonen M, Flego D, Heikinheimo R, Palva ET. A small diffusible molecule is responsible for the global control of virulence and exoenzyme production in the plant pathogen *Erwinia carotovora*. EMBO J 1993; 12:2467-2476.

52. Gambello MJ, Iglewski BH. Cloning and characterization of the *Pseudomonas aeruginosa lasR* gene, a transcriptional activator of elastase expression. J Bacteriol 1991; 173:3000-3009.

53. Passador LC, JM, Gambello MJ, Rust L, Iglewski BH. Expression of *Pseudomonas aeruginosa* virulence genes requires cell-to-cell communication. Science 1993; 260:1127-1130.

54. Stevens AM, Dolan KM, Greenberg EP. Synergistic binding of the *Vibrio fischeri* LuxR transcriptional activator domain and RNA polymerase to the *lux* promoter region. Proc Natl Acad Sci USA 1994; 91:12619-12623.

55. Aldea M, Hernández-Chico C, de la Campa AG, Kushner SR, Vicente M. Identification, cloning, and expression of *bolA*, an *ftsZ*-dependent morphogene of *Escherichia coli*. J Bacteriol 1988; 170:5169-5176.

56. Aldea M, Garrido T, Hernández-Chico C, Vicente M, Kushner SR. Induction of a growth-phase-dependent promoter triggers transcription of *bolA*, an *Escherichia coli*morphogene. EMBO 1989; 8:3923-3931.

56a. Donachie W, Sullivan N, Kenan D, Derbyshire V, Begg K, Kagan-Zur V. Genes and cell division in *Escherichia coli*. In: Cahloupka J, Kotyk A, Streiblova E, eds. Progress in Cell Cycle Controls. Prague: Czechoslovak Acad Sci, 1983:28-33.

57. Fukuda R, Nishimura A, Serizawa H. Genetic mapping of the *Escherichia coli* gene for the stringent starvation protein. Mol Gen Genet 1988; 211.

58. Williams MD, Ouyang TX, Flickinger MC. Starvation-induced expression of SspA and SspB: the effects of a null mutation in *sspA* on *Escherichia coli* protein synthesis and survival during growth and prolonged starvation. Molec Microbiol 1994; 11:1029-1043.

59. Lange R, Hengge-Aronis R. Identification of a central regulator of stationary-phase gene expression in *Escherichia coli*. Molec Microbiol 1991; 5:49-59.

60. Bohannon DE, Connell N, Keener J, Tormo A, Espinosa-Urgel M, Zambranao MM, Kolter R. Stationary-phase-inducible 'gearbox' promoters: differential effects of *katF* mutations and role of Sigma 70. Mol Microbiol 1991; 1:195-201.

61. Lange R, Hengge-Aronis R. Growth phase-regulated expression of bolA and morphology of stationary-phase *Escherichia coli* cells are controlled by the novel sigma factor sigma S. J Bacteriol 1991; 173:4474-4481.

62. Masters M, Paterson T, Popplewell AG, Owen-Hughes T, Pringle JH, Begg KJ. The effect of DnaA protein levels and the rate of initiation at *oriC* on transcription originating in the *ftsQ* and *ftsA* genes: In vivo experiments. Mol Gen Genet 1989; 216:475-483.

63. Bouché F, Bouché J-P. Genetic evidence that DicF, a second division inhibitor encoded by the *Escherichia coli dicB* operon, is probably RNA. Molec Microbiol 1989; 3:991-994.

64. Faubladier M, Cam K, Bouché J-P. *Escherichia coli* cell division inhibitor DicF-RNA of the *dicB* operon. Evidence for its generation in vivo by transcription termination and by RNase III and RNase E-dependent processing. J Mol Biol 1990; 212:461-471.

65. Wachi M, Matsuhashi M. Negative control of cell division by *mreB*, a gene that functions in determining the rod shape of *Escherichia coli* cells. J Bacteriol 1989; 171:3123-3127.

66. Dewar SJ, Kagan-Zur V, Begg KJ, Donachie WD. Transcriptional regulation of cell division genes in *Escherichia coli*. Molec Microbiol 1989; 3:1371-1377.

67. Robin A, Joseleau-Petit D, D'Ari R. Transcription of the *ftsZ* gene and cell division in *Escherichia coli*. J Bacteriol 1990; 172:1392-1399.

68. Mizusawa S, Gottesman S. Protein degradation in *Escherichia coli*: the *lon* gene controls the stability of *sulA* protein. Proc Natl Acad Sci USA 1983; 80:358-62.

69. Bi E, Lutkenhaus J. Cell division inhibitors, SulA and MinCD, prevent localization of FtsZ. J Bacteriol 1993; 175:1118-1125.

70. George J, Castellazzi M, Buttin G. Prophage induction and cell division in *E. coli*. III. Mutations *sfiA* and *sfiB* restore division in tif and lon strains and permit the expression of mutator properties of *tif*. Molec Gen Genetics 1975; 140:309-332.

71. Johnson BF, Greenberg J. Mapping of *sul*, the suppressor of *lon* in *Escherichia coli*. J Bacteriol 1975; 122:570-574.

72. Brody H, Greener A, Hill CW. Excision and reintegration of *Escherichia coli* K-12 chromosomal element *e14*. J Bacteriol 1985; 161:1112-1117.

73. Maguin E, Brody H, Hill CW, D'Ari R. SOS-associated division inhibition gene *sfiC* is part of excisable element *e14* in *Escherichia coli*. J Bacteriol 1986; 168:464-466.

74. van de Putte P, Plasterk R, Kuijpers A. A Mu *gin* complementing function and an invertible DNA region in *Escherichia coli* K-12 are situated in genetic element *e14*. J Bacteriol 1984; 158:517-522.

75. Maguin E, Lutkenhaus J, D'Ari R. Reversibility of SOS-associated division inhibition in *Escherichia coli*. J Bacteriol 1986; 166:733-738.

76. de Boer PAJ, Crossley RE, Rothfield LI. A division inhibitor and a topological specificity factor coded for by the minicell locus determine proper placement of the division septum in *Escherichia coli*. Cell 1989; 56:641-649.

77. Béjar S, Bouché F, Bouché J-P. Cell division inhibition gene *dicB* is regulated by a locus similar to lambdoid bacteriophage immunity loci. Mol Gen Genet 1988; 212:11-19.

78. Faubladier M, Bouché J-P. Division inhibition gene *dicF* of *Escherichia coli* reveals a widespread group of prophage sequences in bacterial genomes. J Bacteriol 1994; 176:1150-1156.

79. Rothfield L. Bacterial chromosome segregation. Cell 1994; 77:963-966.

80. Ogura T, Hiraga S. Mini-F plasmid genes that couple cell division to plasmid proliferation. Proc Natl Acad Sci USA 1983; 80:4784-4788.

80a. Maki S, Takiguchi S, Miki T, Horiuchi T. Modulation of DNA supercoiling activity of *Escherichia coli* DNA gyrase by F plasmid proteins:antagonistic actions of LetA (CcdA) and LetD (CcdB) proteins. J Biol Chem 1992; 267:12244-51.

80b. Bernard P, Kezdy KE, Van Melderen L, Steyaert J, Wyns L, Pato ML, Higgins PN, Couturier M. The F plasmid CcdB protein induces efficient ATP-dependent DNA cleavage by gyrase. J Mol Biol 1993; 234:534-41.

80c. Miki T, Park JA, Nagao K, Murayama N, Horiuchi T. Control of segregation of chromosomal DNA by sex factor F in *Escherichia coli*. Mutants of DNA gyrase subunit A suppress *letD* (*ccdB*) product growth inhibition. J Mol Biol 1992; 225:39-52.

80d. Bernard P, Couturier M. Cell killing by the F plasmid CcdB protein involves poisoning of DNA-topoisomerase II complexes. J Mol Biol 1992; 226:735-45.

81. Van Melderen I, Bernard P, Couturier M. Lon-dependent proteolysis of CcdA is the key control for activation of CcdB in plasmid-free segregant bacteria. Mol Microbiol 1994; 11:1151-1157.

82. Oliver DB, Beckwith J. An *E. coli* mutant pleiotropically defective in the export of secreted proteins. Cell 1982; 25:165-772.

83. Georgopolous CP, Eisen H. Bacterial mutants that block phage assembly. J Supramolec Struct 1974; 2:349-359.

84. Komano T, Utsumi R, Kawamukai M. Functional analysis of the *fic* gene involved in regulation of cell division. Res Microbiol 1991; 142:269-277.

85. Kren B, Fuchs JA. Characterization of the *ftsB* gene as an allele of the *nrdB* gene in *Escherichia coli*. J Bacteriol 1987; 169:14-18.

86. Taschner PEM, Verest JGJ, Woldringh CL. Genetic and morphological characterization of *ftsB* and *nrdB* mutants of *Escherichia coli*. J Bacteriol 1987; 169:19-25.

87. Herman C, Ogura T, Tomoyasu HS, Akiyama Y, Ito K, Thomas R, D'Ari R, Bouloc P. Cell growth and lambda phage development controlled by the same essential *Escherichia coli* gene, *ftsH/hflB*. Proc Natl Acad Sci USA 1993; 90:10861-10865.

88. Miyoshi Y, Yamagata H. Sucrose-dependent spectinomycin-resistant mutants of Escherichia coli. J Bacteriol 1976; 125:142-148.

89. Vinella D, D'Ari R. Thermoinducible filamentation in *Escherichia coli* due to an altered RNA polymerase beta subunit is suppressed by high levels of ppGpp. J Bacteriol 1994; 176:966-72.

90. Leclerc G, Sirard C, Drapeau GR. The *Escherichia coli* cell division mutation *ftsM1* is in *serU*. J Bacteriol 1989; 171:2090-2095.

91. Chen MX, Bouquin V, Norris V, Casaregola S, Seror SJ, Holland IB. A single base change in the acceptor stem of tRNA$_3$^{Leu} confers resistance of *Escherichia coli* to the calmodulin inhibitor 48/80. EMBO J 1991; 10:3113-3122.

92. Meide VD, Borman THPH, Van Kimmenade AMA, Putte Vd, Bosch L. Elongation factor Tu isolated from *Escherichia coli* mutants altered in *tufA* and *tufB*. Proc Natl Acad Sci USA 1980; 77:3922-3926.

93. Vinella D, Joseleau-Petit D, Thevenet D, Bouloc P, D'Ari R. Penicillin-binding protein 2 inactivation in *Escherichia coli* results in cell division inhibition, which is relieved by FtsZ overexpression. J Bacteriol 1993; 175:6704-6710.

94. Kepes F, Kepes A. Postponement of cell division by nutritional shift-up in *Escherichia coli*. J Gen Microbiol 1985; 131:677-685.

95. Kepes F, D'Ari R. Involvement of FtsZ protein in shift-up induced division delay in *Escherichia coli*. J Bacteriol 1987; 169:4036-4040.

96. Miller JH. Experiments in Molecular Genetics. Cold Spring Harbor Harbor, NY: Cold Spring Harbor Laboratory, 1972.

REGULATION OF GENE EXPRESSION IN STATIONARY PHASE

Heidi Goodrich-Blair, María Uría-Nickelsen and Roberto Kolter

INTRODUCTION

Among the first concepts presented to a student of microbiology is the simplified view of the phases of a bacterial culture: lag, exponential, and stationary. Interest in studying the underlying physiology of a bacterial cell as it shifts between these phases has always been high among bacterial physiologists. Despite this interest, molecular genetic approaches were not applied to the study of regulation of gene expression in stationary phase *Escherichia coli* cultures until the last few years. This may have been partially due to the fact that many investigators felt that, in contrast to the highly differentiated spores of starving *Bacillus subtilis*,[1] *Myxococcus xanthus*,[2] or *Streptomyces*,[3] starved *E. coli* cells were simply a nongrowing version of exponential phase cells. The fact that microorganisms in nature must by necessity spend most of their existence in a nongrowing, or an extremely slow growing state, eventually led some investigators to the study of nongrowing *E. coli* in the laboratory. As more and more researchers explore the *E. coli* starvation response we are beginning to understand the remarkable developmental changes this organism undergoes during stationary phase.[4-7] Cells respond to starvation by turning on a reversible program of gene expression enabling them to survive prolonged periods of nutrient deprivation. Although these developmental processes do not give rise to a bacterial spore, starved *E. coli* gain resistance to many different environmental stresses. Yet at the same time they remain metabolically active and primed to reinitiate growth immediately after reencountering nutrients.

The adaptive changes for starvation are summarized in Figure 27.1. The synthesis of trehalose and the induction of glycine-betaine transport allow increased osmotic shock protection.[8,9] Cells become more resistant to oxidative damage[10-12] and heat shock.[13] Storage of the high energy compounds glycogen and polyphosphate occurs, presumably to provide the dormant cell with reserve nutrients and energy sources for survival and future growth.[14-16] A number of changes occur in the cell

Regulation of Gene Expression in Escherichia coli, edited by E. C. C. Lin and A. Simon Lynch. © 1996 R.G. Landes Company.

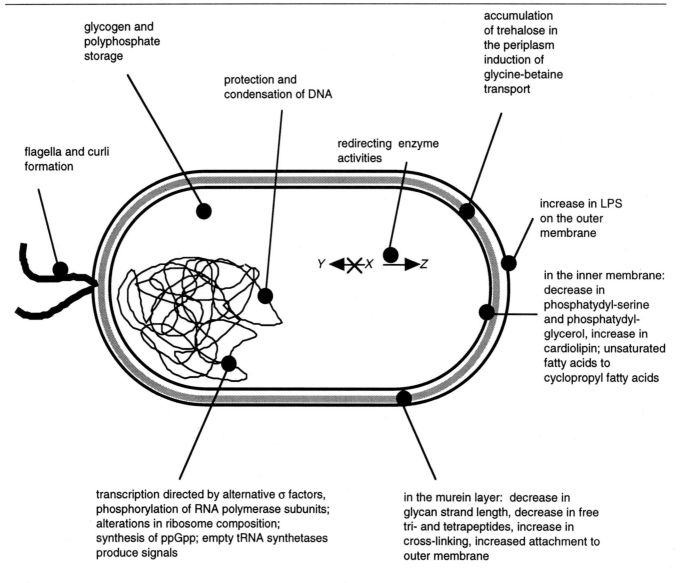

glycogen and
polyphosphate
storage

protection and
condensation of DNA

accumulation
of trehalose in
the periplasm
induction of
glycine-betaine
transport

redirecting enzyme
activities

flagella and curli
formation

increase in LPS
on the outer
membrane

in the inner membrane:
decrease in
phosphatydyl-serine
and phosphatydyl-
glycerol, increase in
cardiolipin; unsaturated
fatty acids to
cyclopropyl fatty acids

$Y \blacktriangleleft\!\!\times\!\!\blacktriangleright X \blacktriangleright Z$

transcription directed by alternative σ factors,
phosphorylation of RNA polymerase subunits;
alterations in ribosome composition;
synthesis of ppGpp; empty tRNA synthetases
produce signals

in the murein layer: decrease in
glycan strand length, decrease in free
tri- and tetrapeptides, increase in
cross-linking, increased attachment to
outer membrane

Fig 27.1. Summary of phenotypic changes in stationary phase Escherichia coli.

wall and membrane, including an increase in lipopolysaccharides (LPS) in the outer membrane[17] and an increase in covalent cross-linking in the murein layer.[18,19] Morphological changes also occur, causing the cells to become compact and spherical. Two-dimensional gel electrophoresis of pulse-labeled cell extracts allowed the identification of at least 30 protein products that are induced at the onset of stationary phase and that presumably are associated with or cause these developmental changes.[12,20,21]

Many of the genes induced at the onset of stationary phase have been identified, and they can be loosely grouped into two types: those that are dependent on the expression of a stationary phase sigma factor, and those that are not. The latter group includes, as an example, those induced at the onset of stationary phase due to increases in cAMP levels: the glycogen synthesis genes *glgCAP*[22,23] and the *cstA* gene involved in the utilization of peptides as carbon and energy sources.[24] The *mcb* genes, involved in the production of the gyrase inhibitor microcin B17, are induced at the onset of stationary phase in an OmpR-dependent fashion.[25] However, the focus of this chapter is to present what is known about changes in gene expression directed by the stationary phase sigma factor, σ^s.

THE σˢ REGULON

The *rpoS* gene encodes σˢ (the "s" standing for stationary phase), and certain mutations in this gene give rise to pleiotropic phenotypes that are indicative of the role of σˢ in stationary phase adaptation. Cells lacking a functional *rpoS* gene show increased sensitivity to a variety of environmental stresses and do not survive as well as their parent strain during prolonged starvation.[21,26] Furthermore, the morphological changes typical of stationary phase cells do not occur in *rpoS* mutants.[27] The gene was originally identified as a regulator of a set of genes with diverse functions and has been known as *katF*,[10] *appR*,[28] *nur*,[29] and *csi-2*.[21] The predicted protein sequence of RpoS displays a high level of similarity to sigma factors, most strikingly to σ⁷⁰.[30] In vitro biochemical analyses have shown that RpoS can interact with core RNA polymerase and in this way direct the transcription of several promoters. It has therefore been designated σ³⁸ or, more commonly, σˢ.[31-33]

Over 20 genes whose expression is σˢ-dependent have been identified and this number continues to increase. A listing of these genes, their map positions and functions is presented in Table 27.1. Many of these genes have obvious relevance to the development of the stress-resistant stationary phase state. The product of the *csiD* gene is involved in thermotolerance[34] as are the products of the *otsBA* genes that encode enzymes required for the production of trehalose.[8,35] Protection from oxidative damage is conferred by a 20 kDa DNA binding protein encoded by the *dps* gene,[12] by a catalase (HPII) encoded by the *katE* gene[36] and by the DNA repair enzyme exonuclease III encoded by the *xthA* gene.[11] The accumulation of glycogen is directed in part by the *glgS* gene product,[37] and the increase in the conversion of unsaturated fatty acids to cyclopropyl fatty acids in the cell membrane is mediated by CFA synthase, the product of the *cfa* gene.[38]

In addition to expression of the primary tier of genes under the control of σˢ, there is a secondary cascade of gene expression. A number of σˢ-dependent genes are themselves regulators of gene expression, although the mechanisms by which they act are not yet well defined. Labeling and two dimensional gel electrophoretic analysis of both *dps*[12] and *csiD* (Hengge-Aronis, personal communication) mutant extracts show marked differences in the proteins expressed when compared to their respective wild-type counterparts. BolA, a 13.5 kDa regulator of cell division genes, acts as a transcriptional activator for the gene encoding PBP6 which is involved in cell wall synthesis at the septum.[39,40] AppY also appears to be a transcriptional activator, affecting the expression of the *cyxAB appA* operon encoding a third cytochrome oxidase and an acid phosphatase, and the *hya* operon encoding hydrogenase 1 (refs. 41-43 and T. Atlung, personal communication).

Attempts to identify a consensus recognition site for this sigma factor in the promoter region of its dependent genes did not reveal any clear distinguishing features. A consensus sequence that appears to be indicative of growth phase regulation, designated a "gearbox" promoter, was identified but its presence does not guarantee transcription by σˢ.[44-46] For example, the σˢ-independent gene *mcbA* is preceded by a gearbox promoter, whereas the σˢ-dependent gene, *katE*, is not. Initial in vitro transcription experiments to test promoter recognition by σˢ did not clarify the situation because the specificity of transcription initiation did not match that observed in vivo. Eσˢ holoenzyme can initiate transcription not only from σˢ promoters but also from some σ⁷⁰ promoters.[32,33] This lack of specificity in vitro suggests that other elements contribute to promoter recognition. Further in vitro studies

Table 27.1. σ^s*-dependent genes*

Gene	Gene Product	Function	Map Position (min)	Reference
aidB	Dehydrogenase	DNA protection	95	73
dps	DNA-binding protein	DNA protection	18	12
katE	Catalase HPII	H_2O_2 resistance	37	10
katG	Catalase HPI	H_2O_2 resistance	89	74
xthA	Exonuclease III	DNA repair, H_2O_2 resistance	38	11
bolA	Regulatory protein	Morphogen, control of PBP6 synthesis	10	75 27
ficA	Regulatory protein	Regulation of cell division genes	74	33 77
csgA	Curli subunit protein	Fibronectin binding	23	51
otsBA	Trehalose-6P-phosphatase Trehalose-6P-synthase	Trehalose synthesis, osmoprotection, thermotolerance	41	8 9 35
csiD	Regulatory protein	Thermotolerance	58	(R. Hengge-Aronis, unpublished)
treA	Trehalase	Trehalose uptake	26	35,77
htrE	Protein with sequence similarity to pilin papC	Thermotolerance, osmoprotection	3	78
glgS	GlgS	Glycogen synthesis	66	37
appY	Regulatory protein	Control of appABC expression	13	41,(T. Atlung, personal communication)
appA BC	Acid phosphatase, third cytochrome oxidase	Anaerobic metabolism	22	42
hyaA BCDEF	Hydrogenase 1	Hydrogen uptake system	22	43
osmB	Lipoprotein	Cell surface alterations	28	35,79
osmY	Periplasmic protein	?	99	21,79
cfa	CFA synthase	CFAs synthesis	36	80
wrbA	WrbA	Enhancement of repression of trp operon	23	81
mccABCDEF	Several proteins	Synthesis and secretion of microcin C7	plasmid	82
csiF	?	?	9	34
csiE	?	?	55	34
poxB	Pyruvate oxidase	?	19	83
proP	Transport protein	Transport of proline and glycine betaine	93	84
cbpA	Analog of DnaJ	Molecular chaperone	?	73
aldB	Aldehyde dehydrogenase	?	81	85

have shown that the conditions of the transcription reactions do indeed have a strong effect on promoter recognition. Specifically, changing the molar ratio between RNA polymerase and the promoter, and/or changing the concentrations and nature of salt used in the buffer can cause a drastic increase in the selectivity of $E\sigma^s$ for promoters known to be dependent on this form of the polymerase.[31] It is therefore likely that, in vivo, changes in the intracellular environment contribute to promoter recognition at the onset of stationary phase, adding complexity to the simple scenario of a switch in sigma factors.

There are many candidates for changes that may alter promoter recognition at the onset of stationary phase. Potassium glutamate increases the promoter specificity of $E\sigma^s$ in vitro, and its accumulation in vivo may have a similar effect. This observation is particularly interesting since glutamate accumulates as an osmolyte when cells are challenged osmotically—a stress that causes the induction of many σ^s-dependent genes.[31]

It has also been proposed that the topology of DNA may be an important factor in promoter recognition by $E\sigma^s$. For instance, the expression of the stationary-phase inducible operon, *proU*, is in part dependent upon DNA supercoiling,[47] and DNA bending may help σ^s recognize and bind to some stationary phase-inducible promoters.[48] Another aspect of modulation of σ^s activity is the action of additional regulators. The *dps* promoter is an excellent example of the regulatory complexity of some σ^s-dependent genes. The σ^s-dependent stationary phase induction of *dps* is facilitated by IHF.[49] In addition, *dps* can be induced in an oxidative stress response in exponential phase utilizing σ^{70} and OxyR.[49] Many σ^s-dependent promoters have additional regulatory proteins acting independently of σ^s. For example, *osmY* is activated by σ^s, but repressed by Lrp, cAMP·CRP, and IHF[50] and *csgA* expression is modulated by both σ^s and H-NS.[51]

The switch from vegetative to stationary phase gene expression might be additionally controlled by changes in the relative concentrations of σ^s and σ^{70} and possibly by changes in the affinity of core polymerase for each sigma factor. The composition of core RNA polymerase does indeed change as a function of growth phase, as detected by altered elution properties from phosphocellulose columns.[52] The stationary phase forms of core polymerase were found to have different promoter recognition properties from the exponential version, based on in vitro transcription assays with σ^{70}-dependent promoters. It is possible that these changes in the core RNA polymerase might alter its affinity for one or the other sigma factor upon entry into stationary phase.[16,52]

RpoS REGULATION

In order to understand more fully how *E. coli* coordinates this complex response of gene expression with changes in environmental conditions, it is essential that we study how the key regulator, σ^s, is itself regulated. How is the depletion of nutrients sensed? At what point in σ^s expression are sensors causing induction? How are individual signals able to produce a subset of responses that are fine-tuned to counteract a particular stress without inducing the entire system (for example, induction of osmoprotectants during osmotic shock vs. stationary phase)? All of the research accumulated to date points to one fact: the regulation of σ^s expression is extremely sophisticated. Controls and sensor inputs are placed at each step of expression: transcription, translation, and post-translation, and the dissection of these stages is just beginning (Fig. 27.2).

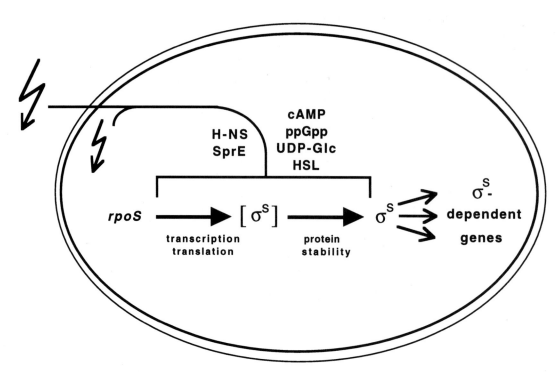

Fig. 27.2. Regulation of rpoS expression.

Analysis of *rpoS::lacZ* transcriptional fusions in cells growing in rich medium show a low basal expression in exponential phase, with induction occurring during late-exponential phase.[21,53,54] The promoter region of *rpoS* has recently been analyzed (refs. 55, 56 and R. Hengge-Aronis, personal communication) and although conflicting data still prohibit definitive interpretations some consistent conclusions can be drawn. Immediately upstream of *rpoS* is a gene, *nlpD*, required for lipopolysaccharide formation, that is not regulated by growth phase.[55,57] One group identified four different promoters within the *nlpD* gene that contribute to *rpoS* transcription.[56] Another group found that two promoters preceding the *nlpD* gene contribute to the basal expression of *rpoS* during exponential growth[55] and that there are at most two promoters within the *nlpD* gene contributing to stationary phase expression (R. Hengge-Aronis, personal communication). Regardless of the discrepancies in the experimental results concerning the precise number of active promoters, the two groups agree that one of the promoters within *nlpD*, designated P2 or *rpoS*p1, is responsible for the majority of *rpoS* transcript. This promoter appears to be σ^{70}-dependent, and in vitro this site can be recognized by both the σ^{70} and σ^{s} holoenzymes.[56] The region harbors two consensus CRP binding sites, which is of particular interest in light of reports indicating some involvement of catabolite repression in the transcription of *rpoS* (see below).

What factors contribute to the induction of transcription from the *rpoS* promoter(s)? One report shows that a dialyzable, heat stable factor in spent LB medium can induce transcription of *rpoS::lacZ* fusions.[53] Attempts to further define the nature of this factor were unsuccessful. However, it was found that the addition of benzoic acid to the medium could induce early transcription of *rpoS*. Other compounds that have been implicated as inducers are weak acids such as acetate and propionate. It is possible that these compounds in some way modulate intracellular proton accumulation coupled to a reduction of intracellular pH, and that *rpoS* may be responding to the resulting changes in ΔpH or $\Delta\varphi$.[53,54]

As alluded to above, the cAMP·CRP complex appears to contribute to the control of transcription of *rpoS*. Unfortunately, these reports are not always consistent. One group found that transcription measured from an *rpoS::lacZ* operon fusion was abolished in a Δcya mutant, suggesting cAMP has a positive regulatory role in *rpoS* expression.[26] However, another group also measured transcription of *rpoS::lacZ* fusions but found an elevated level of transcription in Δcya mutants which could be counteracted by the addition of cAMP, clearly suggesting negative regulation by cAMP.[21] These discrepancies may be partly due to the problems associated with differences between fusion constructions. A later report supported the negative affect of cAMP on σ^s levels detected in immunoblots. The authors of this report suggest that the cAMP/CRP complex may act both indirectly via growth rate changes and directly as a repressor of *rpoS* transcription.[58]

Another signal that appears to affect σ^s levels, possibly at the level of transcription, is ppGpp.[59] Strains harboring $\Delta relA$ $\Delta spoT$ mutations and lacking measurable ppGpp levels (ppGpp^o) have some phenotypic similarities to *rpoS* mutants, including reduced viability, morphological alterations, and salt sensitivity. This indicates that ppGpp might be involved in the expression of *rpoS*. Indeed, immunoblots with monoclonal anti-σ^s antibody show that ppGpp^o strains have delayed and reduced induction of σ^s. Further studies using transcriptional and translational fusions have suggested that depletion of ppGpp affects σ^s levels at the transcriptional stage: a $\Delta relA$ $\Delta spoT$ mutant has a 10-fold reduction in β-galactosidase activity from an *rpoS* transcriptional fusion (R. Hengge-Aronis, personal communication). However, it appears that rather than affecting the initiation of transcription from the *rpoS*p1 promoter, the lack of ppGpp may instead affect the elongation or stability of the nascent mRNA transcript, since even the basal level of transcription from the other upstream promoters was reduced. It was previously reported that ppGpp null mutants have uncoupled transcription and translation and that this leads to increased degradation of *lacZ* mRNA.[60,61]

Another small metabolite, homoserine lactone (HSL), was recently shown to be involved in *rpoS* transcriptional regulation. A connection between HSL and *rpoS* transcription was first suspected because of the action of a gene product, RspA (Regulator of stationary phase) which, when overproduced, suppresses *rpoS* transcription.[62] RspA overproduction also inhibits acylated-HSL mediated luminescence. The *Vibrio fischeri* genes for the regulation and production of luciferase can be expressed in *E. coli*, causing these cells to luminesce in an acylated-HSL dependent manner. If RspA is overproduced however, luminescence is inhibited. This connection between the inhibition of *rpoS* transcription and acylated-HSL mediated luminescence suggested that some form of modified HSL may in fact be an inducer of *rpoS*.[62] Although acylated-HSL mediated induction systems are very well defined,[63] the source of the HSL moeity in these signaling molecules has not yet been determined. One possible source is homoserine, an intermediate in the biosynthetic pathways of threonine and methionine. Mutants in this pathway show a decreased level of σ^s in immunoblots, and exogenous addition of HSL rescues this decrease.[62] Hence there appears to be a connection between *rpoS* and HSL, and it is possible that the source of intracellular HSL is a key metabolic intermediate. Such a molecule would represent an ideal signal for the onset of stationary phase, as it could be derived directly from an intermediate that might build up as protein synthesis and metabolism slow down. However, whether or not the *rpoS* inducer is an acylated form of HSL, how the HSL moeity is formed, and if any regulator protein is required for induction by this signal remain unanswered questions. In all other known acylated-HSL mediated systems a regulatory molecule is required, the prototype of which is LuxR.[64] *E. coli* has a LuxR homolog, SdiA, but SdiA is not required for the induction of *rpoS* transcription.[62,65]

Clearly, there are many signals that could be contributing to the induction of *rpoS* transcription. However, it does not appear that transcription is the major point of regulation in the synthesis of this sigma factor. Instead, there is much evidence indicating that most of the induction that occurs at the onset of stationary phase is at the post-transcriptional level. *rpoS* translational fusions achieve a higher level of β-galactosidase activity than transcriptional fusions.[26,58,66] Thus far, the mechanism by which this post-transcriptional induction occurs has not been defined. It may require some portion of the coding region of *rpoS*, since induction levels are lower in fusions that have less *rpoS* coding region than in those that have more.[58,67] There are several signals that may feed into this tier of regulation. One, UDP-glucose, an intermediate in glucose metabolism, appears to repress translation of σ^s.[68] Blocks in glucose metabolism that decrease UDP-glucose formation cause an increase in σ^s-dependent gene expression. These same mutants have an increase in σ^s levels detected by immunoblotting. According to the authors, preliminary unpublished data suggest UDP-glucose is a post-transcriptional regulator. Again, it is interesting to note that small metabolites that may be accumulating during different phases of growth are affecting the expression of σ^s.

σ^s also seems to rely upon the action of several proteins for its expression. These proteins may be the sensors of the proposed metabolite inducers, or they may act independently. H-NS, a nucleoid protein, was recently shown to negatively regulate σ^s.[69] *hns* mutants have increased cellular contents of σ^s due to both an increase in translation and an increase in stability of the protein. Another negative protein regulator was recently identified: the SprE (<u>s</u>tationary <u>p</u>hase <u>r</u>egulatory <u>e</u>lement) protein inhibits the translation of *rpoS* and is similar to the response regulator family of proteins (T. Silhavy, personal communication). These response regulators are typically part of a two-component regulatory system, the other component being a sensor/kinase.[70] In this case a sensor may recognize external nutrient availability and, when energy sources are available, transduce a signal to inhibit σ^s expression via SprE. This is the first example of a molecule of this type affecting translation rather than transcription, and it is the first example of a regulator that is responding to a specific external condition to regulate σ^s activity.

σ^s and a subset of its dependent genes are induced, in addition to stationary phase, during osmotic upshift. This induction seems to depend upon another form of translational regulation, the mechanism of which requires a portion of the *rpoS* coding region. Varying the length of the *rpoS* sequence preceding the fusion junction to *lacZ* yields differences in the amount of induction measured in translational fusions. Single copy chromosomal "early" translational *rpoS::lacZ* fusions show only a slight induction in response to osmotic upshift, and this induction occurs too late to explain the reported induction of *rpoS*-dependent genes during this stress.[58] However, a translational fusion that carries more *rpoS* sequence displays a much higher level of *rpoS* induction[58] that is consistent with similar kinetics of σ^s-dependent osmotic induction of genes such as *osmY*.[71] These results suggest a novel mechanism of osmotic regulation that operates at the level of translation.[58]

Although the signal that reflects changes in medium osmolarity and also affects *rpoS* translation has not been identified, there is some evidence that ppGpp may have a positive regulatory role in *rpoS* osmotic induction (R. Hengge-Aronis, personal communication). However, this effect may in fact be due to changes in growth rate that are brought about by the increase in ppGpp levels. It remains to be clarified if the induction of *rpoS* continues even after the decrease in growth rate brought about by the induction of ppGpp synthesis has ended.

Post-translational regulation also appears to play a major role in the increase in σ^s activity in stationary phase. It has been observed that the stability of the sigma factor increases dramatically at the onset of stationary phase from a half-life of approximately 2 minutes in exponential phase to 10-25 minutes in stationary phase.[56,58] Mutations within *rpoS* have been obtained that inhibit the ability of σ^s to stabilize upon entry to stationary phase. As a result these mutants have significantly reduced levels of σ^s-dependent expression (ref. 72 and G.W. Huisman, S. Gupta and R. Kolter, unpublished observation). In one case, the mutation has been identified to be a 46-base pair duplication at the 3' end of the *rpoS* gene, causing a frameshift that results in the last four amino acids being replaced by 39 additional residues.[72] How this mutation results in decreased stabilization of the protein is unclear.

It is obvious that complex, well-regulated changes in gene expression are occurring during the stationary phase of *E. coli*. These changes must be directly linked to the sensing of the external and internal environments. Determining how starvation is sensed and transduced to regulate gene expression will help us to define how bacteria coordinate nutrient supply with maximum growth efficiency. The identification of the *rpoS* regulon has provided a rich source of tools to address these questions at the experimental level.

REFERENCES

1. Losick R, Stragier P. Crisscross regulation of cell-type-specific gene expression during development in *B. subtilis*. Nature 1992; 355:601-604.
2. Kim SK, Kaiser D, Kuspa A. Control of cell density and pattern by intercellular signaling in *Myxococcus* development. Annu Rev Microbiol 1992; 46:117-139.
3. Chater KF. Genetics of differentiation in Streptomyces. Annu Rev Microbiol 1993; 47:685-713.
4. Matin A. The molecular basis of carbon-starvation-induced general resistance in *Escherichia coli*. Mol Microbiol 1991; 5:3-10.
5. Hengge-Aronis R. Survival of hunger and stress: the role of *rpoS* in early stationary phase gene regulation in *Escherichia coli*. Cell 1993; 72:165-168.
6. Huisman G, Kolter R. Regulation of gene expression at the onset of stationary phase in *Escherchia coli*. In: Piggot P, Moran C, Youngman P, eds. Regulation of Bacterial Differentiation. Washington, DC: American Society for Microbiology, 1993:21-40.
7. Kolter R, Siegele DA, Tormo A. The Stationary Phase of the Bacterial Life Cycle. Ann Rev Microbiol 1993; 47:855-874.
8. Giaever HM, Styrvold OB, Kaasen I et al. Biochemical and genetic characterization of osmoregulatory trehalose synthesis in *Escherichia coli*. J Bacteriol 1988; 170:2841-2849.
9. Kaasen I, Falkenberg P, Styrvold OB et al. Molecular cloning and physical mapping of the *otsAB* genes, which encode the osmoregulatory trehalose pathway of *Escherichia coli*: Evidence that transcription is activated by KatF (AppR). J Bacteriol 1992; 174:889-898.

10. Loewen PC, Triggs BL. Genetic mapping of *katF*, a locus that with *katE* affects the synthesis of a second catalase species in *Escherichia coli*. J Bacteriol 1984; 160:668-675.

11. Sak BD, Eisenstark A, Touati D. Exonuclease III and the hydroperoxidase II in *Escherichia coli* are both regulated by the *katF* product. Proc Natl Acad Sci USA 1989; 86:3271-3275.

12. Almirón M, Link A, Furlong D et al. A novel DNA binding protein with regulatory and protective roles in starved *E. coli*. Genes Dev 1992; 6:2646-2654.

13. Jenkins DE, Auger EA, Matin A. Role of RpoH, a heat shock regulator protein, in *Escherichia coli* carbon starvation protein synthesis and survival. J Bacteriol 1991; 173:1992-1996.

14. Preiss J. Bacterial glycogen synthesis and its regulation. Ann Rev Microbiol 1984; 38:419-458.

15. Damotte M, Cattaneo J, Sigal N et al. Mutants of *Escherichia coli* K12 altered in their ability to store glycogen. Biochem Biophys Res Comm 1968; 32:916-920.

16. Crooke E, Akiyama M, Rao NN et al. Genetically altered levels of inorganic polyphosphate in *Escherichia coli*. J Biol Chem 1994; 269:6290-6295.

17. Ivanov AI, Fomchenkov VM. Relation between the damaging effect of surface-active compounds on *Escherichia coli* cells and the phase of culture growth. Mikrobiologiia 1989; 58:969-975.

18. Wensink J, Gilden N, Witholt B. Attachment of lipoprotein to the murein of *Escherichia coli*. Eur J Biochem 1982; 122:587-590.

19. Pisabarro AG, de Pedro MA, Vazquez D. Structural modifications in the peptidoglycan of *Escherichia coli* associated with changes in the state of growth of the culture. J Bacteriol 1985; 161:238-242.

20. Groat RG, Schultz JE, Zychlinsky E et al. Starvation proteins in *Escherichia coli*: Kinetics of synthesis and role in starvation survival. J Bacteriol 1986; 168:486-493.

21. Lange R, Hengge-Aronis R. Identification of a central regulator of stationary phase gene expession in *E. coli*. Mol Microbiol 1991; 5:49-59.

22. Romeo T, Preiss J. Genetic regulation of glycogen biosynthesis in *Escherichia coli*: In vitro effects of cyclic AMP and guanosine 5'-diphosphate 3'-diphosphate and analysis of in vivo transcripts. J Bacteriol 1989; 171:2773-2782.

23. Okita T, Rodriguez R, Preiss J. Biosynthesis of bacterial glycogen. J Biol Chem 1981; 256:6944-6952.

24. Schultz JE, Matin A. Molecular and functional characterization of a carbon starvation gene of *Escherichia coli*. J Mol Biol 1991; 218:129-140.

25. Hernandez-Chico C, San Millan JL, Kolter R et al. Growth phase and OmpR regulation of transcription of the Microcin B17 genes. J Bacteriol 1986; 167:1058-1065.

26. McCann MP, Kidwell JP, Matin A. The putative σ factor KatF has a central role in development of starvation-mediated general resistance in *Escherichia coli*. J Bacteriol 1991; 173:4188-4194.

27. Lange R, Hengge-Aronis R. Growth phase-regulated expression of *bolA* and morphology of stationary phase *Escherichia coli* cells is controlled by the novel sigma factor σ^S (*rpoS*). J Bacteriol 1991; 173:4474-4481.

28. Touati E, Dassa E, Boquet PL. Pleiotropic mutations in *appR* reduce pH 2.5 acid phosphatase expression and restore succinate utilization in CRP-deficient strains of *Escherichia coli*. Mol Gen Genet 1986; 202:257-264.

29. Tuveson RW. The interaction of a gene (*nur*) controlling near-UV sensitivity and the *polA1* gene in strains of *E. coli* K-12. Photochem Photobiol 1981; 33:919-923.

30. Mulvey MR, Loewen PC. Nucleotide sequence of *katF* of *Escherichia coli* suggests KatF protein is a novel transcription factor. Nucleic Acids Res 1989; 17:9979-9991.

31. Ding Q, Kusano S, Villarejo M et al. Promoter selectivity control of *Escherichia coli* RNA polymerase by ionic strength: differential recognition of osmoregulated promoters by Eσ^D and Eσ^S holoenzymes. Mol Microbiol 1995; 16:649-656.

32. Nguyen LH, Jensen DB, Thompson NE et al. In vitro functional characterization of overproduced *Escherichia coli katF/rpoS* gene product. Biochemistry 1993; 32:11112-1117.

33. Tanaka K, Takayanagi Y, Fujita N et al. Heterogeneity of the principal sigma factor in *Escherichia coli*: the *rpoS* gene product, σ^{38}, is a principal sigma factor of RNA polymerase in stationary phase *Escherichia coli*. Proc Natl Acad Sci USA 1993; 90:3511-3515.

34. Weichart D, Lange R, Henneberg N et al. Identification and characterization of stationary phase-inducible genes in *Escherichia coli*. Mol Microbiol 1993; 10:407-420.

35. Hengge-Aronis R, Klein W, Lange R et al. Trehalose synthesis genes are controlled by the putative sigma factor encoded by *rpoS* and are involved in stationary phase thermotolerance in *Escherichia coli*. J Bacteriol 1991; 173:7918-7924.

36. Loewen PC, Switala J, Triggs-Raine BL. Catalase HPI and HPII in *Escherichia coli* are induced independently. Arch Biochem Biophys 1985; 243:144-149.

37. Hengge-Aronis R, Fischer D. Identification and molecular analysis of *glgS*, a novel growth-phase-regulated and *rpoS*-dependent gene involved in glycogen synthesis in *Escherichia coli*. Mol Microbiol 1992; 6:1877-1886.

38. Wang A-Y, Cronan JE. The growth phase-dependent synthesis of cyclopropane fatty acids in *Escherichia coli* is the result of an RpoS(KatF)-dependent promoter plus enzyme instability. Mol Microbiol 1994; 11:1009-1017.

39. Begg KJ, Takasuga A, Edwards DH et al. The balance between different peptidoglycan precursors determines whether *Escherichia coli* cells will elongate or divide. J Bacteriol 1990; 172:6697-6703.

40. Dougherty TJ, Pucci MJ. Penicillin-binding proteins are regulated by *rpoS* during transitions in growth states of *Escherichia coli*. Antimicrob Agents Chemother 1994; 1994:205-210.

41. Atlung T, Nielsen A, Hansen FG. Isolation, characterization, and nucleotide sequence of *appY*, a regulatory gene for growth-phase-dependent gene expression in *Escherichia coli*. J Bacteriol 1989; 171:1683-1691.

42. Dassa J, Fsihi H, Marck C et al. A new oxygen-regulated operon in *Escherichia coli* comprises the genes for a putative third cytochrome oxidase and for pH 2.5 acid phosphatase (*appA*). Mol Gen Genet 1992; 229:342-352.

43. Brøndsted L, Atlung T. Anaerobic regulation of the hydrogenase 1 (*hya*) operon of *Escherichia coli*. J Bacteriol 1994; 176:5423-5428.

44. Aldea M, Garrido T, Pla J et al. Division genes in *Escherichia coli* are expressed coordinately to cell septum requirements by gearbox promoters. EMBO J 1990; 9:3787-3794.

45. Bohannon DE, Connell N, Keener J et al. Stationary-phase-inducible "gearbox" promoters: differential effects of *katF* mutations and the role of σ^{70}. J Bacteriol 1991; 173:4482-4492.

46. Vicente M, Kushner SR, Garrido T et al. The role of the "gearbox" in the transcription of essential genes. Mol Microbiol 1991; 5:2085-2091.

47. Higgins CF, Dorman CJ, Stirling DA et al. A physiological role for DNA supercoiling in the osmotic regulation of gene expression in *S. typhimurium* and *E. coli*. Cell 1988; 52:569-584.

48. Espinosa-Urgel M, Tormo A. Sigma σ-dependent promoters in *Escherichia coli* are located in DNA regions with intrinsic curvature. Nuc Acids Res 1993; 21:3667-3670.

49. Altuvia S, Almirón M, Huisman G et al. The *dps* promoter is activated by OxyR during growth and by IHF and σ^s in stationary phase. Mol Microbiol 1994; 13:265-272.

50. Lange R, Barth M, Hengge-Aronis R. Complex transcriptional control of the σ-^{S} dependent stationary-phase-induced and osmotically regulated *osmY* (*csi-5*) gene suggests novel roles for Lrp, cyclic AMP (cAMP) receptor protein-cAMP complex, and integration host factor in the stationary-phase response of *Escherichia coli*. J Bacteriol 1993; 175:7910-7917.

51. Olsén A, Arnqvist A, Sukupolvi S et al. The RpoS sigma factor relieves H-NS mediated transcriptional repression of *csgA*, the subunit gene of fibronectin binding curli in *Escherichia coli*. Mol Microbiol 1993; 7:523-536.

52. Ozaki M, Wada A, Fujita N et al. Growth phase-dependent modification of RNA polymerase in *Escherichia coli*. Mol Gen Genet 1991; 230:17-23.

53. Mulvey MR, Switala J, Borys A et al. Regulation of transcription of *katE* and *katF* in *Escherichia coli*. J Bacteriol 1990; 172:6713-6720.

54. Schellhorn HE, Stones VL. Regulation of *katF* and *katE* in *Escherichia coli* K-12 by weak acids. J Bacteriol 1992; 174:4769-4776.

55. Lange R, Hengge-Aronis R. The *nlpD* gene is located in an operon with *rpoS* on the *Escherichia coli* chromosome and encodes a novel lipoprotein with a potential function in cell wall formation. Mol Microbiol 1994; 13:733-743.

56. Takayanagi Y, Tanaka K, Takahashi H. Structure of the 5' upstream region and the regulation of the *rpoS* gene of *Escherichia coli*. Mol Gen Genet 1994; 243:525-531.

57. Ichikawa JK, Li C, Fu J et al. A gene at 59 minutes on the *Escherichia coli* chromosome encodes a lipoprotein with unusual amino acid repeat sequences. J Bacteriol 1994; 176:1630-1638.

58. Lange R, Hengge-Aronis R. The cellular concentration of the σ^s subunit of RNA polymerase in *Escherichia coli* is controlled at the levels of transcription, translation, and protein stability. Genes Dev 1994; 8:1600-1612.

59. Gentry DR, Hernández VJ, Nguyen LH et al. Synthesis of the stationary-phase sigma factor, σ^s, is positively regulated by ppGpp. J Bacteriol 1993; 175:7982-7989.

60. Vogel U, Sørensen M, Pedersen S et al. Decreasing transcription elongation rate in *Escherichia coli* exposed to amino acid starvation. Mol Microbiol 1992; 6:2191-2200.

61. Faxen M, Isaksson LA. Functional interactions between translation, transcription and ppGpp in growing *Escherichia coli*. Biochim Biophys Acta 1994; 1219:425-434.

62. Huisman GW, Kolter R. Sensing starvation: a homoserine lactone-dependent signaling pathway in *Escherichia coli*. Science 1994; 265:537-539.

63. Fuqua WC, Winans SC, Greenberg EP. Quorum sensing in bacteria: the LuxR-LuxI family of cell density-responsive transcriptional regulators. J Bacteriol 1994; 176:269-275.

64. Engebrecht J, Nealson K, Silverman M. Bacterial bioluminescence: isolation and genetic analysis of functions from *Vibrio fischeri*. Cell 1983; 32:773-781.

65. Wang X, deBoer PAJ, Rothfield LI. A factor that positively regulates cell division by activating transcription of the major cluster of essential cell division genes of *Escherichia coli*. EMBO J 1991; 10:3363-3372.

66. Loewen PC, von Ossowski I, Switala J et al. Regulation of KatF synthesis in *Escherichia coli*. J Bacteriol 1993; 175:2150-2153.

67. McCann MP, Fraley CD, Matin A. The putative σ factor KatF is regu-

lated post-transcriptionally during carbon starvation. J Bacteriol 1993; 175:2143-2149.

68. Bohringer J, Fischer D, Mosler G et al. UDP-glucose is a potential intracellular signal molecule in the control of expression of σ^s and σ^s-dependent genes in *Escherichia coli*. J Bacteriol 1995; 177:413-422.

69. Yamashino T, Ueguchi C, Mizuno T. Quantitative control of the stationary phase-specific sigma factor, σ^s, in *Escherichia coli*: involvement of the nucleoid protein H-NS. EMBO J 1995; 14:594-602.

70. Bosl M, Kersten H. Organization and functions of genes in the upstream region of *tyr*T of *Escherichia coli*: phenotypes of mutants with partial deletion of a new gene (*tgs*). J Bacteriol 1994; 176:221-231.

71. Hengge-Aronis R, Lange R, Henneberg N et al. Osmotic regulation of *rpoS*-dependent genes in *Escherichia coli*. J Bacteriol 1993; 175:259-265.

72. Zambrano MM, Siegele DA, Almirón M et al. Microbial competition: *Escherichia coli* mutants that take over stationary phase cultures. Science 1993; 259:1757-1760:

73. Yamashino T, Kakeda M, Ueguchi C et al. An analog of the DnaJ molecular chaperone whose expression is controlled by σ^s during the stationary phase and phosphate starvation in *Escherichia coli*. Mol Microbiol 1994; 13:475-483.

74. Mukhopadhyay S, Schellhorn HE. Induction of *Escherichia coli* hydroperoxidase I by acetate and other weak acids. J Bacteriol 1994; 176:2300-2307.

75. Aldea M, Garrido T, Hernández-Chico C et al. Induction of a growth-phase-dependent promoter triggers transcription of *bolA*, an *E. coli* morphogene. EMBO J 1989; 8:3923-3931.

76. Kawamukai M, Matsuda H, Fujii W et al. Cloning of the *fic*-1 gene involved in cell filamentation induced by cyclic AMP and construction of a Δ*fic Escherichia coli* strain. J Bacteriol 1988; 170:3864-3869.

77. Boos W, Ehmann U, Bremer E et al. Trehalase of *Escherichia coli*. J Biol Chem 1987; 262:13212-13218.

78. Raina S, Missiakis D, Baird L et al. Identification and transcriptional analysis of the *Escherichia coli htrE* operon which is homologous to *pap* and related pili operons. J Bacteriol 1993; 175:5009-5021.

79. Jung JU, Gutierrez C, Martin F et al. Transcription of *osmB*, a gene encoding an *Escherichia coli* lipoprotein, is regulated by dual signals. J Biol Chem 1990; 265:10574-10581.

80. Magnuson K, Jackowski S, Rock CO et al. Regulation of fatty acid biosynthesis in *Escherichia coli*. Microbiol Rev 1993; 57:522-542.

81. Yang CC, Konisky J. Colicin V-treated *Escherichia coli* does not generate membrane potential. J Bacteriol 1984; 158:757-759.

82. Diaz-Guerra L, Moreno F, SanMillan JL. *app*R gene product activates transcription of microcin C7 plasmid genes. J Bacteriol 1989; 171: 2906-2908.

83. Chang Y-Y, Wang A-Y, Cronan JE. Expression of *Escherichia coli* pyruvate oxidase (PoxB) depends on the sigma factor encoded by the *rpoS*(*katF*) gene. Mol Microbiol 1994; 11:1019-1028.

84. Mellies J, Wise A, Villarejo M. Two different *Escherichia coli pro*P promoters respond to osmotic and growth phase signals. J Bacteriol 1995; 177:144-151.

85. Xu J, Johnson RC. *ald*B, an RpoS-dependent gene in *Escherichia coli* encoding an aldehyde dehydrogenase that is repressed by Fis and activated by Crp. J Bacteriol 1995; 177:3166-3175.

INDEX

Page numbers in italics denote figures (f) or tables (t).

A

acetyl-CoA, 318-320
ackA, 304-305
activating region 1, 266-267
ada promoter, 131, 133
Adhya S, 2
Agrobacterium tumefaciens, 4, 495, 554
ahpCF, 436-437, 443
Aiba H, 242, 256, 269
Alcaligenes eutrophus, 344
Aldea M, 555-556
α operon, 90, 92, 97, 102, 106
    mRNA, 105
Amouyal M, 131
amylomaltase, 209-210
Anfinsen CB, 481
anticodon of tRNA^Met (CAU), 96
anti-sense RNA *micF*, 383, 442
Aoyama T, 393
araBAD promoter, 267-268, 421
araFG promoter, 268
Arc/*arc*
    genes, 361-365
    mutants, 365-366
    system, 361-374
ArcA/*arcA*, 320, 323, 329, 362-374
ArcA/ArcB system, 344, 354, 361
ArcB/*arcB*, 362-369, 372-374
argV, 137
ATP, 10, 33, 239, 241, 245-246, 281, 284-286, 310, 456-457,
    460-461, 484, 486, 511-512, 524
ATPγS, 457, 460-461
attenuation, 34-35, 48-52, 100
attL, 158-160, 166
attP, *159f*, 160-161, 166
Audureau A, 1
Ausubel FM, 284

B

Bacillus, 35
    subtilis, 98, 100, 244, 468-469, 495, 508, 571
        flagellin promoter, 526
bacteriophage(s)
    Com protein, 85
    K20, 387
    γ, 85-86, 88, 108, 132, 149, 181
        cII protein, 167, 561
        cIII gene, 96
        gene
            N protein, 34-35, 37-38
            Q protein, 35-37
        repressor, 462
    M13 gene V, 110
    MS2, 86, 88
    Mu, 149, 160-161, 163-164, 559
        com gene, 96
        mom gene, 85, 443, 445
    P gene, 482
    P1, 513
    P22, 561
    PA-2, 387
    Φ80 repressor, 460-462
    ΦX174, 128
    PM2 DNA, 138-139
    RNA, 85
        MS2, 86, 88
        SP6, 9
    T3, 9
    T4, 85, 110, 512
        DNA polymerase (gene 43), 91
        gene product (gp32). *See* gp32
        gene 44, 89
        RegA protein, 87, 89, 100
    T7, 9, 85, 108
        RNA polymerase, 135
    TuIb, 387
BarA, 366
Barrett C, 386
Bassford PJ, 387-388
Baumgartner JW, 401
Beachem IR, 386
Beckwith J, 292
bgl operon, 35, 100
bglG gene product, 100
Bochkareva ES, 485
bolA, 555
Bonekamp F, 61
Botsford JL, 256
Bouché J-P, 556
Bradyrhizobium, 284
    japonicum, 332, 337, 495
Bremer H, 533-534
Buc H, 131
Bukau B, 487
Bukhari AI, 505-506
Busby S, 2
Buxton RS, 362

C

cI, 505, 510
CII/cII, 69, 74, 504, 510-511
Calvo JM, 420, 428
cAMP·CAP complex, 181, 184-185, 187-193, 195, 202, 205-206,
        232, 235, 237, 242, 245-246, 255-257, 259-264, 267-268,
        330, 560, 575, 577
cAMP·CRP complex. *See* cAMP·CAP complex
carbon catabolite repression, 233-234, 239-241, 245-247
Cashel M, 534
catabolite gene activator protein (CAP), 181, 234, 242-243, 248,
        288, 317-318, 320, 324, 327-328, 330, 331, 335, 337,
        406, 425
    hybrids, 325
    modulon, 255-271
    protein, 15
    synthesis, 245
        See also cyclic-AMP receptor protein (CRP)/*crp*
Caulobacter crescentus, 508
CcdA/*ccdA*, 512-513, 561
CcdB/*ccdB*, 512, 561
cdd promoter, 266, 270
cGMP, 247
Chai TJ, 387
Chamberlin MJ, 11-12, 14, 16, 19, 130
chromatin, 169-170
Clp/*clp*, 486, 511
ClpA/*clpA*, 486, 509-510
ClpP/*clpP*, 486, 509-510
ClpX, 486
ClpXP/*clpXP*, 510-513
Cohn M, 244
ColE1, 68, 72-73, 76-78, 138-139, 330, 459
ColE2, 386
ColIb factor, 107
coliphage, RNA, R17, 89, 91-92, 98
Comeau DE, 390
CopA, 74, 78
CopT, 74, 78
Cox GB, 293
CpxA/*cpxA*, 367-368
cpxC, 362
CpxR, 368
CreC/*creC*, 300, 302-312
Crp/*crp*, 149, 245, 257, 269
crr, 242, 243
CTP, 10, 524
cya, 184, 245, 255-257
    -*lacZ* fusion strains, 242
    P2 promoter, 269
cyaA, 241
cyclic AMP (cAMP), 4, 156, 182, 234, 236, 239-24, 243, 245-248,
        256, 258, 265, 271, 288, 318, 323, 425, 572, 577
cyclic-AMP receptor protein (CRP)/*crp*, 2, 149, 167-168, 184, 190,
        232, 234-235, 242, 245, 255, 257, 269, 318, 335.
        See also catabolite gene activator protein (CAP)
cyd, 323, 372-373
cydAB, 323, 368-369, 371-372
cyo, 323, 372
CytR/*cytR*, 269-270

D

Dairi T, 390
Danese P, 494
Danot O, 205
Davies JK, 386
Defais M, 455
deg, 505-506
degR, 505
degT, 505
Delbrück M, 291
Dewar S, 557
diauxie, 232
dic, 560-561
dithiothreitol (DTT), 443-444
DMSO, 320-321, 343-344, 354
DNA gyrase, 128, 132, 137, 139
DNA topoisomerase, 127-128, 137, 140
DnaA/*dnaA*, 162, 459-460, 487, 557
DnaK/*dnaK*, 219, 362, 482-487, 489-491
DnaJ/*dnaJ*, 482-484, 489-490
dnaN, 459
Doetsch, 10
Doherty P, 386
Donachie W, 557
Doudoroff M, 291
Dps/*dps*, 445
Drolet M, 139
Drosophila, 248
Drury LS, 362
dsg gene, 106

E

Echols H, 292
EcoRI site, 55, 58
Egan SM, 346
electrophoretic mobility shift assays (EMSA), 153, 371
Englesberg E, 2, 181, 395
Enterococcus faecium BM4147, 372
envZ, 202, 205, 210, 383-384, 386, 389-395, 397-406
Enzyme I (EI), 237-239, 242, 244, 248
Enzyme II (EII), 237-238, 242
enzyme II^{Glc}, 210, 215
ε14, 559-560
Erickson JW, 491
Erwinia carotovora, 4, 554
Eσ^{32}, 487, 488t, 490-491, 493
Eσ^{70}, 487, 490-491, 493, 523, 536
Eσ^{E}, 488t, 493
Eσ^{S}, 573-575

F

fdhF, 353
fdn promoter, 328, 330
fdnG operon, 347-349, 351-354, 372
fdnGHI operon, 344-345
Feng GH, 15
FFpmelR promoter, 321, 325-327, 330-331
FinP, 69
Fis/*fis*, 112, 133, 149, 156, 164, 271, 424, 523, 526-527, 530,
        533-536
flagellin promoter, 164-165, 526
FNR/*fnr*, 268, 317-338, 344, 363, 368-369, 373
    apo-, 324, 327, 334
    holo-, 324, 326

formate-hydrogen lyase complex (FHL), 353-354
Forst S, 394, 399
Foulds J, 386-387
frd, 334
frdA operon, 329, 347-349, 351-354
FruR/*fruR*, 246
Fts/*fts*, 547, 549, 552, 561
ftsA, 549, 551-553, 556, 558
FtsB, 561
ftsE, 551
FtsH/*ftsH*, 489, 511, 561
FtsI/*ftsI*, 548-549, 551, 557-558
FtsL, 548
ftsN, 551
FtsQ/*ftsQ*, 548-549, 552-553, 556, 558
*ftsQ*1,2p, 552-556, 558
ftsQAZ, 552-556, 558, 562
FtsW/*ftsW*, 548-549
FtsZ/*ftsZ*, 71, 507, 512, 548-553, 555-560, 562
*ftsZ*2-4p, 556-558
fumarate reductase (FRD/*frdABCD*), 318-319, 344-345, 354
fumC, 442, 446

G

G6PD, 445-446
gal
    operon(s), 181-188, 191-192, 195
    promoters, 185, 187-188, 192-194
    regulon, *235f*
    system, 3
galM, 182
galOP promoter, 136
galP1, 265-266, 269, 328
galP2, 269
GalR/*galR*, 185-187, 192, 193
GalS, *186f*, 187
galU, 217
γ-^{32}P-ATP, 366-367
Garen A, 292
Garen S, 292
Garrett S, 394, 403
Garrido T, 558
gcv, 421, 423, 427
Geiduschek EP, 169
George J, 506-507
Glass RE, 530
glgA, 216
glgC, 216
glgCAP, 572
glgX, 216
glk, 210
glnA, 282
glnALG operon, 282-283, 287-288, 397
glnAp1, 282-283
glnAp2, 282-288, 398
glnHp2, 134, 160-161
glnHPQ operon, 134
GlpR, 234
glp, 3, *235f*, 244-245
gltBDF, 423, 427
glucokinase, 210-211
glyA, 423-424
Goldberg AL, 506
Goldfarb A, 10, 14, 19
Goloubinoff P, 485
Goodman SD, 158

Gorini L, 291
Gottesman ME, 504
Gottesman S, 486
Gourse RL, 534
gp32, 86, 93-95, 99-100, 102-104, 110, 112
Gragerov A, 482
Gralla J, 12-13
GreA, 15, 19-20, 29
GreB, 15, 19-20, 29
GroEL/*groEL*, 482, 484, 487, 495
GroES/*groES*, 484, 487, 495
Gross CA, 487, 491
GrpE/*grpE*, 482-486, 489-490
Gudas LJ, 456
Guest JR, 344
GUG initiation codon, 105
Gunasekera AS, 264
Gutman GA, 47
GyrA/*gyrA*, 128-130, 133, 135-136, 561
gyrase gene promoters, 133
GyrB/*gyrB*, 128-130, 133, 135
gyrB226, 129

H

Hall MN, 388-389, 393-395, 402
Hancock REW, 387
Harman JG, 256
Hartl F-U, 484-485
Hearst, 16
heat-shock, 2, 4-5, 453, 481-495, 510
    σ factor σ^{32}, 509, 511, 561
Hendrick JP, 484
Henning U, 387-388
Herman C, 489
Hfl/*hfl*, 510-512
HflA/*hflA*, 504, 511
HflB/*hflB*, 489-490, 504, 511
himA, 157
Hin, 164, *165f*
hip, 157
Hirota Y, 547
his, 30, 33, 136
HisJ, 208
HisP, 208
hisR promoter, 133
histidine protein (HPr), 237-239, 242-245
hisU, 129, 133
hisW, 129
hix sites, 164, *165f*
HK022, 38, 464
HMG proteins, 165
H-NS/*hns*, 133, 149, 169, 271, 420, 425, 427, 575, 578
Hok/*hok*, 69, 75-76, 78
homoserine lactone (HSL), 577-578
Horibata K, 244
HPI/II, 436-437, 445
Hsu LM, 14, 16, 19
HtrA/*htrA*, 491-494
HU, 99, 149-170, 271, 424
Huang KJ, 392

I

Ichihara S, 388
Igarashi K, 404
Igo MM, 404
IIA^{Glc}, 239-246, 248
Ikenaka K, 391
ilvGMEDA leader, 48, 55
ilvIH, 419-420, 425-428
*ilv*P$_G$1 promoter, 163
inaA, 446-447
initial transcribed sequences (ITS), 8, 10, 13-14
initial transcribing complexes (ITC), 8-12
initiation factors
    IF1, 88, 109
    IF2, 88, 107-108
    IF3, 88-89, 106-108, 115-116
Inokuchi K, 390-391
Inouye M, 400
Int/*int*, 109, 158-160, 166, 504
integration host factor (IHF), 86, 133-134, 149-170, 271, 285-286,
    329, 350-351, 371, 396, 425, 445, 575
IS10, 70-71, 74, 75f, 78, 79f, 162
Iuchi S, 362-363, 367, 372

J

Jacob F, 181
Jacob-Monod operon, 299
Jensen S, 529, 531, 535
Jo YL, 391
Johnson RC, 164

K

Kaguni JM, 487, 491
Kalckar H, 291
Karstrom, 1
katG, 436-437, 443-444
Kato M, 391, 393
Kawaji H, 385, 401
kgl-tdh, 427-428
Klebsiella pneumoniae, 246, 281-282, 288
Koch AL, 532
Kolb A, 2, 269
Krummel B, 11
Kustu S, 160

L

L1-L11 operon, 111
L4 protein, 521
L7-L10/L12 operon, 90
L7/L12 protein, 101
L10, 101
L10-L7/L12 operon, 96, 98
L11-L1 operon, 95, 113
L14, 101
L-leucine, 419, 422
L-serine, 422-424
Lac/*lac*, 3, 194
    fusion(s), 507
        proteins, 510
    mutant, 506
    operon, 2, 55, 181-188, 191-192, 195, 235, 244
    p$_S$ promoter, 133
    promoter, 12-14, 134, 188-191, 265, 270, 524-526
        -*relA* fusion, 530
    repressor, 157, 168, 185, 193, 236, 384, 469
LacI/*lacI*, 184-187, 192-194, 246
lacUV5, 131-134, 404, 523

LacY, 236
LacZ/*lacZ*, 55-56, 58, 110, 219, 388, 390, 397, 438, 459, 491-492,
    524, 553, 578
    -fusions, 201, 212, 321, 323, 332
lamB, 208
lambda
    CI repressor, 473
    N protein 34-35, 37-38
    P$_R$' promoter, 8
    phage P$_L$ promoter(s), 10, 14, 134, 163
    Q protein, 35-37
    receptor, 201, 208, 503
    repressor, 384, 455-457, 460, 462-463, 465, 469, 473
Landick R, 55, 57, 61-62
Landy A, 158
leu, 3
    500, 128-129
    ABCD, 128
    promoter, 129
leuA, 424
leuV promoter, 524, 526, 536
Levinthal C, 292
LexA/*lexA*, 69, 453-474, 504-505, 507, 510, 558-559
Liberek K, 484, 490
Liebke HH, 536
Liljestrom P, 394
Lin ECC, 232, 237, 354, 362-363, 367, 419
lipopolysaccharide(s) (LPS), 492, 572
Little J, 4
lld operon, 354
Lon/*lon*, 485-487, 504-511
Lrp/*lrp*, 149, 206, 271, 419-428, 575
Lugtenberg B, 385
Lutkenhaus J, 549
lux, 554-555
LuxR/*luxR*, 4, 578
Lwoff A, 291
LysR, 443
lysU, 427

M

Maaloe O, 534
Maeda S, 392-393
Magasanik B, 231, 233, 397
mal, 2-3, 202-204, 210, 213-218, *235f*, 244-245, 267
malA, 203
Malamy MH, 292
malB, 203
malE, 267-268
MalF, 201, *204t*, 207-208, 211
MalG/*malG*, *204t*, 207, 211
malI, 203, *204t*, 215-216, 219-220
MalK/*malK*, 202-203, *204t*, 206-210, 212-219, 243, 267
    -*lacZ* fusion(s), 213, 215-217, 219
malP, *204t*, 209, *213f*
malQ, 202, *204t*, 209, 211-213, 216, 218-219, 291
malS, *204t*, 205-206, 212
MalT/*malT*, 15, 135, 201-206, 209, 211-219, 234, 267, 403
maltodextrin phosphorylase, 209-211
maltose-binding protein (MBP), 201, 206-207, 211-212
MalX/*malX*, 203, *204t*, 215, 220
MalY/*malY*, 203, *204t*, 215-216, 219, 220
MalZ/*malZ*, *204t*, 205-206, 209, 211-213, 218
mar, 445-446
MarA/*marA*, 446-447
MarB/*marB*, 446
marC, 71
Margalit H, 428

Markovitz A, 505-506
marR, 446
marRAB, 446-447
Matsuyama SI, 403
Maurizi MR, 486
McClintock, B, 455
McClure WR, 16
McEntee K, 456
MDO, 401
Mecsas J, 491-492
mel, 2, 244
melAB, 258
MelB, 243
MelR/*melR*, 2, 258, 328, 439-440
MetC/*metC*, 215
metK, 424
Meyenburg K, 386
MicF/*micF*, 71, 439, 446-447
MinC/*minC*, 71, 560
MinD/*minD*, 71, 560
minE, 560
Mizuno T, 389-394, 396
Mizushima S, 388, 392-394
Mn-SOD, 442, 444-445
Model P, 495
mok, 75-76
Monod J, 1, 181, 231, 232, 240, 247, 291-292
mreB, 557-558
msp, 362
mtl operon, 238
mtrB gene, 100
Mu*dlac* fusion, 129
Mutoh N, 390
Myxococcus xanthus, 3, 106, 492-493, 508, 571

N

N25/N25₍antiDSR₎ promoter, 14
NADH, 320, 354, 436
NADPH, 281, 435-436, 441-442, 444
Nagai H, 489
Nakae T, 385
Nakashima K, 402
nap-ccm, 346, 352
Nar modulon systems, 343-355
narG, 329, 347-353
  -*lacZ* reporter, 329
narK, 351
NarL/*narL*, 167, 321, 329-330, 343, 345-355, 370, 372
  phospho-, 347-349, 352-353
NarL/NarX system, 361
NarP/*narP*, 321, 343, 346, 349-351, 355
  phospho-, 352
NarP/NarQ system, 361
NarQ/*narQ*, 343, 345-348
NarX/*narX*, 328, 331, 343, 345-349, 352-354
  (H399K), 348
  (H399Q), 348-349
ndh promoter(s), 326-328, 331
Neidhardt FC, 231, 233, 487
nfo, 438
Nierman WC, 12
nif, 282, 285, 288
NifA/*nifA*, 160, 285, 287, 288, 332, 336-337
  -activated promoters, 164
  regulator, 167
nifH, 160, 285
*nif*HDK operon, 167
Nikaido H, 384

Ninfa AJ, 397
nirB operon, 351
nirBDC operon, 344
nitric oxide (NO·), 435, 441
nitrite reductase (Nrf), 346
nitrogen regulator I/II (NR₍I₎/NR₍II₎), 282-288, 306, 397-398
Noble, 511
Nomura M, 94-95, 116, 386
Norioka S, 391
nrfA operon, 351-352
NTP concentration, 16, 31
Ntr, 282, 287, 398
nus, 533
NusA, 27, 34-35, 37, 135, 527
NusB/*nusB*, 37-38, 527-528
NusE (S10)/*nusE*, 37-38, 86, 521, 527
NusG, 34, 37, 527
nut, 35, 37-38, 527
Nystrom T, 372

O

Oka A, 393
OmpA/*ompA*, 113, 269, 385
ompB, 386-389, 394
OmpC/*ompC*, 363, 366, 383-397, 400-406
OmpF/*ompF*, 71, 363, 366, 383-406, 442, 446
OmpR/*ompR* 205, 210, 367, 372, 383-384, 389-400, 402-403, 572
OmpR/EnvZ, 363-364
OmpR-P, 398-400, 403-404, 406
oriC, 139, 161-162, 166, 168, 446
oriK, 139
osmotic pressure, 2, 4, 156
Ostrow KS, 392
OxyR/*oxyR*, 337, 442-445, 575
oxyS, 445

P

P22 Sar RNA, 70
P22 Sas RNA, 70
P1/P2/P3/P4 promoter(s), 185, 188, *189f*, 191-194, 487, 523, 576
P₍i₎ (inorganic phosphate), 210, 214, 246, 286, 297-298, 312
  control(s), 299-302, 308-309
    independent, 303-307
  inhibition, 208
  metabolism, 310
  promoter, 69
  repression, 292
  starvation, 291
P₍L₎ promoter, 109
P₍RE₎ promoter, 69
P₍SIEB₎ promoter, 70
Paigen K, 231
Palleros DR, 484
pap, 2, 268, 419, 426-427
papBA, 426-427
par, 387-388
Paracoccus denitrificans, 335
Pardee AB, 456
Parsell DA, 486
pBR322 RNA I promoter, 131
Pedersen S, 529, 531, 535
PEP, 242, 244-246
pfl promoter, 371

pgm, 210-211, 217
Φ(*arcA-lacZ*), 363
Φ(*arcB-lacZ*), 365
Φ(*cyd-lac*), 368
Φ(*cyo-lac*), 372
Φ(*fdnG-lacZ*), 346, 352
Φ(*frdA-lacZ*), 346, 348, 352-253
Φ(*ilvIH-lac*), 426
Φ(*lctD-lac*), 365-366
Φ(*lld-lac*), 354
Φ(*lrp-LacZ*), 420, 427
Φ(*narG-lacZ*), 352
Φ(*ompF-lacZ*), 366
Φ(*sdh-lac*), 354, 363, 365, 369, 372
Φ(*traY-lacZ*), 369
Pho/*pho*
    regulon, 210, 297-312
    system, 3, 291-293
PhoA/*phoA*, 292, 299, 303, 311, 394, 403
    -fusion, 201, 207
    -*lacZ* fusion, 293
    promoter, 293
PhoB/*phoB*, 210, 292, 297-311
    phospho-, 302, 311-312
phoBR, 299, 303
PhoE, 384, 388, 394, 403
PhoR/*phoR*, 292-293, 298-312
    creC mutant, 304-305
phoR69 allele, 304
phosphoglucomutase, 210-211
phosphotransferase system (PTS), 231-248
    proteins, 237, 244
PhoU/*phoU*, 293, 298, 300-302, 310, 312
pnp, 109
pnt, 423
Postma PW, 240
ppGpp, 112, 529-535, 537, 562, 577, 579
Pratt LA, 402
prfA, 50, 335
prfB, 50
promoter recognition region (PRR), 13
pSC101, 166
Pseudomonas aeruginosa, 4, 492-493, 554
Pst system, 299-302, 310, 312
PstI site, 55, 59
pT181, plasmid, 72-73
pta mutation, 305-306
Pta-AckA pathway, 305, 310
pts operon, 237, 244, 246
ptsG, 215
ptsI, 240, 243
Pugsley AP, 388
pyrB promoter, 10, 15
pyrBI operon, 34-35, 191
pyrE leader, 61
pyruvate dehydrogenase (PDH) complex, 318-320, 323
pyruvate formate-lyase (PFL), 318-320

Q
qut, 35

R
R1, 69, 74, 78
R17/MS2 coat protein, 98
Radman M, 455-456
Rahmouni AR, 138
Raibaud O, 203, 205
Raina S, 491-492
Rampersaud A, 396
Ras, 463-464
RecA/*recA*, 453-465, 468, 470, 472-474, 504-505, 558, 561
recA441, 458
Reeve ECT, 386
Reeves P, 386-388
RegA, 89, 94, 101, 103-104
RegB, 110
Release Factor 1/2, 50
REP (PU) elements, 154
RepA/*repA*, 74, 75*f*, 512
RepC/*repC*, 72
retroviral nucleocapsid proteins (NCPs), 100
Rex JH, 427
Rhizobium meliloti, 337
Rho/*rho*, 136, 528
    -dependent termination, 32-34
ribosomal binding site (RBS), 86, 88-90, 94, 97, 102, 104,
        106-108, mutations, 113
ribosomal RNA (rRNA), 521-537
Richet E, 203
RcsA/*rcsA*, 507-508, 512
RcsB/*rcsB*, 508, 512
RNA recognition motif (RRM), 100, 114
RNA-IN, 74, 78, 79*f*
RNA-OUT, 70-71, 74, 78, 79*f*
RNAP, 490-491
Rob, 446-447
RobA/*robA*, 306
Roberts CW, 456
Roberts DL, 398
Roberts JW, 456, 460, 504
Robinson M, 63
Roesser JR, 50
Roseman S, 232, 237, 240, 247
Rosenberg H, 293
Rothfield LI, 4
Rothman F, 292
Rouvière P, 491
rpoA, 331, 404, 403
RpoB/*rpoB*, 14-15, 535
rpoC, 15, 535
rpoD, 523
rpoE, 489*f*, 491-493
rpoH, 487-489, 491-492
rpoH p3, 491-494
rpoN (*ntrA*), 160, 246, 282, 353
rpoS, 512, 573, 575-579
rpoS::lacZ fusions, 576-578
rpoZ, 531
rrn antitermination system, 527-529
rrn P1/P2 promoter(s), 522-526, 532-534, 536
rrnB operon, 522, 523*f*, 528
rrnB P1/P2 promoter(s), 10, 14, 523-527, 530, 532, 534, 536
Russo FD, 399-400, 402-403

S

S1, r-protein, 107-108
S4, r-protein, 90, 92, 97, 102, 105-106
S6, r-protein, 108, 445
S7, r-protein, 87, 112, 521
S8, r-protein, 95, 101, 111, 521
S10. *See* NusE/*nusE*
S15, r-protein, 89, 92, 97, 105-106, 109, 114
S20, r-protein, 107, 111
Saccharomyces cerevisiae, 437, 508
Saier MH Jr, 240
Salmonella typhimurium, 128-129, 157, 164, 208, 240, 244-245, 288, 297, 331, 443, 548
  cob operon, 89, 115
  hisR gene, 133
Sarma V, 388
Sayre MH, 169
Schmid D, 484
Schmitges CJ, 387-388
Schnaitman C, 385, 387-388
Schwartz M, 201
sdh, 354, 362
SdiA/*sdiA*, 553-555, 578
sec, 201
SecA, 87, 561
SecY, 116
SELEX, 91
serA, 422, 424, 427
SfiA/*sfiA*, 507, 559, 562-563
sfiB, 507, 559
SfiC/*sfiC*, 559-560
Sfr/*sfr*, 369
sfrA, 362, 369
Sharif TR, 404
Shine-Dalgarno sequence, 104, 108, 110, 114
Shultz W, 9
SieB/*sieB*, 70
sigma factor(s), 8, 12
  RpoS, 219
  σ^{32}, 481, 509, 511-512, 536, 561
    -promoted heat shock response, 487-491
  σ^{54}, 246, 283, 495
    -dependent promoters, 160, 285, 288
    -RNA polymerase, 285, 321
    -*glnAp2* complex, 286
  σ^{70}, 36, 536, 573, 575
    -dependent promoters, 283
  σ^{E}, 481, 491-494
  σ^{S}, 512, 555, 572
σ^{S} regulon, 4, 573-575, 578
Silhavy TJ, 388-389, 393-395, 399-400, 402-403
Silverman PM, 369
Slauch JM, 402
SOD, 436-437, 441-442, 444-445
sodA, 323, 371, 437, 439, 442, 445-447
Sok/*sok*, 69, 75-76, 78
SOS
  induction, 71, 157
  regulatory system, 2-5, 453-474
  response, 504-505, 507, 558-561, 563
SoxR/*soxR*, 337, 435-447
SoxS/*soxS*, 437-441, 446-447
soxRS, 435-447
spc operon, 87, 101, 111, 116, 521
Speyer JF, 536
spf promoter, 270
Spiegelman S, 291
sspp, 555-556

Stewart V, 346
Stock A, 397
Storz G, 443
str operon, 87, 111-112, 521
Streptomyces, 571
suA, 32
SulA(SfiA)/*sulA*, 71, 457, 459, 506-507, 512, 559
SulB/*sulB*, 507
superattenuation, 60
SurA/*surA*, 493
Sutherland, 4
Suzuki H, 52

T

T4 gene 32. *See* gp32
T4 RegA protein, 94, 103
T4 RegB endonuclease, 110
T5 N25 promoter, 14-15
T5 N25$_{antiDSR}$ promoter, 11, 14-15, 19
T7 A1/A2 promoter(s), 10, 12, 14-15
T7 gene
  0.3, 108
  0.7, 107
T7 protein kinase, 116
t21, 31
tac promoter(s), 8-10, 133, 216
Tardat B, 370
Taylor RK, 389-390
TBP protein, 155
tdcR regulator, 167
tdh, 427
TetA/*tetA*, 138
Tétart F, 556
threonyl-tRNA synthetase, 101-102, 104
  gene, 111
thrS, 95
tif, 458, 507
TMAO, 343-344
Tn*10*, 162-163, 166, 215
Tnp/*tnp*, 70-71
Tokishita S, 401
tolC, 71
Toledano MB, 444
tolF, 386-388
Tommassen J, 390
topA, 128-130, 136, 138
Touati D, 370
tra, 68
TraJ/*traJ*, 68-69
TRAP protein, 98, 100
traY, 369, 373
trc::lacZ translational fusion plasmid, 54-55, 58
tricarboxylic acid (TCA), 361-362
tRNA^{His} genes, 136
tRNA^{Thr} isoacceptors, 95-96
Trp/*trp*, 30, 35, 47, 50-52, 55, 57, 59*t*, 60-62, 72, 100
  attenuation assay, 57-60
trpG, 100
trpLep leader, 55, 57-58, 61-62
Tsui P, 396
Tsung K, 391

U

UmuC/*umuC*, 459, 464-465
UmuD/*umuD*, 459, 464-465, 473, 504
umuDC operon, 459, 464
UP Element, 523-527, 530, 533-537
upstream activating region (UAR), 527
UTP, 15
Utsumi R, 401
uvrB, 458-459

V

Vaara M, 384
van Alphen W, 385, 388
van Bogelen RA, 487
Varshavsky, 510
Verhoef C, 387-388, 402
Vibrio, 554
   fischeri, 4, 555, 577
   harveyi luciferase, 441, 486
Vinella D, 562
Virtanen AI, 291
von Hippel PH, 16

W

Wagner I, 486
Wang Q, 420, 428, 487, 491
Watson-Crick base pairs, 10
Wawrzynow A, 486
Weisberg RA, 157, 504
Wells RD, 138
Wickner S, 486
Willsky GR, 292
Winkler U, 291
Witkin EM, 455
Witten C, 386

X

Xis, 504-506, 509

Y

Yang Y, 400
Yanofsky C, 47, 50
yeast GAL4 protein, 136
Yura T, 487

Z

Zhou, 10
Zillig W, 9, 130
Zipser D, 505-506
zwf, 439, 442
Zylicz M, 510